Subcellular Biochemistry

Volume 94

The book series SUBCELLULAR BIOCHEMISTRY is a renowned and well recognized forum for disseminating advances of emerging topics in Cell Biology and related subjects. All volumes are edited by established scientists and the individual chapters are written by experts on the relevant topic. The individual chapters of each volume are fully citable and indexed in Medline/Pubmed to ensure maximum visibility of the work.

More information about this series at http://www.springer.com/series/6515

Ulrich Hoeger · J. Robin Harris
Editors

Vertebrate and Invertebrate Respiratory Proteins, Lipoproteins and other Body Fluid Proteins

Springer

Editors
Ulrich Hoeger
Institute of Molecular Physiology
Johannes Gutenberg University
Mainz, Rheinland-Pfalz, Germany

J. Robin Harris
Institute of Molecular Physiology
Johannes Gutenberg University
Mainz, Rheinland-Pfalz, Germany

ISSN 0306-0225 ISSN 2542-8810 (electronic)
Subcellular Biochemistry
ISBN 978-3-030-41768-0 ISBN 978-3-030-41769-7 (eBook)
https://doi.org/10.1007/978-3-030-41769-7

This Springer imprint is published by the registered company Springer Nature Switzerland AG
The registered company address is: Gewerbestrasse 11, 6330 Cham, Switzerland

Preface

Volumes in the long-established *Subcellular Biochemistry* book series have devoted considerable attention to proteins and protein complexes in recent issues, in an almost encyclopaedic manner. The present volume, covering invertebrate coelomic fluid, hemolymph and vertebrate blood proteins, correlates well with this continuing theme. The book editors, Ulrich Hoeger and J. Robin Harris, spent most of their scientific careers researching diverse aspects within this field of study. Indeed, in retirement it has been a pleasure for us to work together, as former colleagues at the University of Mainz, to compile and co-edit this book.

Our initial aim was to cover the subject as thoroughly as possible, taking into account that chapters on some related proteins groups, such as the ferritins, peroxiredoxins, α_2-macroglobulin and frataxin, have already been included in the series. Despite our efforts, some additional interesting topics were inevitably omitted and some chapters have been lost due to commissioned authors failing to deliver their agreed chapter! The remaining 20 submitted chapters are included here and are split into the two predefined sections: i.e. invertebrate and vertebrate topics. Whilst not dealing here specifically with any one chapter topic, it is clear that although much emphasis is placed within the book to the respiratory proteins, the hemoglobins and hemocyanins, numerous other interesting topics are also given due attention (see Chapter list).

We would like to thank our chapter authors for their enthusiasm and considerable efforts. To write book chapters, whilst also performing research, preparing papers for publication, teaching and academic administration, is not these days often given due acknowledgement in terms of the academic performance and advancement. We believe that advanced book series, such as *Subcellular Biochemistry*, fulfil a useful indeed important place, within the broad body of scientific literature. With the availability of Springer e-book and e-chapter downloads, often via institutional libraries, the dispersal of knowledge now goes well beyond the hard-copy books. The previous and recent volumes of the *Subcellular Biochemistry* series provide a ready source of past and present useful information for many.

We hope that the present volume will be of interest and use to advanced students, postgraduates and academic biologists, biomedical scientists and others researching or teaching relevant aspects of this fascinating subject.

Mainz, Germany

Ulrich Hoeger
J. Robin Harris

Contents

Chapter 1
Annelid Coelomic Fluid Proteins

Sven Schenk and Ulrich Hoeger

Abstract The coelomic cavity is part of the main body plan of annelids. This fluid filled space takes up a considerable volume of the body and serves as an important site of exchange of both metabolites and proteins. In addition to low molecular substances such as amino acids and glucose and lactate, the coelomic fluid contains different proteins that can arise through release from adjacent tissues (intestine) or from secretion by coelomic cells. In this chapter, we will review the current knowledge about the proteins in the annelid coelomic fluid. Given the number of more than 20,000 extant annelid species, existing studies are confined to a relatively few species. Most studies on the oligochaetes are confined to the earthworms—clearly because of their important role in soil biology. In the polychaetes (which might represent a paraphyletic group) on the other hand, studies have focused on a few species of the Nereidid family. The proteins present in the coelomic fluid serve different functions and these have been studied in different taxonomic groups. In oligochaetes, proteins involved antibacterial defense such as lysenin and fetidin have received much attention in past and ongoing studies. In polychaetes, in contrast, proteins involved in vitellogenesis and reproduction, and the vitellogenic function of coelomic cells have been investigated in more detail. The metal binding metallothioneins as well as antimicrobial peptides, have been investigated in both oligochaetes and polychaetes. In the light of the literature available, this review will focus on lipoproteins, especially vitellogenin, and proteins involved in defense reactions. Other annelid groups such as the Pogonophora, Echiura, and Sipuncula (now considered polychaetes), have not received much attention and therefore, this overview is far from being complete.

Keywords Annelids · Polychaetes · Coelomic cells · Lipoproteins · Vitellogenin · Reproduction · Defense reactions · Defensive proteins · Metalloproteins · Antimicrobial peptides

---

S. Schenk (✉)
Max F. Perutz Laboratories, Vienna Biocenter (VBC), Dr. Bohr-Gasse 9/4, 1030 Vienna, Austria
e-mail: Sven.Schenk@univie.ac.at

U. Hoeger
Institut für Molekulare Physiologie, Johannes Gutenberg-Universität, 55099 Mainz, Germany
e-mail: uhoeger@uni-mainz.de

U. Hoeger and J. R. Harris (eds.), *Vertebrate and Invertebrate Respiratory Proteins, Lipoproteins and other Body Fluid Proteins*, Subcellular Biochemistry 94,
https://doi.org/10.1007/978-3-030-41769-7_1

The Coelomic Fluid Compartment

The coelomic fluid compartment is a feature of the body plan in most annelids. It is defined by its lining with a thin mesodermal layer, the coelomic epithelium. In contrast, the blood vessel system in annelids is a separate compartment representing the so called primary body cavity, which is present in many, but not all annelids (Westheide 2013). The coelomic fluid (CF) differs from the circulatory system of vertebrates in that it is not circulated by a heart or equivalent organ (unlike the annelid blood vessel system) but only by body movements. However, it takes up a considerable volume of the live animal. In the marine annelid, the polychaete *Nereis virens*, the coelomic fluid makes up 26% of the live weight (Hoeger et al. 1999). The coelomic fluid usually does not serve oxygen transport and does not contain dissolved hemoglobin. Some marine annelid families (Terebellidae, Glyceridae and Capitellidae), however, possess specialized coelomic cells (erythrocytes) containing high concentrations of hemoglobin and in the latter two groups, a blood vessel system is even missing (Mangum et al. 1975; Dales and Dixon 1981). Leeches are another exception; they have a much reduced coelomic system which functions as a closed vascular system and contains large amounts of extracellular hemoglobin (MacMahon et al. 1997). The known annelid hemoglobins are giant (~3.5 MDa) hexagonal bilayer globin structures consisting of 144 heme bearing subunits with a characteristic quaternary structure (Weber and Vinogradov 2001, *see also* the chapter by Costa-Paiva and Coates in this volume).

Via the thin mesodermal epithelium, the coelomic fluid is in contact with different tissues (muscle, intestine, gonads, nerves). In addition to proteins, the coelomic fluid also contains both low molecular weight compounds such as glucose and lactate (Hoeger and Kunz 1993) and especially lactate can accumulate to considerable concentration during times of natural anaerobiosis in marine annelids (Scott 1976; Hoeger and Kunz 1993). The coelomic fluid thus serves as an important compartment for the exchange of both proteins and metabolites between different tissues and the coelomic cells found in this compartment (*see* below).

Coelomic Cells

The coelomic fluid of annelids contains various cell types involved in both reproduction and defense mechanisms. They will be briefly mentioned here since they synthesize and secrete proteins and are thus involved in the exchange and turnover of proteins in the coelomic fluid.

Granulocytes

Granulocytes are part of the innate immune system in annelids and many other invertebrates. Together with the hyalocytes, they belong to the amoeboid coelomic cells (Vetvicka and Sima 2009), and are involved in defense reactions including non-self recognition, cytotoxicity, phagocytosis, encapsulation of foreign objects and secretion of antimicrobial compounds (Ratcliffe and Rowley 1981a, b; Cooper and Stein 1981; Dhainaut and Porchet-Henneré 1988; Porchet-Hennere et al. 1992; Engelmann et al. 2005; de Eguileor et al. 2000; Vetvicka and Sima 2009). They occur in both the annelid coelomic fluid and the hemolymph of e.g., arthropods and mollusks. Granulocytes are characterized by abundant granular inclusions which have been mainly studied on the electron microscopical level; their biochemical composition, however, has not been studied in great detail in annelids. In earthworms, several subtypes of granulocytes have been identified which differ in their ability for phagocytosis and encapsulation using monoclonal antibodies (Engelmann et al. 2005). In the marine annelid *Nereis diversicolor*, different types of granulocytes have been detected using different lectins (M'Beri et al. 1988). Using monoclonal antibodies, three types (G_1 to G_3) have been identified which cooperate during the encapsulation reaction of foreign material (Porchet-Henneré 1990). The G_2 type has been identified as a cytotoxic natural killer cell, while the G_3 type showed a capacity for phenoloxidase synthesis and melanization (Porchet-Hennere et al. 1992).

Chloragocytes and Eleocytes

Chloragocytes are characteristic cells which are formed in the chloragogen tissue of oligochaetes located in a characteristic depression of the dorsal intestine called the typhlosolis. Chloragocytes express at least two important defensive proteins, lysenin and fetidin (*see* below), which are released into the coelomic fluid upon bacterial challenge (Ohta et al. 2000; Opper et al. 2013; Swiderska et al. 2017).

Eleocytes have been described in both oligochaete and polychaete annelids, but they have different functions in these groups. In oligochaetes, eleocytes are involved in defensive reactions and derive from chloragocytes, when released as free cells into the coelomic fluid. Both chloragocytes and especially eleocytes store large amounts of riboflavin (Vitamin B1) in several but not all oligochaete species investigated (Plytycz and Morgan 2011). Riboflavin has been found to be a chemoattractant and it has been hypothesized that release of riboflavin serves to facilitate the migratory recruitment of immune competent coelomocytes to sites of microbial invasion (Mazur et al. 2011).

In marine annelids in contrast, eleocytes have no known defensive function and assume key functions in growth and development of germ cells. They have been especially well studied in the Nereidid polychaetes and have been defined as early as 1895 (Romieu 1921; Racovitza 1895), but their exact role and contribution to the proteins in the coelomic fluid was revealed only much later. They also occur in other polychaete groups such as the Sabellidae (Giangrande et al. 2000) and other families (Dhainaut and Porchet-Henneré 1988), but their biochemical and physiological properties have not been studied outside the Nereidids. During the reproductive phase, eleocytes accumulate in large numbers in Nereidids while granulocytes are present during all phases of the life cycle (Fischer et al. 1996; Dhainaut and Porchet-Henneré 1988). This is related to their reproductive functions since eleocytes synthesize the storage protein, vitellogenin. Eleocytes also synthesize and store large amounts of tri-, and diacylglycerides (Fontaine et al. 1984a, b)—hence their name which is derived from the Greek word for oil, and contain a large vacuole (Fig. 1.1) that acts as a store for purine nucleotides (Hoeger et al. 1996, 1999; Hoeger and Märker 1997), amino acids (Hoeger and Abe 2004) and heme degradation products (Schenk and Hoeger 2011).

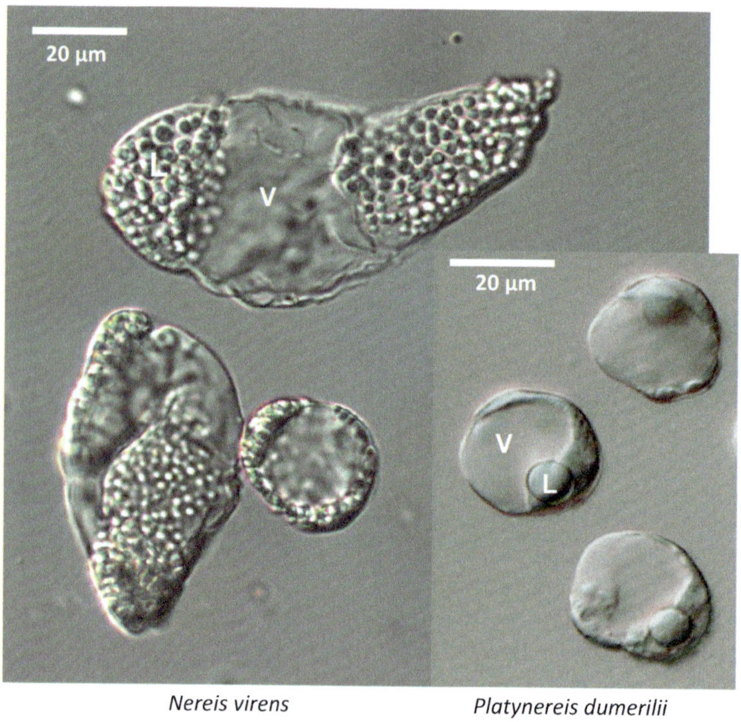

Fig. 1.1 Light microscopical images (DIC) of eleocytes in two polychaetes, *Nereis virens* and *Platynereis dumerilii*. Note the large vacuole (**V**) characteristic of eleocytes and numerous lipid droplets (**L**) in the cytoplasm in *Nereis*; only one lipid droplet is found per cell in *Platynereis*. U. Hoeger, previously unpublished

Lipoproteins

Vitellogenin

Oogenesis requires massive synthesis and storage of the female-specific lipoprotein, vitellogenin (VTG), a key protein for the growth and development of the oocytes. As in other egg laying animals, VTG is usually produced externally (extraovarian synthesis) and is taken up by the oocytes via receptor mediated endocytosis and this is also true in annelids (Fischer et al. 1991; Hafer et al. 1992; Lee et al. 1997). Most (but not all) VTGs belong to the group of large lipid transfer proteins which are present in most oviparous species in both vertebrates and invertebrates. Well known examples of this family include the ApoB type lipoproteins in the blood of vertebrates and the microsomal triglyceride transfer proteins which are essential for the formation of intracellular lipoprotein particles (Hussain et al. 2003). The lipophorins of flying insects also belong to this family (for a review, *see* Smolenaars et al. 2007b). In spite of the diversity of annelids, information on vitellogenin is scarce and at present, only two complete VTG sequences are annotated in the Genbank (*Perinereis aibuhitensis*; AHY02164.1; *Platynereis dumerilii*; APP91162.1). Like other VTGs, the amino acid sequence contains a signal peptide, a lipid binding domain, a domain of unknown function (DUF1943) and a von Willebrand type D domain (Fig. 1.2), which may aid in multimerization and sequestration of VTG once the protein is stored in the oocyte (Journet et al. 1993; Smolenaars et al. 2007a). The molecular mass of Nereidid

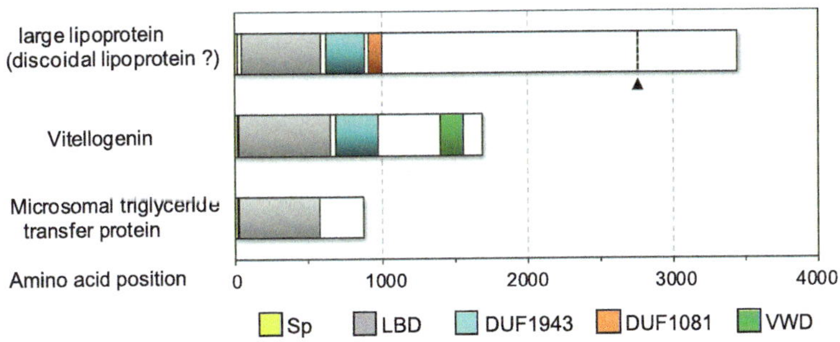

Fig. 1.2 Schematic representation of the domain structure of large lipid transfer proteins (LLTP) in polychaete annelids. The lipid binding domain (LBD) is the characteristic trait of all members of this family. The microsomal triglyceride transfer protein is the simplest LLTP; vitellogenin has a domain of unknown function (DUF1943) and a von Willebrand type D domain (VWD). The putative large discoidal lipoprotein discovered in the polychaete *Nereis virens* has an additional domain of unknown function (DUF1081). A putative propeptide cleavage site is indicated by an arrow. Accession numbers: MTP: GASB01019285.1 TSA (*Glycera dibranchiata*; predicted amino acid sequence); Vitellogenin: APP91162.1 (*Platynereis dumerilii*); putative large discoidal lipoprotein: GBZT01004964.1 TSA (*P. dumerilii*; predicted amino acid sequence). The domain boundaries including signal peptide (Sp) and the propeptide cleavage site were determined using the online tools SMART (https://smart.embl-heidelberg.de/) and ProP 1.0 Server (https://www.cbs.dtu.dk/services/ProP/), respectively

VTG apoproteins is 150–170 kDa and electrophoresis under native conditions show apparent molecular masses between 370 and 530 KDa indicating a homodimeric and -trimeric structures in the coelomic fluid (Baert and Slomianny 1987; Schenk and Hoeger 2011). In the Sabellid, *Pseudopotamilla occelata*, the only non-Nereidid polychaete for which a VTG has been described, VTG is said to be composed of a 200 kDa protein subunit. However, VTG in this species had not been purified and the 200 kDa peptide is the highest molecular mass protein found in yolk purified from oocytes (Lee et al. 1997). Since VTG usually undergoes proteolytic processing upon incorporation in the oocyte (*see* below), a process also demonstrated for *Pseudopotamilla occelata*, it seems likely that the VTG apoprotein is this 200 kDa protein; this notion might also be supported by the fact that a similar molecular mass protein is present in the CF of *Pseudopotamilla occelata* (Lee et al. 1997; Lee and Kim 1993).

In the leech *Theromyzon tessulatum,* VTG and its storage form vitellin (VT) have been characterized as 520 and 490 kDa lipoproteins, present during oogenesis in these hermaphrodite animals. VTG as well as VT are composed of a 165 kDa apoprotein dimer, containing 31% and 24% lipid, respectively (Baert et al. 1991).

In contrast to other annelid groups, the biochemical aspects of vitellogenesis have been studied in Nereidid polychaetes in several species and will now be described. Nereidid polychaetes provide a good model to study gametogenic processes since both coelomic and gamete cells are floating freely in the same compartment and can thus be relatively easy accessed for experiments. In other polychaete families and in Oligochaetes, ovaries are present and oocyte growth is supported by nurse cells or follicle cells (Jamieson 2006; Eckelbarger 2005). In polychaetes, eleocytes are the sole site of vitellogenesis (Fischer 1979; Baert and Slomianny 1987, 1992). They store large amounts of free amino acids with a preference for essential amino acids and their levels fluctuate depending on the nutritional state of the animal (Hoeger and Abe 2004). In female eleocytes, labelled amino acids are rapidly incorporated into VTG (Bonnier and Baert 1992) and possibly other proteins and VTG is released into the coelomic fluid. The lipid moiety of VTG contains mostly phospholipids synthesized by eleocytes (Taki et al. 1989). VTG is then taken up by the oocytes by receptor mediated endocytosis (Hafer et al. 1992). After uptake, VTG is partially cleaved into smaller fragments with a major form of about 100 kDa representing its final storage form VT (Baert 1985, 1986; García-Alonso et al. 2006).

Control of Vitellogenin Synthesis

Vitellogenin synthesis indicates the beginning of the reproductive phase in the life cycle. In polychaetes, this is a dramatic process leading to the production of enormous numbers of oocytes. In *Nereis virens*, oocytes make up 40% of the live weight at the end of oogenesis (Hoeger et al. 1999) and this demand has to be matched by a concomitant synthesis of VTG as the yolk protein precursor. A recent study in *Platynereis dumerilii* (Schenk et al. 2016) has shown that VTG synthesis is under control of the sesquiterpenoid hormone methyl-farnesoate (MF). MF suppresses

VTG synthesis in eleocytes, thus acting as a juvenile hormone. In turn, declining titers of this hormone then lead to the initiation of sexual maturation. The existence of hormonal factors suppressing the processes of sexual maturation have been postulated many years ago (Hauenschild 1956) but the identification of MF as the missing link, including a putative MF receptor, was unsuccessful until recently. The finding that MF titers decrease during the time course of sexual maturation (Schenk et al. 2016) is in agreement with earlier findings in *Perinereis cultrifera* in which eleocytes increase their rate of synthesis during the course of oocyte growth (Baert and Slomianny 1992). As a second control point for the synthesis of VTG, estradiol has been found to enhance VTG release in female vitellogenic eleocytes of *Nereis virens,* but did not have an effect on immature eleocytes (García-Alonso et al. 2006). This could be related to the lack of expression of an eleocyte estradiol receptor which was described in this species (García-Alonso and Rebscher 2005).

Multiple Functions of Vitellogenin

The role of VTG is not restricted to the transport of eleocyte-derived lipids into the oocytes. Recent studies in *Nereis virens* (Schenk and Hoeger 2011) have shown that VTG also carries the hemoglobin degradation product biliverdin. Biliverdin is formed in the eleocytes and is—surprisingly—converted to a novel conjugated form, glutathione biliverdin, which was termed nereioverdin. Heme as the primary degradation product of hemoglobin which originates from the blood system (representing the primary body cavity) is taken up by the eleocytes bound to another lipoprotein, the large discoidal lipoprotein (*see* below) via the coelomic fluid. After its conversion to nereioverdin, the conjugate is stored in the large vacuole of the eleocytes (*see* Fig. 1.1). In female animals, vitellogenin bound nereioverdin is exported into the coelomic fluid and incorporated into the oocytes which become progressively green during growth. In male animals, there is no such export route and nereidoverdin remains in the large vacuole until the death of the animal following the nuptial dance and reproduction, leading to a dark green coloration of eleocytes during the final stages of sexual maturation (Schenk and Hoeger 2011). Hemoglobin is an abundant and common respiratory protein in many annelids (Weber and Vinogradov 2001), but the route of heme degradation as demonstrated in *Nereis* has not been demonstrated in this group before. In other invertebrates (i.e., arthropods) however, other types of biliverdin conjugates resulting from hemoglobin degradation have been found before (Paiva-Silva et al. 2006; Pereira et al. 2007).

The Large Discoidal Lipoprotein

In *Nereis virens*, a non sex-specific discoidal lipoprotein was found in the coelomic fluid (Schenk et al. 2006) after it had been first found in the vascular compartment

(Harris et al. 2001) which, however, most likely resulted from coelomic contaminant. The lipoprotein particle consists of two apoproteins with molecular masses of about 300 and 80 kDa. Although the protein has not been sequenced, a putative lipoprotein was found in the NCBI TSA database. (Accession Nr. GBZT01007058.1). Its predicted sequence (3447 amino acids) has a calculated molecular mass of about 389 kDa and contains a predicted propeptide cleavage site (RHFR) at amino acid position 2760. The cleavage products with calculated molecular masses of 312 and 78 kDa are similar to those calculated from SDS-PAGE gel analysis (Schenk et al. 2006). Under the electron microscope, the lipoprotein particles appear disk-like with a diameter of 40 nm with a thickness of about 10 nm (*see* Fig. 1.3)—much larger than the lipoprotein particles of insects and crustaceans (18–26 nm, Spaziani et al. 1986; Ryan 1990; Weers et al. 1993). Uptake experiments (Schenk and Hoeger 2009) have shown, that the discoidal lipoprotein (dLP) particle is incorporated into the yolk granules of oocytes, and has thus the same target site as VTG. Interestingly, ligand blot experiments showed that the dLp receptor protein has a molecular mass (116 kDa) which is lower than that the VTG receptor (190 kDa, Hafer et al. 1992) suggesting two independent uptake systems. This parallels the situation in insects, in which lipophorin,

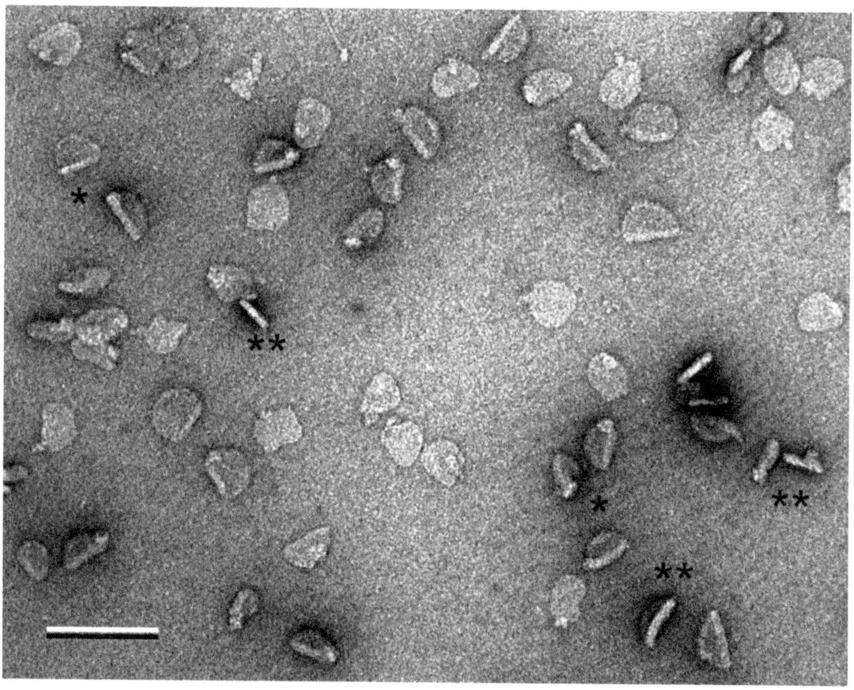

Fig. 1.3 Electron micrograph of discoidal lipoprotein particles isolated from the coelomic fluid of the polychaete, *Nereis virens*. Some of the particles are fixed in a "on edge" position (**); some are partially folded (*). The specimens were prepared for electron microscopy and negative staining as described by Schenk et al. (2006). Scale bar: 100 nm

the main lipid transporting protein in the hemolymph, is also incorporated into the oocyte in addition to vitellogenin (Chino et al. 1977; Kulakosky and Telfer 1990; Telfer et al. 1991). Unlike VTG however, which is partially cleaved after uptake, the large dLp apoprotein is degraded within the oocyte (Schenk, unpublished). Likewise, the lipoprotein cargo does not appear to be stored within the yolk granules. Lipid droplets are found only in the oocyte cytoplasm and contain mainly triacylglycerides (Fontaine et al. 1984a, b), whereas the lipoprotein-bound lipids consist mainly of phospholipids and diacylglycerides (Schenk et al. 2006). This suggests that incorporated lipids are hydrolyzed and transported to the oocyte cytoplasm and are utilized for de novo lipid synthesis.

Further experiments have shown that the dLP has a similar role in male gametogenesis. As in females, males of Nereidid polychaetes produce an enormous number of male gametes, which also develop freely in the coelomic compartment. Uptake experiments (Schenk and Hoeger 2010) have revealed the uptake of labeled dLP by the spermatogonia stages and presumably by the same receptor as found in the oocytes. Lipoprotein uptake and lipid content were found highest in the early spermatogonia stages (i.e. during mitotic proliferation). Lipid analysis by HPLC and mass spectroscopy showed that neutral lipids predominate in early spermatogonia stages (Schenk and Hoeger 2010). This could indicate a metabolic transformation of the lipoprotein-bound phospholipid cargo, as suggested for the oocytes (*see* above) although this view needs to be corroborated by tracer experiments.

A schematic representation of the route of lipoproteins in the coelomic cavity of nereidid polychaetes is shown in Fig. 1.4.

Proteins Involved in Defense Reactions

Like other invertebrates, annelids rely on an innate immune system to defend themselves against pathogens. A key compound in invertebrate immunity are phenoloxidases. Phenoloxidases catalyze the oxidation of phenols to bi-chinons which auto-oxidize and polymerize to form melanin, eventually encapsulating and killing pathogens (Soderhall and Cerenius 1998). In addition to encapsulation and phagocytosis, proteins released into the coelomic fluid may directly interact with the pathogens. These may either be pattern recognition proteins (PRP) binding to specific cell surface glycans, lipopolysaccharide (LPS) recognizing proteins, or antimicrobial peptides and proteins directly involved in killing intruding pathogens, as well as proteinases and lysozyme-like proteins.

In contrast to the reproductive processes and vitellogenesis, molecules involved in defensive reactions have been studied mainly in oligochaetes. The main proteins identified in the coelomic fluid of oligochaetes and the few identified from polychaetes and leeches will be described in the following sections.

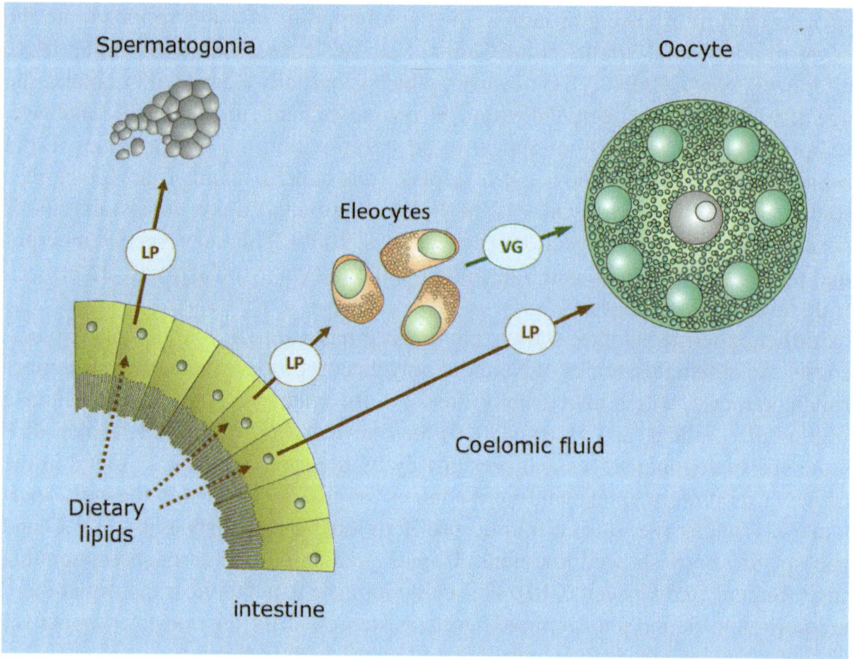

Fig. 1.4 Proposed scheme for the exchange of lipoproteins in the coelomic fluid of a marine annelid, the polychaete *Nereis virens*. Dietary lipids from the intestine are hydrolyzed and used for the de novo synthesis of lipids. The lipids are transported by lipoproteins to both male and female gametes and free coelomic cells, the eleocytes. Eleocytes synthesize and release vitellogenin into the coelomic fluid which is taken up by the oocytes by receptor mediated endocytosis. The large discoidal lipoprotein is taken up by both eleocytes and germ cells. The figure was prepared on the basis of earlier work (Schenk et al. 2006; Schenk and Hoeger 2009, 2010)

Lysenin and Fetidin

Lysenin and fetidin are a group of four highly similar proteins—lysenin, lysenin-related protein 1, lysenin-related protein 2/fetidin and lysenein-related protein 3—with sequence identities between 75 and 89% (Bruhn et al. 2006)—all of which display hemolytic activity and, at least for fetidin also antibacterial properties (Lassegues et al. 1997; Yamaji et al. 1998; Cooper et al. 2001; Kiyokawa et al. 2004; Bruhn et al. 2006). Whilst lysenin and lysenin-related protein 2/fetidin have been shown to be present in both, *Eisenia fetida* and *Eisenia andrei*, lysenin-related protein 1 and 3 seem to be species dependent (Swiderska et al. 2017). All members of this protein family show a strict specificity for sphingomyelin (SM) containing membranes, and at least lysenin specifically requires the phosphorylcholine (POC), sphingosine and the fatty acid moieties of the SM molecule to bind to membranes (Yamaji et al. 1998).

Binding kinetics between lysenin and fetidin are similar, binding SM in the low nano-molar range (~5 nM), lysenin-related protein 1 binds SM only at concentrations roughly ten-times as high (Kiyokawa et al. 2004). Also the hemolytic properties of lysenin-related protein 1 are only a tenth of those of lysenin and fetidin. These features have been linked to the substitution of a phenylalanine residue at position 210 to a isoleucine in lysenin-related protein 1 compared to lysenin and fetidin—the only aromatic to aliphatic amino acid substitution amongst the 30 conserved aromatic amino acid sides between the three proteins (Kiyokawa et al. 2004). Of the four members of the lysenin-family, fetidin and lysenin have received most attention, most probably because fetidin and lysenin are present in both *Eisenia* species, *Eisenia fetida* and *Eisenia andrei* (Swiderska et al. 2017). Therefore, the next section we will deal with what is known of these two members of the lysenin-family.

Fetidin (lysenin-related protein 2) is present as two isoforms with molecular masses of 40 kDa and 45 kDa, representing ~20% of the total CF proteins and is released from coelomocytes (Valembois et al. 1982; Vaillier et al. 1985; Milochau et al. 1997). Both seem to represent different glycoforms of a 34 kDa peptide containing a peroxidase domain and an N-glycosylation consensus sequence, encoded by a 1.44 kBp transcript (Lassegues et al. 1997). Recombinant fetidins were active in biological assays monitoring lysis of sheep red blood cells (Lassegues et al. 1997; Milochau et al. 1997), and since—contrary to the native protein—no peroxidase activity could be detected in the recombinant one (Lassegues et al. 1997), it can be concluded that the peroxidase activity is not needed for hemolysis but rather seems to be needed for antibacterial activity (Cooper 2002). In contrast to hemolytic proteins from many other species, fetidin does not seem to require a co-factor such as Ca^{2+} or Mg^{2+} to be functional, and as hemolysis requires 11,000 bound fetidin molecules per cell, it can be concluded that fetidins bind to specific cell surface receptors, rather than non-specific (Milochau et al. 1997).

While detailed functional studies on fetidins are scarce, much more data are available on lysenin. Lysenin is a 41 kDa protein that was originally isolated from the coelomic fluid of the earthworm *Eisenia foetida* (Sekizawa et al. 1997; Yamaji et al. 1998). Later studies (Ohta et al. 2000; Opper et al. 2013) have shown that lysenin is synthesized by a specific population of coelomocytes, the chloragocytes. While transcription of the lysenin gene occurs in chloragocyctes in the typhosolis (Ohta et al. 2000), translation of the mRNA seems to be restricted to cells floating freely in the coelomic cavity, especially in the highly granular and autofluorescent coelomocytes (called eleocytes; Opper et al. 2013, see above). Exposure of these cells to Gram-positive (*Staphylococcus aureus*) but not Gram-negative (*E. coli*) bacteria led to an increased expression and secretion of lysenin into the coelomic fluid (Opper et al. 2013).

Lysenin acts as a pore forming protein causing cytolysis and represents a distinct clade in among pore forming proteins (Fig. 1.5). It inserts into the membrane after nonamerization, forming a beta barrel channel (Bokori-Brown et al. 2016; Podobnik et al. 2016, see https://www.rcsb.org/3d-view/5GAQ). Insertion into the membrane is initiated by a putative loop structure, characteristic for proteins of the aerolysin family (Bokori-Brown et al. 2016; Podobnik et al. 2016). The pore structure of lysenin has recently been determined to 3.1 Å resolution using single particle cryo-EM (Bokori-Brown et al. 2016) and X-ray crystallography (Podobnik et al. 2016).

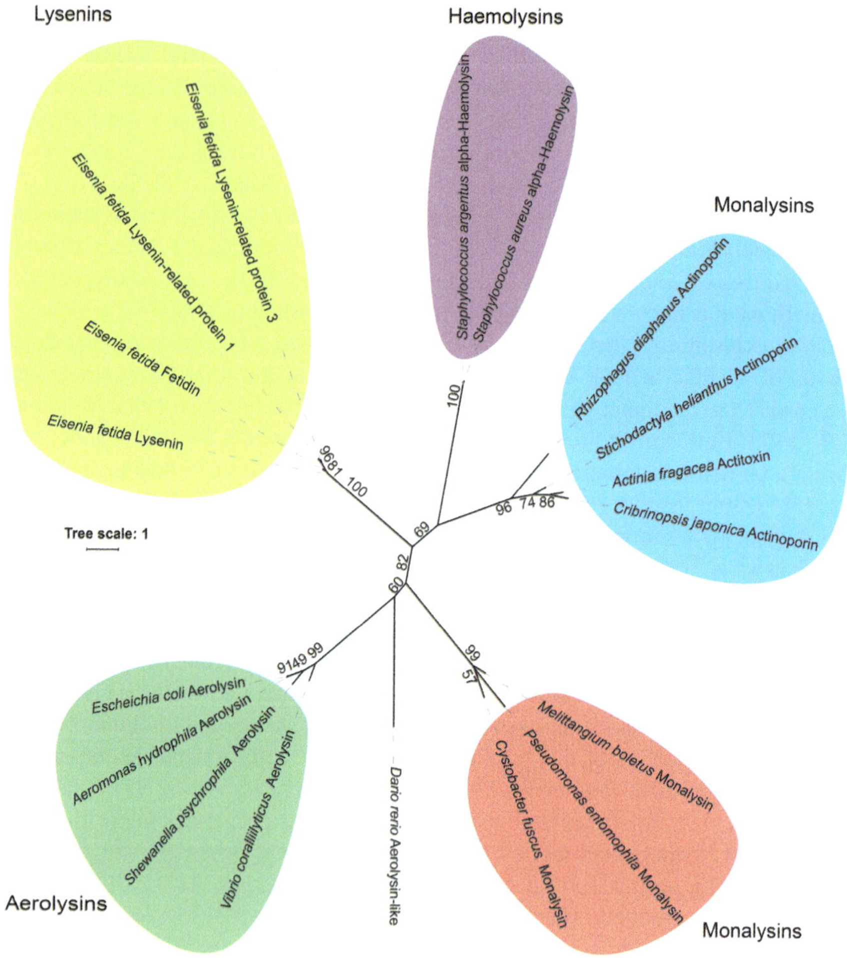

Fig. 1.5 Unrooted maximum likelihood phylogenetic tree of lysenins and representative pore forming proteins. The phylogeny reveals five distinct groups of pore forming proteins. Lysenins form a distinct group next to cnidarian actinoporin/toxins and *Staphlococcus* haemolysins. The alignment was generated using the MUSCLE algorithm (Edgar 2004) implemented in the CLC Main Workbench (v8.1) software suite, the phylogeny was calculated using IQ-tree (v1.63, Minh et al. 2013; Nguyen et al. 2015), ultra fast bootstrapping (Hoang et al. 2018), and the integrated ModelFinder (Kalyaanamoorthy et al. 2017). Accession numbers are: *Eisenia fetida* lysenin BAA21518.1; *Eisenia fetida* fetidin sp|O18425.1 *Eisenia fetida* lysenin-related protein 1 sp|O18424.1, *Eisenia fetida* lysenin related protein 3 sp|Q3LX99.1, *Danio rerio*_aerolysin-like NP_001013322.1, *Staphylococcus argentus* alpha-haemolysin AXH79902.1, *Staphylococcus aureus* alpha-haemolysin SBA32084, *Rhizophagus diaphanus* actinoporin RGB37297.1 *Stichodactyla helianthus* actinoporin AWM11685.1, *Actinia fragacea* actitoxin pdb|5GWF|A, *Cribrinopsis japonica* actinoporin BBC77265.1, *Vibrio coralliilyticus* aerolysin WP_095665830, *Shewanella psychrophila* aerolysin WP_077752031, *Aeromonas hydrophila* aerolysin AEC12846.1, *Escherichia coli* aerolysin MHO04454, *Pseudomonas entomophila* monalysin sp|Q1I8U1.1, *Cystobacter fuscus* monalysin WP_051256506, *Melittangium boletus* monalysin WP_095978255. The tree scale gives the amino acid substitutions per position

Upon formation of the pre-pore complex, the membrane insertion domain unfolds to form the inserted β-hairpin. This unfolding involves a complete re-arrangement of five β-strands and a single 3_{10} α-helix; interestingly, two β-strands previously considered to be flanking the membrane insertion domain, were found to contribute to the inserted β-hairpin. Specifically, the strands β4–β6 and the 3_{10} α-helix form the "core insertion domain" (representing 25% of lysenin's amino acid residues), while strands β3 and β7 represent the flanking strands, leading to insertion of the 73 amino acid residue (V34-I107) β-hairpin between strands β3 and β7 into the membrane (Bokori-Brown et al. 2016; Podobnik et al. 2016). The rest of the lysenin monomer, however, does not undergo major structural changes during pore formation; only the angle between the (N-terminal) β-strand 11 and the C-terminal domain change, tilting the strands β2, β8 and β10 as whole 45° against the C-terminal domain (Podobnik et al. 2016).

X-ray crystallography further confirmed the experimentally observed preference of lysenin for clustered SM over non-clustered SM (Yamaji et al. 1998): The surface of the bottom part of the cap formed by the pore contains cavities harboring positively and negatively charged residues. Molecular modeling has now revealed that the POC heads of two SM molecules fit into one cavity of the membrane interacting part with a third one being adjacent to cap of the pore (Podobnik et al. 2016).

The strict requirement of lysenin for SM is puzzling since sphingolipids are not generally found in bacterial membranes, but seem to be mainly restricted to anaerobe bacteria and the occurrence of SM is very rare (Olsen and Jantzen 2001). Lysenin/fetidin are potent toxins, thus the requirement for SM to lyse target cells may be explained by proving an easy way to prevent autotoxicity of these proteins since SM is virtually absent from the cell membranes of *Eisenia* (McLaughlin 1971). However, SM is present in cell membranes of eukaryotic parasites which might infest the animal (Lange et al. 1997, and cited references see also below—Eiseniapore). Further, it has been speculated that lysenin is secreted through dorsal pores to the outside of the animal to act as protection against predators, or to be secreted along with spermatozoa to prevent them from being fed on (Shakor et al. 2003). Oligomerization of lysenin to form pore-like structures was observed on erythrocytes but not on bacterial membranes (Bruhn et al. 2006). However, bacterial membranes were permeabilized by lysenin, but at a much slower rate. The authors concluded that lysenin displays SM-dependent and -independent activities. In addition, the interaction of lysenin with SM is affected by the presence of other lipids in the plasma membrane (Ishitsuka and Kobayashi 2007). Under in vitro conditions, cholesterol was found to promote the oligomerization of lysenin, while a high SM/lysenin ratio led to a decrease in oligomerization.

Whereas numerous structural and biochemical data exist for lysenin, no detailed information towards its biology is available. This, however, is different for fetidin, for which a mode of action has been put forward: following infection, fetidins are released from the chloragogosomes together with an activating serine-proteinase which is itself de-activated by a 14 kDa serine-proteinase inhibitor present in the coelomic fluid (Roch et al. 1998). Activated fetidins then polymerize on target cells to form pores, leading to cell lysis and inactivation of the pathogen (Roch et al. 1998).

Due to the highly similar primary structure of fetidin and lysenin (89% amino acid identity and 94% similarity, Kiyokawa et al. 2004), a similar mechanism might be assumed for lysenin. However, studies with recombinant lysenin did not need proteolytic activation to form pores and show hemolytic properties (see e.g. Kiyokawa et al. 2004; Bokori-Brown et al. 2016; Podobnik et al. 2016), much in contrast to other aerolysin-type toxins, such as aerolysin from *Aeromonas hydrophila* or monalysin from *Pseudomonas entomophila* that require proteolytic cleavage of a C- or N-terminal pro-peptide, respectively (*see* Podobnik et al. 2016). Interestingly, recombinant fetidins also showed hemolytic activity without proteinase activation (Lassegues et al. 1997).

As both, lysenin and fetidin show antibacterial activity and since both are specific for SM-containing plasma membranes which is absent in bacterial membranes (Olsen and Jantzen 2001) a SM-dependent and a SM-independent cytotoxic mechanism for lysenin (and feditin) action has been proposed (Bruhn et al. 2006, *see* above). While the SM-dependent mechanism clearly involves binding to the cell surface, oligomerization, pore formation and cell leakage, the SM-independent mechanism might be attributed to the peroxidase activity of lysenin/fetidin (Cooper 2002). However, since a proteolytic activation cascade was proposed for in vivo activation of fetidin, it could also be possible that this proteolytic activation is needed to reduce SM specificity and/or to enable oligomerization (and pore-formation) on bacterial cell membranes, which was not observed with un-processed lysenin (Bruhn et al. 2006).

Additional to its hemolytic and cytotoxic properties, lysenin has been shown to bind to (silver) nanoparticles in a specific manner forming so-called "protein coronas" in vitro. The thus coated particles are eventually sequestered by amoebocytes and even by eleocytes, normally not thought of being phagocytotic (Hayashi et al. 2016). Thus, lysenin/fetidins seem to be multifunctional proteins in *Eisenia* host defense.

Eiseniapore

Another cytolytic protein found in the CF of *Eisenia* was termed eiseniapore (Lange et al. 1999) for which no sequence is available to-date so it cannot be related to other pore forming proteins. Eiseniapore is 38 kDa protein which upon reduction of its thiol-groups forms ring-like structures in the target cell membrane, thereby lysing them (Lange et al. 1999). These ring-like structures consist of hexamers and form pores with outer and inner diameters of 10 nm and 3 nm, respectively, as revealed by electron microscopy (Lange et al. 1999). Contrary to the afore mentioned fetidins, metal ions seem to impair pore formation of eiseniapore, depending on the oxidation state; metals with higher oxidation states being stronger inhibitors than metals with lower oxidation states. Similar to lysenin, binding of eiseniapore to cell membranes was shown to be dependent on the presence of sphingomyelin. However, eiseniapore is not solely dependent on sphingomyelin (as are lysenins) as it also recognizes and binds to galactosylceramide (Lange et al. 1997). Sphingomyelin is virtually absent from the Earthworm's cell membranes, thus preventing autotoxicity, as well as from bacterial cell membranes, but seems to be typically present for Earthworm parasite like ciliates, cestodes and nematodes (Lange et al. 1997). This hints at eiseniapore being specific for Eukaryotic parasites while lysenins seem to be responsible for fighting Eukaryotic and bacterial infections.

Coelomocyte Cytolytic Factor 1

The coelomocyte cytolytic factor 1 (CCF-1) was first isolated from the CF of *Eisenia foetida* (Bilej et al. 1995) after cytotoxic properties of *Eisenia* CF had first been described by Kauschke and Mohrig (1987) by assaying the effect of *Eisenia* CF on chicken fibroblasts, insect hemocytes and guinea pig PMN-leukocytes. Purified CCF-1 has a molecular mass of 42 kDa (Bilej et al. 1995), similar to that of fetidins and lysenins. However, contrary to fetidins and lysenins, which seem to be unique for *Eisenia*, CCF-1 has been found in several species of earthworms such as *Aporrectodea caliginosa*, *Aporrectodea icterica*, *Aporrectodea longa*, *Aporrectodea rosea*, *Dendrobaena veneta*, *Lumbricus rubellus* and *Lumbricus terrestris* (Silerova et al. 2006).

Molecular characterization showed that CCF-1 contains a glucanase domain (Pfam PF00722), with specificity for β-1,3-glucans (Beschin et al. 1998; Silerova et al. 2006) and, for *Eisenia*, also a N,N′-diacetylchitobiose-binding capacity (Silerova et al. 2006). These molecular features as well as its cytotoxic activity against TNF-sensitive tumor cell lines have led to the recognition of CCF-1 as a functional TNF analog in Lumbricidae (Olivares Fontt et al. 2002). *Eisenia* CCF-1 binds β-1,3-glucans and lipopolysaccharides (LPS) with high affinity and when bound to either glucans or LPS is a potent activator of the pro-phenoloxidase cascade (Beschin et al. 1998; Olivares Fontt et al. 2002). Thus, CCF-1 serves as pattern recognition protein in Lumbricidid earthworms.

Phenoloxidases

Phenoloxidases are type-3 copper enzymes central to the immune system of many invertebrates. They are present in the body fluids as inactive pro-enzymes requiring activation through proteolytic cascades (Soderhall and Cerenius 1998). The term Phenoloxidase subsumes two distinct enzymes: Tyrosinase (E.C. 1.14.18.1) and catecholoxidase (E.C. 1.10.3.1). Tyrosinases are mono-phenoloxidases catalyzing the *ortho*-hydroxylation of mono-phenols to *ortho*-diphenols and the subsequent oxidation to *ortho*-chinons, whereas catecholoxidases are di-phenoloxidases catalyzing only the oxidation of *ortho*-diphenols to *ortho*-chinons (Sánchez-Ferrer et al. 1995). Even though phenoloxidases are involved in melanization processes in bacteria, plants and animals (Sánchez-Ferrer et al. 1995; Claus and Decker 2006; Marusek et al. 2006), their presence in annelids remained uncertain as no phenoloxidase activity could be detected in the CF of the polychaetes *Arenicola marina* and *Aphrodite aculeata* (Smith and Söderhäll 1991). However, phenoloxidase activity was detected in oligochaetes (Valembois et al. 1991), and later, Porchet-Henneré and Vernet (1992) demonstrated melanin production and phenoloxidase activity in de-granulating G_2 granulocytes of *Nereis diversicolor* encapsulating foreign material (injected latex beads), thus challenging previous findings (Smith and Söderhäll 1991). However, no biochemical characterization such as purification, kinetics, or structural analysis has been undertaken so that detailed information about these enzymes in annelids are unfortunately still lacking.

Lysozyme

Lysozymes are widely spread proteins occurring in bacteria to plants and animals. They are small proteins displaying N-acteylmuramidase activity, i.e. cleaving the β-1,4-glycosidic bond between N-acetylmuraminic acid and N-acetylglucosamine comprising the vast majority of Gram-positive bacteria cell wall polymers (McKenzie and White 1991). Lysozymes have been identified in the coelomic fluid of a wide range of annelids, including both oligochaetes and polychaetes (Périn and Jollès 1972; Dales and Dixon 1980; Çotuk and Dales 1984; Lassalle et al. 1988; Hirigoyenberry et al. 1990; Marcano et al. 1997; Ito et al. 1999; Josková et al. 2009; Fiołka et al. 2012). All annelid lysozymes characterized at the molecular level belong to the invertebrate-type of lysozymes (Marcano et al. 1997; Ito et al. 1999; Josková et al. 2009; Fiołka et al. 2012). Biochemically, annelid lysozymes have been characterized as ~13–20 kDa proteins with a pH optimum between pH 6.0 and 6.8 (Périn and Jollès 1972; Lassalle et al. 1988; Josková et al. 2009; Fiołka et al. 2012).

Even though lysozymes show activity only against Gram-positive bacteria, their expression was also induced by Gram-negative bacteria in *Eisenia andrei*, which seems to be an unspecific induction (Lassalle et al. 1988; Hirigoyenberry et al. 1990; Josková et al. 2009). For *Eisenia andrei* it has been shown that lysozyme activity was highest around 4 h after infection, lasting around 1 h, due to lysozyme de novo synthesis (Hirigoyenberry et al. 1990). It was further demonstrated that upon bacterial infection lysenin/fetidin transcription was strongly enhanced but that in contrast to lysozyme, translation of the lysenin/fetidin mRNA took roughly 3d. However, lysenin/fetidin activity remained high for five more days until declining to normal levels over another five days (Hirigoyenberry et al. 1990). Thus, lysozyme and lysenin/fetidin seem to act in concert to fight bacterial infection with lysozyme being the quick response and lysenin/fetidin a long term anti-bacterial agent, fitting well with its observed slow action against bacterial cell membranes (*see* above).

Proteinases

In invertebrates, proteinases play a vital role in triggering immune responses in such way that they convert pro-phenoloxidases into active phenoloxidases (Schmidt et al. 2010), or in chelicerates activate the oxygen carrier protein hemocyanin into an active phenoloxidase (Decker and Rimke 1998; Nagai and Kawabata 2000).

In the earthworms *Eisenia fetida* and *Lumbricus terrestris,* coelomic fluid proteinases have been described; and for *Eisenia fetida* it was shown that the proteolytic activity is chymotrypsin-like, as it could be inhibited by the serine proteinase inhibitor phenylmethylsulfonyl fluoride (PMSF), but not by the trypsin-specific inhibitor tosyl-L-lysyl-chloromethane hydrochloride (TLCK, Tučková et al. 1986). Later, detailed characterizations of CF proteinases in *Lumbricus terrestris* (Kauschke et al. 1997), *Apporectodea caliginosa* and *Lumbricus rubellus* (Kauschke et al. 2007), showed that a vast array of proteinases is present in the CF of the animals. In *L. terrestris* CF proteolytic activity is highest at a neutral to slightly basic pH and inducible by immunostimulation, reaching maximum activity after 24 h. In naïve animals, four electrophoretically distinct proteinase activities were detected by gelatin-zymography and in animals stimulated by erythrocyte injection this number increased up to 18 times (Kauschke et al. 1997, 2007). When CF from stimulated animals was subjected to gel filtration and anion exchange chromatography, eight electrophoretically distinct proteinases could be shown in native zymography (Kauschke et al. 1997) representing most likely 13 distinct proteinases. Enzymatic analysis using different chromogenic substrates showed that these proteinases are chymotrypsin-like, although weak trypsin-like activity was detected as well (Kauschke et al. 1997). As in *L. terrestris,* proteolytic activity in the CF of *Apporectodea caliginosa* and *Lumbricus rubellus* could be stimulated by injection of erythrocytes and maximum activity was observed 24 h after stimulation of the animals. Studies using proteinase inhibitors demonstrated that these proteinases are serine-proteinase like (Kauschke et al. 2007). Proteolytic activity in *L. rubellus* had a pH optimum from neutral to basic, similar to what was observed in *L. terrestris* (Kauschke et al. 1997); in *A. caliginosa,* the proteolytic pH optimum ranged from acidic to basic (Kauschke et al. 2007). In *L. rubellus*, naïve animals showed 14 proteolytic active bands in zymography which increased to 17 upon stimulation, and in *A. caliginosa* naïve and stimulated animals showed seven and 15 active bands, respectively (Kauschke et al. 2007).

In the polychaete *Nereis virens*, a fibrinolytic serine-proteinase has been detected in the CF (Zhang et al. 2007). This enzyme had a molecular mass of 29 kDa and a pI of ~4.5 and is active between pH 5.0 and 9.0, with a maximum at pH 7.8. Its activity was strongest against the fibrin α-fiber, and lower against the β- and γ-fibers (Zhang et al. 2007).

Mostly, these proteinases are synthesized and stored in coelomic cells and released into the CF upon immune challenge to set in motion defensive reactions (Roch et al. 1998, *see* e.g. above, sections on Defense proteins and Phenoloxidases). In *Eisenia*, a role for activating lysenin/fetidin has been ascribed to a serine-proteinase which is secreted into the coelomic fluid along with the inactive lysenin/fetidin (see above section lysenin/fetidin, Roch et al. 1998). Also in *Eisenia* (*andrei*), the serine proteinase elastase has been shown to participate in extracellular trap formation (Homa et al. 2016). Extracellular traps are aggregates consisting of DNA, histones and heat shock proteins that are formed by coelomocytes (both amoebocytes and eleocytes). These aggregates entangle pathogens, serve as nuclei for melanin encapsulation and mark the trapped pathogens for phagocytosis (Homa et al. 2016).

Antimicrobial Peptides

Antimicrobial peptides are short to medium sized peptides displaying lytic activity against Gram-positive and Gram-negative bacteria, fungi and other parasites. Originally discovered in the innate immune defense of invertebrates they were later also found in mammals and plants. Antimicrobial peptides are classified in four groups according to their secondary structure, namely: α-helical peptides (group I), peptides over-expressing one or more amino acid (group II), peptides with loop structures (group III), and β-sheet peptides (group IV). All peptides in any of these groups display three core features: (i) molecular mass between 1 and 5 kDa, (ii) overall positively charged, or containing a positively charged stretch of 10–25 amino acids, and (iii) an amphipathic structure in lipid bilayers or non-polar solvents (Van't Hof et al. 2001).

Annelids produce a wide range of antimicrobial peptides (for a recent overview, *see* Bruno et al. 2019) which will be discussed in the following section. Lumbricins are antimicrobial peptides initially discovered in the earthworm *Lumbricus rubellus* (Cho et al. 1998) that later were revealed to be more broadly distributed as lumbricins were discovered in the earthworms *Pheretima tschiliensis* (Wang et al. 2003), *Pheretima guillelmi* (Li et al. 2011), *Eisenia andrei* (Bodó et al. 2019) and the Medicinal Leech *Hirudo medicinalis* (Schikorski et al. 2008). Mature lumbricins are extracellular 59–73 amino acid peptides, however, signal peptides have only been described in the lumbricins isolated from *L. rubellus* and *P. guillelmi* (Cho et al. 1998; Wang et al. 2003; Schikorski et al. 2008; Li et al. 2011; Bodó et al. 2019). A characteristic of lumbricins is their high proline content (12–15% of the amino acids, Cho et al. 1998; Wang et al. 2003; Schikorski et al. 2008; Li et al. 2011; Bodó et al. 2019), which promotes a random coil secondary structure rather than an ordered one as was shown for *L. rubellus* lumbricin (Cho et al. 1998). Lumbricins were demonstrated to have activity against Gram-positive and Gram-negative bacteria, as well as fungi (Cho et al. 1998; Wang et al. 2003; Schikorski et al. 2008; Li et al. 2011; Bodó et al. 2019). Hemolytic properties were generally not observed (Cho et al. 1998; Wang et al. 2003; Schikorski et al. 2008; Bodó et al. 2019), however, weak hemolytic activity was described in lumbricin from *Pheretima guillelmi* (Li et al. 2011). Using synthetic lumbricin, Cho et al. (1998) demonstrated that the N-terminal half of the peptide (Lumbricin 6–34) is more active than the full length peptide, suggesting that the N-terminus is responsible for antimicrobial activity. This notion might be supported by the finding that the N-terminal part of lumbricin shows a higher degree of sequence conservation than the C-terminal part (Wang et al. 2003; Schikorski et al. 2008; Li et al. 2011; Bodó et al. 2019). Whereas lumbricin seems to be constitutively expressed in *L. rubellus* (Cho et al. 1998), its expression could be induced by challenging the animals with bacteria in *P. tschiliensis*, *E. andrei* and *H. medicinalis* (Wang et al. 2003; Schikorski et al. 2008; Bodo et al. 2019). Interestingly, these are the lumbricins lacking a signal peptide (Wang et al. 2003; Schikorski et al. 2008; Bodó et al. 2019).

Bodó et al. (2019) also identified a lumbricin-related peptide in *Eisenia andrei*. This peptide showed the same biological activity as *Eisenia* lumbricin but was only 59 amino acids long (*Eisenia* lumbricin: 62). Molecular phyolgenetic analysis revealed that lumbricin-related peptide clustered with *Hirudo* lumbricin whereas *Eisenia* lumbricin clustered with *L. rubellus* and the two identified *Pheretima* lumbricins (Bodó et al. 2019). Whereas lumbricin in Earthworms is rather broadly expressed and seems to be present inside the animal as well as secreted along the mucus to the outside (Cho et al. 1998; Wang et al. 2003; Li et al. 2011; Bodó et al. 2019), *Hirudo* lumbricin appears to be restricted to the nervous system (Schikorski et al. 2008). For this lumbricin Schikorski and coworkers also demonstrated another function: they found that lumbricin is up-regulated during nerve cord regeneration and reduces the time needed to regenerate (Schikorski et al. 2008).

Hedistin is an antimicrobial peptide that was isolated from the CF of the Nereidid *Nereis* (*Hediste*) *diversicolor*. Hedistin is a bromotyrosine containing antibacterial peptide (molecular mass: 2,223 Da) which is constitutively expressed in natural killer like type 3 granulocytes and released upon immune challenge (Tasiemski et al. 2004). In aqueous solutions, hedistin exists as a random coil, however once attached to phospholipid bilayers, secondary structures are induced, leading to the formation of helix-turn-helix structures. This conformational change enables the peptide to insert itself into the target (bacterial) membrane generating a hole leading to cell lysis and death (Xu et al. 2009).

From whole body extracts of the clamworm *Perinereis aibuhitensis,* Pan et al. (2004) purified the antimicrobial peptide perinerin. Perinerin shows activity against Gram-positive and Gram-negative bacteria, as well as against the fungus *Paecilomyces heliothis*. Perinerein is a 51 amino acid peptide with a molecular mass of 5,975 Da stabilized by two intramolecular disulfide bridges. It is constitutively expressed in the animal and could not be induced by bacterial injections. Cleavage of perinerin with *Staphylococcus aureus* proteinase V8 resulted in two main peptides with amino acid length of 26 and 22 amino acids, both of which lack any biological activity.

In the lugworm *Arenicola marina*, two antimicrobial peptides, arenicin-1 and arenicin-2 with masses of 2,758 Da and 2,772 Da, respectively, are present. Both peptides differ only by a VI substitution at position 10 and have been isolated from coelomic cells; they are active against Gram-positive and Gram-negative bacteria, as well as against *Candida albicans*. Interestingly, these 21 amino acid peptides form an 18 amino acid ring structure formed by a disulfide bridge between their only two cysteine-residues (C3–C20, Ovchinnikova et al. 2007). Solution structure analysis of arenicin-2 by NMR spectroscopy revealed that arenicin-2 consists of two β-strands

(C3-I10 and V13-C20) which are right-handed twisted (Ovchinnikova et al. 2007). Structural studies on arenicin-1 showed a similar structure to arenicin-2 which could be expected due to the nearly identical primary structure of the two peptides. However, arenicin-1 seems to be somewhat more compact when compared to arenicin-2 (Andrä et al. 2008). These authors also describe a structural re-arrangement, i.e. an "untwisting" of the β-sheets forming the 18 amino acid loop, forming an amphipatic structure of arenicin-1 upon insertion into lipid vesicles. This was not observed experimentally by Ovchinnikova et al. (2007) using arenicin-2, however, this scenario was proposed by these authors based on the amino acid sequence and the "twisted" β-sheet structures of arenicin-2. Later studies showed that arenicins are broadly expressed in the animal, with expression in the body wall, the central nerve cord, gut and coelomocytes (Maltseva et al. 2014). Interestingly, arenicins could not be detected in the coelomic fluid; expression in granulocytes was found to be colocalized with phagocytozed particles, thus arenicins may be primary involved in killing phagocytozed pathogens, rather than killing them in the coelomic cavity (Maltseva et al. 2014).

In the polychaete *Marphysa sanguinea*, a 1,737 Da antimicrobial peptide has been purified which showed antibacterial activity against Gram-positive and Gram-negative bacteria. This 14 amino acid peptide is N-terminally acetylated (Ac-SVEIPKPFKWNDSF) and secondary structure prediction showed an extended β-strand and unordered structures with a net charge of zero, clearly setting it apart from the generally positively charged antibacterial peptides. Interestingly, this peptide is derived from the N-terminus of myo-hemerythrin of *Marphysa sanguinea,* adding another function to the oxygen carrier, at least in this species (Seo et al. 2016).

Contrary to the polychaete antimicrobial peptides, the antimicrobial peptides from the leech *Theromyzon tessulatum*, termed theromyzin and theromazin, are relatively large molecules with molecular masses of 9,973 Da and 8,518 Da for theromyzin and theromazin, respectively. Both peptides have been shown to be induced after stimulation of the animal's immune system. Theromyzin and theromazin are both exclusively synthesized in so-called Large Fat Cells and released into the CF. Once secreted, the peptides may exert their action directly in the CF, or are taken up into gut skin epithelia to challenge bacterial infections (Tasiemski et al. 2004).

Metal Binding Proteins

Hemerythrins

Hemerythrins (HR) are a group of relatively small non-heme iron-binding proteins exclusive to invertebrates, which can be divided into four major groups. The circulating cHR, myo-HR, neuro-HR and ovo-HR. HR's contain two Fe^{2+} complexed to four histidine residues as well as the carboxylate groups from one glutamic acid and one aspartic acid at their active site and, in contrast to hemoglobins (and other heme containing oxygen binding/utilizing proteins), HR's bind oxygen by oxidation of the iron in their active site, resulting in binding the oxygen as a peroxide ion. HR's are typically ~120 amino acid proteins arranged in a four-helix bundle and form homooctamers, although different stoichiometries have been observed (Klotz and Kurtz 1984; Costa-Paiva et al. 2017). More details on the structure, function and evolution of these proteins can be found in the literature dedicated to HR's (e.g. Klotz and Kurtz 1984; Costa-Paiva et al. 2017; Coates and Decker 2017, and Coates and Costa-Paiva in this volume). In the following section we will only briefly review what is known about HR's and related proteins in annelids.

In the polychaete *Nereis diversicolor*, Nejmeddine et al. (1988) and Dhainaut et al. (1989) found a cadmium binding protein with antibacterial activity. This protein, called Metalloprotein II (MP II), was found in type-1 granulocytes using a monoclonal antibody (Porchet-Henneré et al. 1987). After establishing its amino acid sequence (118 amino acids; UniProtKB/Swiss-Prot: P80255.2), an 80% identity with *Nereis diversicolor* myo-HR was revealed (Demuynck et al. 1991, 1993). Expression of MP II was found in muscle and coelomic cells (Salzet-Raveillon et al. 1993). Later, the protein was also found to be expressed in intestinal cells where MP II levels increased after cadmium exposure. Further experiments showed that the antibacterial activity of MP II was due to its scavenging of iron, which is essential for microbial growth (Deloffre et al. 2003). Binding of either cadmium or iron as the natural ligand therefore confers both metal detoxification and antibacterial defense to this hemerythrin-related protein. A protein of the same molecular mass (14 kDa) was isolated from the coelomic fluid of *Allolobophora caliginosa* (Nejmeddine et al. 1992, 1997). As a difference to the MP II in *Nereis diversicolor*, however, the protein isolated from *Allolobophora* could sequester both iron and cadmium, while in cadmium-exposed *Nereis*, iron was absent in the MP II (Demuynck et al. 1993). Due to its low cysteine content and the high content of aromatic amino acids, the authors concluded that the MP II proteins are distinct from the metallothionins which usually have elevated cysteine content (*see* below).

Whereas HR's generally seem to be intracellular proteins—either packed in hemerythrocytes (cHR), muscle tissue (myo-HR), neuronal tissues (neuro-HR), and eggs (ovo-HR)—ovo-HR is also found extracellular. Ovo-HR, initially isolated as a major constituent of the yolk (~30%) in the leech *Theromyzon tessulatum*, was also present in high concentrations in the CF of *T. tessulatum* (Baert et al. 1992) and seems to be the only extracellular HR characterized up to now. N-terminal sequencing of ovo-HR revealed a higher sequence similarity to Sipunculid myo-HR than to Sipunculid cHR—75% to 50%, respectively—unfortunately no complete sequence information is available and thus detailed phylogeny is lacking. Like vitellogenin, ovo-HR concentrations rise in the CF during the later stages of oogenesis, and drop just before egg-laying (Baert et al. 1992). However, contrary to vitellogenin which appeared in the CF after the 3rd (last) blood meal of the animal, ovo-HR was present in the CF after the 1st and 2nd blood meals. This led to the hypothesis that ovo-HR might be involved in sequestering ingested iron and thus assume a protective (metallothionein-like) role (Baert et al. 1992), similar to MPII from *Nereis diversicolor* (Nejmeddine et al. 1988; Dhainaut et al. 1989), or the 14 kDa HR-like protein found in *A. caliginosa* (Nejmeddine et al. 1992, 1997). However, there may also be additional roles, since the purified protein had a pink color (Baert et al. 1992) which is characteristic for oxy-HR (Klotz and Kurtz 1984). Thus, a role in oxygen transport may be present as well as iron and/or oxygen storage for the developing embryo (Baert et al. 1992).

Metallothioneins

Metallothioneins (MT's) are a heterogeneous group of cationic proteins of low molecular mass (25–150 amino acids; up to 12 kDa) characterized by a high cysteine content (up to 25%), enabling them to coordinate large amounts of metal ions, for example, Zn^{2+}, Cd^{2+}, and Cu^{+}, in the form of metal-thiolate clusters, and in the case of vertebrate MT's virtually devoid of aromatic amino acids and histidine (Fowler et al. 1987).

After the first discovery of a metallothionein in equine kidney (Margoshes and Vallee 1957), MT's have been found in many phyla including some prokaryotes. MT's from different taxonomic groups differ in their sequence length, the number of cysteine residues and the relative positions of the Cys residues in the polypeptide chain (see https://www.uniprot.org/docs/metallo.txt), but within some taxonomic groups, MT sequences are homologous (Stürzenbaum et al. 2001). Invertebrate MT's differ from those of vertebrates by their higher contents in glycine and aromatic amino acids (Nejmeddine et al. 1992, and cited references).

In *Lumbricus rubellus* two MT isoforms, MT1 and MT2 with 79 and 77 amino acids are expressed. Both proteins show 74.7% sequence identity and 91.1% similarity, and contain 20 cysteine-residues which are conserved between the two isoforms. The major difference between the two isoforms is a two amino acid insertion/deletion after amino acid position 48 in MT1 and MT2, respectively (Stürzenbaum et al. 2001). The presence of two MT isoforms has also been reported for other invertebrates and seems to be a common feature (Roesijadi and Fowler 1991; Stürzenbaum et al. 2001). The biological relevance of these two MT isoforms is not clear. However, it has been postulated by Stürzenbaum et al. (2001) that at least in *L. rubellus* the two isoforms could have distinct roles. Whereas MT2 sequesters harmful heavy metals like cadmium, MT1, having a less stable metal-thiolate cluster, acts as a donor for essential bivalent cations.

Gruber et al. (2000) identified and characterized an unusual cadmium inducible MT from *Eisenia foetida*. Transcription of the MT gene leads to an 80 amino acid peptide, which is cleaved near the N-terminus and in the linker domain between the two metal-binding domains, resulting in a mature MT of 41 amino acids containing twelve cysteine residues that bind four cadmium ions (Gruber et al. 2000).

Calisi et al. (2014) demonstrated in *Lumbricus terrestris* that MT concentration is significantly elevated in CF when compared to the whole animal. They further showed that MT levels increase more strongly in the CF compared to the whole animal upon exposure of the animals to Cu^{2+}, Cd^{2+} and Cu^{2+}-Cd^{2+} mixtures (Calisi et al. 2014). As one of the sources for coelomic fluid MT, Stürzenbaum et al. (2004) proposed the synthesis of MT by the cells of the chloragog tissue and the formation of Cd-MT after Cd uptake from the alimentary tract. Cd-MT is then released into the CF and subsequently sequestered by the free coelomocytes; the fate of the Cd-MT is not yet clear.

From the polychaete *Perinereis nuntia* a 68 amino acid MT with 17 cysteine-residues has been cloned and characterized (Won et al. 2011). The MT gene was characterized as a three-exon gene with a total length of 2,124 bp harboring three metal-response elements in the 5′-UTR. Exposing the animals to different divalent cations (Zn^{2+}, Cu^{2+}, Cd^{2+}, As^{2+} and Pb^{2+}) showed that MT expression was enhanced differently by the different metals tested. For instance, while incubation with Cu^{2+} led to a three-fold increase in MT mRNA after 6 h, which returned to normal values after 48 h, incubation with Cd^{2+} induced a gradual increase of the MT transcript from four-fold after 6 h to nine-fold after 48 h of exposure (Won et al. 2011). Challenging animals with different metals also induced expression of MT. Recombinant MT was found to bind the tested metals with different affinities: MT bound Cd^{2+} with the highest affinity. Pb^{2+} was bound with the lowest. Zn^{2+} was bound by recombinant MT with an affinity only slightly lower than that of Cd^{2+}; interestingly, however, Zn^{2+} did not induce MT expression in vivo (Won et al. 2011).

Since the first sequence for MT's in Annelids was published for *Lumbricus rubellus* by Stürzenbaum in 1998 (Stürzenbaum et al. 1998, GenBank CAA06720.1), MT sequences of about 20 annelid species—mostly earthworms—have been added to the NCBI database, due to their role as bioindicators for soil biomonitoring.

Ferritin

Ferritins are iron transport proteins complexing ferric iron (Fe^{3+}), fulfilling a dual role in (i) keeping potential toxic effects of iron under control and (ii) transporting iron through the body (Nichol et al. 2002). The holoparticle has a molecular mass of 400–500 kDa and consists of 24 protein subunits and up to 4,500 complexed iron molecules (Arosio et al. 1984; Nichol et al. 2002; Prochazkova et al. 2011). In annelids, ferritins have been described from the CF as well as intracelluar (Arosio et al. 1984; Schüßler et al. 1995; Harris et al. 2001; Prochazkova et al. 2011). On the sequence level, annelid ferritins show more similarity with the heart-type (H) ferritins than with the liver-type (L) ferritins known from vertebrates, as do the ferritins found in mollusks and crustacea (Jeong et al. 2006; Prochazkova et al. 2011).

The first ferritin purified from annelids was from whole body homogenates of the earthworm *Octolasium complanatum* (Arosio et al. 1984). This ferritin is composed of two subunits of 19.7 and 20.5 kDa present in roughly equal amounts, forming the ~460 kDa ferritin holoparticles. However, isoelectric focussing revealed that the apoprotein stoichiometry in the native ferritin particle seems to vary, thus leading to different isoforms (Arosio et al. 1984).

Ferritin from *Eisenia andrei* has been characterized at the molecular level, after first being discovered in the CF by a MALDI-MS screen (Prochazkova et al. 2011). *Eisenia* ferritin is encoded by a 1.1 kbp transcript with an open reading frame coding for a 172 amino acid protein without signal peptide, and a conserved iron response element (IRE) in the 5′-UTR. The apoferritin has a calculated molecular mass of 19.7 kDa with a pI of 5.2. Typical for annelid ferritins, *Eisenia* apoferritin is more similar to H subunits than to L subunits. The native ferritin particle was found to have a molecular mass of ~400 kDa as demonstrated by subjecting CF samples to native PAGE and staining for iron. Electroelution of the iron-positive band followed by SDS-PAGE showed that *Eisenia* ferritin is composed of 20 kDa subunits as predicted from the cDNA sequence. Transcript level analysis by quantitative real-time PCR (qPCR) showed that stressing the animals with 10 mM $FeCl_3$ increased transcript abundance three-times over that of non-stimulated control animals. However, challenging the animal's immune system by the injection of either *E. coli* or *B. subtilis* led to an eight-fold transcript increase for *B.subtilis* challenged animals and a 96-fold increase for animals challenged with *E. coli* after 4 h. Eight hours after challenge the transcripts levels were 255-fold and 327-fold increased for *E. coli* and *B. subtilis*, respectively, and remained very high for another eight hours until declining after 24 h (Prochazkova et al. 2011). This implies, that ferritin might also function in the animal's immune response (Prochazkova et al. 2011), probably due to its complexing of iron essential for microbial growth, and thus might be functioning as an antibacterial protein in addition to the role in iron transport and storage.

Jeong et al. (2006) characterized a ferritin from the polychaete *Periserrula leucophryna*. Like the *Eisenia* ferritin (Prochazkova et al. 2011), this ferritin is coded by an ~1.1 kbp transcript. The coding region of *Periserrula* ferritin has 525 bp and encodes a 174 amino acid protein containing a putative 17 amino acid signal peptide and an IRE in the 5′-UTR (Jeong et al. 2006). The deduced molecular masses of the apoferritins are 20.2 and 18.3 kDa for the immature and mature apoferritin, respectively. Semi-quantitative Northern blot analysis showed that *Periserrula* ferrtin is expressed in equal levels in all tissues examined (body, head, gut and skin, Jeong et al. 2006).

In the polychaete *Platynereis dumerilii*, ferritins have been implied to have an additional role in immune defense (Altincicek and Vilcinskas 2007) similar to that observed in *Eisenia* (Prochazkova et al. 2011), but detailed studies are lacking.

Hemocyanin

Recently, Costa-Paiva et al. (2018) described the presence of hemocyanin genes in coelomic cells of oligochaetes as well as polychaetes. Interestingly, both types of hemocyanins, arthropod (multimers of hexamers of ~72 kDa subunits) and mollusk (multimers of ~350–400 kDa subunits containing seven to eight functional units, Decker et al. 2007), were found to be expressed in annelids. Whether these hemocyanins are indeed involved in oxygen transport or serve other functions remains to be investigated. However, the presence of hemocyanins in polychaetes makes them the taxonomic group with the greatest diversity of respiratory pigments described so far (Costa-Paiva et al. 2018).

Conclusions

The coelomic fluid space of annelids is in contact with different tissues and the proteins released into and taken up from this compartment serve many different functions. In spite of their large diversity, coelomic fluid proteins have been investigated in a few annelid groups only, the Nereidid polychaetes and the oligochaetes mainly from the lumbricid (earthworm) family.

The main body of research done over the past years has focused mainly on the proteins involved in reproduction and defense and less on those with a more general metabolic function such as the lipoproteins. Other proteins with defensive function, such as the lectins, which have been characterized in crustacean and insect hemolymph, have been largely neglected so far. Studies on other groups emerge only slowly such as the first identification of thymosins in annelids, very recently identified in the coelomic fluid of the echiurid worm *Urechis* (Zhang et al. 2020). Thymosins represent an interesting group of proteins involved in many different cellular functions including defense reactions.

Analysis of the coelomic fluid has also lead to recent discoveries related to medical applications such as the identification of analgesic and anti-inflammatory peptides (Li et al. 2017) and the antitumor activity and apoptotic effects as well as an antifungal (*Candida albicans*) activity of yet unknown coelomic fluid proteins, again found in earthworms (Fiołka et al. 2019a, b).

Since many annelid groups remain largely untouched with regards to their biochemistry, annelids still represent a vastly unexplored area of research.

References

Altincicek B, Vilcinskas A (2007) Analysis of the immune-related transcriptome of a lophotrochozoan model, the marine annelid *Platynereis dumerilii*. Front Zool 4:18. https://doi.org/10.1186/1742-9994-4-18

Andrä J, Jakovkin I, Grötzinger J, Hecht O, Krasnosdembskaya AD, Goldmann T, Gutsmann T, Leippe M (2008) Structure and mode of action of the antimicrobial peptide arenicin. Biochem J 410(1):113–122

Arosio P, Levi S, Gabri E, Stefanini S, Finazziagro A, Chiancone E (1984) Properties of ferritin from the earthworm *Octolasium complanatum*. Biochim Biophys Acta 787(3):264–269. https://doi.org/10.1016/0167-4838(84)90318-2

Baert JL (1985) Multiple forms of vitellin in young oocytes of *Perinereis cultrifera* (Polychaete Annelid): occurrence and relation to vitellin maturation in the oocyte. Comp Biochem Physiol 81B:851–856. https://doi.org/10.1016/0305-0491(85)90078-1

Baert JL (1986) Evidence for vitellin maturation within the oocytes of *Perinereis cultrifera* (Polychaete Annelid). Comp Biochem Physiol 83B:847–853. https://doi.org/10.1016/0305-0491(86)90159-8

Baert JL, Slomianny MC (1987) Heterosynthetic origin of the major yolk protein, vitellin, in a nereid, *Perinereis cultrifera* (Polychaete annelid). Comp Biochem Physiol 88B:1191–1199. https://doi.org/10.1016/0305-0491(87)90023-X

Baert JL, Slomianny MC (1992) Vitellin accumulation and vitellogenin synthesis in relation to oogenesis in *Perinereis cultrifera* (Polychaeta, Annelida). Invertebr Reprod Dev 21:121–128. https://doi.org/10.1080/07924259.1992.9672228

Baert JL, Britel M, Slomianny MC, Delbart C, Fournet B, Sautiere P, Malecha J (1991) Yolk protein in leech. Identification, purification and characterization of vitellin and vitellogenin. Eur J Biochem 201 (1):191–198. https://doi.org/10.1111/j.1432-1033.1991.tb16273.x

Baert JL, Britel M, Sautiere P, Malecha J (1992) Ovohemerythrin, a major 14-kDa yolk protein distinct from vitellogenin in leech. Eur J Biochem 209(2):563–569. https://doi.org/10.1111/j.1432-1033.1992.tb17321.x

Beschin A, Bilej M, Hanssens F, Raymakers J, Van Dyck E, Revets H, Brys L, Gomez J, De Baetselier P, Timmermans M (1998) Identification and cloning of a glucan- and lipopolysaccharide-binding protein from *Eisenia foetida* earthworm involved in the activation of prophenoloxidase cascade. J Biol Chem 273(38):24948–24954. https://doi.org/10.1074/jbc.273.38.24948

Bilej M, Brys L, Beschin A, Lucas R, Vercauteren E, Hanušová R, De Baetselier P (1995) Identification of a cytolytic protein in the coelomic fluid of *Eisenia foetida* earthworms. Immunol Let 45:123–128

Bodo K, Boros A, Rumpler E, Molnar L, Borocz K, Nemeth P, Engelmann P (2019) Identification of novel lumbricin homologues in *Eisenia andrei* earthworms. Dev Comp Immunol 90:41–46. https://doi.org/10.1016/j.dci.2018.09.001

Bokori-Brown M, Martin TG, Naylor CE, Basak AK, Titball RW, Savva CG (2016) Cryo-EM structure of lysenin pore elucidates membrane insertion by an aerolysin family protein. Nat Commun 7:11293. https://doi.org/10.1038/ncomms11293

Bonnier P, Baert J-L (1992) Vitellogenesis in the sand worm, *Nereis diversicolor*. Comp Biochem Physiol B 102(5):785–790. https://doi.org/10.1016/0305-0491(92)90080-B

Bruhn H, Winkelmann J, Andersen C, Andra J, Leippe M (2006) Dissection of the mechanisms of cytolytic and antibacterial activity of lysenin, a defence protein of the annelid Eisenia fetida. Dev Comp Immunol 30(7):597–606. https://doi.org/10.1016/j.dci.2005.09.002

Bruno R, Maresca M, Canaan S, Cavalier JF, Mabrouk K, Boidin-Wichlacz C, Olleik H, Zeppilli D, Brodin P, Massol F, Jollivet D, Jung S, Tasiemski A (2019) Worms' antimicrobial peptides. Mar Drugs 17(9). https://doi.org/10.3390/md17090512

Calisi A, Lionetto MG, De Lorenzis E, Leomanni A, Schettino T (2014) Metallothionein induction in the coelomic fluid of the earthworm *Lumbricus terrestris* following heavy metal exposure: a short report. Biomed Res Int 2014:109386. https://doi.org/10.1155/2014/109386

Chino H, Downer RG, Takahashi K (1977) The role of diacylglycerol-carrying lipoprotein I in lipid transport during insect vitellogenesis. Biochim Biophys Acta 487(3):508–516. https://doi.org/10.1016/0005-2760(77)90220-x

Cho JH, Park CB, Yoon YG, Kim SC (1998) Lumbricin I, a novel proline-rich antimicrobial peptide from the earthworm: purification, cDNA cloning and molecular characterization. Biochim Biophys Acta 1408(1):67–76. https://doi.org/10.1016/s0925-4439(98)00058-1

Claus H, Decker H (2006) Bacterial tyrosinases. Syst Appl Microbiol 29:3–14. https://doi.org/10.1016/j.syapm.2005.07.012

Coates CJ, Decker H (2017) Immunological properties of oxygen-transport proteins: hemoglobin, hemocyanin and hemerythrin. Cell Mol Life Sci 74(2):293–317. https://doi.org/10.1007/s00018-016-2326-7

Cooper EL (2002) Invertebrate immune responses. In: Advances in comparative and environmental physiology, vol 23. Springer, Berlin. https://doi.org/10.1007/978-3-642-79693-7

Cooper EL, Stein EA (1981) Oligochaetes. In: Ratcliffe NA, Rowley AF (eds) Invertebrate blood cells, vol 1. Academic Press, New York, pp 76–140

Cooper EL, Kauschke E, Cossarizza A (2001) Annelid humoral immunity: cell lysis in earthworms. In: Beck G, Sugumaran M, Cooper EL (eds) Phylogenetic perspectives on the vertebrate immune system. Adv. Exp. Med. Biol. Springer, Berlin,

Costa-Paiva EM, Schrago CG, Halanych KM (2017) Broad phylogenetic occurrence of the oxygen-binding hemerythrins in bilaterians. Genome Biol Evol 9(10):2580–2591. https://doi.org/10.1093/gbe/evx181

Costa-Paiva EM, Schrago CG, Coates CJ, Halanych KM (2018) Discovery of novel hemocyanin-like genes in metazoans. Biol Bull 235:134–151. https://doi.org/10.1086/700181

Çotuk A, Dales RP (1984) Lysozyme activity in the coelomic fluid and coelomocytes of the earthworm *Eisenia foetida* Sav. in relation to bacterial infection. Comp Biochem Physiol A 78:469–474. https://doi.org/10.1016/0300-9629(84)90580-2

Dales RP, Dixon LJR (1980) Responses of polychaete annelids to bacterial infection. Comp Biochem Physiol A 67:391–396. https://doi.org/10.1016/S0300-9629(80)80014-4

Dales RP, Dixon RJ (1981) Polychaetes. In: Ratcliffe NA, Rowley AF (eds) Invertebrate blood cells, vol 1. Academic Press, New York, pp 35–74

de Eguileor M, Grimaldi A, Tettamanti G, Valvassori R, Cooper EL, Lanzavecchia G (2000) Lipopolysaccharide-dependent induction of leech leukocytes that cross-react with vertebrate cellular differentiation markers. Tissue Cell 32(5):437–445

Decker H, Rimke T (1998) Tarantula hemocyanin shows phenoloxidase activity. J Biol Chem 273(40):25889–25892. https://doi.org/10.1074/jbc.273.40.25889

Deloffre L, Salzet B, Vieau D, Andries JC, Salzet M (2003) Antibacterial properties of hemerythrin of the sand worm Nereis diversicolor. Neuro Endocrinol Lett 24(1–2):39–45

Demuynck S, Sautiere P, van Beeumen J, Dhainaut-Courtois N (1991) Homologies between hemerythrins of sipunculids and cadmium-binding metalloprotein (MP II) from a polychaete annelid, *Nereis diversicolor*. C R Acad Sci III 312(7):317–322

Demuynck S, Li KW, Van der Schors R, Dhainaut-Courtois N (1993) Amino acid sequence of the small cadmium-binding protein (MP II) from Nereis diversicolor (annelida, polychaeta). Evidence for a myohemerythrin structure. Eur J Biochem 217 (1):151–156. https://doi.org/10.1111/j.1432-1033.1993.tb18230.x

Dhainaut A, Porchet-Henneré E (1988) Haemocytes and coelomocytes. In: Westheide W, Hermans CO (eds) Microfauna marina, vol 4. Gustav Fischer Verlag, Stuttgart, pp 216–230

Dhainaut A, Raveillon B, Mberi M, Porchet-Henneré E, Demuynck S (1989) Purification of an antibacterial protein in the coelomic fluid of *Nereis diversicolor* (Annelida, Polychaeta)—similitude with a cadmium-binding protein. Comp Biochem Phys C 94(2):555–560. https://doi.org/10.1016/0742-8413(89)90112-6

Eckelbarger KJ (2005) Oogenesis and oocytes. Hydrobiologia 535:179–198. https://doi.org/10.1007/s10750-004-4397-y

Edgar RC (2004) MUSCLE: multiple sequence alignment with high accuracy and high throughput. Nucleic Acids Res 32(5):1792–1797. https://doi.org/10.1093/nar/gkh340

Engelmann P, Palinkas L, Cooper EL, Nemeth P (2005) Monoclonal antibodies identify four distinct annelid leukocyte markers. Dev Comp Immunol 29(7):599–614. https://doi.org/10.1016/j.dci.2004.10.008

Fiołka MJ, Zagaja MP, Hułas-Stasiak M, Wielbo J (2012) Activity and immunodetection of lysozyme in earthworm *Dendrobaena veneta* (Annelida). J Invertebr Pathol 109:83–90. https://doi.org/10.1016/j.jip.2011.10.002

Fiołka MJ, Czaplewska P, Macur K, Buchwald T, Kutkowska J, Paduch R, Kaczynski Z, Wydrych J, Urbanik-Sypniewska T (2019a) Anti-Candida albicans effect of the protein-carbohydrate fraction obtained from the coelomic fluid of earthworm *Dendrobaena veneta*. PLoS ONE 14(3):e0212869. https://doi.org/10.1371/journal.pone.0212869

Fiołka MJ, Rzymowska J, Bilska S, Lewtak K, Dmoszynska-Graniczka M, Grzywnowicz K, Kazmierski W, Urbanik-Sypniewska T (2019b) Antitumor activity and apoptotic action of coelomic fluid from the earthworm *Dendrobaena veneta* against A549 human lung cancer cells. APMIS 127(6):435–448. https://doi.org/10.1111/apm.12941

Fischer A (1979) A vitellin-like antigen in the coelomic fluid of maturing *Nereis virens* females. Naturwissenschaften 66:316

Fischer A, Rabien H, Heacox AE (1991) Specific, concentration-dependent uptake of vitellin by the oocytes of *Nereis virens* (Annelida: Polychaeta) in vitro. J Exp Zool 260:106–115. https://doi.org/10.1002/jez.1402600114

Fischer A, Dorresteijn AWC, Hoeger U (1996) Metabolism of oocyte construction and the origin of histospecificity in the cleaving egg. Lessons from nereid annelids. Int J Develop 40:421–430

Fontaine F, Gevaert MH, Porchet M (1984a) Acylglycerol metabolism in the coelomic constituents during vitellogenesis of *Perinereis cultrifera* (Annelida Polychaeta). Comp Biochem Physiol B 78:581–584. https://doi.org/10.1016/0305-0491(84)90101-9

Fontaine F, Gevaert MH, Porchet M (1984b) Distribution of neutral lipids in coelomic constituents during oogenesis of *Perinereis cultrifera* (Annelida Polychaeta). Comp Biochem Physiol A 77:45–50. https://doi.org/10.1016/0300-9629(84)90009-4

Fowler BA, Hildebrand CE, Kojima Y, Webb M (1987) Nomenclature of metallothionein. Experientia Suppl 52:19–22. https://doi.org/10.1007/978-3-0348-6784-9_2

García-Alonso J, Rebscher N (2005) Estradiol signalling in *Nereis virens* reproduction. Invert Repr Develop 48(1–3):95–100. https://doi.org/10.1080/07924259.2005.9652175

García-Alonso J, Hoeger U, Rebscher N (2006) Regulation of vitellogenesis in *Nereis virens* (Annelida: Polychaeta): Effect of estradiol-17b on eleocytes. Comp Biochem Physiol A 143(1):55–61. https://doi.org/10.1016/j.cbpa.2005.10.022

Giangrande A, Licciano M, Pagliara P, Gambi MC (2000) Gametogenesis and larval development in *Sabella spallanzanii* (Polychaeta: Sabellidae) from the Mediterranean Sea. Mar Biol 136(5):847–861. https://doi.org/10.1007/s002279900251

Gruber C, Stürzenbaum SR, Gehring P, Sack R, Hunziker P, Berger B, Dallinger R (2000) Isolation and characterization of a self-sufficient one-domain protein (Cd)-Metallothionein from *Eisenia foetida*. Eur J Biochem 267:573–582. https://doi.org/10.1046/j.1432-1327.2000.01035.x

Hafer J, Fischer A, Ferenz HJ (1992) Identification of the yolk receptor protein in oocytes of *Nereis virens* (Annelida, Polychaeta) and comparison with the locust vitellogenin receptor. J Comp Physiol B 162:148–152. https://doi.org/10.1007/BF00398340

Harris JR, Hoeger U, Adrian M (2001) Transmission electron microscopical studies on some haemolymph proteins from the marine polychaete Nereis virens. Micron 32(6):599–613. https://doi.org/10.1016/S0968-4328(00)00051-2

Hauenschild C (1956) Hormonale Hemmung der Geschlechtsreife und Metamorphose bei dem Polychaeten *Platynereis dumerilii*. Zt Naturforsch B 11:125–132

Hayashi Y, Miclaus T, Engelmann P, Autrup H, Sutherland DS, Scott-Fordsmand JJ (2016) Nanosilver pathophysiology in earthworms: Transcriptional profiling of secretory proteins and the implication for the protein corona. Nanotoxicology 10:303–311. https://doi.org/10.3109/17435390.2015.1054909

Hirigoyenberry F, Lassalle F, Lassegues M (1990) Antibacterial activity of *Eisenia fetida andrei* coelomic fluid: Transcription and translation of lysozyme and proteins evidenced after bacterial infestation. Comp Biochem Physiol B 95:71–75. https://doi.org/10.1016/0305-0491(90)90250-w

Hoang DT, Chernomor O, von Haeseler A, Minh BQ, Vinh LS (2018) UFBoot2: Improving the Ultrafast Bootstrap Approximation. Mol Biol Evol 35(2):518–522. https://doi.org/10.1093/molbev/msx281

Hoeger U, Abe H (2004) ß-Alanine and other free amino acids during salinity adaption of the polychaete *Nereis japonica*. Comp Biochem Physiol A 137(1):161–171. https://doi.org/10.1016/S1095-6433(03)00286-1

Hoeger U, Kunz I (1993) Metabolic enzymes in coelomic cells (eleocytes) of the polychaete *Nereis virens*: sex specific changes during sexual maturation. Mar Biol 115:653–660. https://doi.org/10.1016/j.cbpb.2010.06.001

Hoeger U, Märker C (1997) Nucleotide pool changes in coelomic cells (eleocytes) of the polychaete *Nereis virens* during sexual maturation. Exp Biol Online 2–12 (12)

Hoeger U, Märker C, Geier G (1996) Adenylate storage, metabolism and utilization of the polychaete *Nereis virens* (Annelida, Polychaeta). Experientia 52:481–486

Hoeger U, Rebscher N, Geier G (1999) Metabolite supply in growing oocytes of *Nereis virens*: role of nucleosides. Hydrobiologia 402:163–174. https://doi.org/10.1023/A:1003792525942

Homa J, Ortmann W, Kolaczkowska E (2016) Conservative mechanisms of extracellular trap formation by annelida *Eisenia andrei*: serine protease activity requirement. PLoS ONE 11:e0159031. https://doi.org/10.1371/journal.pone.0159031

Hussain MM, Shi J, Dreizen P (2003) Microsomal triglyceride transfer protein and its role in apoB-lipoprotein assembly. J Lipid Res 44(1):22–32. https://doi.org/10.1194/jlr.r200014-jlr200

Ishitsuka R, Kobayashi T (2007) Cholesterol and lipid/protein ratio control the oligomerization of a sphingomyelin-specific toxin, lysenin. Biochemistry 46(6):1495–1502. https://doi.org/10.1021/bi061290k

Ito Y, Ayako Y, Hotani T, Fukuda S, Sugimura K, Imoto T (1999) Amino acid sequences of lysozymes newly purified from invertebrates imply wide distribution of a novel class in the lysozyme family. Eur J Biochem 259:456–461. https://doi.org/10.1046/j.1432-1327.1999.00064.x

Jamieson BGM (2006) Reproductive biology and phylogeny of annelida. In: Reproductive biology and phylogeny, vol 4, 1st edn. CRC Press, Boca Raton. https://doi.org/10.1201/9781482280159

Jeong BR, Chung SM, Baek NJ, Koo KB, Baik HS, Joo HS, Chang CS, Choi JW (2006) Characterization, cloning and expression of the ferritin gene from the Korean polychaete, *Periserrula leucophryna*. J Microbiol 44(1):54–63

Josková R, Šilerová M, Procházková P, Bilej M (2009) Identification and cloning of an invertebrate-type lysozyme from *Eisenia andrei*. Dev Comp Immunol 33:932–938. https://doi.org/10.1016/j.dci.2009.03.002

Journet AM, Saffaripour S, Wagner DD (1993) Requirement for both D domains of the propolypeptide in von Willebrand factor multimerization and storage. Thromb Haemost 70(6):1053–1057

Kalyaanamoorthy S, Minh BQ, Wong TKF, von Haeseler A, Jermiin LS (2017) ModelFinder: fast model selection for accurate phylogenetic estimates. Nat Methods 14(6):587–589. https://doi.org/10.1038/nmeth.4285

Kauschke E, Mohrig W (1987) Cytotoxic activity in the coelomic fluid of the annelid *Eisenia foetida* Sav. J Comp Physiol B 157:77–83. https://doi.org/10.1007/bf00702731

Kauschke E, Pagliara P, Stabili L, Cooper EL (1997) Characterization of proteolytic activity in coelomic fluid of *Lumbricus terrestris* L (Annelida, Lumbricidae). Comp Biochem Physiol B 116(2):235–242. https://doi.org/10.1016/S0305-0491(96)00248-9

Kauschke E, Mohrig W, Cooper EL (2007) Coelomic fluid proteins as basic components of innate immunity in earthworms. Eur J Soil Biol 43:S110–S115. https://doi.org/10.1016/j.ejsobi.2007.08.043

Kiyokawa E, Makino A, Ishii K, Otsuka N, Yamaji-Hasegawa A, Kobayashi T (2004) Recognition of sphigomyelin by lysenin and lysenin-related protein. Biochemistry 43:9766–9773. https://doi.org/10.1021/bi049561j

Klotz IM, Kurtz DM (1984) Binuclear oxygen carriers: hemerythrin. Accounts Chem Res 17(1):16–22. https://doi.org/10.1021/ar00097a003

Kulakosky PC, Telfer WH (1990) Lipophorin as a yolk precursor in *Hyalophora cecropia*: Uptake kinetics and competition with vitellogenin. Arch Biochem Biophys 14:269–285. https://doi.org/10.1002/arch.940140406

Lange S, Nussler F, Kauschke E, Lutsch G, Cooper EL, Herrmann A (1997) Interaction of earthworm hemolysin with lipid membranes requires sphingolipids. J Biol Chem 272(33):20884–20892. https://doi.org/10.1074/jbc.272.33.20884

Lange S, Kauschke E, Mohrig W, Cooper EL (1999) Biochemical characteristics of Eiseniapore, a pore-forming protein in the coelomic fluid of earthworms. Eur J Biochem 262(2):547–556. https://doi.org/10.1046/j.1432-1327.1999.00407.x

Lassalle F, Lassegues M, Roch P (1988) Protein analysis of earthworm coelomic fluid - IV. Evidence, activity induction and purification of *Eisenia fetida andrei* lysozyme (Annelidae). Comp Biochem Physiol B 91:187–192. https://doi.org/10.1016/0305-0491(88)90133-2

Lassegues M, Milochau A, Doignon F, Du Pasquier L, Valembois P (1997) Sequence and expression of an *Eisenia fetida*-derived cDNA clone that encodes the 40-kDa fetidin antibacterial protein. Eur J Biochem 246:756–762. https://doi.org/10.1111/j.1432-1033.1997.00756.x

Lee YR, Kim YN (1993) Selective transport of coelomic fluid protein into oocytes of a tubicolous polychaete, *Pseudopotamilla occelata*. Invert Reprod Develop 24(2):119–126. https://doi.org/10.1080/07924259.1993.9672341

Lee BK, Nam HJ, Lee YR (1997) Receptor-mediated transport of vitellin during oogenesis of a polychaete, *Pseudopotamilla ocelata*. Kor J Biol Sci 1:341–344

Li W, Li S, Zhong J, Zhu Z, Liu J, Wang W (2011) A novel antimicrobial peptide from skin secretions of the earthworm *Pheretima guillelmi* (Michaelsen). Peptides 32:1146–1150. https://doi.org/10.1016/j.peptides.2011.04.015

Li C, Chen M, Li X, Yang M, Wang Y, Yang X (2017) Purification and function of two analgesic and anti-inflammatory peptides from coelomic fluid of the earthworm, Eisenia foetida. Peptides 89:71–81. https://doi.org/10.1016/j.peptides.2017.01.016

MacMahon BR, Wikens JL, Smith PJS (1997) Invertebrate circulatory systems. In: Dantzler WH (ed) Comparative Physiology. Oxford Univ. Press, New York, pp 931–1008

Maltseva, AL, Kotenko, ON, Kokryakov, VN, Starunov, VV, Krasnodembskaya, AD (2014) Expression pattern of arenicins—the antimicrobial peptides of polychaete *Arenicola marina*. Front Physiol 5. https://doi.org/10.3389/fphys.2014.00497

Mangum CP, Woodin BR, Bonaventura C, Sullivan B, Bonaventura J (1975) Role of coelomic and vascular hemoglobin in annelid family terebellidae. Comp Biochem Phys A 51(2):281–294. https://doi.org/10.1016/0300-9629(75)90372-2

Marcano L, Nusetti O, Rodriguez-Grau J, Briceño J, Vilas J (1997) Coelomic fluid lysozyme activity induction in the polychaete *Eurythoe complanata* as a biomarker of heavy metal toxicity. Bull Environ Contam Toxicol 59:22–28. https://doi.org/10.1007/s001289900438

Margoshes M, Vallee BL (1957) A cadmium protein from equine kidney cortex. J Am Chem Soc 79(17):4813–4814. https://doi.org/10.1021/ja01574a064

Marusek CM, Trobaugh NM, Flurkey WH, Inlow JK (2006) Comparative analysis of polyphenol oxidase from plant and fungal species. J Inorg Biochem 100:108–123. https://doi.org/10.1016/j.jinorgbio.2005.10.008

Mazur AI, Klimek M, Morgan AJ, Plytycz B (2011) Riboflavin storage in earthworm chloragocytes and chloragocyte-derived eleocytes and its putative role as chemoattractant for immunocompetent cells. Pedobiologia 54:S37–S42. https://doi.org/10.1016/j.pedobi.2011.09.008

M'Beri M, Debray H, Dhainaut A (1988) Separation of two different populations of granulocytes of *Nereis diversicolor* (Annelida) by selective agglutination with lectins. Dev Comp Immunol 12(2):279–285. https://doi.org/10.1016/0145-305x(88)90004-3

McKenzie HA, White FH Jr (1991) Lysozyme and α-lactalbumin: structure, function and interrelationships. Adv Prot Chem 41:173–315. https://doi.org/10.1016/s0065-3233(08)60198-9

McLaughlin J (1971) Biochemical studies on *Eisenia foetida* (Savigny, 1826), the brandling worm - I. Tissue lipids and sterols. Comp Biochem Physiol B 38:147–163. https://doi.org/10.1016/0305-0491(71)90294-X

Milochau A, Lassègues M, Valembois P (1997) Purification, characterization and activities of two hemolytic and antibacterial proteins from coelomic fluid of the annelid *Eisenia fetida andrei*. Biochim Biophys Acta 1137:123–132. https://doi.org/10.1016/s0167-4838(96)00160-4

Minh B, Nguyen M, von Haeseler A (2013) Ultrafast approximation for phylogenetic bootstrap. Mol Biol Evol 30:1188–1195. https://doi.org/10.1093/molbev/mst024

Nagai T, Kawabata S (2000) A link between blood coagulation and prophenol oxidase activation in arthropod host defense. J Biol Chem 275(38):29264–29267. https://doi.org/10.1074/jbc.M002556200

Nejmeddine A, Dhainautcourtois N, Baert JL, Sautiere P, Fournet B, Boulenguer P (1988) Purification and characterization of a cadmium-binding protein from *Nereis diversicolor* (Annelida, Polychaeta). Comp Biochem Phys C 89(2):321–326. https://doi.org/10.1016/0742-8413(88)90231-9

Nejmeddine A, Sautiere P, Dhainaut-Courtois N, Baert JL (1992) Isolation and characterization of a Cd-binding protein from *Allolobophora caliginosa* (Annelida, Oligochaeta): distinction from metallothioneins. Comp Biochem Physiol C 101(3):601–605. https://doi.org/10.1016/0742-8413(92)90093-M

Nejmeddine A, Wouters-Tyrou D, Baert JL, Sautiere P (1997) Primary structure of a myohemerythrin-like cadmium-binding protein, isolated from a terrestrial annelid oligochaete. C R Acad Sci III 320(6):459–468

Nguyen LT, Schmidt HA, von Haeseler A, Minh BQ (2015) IQ-TREE: a fast and effective stochastic algorithm for estimating maximum-likelihood phylogenies. Mol Biol Evol 32(1):268–274. https://doi.org/10.1093/molbev/msu300

Nichol H, Law JH, Winzerling JJ (2002) Iron metabolism in insects. Annu Rev Entomol 47:535–559. https://doi.org/10.1146/annurev.ento.47.091201.145237

Ohta N, Shioda S, Sekizawa Y, Nakai Y, Kobayashi H (2000) Sites of expression of mRNA for lysenin, a protein isolated from the coelomic fluid of the earthworm *Eisenia foetida*. Cell Tissue Res 302(2):263–270. https://doi.org/10.1007/s004410000284

Olivares Fontt E, Beschin A, van Dijck E, Vercruysse V, Bilej M, Lucas R, de Baetselier P, Vray B (2002) *Trypanosoma cruzi* is lysed by coelomic cytolytic factor-1, an invertebrate analogue of tumor necrosis factor, and induces phenoloxidase activity in the coelomic fluid of *Eisenia foetida foetida*. Dev Comp Immunol 26:27–34. https://doi.org/10.1016/s0145-305x(01)00048-9

Olsen I, Jantzen E (2001) Sphingolipids in bacteria and fungi. Anaerobe 7:103–112. https://doi.org/10.1006/anae.2001.0376

Opper B, Bognar A, Heidt D, Nemeth P, Engelmann P (2013) Revising lysenin expression of earthworm coelomocytes. Dev Comp Immunol 39(3):214–218. https://doi.org/10.1016/j.dci.2012.11.006

Ovchinnikova, TV, Shenkarev, ZO, Nadezhdin, KD, Balandin, SV, Zhmak, MN, Kudelina, IA, Finkina, EI, Kokryakov, VN, Arseniev, AS (2007) Recombinant expression, synthesis, purification, and solution structure of arenicin. Biochem Biophys Res Com 360(1):156–162. https://doi.org/10.1016/j.bbrc.2007.06.029

Paiva-Silva GO, Cruz-Oliveira C, Nakayasu ES, Maya-Monteiro CM, Dunkov BC, Masuda H, Almeida IC, Oliveira PL (2006) A heme-degradation pathway in a blood-sucking insect. Proc Natl Acad Sci U S A 103(21):8030–8035. https://doi.org/10.1073/pnas.0602224103

Pan, W, Liu, X, Ge, F, Han, J, Zheng, T (2004) Perinerin, a novel antimicrobial peptide purified from the clamworm *Perinereis aibuhitensis* Grube and its partial characterization. J Biochem 135(3):297–304. https://doi.org/10.1093/jb/mvh036

Pereira LOR, Oliveira PL, Almeida IC, Paiva-Silva GO (2007) Biglutaminyl-biliverdin IX alpha as a heme degradation product in the dengue fever insect-vector *Aedes aegypti*. Biochemistry 46(23):6822–6829. https://doi.org/10.1021/bi700011d

Périn J-P, Jollès P (1972) The lysozyme from *Nephthys hombergi* (Annelid). Biochim Biophys Acta 263:683–689. https://doi.org/10.1016/0005-2795(72)90051-7

Plytycz B, Morgan AJ (2011) Riboflavin storage in earthworm chloragocytes/eleocytes in an eco-immunology perspective. Isj-Invert Surviv J 8(2):199–209

Podobnik M, Savory P, Rojko N, Kisovec M, Wood N, Hambley R, Pugh J, Wallace EJ, McNeill L, Bruce M, Liko I, Allison TM, Mehmood S, Yilmaz N, Kobayashi T, Gilbert RJ, Robinson CV, Jayasinghe L, Anderluh G (2016) Crystal structure of an invertebrate cytolysin pore reveals unique properties and mechanism of assembly. Nat Commun 7:11598. https://doi.org/10.1038/ncomms11598

Porchet-Henneré E (1990) Cooperation between different coelomocyte populations during encapsulation response of *Nereis diversicolor* demonstrated by using monoclonal antibodies. J Invert Pathol 56:353–361. https://doi.org/10.1016/0022-2011(90)90122-M

Porchet-Henneré E, Nejmeddine A, Baert JL, Dhainaut-Courtois N (1987) Selective immunostaining of type-1 granulocytes of the polychaete annelid *Nereis diversicolor* by a monoclonal-antibody against acadmium-binding protein (Mp-II). Biol Cell 60(3):259–261

Porchet-Hennere E, Vernet G (1992) Cellular immunity in an annelid (*Nereis diversicolor*, Polychaeta): production of melanin by a subpopulation of granulocytes. Cell Tissue Res 269(1):167–174. https://doi.org/10.1007/bf00384737

Porchet-Hennere E, Dugimont T, Fischer A (1992) Natural killer cells in a lower invertebrate, *Nereis diversicolor*. Eur J Cell Biol 58(1):99–107

Prochazkova P, Dvorak J, Silerova M, Roubalova R, Skanta F, Halada P, Bilej M (2011) Molecular characterization of the iron binding protein ferritin in Eisenia andrei earthworms. Gene 485(2):73–80. https://doi.org/10.1016/j.gene.2011.06.010

Racovitza E (1895) Sur le role des amibocytes chez les annélides polychètes. C R Acad Sci, Paris 120:464–467

Ratcliffe NA, Rowley AF (1981a) Arthropods to urochordates. In: Ratcliffe NA, Rowley AF (eds) Invertebrate blood cells, vol 2. Academic Press, New York

Ratcliffe NA, Rowley AF (1981b) General aspects. In: Ratcliffe NA, Rowley AF (eds) Invertebrate blood cells, vol 1. Academic Press, New York

Roch P, Ville P, Cooper EL (1998) Characterization of a 14 kDa plant-related serine protease inhibitor and regulation of cytotoxic activity in earthworm coelomic fluid. Dev Comp Immunol 22(1):1–12. https://doi.org/10.1016/s0145-305x(97)00047-5

Roesijadi G, Fowler BA (1991) Purification of invertebrate metallothioneins. Meth Enzymol 205:263–273. https://doi.org/10.1016/0076-6879(91)05106-6

Romieu M (1921) Sur les éléocytes de *Perinereis cultrifer*a (Grube). C R Acad Sci Paris 173:246–249

Ryan RO (1990) Dynamics of insect lipophorin metabolism. J Lipid Res 31:1725–1739

Salzet-Raveillon B, Rentier-Delrue F, Dhainaut A (1993) Detection of mRNA encoding an antibacterial-metalloprotein (MPII) by in situ hybridization with a cDNA probe generated by polymerase chain reaction in the worm *Nereis diversicolor*. Cell Mol Biol (Noisy-le-grand) 39 (1):105–114

Sánchez-Ferrer Á, Rodríguez-López JN, García-Cánovas F, García-Carmona F (1995) Tyrosinase: a comprehensive review of its mechanism. Biochim Biophys Acta 1247:1–11. https://doi.org/10.1016/0167-4838(94)00204-t

Schenk S, Hoeger U (2009) Lipoprotein mediated lipid uptake in oocytes of polychaetes (Annelida). Cell Tiss Res 337(2):341–348. https://doi.org/10.1007/s00441-009-0817-7

Schenk S, Hoeger U (2010) Lipid accumulation and metabolism in polychaete spermatogenesis: Role of the large discoidal lipoprotein. Mol Reprod Dev 77:710–719. https://doi.org/10.1002/mrd.21208

Schenk S, Hoeger U (2011) Glutathionyl-biliverdin IXa, a new heme catabolite in a marine annelid: sex and cell specific accumulation. Biochimie 93(2):302–316

Schenk S, Harris JR, Hoeger U (2006) A discoidal lipoprotein from the coelomic fluid of the polychaete *Nereis virens*. Comp Biochem Physiol B 143(2):236–243. https://doi.org/10.1016/j.cbpb.2005.11.012

Schenk S, Krauditsch C, Fruhauf P, Gerner C, Raible F (2016) Discovery of methylfarnesoate as the annelid brain hormone reveals an ancient role of sesquiterpenoids in reproduction. Elife 5. https://doi.org/10.7554/eLife.17126

Schikorski D, Cuvillier-Hot V, Leippe M, Boidin-Wichlacz C, Slomianny C, Macagno ER, Salzet M, Tasiemski A (2008) Microbial challenge promotes the regenerative process of the injured central nervous system of the medicinal leech by inducing the synthesis of antimicrobial peptides in neurons and microglia. J Immunol 181 (1083–1095). https://doi.org/10.4049/jimmunol.181.2.1083

Schmidt O, Soderhall K, Theopold U, Faye I (2010) Role of adhesion in arthropod immune recognition. Ann Rev Entomol 55:485–504. https://doi.org/10.1146/annurev.ento.54.110807.090618

Schüßler P, Potters E, Winnen R, Bottke W, Kunz W (1995) An isoform of ferritin as a component of protein yolk platelets in *Schistosoma mansoni*. Mol Reprod Dev 41(3):325–330. https://doi.org/10.1002/mrd.1080410307

Scott DM (1976) Circadian rhythm of anaerobiosis in a polychaete annelid. Nature 262:811–813. https://doi.org/10.1038/262811a0

Sekizawa Y, Kubo T, Kobayashi H, Nakajima T, Natori S (1997) Molecular cloning of cDNA for lysenin, a novel protein in the earthworm Eisenia foetida that causes contraction of rat vascular smooth muscle. Gene 191(1):97–102. https://doi.org/10.1016/s0378-1119(97)00047-4

Seo J-K, Nam B-H, Go H-J, Jeong M, Lee K-Y, Cho S-M, Lee I-A, Park NG (2016) Hemerythrin-related antimicrobial peptide, msHemerycin, purified from the body of the lugworm, *Marphysa sanguinea*. Fish Shellfish Immunol 57:49–59. https://doi.org/10.1016/j.fsi.2016.08.018

Shakor AB, Czurylo EA, Sobota A (2003) Lysenin, a unique sphingomyelin-binding protein. FEBS Lett 542(1–3):1–6. https://doi.org/10.1016/s0014-5793(03)00330-2

Silerova M, Prochazkova P, Joskova R, Josens G, Beschin A, De Baetselier P, Bilej M (2006) Comparative study of the CCF-like pattern recognition protein in different Lumbricid species. Dev Comp Immunol 30(9):765–771. https://doi.org/10.1016/j.dci.2005.11.002

Smith VJ, Söderhäll K (1991) A comparison of phenoloxidase activity in the blood of marine invertebrates. Dev Comp Immunol 15:251–261. https://doi.org/10.1016/0145-305x(91)90018-t

Smolenaars MMW, De Morreé A, Kerver J, Van der Horst DJ, Rodenburg KW (2007a) Insect lipoprotein biogenesis depends on an amphipatic ß cluster in apolipophorin II/I and is stimulated by microsomal triglyceride transfer protein. J Lipid Res 48:1955–1965. https://doi.org/10.1194/jlr.M600434-JLR200

Smolenaars MMW, Madsen O, Rodenburg KW, Van der Horst DJ (2007b) Molecular diversity of the large lipid transfer protein superfamily. J Lipid Res 48:489–502. https://doi.org/10.1194/jlr.R600028-JLR200

Soderhall K, Cerenius L (1998) Role of the prophenoloxidase-activating system in invertebrate immunity. Curr Opin Immunol 10(1):23–28. https://doi.org/10.1016/s0952-7915(98)80026-5

Spaziani E, Havel RJ, Hamilton RL, Hardman DA, Stoudemire JB, Watson RD (1986) Properties of serum high-density lipoproteins in the crab, *Cancer antennarius* Stimpson. Comp Biochem Physiol B 85(2):307–314. https://doi.org/10.1016/0305-0491(86)90005-2

Stürzenbaum SR, Kille P, Morgan AJ (1998) The identification, cloning and characterization of earthworm metallothionein. FEBS Lett 431(3):437–442. https://doi.org/10.1016/s0014-5793(98)00809-6

Stürzenbaum SR, Winters C, Galay M, Morgan AJ, Kille P (2001) Metal ion trafficking in earthworms. Identification of a cadmium-specific metallothionein. J Biol Chem 276 (36):34013–34018. 10.1074/jbc.M103605200

Sturzenbaum SR, Georgiev O, Morgan AJ, Kille P (2004) Cadmium detoxification in earthworms: from genes to cells. Environ Sci Technol 38(23):6283–6289. https://doi.org/10.1021/es049822c

Swiderska B, Kedracka-Krok S, Panz T, Morgan AJ, Falniowski A, Grzmil P, Plytycz B (2017) Lysenin family proteins in earthworm coelomocytes - Comparative approach. Dev Comp Immunol 67:404–412. https://doi.org/10.1016/j.dci.2016.08.011

Taki H, Baert JL, Dhainaut A (1989) Synthesis and release of vitellogenin-associated phospholipids by the coelomocytes of *Perinereis cultifera* (Annelida, Polychaeta). Comp Biochem Physiol B 92(1):167–173. https://doi.org/10.1016/0305-0491(89)90330-1

Tasiemski A, Vandenbulcke F, Mitta G, Lemoine J, Lefebvre C, Sautiere PE, Salzet M (2004) Molecular characterization of two novel antibacterial peptides inducible upon bacterial challenge in an annelid, the leech *Theromyzon tessulatum*. J Biol Chem 279(30):30973–30982. https://doi.org/10.1074/jbc.M312156200

Telfer WH, Pan M-L, Law JH (1991) Lipophorin in developing adults of Hyalophora cecropia: support of yolk formation and preparation for flight. Insect Biochem Mol Biol 21(-):653–663

Tučková L, Rejnek J, Šíma P, Ondřejová R (1986) Lytic activities in coelomic fluid of *Eisenia foetida* and *Lumbricus terrestris*. Dev Comp Immunol 10(2):181–189. https://doi.org/10.1016/0145-305X(86)90002-9

Vaillier J, Cadoret MA, Roch P, Valembois P (1985) Protein analysis of earthworm coelomic fluid. III. Isolation and characterization of several bacteriostatic molecules from Eisenia fetida andrei. Dev Comp Immunol 9(1):11–20. https://doi.org/10.1016/0145-305x(85)90055-2

Valembois P, Roch P, Lassegues M, Davant N (1982) Bacteriostatic activity of a chloragogen cell secretion. Pedobiologia 24(4):191–195

Valembois P, Seymour J, Roch P (1991) Evidence and cellular localization of an oxidase activity in the coelomic fluid of the earthworm *Eisenia fetida andrei*. J Invertebr Pathol 57:177–183

Vetvicka V, Sima P (2009) Origins and functions of annelide immune cells: the concise survey. Isj-Invert Surviv J 6(2):138–143

Wang X, Wang X, Zhang Y, Qu X, Yang S (2003) An antimicrobial peptide of the earthworm *Pheretima tschiliensis*: cDNA cloning, expression and immunolocalization. Biotechnol Lett 25:1317–1323

Weber RE, Vinogradov SN (2001) Nonvertebrate hemoglobins: functions and molecular adaptations. Physiol Rev 81(2):569–628. https://doi.org/10.1152/physrev.2001.81.2.569

Weers PM, Van Marrewijk WJ, Beenakkers AM, Van der Horst DJ (1993) Biosynthesis of locust lipophorin. Apolipophorins I and II originate from a common precursor. J Biol Chem 268 (6):4300–4303

Westheide W (2013) Spezielle Zoologie, vol 1. Einzeller und wirbellose Tiere, 3rd edn. Springer-Spektrum, Berlin

Won EJ, Rhee JS, Ra K, Kim KT, Au DW, Shin KH, Lee JS (2011) Molecular cloning and expression of novel metallothionein (MT) gene in the polychaete *Perinereis nuntia* exposed to metals. Environ Sci Pollut Res Int 19(7):2606–2618. https://doi.org/10.1007/s11356-012-0905-1

Yamaji A, Sekizawa Y, Emoto K, Sakuraba H, Inoue K, Kobayashi H, Umeda M (1998) Lysenin, a novel sphingomyelin-specific binding protein. J Biol Chem 273(9):5300–5306. https://doi.org/10.1074/jbc.273.9.5300

Zhang Y, Cui J, Zhang R, Wang Y, Hong M (2007) A novel fibrinolytic serine protease from the polychaete *Nereis* (*Neanthes*) *virens* (Sars): purification and characterization. Biochimie 89(1):93–103. https://doi.org/10.1016/j.biochi.2006.07.023

Zhang L, Cui W, Chen Q, Ren Q, Zhu Y, Zhang Y (2020) Thymosin-beta12 characteristics and function in *Urechis unicinctus*. Comp Biochem Physiol B 239:110366. https://doi.org/10.1016/j.cbpb.2019.110366

Chapter 2
Crustacean Hemolymph Lipoproteins

Ulrich Hoeger and Sven Schenk

Abstract Lipoproteins mediate the transport of apolar lipids in the hydrophilic environment of physiological fluids such as the vertebrate blood and the arthropod hemolymph. In this overview, we will focus on the hemolymph lipoproteins in Crustacea that have received most attention during the last years: the high density lipoprotein/β-glucan binding proteins (HDL-BGBPs), the vitellogenins (VGs), the clotting proteins (CPs) and the more recently discovered large discoidal lipoproteins (dLPs). VGs are female specific lipoproteins which supply both proteins and lipids as storage material for the oocyte for later use by the developing embryo. Unusual within the invertebrates, the crustacean yolk proteins—formerly designated VGs—are more related to the ApoB type lipoproteins of vertebrates and are now termed apolipocrustaceins. The CPs on the other hand, which are present in both sexes, are related to the (sex specific) VGs of insects and vertebrates. CPs serve in hemostasis and wound closure but also as storage proteins in the oocyte. The HDL-BGBPs are the main lipid transporters, but are also involved in immune defense. Most crustacean lipoproteins belong to the family of the large lipid transfer proteins (LLTPs) such as the intracellular microsomal triglyceride transfer protein, the VGs, CPs and the dLPs. In contrast, the HDL-BGBPs do not belong to the LLTPs and their relationship with other lipoproteins is unknown. However, they originate from a common precursor with the dLPs, whose functions are as yet unknown. The majority of lipoprotein studies have focused on decapod crustaceans, especially shrimps, due to their economic importance. However, we will present evidence that the HDL-BGBPs are restricted to the decapod crustaceans which raises the question as to the main lipid transporting proteins of the other crustacean groups. The diversity of crustaceans lipoproteins thus appears to be more complex than reflected by the present state of knowledge.

U. Hoeger (✉)
Institut für Molekulare Physiologie, Johannes Gutenberg-Universität, 55099 Mainz, Germany
e-mail: uhoeger@uni-mainz.de

S. Schenk
MAX F. PERUTZ LABORATORIES, Vienna Biocenter (VBC), Dr. Bohr-Gasse 9/4, 1030 Vienna, Austria
e-mail: Sven.Schenk@univie.ac.at

U. Hoeger and J. R. Harris (eds.), *Vertebrate and Invertebrate Respiratory Proteins, Lipoproteins and other Body Fluid Proteins*, Subcellular Biochemistry 94,
https://doi.org/10.1007/978-3-030-41769-7_2

Keywords Crustacea · Hemolymph · High density lipoprotein/β-glucan binding protein · Discoidal lipoprotein · Vitellogenin · Clotting protein · Lipids · Coagulation · Immune defense

Introduction

The hemolymph of arthropods circulates in an open circulatory system which represents a large fluid-filled space. In crustaceans, the hemolymph volume comprises a relatively large volume and varies between 19 and 33% of the whole body, depending on the species and the methodology used (measured in decapods; see Hoeger and Florey 1989 and cited references). The hemolymph protein concentrations are high, ranging between 30 and >100 mg/ml, within which the respiratory protein hemocyanin constitutes between 25 and 90% depending on the physiological state of the animal (Depledge and Bjerregaard 1989). As outlined above, hemocyanin and—in some species—hemoglobin (see the chapter by Zeis et al. in this volume) is the dominant protein in the hemolymph; and during the premolt phase, a hemocyanin related non respiratory protein, cryptocyanin, arises in the hemolymph becoming the most abundant protein in the hemolymph, as found in the crab, *Cancer magister* and proposed for the lobster, *Homarus americanus* (Burmester 1999; Terwilliger et al. 1999). Nevertheless, the hemolymph also contains significant concentrations of lipoproteins in the range of several mg/ml. These lipoproteins are essential in mediating the transport of apolar lipids in the polar environment of the hemolymph. Hemocyanin itself has little capacity for lipid transport, but due to its high concentration, it contributes significantly to the amount of transported lipids. This aspect is discussed for chelicerate hemocyanins in the chapter by Cunningham et al. in this volume.

Most invertebrates possess only one major type of sex independent lipoprotein such as the lipophorins in flying insects (Soulages and Wells 1994; Rodenburg and van der Horst 2005), chelicerates (Haunerland and Bowers 1989; Cunningham et al. 2007) or the HDL-type lipoprotein in a marine annelid (see Schenk and Hoeger, this volume) while in vertebrates, lipoprotein evolution has led to a multitude of lipoproteins with various lipid and apoprotein composition (Khovidhunkit et al. 2004). Lipoproteins are also involved in functions other than lipid transport such as the participation of insect lipophorins in clot formation and binding of bacterial lipopolysacharides (Schmidt et al. 2010) and the function of insect apolipophorin III as pattern recognition protein (Whitten et al. 2004). Likewise, the participation of lipoproteins in clot formation has been shown in the tarantula, *Acanthoscurria geniculata* (Sanggaard et al. 2016). Furthermore, the evolution of lipoproteins has led to new functions such as the crustacean clotting proteins (see below) or the activation of chelicerate hemocyanin to a phenoloxidase by lipoprotein (Schenk et al. 2015). Similarly, the vitellogenin (VTG)-like melanin engaging protein works in concert with the phenoloxidase system enhancing melanin synthesis (Lee et al. 2000). This protein was first described in the flour beetle *Tenebrio*, but related proteins

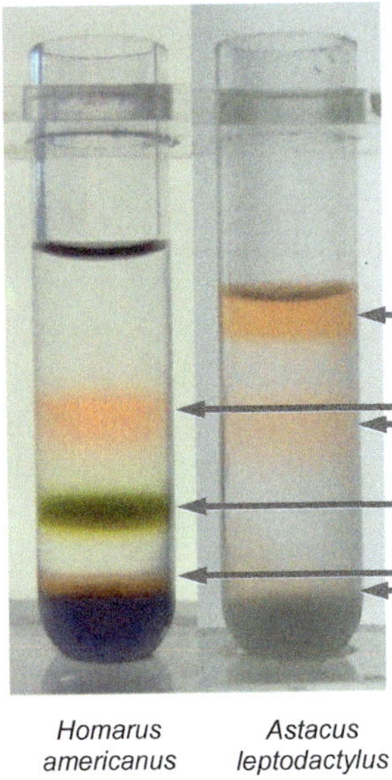

Density class/ Lipoprotein	Density (lipid content)	Protein family
HDL/Large discoidal lipoprotein	1,10 (70%)	LLT
HDL/Beta-glucan binding protein	1,14-1,16 (25-55%)	?
HDL/„Vitellogenin“ (Apolipocrustacein)	1,19-1,20 (23,5-28%)	LLT, Apo B related
VHDL/ Clotting protein	1,20-1,30 (6-28%)	LLT, Vitellogenin related

Fig. 2.1 Density gradient centrifugation (KBr) of the hemolymph of a female lobster (*Homarus americanus*) and male crayfish (*Astacus leptodactylus*). The separated lipoproteins and their corresponding density class is given in the table. The coloration is due to carotenoids which form part of the lipid moiety. For methods see Schenk et al. (2009)

were found recently in copepod crustaceans (Order Maxillopoda, Lee et al. 2016). Finally, another vitellogenin related protein, Crossveinless, which is required for BMP signaling in the wing was found in *Drosophila* (Chen et al. 2012).

In this overview we will focus on the major lipoproteins circulating in the hemolymph space of crustaceans, the high-density lipoprotein/β-glucan binding protein, the large discoidal lipoprotein, the vitellogenin and the clotting protein. Most invertebrate lipoproteins belong to the high density lipoproteins with buoyant densities between 1.064 and 1.210 g/ml (Chapman 1980) and this is also true for crustaceans (Yepiz-Plascencia et al. 1995 and cited references). These differences in the buoyant densities are used to separate the lipoproteins from other hemolymph proteins in KBr gradients as shown for the isolation of lipoproteins from lobster and crayfish hemolymph (Fig. 2.1).

Large Discoidal Lipoprotein

The large discoidal lipoprotein (dLP) is the most recent lipoprotein discovered in the Crustacea. It was first isolated from the hemolymph of the crayfish *Astacus leptodactylus* (Stieb et al. 2008). Its concentration (2.5–6 mg/ml hemolymph) was in the same range as found for the HDL/BGBP (see below). Electron microscopical images (Fig. 2.2) showed that it is an unusually large lipoprotein particle with a diameter of about 40 nm and a thickness of about 7 nm having the same size as the large lipoprotein particle found in the coelomic fluid of the marine annelid *Nereis virens* (see chapter by Schenk and Hoeger in this volume).

The dLp has an extremely high lipid binding capacity (about 70% of the particle mass) which exceeds by far that of the other known lipoproteins in the hemolymph (see Table 1). The first analysis subjecting excised protein bands to MALDI-ToF mass spectrometry revealed that the larger dLp subunit contained a peptide that showed sequence similarities with another well known lipoprotein, the high density lipoprotein/β-glucan binding protein (HDL-BGBP; Stieb et al. 2008) which will be described below. Further analysis by mass spectroscopy and sequencing (Stieb et al. 2014) then revealed that the dLp and the HDL-BGBP originate from a common precursor, which contains the sequences of the two subunits of the dLP, and the sequence of the HDL-BGBP. Two potential dibasic furin cleavage sites with the motive K/R-X-K/R-R separate the sequences (Fig. 2.3) Domain architecture analysis of the dLp precursor (Stieb et al. 2014) showed that the larger subunit of dLp represents a member of the large lipid transfer protein family (LLTP, Fig. 2.2). This finding was somewhat unexpected as it is in contrast to the architecture of the well

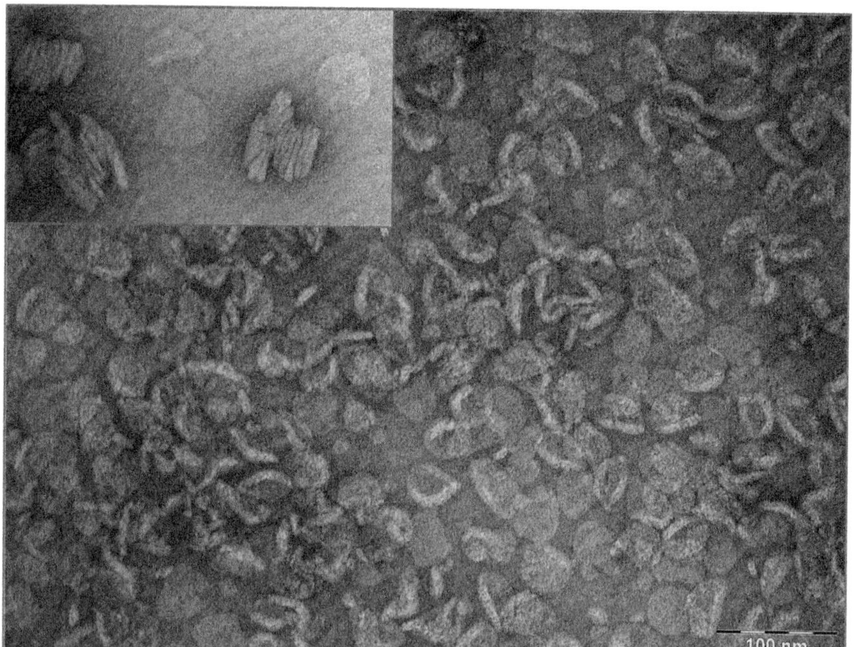

Fig. 2.2 Electron micrograph of lipoprotein particles isolated from the coelomic fluid of *Astacus leptodactylus*. Some of the particles are partially folded and display rim like structures at their edges. At higher concentrations, the particles form disk stacks (insert). The specimens were prepared for electron microscopy by negative staining, as described by Schenk et al. (2006)

Table 2.1 Comparison of the lipid content of different lipoproteins isolated from crustacean hemolymph

Lipoprotein/species	Lipid content (%)	References
High density lipoprotein/β-glucan binding protein		
Pacifastacus leniusculus	47.6	Hall et al. (1995)
Astacus leptodactylus	25	Stieb et al. (2008)
Astacus astacus	31	Stieb et al. (2008)
Litopenaeus vanamei	50	Ruiz-Verdugo et al. (1997)
Penaeus semisulcatus	50, 35	Lubzens et al. (1997), Ravid et al. (1999)
Large discoidal lipoprotein		
Astacus leptodactylus	67	Stieb et al. (2008)
Apolipocrustaceins/Vitellogenins		
Cancer antennarius	45	Spaziani et al. (1986)
Penaeus semisulcatus	50	Lubzens et al. (1997)
Callinectes sapidus	47	Lee and Puppione (1988)
Emerita asiatica	27.9	Tirumalai and Subramoniam (1992)
Triops longicaudatus	19	Schilz and Hoeger, unpubl
Apolipocrustaceins/(Lipo-)Vitellins		
Pagurus pollicaris	33	Wallace et al. (1967)
Uca pugilator	32	Wallace et al. (1967)
Sesarma reticulatum	29	Wallace et al. (1967)
Libinia emarginata	30	Wallace et al. (1967)
Homarus americanus	29	Wallace et al. (1967)
Cancer irroratus	28	Wallace et al. (1967)
Procambarus spec.	35	Fyffe and O'Connor (1974)
Triops longicaudatus	3	Schilz and Hoeger, unpubl.
Clotting protein		
Pacifastacus leniusculus	11.4	Hall et al. (1995)
Astacus leptodactylus	8.5	Stieb et al. (2008)
Astacus astacus	6	Stieb et al. (2008)
Penaeus vannamei	9–15	Yepiz-Plascencia et al. (2009)
Panulirus interuptus	5	Kollman and Quispe (2005)
Penaeus semisulcatus	6.1–8.2	Ravid et al. (1999)
Ibacus ciliatus	5.9–6.3	Komatsu and Ando (1998)

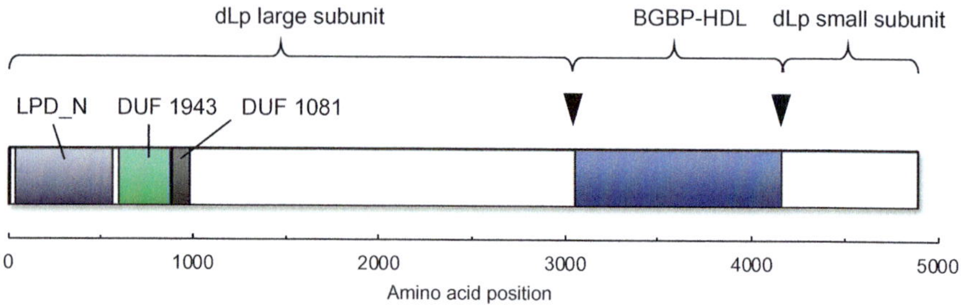

Fig. 2.3 Schematic representation of the domain structure for the common precursor of the large discoidal lipoprotein (dLP) and high density lipoprotein/β-glucan binding protein (HDL-BGBP) using the sequence of *Astacus leptodactylus* (AHJ78589.1) and *Astacus astacus* (AHK23026.1) as example. LPD_N, lipoprotein N terminal domain; DUF 1943, DUF 1081, domains of unknown function. Arrows indicate two furin type cleavage sites enclosing the high density lipoprotein/β-glucan binding protein. After Stieb et al. (2014)

characterized insect lipophorins in which the smaller of the two lipoprotein particle-associated subunits possess LLTP family characteristics (Smolenaars et al. 2007a, b).

Although tissue specific expression of the dLp has not yet been studied, the observation that dLp is derived from a precursor in common with the HDL-BGBP (see below) indicates that both proteins have the same expression sites. Thus, dLp would primarily be expressed in the hepatopancreas, intestine, muscle tissues and hemocytes (see section on HDL-BGBP for details). At the protein level, antibody staining for dLP reveals the same staining pattern in the hepatopancreas as HDL-BGBP (Dal Magro, Lehmann und Hoeger, unpublished observation). However, the staining could as well arise from protein uptake. A striking staining pattern was found in the blood cells of several decapod species which revealed a fine granule staining pattern which extends in the fine filopodia characteristic of these cells. An example of lobster hemocytes is shown in Fig. 2.4. Again, it is unclear if the dLP-like proteins are released into the coelomic fluid from hemocytes or if the dLP could perhaps have a function in mediating the contact with foreign material and could thus be involved in antimicrobial defense.

High Density Lipoprotein/β-Glucan Binding Protein

High density lipoproteins isolated from crustacean hemolymph have been characterized by various authors and studied with respect to their role in general lipid metabolism and their relation to vitellogenesis and vitellogenins (Lee and Puppione 1978; Puppione et al. 1986; Spaziani et al. 1986; Lee and Puppione 1988). β-glucan binding proteins were originally isolated from the hemolymph of the Signal Crayfish *Pacifastacus leniusculus* by precipitation with the β-1,3-glucan laminarin and were characterized as proteins involved in pattern recognition (Duvic and Söderhäll 1990).

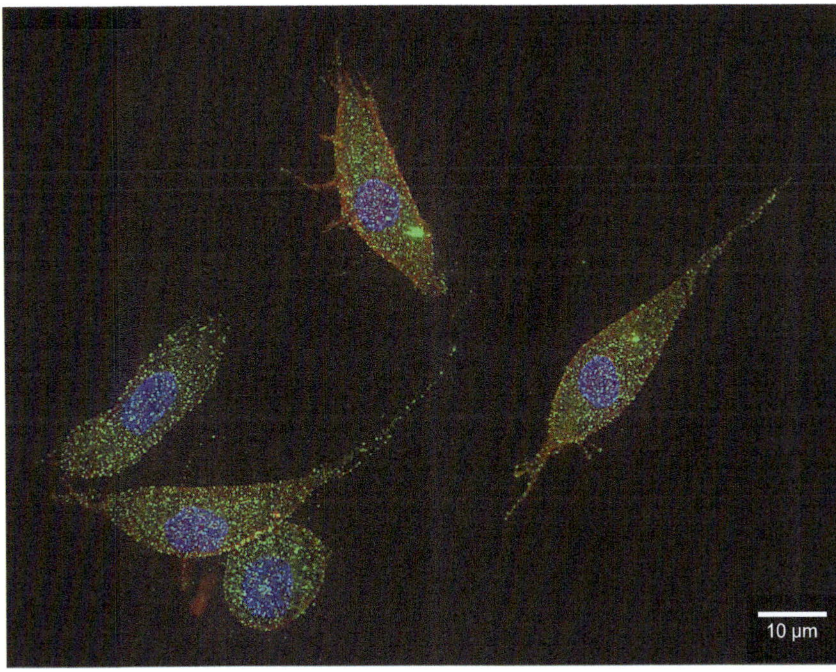

Fig. 2.4 Immunohistochemical staining of lobster (*Homarus americanus*) hemocytes using antiserum against the discoidal lipoprotein (green fluorescence). The red fluorescence indicates labeling of the actin cytoskeleton using TRITC-phalloidin. Nuclei are shown in blue (DAPI staining). The cells were prepared as described in Stieb et al. (2014)

Later however, it was shown, that the β-glucan binding proteins are also involved in lipid transport and represent high-density lipoproteins identical to the initially characterized lipoproteins (Hall et al. 1995; Ruiz-Verdugo et al. 1997; Yepiz-Plascencia et al. 1998). Thus, these proteins are fulfilling a dual role in the animal and the term high density lipoprotein/β-glucan binding protein (HDL-BGBP) was coined to reflect this dual role and is now used to describe these special proteins. In the hemolymph of the crayfish *Astacus leptodactylus*, HDL-BGBP is a rather abundant protein present in concentration ranges between 1 and 6 mg/ml (Stieb et al. 2014), whereas in marine crustaceans HDL-BGBP seems to be present in even broader concentration ranges, ranging between 0.3 and 6.8 mg/mL (Lee and Puppione 1978, 1988; Spaziani and Wang 1991).

Structure of HDL-BGBP

One of the interesting features of HDL-BGBP is the complete absence of cysteine residues as can be seen in the sequences (Accession No.'s in parentheses) of *Litopenaeus vannamei*, (P81182), *Fenneropenaeus chinensis* (ADW10720), *Pacifastacus leniusculus* (CAA56703), *Pontastacus leptodactylus* (AHJ78589) and *Astacus astacus* (AHK23026). This results in a lack of disulfide bridges that might be necessary to maintain an ordered 3D structure. On the other hand, the lack of cysteine residues indicates a flexible peptide backbone which might be advantageous for the protein's defensive function (see below) when docking to pathogen glucans, as has been suggested for some hemocyanin derived antimicrobial peptides, likewise lacking cysteine residues (Coates and Decker 2017). For the HDL-BGBP's, this trait is striking considering a peptide chain of about 1,100 amino acids and indicates a strong evolutionary selection.

As yet another trait, the sequences of the HDL-BGBPs contain one or two integrin binding motives (Arg-Gly-Asp, RGD, Ruoslahti 1996) which may aid in binding to cellular membranes such as the hemocytes (Johansson 1999). The RGD motive is also present in the sequences of the so-called lipopolysaccharide and glucan binding proteins (see below, Liu et al. 2009; Zhao et al. 2009).

Lipid Content

Unlike the dLP, HDL-BGBPs do not belong to the LLTP family of proteins, to which most lipoproteins belong. Whereas the lipidation process is well understood in LLTP (Mann et al. 1999; Richardson et al. 2005; Smolenaars et al. 2007a), nothing is currently known as to how HDL-BGBPs are lipidated. One possibility is that they are lipidated alongside the discoidal lipoprotein (dLP, see above) which is a LLTP member and that they are then excised from the lipoprotein particle. However (lipidated) dLP has so far only been isolated from the crayfish *Astacus leptodactylus,* although very similar sequences can be found in the transcriptomes of many other decapod crustaceans as will be discussed in the next section. This would raise the question as to how HDL-BGBPs are lipidated in species with a non-lipidated dLP and what is the fate of dLP in these species after lipidation and excision of HDL-BGBP? Interestingly, HDL-BGBPs are rather hydrophilic proteins (Kyle-Doolittle hydrophobicity −0.41 to −0.53, average of all 20 amino acids: −0.49) compared to LLTP family members (Kyle-Doolittle hydrophopacity: ~ −0.20 for the LLTP part; unpublished data) Nevertheless, HDL-BGBPs have a lipid carrying capacity between 30 and 50% depending on the species (see Table 1).

The exact lipid binding motives of HDL-BGBPs are at not yet known. Based on analogy to mammalian lipoproteins and insect lipophorins the lipid binding region has been associated with the β-sheet portion of the protein (Vargas-Requena et al. 2002), as is true for the beta-sheets in the N-terminal lipid binding domain of the LLT proteins (Segrest et al. 2001). Indeed, native HDL-BGBPs have a high content in β-sheets, comprising 50–53% of the secondary structures in *P. vannamei* HDL-BGBP (Vargas-Requena et al. 2002; Romo-Figueroa et al. 2004), 39% for *A. leptodactylus* and 41% for *A. astacus* HDL-BGBP, respectively (Stieb, Schenk and Hoeger, unpublished), 36% for *Carcinus maenas* and *Cancer pagurus* HDL-BGBP (Schenk and Hoeger, unpublished) as determined by far-UV circular dichroism spectroscopy.

The transport and uptake of HDL-BGBP bound lipids has been neglected so far. There is only one uptake study on HDL-BGBP showing that the protein binds to isolated membranes and is taken up into the Y-organs of the crab *Cancer antennarius* and contributes to cholesterol supply of these organs necessary for ecdysteroid synthesis (Kang and Spaziani 1995a, b, c). Likewise, no studies on the putative receptors for this protein have been carried out.

Expression of HDL-BGBP

Most of the HDL-BGBP circulating in the in the hemolymph is thought to originate in the hepatopancreas, although it seems to be ubiquitously expressed. The expression levels vary with species and tissue. In *Penaeus vannamei*, Romo-Figueroa et al. (2004) found highest expression in the hepatopancreas and less in muscle and gill tissue. No expression was found in hemocytes. In contrast in *Fenneropenaeus chinensis*, Lai et al. (2011) found highest expression in muscle, less in hepatopancreas, gill and intestine and a clear expression also in hemocytes. Animals challenged with the white spot syndrome virus showed, after an initial (6 h) decrease, upregulation of HDL-BGBP over a four day period in hepatopancreas, but downregulation in muscle, intestine and gills (Lai et al. 2011). In *L. vannamei*, mRNA levels of HDL-BGBP were decreased in the hepatopancreas under short term starvation (Muhlia-Almazan et al. 2005).

Role of the HDL-BGBP in the Innate Immune Defense

HDL-BGBPs represent pattern recognition proteins which are wide spread in many vertebrate and invertebrate taxa. HDL-BGBPs are specific for microbial β-1,3-glucans and are thus part of the innate immune system and act in concert with the hemocytes. In crustaceans and other invertebrates, another group of glucan binding proteins exist, the lipopolysaccharide and glucan binding proteins (LGBPs) containing a glucanase domain and, like the HDL-BGBPs, a RGD motive (Liu et al. 2009; Zhao et al. 2009). Both groups are sometimes simply termed β-glucan binding proteins although they represent different groups of proteins. While the relationship of LGBPs with glucanases is well established (Lin et al. 2008; Padhi and Verghese 2008; Pauchet et al. 2009) the relationship of HDL-BGBPs with known protein families is not clear. The binding of β-1,3-glucans in vitro has been demonstrated by several authors (Duvic and Söderhäll 1990; Jimenez-Vega et al. 2002; Vargas-Requena et al. 2002) and two glucanase like domains have been postulated as possible binding sites in *Pacifastacus* and *Fenneropenaeus* HDL-BGBP (Cerenius et al. 1994; Lai et al. 2011), but a domain search using the SMART online server (http://smart.embl-heidelberg.de/) has not revealed any clues as to the presence of such domains in *Pacifastacus*, *Pontastacus* and *Astacus* (unpublished observations). LGBP's have a lower molecular mass 40–50 kDa and do not bind lipids, while HDL-BGBPs contain about 30–50% lipids (mostly phospholipids, Hall et al. 1995; Stieb et al. 2008; Vargas-Requena et al. 2002, see Table 1) and their apoproteins have a molecular mass around 100 kDa (Puppione et al. 1986; Spaziani et al. 1986; Lee and Puppione 1988; Hall et al. 1995; Jimenez-Vega et al. 2002; Yepiz-Plascencia et al. 2002; Stieb et al. 2008; Goncalves et al. 2012).

As found already many years ago (Duvic and Söderhäll 1990), HDL-BGBP enhances the activity of phenoloxidase (PO) and a peptidase from the hemocytes only when bound to β-1,3-glucans. These early findings were later supported by the finding of a corresponding receptor for HDL-BGBP in the hemocyte membrane, which however could only bind to HDL-BGBP when associated with β-1,3-glucan (Duvic and Soderhall 1992). Both HDL-BGBP and LGBP thus act as opsonizing factors for microbial components. This finding was confirmed more recently for two shrimp species (Goncalves et al. 2012). However, the question remains as to how HDL-BGBP is taken up into target tissues when not associated with β-1,3-glucan as shown for the uptake into Y-organs (Kang and Spaziani 1995a, b, c; see above).

The glucan/HDL-BGBP complex leads to the exocytic release of the components of the PO system from a subpopulation of the hemocytes (the semigranular cells; Johansson and Söderhäll 1985; Barracco et al. 1991) and triggers a cascade of serine proteases leading to the conversion of prophenoloxidase to its active form, phenoloxidase. Phenoloxidases catalyze the early steps of melanization by oxidation of monophenols such as tyrosine di-chinones followed by auto-oxidation to melanin, eventually leading to the encapsulation of the invading pathogens (Cerenius and Soderhall 2004). As an important side effect, toxic chinones are produced during this reaction. For more details on this complex system, the reader is referred to several reviews on this subject (Cerenius and Soderhall 2018; Cerenius and Soderhall 2004; Cerenius et al. 2010a, b). This aspect is also discussed in the chapter by Coates and Costa-Paiva in this volume.

A direct antimicrobial effect of the HDL-BGBPs has not yet been studied in great detail. In two shrimp species (*Fenneropenaeus chinensis* and *Litopenaeus schmittii*), HDL-BGBP was reported to agglutinate fungal (yeast) cells (Goncalves et al. 2012) but no effect on cell viability was studied. In a preliminary study from our laboratory (Mogler 2017), the inhibitory effect of HDL-BGBP on the growth of *E. coli* (gram negative), *Micrococcus luteus* (gram positive) and two yeasts (*Saccharomyces cerevisiae* and *Pichia jardinii*) was tested. At concentrations up to 200 μg/ml, no inhibitory effect was found except for a slight growth inhibition in *Saccharomyces*.

Are dLP and HDL-BGBP Proteins Restricted to the Malacostraca?

An enigmatic finding is that the dLP could be isolated as a lipoprotein particle only in *Astacus leptodactylus*, while in its congener *Astacus astacus*, a lipidated dLP was not found in the hemolymph, although the corresponding gene has a sequence identity of 97% (Dal Magro, Lehmann and Hoeger, unpublished). At present, complete sequences for the dLP/HDL-BGBP precursor in the NCBI database are from these two species only. Therefore, we searched the NCBI Transcriptome Shotgun Assembly (TSA) sequence database (https://www.ncbi.nlm.nih.gov/genbank/tsa/) for related proteins in the presently available transcriptomes of Crustacea. Using the *Astacus leptodactylus* dLP/HDL-BGBP precursor sequence (Accession Nr. AHJ78589.1) as query, a tblastn-search revealed more than 50 hits with sequence coverages of 99–100% and E-values close to zero indicating a high degree of sequence similarity. Fifteen of these sequences were analyzed for the presence of domains and motives characteristic for the dLP/HDL-BGBP precursor as shown in Fig. 2.3. Highly similar sequences were found in representatives of different malacostracan groups, i.e., decapods (amino acid sequence identity 43–85%), euphausiids (40–41%), isopods and amphipods (29–32%). In contrast, no database hits were obtained for representatives of the lower crustaceans: branchiopods (*Daphnia*, *Triops*) and copepods (8 species). It seems unlikely that the absence of the dLP/HDL-BGBP precursor is caused by the lack of transcriptome coverage, but rather reflects taxonomic differences. This finding has further implications. Given our previous observation that both the dLP and the HDL-BGBP originate from a common precursor (see above), this also would indicate that by the absence of the dLP/HDL-BGBP precursor a functional HDL-BGBP is lacking in these groups. This is a surprising result, since in decapod crustaceans the HDL-BGBP has two important functions: first, it has been considered as the main lipid transporting protein exhibiting a lipid binding capacity of 40–50% (see Table 1) and second, it serves as a pattern recognition protein triggering the innate immune response as described above. In line with this, no genes encoding for peptidoglycan recognition proteins could be found in the genome of *Daphnia pulex* (McTaggart et al. 2009). Therefore, if these findings are correct, the question arises as to how the non-malacostracan crustaceans carry out their major lipid transport and which alternative mechanisms are realized in these other groups to support the innate immune system.

Apolipocrustacein/Vitellogenin

Vitellogenin (VTG) is the precursor of the major yolk protein, vitellin, which accumulates in the oocytes of oviparous animals. The term "vitellogenin" was introduced by Pan et al. (1969) for insect yolk protein precursors and later used to describe the yolk precursor proteins of other animals, such as vertebrates and also crustaceans. In crustaceans, the term "Lipo-vitellin" was used to describe the major egg yolk protein isolated from crustacean eggs (Wallace et al. 1967). VTGs have native molecular masses between 325 and 700 KDa (see Wilder et al. 2010, Table 2; and Lee 1991, Table 5, for an overview) with lipid contents between 19 and 50% (see Table 1). In addition to neutral- and phospholipids, crustacean VTGs contain various amounts of carotenoids, mainly astaxanthin, canthaxanthine and β-carotene (for an overview, see Lee 1991). Carotenoids may protect from excess radiation (Sagi et al. 1995 and cited references). More recently, Babin et al. (2010) found that dietary carotenoids enhanced phenoloxidase activity and resistance to bacterial infection in gammarid amphipods.

VTGs belong to the group of the non-exchangeable large lipid transfer proteins, which include the ApoB type lipoproteins of vertebrates, the lipophorins of flying insects (for a review, see Smolenaars et al. 2007b) and the microsomal triglyceride transfer proteins (MTPs) which are essential for the formation of intracellular lipoprotein particles (reviewed by Hussain et al. 2003, 2012). For crustacean MTP, this has recently been demonstrated for the first time in the ectoparasitic *Lepeophtheirus salmonis* (Khan et al. 2017). Molecular analysis has revealed that these three protein groups derive from a common ancestor apoprotein (Babin et al. 1999). At present, most of the complete crustacean VTG sequences listed in the NCBI nr database are derived from decapods. The sequences show a length between 2500 and 2600 amino acids whereas available sequences for other groups (amphipods, copepods) are generally shorter (1500–1700 amino acids).

Recent phylogenetic analysis of the VTG genes of crustaceans has revealed that they are related to the vertebrate ApoB type lipoproteins and not to the "classical" VTGs of vertebrates and insects (Avarre et al. 2007). Therefore, the term "apolipocrustacein" has been suggested for the former crustacean VTGs to avoid confusion (Avarre et al. 2007). In contrast, the crustacean clotting proteins (see below) do show a relationship with these classical vitellogenins, a finding supported by the absence of a domain of unknown function (DUF 1081) in the clotting proteins and the classical VTGs (see Fig. 2.5), while this domain is present in the apolipocrustaceins. In contrast, the yolk protein of the branchiopod *Daphnia* clusters with the classical VTGs (Avarre et al. 2007) and a more recent paper on the lipoproteins of the copepod *Tigriopus* (Lee et al. 2016) indicates that the VTG-like proteins in this and four other copepod species are also more related to the true VTGs and decapod crustacean clotting proteins. This could indicate that the evolution of apolipocrustaceins took place only on the level of the higher Crustacea (Malacostraca) but this hypothesis should be corroborated by more extensive analysis of the corresponding lipoproteins of additional lower crustacean groups.

The site of VTG synthesis has been considered different in the two major groups of the decapod crustaceans, the Pleocyemata (comprising shrimps and prawns) and the Dendrobranchiata (most other decapod crustaceans). In the former group, expression has been found both in the hepatopancreas and the ovaries (where it is synthesized by the follicle cells; Tsutsui et al. 2000), while in the latter group, VTG expression is restricted to the hepatopancreas (Avarre et al. 2007; Jeon et al. 2010). However, this view has been challenged by the expression of VTG in both tissues in some

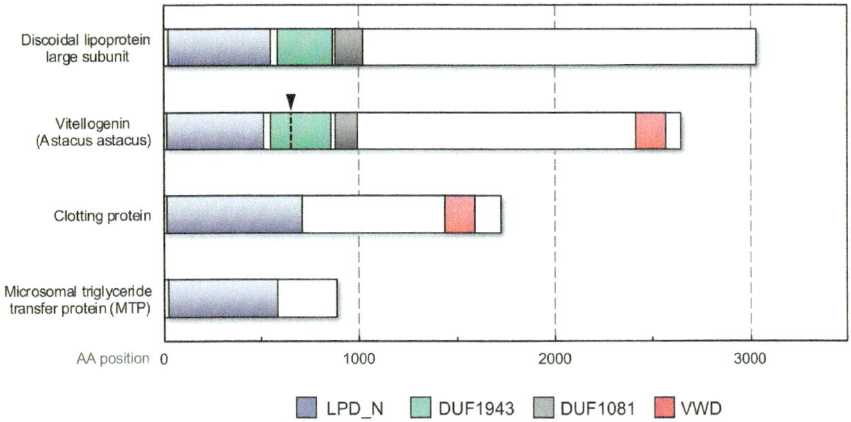

Fig. 2.5 Schematic representation of the domain structure of Crustacean large lipid transfer proteins (LLTP) using the gene sequences of *Astacus leptodactylus* and *A. astacus* as an example. The lipid binding domain (LPD_N) is the characteristic trait of all members of this family. The microsomal triglyceride transfer protein (MTP) is the simplest LLTP; the clotting protein has a von Willebrand type D domain (VWD). The large subunit of the discoidal lipoprotein has two additional domains of unknown function (DUF1943, DUF1081); Vitellogenin has all the latter three domains. Accession numbers: MTP: GAFS01003576.1 TSA; clotting protein: GBEI01001282.1 TSA; vitellogenin: GEDF01024702.1 (*A. astacus;*); large discoidal lipoprotein: KF896205.1. For accession numbers from the TSA database, the predicted amino acid sequence was determined from the open reading frames using the online tool Translate (http://web.expasy.org/translate/); the domain boundaries were determined using SMART (http://smart.embl-heidelberg.de/). The arrow indicates a furin cleavage site in the vitellogenin sequence. Signal peptides sequences have been omitted

species of the Dendrobranchiata In the hepatopancreas, two different VTG genes were found, while the ovaries expressed only one of these genes (Jeon et al. 2010 and cited references).

A difference between these two groups can also be found in the degree of VTG glycosylation. Roth et al. (2010) have shown that the Pleocyemata group VTG possess multiple putative N-glycosylation sites (8 species analyzed), while in the Dendrobranchiata (8 species) VTG showed only O-glycosylation sites and were largely deficient in N-glycosylation sites. According to the authors, this could have to do with differences in the reproductive mode, since members of the Pleocyemata carry their eggs until completion of the development, while the Dendrobranchiata release their eggs which develop into a lecitotrophic nauplius freely floating in the water column.

Vitellogenin synthesis proceeds in coordination with the molting cycle and is under control of inhibitory neuropeptides produced in the X-organ/sinus gland complex (vitellogenesis-inhibiting hormone and molt-inhibiting hormone) and other compounds like the ecdysteroids (produced in the Y-organs) and the sesquiterpenoid methyl-farnesoate (produced in the mandibular organs). In addition, in gonochoristic species, the sex specific expression is also controlled by the androgenic gland

(Okumura et al. 2005; Kang et al. 2008). These aspects will not be dealt with here and the reader is referred to appropriate reviews (e.g., Nagaraju 2011).

Processing of Apolipocrustacein

It is well known that VTG is partially cleaved after uptake into the oocyte leading to its final storage form, vitellin (VT). In crustaceans, the situation is more complicated that in many species the VTG precursor is already cleaved upon release into the hemolymph and upon uptake into the oocytes further cleavage can take place. In Crustacea, this has been well documented but mostly for decapods as the economically most important group (for examples, see Wilder et al. 2010, Table 2). However, VTG processing has also been observed in other groups such as isopods (Okuno et al. 2000).

Detailed studies by Okuno et al. (2002) have shown that the VTG of *Macrobrachium rosenbergii* is cleaved first at a furin-type cleavage site before its release into the hemolymph. A second cleavage at a furin-type consensus sequence (RDRR) occurs when VTG circulates in the hemolymph leading to three apolipoprotein subunits, presumably associated with one VTG particle. However, the existence of an extracellular furin-type protease needs to be established. In *Penaeus semisulcatus*, Avarre et al. (2003) likewise found a primary cleavage of VTG at amino acid position 710 while the second cleavage at position 1796 took place only after uptake into the ovary. Later, Jeon et al. (2010) have compared the furin cleavage sites in 20 decapod crustaceans of different suborders and found one highly conserved cleavage site located N-terminally at amino acid positions 725–730 depending on the species. Cleavage of the apolipocrustacein at position 725–730 would affect the DUF 1943 domain which is located next to the N-terminal lipid binding domain at amino acid positions between 550 and 850 (see Fig. 2.2). Although still designated a domain of unknown functions in the NCBI database, this domain is part of a series of 17 helices forming a "lipid pocket" which serves for the formation of the nascent lipoprotein particle (Richardson et al. 2005). Homology modelling on the basis of the lamprey vitellin showed a high structural similarity with vertebrate ApoB lipoproteins and also the apolipocrustacein (formerly Vitellogenin) of *Macrobrachium rosenbergii* (Roth et al. 2010). The cleavage in the central lipid binding region does not necessarily impair the lipid transport capacity of VTG, since it might be stabilized by interaction with the lipid moiety or by the α-Helix region surrounding the lipid pocket (Richardson et al. 2005). Likewise in insects, furin cleavage of the apolipophorin II/I takes place within the N-terminal lipid binding domain (at position 720, Smolenaars et al. 2005; Marinotti et al. 2006; Wen et al. 2017). The former authors showed that inhibition of cleavage does not prevent lipidation and secretion of the lipophorin particle. As in other oviparous species, uptake of VTG is mediated by a member of the LDL-receptor family. This crustacean VTG receptor is a 210 kDa membrane-bound protein and has been identified and studied in *Scylla serrata, Macrobrachium rosenbergii*, *Penaeus monodon* and *Marsupenaeus japonicus* (Warrier and Subramoniam

2002; Mekuchi et al. 2008; Tiu et al. 2008; Roth and Khalaila 2012). The receptor binding region of VTG has been recently identified in *Macrobrachium rosenbergii* and was found located in the beta-sheet domain of the outer core of the N-terminal region (residues 237–260 Roth et al. 2013). Interestingly, three different regions were identified where interaction with VTG could take place. Thus, the reason for the cleavage within the lipid binding region could change its conformation to improve the interaction with the VTG receptor.

Clotting Protein

Clotting Systems and Clotting Proteins of Invertebrates

In both vertebrates and invertebrates, clotting of blood or hemolymph is essential to prevent blood/hemolymph loss and at the same time to block the invasion of foreign microorganisms. With an open circulatory system, arthropods are especially vulnerable to the loss of hemolymph and bacterial infections. Different mechanisms have evolved in vertebrates and invertebrates. In mammalian blood, coagulation starts with fibrinogens which are first proteolytically processed thus allowing the interaction of the resulting fibrin to form polymers. These polymers are further stabilized by crosslinking catalyzed by the enzyme, plasma transglutaminase. Transglutaminase (TGase; Protein-glutamine gamma-glutamyltransferase; EC 2.2.13) catalyzes thiol- and calcium dependent transamidation reactions. Transamidation leads to the formation of a covalent bond between the γ-carboxamide group of a peptide bound glutamine residue and a primary amine group of another peptide such as for example the ε-amino group of a lysine residue. This leads to a covalent crosslink between the γ-carboxamide group of a peptide bound glutamine and the primary amino group (Griffin et al. 2002).

Among invertebrates, the known clotting mechanisms are diverse but again, crosslinking is an important process in clot formation, as will be shown below. The biochemical and molecular basis of clotting and coagulation has been mainly studied in the chelicerate *Limulus*, in crustaceans and in insects. In the well-studied *Limulus* system, the clotting components are all stored in inactive form in the hemocytes. Elicitors such as bacterial cell wall lipopolysaccharides and fungal β-1,3-glucans are recognized by pattern recognition proteins leading to a release of the granules and a proteolytic cascade is initiated in the hemolymph. The *Limulus* clotting protein, termed coagulogen, is a 175 amino acid protein of 19.6 kDa mass and belongs to the cystine knot superfamily which includes the nerve growth factor Neurotrophin and the Toll receptor ligand Spätzle (Bergner et al. 1996). Clotting is initiated by the clotting enzyme, a protease which converts the clotting protein coagulogen into an insoluble form, coagulin. Coagulin self-assembles to form large aggregates. These aggregates are further stabilized by proxins, proline-rich proteins located on the hemocyte surface. Proxins are then crosslinked with coagulin by the action of transglutaminase thus stabilizing the clot (Osaki et al. 2002). For an overview on this subject, see Kawabata and Muta (2010) and Cerenius and Söderhäll (2011).

In insects, clotting is likewise initiated by the activation of hemocytes mediated by different pattern recognition receptors which activate different signaling pathways such as the Toll pathway (Strand 2008). Both extracellular and intracellular (i.e. hemocytes) proteins are recruited for clot formation. The large number of different proteins identified to-date belong to quite different protein families. These include proteins that are also involved in other cellular functions such as the lipophorins, members of the large lipid transfer proteins, hexamerins, which are larval storage proteins related to hemocyanins and a secreted version of gelsolin, an actin binding protein. In addition, hemolectin and hemocytin, large (340 and 430 kDa, respectively) proteins are released from hemocytes along with TGase and constitute the most abundant proteins in the clot. Hemolectin and hemocytin carry domains which are also present in the human von Willebrand factor. Some of the clotting factors are species-specific, such as the Fondue and Glutactin protein in *Drosophila* (Scherfer et al. 2004; Schmid et al. 2019). Characteristic for insects, coagulation involves the activation of the prophenoloxidase-activating system leading to further stabilization of the clot by melanization (Schmid et al. 2019). For further details on insect clotting, see the reviews by Dushay (2009), Loof et al. (2011) and the chapter by K. Clark in this issue.

In contrast to insects, the clotting system and the known components in crustaceans appear to be less complex. TGase is also a key component, but unlike in the *Limulus* system, where the enzyme stabilizes the initial clot produced by coagulin auto-polymerization, TGase is essential for the polymer formation. TGase is stored in the hemocytes, while its substrate, the clotting protein (see below), is an extracellular protein circulating in the hemolymph. Clotting is initiated by exocytosis of granules containing TGase triggered by microbial compounds and the LGBP and the HDL/BGBP acting as pattern recognizing proteins as described above. As in insects, Toll like receptors are likely to be involved in the signaling pathway. Several Toll like receptors have been described in crustaceans (see Wang and Wang 2013 and cited references) in several tissues including hemocytes (Wang et al. 2012).

The Crustacean Clotting Proteins

So far, clotting reaction and clotting proteins have been studied only in the decapod crustaceans (for a recent overview, see Perdomo-Morales et al. 2019). In earlier studies, CP was isolated from the hemolymph as a 170–220 kDa protein in a homodimeric form with the subunits linked by disulfide bridges (Fuller and Doolittle 1971; Kopacek et al. 1993; Hall et al. 1995, 1999; Komatsu and Ando 1998; Montano-Perez et al. 1999; Yeh et al. 1999; Yepiz-Plascencia et al. 2002; Perazzolo et al. 2005; Cheng et al. 2008b; Wang et al. 2013). In *Pacifastacus leniusculus* and

Penaeus monodon, the transglutaminase-mediated polymerization of CP was visualized by electron microscopy using purified clotting protein (Hall et al. 1999; Chen et al. 2005). Clotting protein molecules rapidly assembled into long, flexible and occasionally branching chains. Silencing of either the transglutaminase or clotting protein gene in *Marsupenaeus japonicus* using double stranded RNA injection led to a loss of hemolymph clotting over several days (Maningas et al. 2008).

CPs belong to the very high-density lipoproteins (see Fig. 2.1) and their lipid carrying capacity is rather low ranging between 6 and 12% (Table 2.1). In *Pacifastacus leniusculus*, most of the CP-bound lipids consist of diacylglycerols and phosphatidylcholine (Cerenius et al. 1994) and considering its concentration in the hemolymph, it has been calculated that in *P. leniusculus,* more than 50% of the total hemolymph lipids are bound to CP while about 30% are bound to the HDL-BGBP (Cerenius et al. 1994). Although CP lipid content is low, the clotting proteins may also have a defensive function since lipoprotein particles can bind bacterial lipopolysaccharides (see Schmidt et al. 2010 and cited references). More recently, the lipid moiety of the locust lipophorin was found to bind and aggregate (gram-positive) bacteria and fungi (Wynant et al. 2014). In the light of this antimicrobial function, this view may also be true for the lipid moiety of HDL-BGBP discussed above.

After the first N-terminal sequences were published, the relationship of CP with vitellogenin and the large lipid transfer family of proteins became apparent (Doolittle and Riley 1990; Hall et al. 1995, 1999; see Fig. 2.5). As already pointed out above, it is one of the peculiarities of the higher crustaceans that their yolk proteins (the former vitellogenins, now termed apolipocrustaceins) are more related to the ApoB type lipoproteins while the CPs (which are present in both sexes) are related to the (sex specific) vitellogenins of insects and vertebrates.

The sequences of several CPs show an enrichment of both glutamine and lysine residues which make them preferred substrates for the TGases and lysine- and glutamine-rich stretches have been found in several malacostracan crustaceans (Hall et al. 1999; Yeh et al. 1999; Cheng et al. 2008b). We have expanded the search for glutamine- and lysine-rich regions by including putative CPs predicted from transcriptomes of a number of crustaceans available in the NCBI TSA database (https://www.ncbi.nlm.nih.gov/Traces/wgs/?page=1&view=tsa). A tblastn search was performed using the clotting protein of *Pacifastacus leniusculus* (gb AAD16454.1) as query. The resulting hits with high query sequence cover (>95%) and low E-values ($<1^{-150}$) were tested for the presence (N-terminal lipid binding domain, von Willebrand type D domain) and the absence (DUF1943 domain) of the motives characteristic for the annotated CPs and grouped in different taxonomic groups. Our search indicated that glutamine-rich regions were highly variable and appeared to be characteristic for different taxonomic groups (Fig. 2.6). Polyglutamine (polyQ) stretches were especially evident in some amphipod species and even more so in some copepods, were up to 20 polyQ residues were present (see Fig. 2.6).

Characteristic lysine-rich stretches arranged in Ser-Lys-Thr or Thr-Lys-Thr repeats as described before (Hall et al. 1999; Yeh et al. 1999; Cheng et al. 2008b) were found only in predicted clotting protein sequences of six crayfish species, five species of shrimp, as listed in Fig. 2.6 and two species of crabs (data not shown)

```
Astacidea (crayfish)
Pac. len.      1152 | GALEFQ-FGQGETESFGQE-----QQQETQSRQRIVQLQQDFMPITGGIVRTVSMGVELKGATKK | 1210
Ast. ast._     1152 | GALEFQ-FGQGEAETLGQEQQQQQQQQETQSRLRIVKLQEEFMPITGGIVRTVSIGVELQGATKR | 1215
Ast. lep.CP1   1120 | GALEFQ-FGQGEAETFGQDQQQQQQQQETQSRLRIVKLQEEFMPITGGIVRTISIGVELQGATKR | 1183
Ast. lep.CP2   1141 | GSLAIQKFGQGEAEPFGLL---QNEQQENQSRQRIVKLLEDVMPITGGRVNGVSLGVELDG-SPS | 1201
Proc. clar.    1145 | GSLEFQ-LGQGETDSFGQQ-----QQLESQSRQRIVKLQEELMPVTGGIVRTVSVGVELQAATKR | 1203
Hom. am.       1131 | ---DYQEFG--ETEIFA-------QEQQSESRQRIAKLQEQVMPISGGYVQSVSVGVELQGSSTS | 1183
Cher. quad.    1151 | GSRDFQEIGDLELETFGQ------QQQQTNYRQRIEKLQEEVMPISGGYVKSISIGLDLQG-STS | 1208
                           *  *  *                   * ** *    **  **     * ** *
Penaeoidea (Penaeid shrimp)
Pen. mon.      1597 | EGQYHCKALSPMPVCRPECVASHPITTSVEYECRFSGHQQQAQQQQQQVQGTQWEECFPHFYTL | 1661
Lit. vann.     1594 | EGQYHCKALSPMPVCRPECVANPPITTSVEYECKFSEHRQQAQ-QLQQQVQGTKWVECFPHSYTL | 1657
Mar. jap.      1599 | EGHYHCKALTPMPVCRPECVANSPITTSVEYECKFTGHHQQAQ-QQQQQVEGTQWDECFPHSYTL | 1662
Fen. pen.      1596 | EGQYHCKALSPMPVCRPGCIANPPITTSVEYECKFSGRQQQA--QQQQQVQGTQWEECFPHSYTL | 1658
Farf. azt.     1590 | EGQYHCKALSPMPVCRPECVANHPITTSVEYECKFVGHQQQGQKQQQQQVQGRQWVECYPHSYTL | 1654
                      **  ***** ******* * ** ************             *      ** ** ***
Gammaroidea (Gammarid amphipods)
Gam. fos.      1868 | EQQQ------QQHHQRHHQSVEGQQNWEQQQQQQQRRQQEQ------EQSQQQQERRQQEQQQGR | 1920
Het. sop.      1868 | EQQQQRGQDEQQHHHRRHQSVEGQQNRDQQQRRQQ------------EQVQSQQERWQQEQQQNR | 1920
Mac. wag.      1867 | EQQQQRGQDEQQHYHRGHESVEGQQNHDQQQRRQHEQ----------EQEQSQQQRWQQEQQQNR | 1921
Pen. daw.      1868 | EQQQQRGQGEQQHHHRRHQSVKGQQNHDQQQRRQQ------------EQEQSQQERWQQEQQQNR | 1920
Brac. gr.      1863 | EQQQQRGQDEQQHHHRRHQSVEGQQNRDQQQPRQQEQEQSQQQGRQHEQEQSHQERWQQEQQQNR | 1927
                      ****      ***  * * *******  ***                ** *  * ******** *
Calanoida (Calanoid Copepods)
Euc. ser.CP1   1171 | -EKVQGEHKIKYHMV----------------QLKNDQQYQDESRQNNQRQQQQQQLKNVLLKL-- | 1216
Euc. ser.CP2   1162 | -KKEQKQDQDQQQQQQQQQR-----------QRFNSYQDEDEEQQEQEQQQQQNGIKYQTIKVKD | 1214
Lab. mad.      1187 | QQKWQTQQQQQQQQQAQQQELQQLLQQQSLQQQQPLQLQQLQKLKQAQQLQQREQIQQNQQQL-- | 1249
Tem. lon.      1212 | IENQQRQQQLQQQQQEQVQKAQDKFQQGQRTQLQQQQQQQQQQQQQQQQQQQQQQWQQQQQQLQQ | 1276
Eur. aff.      1223 | NQRQQQQQLMQQQQQQQKQKIQQAHQQ----QLNQANQQAEQEQQEEQQQKQQQQQQQQEQLFSQ | 1283
                                                     *                   *
```

Fig. 2.6 Comparison of glutamine-rich regions in the sequences of known and putative clotting proteins of different crustacean groups after multiple partial alignment. The sequences were retrieved by blast searches of the NCBI non-redundant protein sequences database or by tblastn searches in the NCBI TSA database using the *Pacifastacus leniusculus* clotting protein (Accession no. gb_AAD16454.1) as query. The open reading frames and the domain boundaries were determined using the online tools Translate (http://web.expasy.org/translate/) and SMART (http://smart.embl-heidelberg.de/), respectively. Flanking numbers indicate amino acid positions of the ungapped sequences. The following sequences were used (GenBank accession no. in parentheses): Pac. len. (*Pacifastacus leniusculus*, AAD16454.1); Ast. ast. (*Astacus astacus*, GEDF0106568.1 TSA); Ast. lep. cp1 (*Astacus leptodactylus* type 1), GEI01001282.1 TSA); Ast. lep. cp2 (*A. leptodactylus* type 2, GBEI01062822.1 TSA); Proc. clar. (*Procambarus clarkii*, GBEV01000942.1 TSA); Hom. am. (*Homarus americanus*, GEBG01001431.1 TSA); Cher. quad. (*Cherax quadricarinatus*, HACK01011849.1 TSA); Pen. mon. (*Penaeus monodon*, AAF19002.1 TSA); Lit. vann. (*Litopenaeus vannamei*, ABI9536.1 TSA); Mar. jap. (*Marsupenaeus japonicus*, ABK59925.1 TSA); Fen. pen. (*Feneropenaeus penicillatus*, GFRT01002262.1 TSA); Farf. azt. (*Farfantepenaeus aztecus*, GEUA91959419.1 TSA); Gam. fos. (*Gammarus fossarum*, GHDC01023472.1 TSA); Het. sop. (*Heterogammarus sophianosii*, TSA gb GEPR01048044.1); Mac. wag. (*Macropereiopus wagneri*, GEQL01039548.1 TSA); Pen. daw. (*Pentagonurus dawydowi*, GEQV01048596.1 TSA); Brac. gr. (*Brachyuropus grewingki*i, GEPD01012907.1 TSA); Euc. ser. cp1 (*Eucyclops serrulatus* type 1, CP1GARW01000128.1 TSA); Euc. ser. cp2 (*Eucyclops serrulatus* type 2, CP1GARW01001931.1 TSA); Lab. mad. (*Labidocera madurae*, GFWO01074197.1 TSA); Tem. lon. (*Temora longicornis*, GGQN01004610.1 TSA); Eur. aff. (*Eurytemora affinis*, GBGO01109147.1 TSA)

but not in amphipods, copepods (listed in Fig. 2.6) and isopods (data not shown). This could indicate different susceptibilities of the different type of CPs against the crosslinking activity of TGase. PolyQ stretches are excellent substrates for (human) TGase and the reactivity of TGase has been found to greatly increase with increasing number of glutamine residues (Kahlem et al. 1996).

In humans, polyQ stretches can arise from genetic disorders caused by repeats in the codon CAG encoding for glutamine, which results in an expanded polyQ tract in the translated protein. PolyQs promote protein aggregation and are the cause of neurodegenerative diseases such as Huntington's disease. However, there is a threshold of the polyQ length critical for the onset of disease (see Adegbuyiro et al. 2017 for a recent review) and in the human proteome Ramazzotti et al. (2012) found that polyQs in proteins associated with genetic diseases were longer than polyQ's in disease-unrelated proteins. PolyQ tracts seem to be a normal trait of proteins.

In addition to serving as substrates for TGase, polyQ stretches might also aid in the interaction with proteins other than clotting proteins which are also recruited for clot formation. This might be supported by a bioinformatic study on different taxa (plants, fungi, Nematoda and Chordata) in which Schaefer et al. (2012) found polyQ stretches in many unrelated protein families. This led the authors to conclude that polyQ tracts are involved in protein interactions due to their enrichment in protein complexes and their association with coiled-coil regions which are typical motives for protein interactions. In crustaceans, these additional proteins include the protease inhibitor alpha-2-macroglobulin which is also a substrate of transglutaminase (Hall and Söderhäll 1994; Chaikeeratisak et al. 2012). Alpha-2-macroglobulin is likely to prevent the degradation of the CP by bacterial proteases (Chaikeeratisak et al. 2012) as found during infection of the shrimp *Penaeus vannamei* with the bacterium *Vibrio harveyi* (Chuang et al. 2016). In the same species, Yao et al. (2019) have identified hemocyanin, HDL-BGBP, alpha-macroglobulin and some other proteins, in addition to CP, as clot components. Hemocyanin was also identified as a substrate of transglutaminase. In the tarantula *Acanthoscurria geniculata*, mass spectroscopy of the clot proteins by Sanggaard et al. (2016) revealed hemocyanin as the major component accounting for more than 50% of the clot proteins, followed by two hemolectins (15%) and vitellogenin (6%) and smaller amounts of a large number of additional proteins. In the light of the complex clot proteins found in insects (see above), recruitment of proteins other than those involved in the actual clotting process seems to be a common theme. Considering these differences found between the *Limulus* system, tarantula, the insects and the crustaceans, we can expect many more clotting mechanisms among the invertebrate taxa that have not been studied in great detail, such as in the mollusks.

Tissue Expression and Additional Functions of Crustacean Clotting Proteins

The concentration of CP in the hemolymph has been measured in the shrimp *Penaeus monodon* (3 mg/ml, Yeh et al. 2007). Experimental injury (i.e., withdrawal of hemolymph) led to a four-fold increase of CP in the hemolymph suggesting resynthesis. The protein was found in many tissues with highest expression in the lymphoid (hematopoietic) organ, gill and nervous system, and less expression in hepatopancreas and muscle tissue. No expression was found in hemocytes (Yeh et al. 1999; 2007). Cheng et al. (2008a) found CP expression in mature ovaries (but not in immature ovaries and testes) and eye stalk ablation led to a 40-fold increase of its expression (measured by RT-PCR) in the ovaries. Furthermore, CP was consumed during embryonic development suggesting a role as a storage protein—an interesting finding highlighting the relationship of CP with the vitellogenins.

The same authors found that purified CP reacted with anti-phosphoserine, anti-phosphothreonine, and anti-phosphotyrosine antibodies suggesting that CP is subjected to phosphorylation. So far, this property has not been studied in CPs from tissues other than ovaries, but given the occurrence of the proteins in many different tissues this finding suggests intracellular functions in addition to clotting.

A novel function of CP in hematopoiesis was recently described in the crayfish *Pacifastacus* (Junkunlo et al. 2018). Hematopoietic progenitor cells were found to highly express both CP and TGase while in differentiated hemocytes, their expression levels were low. CP was secreted by these cells along with TGase and produced a filamentous extracellular network as a result of TGase crosslinking activity. Thus, CP participates in the formation of an extra cellular matrix in the hematopoietic tissue, together with collagen. This matrix was found to influence the morphology, spreading and migration of hematopoietic cells. The progenitor cell behavior was also stimulated by the TGase inhibitor, Astakine 1 (Soderhall et al. 2005; Lin et al. 2010), whose function was thought to modulate the structure of the extracellular matrix by controlling the degree of crosslinking by the TGase. Another TGase inhibitor, cystamine, led to a release of progenitor cells into the hemolymph (Junkunlo et al. 2019).

Some properties of the CP are not yet clear and need to be further investigated, such as the presence of integrin binding (RGD) motives in shrimp CPs (Cheng et al. 2008b), the significance of the glycosylation (Yeh et al. 1998) and a latent protease activity found in the CP of *Litopenaeus vanamei* (Reyes-Izquierdo and Vargas-Albores 2001), which appeared when the protein was reduced.

References

Adegbuyiro A, Sedighi F, Pilkington AWt, Groover S, Legleiter J (2017) Proteins containing expanded polyglutamine tracts and neurodegenerative disease. Biochemistry 56(9):1199–1217. https://doi.org/10.1021/acs.biochem.6b00936

Avarre JC, Lubzens E, Babin PJ (2007) Apolipocrustacein, formerly vitellogenin, is the major egg yolk precursor protein in decapod crustaceans and is homologous to insect apolipophorin II/I and vertebrate apolipoprotein B. BMC Evol Biol 7:3. https://doi.org/10.1186/1471-2148-7-3

Avarre JC, Michelis R, Tietz A, Lubzens E (2003) Relationship between vitellogenin and vitellin in a marine shrimp (*Penaeus semisulcatus*) and molecular characterization of vitellogenin complementary DNAs. Biol Reprod 69(1):355–364. https://doi.org/10.1095/biolreprod.102.011627

Babin A, Biard C, Moret Y (2010) Dietary supplementation with carotenoids improves immunity without increasing its cost in a crustacean. Am Nat 176(2):234–241. https://doi.org/10.1086/653670

Babin PJ, Bogerd J, Kooiman FP, Van Marrewijk WJ, Van der Horst DJ (1999) Apolipophorin II/I, apolipoprotein B, vitellogenin, and microsomal triglyceride transfer protein genes are derived from a common ancestor. J Mol Evol 49(1):150–160. https://doi.org/10.1007/pl00006528

Barracco MA, Duvic B, Soderhall K (1991) The beta-1,3-glucan-binding protein from the crayfish *Pacifastacus leniusculus*, when reacted with a beta-1,3-glucan, induces spreading and degranulation of crayfish granular cells. Cell Tissue Res 266(3):491–497. https://doi.org/10.1007/Bf00318590

Bergner A, Oganessyan V, Muta T, Iwanaga S, Typke D, Huber R, Bode W (1996) Crystal structure of a coagulogen, the clotting protein from horseshoe crab: a structural homologue of nerve growth factor. EMBO J 15(24):6789–6797

Burmester T (1999) Identification, molecular cloning, and phylogenetic analysis of a non-respiratory pseudo-hemocyanin of *Homarus americanus*. J Biol Chem 274:13217–13222. https://doi.org/10.1074/jbc.274.19.13217

Cerenius L, Jiravanichpaisal P, Liu HP, Soderhall I (2010a) Crustacean immunity. Adv Exp Med Biol 708:239–259. https://doi.org/10.1007/978-1-4419-8059-5_13

Cerenius L, Kawabata S, Lee BL, Nonaka M, Soderhall K (2010b) Proteolytic cascades and their involvement in invertebrate immunity. Trends Biochem Sci 35(10):575–583. https://doi.org/10.1016/j.tibs.2010.04.006

Cerenius L, Liang Z, Duvic B, Keyser P, Hellman U, Palva ET, Iwananga S, Söderhäll K (1994) Structure and biological activity of a 1,3-ß-D-glucan-binding protein in crustacean blood. J Biol Chem 269(47):29462–29467

Cerenius L, Soderhall K (2004) The prophenoloxidase-activating system in invertebrates. Immunol Rev 198:116–126. https://doi.org/10.1111/j.0105-2896.2004.00116.x

Cerenius L, Soderhall K (2011) Coagulation in Invertebrates. J Innate Immun 3(1):3–8. https://doi.org/10.1159/000322066

Cerenius L, Soderhall K (2018) Crayfish immunity—recent findings. Dev Comp Immunol 80:94–98. https://doi.org/10.1016/j.dci.2017.05.010

Chaikeeratisak V, Somboonwiwat K, Tassanakajon A (2012) Shrimp alpha-2-macroglobulin prevents the bacterial escape by inhibiting fibrinolysis of blood clots. PLoS ONE 7(10):e47384. https://doi.org/10.1371/journal.pone.0047384

Chapman MJ (1980) Animal lipoproteins: chemistry, structure, and comparative aspects. J Lipid Res 21:789–853

Chen J, Honeyager SM, Schleede J, Avanesov A, Laughon A, Blair SS (2012) Crossveinless d is a vitellogenin-like lipoprotein that binds BMPs and HSPGs, and is required for normal BMP signaling in the *Drosophila* wing. Development 139(12):2170–2176. https://doi.org/10.1242/dev.073817

Chen MY, Hu KY, Huang CC, Song YL (2005) More than one type of transglutaminase in invertebrates? A second type of transglutaminase is involved in shrimp coagulation. Dev Comp Immunol 29(12):1003–1016. https://doi.org/10.1016/j.dci.2005.03.012

Cheng W, Chiang PC, Lai CY, Yeh MS (2008a) Expression of clottable protein of tiger shrimp (*Penaeus monodon*) in gonads and its possible role as nutrient source for the embryo. Dev Comp Immunol 32(12):1422–1429. https://doi.org/10.1016/j.dci.2008.05.011

Cheng W, Tsai IH, Huang CJ, Chiang PC, Cheng CH, Yeh MS (2008b) Cloning and characterization of hemolymph clottable proteins of kuruma prawn (*Marsupenaeus japonicus*) and white shrimp (*Litopenaeus vannamei*). Dev Comp Immunol 32(3):265–274. https://doi.org/10.1016/j.dci.2007.05.009

Chuang WH, Liu PC, Chien YC, Wu YC, Lee KK (2016) Effects of Vibrio harveyi on plasma clotting protein (coagulogen) of white shrimp *Litopenaeus vannamei*. Aquac Res 47(11):3464–3476. https://doi.org/10.1111/are.12796

Coates CJ, Decker H (2017) Immunological properties of oxygen-transport proteins: hemoglobin, hemocyanin and hemerythrin. Cell Mol Life Sci 74(2):293–317. https://doi.org/10.1007/s00018-016-2326-7

Cunningham M, Garcia F, Pollero RJ (2007) Arachnid lipoproteins: Comparative aspects. Comp Biochem Physiol C 146:79–87. https://doi.org/10.1016/j.cbpc.2006.06.011

Depledge MH, Bjerregaard P (1989) Hemolymph protein-composition and copper levels in decapod crustaceans. Helgolander Meeresun 43(2):207–223. https://doi.org/10.1007/Bf02367900

Doolittle RF, Riley M (1990) The amino-terminal sequence of lobster fibrinogen reveals common ancestry with vitellogenins. Biochem Biophys Res Commun 167(1):16–19. https://doi.org/10.1016/0006-291x(90)91723-6

Dushay MS (2009) Insect hemolymph clotting. Cell Mol Life Sci 66(16):2643–2650. https://doi.org/10.1007/s00018-009-0036-0

Duvic B, Soderhall K (1992) Purification and partial characterization of a beta-1,3-glucan-binding-protein membrane receptor from blood cells of the crayfish *Pacifastacus leniusculus*. Eur J Biochem 207(1):223–228. https://doi.org/10.1111/j.1432-1033.1992.tb17041.x

Duvic B, Söderhäll K (1990) Purification and characterization of a ß-1,3-glucan binding protein from plasma of the crayfish *Pacifastacus leniusculus*. J Biol Chem 265(16):9327–9332

Fuller GM, Doolittle RF (1971) Studies of invertebrate fibrinogen. I. Purification and characterization of fibrinogen from the spiny lobster. Biochemistry 10(8):1305–1311. https://doi.org/10.1021/bi00784a005

Fyffe WE, O'Connor JD (1974) Characterization and quantification of a crustacean lipovitellin. Comp Biochem Physiol B 47(4):851–867. https://doi.org/10.1016/0305-0491(74)90030-3

Goncalves P, Vernal J, Rosa RD, Yepiz-Plascencia G, de Souza CR, Barracco MA, Perazzolo LM (2012) Evidence for a novel biological role for the multifunctional β-1,3-glucan binding protein in shrimp. Mol Immunol 51(3–4):363–367. https://doi.org/10.1016/j.molimm.2012.03.032

Griffin M, Casadio R, Bergamini CM (2002) Transglutaminases: nature's biological glues. Biochem J 368(Pt 2):377–396. https://doi.org/10.1042/BJ20021234

Hall M, Söderhäll K (1994) Crayfish [alpha]-macroglobulin as a substrate for transglutaminases. Comp Biochem Physiol B 108(1):65–72. https://doi.org/10.1016/0305-0491(94)90166-X

Hall M, van Heusden MC, Soderhall K (1995) Identification of the major lipoproteins in crayfish hemolymph as proteins involved in immune recognition and clotting. Biochem Biophys Res Commun 216(3):939–946. https://doi.org/10.1006/bbrc.1995.2711

Hall M, Wang R, van Antwerpen R, Sottrup-Jensen L, Soderhall K (1999) The crayfish plasma clotting protein: a vitellogenin-related protein responsible for clot formation in crustacean blood. Proc Natl Acad Sci U S A 96(5):1965–1970. https://doi.org/10.1073/pnas.96.5.1965

Haunerland NH, Bowers WS (1989) Comparative studies on arthropod lipoproteins. Comp Biochem Physiol B 92(1):137–141. https://doi.org/10.1016/0305-0491(89)90326-X

Hoeger U, Florey E (1989) Catecholamine Degradation in the Hemolymph of the Chinese Crab Eriocheir sinensis. Comp Biochem Phys C 92(2):323–327. https://doi.org/10.1016/0742-8413(89)90062-5

Hussain MM, Rava P, Walsh M, Rana M, Iqbal J (2012) Multiple functions of microsomal triglyceride transfer protein. Nutr Metab (Lond) 9:14. https://doi.org/10.1186/1743-7075-9-14

Hussain MM, Shi J, Dreizen P (2003) Microsomal triglyceride transfer protein and its role in apoB-lipoprotein assembly. J Lipid Res 44(1):22–32. https://doi.org/10.1194/jlr.r200014-jlr200

Jeon JM, Lee SO, Kim KS, Baek HJ, Kim S, Kim IK, Mykles DL, Kim HW (2010) Characterization of two vitellogenin cDNAs from a Pandalus shrimp (*Pandalopsis japonica*): expression in hepatopancreas is down-regulated by endosulfan exposure. Comp Biochem Physiol 157(1):102–112. https://doi.org/10.1016/j.cbpb.2010.05.006

Jimenez-Vega F, Sotelo-Mundo RR, Ascencio F, Vargas-Albores F (2002) 1,3-beta-D glucan binding protein (BGBP) from the white shrimp, *Penaeus vannamei*, is also a heparin binding protein. Fish Shellfish Immunol 13(3):171–181. https://doi.org/10.1006/fsim.2001.0392

Johansson MW (1999) Cell adhesion molecules in invertebrate immunity. Dev Comp Immunol 23(4–5):303–315. https://doi.org/10.1016/S0145-305X(99)00013-0

Johansson MW, Söderhäll K (1985) Exocytosis of the prophenoloxidase activating system from crayfish haemocytes. J Comp Physiol B 156(2):175–181. https://doi.org/10.1007/bf00695771

Junkunlo K, Soderhall K, Soderhall I (2018) Clotting protein—an extracellular matrix (ECM) protein involved in crustacean hematopoiesis. Dev Comp Immunol 78:132–140. https://doi.org/10.1016/j.dci.2017.09.017

Junkunlo K, Soderhall K, Soderhall I (2019) Transglutaminase inhibition stimulates hematopoiesis and reduces aggressive behavior of crayfish. Pacifastacus leniusculus. J Biol Chem 294(2):708–715. https://doi.org/10.1074/jbc.RA118.005489

Kahlem P, Terre C, Green H, Djian P (1996) Peptides containing glutamine repeats as substrates for transglutaminase-catalyzed cross-linking: relevance to diseases of the nervous system. Proc Natl Acad Sci U S A 93(25):14580–14585. https://doi.org/10.1073/pnas.93.25.14580

Kang BJ, Nanri T, Lee JM, Saito H, Han CH, Hatakeyama M, Saigusa M (2008) Vitellogenesis in both sexes of gonochoristic mud shrimp, *Upogebia major* (Crustacea): analyses of vitellogenin gene expression and vitellogenin processing. Comp Biochem Physiol B 149(4):589–598. https://doi.org/10.1016/j.cbpb.2007.12.003

Kang BK, Spaziani E (1995a) Uptake of high-density lipoprotein by Y-organs of the crab *Cancer antennarius*. 3. Evidence for adsorptive endocytosis and the absence of lysosomal processing. J Exp Zool 273(5):425–433. https://doi.org/10.1002/jez.1402730506

Kang BK, Spaziani E (1995b) Uptake of high-density-lipoprotein by y-organs of the crab, *Cancer antennarius*. 1. characterization in-vitro and effects of stimulators and inhibitors. Arch Insect Biochem 30(1):61–75. https://doi.org/10.1002/arch.940300105

Kang BK, Spaziani E (1995c) Uptake of high-density-lipoprotein by y-organs of the crab, *Cancer antennarius*. 2. Formal characterization of receptor-mediation with isolated membranes. Arch Insect Biochem 30(1):77–91. https://doi.org/10.1002/arch.940300106

Kawabata S, Muta T (2010) Sadaaki Iwanaga: Discovery of the lipopolysaccharide- and beta-1,3-D-glucan-mediated proteolytic cascade and unique proteins in invertebrate immunity. J Biochem 147(5):611–618. https://doi.org/10.1093/jb/mvq026

Khan MT, Dalvin S, Nilsen F, Male R (2017) Microsomal triglyceride transfer protein in the ectoparasitic crustacean salmon louse (*Lepeophtheirus salmonis*). J Lipid Res 58(8):1613–1623. https://doi.org/10.1194/jlr.M076430

Khovidhunkit W, Kim M-S, Memon RA, Shigenaga JK, Moser AH, Feingold KR, Grunfeld C (2004) Effects of infection and inflammation on lipid and lipoprotein metabolism: mechanisms and consequences to the host. J Lipid Res 45:1169–1196. https://doi.org/10.1194/jlr.R300019-JLR200

Kollman JM, Quispe J (2005) The 17 Å structure of the 420 kDa lobster clottable protein by single particle reconstruction from cryoelectron micrographs. J Struct Biol 151:306–314

Komatsu M, Ando S (1998) A very-high-density lipoprotein with clotting ability from hemolymph of sand crayfish Ibacus ciliatus. Biosci Biotechnol Biochem 62(3):459–463. https://doi.org/10.1271/bbb.62.459

Kopacek P, Hall M, Soderhall K (1993) Characterization of a clotting protein, isolated from plasma of the freshwater crayfish *Pacifastacus leniusculus*. Eur J Biochem 213(1):591–597. https://doi.org/10.1111/j.1432-1033.1993.tb17798.x

Lai X, Kong J, Wang Q, Wang W, Meng X (2011) Cloning and characterization of a beta-1,3-glucan-binding protein from shrimp *Fenneropenaeus chinensis*. Mol Biol Rep 38(7):4527–4535. https://doi.org/10.1007/s11033-010-0583-3

Lee KM, Lee KY, Choi HW, Cho MY, Kwon TH, Kawabata S, Lee BL (2000) Activated phenoloxidase from *Tenebrio molitor* larvae enhances the synthesis of melanin by using a vitellogenin-like protein in the presence of dopamine. Eur J Biochem 267(12):3695–3703. https://doi.org/10.1046/j.1432-1327.2000.01402.x

Lee RF (1991) Lipoproteins from the hemolymph and ovaries of marine invertebrates. In: Gilles R (ed) Advances in comparative and environmental physiology. Springer, Berlin, pp 187–207

Lee RF, Puppione DL (1978) Serum lipoproteins in the spiny lobster *Panulirus interruptus*. Comp Biochem Physiol B 59:239–243. https://doi.org/10.1016/0305-0491(78)90253-5

Lee RF, Puppione DL (1988) Lipoproteins I and II from the hemolymph of the blue crab *Callinectes sapidus*: Lipoprotein II associated with vitellogenesis. J Exp Zool 248:278–289. https://doi.org/10.1002/jez.1402480306

Lee SR, Lee JH, Kim AR, Kim S, Park H, Baek HJ, Kim HW (2016) Three cDNAs encoding vitellogenin homologs from Antarctic copepod, *Tigriopus kingsejongensis*: Cloning and transcriptional analysis in different maturation stages, temperatures, and putative reproductive hormones. Comp Biochem Physiol B 192:38–48. https://doi.org/10.1016/j.cbpb.2015.11.008

Lin X, Novotny M, Soderhall K, Soderhall I (2010) Ancient cytokines, the role of astakines as hematopoietic growth factors. J Biol Chem 285(37):28577–28586. https://doi.org/10.1074/jbc.M110.138560

Lin YC, Vaseeharan B, Chen JC (2008) Identification and phylogenetic analysis on lipopolysaccharide and beta-1,3-glucan binding protein (LGBP) of kuruma shrimp *Marsupenaeus japonicus*. Dev Comp Immunol 32(11):1260–1269. https://doi.org/10.1016/j.dci.2008.05.003

Liu F, Li F, Dong B, Wang X, Xiang J (2009) Molecular cloning and characterisation of a pattern recognition protein, lipopolysaccharide and beta-1,3-glucan binding protein (LGBP) from Chinese shrimp *Fenneropenaeus chinensis*. Mol Biol Rep 36(3):471–477. https://doi.org/10.1007/s11033-007-9203-2

Loof TG, Schmidt O, Herwald H, Theopold U (2011) Coagulation Systems of Invertebrates and Vertebrates and Their Roles in Innate Immunity: The Same Side of Two Coins? J Innate Immun 3(1):34–40. https://doi.org/10.1159/000321641

Lubzens E, Ravid T, Khayat M, Daube N, Tietz A (1997) Isolation and characterization of the high-density lipoproteins from the hemolymph and ovary of the penaeid shrimp *Penaeus semisulcatus* (de Haan): Apoproteins and lipids. J Exp Zool 278(6):339–348

Maningas MB, Kondo H, Hirono I, Saito-Taki T, Aoki T (2008) Essential function of transglutaminase and clotting protein in shrimp immunity. Mol Immunol 45(5):1269–1275. https://doi.org/10.1016/j.molimm.2007.09.016

Mann CJ, Anderson TA, Read J, Chester SA, Harrison GB, Köchl S, Ritchie PJ, Bradbury P, Hussain FS, Amey J, Vanloo B, Rosseneu M, Infante R, Hancock JM, Levitt DG, Banaszak LJ, Scott J, Shoulders CC (1999) The structure of vitellogenein provides a molecular model for the assembly and secretion of of atherogenic lipoproteins. J Mol Biol 285:391–408. https://doi.org/10.1006/jmbi.1998.2298

Marinotti O, Cappurro MdL, Nirmala X, Calvo E, James AA (2006) Structure and expression of the lipophorin-encoding gene of the malaria vector. Anopheles gambiae. Comp Biochem Physiol B 144(1):101–109

McTaggart SJ, Conlon C, Colbourne JK, Blaxter ML, Little TJ (2009) The components of the *Daphnia pulex* immune system as revealed by complete genome sequencing. BMC Genom 10:175. https://doi.org/10.1186/1471-2164-10-175

Mekuchi M, Ohira T, Kawazoe I, Jasmani S, Suitoh K, Kim YK, Jayasankar V, Nagasawa H, Wilder MN (2008) Characterization and expression of the putative ovarian lipoprotein receptor in the Kuruma prawn *Marsupenaeus japonicus*. Zoolog Sci 25(4):428–437. https://doi.org/10.2108/zsj.25.428

Mogler M (2017) Interaktionen von Lipoproteinen aus der Hämolymphe zweier Crustaceen mit verschiedenen Mikroorganismen. Masters Thesis, Institute of Molecular Physiology, Johannes Gutenberg-University, Mainz

Montano-Perez K, Yepiz-Plascencia G, Higuera-Ciapara I, Vargas-Albores F (1999) Purification and characterization of the clotting protein from the white shrimp *Penaeus vannamei*. Comp Biochem Phys B 122(4):381–387. https://doi.org/10.1016/S0305-0491(99)00011-5

Muhlia-Almazan A, Sanchez-Paz A, Garcia-Carreno F, Peregrino-Uriarte AB, Yepiz-Plascencia G (2005) Starvation and diet composition affect mRNA levels of the high density-lipoprotein-beta glucan binding protein in the shrimp *Litopenaeus vannamei*. Comp Biochem Physiol B 142(2):209–216. https://doi.org/10.1016/j.cbpc.2005.07.005

Nagaraju GP (2011) Reproductive regulators in decapod crustaceans: an overview. J Exp Biol 214(Pt 1):3–16. https://doi.org/10.1242/jeb.047183

Okumura T, Nikaido H, Yoshida K, Kotaniguchi M, Tsuno Y, Seto Y, Watanabe T (2005) Changes in gonadal development, androgenic gland cell structure, and hemolymph vitellogenin levels during male phase and sex change in laboratory-maintained protandric shrimp, *Pandalus hypsinotus* (Crustacea: Caridea: Pandalidae). Mar Biol 148(2):347–361. https://doi.org/10.1007/s00227-005-0073-7

Okuno A, Katayama H, Nagasawa H (2000) Partial characterization of vitellin and localization of vitellogenin production in the terrestrial isopod Armadillidium vulgare. Comp Biochem Physiol B 126(3):397–407. https://doi.org/10.1016/s0305-0491(00)00214-5

Okuno A, Yang WJ, Jayasankar V, Saido-Sakanaka H, Huong DT, Jasmani S, Atmomarsono M, Subramoniam T, Tsutsui N, Ohira T, Kawazoe I, Aida K, Wilder MN (2002) Deduced primary structure of vitellogenin in the giant freshwater prawn, *Macrobrachium rosenbergii*, and yolk processing during ovarian maturation. J Exp Zool 292(5):417–429. https://doi.org/10.1002/jez.10083

Osaki T, Okino N, Tokunaga F, Iwanaga S, Kawabata S (2002) Proline-rich cell surface antigens of horseshoe crab hemocytes are substrates for protein cross-linking with a clotting protein coagulin. J Biol Chem 277(42):40084–40090. https://doi.org/10.1074/jbc.M206773200

Padhi A, Verghese B (2008) Detecting molecular adaptation at individual codons in the pattern recognition protein, lipopolysaccharide- and beta-1,3-glucan-binding protein of decapods. Fish Shellfish Immunol 24(5):638–648. https://doi.org/10.1016/j.fsi.2008.02.002

Pan ML, Bell WJ, Telfer WH (1969) Vitellogenic blood protein synthesis by insect fat body. Science 165(3891):393–394. https://doi.org/10.1126/science.165.3891.393

Pauchet Y, Freitak D, Heidel-Fischer HM, Heckel DG, Vogel H (2009) Immunity or digestion: glucanase activity in a glucan-binding protein family from lepidoptera. J Biol Chem 284(4):2214–2224. https://doi.org/10.1074/jbc.M806204200

Perazzolo LM, Lorenzini DM, Daffre S, Barracco MA (2005) Purification and partial characterization of the plasma clotting protein from the pink shrimp *Farfantepenaeus paulensis*. Comp Biochem Physiol B 142:302–307. https://doi.org/10.1016/j.cbpb.2005.07.015

Perdomo-Morales R, Montero-Alejo V, Perera E (2019) The clotting system in decapod crustaceans: History, current knowledge and what we need to know beyond the models. Fish Shellfish Immunol 84(-):204–212. https://doi.org/10.1016/j.fsi.2018.09.060

Puppione DL, Jensen DF, O'Connor JD (1986) Physicochemical study of rock crab lipoproteins. Biochim Biophys Acta 875:563–568. https://doi.org/10.1016/0005-2760(86)90078-0

Ramazzotti M, Monsellier E, Kamoun C, Degl'Innocenti D, Melki R (2012) Polyglutamine repeats are associated to specific sequence biases that are conserved among eukaryotes. PLoS ONE 7(2):e30824. https://doi.org/10.1371/journal.pone.0030824

Ravid T, Tietz A, Khayat M, Boehm E, Michelis R, Lubzens E (1999) Lipid accumulation in the ovaries of a marine shrimp penaeus semisulcatus (de haan). J Exp Biol 202(Pt 13):1819–1829

Reyes-Izquierdo T, Vargas-Albores F (2001) Proteinase activity in the white shrimp (*Penaeus vannamei*) clotting protein. Biochem Biophys Res Commun 287(2):332–336. https://doi.org/10.1006/bbrc.2001.5595

Richardson PE, Manchekar M, Dashti N, Jones MK, Beigneux A, Young SG, Harvey SC, Segrest JP (2005) Assembly of lipoprotein particles containing apolipoprotein-B: structural model for the nascent lipoprotein particle. Biophys J 88:2789–2800. https://doi.org/10.1529/biophysj.104.046235

Rodenburg KW, van der Horst DJ (2005) Lipoprotein-mediated lipid transport in insects: analogy to the mammalian lipid carrier system and novel concepts for the functioning of LDL receptor family members. Biochim Biophys Acta 1736:10–29. https://doi.org/10.1016/j.bbalip.2005.07.002

Romo-Figueroa MG, Vargas-Requena C, Sotelo-Mundo RR, Vargas-Albores F, Higuera-Ciapara I, Soderhall K, Yepiz-Plascencia G (2004) Molecular cloning of a beta-glucan pattern-recognition lipoprotein from the white shrimp *Penaeus* (*Litopenaeus*) *vannamei*: correlations between the deduced amino acid sequence and the native protein structure. Dev Comp Immunol 28(7–8):713–726. https://doi.org/10.1016/j.dci.2003.11.008

Roth Z, Khalaila I (2012) Identification and characterization of the vitellogenin receptor in *Macrobrachium rosenbergii* and its expression during vitellogenesis. Mol Reprod Dev 79(7):478–487. https://doi.org/10.1002/mrd.22055

Roth Z, Parnes S, Wiel S, Sagi A, Zmora N, Chung JS, Khalaila I (2010) N-glycan moieties of the crustacean egg yolk protein and their glycosylation sites. Glycoconj J 27(1):159–169. https://doi.org/10.1007/s10719-009-9268-3

Roth Z, Weil S, Aflalo ED, Manor R, Sagi A, Khalaila I (2013) Identification of receptor-interacting regions of vitellogenin within evolutionarily conserved beta-sheet structures by using a peptide array. ChemBioChem 14(9):1116–1122. https://doi.org/10.1002/cbic.201300152

Ruiz-Verdugo LM, García-Bañuelos ML, Vargas-Albores F, Inocencio H-C, Yepiz-Plascencia GM (1997) Amino acids and lipids of the plasma HDL from the white shrimp *Penaeus vannamei* Boone. Comp Biochem Physiol B 118(1):91–96. https://doi.org/10.1016/S0305-0491(97)00025-4

Ruoslahti E (1996) RGD and other recognition sequences for integrins. Ann Rev Cell Dev Biol 12:697–715. https://doi.org/10.1146/annurev.cellbio.12.1.697

Sagi A, Rise M, Isam K, Arad SM (1995) Carotenoids and their derivatives in organs of the maturing female crayfish *Cherax-quadricarinatus*. Comp Biochem Phys B 112(2):309–313. https://doi.org/10.1016/0305-0491(95)00069-0

Sanggaard KW, Dyrlund TF, Bechsgaard JS, Scavenius C, Wang T, Bilde T, Enghild JJ (2016) The spider hemolymph clot proteome reveals high concentrations of hemocyanin and von Willebrand factor-like proteins. Biochim Biophys Acta 1864(2):233–241. https://doi.org/10.1016/j.bbapap.2015.11.004

Schaefer MH, Wanker EE, Andrade-Navarro MA (2012) Evolution and function of CAG/polyglutamine repeats in protein-protein interaction networks. Nucleic Acids Res 40(10):4273–4287. https://doi.org/10.1093/nar/gks011

Schenk S, Gras H, Marksteiner D, Patasic L, Prommnitz B, Hoeger U (2009) The *Pandinus imperator* haemolymph lipoprotein, an unusual phosphatidylserine carrying lipoprotein. Insect Biochem Mol Biol 39(10):735–744. https://doi.org/10.1016/j.ibmb.2009.08.009

Schenk S, Harris JR, Hoeger U (2006) A discoidal lipoprotein from the coelomic fluid of the polychaete *Nereis virens*. Comp Biochem Physiol B 143(2):236–243. https://doi.org/10.1016/j.cbpb.2005.11.012

Schenk S, Schmidt J, Hoeger U, Decker H (2015) Lipoprotein-induced phenoloxidase-activity in tarantula hemocyanin. Biochim Biophys Acta 1854:939–949. https://doi.org/10.1016/j.bbapap.2015.03.006

Scherfer C, Karlsson C, Loseva O, Bidla G, Goto A, Havemann J, Dushay MS, Theopold U (2004) Isolation and characterization of hemolymph clotting factors in *Drosophila melanogaster* by a pullout method. Curr Biol 14(7):625–629. https://doi.org/10.1016/j.cub.2004.03.030

Schmid MR, Dziedziech A, Arefin B, Kienzle T, Wang Z, Akhter M, Berka J, Theopold U (2019) Insect hemolymph coagulation: Kinetics of classically and non-classically secreted clotting factors. Insect Biochem Mol Biol 109:63–71. https://doi.org/10.1016/j.ibmb.2019.04.007

Schmidt O, Soderhall K, Theopold U, Faye I (2010) Role of adhesion in arthropod immune recognition. Ann Rev Entomol 55:485–504. https://doi.org/10.1146/annurev.ento.54.110807.090618

Segrest JP, Jones MK, De Loof H, Dashti N (2001) Structure of apolipoprotein B-100 in low density lipoproteins. J Lipid Res 42(9):1346–1367

Smolenaars MM, Kasperaitis MA, Richardson PE, Rodenburg KW, Van der Horst DJ (2005) Biosynthesis and secretion of insect lipoprotein: involvement of furin in cleavage of the apoB homolog, apolipophorin-II/I. J Lipid Res 46(3):412–421. https://doi.org/10.1194/jlr.M400374-JLR200

Smolenaars MMW, De Morreé A, Kerver J, Van der Horst DJ, Rodenburg KW (2007a) Insect lipoprotein biogenesis depends on an amphipatic ß cluster in apolipophorin II/I and is stimulated by microsomal triglyceride transfer protein. J Lipid Res 48:1955–1965. https://doi.org/10.1194/jlr.M600434-JLR200

Smolenaars MMW, Madsen O, Rodenburg KW, Van der Horst DJ (2007b) Molecular diversity of the large lipid transfer protein superfamily. J Lipid Res 48:489–502. https://doi.org/10.1194/jlr.R600028-JLR200

Soderhall I, Kim YA, Jiravanichpaisal P, Lee SY, Soderhall K (2005) An ancient role for a prokineticin domain in invertebrate hematopoiesis. J Immunol 174(10):6153–6160. https://doi.org/10.4049/jimmunol.174.10.6153

Soulages JL, Wells MA (1994) Lipophorin: the structure of an insect lipoprotein and its role in lipid transport in insects. Adv Protein Chem 45:371–415. https://doi.org/10.1016/S0065-3233(08)60644-0

Spaziani E, Havel RJ, Hamilton RL, Hardman DA, Stoudemire JB, Watson RD (1986) Properties of serum high-density lipoproteins in the crab, *Cancer antennarius* Stimpson. Comp Biochem Physiol B 85(2):307–314

Spaziani E, Wang WL (1991) Serum high-density lipoproteins in the crab, *Cancer antennarius*—III. Density-gradient profiles and lipid composition of subclasses. Comp Biochem Physiol B 98(4):555–561. https://doi.org/10.1016/0305-0491(86)90005-2

Stieb S, Hoeger U, Schenk S (2008) A large discoidal lipoprotein present in only one of two closely related crayfish. J Comp Physiol B 178(6):755–765. https://doi.org/10.1007/s00360-008-0266-8

Stieb S, Roth Z, Dal Magro C, Fischer S, Butz E, Sagi A, Khalaila I, Lieb B, Schenk S, Hoeger U (2014) One precursor, three apolipoproteins: the relationship between two crustacean lipoproteins, the large discoidal lipoprotein and the high density lipoprotein/beta-glucan binding protein. Biochim Biophys Acta 1841(12):1700–1708. https://doi.org/10.1016/j.bbalip.2014.09.020

Strand MR (2008) The insect cellular immune response. Insect Sci 15(1):1–14. https://doi.org/10.1111/j.1744-7917.2008.00183.x

Terwilliger NB, Dangott L, Ryan M (1999) Cryptocyanin, a crustacean molting protein: evolutionary link with arthropod hemocyanins and insect hexamerins. Proc Natl Acad Sci U S A 96(5):2013–2018. https://doi.org/10.1073/pnas.96.5.2013

Tirumalai R, Subramoniam T (1992) Purification and characterization of vitellogenin and lipovitellins of the sand crab *Emerita asiatica*: molecular aspects of crab yolk proteins. Mol Reprod Dev 33(1):16–26. https://doi.org/10.1002/mrd.1080330104

Tiu SH, Benzie J, Chan SM (2008) From hepatopancreas to ovary: molecular characterization of a shrimp vitellogenin receptor involved in the processing of vitellogenin. Biol Reprod 79(1):66–74. https://doi.org/10.1095/biolreprod.107.066258

Tsutsui N, Kawazoe I, Ohira T, Jasmani S, Yang WJ, Wilder MN, Aida K (2000) Molecular characterization of a cDNA encoding vitellogenin and its expression in the hepatopancreas and ovary during vitellogenesis in the Kuruma prawn Penaeus japonicus. Zoolog Sci 17(5):651–660. https://doi.org/10.2108/zsj.15.651

Vargas-Requena CL, Hernandez-Santoyo A, Yepiz-Plascencia G, Vargas-Albores F, Higuera-Ciapara I, Rodriguez-Romero A, Soulages JL, Sotelo-Mundo RR (2002) Biophysical evidence of lipid and carbohydrate binding activities of shrimp high density lipoprotein/beta glucan binding protein. Protein Pept Lett 9(4):337–344. https://doi.org/10.2174/0929866023408706

Wallace RA, Walker SL, Hauschka PV (1967) Crustacean lipovitellin. Isolation and characterization of the major high-density lipoprotein from the eggs of decapods. Biochemistry 6(6):1582–1590. https://doi.org/10.1021/bi00858a003

Wang BJ, Peng HN, Liu M, Jiang KY, Zhang GF, Wang L (2013) Purification and identification of a clotting protein from the hemolymph of Chinese shrimp (*Fenneropenaeus chinensis*). J Ocean U China 12(3):477–483. https://doi.org/10.1007/s11802-013-2026-y

Wang PH, Liang JP, Gu ZH, Wan DH, Weng SP, Yu XQ, He JG (2012) Molecular cloning, characterization and expression analysis of two novel Tolls (LvToll2 and LvToll3) and three putative Spatzle-like Toll ligands (LvSpz1-3) from *Litopenaeus vannamei*. Dev Comp Immunol 36(2):359–371. https://doi.org/10.1016/j.dci.2011.07.007

Wang XW, Wang JX (2013) Pattern recognition receptors acting in innate immune system of shrimp against pathogen infections. Fish Shellfish Immunol 34(4):981–989. https://doi.org/10.1016/j.fsi.2012.08.008

Warrier SA, Subramoniam T (2002) Receptor mediated yolk protein uptake in the crab *Scylla serrata*: crustacean vitellogenin receptor recognizes related mammalian serum lipoproteins. Mol Reprod Dev 61:536–548. https://doi.org/10.1002/mrd.10106

Wen D, Luo H, Li T, Wu C, Zhang J, Wang X, Zhang R (2017) Cloning and characterization of an insect apolipoprotein (apolipophorin-II/I) involved in the host immune response of *Antheraea pernyi*. Dev Comp Immunol 77:221–228

Whitten MM, Tew IF, Lee BL, Ratcliffe NA (2004) A novel role for an insect apolipoprotein (apolipophorin III) in beta-1,3-glucan pattern recognition and cellular encapsulation reactions. J Immunol 172(4):2177–2185. https://doi.org/10.4049/jimmunol.172.4.2177

Wilder MN, Okumura T, Tsutsui N (2010) Reproductive mechanisms in Crustacea focusing on selected prawn species: vitellogenin structure, processing and synthetic control. Aqua-Bio Sci Monogr 3(3):73–110. https://doi.org/10.5047/absm.2010.00303.0073

Wynant N, Duressa TF, Santos D, Van Duppen J, Proost P, Huybrechts R, Vanden Broeck J (2014) Lipophorins can adhere to dsRNA, bacteria and fungi present in the hemolymph of the desert locust: a role as general scavenger for pathogens in the open body cavity. J Insect Physiol 64:7–13. https://doi.org/10.1016/j.jinsphys.2014.02.010

Yao D, Wang Z, Wei M, Zhao X, Aweya JJ, Zhong M, Li S, Zhang Y (2019) Analysis of *Litopenaeus vannamei* hemocyanin interacting proteins reveals its role in hemolymph clotting. J Proteomics 201:57–64. https://doi.org/10.1016/j.jprot.2019.04.013

Yeh MS, Chen YL, Tsai IH (1998) The hemolymph clottable proteins of tiger shrimp, *Penaeus monodon*, and related species. Comp Biochem Physiol B 121(2):169–176. https://doi.org/10.1016/j.fsi.2006.10.007

Yeh MS, Huang CJ, Cheng JH, Tsai IH (2007) Tissue-specific expression and regulation of the haemolymph clottable protein of tiger shrimp (*Penaeus monodon*). Fish Shellfish Immunol 23(2):272–279. https://doi.org/10.1016/j.fsi.2006.10.007

Yeh MS, Huang CJ, Leu JH, Lee YC, Tsai IH (1999) Molecular cloning and characterization of a hemolymph clottable protein from tiger shrimp (*Penaeus monodon*). Eur J Biochem 266(2):624–633. https://doi.org/10.1016/S0305-0491(98)10085-8

Yepiz-Plascencia G, Jimenez-Vega F, Romo-Figueroa MG, Sotelo-Mundo RR, Vargas-Albores F (2002) Molecular characterization of the bifunctional VHDL-CP from the hemolymph of white shrimp *Penaeus vannamei*. Comp Biochem Physiol B 132(3):585–592. https://doi.org/10.1016/S1096-4959(02)00074-X

Yepiz-Plascencia G, Vargas-Albores F, Jimenez-Vega F, Ruiz-Verdugo LM, Romo-Figueroa G (1998) Shrimp plasma HDL and beta-glucan binding protein (BGBP): comparison of biochemical characteristics. Comp Biochem Physiol B 121(3):309–314. https://doi.org/10.1016/S0305-0491(98)10104-9

Yepiz-Plascencia GM, Sotelo-Mundo R, Vazquez-Moreno L, Ziegler R, Higuera-Ciapara I (1995) A non-sex-specific hemolymph lipoprotein from the white shrimp *Penaeus vannamei* Boone. Isolation and partial characterization. Comp Biochem Physiol 111:181–187. https://doi.org/10.1016/0305-0491(94)00254-R

Zhao D, Chen L, Qin C, Zhang H, Wu P, Li E, Chen L, Qin J (2009) Molecular cloning and characterization of the lipopolysaccharide and beta-1, 3-glucan binding protein in Chinese mitten crab (*Eriocheir sinensis*). Comp Biochem Physiol B 154(1):17–24. https://doi.org/10.1016/j.cbpb.2009.04.014

Chapter 3
The Anti-lipopolysaccharide Factors in Crustaceans

Shihao Li and Fuhua Li

Abstract Anti-lipopolysaccharide factors (ALFs) are a type of antimicrobial peptide (AMP) which show broad-spectrum antimicrobial activity against Gram-positive bacteria, Gram-negative bacteria, fungi and viruses. In this chapter, we review the discovery and classification of this kind of antimicrobial peptide in crustaceans. The structure and function, as well as the mechanism of antibacterial and antiviral activities of ALFs will be summarized and discussed. We will then describe the expression and regulation of various ALF genes in different crustacean species. Finally, the application prospects of ALFs in drug development and disease-resistant genetic breeding will be pointed out and discussed. The review will also discuss several key questions such as the systematic classification and expression regulation of the ALF genes, as well as the future application of ALFs and ALF-derived peptides.

Keywords Anti-lipopolysaccharide factors · LPS-binding domain · Innate Immunity · Structure · Function · Antimicrobial, antiviral activity · Regulation mechanisms · Disease control

Abbreviations

ALF	Anti-lipopolysaccharide factor
AMP	Antimicrobial peptide
EST	Expressed sequence tag

S. Li · F. Li (✉)
Key Laboratory of Experimental Marine Biology, Institute of Oceanology, Chinese Academy of Sciences, Qingdao, China
e-mail: fhli@qdio.ac.cn

S. Li
e-mail: lishihao@qdio.ac.cn

Laboratory for Marine Biology and Biotechnology, Qingdao National Laboratory for Marine Science and Technology, Qingdao, China

Center for Ocean Mega-Science, Chinese Academy of Sciences, Qingdao, China

U. Hoeger and J. R. Harris (eds.), *Vertebrate and Invertebrate Respiratory Proteins, Lipoproteins and other Body Fluid Proteins*, Subcellular Biochemistry 94,
https://doi.org/10.1007/978-3-030-41769-7_3

RACE	Rapid amplification of cDNA end
LPS	Lipopolysaccharide
cDNA	Complementary deoxyribonucleic acid
LPS-BD/LBD	LPS-binding domain
*p*I	Isoelectric point
CDS	Coding sequence
NMR	Nuclear magnetic resonance
LTA	Lipoteichoic acid
RNAi	RNA interference
dsRNA	Double-strand RNA
qPCR	Quantitative PCR
HPT	Hematopoietic tissue
WSSV	White spot syndrome virus
IC50	Half maximal inhibitory concentration
TCBS	Thiosulphate citrate bile salts
SNP	Single nucleotide polymorphism
NF-κB	Nuclear factor-kappaB
IMD	Immune deficiency
JAK/STAT	Janus kinase/signal transducer and activator of transcription
SCOS2	Suppressor of cytokine signaling 2
MyD88	Myeloid differentiation factor 88
TRAF6	Tumor necrosis factor-associated factor 6
CFU	Colony forming unit

Discovery of Anti-lipopolysaccharide Factors

ALFs are a kind of AMP which exhibit a broad-spectrum antimicrobial activity in marine chelicerates (horseshoe crabs) and crustaceans. ALFs were originally purified from the amoebocytes of the chelicerate horseshoe crab, *Limulus polyphemus* (ALF-L) and *Tachypleus tridentatus* (ALF-T) (Ohashi et al. 1984; Tanaka et al. 1982). They were initially discovered in crustaceans from signature sequences in the EST libraries of the shrimps *Litopenaeus vannamei* and *Litopenaeus setiferus* (Gross et al. 2001). Thereafter more and more ALFs were identified from different crustacean species, in which ALFs always showed a large sequence diversity. In *Penaeus monodon*, six different ALF variants (ALFPm1 to 6) have been identified using an EST approach (Ponprateep et al. 2012; Supungul et al. 2004). In *Portunus trituberculatus*, seven anti-lipopolysaccharide factor (PtALF1-7) isoforms, which are encoded by multiple genomic loci, were derived from two full-length enriched cDNA libraries constructed from hemocytes and eyestalks (Liu et al. 2011). In the Chinese shrimp *Fenneropenaeus chinensis*, seven ALF isoforms (FcALF1-7) were identified from the transcriptome database (Li et al. 2013a, b) and one new ALF, FcALF8, was reported recently (Li et al. 2019). In *Macrobrachium nipponense*, five ALF ESTs were discovered through analysis of the established *M. nipponense* transcriptome, and their full-length cDNA sequences were cloned using the RACE method (Wang et al. 2015a). ALF was also identified from the deep-sea shrimp *Rimicaris* sp. from the hydrothermal vent in the Desmos, Manus Basin (Gu et al. 2018). To date, with the powerful deep sequencing technology, ALFs have been reported in many crustacean species including crab, shrimp, prawn and lobster, which provides a basis for the study of their functional diversity.

Classification

As progressively more ALF isoforms were identified, they have been classified into different groups. Initially, five ALF variants from *P. monodon* were divided into two groups, group A (ALFPm1 and ALFPm2) and group B (ALFPm3-5) (Tharntada et al. 2008). Group A and group B ALFs were encoded by two distinct genomic loci. The gene encoding group A ALF consisted of three exons interrupted by two introns whereas the gene encoding group B ALF contained four exons interrupted by three introns. The alignment of genomic sequences with the ALF cDNA sequences revealed that different transcripts in both groups were generated by alternative RNA splicing of the pre-mRNA transcripts, which generated ALF isoforms with the same LPS-BD in the same group (Tharntada et al. 2008). In 2012, a new ALF, ALFPm6, was isolated from *P. monodon* and classified as group C since the putative LPS-BD of ALFPm6 was different from those of groups A and B (Ponprateep et al. 2012). Actually, these three groups represent three distinct genomic loci that encode ALF genes.

A further study by performing phylogenetic analysis on 42 ALF sequences from penaeid shrimp and from the horseshoe crabs identified a new group ALF, group D (Rosa et al. 2013), which contained sequences similar to ALFs found in *L. setiferus* (Gross et al. 2001) and *Litopenaaeus stylirostris* (de Lorgeril et al. 2008). ALFs from different groups have significant differences in their theoretical *p*I. ALFs from group B and C were highly cationic (calculated *p*I = 9.70–10.29) or cationic (calculated *p*I = 7.77–9.77). ALFs from group A ranged from anionic to cationic (calculated *p*I = 5.88–8.06), whereas peptides from group D were high anionic (calculated *p*I = 5.58–6.10) (Rosa et al. 2013). Furthermore, group D ALFs exhibited an incomplete LPS-binding site, which lacked most of the residues proposed to be involved in LPS-binding of group B ALFs (Rosa et al. 2013; Yang et al. 2009). In *Marsupenaeus japonicus*, two ALF members (*Mj*ALF-E1 and *Mj*ALF-E2) with low sequence similarities to groups A to D ALFs were designated as group E (Jiang et al. 2015). *Mj*ALF-E1 and *Mj*ALF-E2 also showed a low sequence similarity with each other. The theoretical *p*I of mature peptide and the LPS-BD of *Mj*ALF-E1 were 9.27 and 5.82, while those of *Mj*ALF-E2 were 5.33 and 9.31, respectively (Jiang et al. 2015).

Recently, more new groups of ALFs were defined based on multiple sequence alignments and phylogenetic analysis. In a recent study, multiple alignments on 47 unique ALF sequences with complete CDS from 10 penaeid shrimp species were performed. The results showed that they were clustered into seven distinct groups, which included the previously described groups A to E (Matos et al. 2018). In silico mining analysis revealed that the seven distinct groups represented seven unique genomic sequences identified in *P. monodon* and each genomic sequence corresponded to a specific ALF member. In addition, all these genes shared a similar structural organization in their coding regions: three exons interrupted by two introns. The second exon encodes the four stranded β-sheets, with the two cysteines delimiting the central β-hairpin (Matos et al. 2018). Phylogenetic analysis indicated that the seven ALF groups were clustered into two main clades: one clade containing ALFs from group A and another clade containing ALFs from groups B to G. The second clade could be divided into three branches. All cationic shrimp ALFs (groups B, C and F) were in the first branch, the anionic ALFs from groups E and G were in the second branch whereas ALFs from group D were in the third branch (Matos et al. 2018).

Currently, the classification of ALFs is mainly performed on genes derived from the shrimp. However, the classification results were not completely consistent in different publications. For instance, the sequence MjALF-A2 regarded as group A member in a previous study (Jiang et al. 2015) was clustered into Group G in the recent report (Matos et al. 2018). The sequence MjALF-E1, previously classified as a cationic member of Group E (Jiang et al. 2015; Smith and Dyrynda 2015), even did not fit in any ALF Group (Matos et al. 2018). Taking all ALFs from prawn, crab and lobster into consideration, the classification of ALFs would become more complicated. A phylogenetic analysis on 39 ALFs with complete CDS from 12 crustacean species including shrimp, prawn and crab, revealed that ALFPm3 (previously clustered into group B) and ALFPm6 (previously clustered into group C) were classified into one same category (Li et al. 2019). Therefore, the classification of all crustacean ALFs still needs a comprehensive analysis in combination with the evolution of ALF genes among different species.

Structure and Function

The tertiary structure of ALF (Fig. 3.1) from the horseshoe crab *L. polyphemus* (ALF-L) and the shrimp *P. monodon* (ALFPm3) has been clearly resolved using X-ray and NMR strategies, respectively. ALF-L and ALFPm3 share a similar three-dimensional structure, consisting of three α-helices packed against a four-stranded β-sheet. Two conserved cysteine residues are involved in an intramolecular disulfide bridge which delimits the central β-hairpin (Rosa et al. 2013). In the horseshoe crab ALF, positively-charged residues recognizing the lipid A moiety of LPS are located in the β-hairpin, stabilized by the disulfide bridge (Hoess et al. 1993). Such a β-hairpin structure is preserved in cyclic synthetic peptides designed on the central-most residues of the β-hairpin, hydrophobic residues at position 44 and 46 of the horseshoe crab ALF sequence playing a crucial role in stabilizing the hydrophobic face of the β-hairpin (Mora et al. 2008). In such cyclic synthetic peptides, the positively-charged

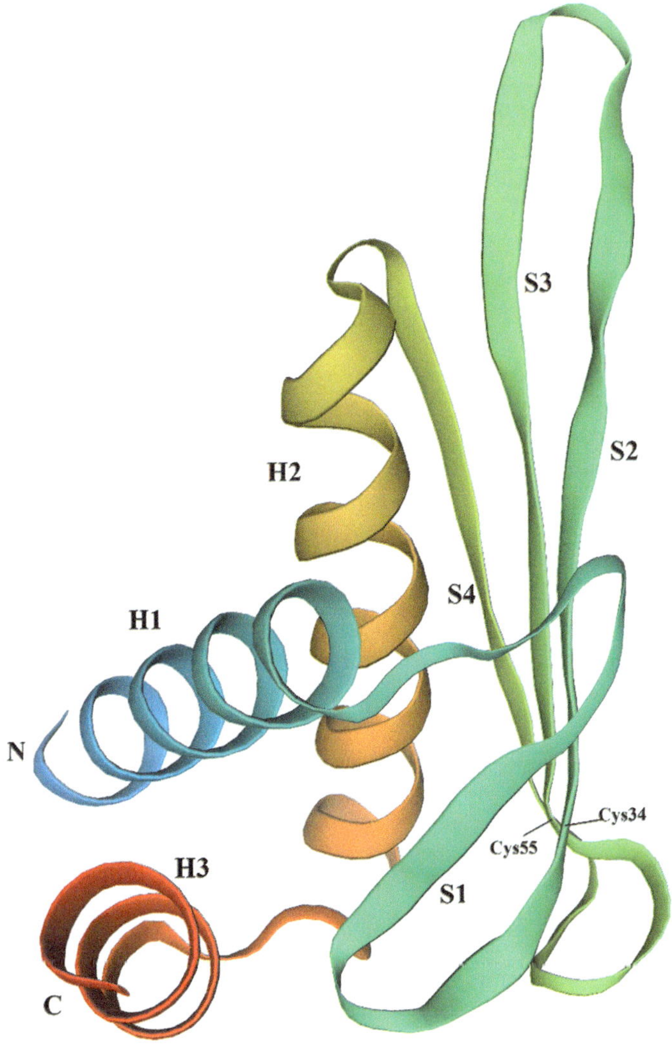

Fig. 3.1 A predicted tertiary structure of the ALF (NCBI accession number: AAX63831) from *F. chinensis*. The three-dimensional structure of the ALF was predicted by the online SWISS-MODEL software (https://swissmodel.expasy.org/) using the default parameters. N and C represent amino and carboxyl terminals, respectively. The predicted helixes (H1-H3) and β-strands (S1-S4) were numbered from N to C terminals. Two conserved cysteine residues forming the disulfide bond were marked as Cys34 and Cys55, respectively

residues bind in an exothermic reaction with the negative charges of LPS and Lipid A, which further shows the crucial role of the β-hairpin in its interaction with the LPS acyl chains (Andra et al. 2007).

Like the native horseshoe crab ALF, the recombinant shrimp ALFPm3 binds to Lipid A and the most probable LPS-binding site involves six positively-charged residues and one negatively-charged amino acid located in the cysteine-stabilized β-hairpin and in the two neighboring β-strands (Yang et al. 2009). Recombinant ALFPm3 was also shown to bind to LTA, a major cell wall component of Gram-positive bacteria (Somboonwiwat et al. 2008). However, the amino acid residues involved in LTA-binding are still unknown. The ability of shrimp ALFs binding to bacterial cell wall components, such as LPS and LTA, was suggested to be associated with their antibacterial activities (Somboonwiwat et al. 2008). Structure resolution of ALFs revealed their binding abilities to the major cell wall components LPS and LTA of bacteria through the LPS-binding sites.

ALFs were first recognized as anticoagulant or anti-LPS factors since they could inhibit the activation of the coagulation factor B and factor C through the binding of ALFs with LPS (Morita et al. 1985; Tanaka et al. 1982). Currently, it is well known that ALF, as well as its functional domain LPS-BD or LBD, exhibits broad-spectrum antimicrobiall activity against microbes that cover a large number of Gram-positive and Gram-negative bacteria, fungi as well as viruses. In *P. monodon*, the antimicrobiall activity spectrum of the recombinant ALFPm3 (rALFPm3) expressed in the yeast *Pichia pastoris*, was tested against a series of Gram-positive and Gram-negative bacteria and filamentous fungi strains (Somboonwiwat et al. 2005). rALFPm3 showed high activities against most of the tested Gram-positive and Gram-negative bacteria as well as the filamentous fungi. A synthetic peptide corresponding to the putative LPS-BD of ALFPm3 displayed activities mainly against Gram-positive bacteria (Somboonwiwat et al. 2005). A further immunohistochemistry assay using anti-ALFPm3 antibody showed that ALFPm3 protein was primarily localized in hemocytes and a rapid increase in the number of hemocytes producing ALFPm3 was observed in *V. harveyi*-injected shrimp, which also suggested that ALFPm3 played important function in defense against invading pathogens (Somboonwiwat et al. 2008).

The antiviral activity of ALF was first reported in *Pacifastacus leniusculus* (Liu et al. 2006). Knockdown of ALF by RNAi specifically resulted in higher rates of viral propagation in a cell culture of HPT from *P. leniusculus*. After a pre-stimulation of ALF transcription level by the administration of UV-treated WSSV, a partial protection against a subsequent challenge with the active WSSV was conferred to the host. The data indicated that ALF can protect the crayfish against WSSV infection. The rALFPm3 protein was also examined for its role in the defense against WSSV infection in Hpt cell cultures of the crayfish *P. leniusculus*, as well as in the live shrimp *P. monodon* (Tharntada et al. 2009). Incubation of Hpt cell cultures with a mixture of WSSV and rALFPm3 led to a dose-dependent decrease of VP28 gene expression level, compared with those incubated with WSSV alone, with an rALFPm3 IC50 value lower than 2.5 μM. The synthetic peptide corresponding to the crayfish ALF LPS-BD could protect cells at a higher concentration (10 μM). Injection of rALFPm3 pre-treated WSSV was able to reduce the in vivo WSSV propagation and prolong the survival of *P. monodon* (Tharntada et al. 2009).

In the last decade, several publications reported the broad-spectrum antimicrobial activity of ALFs from different crustacean species. In *L. vannamei*, knockdown of *LvALF1* through dsRNA mediated RNAi resulted in a higher mortality of shrimps after *Vibrio penaeicida* or *Fusarium oxysporum* infection when compared with controls, showing the protective role of *LvALF1* in shrimps against both bacterial and fungal infections (de la Vega et al. 2008). In *Eriocheir sinensis*, the recombinant protein EsALF-1 displayed bactericidal activity against both G^{-}- and G^{+}-bacteria including *Vibrio anguillarum*, *Escherichia coli* Top10F', and *Micrococcus luteus* (Li et al. 2008) and the recombinant protein EsALF-2 showed antibacterial effects on the growth of *V. anguillarum* and *Pichia pastoris* (Zhang et al. 2010). In the crayfish *Procambarus clarkii*, the recombinant protein PcALF1 could bind to all tested bacteria and fungi as well as the microbial polysaccharides LPS, LTA, and beta-glucan. It also showed a broad spectrum against Gram-positive and Gram-negative bacteria and a fungus, as well as an in vivo clearance activity on *V. anguillarum* in a dose-dependent manner (Sun et al. 2011). In the mud crab *Scylla paramamosain*, both the recombinant protein and the synthetic peptides corresponding to two Sp-ALFs exhibited strong effects against most of the tested Gram-positive and Gram-negative bacteria, as well as the WSSV virus (Liu et al. 2012a). The synthetic peptide of the LPS-BD of MrALF from the prawn *Macrobrachium rosenbergii* could recognize and bind to the LPS of bacterial cell walls and showed an effective antimicrobial activity (Arockiaraj et al. 2014). The antimicrobial activities of ALFs from additional species, *P. trituberculatus* (Liu et al. 2013), *F. chinensis* (Li et al. 2014, 2019; Yang et al. 2015), *M. nipponense* (Wang et al. 2015a), *E. carinicauda* (Lv et al. 2017, 2018), and the deep-sea shrimp *Rimicaris* sp. (Gu et al. 2018), were also characterized in recent years. Taken together, although ALFs show broad-spectrum antimicrobial properties, distinct ALFs exhibit activities against different microbes, which might contribute to their important roles in defending against infections from various pathogens.

Mechanism of Antibacterial Activity

The functional mechanisms of ALFs against bacteria have been widely investigated using the LBD domain, the functional domain of ALFs. In *F. chinensis* a synthetic LBD, FcALF-LBDc, exhibited apparent antibacterial activity against both Gram-negative (*E. coli* and *V. anguillarum*) and Gram-positive bacteria (*M. luteus* and *M. lysodeikticus*) with MIC ranges of 32–64, 2–4, 1–2, and 32–64 μM, respectively. However, both opening the disulfide bridge and replacing the basic amino acids (lysine or arginine) with neutral ones impaired the antibacterial activities of FcALF-LBDc (Guo et al. 2014). A further study showed that the antibacterial activities of the synthetic LBD peptides, LBDb and LBDv, were apparently enhanced and broadened after the number of basic amino acids was increased in these peptide sequences (Li et al. 2014). A modified FcALF2 (mFcALF2), in which the LBD was substituted by LBDv, was recombinantly expressed in yeast *Pichia pastoris* GS115, and it also exhibited strong antibacterial activity against Gram-negative bacteria, including *E. coli*, *V. alginolyticus*, *V. harveyi*, and *V. parahaemolyticus*, as well as against Gram-positive bacteria, including *Bacillus licheniformis* and *Staphylococcus epidermidis* (Yang et al. 2016b). These findings suggest that the disulfide loop and the basic amino acids in the LBD play key roles in its antibacterial activities.

A previous review summarized the antimicrobial activity of different types ALFs and found a relationship between the activity and the *p*I of the LPS-BD (Tassanakajon et al. 2015). Highly cationic ALFs like ALFPm3 exhibit a broad spectrum of activities against Gram-negative and Gram-positive bacteria, fungi and virus (Carriel-Gomes et al. 2007; de la Vega et al. 2008; Liu et al. 2006; Somboonwiwat et al. 2005; Tassanakajon et al. 2015). Cationic ALFs, such as PcALF1, SpALF1 and SpALF2, and anionic ALFs, such as EsALF2, PtALF6, MrALF and SpALF4, exhibit a lower strength antibacterial activity against Gram-negative and Gram-positive bacteria and in some cases against Gram-negative bacteria only (Liu et al. 2013, 2014a; Sun et al. 2011; Zhang et al. 2010; Zhu et al. 2014). Moreover, highly anionic ALFs have an impaired LPS-BD and are devoid of antimicrobial activity (Rosa et al. 2013). The absence of a positively charged amino acid cluster in the anionic LPS-BD causes the deficient LPS binding ability and the lack of antimicrobial activity of the ALF. These data support the view that basic amino acids in the LBD peptides play key roles in their antibacterial activities.

In addition to the LBD region, positively charged amino acids outside the LBD but close to this region also contribute to the binding and antimicrobial activity of the peptide. A recent study reported an ALF, SpALF6, and its variant, SpALF6-V, from the mud crab *S. paramamosain* (Hou et al. 2017). The SpALF6-V had two amino acid mutations (H46 to R and A110 to P) due to the presence of SNPs. The mutation H46 to R was located outside but close to LBD region whereas another mutation was distant from this region. The recombinant protein corresponding to SpALF6-V exhibited more potent antimicrobial and cell-binding activity on Gram-positive bacteria, as well as stronger LTA-binding activity than SpALF6, indicating the role of positively charged amino acids outside the LBD region on its antibacterial activity.

The above-mentioned features of LBD peptides enable them to directly bind to the major cell wall components, LPS and LTA, of bacteria cells, as shown in the structures of ALF-L and ALFPm3 (Andra et al. 2007; Hoess et al. 1993; Mora et al. 2008; Yang et al. 2009). The binding of LBD to the bacterial cell led to a series of subsequent effects. Upon binding, the LBDv peptide was found to quickly cause bacterial agglutination, and kill bacteria by damaging their membrane integrity (Fig. 3.2) (Yang et al. 2016a). A previous study on the recombinant ALF protein rALFPm3 also exhibited the same effect. The rALFPm3 protein could attach to *V. harveyi* cells in vivo and then induce membrane permeabilization and leakage of cytoplasmic components. In vitro rALFPm3 treatment of the bacteria cells also led to membrane disruption and damage, bleb and pore formation, and leakage of cytoplasmic contents (Jaree et al. 2012).

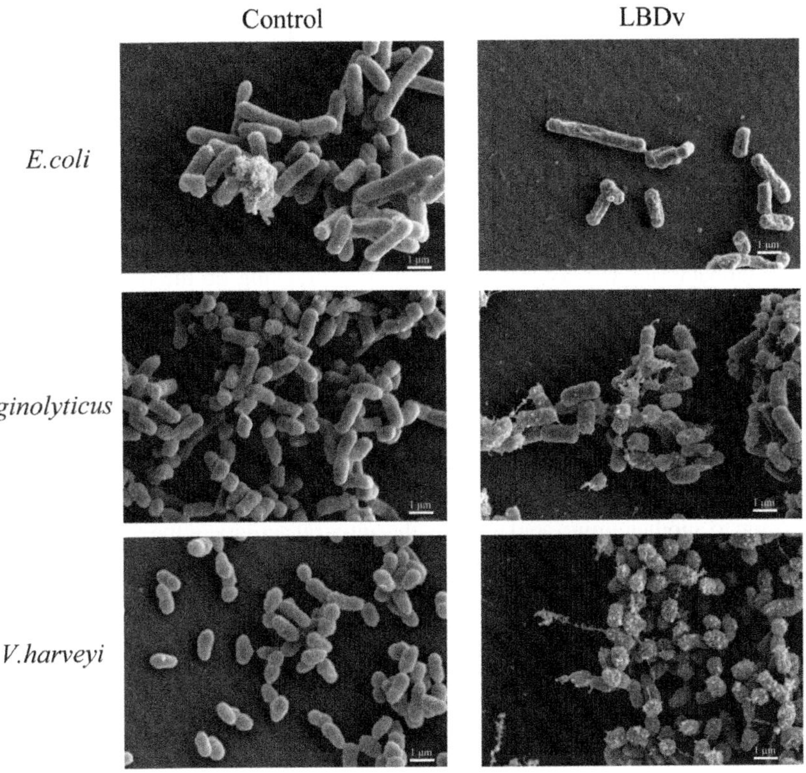

Fig. 3.2 Morphology of the bacteria before and after treatment with LBDv peptide. The 10^8 cfu/mL different bacteria were incubated with 64 μM LBDv peptide for 2 h. The bacteria treated with same concentration pGFP peptide were used as control (Yang et al. 2016a)

Apart from the direct killing of microbes, ALFs also perform immunological roles through regulating host immunity and microbial balance. In *M. rosenbergii*, the phagocytic activity of hemocytes was enhanced after treated with MrALF at a lower concentration than the minimum inhibitory concentration, suggesting that MrALF functions as an opsonin (Liu et al. 2014a). In addition, silencing of *EcALF1* in healthy individuals caused a lesion of the hepatopancreas in *E. carinicauda* and finally led to the death of the prawn (Lv et al. 2017). Subsequent detection of bacteria in the hepatopancreas showed that the colonies of TCBS cultured bacteria had increased dramatically in the *EcALF1*-silenced prawn. The dominant bacteria were *V. alginolyticus* and *V. parahaemolyticus*, which are the main pathogenic bacteria species in crustaceans. In vitro antimicrobial detection showed that the synthetic LBD peptide of EcALF1 displayed strong antibacterial activity against *V. alginolyticus*. These data showed the important role of the ALF in modulating the in vivo microbial balance in *E. carinicauda*. Taken together, ALFs perform their antibacterial activities through direct binding to bacterial cells, based on the sequence features, as well as their in vivo roles in regulating host immunity and microbial balance, which contribute to their important roles in defending against infections from various bacteria.

Mechanism of Antiviral Activity

The functional mechanism of ALFs against viruses is apparently different from their antibacterial activity. Linearization (i.e., breaking up the disulfide loop) and replacement of arginine residues in the FcALF-LBDc peptide did not affect its anti-white spot syndrome virus (WSSV) activity. However, its anti-WSSV activity did become deficient when lysine residues were replaced demonstrating that some lysine residues were indispensable in the anti-WSSV activity of FcALF-LBDc peptide (Guo et al. 2014). Sequence analysis on different LBD peptides with or without anti-WSSV activity revealed that an identical lysine residue site was specifically conserved in peptides with anti-WSSV activity. When changing the lysine residue at this specific position to another amino acid residue, the antiviral activity of the peptide apparently decreased. On the contrary, replacing the residue with a lysine residue at this site in a LBD peptide without anti-WSSV activity, the antiviral activity to WSSV of the peptide was restored (Li et al. 2015). These data suggested the important role of certain lysine residues in the LBD peptide for its antiviral activity.

ALF or its functional domain LBD also shows direct interaction with the envelope protein of WSSV. Under the yeast two-hybrid screening system, five WSSV proteins encoded by the transcripts WSSV186, WSSV189, WSSV395, WSSV458 and WSSV471 showed interaction with ALFPm3 protein. All five WSSV gene transcripts were expressed in the late phase of infection (24 and 48 h post-infection). Of these genes, WSSV189 was found to encode an envelope protein as shown by western blot and immunoelectron microscopy analyses. The recombinant WSSV189 protein (rWSSV189) could interact with ALFPm3 in an in vitro pulldown assay. Injection of rALFPm3 pre-treated WSSV reduced the cumulative mortality of shrimp, whereas injection of rWSSV189 and rALFPm3 pre-treated WSSV increased the cumulative mortality of shrimp compared to animals treated with rALFPm3 pre-treated WSSV only (Suraprasit et al. 2014).

Another study found that the recombinant protein rFcALF5 could inhibit the in vivo WSSV propagation when rFcALF5 was used to pre-treat WSSV or injected within a short time (2 h) before or after WSSV injection (Yang et al. 2015). However, the inhibition effect was lost when rFcALF5 was injected into the shrimp 15 h after the WSSV injection, which indicated a direct interaction of rFcALF5 with WSSV. Further investigations revealed that rFcALF5, as well as its functional domain LBD-FcALF5, could interact directly with the envelope protein VP24 of WSSV. These data suggested that ALF played a neutralization effect on WSSV. However, the antiviral mechanism of ALFs still needs to be further elucidated with more evidence.

Expression and Regulation

The distribution of distinct ALF transcripts show significant differences. The transcriptional expression of some ALF isoforms is tissue specific while other ALF transcripts are widely expressed in many tissues. Many ALFs are mainly expressed in hemocytes, the major immune-related cells of crustaceans (Arockiaraj et al. 2014; Liu et al. 2005, 2012a, b; Lv et al. 2018; Nagoshi et al. 2006; Ponprateep et al. 2012; Somboonwiwat et al. 2008). Some ALFs are predominantly expressed in other immune-related tissues, including the gills (Liu et al. 2013), lymphoid organ (Li et al. 2019), intestine and hepatopancreas (Zhou et al. 2019). Some other ALFs are widely expressed in several tissues at different levels (Lv et al. 2018; Mekata et al. 2010; Ponprateep et al. 2012; Sun et al. 2015; Wang et al. 2015a). An in vivo expression study showed that the MrALF protein concentration in prawn plasma significantly increased after LPS treatment for 6 h, whereas the MrALF protein concentration in the hemocytes decreased (Liu et al. 2014a), suggesting that MrALF was released from hemocytes to plasma upon challenge. ALFs that were widely distributed among tissues might also be released into the hemolymph, the open circulatory system of crustacean, to encounter invasive pathogens. The distinct expression patterns of ALFs, as well as their various isoforms and different antimicrobial functions, might endow ALFs with a complementary function in the host against diverse pathogen infections.

Appearance of distinct expression patterns might be related to the regulation of ALFs. Initially, the promoter region of ALF was identified and characterized to study the expression regulation of the gene. In *P. monodon*, the transcription factor-binding sites, including Octamer, GATA, CCAAT box and GAAA motifs, were predicted in the promoter region of both the ALF group A and B genes, whereas the putative binding site for activator protein 1 and NF-κB were only predicted at the promoter region of the ALF group A genes (Tharntada et al. 2008). The binding sites for NF-κB, Octamer binding protein 1 and GATA binding factor 1 were also predicted in the promoter region from +1 to −702 of the *ALFFc* gene from *F. chinensis*. In the insect sf9 cell lines, the putative promoter sequence of pALF-702 could successfully drive the expression of the EGFP reporter gene by adding LPS or (1,3)-β-D-glucan, while the shorter promoter sequence pALF-318 was only effective by adding (1,3)-β-D-glucan (Tang et al. 2014). This indicated that different transcription elements might contribute to the transcriptional regulation of the *ALFFc* gene.

Recent studies have revealed that the transcription of *ALFs* are regulated by several classical signaling pathways, including the IMD, Toll and JAK/STAT pathways. The expression of *ALF1* and *ALF2* in *M. japonicus* was regarded to be regulated by the IMD signaling pathway since knockdown of MjIMD expression inhibited transcription of the *ALF1* and *ALF2* in shrimp post bacterial challenge, when compared with the control group (Lan et al. 2013). In *L. vannamei*, the transcription of one *ALF* gene was also considered to be regulated by LvIMD because of a similar phenomenon as in *M. japonicus* (Hou et al. 2014). In *M. japonicus*, SOCS2, a negative regulator of the JAK/STAT signaling pathway, could regulate the expression of several other ALFs including ALF-C1, C2 and D1. Knockdown of SCOS2 in shrimps post *V. anguillarum* infection up-regulated the expression level of these ALFs, while injection of recombinant SOCS2 down-regulated the expression (Sun et al. 2016). The data indicated that the expression of some ALFs might be regulated by the JAK/STAT signaling pathway.

Further studies revealed that the expression of some ALFs was influenced by the Toll signaling pathway. In *P. clarkii*, knockdown of *PcToll* decreased the expression of *PcALF2* in the crayfish after *V. anguillarum* injection (Wang et al. 2015b). Knockdown of *PcToll3* influenced the expression of *PcMyD88*, *PcTRAF6*, *PcDorsal* and *PcALF1*, and inhibited translocation of PcDorsal from cytoplasm to the nucleus (Lan et al. 2016). Another Toll gene from *P. clarkii*, *PcToll4*, seems to regulate the expression of several ALFs. The expression of *PcALF1*, *PcALF2*, *PcALF4*, *PcALF7*, and *PcALF10* was significantly up-regulated after WSSV infection, while knockdown of *PcToll4* could inhibit the up-regulation of these ALFs in the crayfish after WSSV infection (Ying et al. 2017). In *Cherax quadricarinatus,* the expression of *CqALF* was also inhibited after gene silencing of *CqToll* in HPT cell cultures (Li et al. 2017). As the key cytoplasm signal adaptor that mediates signals activated by Toll-like receptors, TRAF6 from the mud crab was found to regulate the expression of four ALFs. Silencing of *Sp-TRAF6* gene could inhibit expression of *SpALF1*, *SpALF2*, *SpALF5* and *SpALF6* in hemocytes, and the expression of *SpALF1*, *SpALF3*, *SpALF4*, *SpALF5* and *SpALF6* in hepatopancreas (Sun et al. 2017). Two serine proteases from the mud crab, SP3 and SP5, interact with TRAF6 and regulate the expression of different ALFs, which was confirmed by gene silencing and over-expression (Wei et al. 2019). In *L. vannamei*, the expression of *ALF1-4* was influenced by the Toll4-Dorsal cascade under WSSV infection. Silencing of *Toll4* and *Dorsal* impaired the expression of these ALF genes (Li et al. 2018). These data indicate that the Toll signaling pathway plays important roles in regulating the expression of various ALFs.

As evident from the literature, considerable work has been performed to explain the expression regulation of ALFs in crustaceans. However, more direct evidence is still necessary to elucidate this question. The existing research results suggest that different transcription factors might bind to the promoter region of ALF genes and that the three classical signaling pathways in humoral immunity participate in regulating the expression of different ALFs. However, whether the transcription factors, Relish, Dorsal and STAT, regulate ALFs expression directly still needs to be clarified. Future studies are awaited to clearly interpret the direct interaction between these transcription factors and specific promoter regions of different ALF genes.

Applications and Prospects

Drug Development

ALF exhibits a variety of antimicrobial and immune regulatory activities. Following its identification and functional characterization, researchers attempted to apply it as an anti-disease component in aquaculture. However, it was gradually recognized to be not easily feasible due to the high cost of preparing the agent. On the other hand, as the functional domain of ALF, LBD, consists of only 22 amino acid residues, the accrued knowledge about this now well characterized peptide should greatly facilitate its preparation and future applications.

In recent years, researchers have developed the perspective of possible medical usage of LBD or LBD-derived peptides. In conjunction with inactivated MBT-2 lysate, the synthetic LBD peptide of SALF promoted innate IL-1β production in mouse macrophages, then further modulated tumor-associated chemokines and cytokines, enhanced infiltration of immune cells into the tumor site and activated tumor-associated immune cells, which finally prevented murine bladder-associated tumors in mice (Huang et al. 2015). A looped peptide CLP-19, derived from the LBD of LALF, attenuated the LPS-stimulated proinflammatory factor secretion while enhancing anti-inflammatory factor production. This finding revealed that CLP-19 seems to be a potential therapeutic agent for clinical treatment of septic shock (Yang et al. 2016c). An LBD-containing PT-peptide, derived from ALF of *P. trituberculatus*, was encapsulated in raw milk-derived extracellular vesicles (EVs) and could be delivered to monocytes. This resulted in elevated immune activity, including reactive oxygen species level, superoxide anion production and phagocytosis, as well as the secretion of cytokines, such as interferon-γ, interleukin-6 and tumor necrosis factor-α in human monocytic THP-1 cells (Lee et al. 2019). These data reveal the prospects of medical applications of LBD or LBD-derived peptides for their activities in anti-tumor growth and immune regulation.

A very recent study found that an LBD-derived peptide even as short as ten amino acids in length without disulfide bonds, named crab-ALF2A, also showed strong antimicrobial activities against Gram-positive bacteria and Gram-negative bacteria; it disrupted the bacterial cell membrane and led to the leakage of cytoplasmic contents (Nam et al. 2019). This finding provided further evidence for the application prospects of LBD-derived peptide with lower preparation cost.

Disease-Resistant Genetic Breeding

Diseases caused by various pathogens are still the biggest threat in crustacean aquaculture. Considering the multiple antimicrobial activities and sequence diversity of ALFs in crustaceans, researchers also tested the applicability of ALF genes in disease-resistant genetic breeding. Several SNPs located in ALF genes have been reported to be related to the resistance/susceptibility of the host to *V. alginolyticus* (Li et al. 2013b) or WSSV (Liu et al. 2014b). ALFs, as well as other AMPs, are the most crucial immune factors playing roles in direct killing of pathogens or indirect regulation of host immunity. Therefore, these genes are suggested as the important target genes for marker selection related to disease-resistance.

Acknowledgements This work was supported by the Key Program of National Natural Science Foundation of China (31830100) and the China Agriculture Research system-48 (CARS-48).

References

Andra J, Howe J, Garidel P, Rossle M, Richter W, Leiva-Leon J, Moriyon I, Bartels R, Gutsmann T, Brandenburg K (2007) Mechanism of interaction of optimized Limulus-derived cyclic peptides with endotoxins: thermodynamic, biophysical and microbiological analysis. Biochem J 406:297–307

Arockiaraj J, Kumaresan V, Bhatt P, Palanisamy R, Gnanam AJ, Pasupuleti M, Kasi M, Chaurasia MK (2014) A novel single-domain peptide, anti-LPS factor from prawn: synthesis of peptide, antimicrobial properties and complete molecular characterization. Peptides 53:79–88

Carriel-Gomes MC, Kratz JM, Barracco MA, Bachere E, Barardi CR, Simoes CM (2007) In vitro antiviral activity of antimicrobial peptides against herpes simplex virus 1, adenovirus, and rotavirus. Mem Inst Oswaldo Cruz 102:469–472

de la Vega E, O'Leary NA, Shockey JE, Robalino J, Payne C, Browdy CL, Warr GW, Gross PS (2008) Anti-lipopolysaccharide factor in Litopenaeus vannamei (LvALF): a broad spectrum antimicrobial peptide essential for shrimp immunity against bacterial and fungal infection. Mol Immunol 45:1916–1925

de Lorgeril J, Gueguen Y, Goarant C, Goyard E, Mugnier C, Fievet J, Piquemal D, Bachere E (2008) A relationship between antimicrobial peptide gene expression and capacity of a selected shrimp line to survive a Vibrio infection. Mol Immunol 45:3438–3445

Gross PS, Bartlett TC, Browdy CL, Chapman RW, Warr GW (2001) Immune gene discovery by expressed sequence tag analysis of hemocytes and hepatopancreas in the Pacific White Shrimp, Litopenaeus vannamei, and the Atlantic White Shrimp. L. setiferus. Dev Comp Immunol 25:565–577

Gu HJ, Sun QL, Jiang S, Zhang J, Sun L (2018) First characterization of an anti-lipopolysaccharide factor (ALF) from hydrothermal vent shrimp: Insights into the immune function of deep-sea crustacean ALF. Dev Comp Immunol 84:382–395

Guo SY, Li SH, Li FH, Zhang XJ, Xiang JH (2014) Modification of a synthetic LPS-binding domain of anti-lipopolysaccharide factor from shrimp reveals strong structure-activity relationship in their antimicrobial characteristics. Dev Comp Immunol 45:227–232

Hoess A, Watson S, Siber GR, Liddington R (1993) Crystal-Structure of an Endotoxin-Neutralizing Protein from the Horseshoe-Crab, Limulus Anti-Lps Factor, at 1.5 Angstrom Resolution. Embo J 12:3351–3356

Hou FJ, He SL, Liu YJ, Zhu XW, Sun CB, Liu XL (2014) RNAi knock-down of shrimp Litopenaeus vannamei Toll gene and immune deficiency gene reveals their difference in regulating antimicrobial peptides transcription. Dev Comp Immunol 44:255–260

Hou ZG, Wang Y, Hui KM, Fang WH, Zhao S, Zhang JX, Ma HY, Li XC (2017) A novel anti-lipopolysaccharide factor SpALF6 in mud crab Scylla paramamosain exhibiting different antimicrobial activity from its single amino acid mutant. Dev Comp Immunol 72:44–56

Huang HN, Rajanbabu V, Pan CY, Chan YL, Chen JY, Wu CJ (2015) Enhanced Control of Bladder-Associated Tumors Using Shrimp Anti-Lipopolysaccharide Factor (SALF) Antimicrobial Peptide as a Cancer Vaccine Adjuvant in Mice. Mar Drugs 13:3241–3258
Jaree P, Tassanakajon A, Somboonwiwat K (2012) Effect of the anti-lipopolysaccharide factor isoform 3 (ALFPm3) from Penaeus monodon on Vibrio harveyi cells. Dev Comp Immunol 38:554–560
Jiang HS, Zhang Q, Zhao YR, Jia WM, Zhao XF, Wang JX (2015) A new group of anti-lipopolysaccharide factors from Marsupenaeus japonicus functions in antibacterial response. Dev Comp Immunol 48:33–42
Lan JF, Zhou J, Zhang XW, Wang ZH, Zhao XF, Ren Q, Wang JX (2013) Characterization of an immune deficiency homolog (IMD) in shrimp (Fenneropenaeus chinensis) and crayfish (Procambarus clarkii). Dev Comp Immunol 41:608–617
Lan JF, Wei S, Wang YQ, Dai YJ, Tu JG, Zhao LJ, Li XC, Qin QW, Chen N, Lin L (2016) PcToll3 was involved in anti-Vibrio response by regulating the expression of antimicrobial peptides in red swamp crayfish, Procambarus clarkii. Fish Shellfish Immun 57:17–24
Lee BH, Chen BR, Huang CT, Lin CH (2019) The immune activity of PT-Peptide derived from Anti-Lipopolysaccharide factor of the swimming crab portunus trituberculatus is enhanced when encapsulated in milk-derived extracellular vesicles. Mar Drugs 17(5)
Li CH, Zhao JM, Song LS, Mu CK, Zhang H, Gai YC, Qiu LM, Yu YD, Ni DJ, Xing KZ (2008) Molecular cloning, genomic organization and functional analysis of an anti-lipopolysaccharide factor from Chinese mitten crab Eriocheir sinensis. Dev Comp Immunol 32:784–794
Li SH, Zhang XJ, Sun Z, Li FH, Xiang JH (2013a) Transcriptome analysis on Chinese Shrimp Fenneropenaeus chinensis during WSSV Acute Infection. Plos One 8(3)
Li XH, Cui ZX, Liu Y, Song CW, Shi GH, Wang CL (2013b) Polymorphisms of anti-lipopolysaccharide factors in the swimming crab Portunus trituberculatus and their association with resistance/susceptibility to Vibrio alginolyticus. Fish Shellfish Immunol 34:1560–1568
Li SH, Gu SY, Li FH, Xiang JH (2014) Characterization and function analysis of an anti-lipopolysaccharide factor (ALF) from the Chinese shrimp Fenneropenaeus chinensis. Dev Comp Immunol 46:349–355
Li SH, Guo SY, Li FH, Xiang JH (2015) Functional Diversity of Anti-Lipopolysaccharide Factor Isoforms in Shrimp and Their Characters Related to Antiviral Activity. Mar Drugs 13(5):2602–2616
Li YY, Chen XX, Lin FY, Chen QF, Ma XY, Liu HP (2017) CqToll participates in antiviral response against white spot syndrome virus via induction of anti-lipopolysaccharide factor in red claw crayfish Cherax quadricarinatus. Dev Comp Immunol 74:217–226
Li HY, Yin B, Wang S, Fu QH, Xiao B, Lu K, He JG, Li CZ (2018) RNAi screening identifies a new Toll from shrimp Litopenaeus vannamei that restricts WSSV infection through activating Dorsal to induce antimicrobial peptides. Plos Pathog 14(9)
Li SH, Lv XJ, Li FH, Xiang JH (2019) Characterization of a lymphoid organ specific anti-lipopolysaccharide factor from shrimp reveals structure-activity relationship of the LPS-binding domain. Front Immunol 10
Liu FS, Liu YC, Li FH, Dong B, Xiang JH (2005) Molecular cloning and expression profile of putative antilipopolysaccharide factor in Chinese shrimp (Fenneropenaeus chinensis). Mar Biotechnol 7:600–608
Liu H, Jiravanichpaisal P, Soderhall I, Cerenius L, Soderhall K (2006) Antilipopolysaccharide factor interferes with white spot syndrome virus replication in vitro and in vivo in the crayfish Pacifastacus leniusculus. J Virol 80:10365–10371
Liu Y, Cui ZX, Song CW, Wang SY, Li QQ (2011) Multiple isoforms of immune-related genes from hemocytes and eyestalk cDNA libraries of swimming crab Portunus trituberculatus. Fish Shellfish Immunol 31:29–42
Liu HP, Chen RY, Zhang QX, Wang QY, Li CR, Peng H, Cai L, Zheng CQ, Wang KJ (2012a) Characterization of two isoforms of antiliopolysacchride factors (Sp ALFs) from the mud crab Scylla paramamosain. Fish Shellfish Immunol 33:1–10

Liu Y, Cui Z, Li X, Song C, Li Q, Wang S (2012b) Molecular cloning, expression pattern and antimicrobial activity of a new isoform of anti-lipopolysaccharide factor from the swimming crab Portunus trituberculatus. Fish Shellfish Immunol 33:85–91

Liu Y, Cui ZX, Li XH, Song CW, Shi GH, Wang CL (2013) Molecular cloning, genomic structure and antimicrobial activity of PtALF7, a unique isoform of anti-lipopolysaccharide factor from the swimming crab Portunus trituberculatus. Fish Shellfish Immunol 34:652–659

Liu CC, Chung CP, Lin CY, Sung HH (2014a) Function of an anti-lipopolysaccharide factor (ALF) isoform isolated from the hemocytes of the giant freshwater prawn Macrobrachium rosenbergii in protecting against bacterial infection. J Invertebr Pathol 116:1–7

Liu JW, Yu Y, Li FH, Zhang XJ, Xiang JH (2014b) A new anti-lipopolysaccharide factor (ALF) gene with its SNP polymorphisms related to WSSV-resistance of Litopenaeus vannamei. Fish Shellfish Immunol 39:24–33

Lv X, Li S, Liu F, Li F, Xiang J (2017) Identification and function analysis of an anti-lipopolysaccharide factor from the ridgetail prawn Exopalaemon carinicauda. Dev Comp Immunol 70:128–134

Lv XJ, Li SH, Zhang CS, Xiang JH, Li FH (2018) Multiple isoforms of anti-lipopolysaccharide factors and their antimicrobial functions in the ridgetail prawn Exopalaemon carinicauda. Mar Drugs 16(5)

Matos GM, Schmitt P, Barreto C, Farias ND, Toledo-Silva G, Guzman F, Destoumieux-Garzon D, Perazzolo LM, Rosa RD (2018) Massive gene expansion and sequence diversification is associated with diverse tissue distribution, regulation and antimicrobial properties of anti-lipopolysaccharide factors in shrimp. Mar Drugs 16(10)

Mekata T, Sudhakaran R, Okugawa S, Kono T, Sakai M, Itami T (2010) Molecular cloning and transcriptional analysis of a newly identified anti-lipopolysaccharide factor gene in kuruma shrimp, Marsupenaeus japonicus. Lett Appl Microbiol 50:112–119

Mora P, De la Paz ML, Perez-Paya E (2008) Bioactive peptides derived from the Limulus anti-lipopolysacchari de factor: structure-activity relationships and formation of mixed peptide/lipid complexes. J Pept Sci 14:963–971

Morita T, Ohtsubo S, Nakamura T, Tanaka S, Iwanaga S, Ohashi K, Niwa M (1985) Isolation and biological activities of limulus anticoagulant (anti-LPS factor) which interacts with lipopolysaccharide (LPS). J Biochem 97:1611–1620

Nagoshi H, Inagawa H, Morii K, Harada H, Kohchi C, Nishizawa T, Taniguchi Y, Uenobe M, Honda T, Kondoh M, Takahashi Y, Soma GI (2006) Cloning and characterization of a LPS-regulatory gene having an LPS binding domain in kuruma prawn Marsupenaeus japonicus. Mol Immunol 43:2061–2069

Nam BH, Park EH, Shin EH, Kim YO, Kim DG, Kong HJ, Park JY, Seo JK (2019) Development of novel antimicrobial peptides derived from anti-lipopolysaccharide factor of the swimming crab, Portunus trituberculatus. Fish Shellfish Immunol 84:664–672

Ohashi K, Niwa M, Nakamura T, Morita T, Iwanaga S (1984) Anti-LPS Factor in the Horseshoe Crab, Tachypleus tridentatus - Its Hemolytic Activity on the Red Blood Cell Sensitized with Lipopolysaccharide. Febs Lett 176:207–210

Ponprateep S, Tharntada S, Somboonwiwat K, Tassanakajon A (2012) Gene silencing reveals a crucial role for anti-lipopolysaccharide factors from Penaeus monodon in the protection against microbial infections. Fish Shellfish Immunol. 32:26–34

Rosa RD, Vergnes A, de Lorgeril J, Goncalves P, Perazzolo LM, Saune L, Romestand B, Fievet J, Gueguen Y, Bachere E, Destoumieux-Garzon D (2013) Functional divergence in shrimp anti-lipopolysaccharide factors (ALFs): from recognition of cell wall components to antimicrobial activity. PLoS One 8(7):e67937

Smith VJ, Dyrynda EA (2015) Antimicrobial proteins: From old proteins, new tricks. Mol Immunol 68:383–398

Somboonwiwat K, Marcos M, Tassanakajon A, Klinbunga S, Aumelas A, Romestand B, Gueguen Y, Boze H, Moulin G, Bachere E (2005) Recombinant expression and antimicrobiall activity of

anti-lipopolysaccharide factor (ALF) from the black tiger shrimp Penaeus monodon. Dev Comp Immunol 29:841–851
Somboonwiwat K, Bachere E, Rimphanitchayakit V, Tassanakajon A (2008) Localization of anti-lipopolysaccharide factor (ALFPm3) in tissues of the black tiger shrimp, Penaeus monodon, and characterization of its binding properties. Dev Comp Immunol 32:1170–1176
Sun C, Xu WT, Zhang HW, Dong LP, Zhang T, Zhao XF, Wang JX (2011) An anti-lipopolysaccharide factor from red swamp crayfish, Procambarus clarkii, exhibited antimicrobial activities in vitro and in vivo. Fish Shellfish Immunol 30:295–303
Sun WW, Wan WS, Zhu SO, Wang SS, Wang SQ, Wen XB, Zheng HP, Zhang YL, Li SK (2015) Characterization of a novel anti-lipopolysaccharide factor isoform (SpALF5) in mud crab, Scylla paramamosain. Mol Immunol 64:262–275
Sun JJ, Lan JF, Xu JD, Niu GJ, Wang JX (2016) Suppressor of cytokine signaling 2 (SOCS2) negatively regulates the expression of antimicrobial peptides by affecting the Stat transcriptional activity in shrimp Marsupenaeus japonicus. Fish Shellfish Immunol 56:473–482
Sun WW, Zhang XX, Wan WS, Wang SQ, Wen XB, Zheng HP, Zhang YL, Li SK (2017) Tumor necrosis factor receptor-associated factor 6 (TRAF6) participates in anti-lipopolysaccharide factors (ALFs) gene expression in mud crab. Dev Comp Immunol 67:361–376
Supungul P, Klinbunga S, Pichyangkura R, Hirono I, Aoki T, Tassanakajon A (2004) Antimicrobial peptides discovered in the black tiger shrimp Penaeus monodon using the EST approach. Dis Aquat Organ 61:123–135
Suraprasit S, Methatham T, Jaree P, Phiwsaiya K, Senapin S, Hirono I, Lo CF, Tassanakajon A, Somboonwiwat K (2014) Anti-lipopolysaccharide factor isoform 3 from Penaeus monodon (ALFPm3) exhibits antiviral activity by interacting with WSSV structural proteins. Antivir Res 110:142–150
Tanaka S, Nakamura T, Morita T, Iwanaga S (1982) Limulus Anti-Lps Factor: an Anticoagulant Which Inhibits the Endotoxin-Mediated Activation of Limulus Coagulation System. Biochem Bioph Res Co 105:717–723
Tang T, Li LX, Sun LL, Bu JC, Xie S, Liu FS (2014) Functional analysis of Fenneropenaeus chinensis anti-lipopolysaccharide factor promoter regulated by lipopolysaccharide and (1,3)-beta-D-glucan. Fish Shellfish Immun 38:348–353
Tassanakajon A, Somboonwiwat K, Amparyup P (2015) Sequence diversity and evolution of antimicrobial peptides in invertebrates. Dev Comp Immunol 48:324–341
Tharntada S, Somboonwiwat K, Rimphanitchayakit V, Tassanakajon A (2008) Anti-lipopolysaccharide factors from the black tiger shrimp, Penaeus monodon, are encoded by two genomic loci. Fish Shellfish Immun 24:46–54
Tharntada S, Ponprateep S, Somboonwiwat K, Liu H, Soderhall I, Soderhall K, Tassanakajon A (2009) Role of anti-lipopolysaccharide factor from the black tiger shrimp, Penaeus monodon, in protection from white spot syndrome virus infection. J Gen Virol 90:1491–1498
Wang Y, Tang T, Gu J, Li X, Yang X, Gao X, Liu F, Wang J (2015a) Identification of five anti-lipopolysaccharide factors in oriental river prawn, Macrobrachium nipponense. Fish Shellfish Immunol 46:252–260
Wang Z, Chen YH, Dai YJ, Tan JM, Huang Y, Lan JF, Ren Q (2015b) A novel vertebrates Toll-like receptor counterpart regulating the antimicrobiall peptides expression in the freshwater crayfish, Procambarus clarkii. Fish Shellfish Immunol 43:219–229
Wei ZB, Sun WW, Tran NT, Gong Y, Ma HY, Zheng HP, Zhang YL, Li SK (2019) Two novel serine proteases from Scylla paramamosain involved in the synthesis of anti-lipopolysaccharide factors and activation of prophenoloxidase system. Fish Shellfish Immunol 84:322–332
Yang Y, Boze H, Chemardin P, Padilla A, Moulin G, Tassanakajon A, Pugniere M, Roquet F, Destoumieux-Garzon D, Gueguen Y, Bachere E, Aumelas A (2009) NMR Structure of rALF-Pm3, an Anti-Lipopolysaccharide Factor from Shrimp: Model of the Possible Lipid A-Binding Site. Biopolymers 91:207–220
Yang H, Li SH, Li FH, Lv XJ, Xiang JH (2015) Recombinant expression and functional analysis of an isoform of anti-lipopolysaccharide factors (FcALF5) from Chinese shrimp Fenneropenaeus chinensis. Dev Comp Immunol 53:47–54

Yang H, Li SH, Li FH, Xiang JH (2016a) Structure and bioactivity of a modified peptide derived from the LPS-binding domain of an Anti-Lipopolysaccharide Factor (ALF) of Shrimp. Mar Drugs 14(5)

Yang H, Li SH, Li FH, Yu KJ, Yang FS, Xiang JH (2016b) Recombinant expression of a modified shrimp anti-lipopolysaccharide factor gene in pichia pastoris GS115 and its characteristic analysis. Mar Drugs 14(5)

Yang Y, Li D, Tian ZQ, Lv J, Sun FJ, Wang Q, Liu Y, Xia PY (2016c) Looped limulus anti-lipopolysaccharide derived peptide CLP-19 induces endotoxin tolerance involved inhibition of NF-kappa B activation. Biochem Biophys Res Comm 480:486–491

Ying H, Li TT, Min J, Yin S, Hui KM, Qian R (2017) Newly identified PcToll4 regulates antimicrobial peptide expression in intestine of red swamp crayfish Procambarus clarkii. Gene 610:140–147

Zhang Y, Wang L, Wang L, Yang J, Gai Y, Qiu L, Song L (2010) The second anti-lipopolysaccharide factor (EsALF-2) with antimicrobial activity from Eriocheir sinensis. Dev Comp Immunol 34:945–952

Zhou L, Li G, Jiao Y, Huang D, Li A, Chen H, Liu Y, Li S, Li H, Wang C (2019) Molecular and antimicrobial characterization of a group G anti-lipopolysaccharide factor (ALF) from Penaeus monodon. Fish Shellfish Immunol 94:149–156

Zhu L, Lan JF, Huang YQ, Zhang C, Zhou JF, Fang WH, Yao XJ, Wang H, Li XC (2014) SpALF4: a newly identified anti-lipopolysaccharide factor from the mud crab Scylla paramamosain with broad spectrum antimicrobial activity. Fish Shellfish Immunol 36:172–180

Chapter 4
Insect Defense Proteins and Peptides

Iwona Wojda, Małgorzata Cytryńska, Agnieszka Zdybicka-Barabas and Jakub Kordaczuk

Abstract The composition of insect hemolymph can change depending on many factors, e.g. access to nutrients, stress conditions, and current needs of the insect. In this chapter, insect immune-related polypeptides, which can be permanently or occasionally present in the hemolymph, are described. Their division into peptides or low-molecular weight proteins is not always determined by the length or secondary structure of a given molecule but also depends on the mode of action in insect immunity and, therefore, it is rather arbitrary. Antimicrobial peptides (AMPs) with their role in immunity, modes of action, and classification are presented in the chapter, followed by a short description of some examples: cecropins, moricins, defensins, proline- and glycine-rich peptides. Further, we will describe selected immune-related proteins that may participate in immune recognition, may possess direct antimicrobial properties, or can be involved in the modulation of insect immunity by both abiotic and biotic factors. We briefly cover Fibrinogen-Related Proteins (FREPs), Down Syndrome Cell Adhesion Molecules (Dscam), Hemolin, Lipophorins, Lysozyme, Insect Metalloproteinase Inhibitor (IMPI), and Heat Shock Proteins. The reader will obtain a partial picture presenting molecules participating in one of the most efficient immune strategies found in the animal world, which allow insects to inhabit all ecological land niches in the world.

Keywords Antimicrobial peptides · Defense activity · Insect immunity · Immune proteins · Immunomodulation

I. Wojda (✉) · M. Cytryńska · A. Zdybicka-Barabas · J. Kordaczuk
Department of Immunobiology, Institute of Biological Sciences, Maria Curie-Skłodowska University, Lublin, Poland
e-mail: wojda@poczta.umcs.lublin.pl

M. Cytryńska
e-mail: cytryna@poczta.umcs.lublin.pl

A. Zdybicka-Barabas
e-mail: barabas@poczta.umcs.lublin.pl

J. Kordaczuk
e-mail: jakub.kordaczuk@o2.pl

U. Hoeger and J. R. Harris (eds.), *Vertebrate and Invertebrate Respiratory Proteins, Lipoproteins and other Body Fluid Proteins*, Subcellular Biochemistry 94,
https://doi.org/10.1007/978-3-030-41769-7_4

Introduction

The role of the immune system is to protect a given organism from infecting intruders. There are two main branches of immunity: innate and acquired. Innate immunity relies on mechanisms with which every organism is equipped, irrespective of its earlier immune life history, and reacts immediately to infection. This very quick reaction is possible because it is characterized by high universality. Acquired immunity is an evolutionarily younger type, which has developed in vertebrates and relies on somatic gene rearrangement. Vertebrates posses both innate and acquired immunity, while invertebrate organisms rely only on innate immunity (Buchmann 2014). The acquired branch is characterized by great flexibility depending on the immune life history of individuals. It ensures better protection upon repeated infection with a given pathogen. It needs a so-called lag time to develop, but is highly specific. Acquired immunity is based on memory T and B cells, antibodies, and the major histocompatibility (MHC) complex and co-operates with innate immunity (Kaufmann 2019). Over 95% of animals are invertebrates with only the mechanisms of innate immunity (Cooper 2008). Among them are insects, whose defense system is very efficient and able to cope with different types of infections, which allows these arthropods to successfully inhabit all ecological land niches (Cytryńska et al. 2016).

There are three main components of the insect immune system. The first is represented by anatomical and physiological barriers protecting the organism from intruders. These include the chitin-containing exoskeleton and the chitin lining of the foregut, hindgut, and tracheas. The physiological barrier is created by conditions in the gut and trachea, which are hostile for the development of intruders (Hillyer 2016; Wojda 2017a). When these barriers are broken, microorganisms can enter the hemocoel, where the mechanisms of second and third components are activated. These are cellular and humoral mechanisms. The hemolymph cells, i.e. hemocytes, engulf microorganisms via phagocytosis. When a group of many microorganisms or a very large foreign body, such as an egg of a parasitic wasp, enter the hemocel, they can be entrapped in structures assembled by many hemocytes called nodules and capsules, respectively (Marmaras and Lampropoulou 2009). At the same time, pathogen associated molecular patterns (PAMPs), like peptidoglycan (PGN), lipopolysaccharide (LPS), and β-1,3-glucan or their virulence factors are recognized by pattern recognition receptors. This leads to the activation of pathways regulating the expression of antimicrobial peptides (Ganesan et al. 2011). These peptides are synthesized mainly in the fat body—a metabolically highly active organ, which is functionally analogous to mammalian liver. They are then secreted into the hemolymph, where they act in synergy with other components against the intruder (Wu et al. 2018).

The last twenty years appeared to have been very fruitful in the area of insect immunity. The sequencing of insect genomes, transcriptomes, and proteomes as well as advanced research techniques led to identification of many components of the insect immunity system and mechanisms of its regulation including modulation by abiotic and biotic factors. Abiotic factors are *inter alia* changes in ambient temperature as well as mechanical and oxidative stress, whereas biotic ones include the experience of previous infection with the same or other pathogens. For example, insects are able to improve their resistance upon repeated infection. This suggests the presence of a kind of immune memory in the area of innate immunity (Cooper and Eleftherianos 2017). This phenomenon has been called "immune priming". This is especially important in the light of the fact that insects possess only innate immunity. Reports coming from different laboratories have tried to elucidate the insect immune plasticity at the molecular and biochemical levels. The proposals include:

- diversity of receptors resulting in faster recognition of a given pathogen after repeated infection and, therefore, faster bacterial clearance (Watson et al. 2005; Doolittle et al. 2012),
- infection-induced epigenetic changes in the body of infected insects as a result of modulation of DNA methylases and histone acetylase/deacetylase activities leading to changes in gene expressions (Mukherjee et al. 2015; Vilcinskas 2016; Gegner et al. 2019),
- cross-talk and cross-protection mechanisms. In the first case, two different types of stress can activate the same component(s), for instance a member of a signaling pathway leading to amplification of the signal. The latter mechanism means that protective substances synthesized in response to one stress may protect against the effects of another stress (Sinclair et al. 2013). These two phenomena will also be mentioned below, where the role of Hsp proteins is discussed,
- more efficient immune response in re-infected insects (Taszłow et al. 2017; Vertyporokh et al. 2019).

The insect body cavity, in contrast to that of mammals, has an open space filled with hemolymph. The content of this body fluid depends on nutrients ingested by the insect, environmental conditions including abiotic and biotic factors, and finally, the immune state of the individual (Pendar et al. 2019).

There are many hemolymph polypeptide components contributing to insect defense and participating in the above-mentioned modulation of insect immunity. Some of them are proteins involved in recognition of infection, activation of signaling pathways, and stimulation of phagocytosis or humoral factors, while the others are polypeptides with direct antimicrobial activity (Fig. 4.1).

In the first part of this chapter, we will describe antimicrobial peptides. We will demonstrate their general properties and then focus of some representatives of each group. Further, we will describe hemolymph immune-related polypeptides, yet omitting these whose role in insect immunity is well known, for example peptidoglycan recognition proteins (PGRP) or β-1,3-glucan recognition proteins. They have been described in many excellent review papers (examples: Hughes 2012; Buchon et al. 2014; Kurata 2014; Hillyer 2016). We will focus our attention on proteins that are the most interesting from the point of view of comparative immunobiology and these that can be responsible in insect immune plasticity.

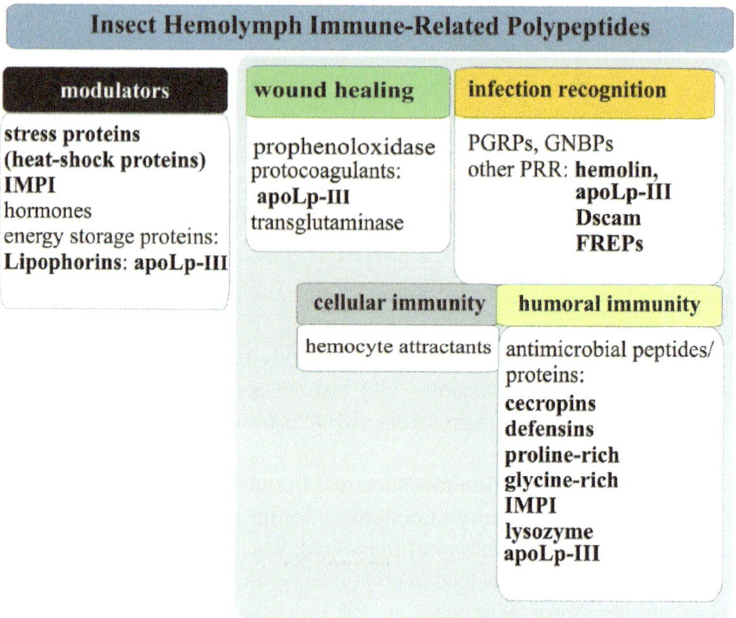

Fig. 4.1 Groups of selected hemolymph proteins and peptides involved in different steps of insect immunity. Polypeptides described in this chapter are indicated in bold. Note that some polypeptides may have more than one function

Insect Immune-Related Peptides

Antimicrobial Peptides as Effectors of Insect Immunity

Antimicrobial peptides are main effectors of humoral immune response in insects. They can be synthesized by different tissues, but the main source of peptides present in hemolymph is the fat body (Li et al. 2019). It is an organ with very high metabolic activity considered as a functional analogue of vertebrate liver. Some antimicrobial peptides can be constitutively present in insect hemolymph; however, their concentration may be immune-modulated (Wojda 2017a). Most of them are synthesized and secreted into the hemolymph in response to infection, where they act against intruders. The concentration of a given antimicrobial peptide in insect hemolymph depends on the infecting organism, severity of infection, and the concentration of other defense proteins and peptides. The defense activity of hemolymph is a resultant of the action of all its components. For example, even a peptide with antifungal activity like gallerimycin can sensitize Gram-negative bacteria against the action of cecropin, as shown in *Galleria mellonella* (Bolouri Moghaddam et al. 2016). The synergistic action of antimicrobial peptides is beneficial to the insect, which can then use less energy for synthesis of defense molecules to achieve a relatively high level of antimicrobial activity.

Antimicrobial peptides are amphipathic 7–100 amino acid molecules (Wu et al. 2018). The presence of hydrophilic and hydrophobic regions determines their antimicrobial properties, because the great majority of these compounds act against microbial membranes interacting with phospholipids. These membranes are negatively charged and most antimicrobial peptides are cationic, which facilitates their mutual interaction; however, there are some exceptional anionic defense molecules. Insect antimicrobial peptides can efficiently serve their defense functions due to the following features (Matsuzaki 1999):

- selective toxicity—this means that they act only against microbial, but not host's membranes at concentrations harmful for microorganisms,
- relatively quick action—quicker than bacterial doubling time, otherwise the bacterial clearance would not be possible,
- they do not cause development of resistance in pathogens like antibiotics do,
- most of them have a broad spectrum of activity.

Unlike in the case of the very specific antigen-antibody interaction, a given peptide is not specific for only one pathogenic species. For example, there are peptides acting only against fungi, against Gram-positive and Gram-negative bacteria, or against a specific group of bacteria. This is probably associated with the fact that there are only two main signaling pathways Toll and Imd regulating their expression. However, the so-called repertoire of defense molecules in insect hemolymph depends on the infecting microorganism (Mak et al. 2010). Taking into account the function of insect antimicrobial peptides, they can be divided into antibacterial (anti-Gram-positive, anti-Gram-negative), antifungal, anti-viral, and anti-parasitic, depending on the target microorganisms. It is also worth mentioning that, besides direct antimicrobial activity, insect defense peptides can neutralize pathogen endotoxins, modulate host defense response, and control the presence of endosymbionts (Brogden 2005; Makarova et al. 2016). In terms of their structure, antimicrobial peptides are divided into three groups: (i) linear; (ii) containing cysteine residues, which stabilize their structure, and, (iii) containing many residues of a given amino acid (Fig. 4.2) (Cytryńska 2009; Cytryńska and Zdybicka-Barabas 2015).

Insect antimicrobial peptides disrupt pathogen homeostasis via a variety of mechanisms, but most of them are targeted against microbial membranes. Due to their amphipathic properties and, in most cases, positive charge, they can easily attach to a negatively charged phospholipid bilayer. When the threshold concentration of antimicrobial peptides is achieved, they interfere with membrane functionality. There are a few models elucidating the action of insect antimicrobial peptides; however, they are not mutually exclusive. In the barrel-stave model, peptides insert themselves across the membranes and form a peptide-lined pore. They line the interior of the pore, building its wall like staves in a barrel. In another model—torroidal, the inserting peptide changes the position of membrane phospholipids, which bend and form a half-ring, so their hydrophilic "heads" are directed towards the channel light. The

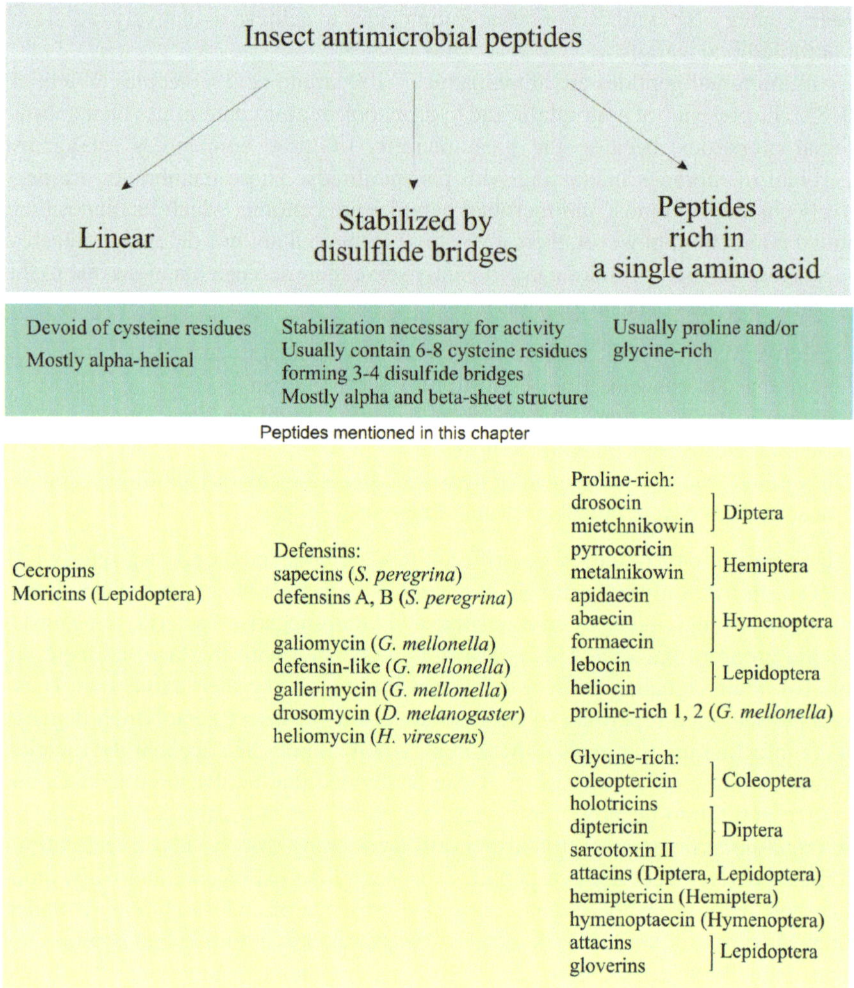

Fig. 4.2 Insect antimicrobial peptides

created pore is lined with both peptides and lipids. In the disordered aggregate model, the structure of the pore is less regular with lower amounts of randomly spread peptides, oriented in different ways. In the carpet model, antimicrobial peptides lie on the surface of the membrane (like a carpet on the floor) and, after reaching the threshold concentration, they solubilize the membrane into micelles (Wiesner and Vilcinskas 2010). These models are shown in Fig. 4.3.

There are also other ways in which the defense peptide can disturb microbial membranes, e.g. an impact on its thickness, membrane depolarization, etc. (Nguyen

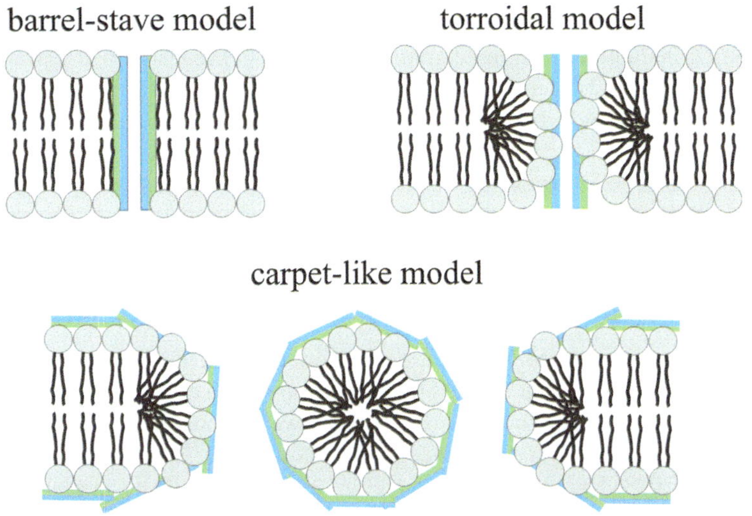

Fig. 4.3 Simplified models of AMP action. In the barrel-stave model, pores in the phospholipid bilayer are lined with AMPs (the blue and green parts reflect the hydrophobic and hydrophilic domains of the peptide, respectively). In the torroidal model, pores are created with AMPs and heads of phospholipids, while the membranes in the carpet model are torn into micelles and small fragments surrounded by the peptides

et al. 2011). Some antimicrobial peptides can also penetrate pathogen cells and interfere with such physiological processes as the synthesis of nucleic acids or proteins, activation of autolysis, and others.

Cecropins

Cecropin was the first insect antimicrobial peptide to be found. This linear peptide was isolated in 1980 from the hemolymph of immunized *Hyalophora cecropia* pupae. The peptides characterized in *H. cecropia* were classified into subfamilies A, B, C, D, and E (Hultmark et al. 1982). Interestingly, whereas most cecropin subfamilies were identified in other insect orders (e.g. Diptera, Coleoptera), cecropins D were described only in lepidopteran insects. The defense peptides classified as cecropins in different insect species often have different names, e.g. sarcotoxin-I in *Sarcophaga peregrina*, papilocin in *Papilio xuthus*, stomoxy in *Stomoxys calcitrans*, and hinnawin in *Artogeia rapae*. Cecropin molecules contain 31–39 amino acid residues and, in a hydrophobic environment, they adopt an α-helical conformation, in which two α-helices are joined by a flexible hinge region. For example, in *H. cecropia* cecropin A, amino acid residues 5–21 and 24–37 form the N-terminal and C-terminal α-helix, respectively. Usually, the N-terminal part of the cecropin molecule is strongly basic and the C-terminal part is hydrophobic. However, in contrast to the highly basic cecropins A and B, cecropins D are less basic and more hydrophobic. In addition, most cecropins have an amide group attached to the C terminus, a modification that increases peptide stability and facilitates interaction with phospholipid membranes. These peptides exhibit a broad spectrum of bactericidal and fungicidal activity (Sidén and Boman 1983; Otvos 2000; Bulet and Stöcklin 2005; Yi et al. 2014). Our recent study on *G. mellonella* neutral cecropin D (p*I* = 6.47) revealed that anti-*E. coli* activity of this peptide is initiated by interaction with lipopolysaccharide (LPS) in the outer membrane and other lipid components in the cell envelope. Effective binding of this neutral defense peptide to *E. coli* cells is possible thanks to the amphipathic character of α-helices, exposure of small positively charged patches on their polar surfaces, and hydrophobic interactions (Zdybicka-Barabas et al. 2019). The changes in the topography of the pathogen cell surface caused by cecropins and other defense factors can be imaged by Atomic Force Microscopy (Fig. 4.4).

Moricins
Among insect α-helical AMPs, there are also moricins, described so far only in Lepidoptera. Moricins were first isolated from the hemolymph of immunized silkworm *Bombyx mori* (Hara and Yamakawa 1995). They were characterized in *Spodoptera litura*, *Manduca sexta*, *G. mellonella*, and *Plutella xylostella* (Oizumi et al. 2005; Brown et al. 2008; Dai et al. 2008; Xia et al. 2017). In contrast to cecropins, the moricin molecule composed of 39–43 amino acid residues forms a single α-helix structure. Moricins have bactericidal activity, especially against Gram-positive *B. cereus*, *S. aureus*, and *Streptococcus pyogenes* and against Gram-negative *E. coli*. *G. mellonella* moricins also exhibit high antifungal activity, including action against filamentous fungi, e.g. *Fusarium graminearum* (Brown et al. 2008). Crucial for moricin antimicrobial activity is the amphipathic N-terminus and the hydrophobic C-terminus. It was demonstrated that the *P. xylostella* moricins, active against *S. aureus* and *E. coli*, led to disruption of the bacterial membrane (Dai et al. 2008; Yi et al. 2014; Xia et al. 2017). Interestingly, two moricins (B and C4) and cecropin D were found to be up-regulated in the hemolymph of *Leishmania panamensis*-challenged *G. mellonella* larvae (Patiño-Márquez et al. 2018).

Defensins
Insect defensins constitute a very large group of hemolymph defense peptides, stabilized by disulfide bridges. They have been described in species belonging to Coleoptera, Diptera, Hemiptera, Hymenoptera, Lepidoptera, and Odonata. Defensin genes are expressed mainly in the insect fat body as well as the intestine in hematophages (bloodsucking insects), e.g. *S. calcitrans* or *Anopheles gambiae*. The first insect defensins, i.e. sapecin and defensins A and B, were obtained from a *S. peregrina* (Diptera) cell line and from the hemolymph of immunized larvae of *Phormia terranovae* (Diptera), respectively. Insect defensins contain 34–51 amino acid residues adopting a characteristic cysteine-stabilized αβ motif (CSαβ). They contain 6 cysteine residues forming disulfide bridges according to the formula C1-C4, C2-C5, and C3-C6. The number of amino acids in the α-helix and two β-sheets is highly preserved in all known insect defensins, while the length of the loop at the N-terminus is variable. The sequences of insect defensins in species classified in the same order have a high degree of homology, which is not found when comparing peptides between different orders. Defensins change the permeability of the cell membrane of Gram-positive bacteria by formation of membrane channels but usually do not show activity against Gram-negative bacteria (Otvos 2000; Bulet and Stöcklin

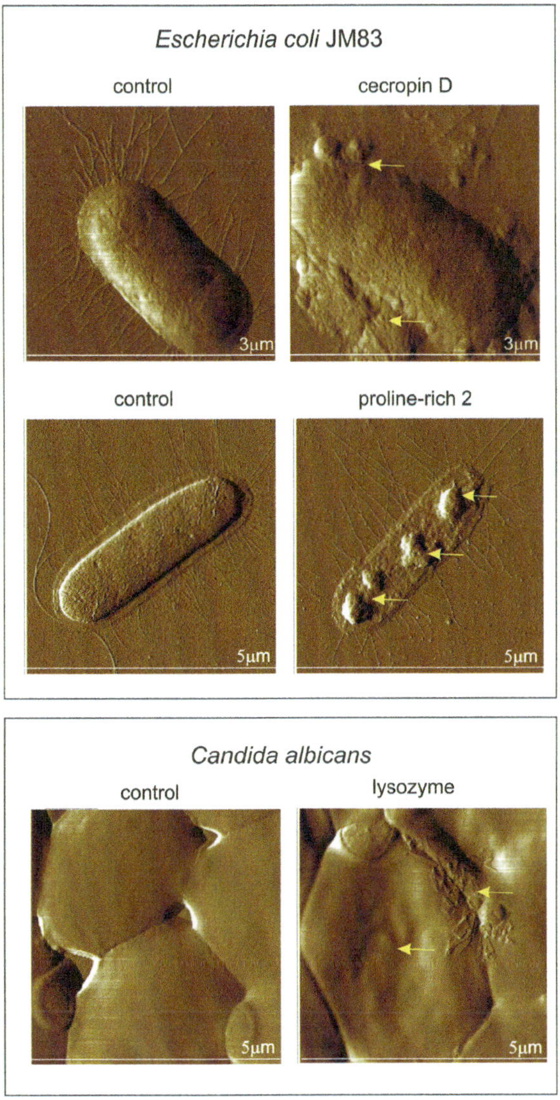

Fig. 4.4 Atomic force microscopy images of alterations in *Escherichia coli* JM83 and *Candida albicans* cell surface topography after incubation with *Galleria mellonella* antimicrobial peptides and proteins. The changes are indicated by the yellow arrows

2005; Gao and Zhu 2014). It has been demonstrated that *P. terranovae* defensin A forms oligomers in the bacterial membrane, which leads to partial depolarization, loss of potassium ions from the cell, a decrease in the cytoplasmic ATP level, and inhibition of respiratory processes (Cociancich et al. 1993).

The first lepidopteran defensin, named galiomycin, was purified from the hemolymph of immunized *G. mellonella* larvae (Lee et al. 2004). This 43 amino acid peptide is active against filamentous fungi and yeasts, but has no antibacterial activity. Another *G. mellonella* defensin, i.e. the Gm defensin-like peptide, contains

44 amino acid residues. Although its sequence relative to galiomycin differs by only two amino acid residues, this peptide exhibits higher antifungal activity and, additionally, antibacterial activity against some Gram-positive bacteria (Cytryńska et al. 2007). In addition to typical insect defensins with spatial conformation stabilized by three disulfide bridges, the structure of some antifungal CSαβ motif-containing peptides is stabilized by four disulfide bridges thanks to the presence of 8 cysteine residues. This group is represented by *Drosophila melanogaster* drosomycin, the first described inducible insect antifungal peptide. Drosomycin strongly inhibits the growth of filamentous fungi and leads to lysis of *Botrytis cinerea* and *Neurospora crassa* cells (Fehlbaum et al. 1994; Zhang and Zhu 2009). Drosomycin-like antifungal peptides have been described in the representatives of Lepidoptera, e.g. heliomicin in *Heliothis virescens* and gallerimycin in *G. mellonella* (Lamberty et al. 1999; Schumann et al. 2003). The *G. mellonella* gallerimycin is active against the entomopathogenic fungus *Metarhizium anisopliae*, but not against yeasts (Schumann et al. 2003). Interestingly, transgenic tobacco expressing gallerimycin was resistant to plant pathogenic fungi *Erysiphe cichoracearum* and *Sclerotinia minor* (Langen et al. 2006).

Proline- and glycine-rich peptides

Insect proline-rich defense peptides have been characterized in the representatives of Diptera, e.g. drosocin and mietchnikowin, Hemiptera, e.g. pyrrocoricin and metalnikowin, Hymenoptera, e.g. apidaecin, abaecin, and formaecin, and Lepidoptera, e.g. lebocin and heliocin. These peptides are composed of 15–39 amino acid residues, where proline residues, usually organized into repetitive Pro-Arg-Pro or Arg-Pro motifs, account for 25–44%. Some insect proline-rich peptides, e.g. *G. mellonella* proline-rich peptide 1 and 2 as well as *Apis mellifera* abaecin, do not contain the characteristic Pro-Arg-Pro motif (Bulet et al. 1999; Cytryńska et al. 2007). They contain two domains: one that is highly conservative and responsible for antimicrobial activity and the other that is more variable and confers specificity. Short-chain peptides (up to 20 amino acid residues) show higher activity against Gram-negative bacteria, whereas long-chain peptides (more than 20 amino acid residues) are highly active against Gram-positive bacteria and fungi. It should be noted that many proline-rich peptides target intracellular components, e.g. the translational apparatus and chaperones in pathogen cells (Kragal et al. 2001; Scocchi et al. 2011; Graf et al. 2017).

Glycine-rich peptides have been characterized in the representatives of many insect orders, e.g. Coleoptera (coleoptericin, holotricins), Diptera (diptericin, attacins, sarcotoxin II), Hemiptera (hemiptericin), Hymenoptera (hymenoptaecin), and Lepidoptera (attacins, gloverins). Some molecules classified in this group are in fact polypeptides (e.g. the 93-amino acid long *A. mellifera* hymenoptaecin) (Casteels et al. 1993) or even small proteins (e.g. approx. 20 kDa attacins). Some glycine-rich peptides exhibit high activity against specific Gram-negative bacteria. *H. cecropia* attacin inhibits synthesis of the *E. coli* outer membrane, while *Manduca sexta* gloverin binds to the cell wall components of the bacteria (Scocchi et al. 2011; Mylonakis et al. 2016). Interestingly, anti-trypanosomal activity against *Critidia bombi* has been demonstrated for *Bombus terrestris* hymenoptaecin (Marxer et al. 2016).

Other peptides
It is important to note that although the vast majority of insect antimicrobial peptides are cationic molecules, there are also anionic defense peptides. Some of these peptides represent insect defensins, which have been detected in *Spodoptera littoralis* (Seufi et al. 2011) and *B. mori* (Wen et al. 2009). Two unique anionic defense peptides without homology to known peptides or proteins have been identified in *G. mellonella* hemolymph (Cytryńska et al. 2007). Interestingly, anionic peptide 2 is present constitutively in *G. mellonella* hemolymph at a relatively high concentration of 12 μM (Mak et al. 2010). It has been demonstrated that this peptide acts synergistically with lysozyme and apolipophorin III against Gram-negative bacteria *E. coli* and yeast-like fungus *C. albicans* (Zdybicka-Barabas et al. 2012a; Sowa-Jasiłek et al. 2014).

Peptide Inhibitors of Metalloproteinases

Two types of metalloproteinase inhibitors have been identified in insects. The first one, detected in *D. melanogaster*, is a homologue of vertebrate tissue inhibitors of metalloproteinases (TIMP), which may regulate the corresponding matrix metalloproteinase (MMP) (Pohar et al. 1999). The other type has been identified in *G. mellonella* larvae and is the first peptide capable to inhibit specifically the activity of metalloproteinases; hence, it was given the name IMPI (*insect metalloproteinase inhibitor*) (Wedde et al. 1998).

IMPI has no similarity with TIMPs. Yet, a database search revealed similarity of the IMPI sequence to a trypsin inhibitor-like cysteine-rich (TIL) domain and a cysteine-rich trypsin inhibitor-like family (Clermont et al. 2004; Rawlings et al. 2004). The *G. mellonella* IMPI is a heat-stable glycosylated peptide with five intramolecular disulfide bonds and molecular mass of 8.6 kDa (Wedde et al. 1998). It is induced and secreted to hemolymph in response to bacterial or fungal challenge. This inhibitor has been detected in the hemolymph of immunized *G. mellonella* larvae, which may suggest its important role in immunity similar to that of AMPs. It has been demonstrated that IMPI inhibits the activation of phenoloxidase by thermolysin in vitro. This can imply that IMPI can probably inhibit the melanization of *G. mellonella* larvae after immunization. This can allow avoidance of excessive activation of PPO by microbial metalloproteinases (Wedde et al. 1998). It has been demonstrated that *G. mellonella* IMPI can inhibit development of fungal entomopathogens (Vilcinskas and Wedde 1997; Wedde et al. 1998). In addition, the results presented by Vertyporokh and Wojda (2017) showed expression of genes encoding IMPI as well as gallerimycin and galiomycin in *G. mellonella* larvae infected with *Beauveria bassiana*, which may suggest its important role in humoral response during fungal infection. IMPI is encoded by the *impi* gene, which contains 510 nucleotides corresponding to 170 amino acids. The calculated molecular mass of the protein without a signal sequence (19 aa) is 16.549 kDa (Wedde et al. 2007). This is almost twice the value for purified IMPI (8.364 kDa) from hemolymph of the greater wax moth *G. mellonella* (Clermont et al. 2004). It has been suggested that the identified putative furin cleavage site in the full-length protein implies that IMPI isolated from hemolymph is a product of post-translational processing and that there are two products of the *impi* gene. The N-terminal peptide (rIMPI-1) corresponds to native IMPI purified from hemolymph, whereas the C-terminal half of the *impi* sequence encodes rIMPI-2. The results regarding two recombinant proteins rIMPI-1 and rIMPI-2 obtained in an analysis of *impi* cDNA either upstream or downstream of the furin cleavage site showed two peptides with distinct activities but some sequence similarity. It has been found that rIMPI-1 inhibits the thermolysin-like metalloproteinase of *Vibrio vulinficus*. In contrast, rIMPI-2 is inactive against this microbial enzyme but moderately inhibits human recombinant MMP-1, MMP-2, MMP-3, and MMP-7 (Wedde et al. 2007). It is known that the *impi* gene is activated during *G. mellonella* metamorphosis, indicating that endogenous matrix metalloproteinases may regulate the expression of this gene (Altincicek and Vilcinskas 2006). In addition, it has been found that the *impi* gene is expressed after injection of both microbial metalloproteinases and MMPs into *G. mellonella* (Wedde et al. 2007). Therefore, it seems that the *impi* gene is expressed during metamorphosis and immune challenge, and two distinct inhibitors of metalloproteinases occur after post-translational processing of the primary protein. One IMPI regulates microbial metalloproteinases and contributes to innate immunity, and the other possibly regulates endogenous metalloproteinases during metamorphosis (Vilcinskas and Wedde 2002; Wedde et al. 2007). It has been proposed that microbial metalloproteinases released into the hemolymph nonspecifically hydrolyze the hemolymph proteins to a number of small *ca.* 3 kDa molecules. These fragments called *protfrags* are recognized probably by a receptor that activates a signaling pathway in which Rel proteins are involved, and this induces transcription and translation of IMPI and AMPs (Fig. 4.5). IMPI secreted into hemolymph inactivates metalloproteinases (Vilcinskas and Wedde 2002; Altincicek and Vilcinskas 2006).

Additionally, inducible serine proteinase inhibitors, i.e. ISPI-1, -2, -3, have been discovered in hemolymph of immunized *G. mellonella* larvae. They can inhibit toxic serine proteinases Pr1 and Pr2 produced by *M. anisopliae* (Fröbius et al. 2000). Serine protease inhibitors (SPIs), which can protect insects against fungal infection, have been identified also in *B. mori* hemolymph. It has been found that the silkworm fungal protease inhibitor-F (FPI-F) from the TIL family inhibits subtilisin and proteases produced by *B. bassiana* and *Aspergillus melleus*. It also inhibits *B. bassiana* growth by affecting conidial germination and germ tube development (Eguchi et al. 1994; Zhao et al. 2012). Li et al. (2012) provided evidence for the presence of BmSPI38 in the hemolymph, midgut, fat body, and integument of *B. mori* after infection with *B. bassiana*. It was demonstrated that BmSPI38 inhibited cuticle-degrading protease CDEP-1 such as Pr1 secreted by *B. bassiana* as well. It has been found that CDEP-1 may trigger melanization in insects. Therefore, it was supposed that the use of

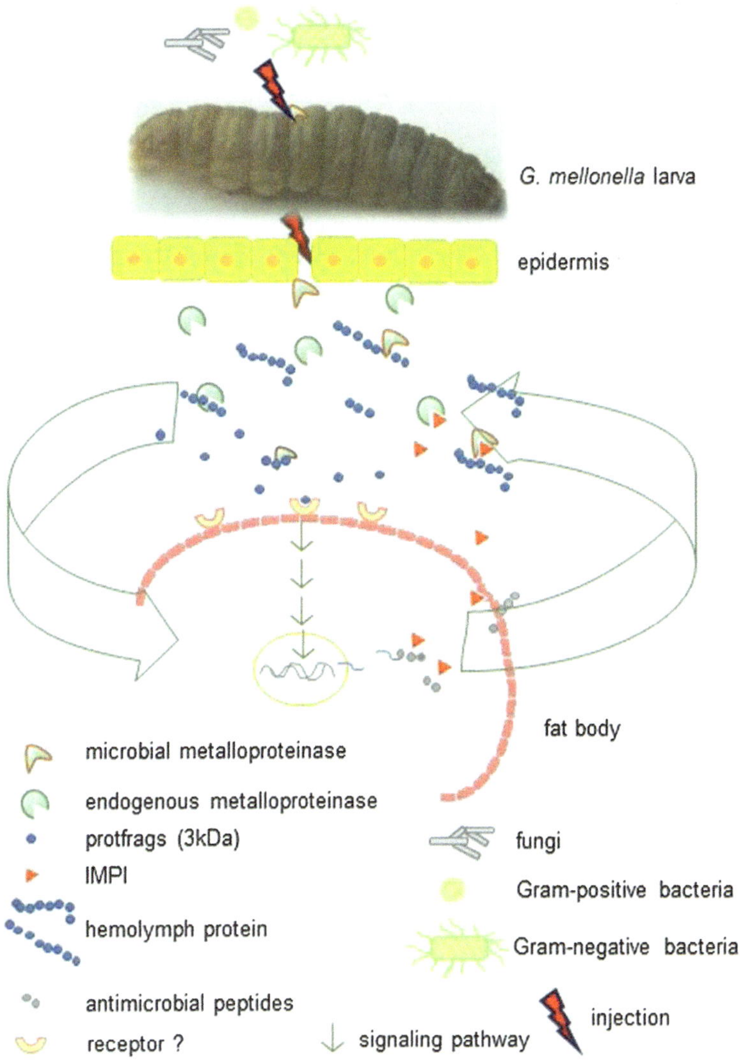

Fig. 4.5 Regulation of IMPI and AMPs expression in *G. mellonella* by fragments of digested hemolymph proteins (protfrags). Description in the text (Vilcinskas and Wedde 2002)

BmSPI38 can prevent CDEP-1 activity. This suggests that SPIs, such as BmSPI38, may suppress fungal infection and enhance the anti-fungal resistance of insects (Li et al. 2012).

Immune-Related Hemolymph Proteins

In this section, we will describe proteins that are components of the insect defense system. We will start with lysozyme, which has antimicrobial activity; therefore, it can be regarded as an antimicrobial peptide. Further, we will go on to present proteins involved in immune recognition: Dscam, FREPs, hemolin, and apoLp-III. Some of these proteins are characterized by high diversity, allowing the insect to recognize different pathogens efficiently. They are very interesting from the evolutionary point of view, since insects have not developed immunoglobulins like vertebrates, and have been considered (as all invertebrates) as organisms devoid of immune memory for many years. We will see that many immune-related proteins have multiple functions of pattern recognition receptors (PRR), modulators of response, and agents with direct antimicrobial activity. In this light, we will also draw the attention of the reader to heat shock proteins, which may modulate insect resistance.

Lysozyme—A Basal Factor of Immunity

Lysozymes (EC 3.2.1.17) are muramidases hydrolysing the β-1,4-glycosidic linkage between N-acetylmuramic acid and N-acetylglucosamine in bacterial peptidoglycan (PGN) (Jolles and Jolles 1984). They can also hydrolyze the β-1,4-linkage of chitooligosaccharides in fungal cell walls (Ren et al. 2009). Lysozymes are well-known immune effectors widespread in many tissues and secretions in a huge variety of organisms from viruses to animals and plants. Several types, e.g. c (chicken), g (goose), i (invertebrate), T4 (phage), bacterial, and plant lysozymes, have been described. Insect lysozyme was described for the first time as the main antimicrobial factor by Mohrig and Messner (1968). The vast majority of known insect lysozymes are chicken type, but both c-type and i-type lysozymes may occur as well (Hultmark 1996). The c-type lysozymes typically contain 120–130 amino acids and their molecular masses range from 14 to 16 kDa. It has been described that lysozymes exhibit enzymatic (muramidase) activity as well as antimicrobial activity typical for cationic defense peptides. The c-type lysozymes are mainly active against Gram-positive bacteria, e.g. *Micrococcus*, through peptidoglycan degradation (Wang et al. 2018). Although Gram-negative bacteria are mostly resistant to enzymatic activity of lysozyme (Kajla et al. 2010; Beckert et al. 2015), lysozymes purified from lepidopteran species *Agrius convolvuli*, *B. mori*, and *G. mellonella* have been shown to have low antimicrobial activity against these bacteria (Abraham et al. 1995; Yu et al. 2002). Surprisingly, the orthopteran *Gryllotalpa orientalis* lysozyme showed similar antimicrobial activity against Gram-positive and Gram-negative bacteria (Kwon et al. 2014). In addition, it has been demonstrated that *E. coli* becomes more sensitive when lysozyme acts synergistically with antimicrobial peptides/proteins such as cecropin (Chalk et al. 1994), attacins (Engström et al. 1984), and *G. mellonella* anionic peptide

2 (Zdybicka-Barabas et al. 2012b) or with *G. mellonella* apolipophorin III (Zdybicka-Barabas et al. 2013). In addition, an important role of lysozyme is to prevent excessive melanization, which is correlated with inhibition of phenoloxidase activity (Li and Paskewitz 2006; Zdybicka-Barabas et al. 2014a, b).

The main site of lysozyme synthesis is the fat body; it may also be present in hemocytes, the gut, and the salivary glands (Hultmark 1996). The lysozyme level is dependent on the stage of insect development (Morishima et al. 1995). It is known that lysozyme is present constitutively at a low level in hemolymph, but injury or injection of bacteria or their cell wall components instantaneously induces an increase in the lysozyme level and activity (Jarosz 1979; Tanaka et al. 2008; Beckert et al. 2015; Mohamed et al. 2016). Similarly, after fungal challenge, the level of c-type lysozyme increases (Kong et al. 2013, 2016; Shen et al. 2014). In *B. mori*, overexpression of c-type lysozyme was found to inhibit significantly viral infection (Chen et al. 2018). Noteworthy, lysozyme c-1 in *Anopheles* has two roles: on the one hand, it protects the mosquito midgut against invasion of an excessive number of *Plasmodium*; on the other hand, the interaction of lysozyme with the parasite facilitates development of *Plasmodium* oocysts in the mosquito (Kajla et al. 2011). It is known that c-type lysozymes have muramidase activity but the i-type insect lysozyme has no such activity (Cancado et al. 2008; Paskewitz et al. 2008; Kajla et al. 2010; Beckert et al. 2016). Multiple i-type lysozyme genes have been identified in coleopteran species *Nicrophorus vespilloides* and *Meligethes aeneus* (Vogel et al. 2011, 2014). In the hemipteran whitefly *Bemisia tabaci*, two types of lysozymes representing c-type and i-type have been identified. The whitefly has three lysozyme genes: one c-type (*Btlysc*) and two i-type genes (*Btlysi1; Btlysi2*). All are constitutively expressed during developmental stages in different tissues (midgut, ovary, fat body), but *Btlysi1* and *Btlysi2* are expressed at a lower level than *Btlysc* (Wang et al. 2018). Also, three different genes of two types of lysozymes have been found in *A. gambiae* (Li et al. 2005; Paskewitz et al. 2008). The levels of Lys i-1 and Lys i-2 in *A. gambiae* and i-lys2 in *Harmonia axyridis* were not induced by immune challenge with bacteria or wounding, which suggested an immunity-unrelated function (Paskewitz et al. 2008; Beckert et al. 2016). Since lysozyme was found in the midgut of some insects (*D. melanogaster*, *M. sexta*), its role in digestion of bacteria was postulated (Daffre et al. 1994).

Besides the typical lysozyme with muramidase activity, lysozyme devoid of one or both residues (Glu^{32}, Asp^{50}) necessary for muramidase activity has been detected in some insects (e.g. *Antheraea mylitta*, *B. mori*, *Spodoptera exigua*). This new class of non-catalytic protein has been named lysozyme-like protein (LLP), whose action relies on peptidoglycan binding but not hydrolysis. LLPs are members of the glycoside hydrolase family. Their expression is regulated by the Imd pathway and was found to increase after bacterial infection (Gandhe et al. 2007; Chapelle et al. 2009; Satyavathi et al. 2018). Summarizing, lysozyme may act enzymatically as a muramidase, but there are also reports on its non-enzymatic antifungal action (Sowa-Jasiłek et al. 2014, 2016). The effect of lysozyme action on the topography of fungal cell surface imaged by AFM is shown in Fig. 4.4.

Fibrinogen-Domain Containing Proteins

The fibrinogen (FBG) domain refers to a conserved region composed of approximately 200 amino acid residues with high sequence similarity to the C-terminus of β and γ fibrinogen chains (Doolittle et al. 2012). These domains are important components of evolutionarily conserved invertebrate and vertebrate molecules with immune functions. The vertebrate FBG domain-containing proteins, e.g. mammalian ficolins, are involved in pattern recognition and complement system activation. The well-recognized engagement of fibrinogen in the coagulation process in vertebrates is phylogenetically relatively recent (Hanington and Zhang 2011). The invertebrate FBG domain-containing proteins or fibrinogen-related proteins (FREPs) function primarily in innate immune response as pattern recognition receptors. For example, tachylectins TL5A and TL5B characterized in the plasma of horseshoe crab *Tachypleus tridentatus* specifically recognize acetyl groups on the surface of Gram-positive and Gram-negative bacteria (Gokudan et al. 1999). Despite the presence of evolutionarily conserved FBG domains in the C-terminal region, the invertebrate representatives of this group differ considerably in the N-terminal domains. FREPs described in mollusks (e.g. *Biomphalaria glabrata*) contain one or two variable immunoglobulin domains that are absent in other invertebrate FBG-containing molecules. With their lectin-like properties, FREPs are involved in agglutination of pathogens, bacterial clearance, and anti-parasite defense (Hanington and Zhang 2011). Not surprisingly, genes encoding FBG-containing proteins have been identified in the genomes of various insect species, including *A. mellifera*, *B. mori*, *D. melanogaster*, *M. sexta*, and *Tribolium castaneum* (Hanington and Zhang 2011; Zhang et al. 2015b). The number of identified genes encoding the proteins of the FBG domain varies from 1 in *Nasonia vitripennis*, 3 in *A. mellifera* and *B. mori*, and up to several or even several dozen in different *Drosophila* species. Genome analysis of twelve *Drosophila* species revealed that the number of genes encoding FBG-related proteins varies from 14 genes in *Drosophila yakuba* and *Drosophila simulans* to 43 genes in *Drosophila grimshawi*. The sequence and structural comparisons allowed researchers to postulate that some of the *D. melanogaster* FREPs can recognize carbohydrates and their derivatives, thus being involved in recognition of microorganisms (Middha and Wang 2008; Hanington and Zhang 2011). However, mosquito genomes appeared to encode the largest number of FREPs, alternatively named fibrinogen-domain immunolectins (FBNs). They are the largest families of pattern recognition receptors in *Culex pipiens quinquefasciatus* and *A. gambiae*, composed of 93 and 59 putative members, respectively (Hanington and Zhang 2011).

Research on the FBG domain-containing protein in *Armigeres subalbatus*, named aslectin (AL-1), revealed its binding to bacterial cells (*E. coli* and *M. luteus*) and specific interaction with N-acetylglucosamine (GlcNAc). Challenge of mosquitoes with *E. coli* or *M. luteus* resulted in up-regulation of *AL-1* expression in hemocytes followed by an increase in the protein level in hemolymph. Hence, involvement of aslectin in pathogen recognition has been postulated (Wang et al. 2004). The function as pattern recognition receptors of bacteria has also been ascribed to some *A.*

gambiae FBNs, but the most interesting role of these molecules in the insect vectors is recognition of malaria parasites and involvement in anti-*Plasmodium* defense (Dong and Dimopoulos 2009; Cirimotich et al. 2010). Since the discovery of engagement of FREPs/FBNs in *A. gambiae-Plasmodium* interactions, mosquito FREPs have become the best characterized representatives of this family of proteins among insects.

Using transcriptomic analysis, it has been shown that some *FREP* genes, i.e. *FBN9* (*FREP13*) and *FBN25* (*FREP63*), can be up-regulated after *E. coli* challenge in *A. gambiae*. FBN9 binding to bacterial cells, including Gram-negative *E. coli* and *Pseudomonas veronii* as well as Gram-positive *Bacillus subtilis*, was demonstrated. The binding assays revealed different levels of affinity for Gram-negative and Gram-positive bacteria and a possibility of FBN9 homodimer formation. Multimerization through intra- and interchain disulfide linkages has been described for horseshoe crab tachylectins (Gokudan et al. 1999), and a similar process may provide recognition of a broader repertoire of compounds by mosquito FREPs. On the other hand, RNAi-mediated gene silencing indicated two other FREPs, i.e. FBN22 (FREP26) and FBN39 (FREP40), as essential participants in bacterial clearance and *A. gambiae* immune defense against bacteria (Dong and Dimopoulos 2009). Notably, *FBN8* (*FREP57*), *FBN9*, and *FBN39* were strongly induced during *Plasmodium* midgut invasion. FBN9 is considered as one of the most potent *Anopheles* anti-*Plasmodium* immune factors. It has been demonstrated to co-localize with ookinetes of the human pathogen *Plasmodium falciparum* and the rodent experimental model pathogen *Plasmodium berghei*. The *FBN9* silencing resulted in increased susceptibility of *A. gambiae* to both pathogens (Dong and Dimopoulos 2009). However, a transgenic *A. gambiae* line specifically overexpressing *FBN9* in the fat body showed increased resistance only to the rodent pathogen *P. berghei*, suggesting differences in the mosquito immune defense against these two parasites (Simões et al. 2017). The transgenic mosquitoes exhibited higher resistance to bacterial infections as well, confirming the involvement of FBN9 in anti-bacterial defense. On the other hand, FBN39 specifically protected the mosquitoes against *P. falciparum*, and *FBN39* silencing exerted a more pronounced effect on the development of this parasite, indicating specificity in regulation of the mosquito resistance to the human pathogen (Dong et al. 2006; Dong and Dimopoulos 2009; Simões et al. 2017).

In addition to the above-mentioned FBNs, two other members of this family, FREP1 and FBN30 (FREP8), have been implicated in *Anopheles-Plasmodium* interactions (Zhang et al. 2015a; Dong et al. 2018). It has been demonstrated that FBN30 is a secreted protein forming an octamer able to inhibit *Plasmodium* infections. Recent studies have revealed specific interactions of FBN30 with *P. berghei* and clinical isolates of *P. falciparum* from natural *A. gambiae* populations in Kenya; however, the ligand of FBN30 has not been identified yet (Niu et al. 2017b). Interestingly, in contrast to other known mosquito FREPs, the FREP1, acting as a tetramer, mediates *P. falciparum* and *P. berghei* invasion by anchoring ookinetes to the peritrophix matrix (Zhang et al. 2015a, b). *FREP1* inactivation in *A. gambiae* through CRISPR/Cas9-mediated gene knockout resulted in suppression of infection with both *Plasmodium* species (Dong et al. 2018). FREP1 is considered as a potential transmission-blocking target for control of malaria (Niu et al. 2017a). Interestingly, although the midgut is the primary site of immune response against invading *Plasmodium* ookinetes, most of the analyzed FREPs/FBNs were detected in the abdominal parts of the mosquito body containing hemolymph and hemocytes (Dong and Dimopoulos 2009). Moreover, the highest expression of *FBN9* was found in the fat body and carcass tissues, suggesting that the hemocoel is a source of FBN molecules that target ookinetes (Baker et al. 2011; Simões et al. 2017).

Down Syndrome Cell Adhesion Molecules

Down syndrome cell adhesion molecules (Dscam) have been described in vertebrates and invertebrates. Their characteristic feature is the presence of two kinds of domains: fibronectin type III (FNIII) and immunoglobulin (Ig) domains. Because of this, Dscam are classified as members of the immunoglobulin (Ig) superfamily (Ng et al. 2014). The human DSCAM proteins are encoded by the *DSCAM* gene mapping to the Down syndrome region of chromosome band 21q22.2-22.3 (Yamakawa et al. 1998). The first invertebrate homolog of human DSCAM, Dscam1, was characterized in *D. melanogaster* (Schmucker et al. 2000). The structure of DSCAM proteins is highly conserved; they are composed of an extracellular region with ten Ig domains and six FNIII domains, a transmembrane region, and a C-terminal cytoplasmic endodomain. However, the insect and crustacean *Dscam1* gene, in contrast to its human counterpart, is a source of the extraordinary level of Dscam protein diversity. The *Dscam1* gene has a complex structure, in which three hypervariable exon clusters (exons 4, 6, 9 in *Drosophila*) encode three Ig domains of the extracellular region of the Dscam molecule. In *Drosophila*, exons 4, 6, and 9 have evolved 12, 48, and 33 alternative variants, respectively (Fig. 4.6; Ghosh et al. 2011), while the corresponding exons in *A. gambiae* encode 14, 30, and 38 alternative sequences, respectively (Xu et al. 2019). Across insect species, these hypervariable exons encode the N-terminal parts of the Ig2 and Ig3 domains and the whole Ig7 domain. Additionally, a cluster of exon 17 encodes two possible forms of a transmembrane region.

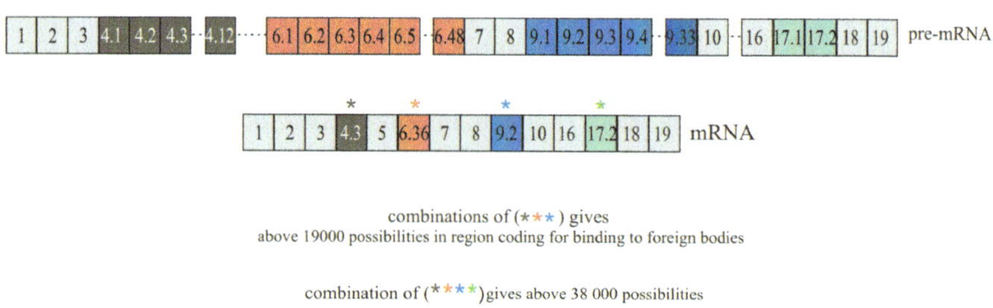

Fig. 4.6 Simplified scheme of *D. melanogaster Dscam* pre-mRNA and an example of mature mRNA. The numbers on the grey background represent the exons and the numbers on the color background signify alternative versions of a given region

Dscam pre-mRNA undergoes mutually exclusive alternative splicing, in which alternative versions of the exons mentioned above can be built in the mature mRNA (Graveley 2005), resulting in a total number of 38,016 (19,008 ectodomains) and 31,920 (15,960 ectodomains) of potential Dscam1 isoforms in *D. melanogaster* and *A. gambiae*, respectively.

Noteworthy, according to the number of alternative variants of the variable exons, *Musca domestica* can produce as many as 69,120 putative Dscam isoforms (Xu et al. 2019). Since exons 19 and 23 encoding for the cytoplasmic tail in *Drosophila* can be skipped or not, the number of potential splice variants increases up to approx. 150,000 (Armitage et al. 2017). Similarly to human DSCAM, *Drosophila* Dscam was found to be involved in normal formation of connections in the neuronal network by mediating self-avoidance process. In *Drosophila*, Dscam hypervariability is essential for obtaining and maintaining the identity of individual neurons through distinguishing between its own and other neuron neurites (Forbes et al. 2011; Zipursky and Grueber 2013; Armitage et al. 2015).

It has been suggested that, thanks to hypervariable Ig domains, the diversity in Dscam isoforms in insects (and crustaceans) could be used by the invertebrate immune system to reach a functional endpoint similar to vertebrate antibodies (Cherry and Silverman 2006; Stuart and Ezekowitz 2008; Ng et al. 2014; Li et al. 2018). In accordance with this idea, it has been demonstrated that different *Dscam* isoforms can be expressed by individual *Drosophila* hemocytes collected from hemolymph of third instar wandering larvae (Neves et al. 2004), and potentially 18,000 extracellular domain isoforms can be produced in the *Drosophila* fat body and hemocytes (Watson et al. 2005). A shorter form of the Dscam protein, corresponding to the extracellular domain, was also found in hemolymph, suggesting a possibility of secretion of a soluble form from the fat body/hemocytes or proteolytic shedding of extracellular domains from hemocytes. Moreover, binding of recombinant Dscam to *E. coli* cells has been demonstrated (Watson et al. 2005). Research on *A. gambiae Dscam, i.e. AgDscam*, revealed that challenge of hemocyte-like immunocompetent cells Sua5B with Gram-positive bacteria *Staphylococcus aureus* and Gram-negative bacteria *E. coli* and *Pseudomonas veronii* resulted in production of distinct mRNA splice variants of exon 4 encoding the Ig2 domain, suggesting synthesis of AgDscam isoforms with different adhesive preferences and interaction specificity. Immunization of adult *A. gambiae* with these bacteria also caused alternative splicing of *AgDscam* (Dong et al. 2006). The AgDscam exposed on the surface of Sua5B hemocyte-like cells concentrated at a site of interaction with *E. coli* and *S. aureus* cells, but not with fungal cells *Saccharomyces cerevisiae*, indicating some level of interaction specificity. Because the AgDscam-depleted cells were compromised in phagocytosis of the bacteria, the AgDscam role in phagocytosis of the bacteria but not the fungus was postulated (Dong et al. 2006). Exposure of *A. gambiae* to the malaria parasite *P. falciparum* caused an increase in the diversity of the AgDscam splice-form, with exon 4 variants 4.11, 4.12, and 4.13 and exon 6 variants 6.1 and 6.9 being particularly important for immune response against malaria parasites in the field (Smith et al. 2011). An invasion of adult mosquitoes with *P. falciparum* and *P. berghei* resulted in different AgDscam splice form repertoires in the midgut epithelium. In addition, silencing of *AgDscam* in adult mosquitoes decreased the survival rate after bacterial infection and increased the number of *P. berghei* oocysts. It was also suggested that the repertoire of AgDscam isoforms induced by a pathogen increased protection against this pathogen (Dong et al. 2006).

Summarizing, it is hypothesized that Dscam proteins in the insect immune system may function as hypervariable pattern recognition receptors. On the other hand, profiling of Dscam1 isoforms in adult *D. melanogaster* exposed to *E. coli* revealed no significant changes in the expression of individual exons and in splicing patterns (Armitage et al. 2014). In addition, the *Dscam1* gene expression was not affected by hemocoelic exposure of *D. melanogaster* to *Bacillus thuringiensis*. Similarly, no changes in *Dscam1* gene expression were detected in *T. castaneum* larvae challenged with *B. thuringiensis*, *Pseudomonas fluorescens*, and *E. coli*. RNAi knockdown of *Dscam1* did not decrease the survival of bacteria-exposed *T. castaneum* (Peuß et al. 2016). Hence, the role of Dscam in insect immunity is still unclear and controversial (Brites and Du Pasquier 2015; Armitage et al. 2017).

Hemolin—A Bifunctional Protein

Hemolin is a member of the Ig superfamily. It has been described in Lepidoptera insects such as *H. cecropia* (Bettencourt et al. 2002), *M. sexta* (Eleftherianos et al. 2007), *Hyphantria cunea* (Shin et al. 1998), *Helicoverpa zea*, *Heliothis virescens* (Terenius et al. 2009), *Plodia interpunctela* (Orozco-Flores et al. 2017), *G. mellonella* (Cytryńska el al. 2002; Shaik and Sehnal 2009), *B. mori* (Aathmanathan et al. 2018), *Lymantria dispar* (Lee et al. 2002), *Samia cynthia ricini* (Bao et al. 2007), *Actis selene* (Qian et al. 2017), and *Antheraea pernyi* (Sun et al. 2016). In 1979, Rasmuson and Boman were the first to purify hemolin from hemolymph of *H. cecropia* diapausing pupae. The purified protein had a molecular weight of 48 kDa and an isoelectric point of 8.2 (Rasmuson and Boman 1979). It is known that hemolin is synthesized in the fat body and the size of this protein shows varied dependence on the insect species, i.e. it can range from 46 to 55 kDa (examples: Lindström-Dinnetz et al. 1995; Ladendorff and Kanost 1999; Terenius 2008). Hemolin has a conservative structure composed of immunoglobulin domains in the shape resembling a horseshoe. *M. sexta* hemolin contains four Ig domains (Ladendorff and Kanost 1991). The number of glycosylation sites is dependent on the insect species, e.g. there are two such sites in *H. cunea* and *H. cecropia*, one in *M. sexta*, and three in *L. dispar* hemolin (Ladendorff and Kanost 1990; Sun et al. 1990; Wang et al. 1995; Shin et al. 1998).

Hemolin is encoded by a single gene in *H. cecropia*. The length of *Hemolin* genes varies between different insect species but they have the same structure, i.e. a regulatory sequence, a promoter, 6 exons, and 5 introns. The introns are located both within and between the Ig-like domains. The *Hemolin* gene has structural similarities to genes coding for cell-adhesion molecules (Lindström-Dinnetz et al. 1995). Based on the comparison of hemolin amino acid sequence with other molecules containing immunoglobulin domains, hemolin was shown to have similarity to neural cell adhesion molecules from insects and vertebrates, e.g. to *Drosophila* neuroglian. This suggests that hemolin is a type of an adhesion molecule and can regulate cellular response. Although the structure of this protein in Lepidoptera is similar, its distribution in insect tissues varies. High levels of hemolin have been detected in the fat body, hemolymph, or midgut (Shaik and Sehnal 2009; Sun et al. 2016; Qian et al. 2017). Hemolin is a soluble extracellular protein, which lacks a transmembrane domain. However, it has been shown that certain amounts of hemolin may be associated with the surfaces of the fat body and some hemocytes (Schmidt et al. 1993). In addition, the presence of hemolin has been shown in Malpighian tubules, gonads, and hemocytes (Sun et al. 2016; Qian et al. 2017). It was revealed that hemolin could be expressed in epidermal and neural tissues of the embryo, in yolk cells, nurse cells, and follicle cells of developing oocytes (Bettencourt et al. 2002; Aye et al. 2008).

It was found that the hemolin levels in the hemolymph of *H. cecropia* pupae increased 18-fold after *Enterobacter cloacae* infection. The hemolin gene is induced by bacteria and fungi, and by viruses in some insects, e.g. *A. pernyi, H. cecropia, B. mori,* and *M. sexta* (Ladendorff and Kanost 1991; Bao et al. 2007; Terenius 2008; Shaik and Sehnal 2009; Terenius et al. 2009; Qian et al. 2017). In contrast, virus infection did not induce hemolin in *Trichoplusia ni*, *Pseudoplusia includens*, or *Spodoptera frugiperda* (Terenius et al. 2009). It was also demonstrated that bacterial cell wall components induced hemolin gene expression or elevated the level of the protein in hemolymph (Cytryńska et al. 2002; Bao et al. 2007). Although it has no direct antibacterial activity, hemolin can modulate insect immune response through binding to LPS of Gram-negative bacteria or to lipoteichoic acid (LTA) of Gram-positive bacteria. Daffre and Faye (1997) have shown that hemolin from *H. cecropia* binds specifically to the lipid A moiety of LPS. Yu and Kanost (2002) have suggested that hemolin contains two binding sites for LPS: one site binds to lipid A and the other interacts with the carbohydrate components from the O-antigen and/or the outer core. They showed that hemolin binds more effectively to smooth LPS than to rough LPS. Competitive binding experiments have indicated that LTA has a higher affinity for hemolin than LPS and both react with the same site in the hemolin molecule (Yu and Kanost 2002). Therefore, since hemolin recognizes the structural patterns of bacteria, it can be classified as a pattern recognition receptor (Bettencourt et al. 2002). Hemolin can bind to hemocytes, changing their adhesive properties and their intracellular phosphorylation pattern. The protein stimulates hemocytes to phagocytosis and this is enhanced by LPS. It has been shown that the hemocytes of *M. sexta* and *H. cecropia* do not aggregate in the presence of hemolin. Probably, hemolin binds to the hemocyte surface via a cell adhesion receptor (Ladendorff and Kanost 1991; Lanz-Mendoza et al. 1996).

Given its ability to bind to hemocytes, hemolin may be regarded as an opsonin, which mediates the interaction between microorganisms and hemocytes. In this way, hemolin can stimulate phagocytosis (Terenius 2008). It was shown that knock-down of hemolin significantly reduced the ability of *M. sexta* hemocytes to phagocytose *E. coli* and suppressed nodule melanization (Eleftherianos et al. 2007). Hemolin can agglutinate bacteria (Yu and Kanost 2002; Qian et al. 2017); yet, hemolin silencing by RNAi was reported to increase insect susceptibility to pathogens and to reduce the ability to form microaggregates around bacteria in vitro (Eleftherianos et al. 2006, 2007). These results confirmed the key role of hemolin in insect cellular immune response. It can be assumed that hemolin plays an important role in regulation of the prophenoloxidase cascade in response to bacterial infections (Terenius et al. 2007). It has been found that the level of hemolin during activation of immune response depends on the Yippee protein. The Yippee protein is highly conserved in eukaryotes, including fungi, plants, and animals. It is expressed constitutively in many tissues, but increases rapidly after challenge with pathogens. This intracellular protein is important due to its involvement in cell proliferation and growth. Knock-down of the *Yippee* gene down-regulated the level of hemolin expression in *A. pernyi* (Sun et al. 2016).

Hemolin is a bifunctional protein playing an important role in insect immunity, embryonic development, and metamorphosis (Bettencourt et al. 2002; Qian et al. 2017). It has been suggested that *Hemolin* expression is upregulated by 20-hydroxyecdysone in *L. dispar*. A 55-kDa protein identified as hemolin was found in the gut of a diapausing *L. dispar* moth and during embryonic development of the insect. This may suggest that hemolin protects the gut of a diapausing gypsy moth from bacterial infection. The *L. dispar* hemolin has higher molecular weight than its *M. sexta* and *H. cecropia* counterparts (48 kDa), because it contains more N-glycosylation sites (Lee et al. 2002). It is known that hemolin is present in oocytes and embryos of *H. cecropia* as well (Bettencourt et al. 2002). It has been demonstrated that the hemolin gene knock-down in females can exert effects on embryo lethality in *H. cecropia*. When the expression of hemolin was down-regulated by RNA interference in *H. cecropia* pupae, the insects developed normally and adult moths mated normally but the next-generation embryos were malformed (Bettencourt et al. 2002).

Lipophorins

Lipophorins are particles circulating in insect hemolymph composed of apolipoproteins (in insects called apolipophorins) and lipids. Their main function consists in transport of lipids and related molecules from the fat body to muscles, which is especially important for energy supply during flight (Ryan and Van der Horst 2000). However, beside this role, they are involved in insect immunity, e.g. serving as so-called "hemolymph coagulogens" in the process of hemolymph clotting. This process is described in another chapter in this book, to which the reader is referred. In addition, a role of lipophorins in immune signaling in hemolymph has been postulated (Dettloff et al. 2001a). The protein component of insect lipophorin particles comprises three apolipophorins: apolipophorin I (apoLp-I), apolipophorin II (apoLp-II), and apolipophorin III (apoLp-III) with molecular mass approx. 250 kDa, 80 kDa, and 18 kDa, respectively. While apoLp-I and apoLp-II are structural non-exchangeable components of the lipophorin particle forming high density lipophorin (HDL), apoLp-III joins transiently the HDL, which leads to formation of low density lipophorin (LDL) facilitating attachment and transport of lipid molecules in the hemolymph. Up to sixteen apoLp-III molecules can bind to one HDL particle (Weers and Ryan 2006). In the muscles, the lipid cargo is released and apoLp-III molecules dissociate from the particle becoming ready to join the next HDL. The apoLp-III involvement in lipid transport is possible thanks to the amphipathic properties of this protein. The apoLp-III molecule is composed of five anti-parallel α-helices forming a bundle. The hydrophobic side chains are hidden inside the bundle, whereas hydrophilic groups are exposed on the bundle surface. Upon lipid binding, the apoLp-III molecule undergoes dramatic conformational rearrangement resulting in opening of the bundle. The capability of such conformational change facilitates transport of large amounts of neutral lipids by lipophorin particles and determines LDL solubility in hemolymph, because lipophorin-bound apoLp-III exposes hydrophilic side chains on the LDL surface. Hence, at least three main fractions of apoLp-III exist in insect hemolymph: free apoLp-III, lipid-bound apoLp-III, and lipophorin-bound apoLp-III. The structural aspects of apoLp-III interaction with lipid molecules have been reviewed in detail elsewhere (Weers and Ryan 2006; Narajanaswami et al. 2010); the role of lipophorins in lipid transport in insects has been presented as well (Van der Horst and Rodenburg 2010).

Whole lipophorin particles as well as free and lipid-bound apoLp-III are engaged in insect immune response. The role of lipophorins in insect immunity has been postulated by Dettloff et al. (2001b), who demonstrated conversion of some of hemolymph HDL particles into LDL ones via interaction with free apoLp-III shortly after immune challenge of *G. mellonella* larvae with heat-inactivated *E. cloacae*. In vitro experiments demonstrated that the LDLs, but not HDLs, were specifically taken up by a subclass of granulocytes (Dettloff et al. 2001b). It was also shown that immune challenge of *G. mellonella* induced apoLp-III binding with lipid molecules and incorporation of such lipid-bound apoLp-III into some granulocytes via endocytosis (Dettloff et al. 2001a), supporting an idea of the function of apolipophorins in insect immunity as signaling particles in hemolymph. Later studies on *G. mellonella* and *Ephestia kuehniella* indicated involvement of lipophorin particles in hemolymph clotting (Li et al. 2002), LPS-mediated immune response, and activation of prophenoloxidase (Rahman et al. 2006). Apolipophorins, together with pattern recognition proteins, phenoloxidase, serine protease homologs, and a serine protease involved in hemocyte recruitment, were found as essential components of multi-functional immune complexes formed in *B. mori* and *Aedes aegypti* hemolymph (Phillips and Clark 2017). The authors postulated functional integration of melanization, coagulation, and encapsulation through incorporation of these proteins into a large protein array. Hanada et al. (2011) reported on binding of *B. mori* apoLp-I/II to the *S. aureus* cell surface, which resulted in reduced expression of hemolysin, i.e. one of the bacterial virulence factors. In further studies, *S. aureus* lipoteichoic acid was identified as the cell wall component responsible for the interaction with apoLp-I/II (Omae et al. 2013). These findings have broadened the range of possible roles of lipophorins in insect immunity by including their direct participation in antibacterial defense. Lipophorins may also play an opposite role interfering with antiparasitic response in mosquitoes. It has been demonstrated that depletion of lipophorins in *A. gambiae* caused reduction of developing *P. berghei* oocysts in the mosquito midgut, an effect mediated directly by a decreased level of vitellogenin, whose proper expression requires lipophorins (Rono et al. 2010). In this study, lipophorins were also recognized as negative regulators of the TEP1 (thioester-containing protein 1) antiparasitic factor reducing its binding to the surface of *Plasmodium* ookinetes (Rono et al. 2010). In a more recent study, apoLp-I/II was characterized as a negative regulator of TEP1-mediated immune response during systemic bacterial and fungal infections in *A. gambiae* (Kamareddine et al. 2016). *ApoLp-I/II* silencing led to a TEP1-dependent increase in mosquito resistance to *E. coli* and *B. bassiana*, indicating that apoLp-I/II plays an important regulatory role not only during parasite infections.

The involvement of apolipophorins in insect immunity has been reviewed relatively recently (Zdybicka-Barabas and Cytryńska 2013); however, new information has become available since then. Our recent results have indicated considerable changes in the content of apoLp-I and apoLp-II in complexes with apoLp-III recovered from *G. mellonella* hemolymph, depending on the pathogen used for immunization (Gram-negative bacteria *E. coli*, Gram-positive bacteria *Micrococcus luteus*, yeast-like fungus *Candida albicans*, filamentous fungus *Fusarium oxysporum*) and time after challenge (Stączek et al. 2018). In addition, the different native-PAGE behavior of apolipophorins and lipids from hemolymph as well as from hemocytes of insects immunized with different pathogens suggested varied pathogen- and time-dependent involvement of these lipophorin components in the insect immune response (Stączek et al. 2018). Research on apolipophorins in *A. pernyi* revealed increased expression of apoLp-I/II after *S. aureus* injection at the mRNA and protein levels (Wen et al. 2017). Interestingly, in non-immunized insects, apoLp-I/II mRNA was detected mainly in the fat body and epidermis but not in hemocytes, whereas high amounts of these proteins were detected especially in cell-free plasma and hemocytes, implying that a portion of apoLp-I/II underwent translocation from hemolymph into hemocytes. After bacterial injection, the apoLp-I/II mRNA level increased significantly in the fat body and epidermis, while the protein level in hemocytes decreased, suggesting release of apoLp-I/II into hemolymph (Wen et al. 2017). In addition, depletion of apoLp-I/II expression by RNAi resulted in increased activity of phenoloxidase (PO) in response to laminarin and peptidoglycan, suggesting an inhibitory role of apoLp-I/II in the PO system (Wen et al. 2017).

The role of apoLp-III in insect immune response has been well documented. This protein functions as a versatile pattern recognition receptor engaged in (i) recognition of bacterial LPS and LTA as well as fungal β-1,3-glucan (Whitten et al. 2004; Wen et al. 2016), (ii) modulation of adhesive properties of hemocytes (Zakarian et al. 2002), (iii) phagocytosis and nodulation of fungal cells (Gotz et al. 1997; Dunphy et al. 2003), (iv) direct bacterial killing (Zdybicka-Barabas et al. 2011; Zdybicka-Barabas and Cytryńska 2011; Kim and Jin 2015), (v) synergistic antimicrobial action with other defense peptides and proteins present in hemolymph (Zdybicka-Barabas et al. 2013; Niere et al. 2001), (vi) modulation of phenoloxidase system activity (Contreras et al. 2013), and (vii) detoxification of microbial toxins together with apoLp-I and apoLp-II (Kato et al. 1994; Vilcinskas and Wedde 1997; Ma et al. 2006). The importance of apoLp-III in insect immune response is underlined by the fact that it is the target of proteolytic enzymes of entomopathogens, e.g. *Pseudomonas aeruginosa*

(Andrejko et al. 2005, 2014). In addition, apoLp-III in *S. exigua* hemolymph was identified as the target of an insulin–like peptide-binding protein from the venom of the solitary wasp *Eumenes pomiformis* (Baek et al. 2016). Binding of apoLp-III with this venom component altered the catepillar metabolism, thus facilitating the development of the parasite by extending the larval stage and suppression of body loss. ApoLp-III has also been implicated in anti-*Plasmodium* response in mosquitoes. It has been recognized as a positive regulator of *P. berghei* development in *Anopheles stephensi*. However, while silencing of the apoLp-III gene in an Indian susceptible *A. stephensi* strain decreased the development of *P. berghei* oocysts, it showed no effect on the *P. falciparum* development in *A. gambiae* (Mendes et al. 2008; Dhawan et al. 2017). The involvement of apoLp-III during *G. mellonella* infection with *B. thuringiensis* and *B. bassiana* has also been reported (Taszłow and Wojda 2015; Vertyporokh et al. 2015; Taszłow et al. 2017).

It seems that apoLp-III plays an opposite role to apoLp-I/II in regulating the activity of the PO system. Wen et al. (2016) reported on activation of the PO system in *A. pernyi* by apoLp-III in the presence of different PAMPs, including laminarin, LPS, LTA, PGN, and mannan. Moreover, RNAi-mediated silencing of apoLp-III resulted in lower expression of cecropin B, attacin, and lysozyme in *S. aureus*-injected *A. pernyi* larvae, suggesting involvement of apoLp-III in regulation of the expression of antimicrobial peptides/proteins in this insect species (Wen et al. 2016).

Interaction of apoLp-III with microbial cell wall components and binding to the cell surface of different bacteria and fungi implies its antimicrobial activity. Indeed, *G. mellonella* apoLp-III exhibits in vitro activity against Gram-positive bacteria *Bacillus circulans* and Gram-negative bacteria *Salmonella typhimurium*, *Klebsiella pneumoniae*, *Legionella pneumophila*, and *Legionella dumoffii* (Zdybicka-Barabas and Cytryńska 2011; Palusińska-Szysz et al. 2012; Zdybicka-Barabas et al. 2014a, b). Our analyses of apoLp-III interactions with both *Legionella* species revealed that the extremely hydrophobic LPS and the increased content of phosphatidylcholine in the cell membrane were responsible for effective apoLp-III binding to *L. pneumophila* and *L. dumoffii*, respectively (Zdybicka-Barabas et al. 2014a, b; Palusińska-Szysz et al. 2016). In addition, apoLp-III binds to yeast cells and conidia of filamentous fungi and affects cell morphology, metabolic activity, and biophysical properties of the cell surface (Zdybicka-Barabas et al. 2012a, b). In *G. mellonella*, apoLp-III can be a source of antimicrobial peptides, e.g. cationic apolipophoricin, which comprises a C-terminal apoLp-III fragment of 51 amino acids (Cytryńska et al. 2007) and can act synergistically with lysozyme against Gram-negative bacteria by enhancing its muramidase activity (Zdybicka-Barabas and Cytryńska 2013).

LPS binding induces changes in the apoLp-III conformation leading to opening of the helix bundle, similar to reorganization during apoLp-III interactions with lipids and lipophorin particles (Leon et al. 2006). It was shown that apoLp-III can disaggregate LPS aggregates interacting initially with exposed carbohydrate regions of LPS molecules and then with the lipid A region. Such interaction results in formation of LPS/apoLp-III complexes (Oztug et al. 2012). Recently, it has been reported that the lipid-bound conformation negatively influences apoLp-III effectiveness of LPS binding (Wijeratne and Weers 2019). This finding implies that, to serve as an effective antimicrobial protein, apoLp-III should be in its lipid-free form. This also suggests that, in order to play such different roles in insect metabolism, including lipid transport and many functions in immune response, different fractions of apoLp-III need to be present simultaneously in hemolymph. The presence of lipid-free and lipophorin-bound apoLp-III fractions was demonstrated in *G. mellonella* larvae (Stączek et al. 2018). In addition, two isoforms of the apoLp-III protein were found in hemolymph of naive *G. mellonella* larvae. Immunization with different pathogens resulted in appearance of additional apoLp-III isoforms and apoLp-III-derived polypeptides differing in p*I* values in a pathogen- and time-dependent manner (Zdybicka-Barabas et al. 2015).

Heat Shock Proteins

Before we summarize the possible involvement of heat shock proteins in insect defense, we will shortly describe the role of ambient temperature in insect immunity. Insects are cold-blooded animals that cannot regulate body temperature by changing metabolism. In this situation, the easiest way to change the temperature is to expose their body to sunlight or formation of clusters with other insects, increasing the temperature of individuals. The phenomenon of temperature regulation by changing the environment is called behavioral fever (Hasday et al. 2000). This behavior can increase immune abilities. On the other hand, insects may be sometimes exposed to temperature stress. Recent research has demonstrated that such abiotic stress may influence insect resistance (Adamo 1998; Kozak et al. 2000; Wojda et al. 2009, 2017b).

Phenomenon of heat shock
Sudden temperature changes greater than 5 °C may induce heat shock. This leads to a number of changes in the cell—*inter alia* changes in the structure and fluidity of biological membranes, denaturation of proteins, or formation of protein aggregates. Heat shock can adversely influence basic molecular processes such as transcription and translation and may sometimes lead to death (Wojda 2010). It is worth emphasizing here that the heat shock phenomenon was first found in *D. melanogaster* in 1962 by Ferrucio Ritossa. He observed that fruit flies grown at high temperature show increased activity of selected *loci* in the polytene chromosomes of the larvae. Subsequent studies have shown that high temperature activates a number of genes coding for heat shock proteins (Ritossa 1962; Wojda 2010). The expression of genes encoding heat shock proteins (Hsps) requires the presence of heat shock factors (Hsf) in all groups of organisms. In studies of *D. melanogaster*, it was shown that the insects have a single *Hsf* gene. Activation of Hsf in insects is similar to that in humans. In optimal conditions, Hsf is found in the cytoplasm as a monomer (Singh and Aballay 2006). Stress conditions lead to phosphorylation and trimerisation of Hsf resulting in its transport to the nucleus, where it binds to the promoters of heat-regulated genes (Westwood and Wu 1993). All promoters of heat shock genes have an inverted,

complementary repetition motif nGAAn (Wojda 2017b). Distances between individual motifs as well as the number of characteristic repetitions are determinants of the expression of genes encoding particular heat shock proteins. The action of Hsf is regulated by the level of Hsp proteins, and their excess results in inhibition of the transcription factor. Although the name of the group of heat shock proteins suggests that they are mainly involved in response to thermal shock, they protect cellular components against different types of stress; hence, they are called stress proteins. Also, in optimal conditions, heat-shock proteins participate in a number of processes important for the maintenance the homeostasis of the organism (Jee 2016). For example, they "help" folding of native proteins, formation of protein complexes, and membrane protein transport. Hsps induce the synthesis of the disaccharide trehalose, which protects proteins against denaturation (Singer and Lindquist 1998; Pirkalla et al. 2001; Jee 2016). The following families of Hsp proteins are distinguished: small Hsp, Hsp40, Hsp60, Hsp70, Hsp90, and Hsp100. The names of the individual groups are derived from the molecular mass of proteins expressed in kDa. Each group possesses a specific set of properties and the ability to create specific functional complexes with other proteins, but these issues are not the subject of this work. Readers interested in this issue are referred to papers that thoroughly discuss the particular Hsp groups (Jee 2016).

Heat shock proteins and immunity
From the immunobiology point of view, heat shock proteins are important elements contributing to the resistance of insects to infection (Adamo 2010). The role of Hsp proteins in the defense of invertebrates has not been fully understood (Singh and Aballay 2006; Wojda 2010; 2017b). It has been proven that the infection of an insect even at optimal temperature leads to an increase in Hsp90, suggesting some input into pathways regulating its expression (Wojda and Jakubowicz 2007). Infection with a pathogen leads to homeostasis disorder and constitutes a biotic stress. Studies of *G. mellonella* larvae proved that insects exposed to thermal shock immediately before infection with entomopathogens showed better resistance than larvae that were not exposed to elevated temperature. The molecular basis of this phenomenon is seen in the action of Hsp90 proteins. The Hsp90 family of proteins is responsible for maintaining the correct conformation of a number of MAP kinases, which may participate in the activation of a number of elements of the insect immune system (Wojda and Taszłow 2013). For example, the JNK MAP pathway is activated by the Imd signaling pathway, i.e. one of the main regulators of the defense system. It is likely that Hsp90 proteins or their degradation products can directly stimulate the immune system as a so-called danger signal (Wojda and Jakubowicz 2007; Wojda and Kowalski 2013).

On the other hand, Hsp proteins influence the full activation of antimicrobial peptides, giving them the correct conformation in the process of folding, and protect against the harmful effects of pathogen's virulence agents (Adamo 2010; Jee 2016). The increase in the heat shock proteins in the fat body after immune stimulation at optimal temperature indicates their involvement in insect defense. It is also suggested that Hsp proteins are involved in the protection of cellular structures and tissues against the harmful effects of reactive oxygen species (ROS), which appear in the body as a result of infection with pathogens (Wojda 2010; Wojda and Kowalski 2013). To determine the function of Hsp proteins and the Hsf transcription factor, geldanamycin, i.e. an inhibitor of heat shock protein Hsp90, was injected to *G. mellonella* larvae. By binding to Hsp90, this compound pulls it away from the Hsf factor, mimicking the heat shock conditions. Its injection caused increased immune response at the optimal growth temperature, as during exposure of larvae to elevated temperature (Wojda and Kowalski 2013). There are suggestions that heat shock proteins are an integral part of the immune response (Vallet–Gely et al. 2008). It has been demonstrated that apoLp-III shows greater resistance to proteolytic enzymes in insects exposed to elevated temperature than in organisms that were kept at the optimal temperature (Vertyporokh et al. 2015). Another report indicates the relationship between the Hsf factor and *D. melanogaster* viral infection. Inhibition of the transcription factor led to greater susceptibility of insects to virus infection, while the increased expression of Hsf led to a more effective fight against the virulence factor (Merkling et al. 2015). Besides the reports mentioned above, the current state of knowledge does not demonstrate how exactly heat shock proteins influence the defense abilities of insects.

Phenomenon of cross- talk and cross-protection
In the context of the effect of heat shock proteins on the immune system, one should consider the phenomenon of cross-talk and cross-protection. The cross-talk mechanism consists in activation of signals from various signaling pathways, resulting in activation of a number of defense mechanisms. Both elevated temperature and infection regulate the expression of Hsp proteins. The temperature-induced increase in Hsp proteins may then increase one of the threat signals (danger signal), giving an additional input into activation of defense response (Sinclair et al. 2013). Cross-protection occurs when a given component synthesized in response to one type of stress protects the organisms from another type of stress. Exposure of insects to heat stress leads to induction of the expression of genes encoding Hsp, which may provide additional protection to host tissue against destruction caused by the invading pathogen (Mattson and Calabrese 2010; Sinclair et al. 2013). The participation both in cross-talk as members of signaling pathways and in cross-protection as effector molecules with protective functions is the unusual feature of heat shock proteins. Perhaps such processes as cross-talk and cross-protection as well as the phenomenon of behavioral fever are the key to full understanding of the role of heat-shock proteins in insect resistance. The answer to this question, however, requires further research focusing on heat-shock proteins in insect organisms. At the end of this part, we should mention that the positive effect of heat stress on insect immunity is not a general rule. Sometimes stressors, including temperature change, may have a negative effect on animal fitness due to the loss of energy in the fight against two different types of stressful conditions. This depends on the insect species and its living conditions, the life history of individuals, and finally on the pathogen and the action of its virulence factors (Stacey et al. 2003).

Conclusions

As we described above, insect hemolymph is very rich in immune polypeptides: they can have direct defense antimicrobial activity or possess regulatory function. Since insects possess an open circulatory system, this body fluid is in direct contact with all internal organs. They may secrete some polypeptides to hemolymph but also take nutrients and hormones for their needs. Some hemolymph proteins, e.g. apoLp-III and heat-shock proteins belong to so-called "moonlighting proteins" i.e. proteins with many functions (Jeffery 1999). The amount of a particular hemolymph immune component is a trade-off of all current immunological, nutritional, motor, and procreative needs of the organism, since all of them require energy expenditure (Adamo et al. 2008; Galenza and Foley 2019). The evolutionary pressure on the diversity of some receptors like Dscam or FREPs has led to enhancement of the level of specificity correlated with achievement of a kind of immune memory but—unlike in vertebrates—using solely the tools of innate immunity. From the biotechnological point of view, hemolymph is a rich source of bioactive molecules. For many years, great hopes have been placed on defense peptides that may supplement or even replace antibiotics in the future, to which pathogens are becoming increasingly resistant (Torres et al. 2019). The interest in bioactive molecules is reflected by the fact that there is a database dedicated to natural antimicrobial peptides (APD, antimicrobial peptide database: http://aps.unmc.edu/AP/main.php; Wang et al. 2016). The defense peptides not only possess antimicrobial activity but also antiparasitic, anticancer, spermicidal, insecticidal effects. They can function as peptide ion channel inhibitors, protease inhibitors, and antioxidant peptides. The bioactive molecules of animals, including insects, may be used in the future for protection of animals, including humans, and plants against different diseases.

References

Aathmanathan VS, Jothi N, Prajapati VK, Krishnan M (2018) Investigation of immunogenic properties of hemolin from silkworm, Bombyx mori as carrier protein: an immunoinformatic approach. Sci Rep 8(1):6957. https://doi.org/10.1038/s41598-018-25374-z

Abraham EG, Nagaraju J, Salunke D, Gupta HM, Datta RK (1995) Purification and partial characterization of an induced antibacterial protein in the silkworm Bombyx mori. J Invertebr Phathol 65(1):17–24

Adamo SA (1998) The Specificity of Behavioral Fever in the Cricket Acheta domesticus. J Parasitol 84(3):529

Adamo SA (2010) Why should an immune response activate the stress response? Insights from the insects (the cricket Gryllus texensis). Brain Behav Immun 24:194–200

Adamo SA, Roberts JL, Easy RH, Ross NW (2008) Competition between immune function and lipid transport for the protein apolipophorin III leads to stress–induced immunosuppression in crickets. J Exp Biol 211:531–538

Altincicek B, Vilcinskas A (2006) Metamorphosis and collagen-IV-fragments stimulate innate immune response in the greater wax moth Galleria mellonella. Dev Comp Immunol 30(12):1108–1118

Andrejko M, Cytryńska M, Jakubowicz T (2005) Apolipophorin III is a substrate for protease IV from Pseudomonas aeruginosa. FEMS Microbiol Lett 243(2):331–337

Andrejko M, Zdybicka-Barabas A, Cytryńska M (2014) Diverse effects of Galleria mellonella infection with entomopathogenic and clinical strains of Pseudomonas aeruginosa. J Invertebr Pathol 115:14–25

Armitage SA, Sun W, You X, Kurtz J, Schmucker D, Chen W (2014) Quantitative profiling of Drosophila melanogaster Dscam1 isoforms reveals no changes in splicing after bacterial exposure. PLoS ONE 9(10):e108660. https://doi.org/10.1371/journal.pone.0108660

Armitage SA, Peuss R, Kurtz J (2015) Dscam and pancrustacean immune memory—a review of the evidence. Dev Comp Immunol 48(2):315–323. https://doi.org/10.1016/j.dci.2014.03.004

Armitage SAO, Kurtz J, Brites D, Dong Y, Du Pasquier L, Wang HC (2017) Dscam1 in pancrustacean immunity: current status and a look to the future. Front Immunol 8:662. https://doi.org/10.3389/fimmu.2017.00662

Aye TT, Shim JK, Rhee IK, Lee KY (2008) Upregulation of the immune protein gene hemolin in the epidermis during the wandering larval stage of the Indian meal moth, Plodia interpunctella. J Insect Physiol 54:1301–1305

Baek JH, Lee SH, Kim WY, Kim MG (2016) An insulin–binding protein from the venom of a solitary wasp Eumenes pomiformis binds to apolipophorin III in lepidopteran hemolymph. Toxicon 111:62–64. https://doi.org/10.1016/j.toxicon.2015.12.019

Baker DA, Nolan T, Fischer B, Pinder A, Crisanti A, Russell S (2011) A comprehensive gene expression atlas of sex– and tissue–specificity in the malaria vector Anopheles gambiae. BMC Genomics 12:296. https://doi.org/10.1186/1471-2164-12-296

Bao Y, Yamano Y, Morishima I (2007) Induction of hemolin gene expression by bacterial cell wall components in eri–silkworm, Samia cynthia ricini. Comp Biochem Physiol B 146:147–151

Beckert A, Wiesner J, Baumann A, Poppel A-K, Vogel H, Vilcinskas A (2015) Two c–type lysozyme boost the innate system of the invasive ladybird Harmonia axyridis. Dev Comp Immunol 49:303–312

Beckert A, Wiesner J, Schmidtberg R, Lehmann R, Baumann A, Vogel H, Vilcinskas A (2016) Expression and characterization of a recombinant i–type lysozyme from the harlequin ladybird beetle Harmonia axyridis. Insect Mol Biol 25(3):202–215

Bettencourt R, Terenius O, Faye I (2002) Hemolin gene silencing by ds–RNA injected into Cecropia pupae is lethal to next generation embryos. Insect Mol Biol 11(3):267–271

Bolouri Moghaddam MR, Tonk M, Schreiber C, Salzig D, Czermak P, Vilcinskas A, Rahnamaeian M (2016) The potential of the Galleria mellonella innate immune system is maximized by the co–presentation of diverse antimicrobial peptides. Biol Chem 397:939–945. https://doi.org/10.1515/hsz-2016-0157

Brites D, Du Pasquier L (2015) Somatic and germline diversification of a putative immunoreceptor within one phylum: Dscam in arthropods. In: Hsu E, Du Pasquier L (eds) Pathogen–host interactions: antigenic variation v. somatic adaptations. Results and problems in cell differentiation, vol 57, pp 131–158. https://doi.org/10.1007/978-3-319-20819-0_6

Brogden KA (2005) Antimicrobial peptides: pore formers or metabolic inhibitors in bacteria? Nat Rev Microbiol 3:238–250. https://doi.org/10.1038/nrmicro1098

Brown SE, Howard A, Kasprzak AB, Gordon KH, East PD (2008) The discovery and analysis of a diverged family of novel antifungal moricin–like peptides in the wax moth Galleria mellonella. Insect Biochem Mol Biol 38(2):201–212. https://doi.org/10.1016/j.ibmb.2007.10.009

Buchmann K (2014) Evolution of innate immunity: clues from invertebrates via fish to mammals. Front Immunol 5:459. https://doi.org/10.3389/fimmu.2014.00459

Buchon N, Silverman N, Cherry S (2014) Immunity in Drosophila melanogaster–from microbial recognition to whole–organism physiology. Nat Rev Immunol 14(12):796–810. https://doi.org/10.1038/nri3763

Bulet P, Stöcklin R (2005) Insect antimicrobial peptides: structures, properties and gene regulation. Protein Pept Lett 12:3–11

Bulet P, Hetru C, Dimarcq J-L, Hoffmann D (1999) Antimicrobial peptides in insects; structure and function. Dev Comp Immunol 23:329–344

Cancado F, Chimoy EP, Terra W, Marana S (2008) Cloning lysozymes from Musca domestica larvae. Braz J Med Biol Res 41:969–977

Casteels P, Ampe C, Jacobs F, Tempst P (1993) Functional and chemical characterization of Hymenoptaecin, an antibacterial polypeptide that is infection-inducible in the honeybee (Apis mellifera). J Biol Chem 268(10):7044–7054

Chalk R, Townson H, Natori S, Desmond H, Ham PJ (1994) Purification of an insect defensin from the mosquito, Aedes aegypti. Insect Biochem Mol Biol 24:403–410

Chapelle M, Girard PA, Cousserans F, Volkoff NA, Duvic B (2009) Lysozymes and lysozyme-like proteins from the fall armyworm Spodoptera frugiperda. Mol Immunol 47(2–3):261–269

Chen TT, Tan LR, Hu N, Dong ZQ, Hu ZG, Jiang YM, Che P, Pan MH, Lu Ch (2018) C–lysozyme to antiviral immunity in Bombyx mori against nucleopolyhedrovirus infection. J Insect Physiol 108:54–60

Cherry S, Silverman N (2006) Host–pathogen interactions in Drosophila: new tricks from an old friend. Nat Immunol 7(9):911–917. https://doi.org/10.1038/ni1388

Cirimotich CM, Dong Y, Garver LS, Sim S, Dimopoulos G (2010) Mosquito immune defenses against Plasmodium infection. Dev Comp Immunol 34(4):387–395. https://doi.org/10.1016/j.dci.2009.12.005

Clermont A, Wedde M, Seitz V, Podsiadlowski L, Lenze D, Hummel M, Vilcinskas A (2004) Cloning and expression of an inhibitor of microbial metalloproteinases from insects contributing to innate immunity. Biochem J 382(Pt 1):315–322

Cociancich S, Ghazi A, Hetru C, Hoffmann JA, Latellier L (1993) Insect defensin, an inducible antibacterial peptide, forms voltage–dependent channels in Micrococcus luteus. J Biol Chem 268:19239–19245

Contreras E, Rausell C, Real MD (2013) Tribolium castaneum apolipophorin–III acts as an immune response protein against Bacillus thuringiensis Cry3Ba toxic activity. J Invertebr Pathol 113(3):209–213. https://doi.org/10.1016/j.jip.2013.04.002

Cooper EL (2008) From Darwin and Metchnikoff to Burnet and beyond. Contrib Microbiol 15:1–11. https://doi.org/10.1159/000135680

Cooper D, Eleftherianos I (2017) Memory and specificity in the insect immune system: current perspectives and future challenges. Front Immunol 8:539. https://doi.org/10.3389/fimmu.2017.00539

Cytryńska M (2009) Immunity without antibodies. Postepy Biol Komorki 36(2):309–324

Cytryńska M, Zdybicka-Barabas A (2015) Defense peptides: recent developments. Biomol Concepts 115(1):397–406. https://doi.org/10.1007/s00436-015-4761-z

Cytryńska M, Zdybicka-Barabas A, Andrejko M, Jabłoński P, Jakubowicz T (2002) Antimicrobial peptides and proteins in Galleria mellonella hemolymph. In: 2001 3rd International conference on arthropods: chemical, physiological and environmental aspects, Lądek Zdrój, Publisher of the University of Wroclaw, Wrocław, pp 193–197

Cytryńska M, Mak P, Zdybicka-Barabas A, Suder P, Jakubowicz T (2007) Purification and characterization of eight peptides from Galleria mellonella immune hemolymph. Peptides 28:533–546

Cytryńska M, Wojda I, Jakubowicz T (2016) How insect combat infections. In: Ballarin L, Cammarata M (eds) Lessons in immunity: from single-cell organisms to mammals. Elsevier, Amsterdam, pp 117–128. https://doi.org/10.1016/B978-0-12-803252-7.00009-6

Daffre S, Kylsten P, Samakovlis C, Hultmark D (1994) The lysozyme locus in Drosophila melanogaster: an expanded gene family adapted for expression in the digestive tract. Mol Gen Genet 242(2):152–162. https://doi.org/10.1007/bf00391008

Daffre S, Faye I (1997) Lipopolysaccharide interaction with hemolin, an insect member of the Ig–superfamily. FEBS Lett 408:127–130

Dai H, Rayaprolu S, Gong Y, Huang R, Prakash O, Jiang H (2008) Solution structure, antibacterial activity, and expression profile of Manduca sexta moricin. J Pept Sci 14(7):855–863. https://doi.org/10.1002/psc.1016

Dettloff M, Kaiser B, Wiesner A (2001a) Localization of injected apolipophorin III in vivo—new insights into the immune activation process directed by this protein. J Insect Physiol 47:789–797

Dettloff M, Wittwer D, Weise C, Wiesner A (2001b) Lipophorin of a lower density is formed during immune responses in the lepidopteran insect Galleria mellonella. Cell Tissue Res 306:449–458

Dhawan R, Gupta K, Kajla M, Kakani P, Choudhury TP, Kumar S, Kumar V, Gupta L (2017) Apolipophorin–III acts as a positive regulator of Plasmodium development in Anopheles stephensi. Front Physiol 8:185. https://doi.org/10.3389/fphys.2017.00185

Dong Y, Dimopoulos G (2009) Anopheles fibrinogen–related proteins provide expanded pattern recognition capacity against bacteria and malaria parasites. J Biol Chem 284(15):9835–9844. https://doi.org/10.1074/jbc.M807084200

Dong Y, Aguilar R, Xi Z, Warr E, Mongin E, Dimopoulos G (2006) Anopheles gambiae immune responses to human and rodent Plasmodium parasite species. PLoS Pathog 2(6):e52

Dong Y, Simões ML, Marois E, Dimopoulos G (2018) CRISPR/Cas9–mediated gene knockout of Anopheles gambiae FREP1 suppresses malaria parasite infection. PLoS Pathog 14(3):e1006898. https://doi.org/10.1371/journal.ppat.1006898

Doolittle RF, McNamara K, Lin K (2012) Correlating structure and function during the evolution of fibrinogen–related domains. Protein Sci 21(12):1808–1823. https://doi.org/10.1002/pro.2177

Dunphy GB, Oberholzer U, Whiteway M, Zakarian RJ, Boomer I (2003) Virulence of Candida albicans mutants toward larval Galleria mellonella (Insecta, Lepidoptera, Galleridae). Can J Microbiol 49(8):514–524

Eguchi M, Itoh M, Nishino K, Shibata H, Tanaka T, Kamei-Hayashi K, Hara S (1994) Amino acid sequence of an inhibitor from the silkworm (Bombyx mori) hemolymph against fungal protease. J Biochem 115(5):881–884

Eleftherianos I, Millichap PJ, Ffrench-Constant RH, Reynolds SE (2006) RNAi suppression of recognition protein mediated immune response in the tabacco homworm Manduca sexta causes increased susceptibility to the insect pathogen Photorhabdus. Dev Comp Immunol 30:1099–1107

Eleftherianos I, Gökçen F, Felföldi G, Millichap PJ, Trenczek TE, Ffrench-Constant RH, Reynolds SE (2007) The immunoglobulin family protein hemolin mediates cellular immune responses to bacteria in the insect Manduca sexta. Cell Microbiol 9(5):1137–1147

Engström P, Carlsson A, Engström Å, Tao ZJ, Bennich H (1984) The antimicrobial effect of attacins from the silk moth Hyalophora cecropia is directed against the outer membrane of Escherichia coli. EMBO J 3:3347–3351

Fehlbaum P, Bulet P, Michaut L, Lagueux M, Broekaert WF, Hetru C, Hoffmann JA (1994) Insect immunity. Septic injury of Drosophila induces the synthesis of a potent antifungal peptide with sequence homology to plant antifungal peptides. J Biol Chem 269(52):33159–33163

Forbes EM, Hunt JJ, Goodhill GJ (2011) The combinatorics of neurite self–avoidance. Neural Comput 23:2746–2769

Fröbius AC, Kanost MR, Götz P, Vilcinskas A (2000) Isolation and characterization of novel inducible serine protease inhibitors from larval hemolymph of the greater wax moth Galleria mellonella. Eur J Biochem 267(7):2046–2053

Galenza A, Foley E (2019) Immunometabolism: insights from the drosophila model. Dev Comp Immunol 94:22–34

Gandhe AS, Janardhan G, Nagaraju J (2007) Immune upregulation of novel antimicrobial proteins from silkmoths (Lepidoptera) that resamble lysozymes but lack muramidase activity. Insect Biochem Mol Biol 37:655–666

Ganesan S, Aggarwal K, Paquette N, Silverman N (2011) NF–κB/Rel proteins and the humoral immune responses of Drosophila melanogaster. Curr Top Microbiol Immunol 349:25–60. https://doi.org/10.1007/82_2010_107

Gao B, Zhu S (2014) An insect defensin–derived β–hairpin peptide with enhanced antibacterial activity. ACS Chem Biol 9(2):405–413. https://doi.org/10.1021/cb400591d

Gegner J, Baudach A, Mukherjee K, Halitschke R, Vogel H, Vilcinskas A (2019) Epigenetic mechanisms are involved in sex–specific trans–generational immune priming in the lepidopteran model host Manduca sexta. Front Physiol 10:137. https://doi.org/10.3389/fphys.2019.00137
Ghosh J, Lun CM, Majeske AJ, Sacchi S, Schrankel CS, Smith LC (2011) Invertebrate immune diversity. Dev Comp Immunol 35(9):959–974. https://doi.org/10.1016/j.dci.2010.12.009
Gokudan S, Muta T, Tsuda R, Koori K, Kawahara T, Seki N, Mizunoe Y, Wai SN, Iwanaga S, Kawabata S (1999) Horseshoe crab acetyl group–recognizing lectins involved in innate immunity are structurally related to fibrinogen. Proc Natl Acad Sci USA 96(18):10086–10091
Gotz P, Weise C, Kopacek P, Losen S, Wiesner A (1997) Isolated apolipophorin III from Galleria mellonella stimulates the immune reactions of this insect. J Insect Physiol 43(4):383–391
Graf M, Mardirossian M, Nguyen F, Carolin Seefeldt A, Gilles G, Marco S, Axel Innis C, Wilson DN, (2017) Proline-rich antimicrobial peptides targeting protein synthesis. Nat Prod Rep 34(7):702–711
Graveley BR (2005) Mutually exclusive splicing of the insect Dscam pre–mRNA directed by competing intronic RNA secondary structures. Cell 123(1):65–73
Hanada Y, Sekimizu K, Kaito C (2011) Silkworm apolipophorin protein inhibits Staphylococcus aureus virulence. J Biol Chem 286(45):39360–39369. https://doi.org/10.1074/jbc.M111.278416
Hanington PC, Zhang S-M (2011) The primary role of fibrinogen–related proteins in invertebrates is defense, not coagulation. J Innate Immun 3:17–27
Hara S, Yamakawa M (1995) Moricin, a novel type of antibacterial peptide isolated from the silkworm Bombyx mori. J Biol Chem 270(50):29923–29927
Hasday JD, Fairchild KD, Shanholtz C (2000) The role of fever in the infected host. Microbes Infect 2(15):1891–1904
Hillyer JF (2016) Insect immunology and hematopoiesis. Dev Comp Immunol 58:102–118
Hughes AL (2012) Evolution of the βGRP/GNBP/β–1,3–glucanase family of insects. Immunogenetics 64(7):549–558. https://doi.org/10.1007/s00251-012-0610-8
Hultmark D (1996) Insect lysozymes. In: Jollès P (ed) Lysozymes: model enzymes in biochemistry and biology. Birkhäuser Verlag, Basel, Switzerland, pp 87–102
Hultmark D, Engström A, Bennich H, Kapur R, Boman HG (1982) Insect immunity: isolation and structure of cecropin D and four minor antibacterial components from Cecropia pupae. Eur J Biochem 127:207–217
Jarosz J (1979) Simultaneous induction of protective immunity and selective synthesis of hemolymph lysozyme protein in larvae of Galleria mellonella. Biol Zbl 98:459–471
Jee H (2016) Size–dependent clasification of heat–shock proteins. A mini–review. J Exerc Rehabil 12:255–259. http://dx.doi.org/10.12965/jer.1632642.321
Jeffery CJ (1999) Moonlighting proteins. TIBS 24:8–11
Jolles P, Jolles J (1984) What's new in lysozyme research? Always a model system, today as yesterday. Mol Cell Biochem 63:165–189
Kajla MK, Andreeva O, Gilbreath TM, Paskewitz SM (2010) Characterization of expression, activity and role in antibacterial immunity of Anopheles gambiae lysozyme c–1. Comp Biochem Physiol B Biochem Mol Biol 155:201–209
Kajla MK, Shi L, Li B, Luckhart S, Li J, Paskewitz SM (2011) A new role for an old antimicrobial: lysozyme c–1 can function to protect malaria parasites in Anopheles mosquitoes. PLoS ONE 6(5):e19649. https://doi.org/10.1371/journal.pone.0019649
Kamareddine L, Nakhleh J, Osta MA (2016) Functional interaction between apolipophorins and complement regulate the mosquito immune response to systemic infections. J Innate Immun 8(3):314–326. https://doi.org/10.1159/000443883
Kato Y, Motoi Y, Taniai K, Kadono-Okuda K, Yamamoto M, Higashino Y, Shimabukuro M, Chowdhury S, Xu J, Sugiyama M, Hiramatsu M, Yamakawa M (1994) Lipopolysaccharide–lipophorin complex formation in insect hemolymph: a common pathway of lipopolysaccharide detoxification both in insects and in mammals. Insect Biochem Mol Biol 24(6):547–555
Kaufmann SHE (2019) Immunology's coming of age. Front Immunol 10:684 https://doi.org/10.3389/fimmu.2019.00684

Kim BY, Jin BR (2015) Apolipophorin III from honeybees (Apis cerana) exhibits antibacterial activity. Comp Biochem Physiol B Biochem Mol Biol 182:6–13. https://doi.org/10.1016/j.cbpb.2014.11.010

Kong HL, Cheng YX, Luo LZ, Sappington TW, Jiang XF, Zhang L (2013) Density–dependent prophylaxis in crowded beet webworm, Loxostege sticticalis (Lepidoptera: Pyralidae) larvae to a parasitoid and a fungal pathogen. Int J Pest Manage 59:174–179

Kong H, Lv M, Mao N, Wang C, Cheng Y, Zhang L, Jiang X, Luo L (2016) Molecular characterization of a lysozyme gene and its altered expression profile in crowded beet webworm (Loxostege sticticalis). PLoS ONE 11(8):e0161384. https://doi.org/10.1371/journal.pone.0161384

Kozak W, Kluger MJ, Tesfaigzi J, Kozak A, Mayfield KP, Wachulec M, Dokladny K (2000) Molecular mechanisms of fever and endogenous antipyresis. Ann NY Acad Sci 917:121–134

Kragol G, Lovas S, Varadi G, Condie BA, Hoffmann R, Otvos L (2001) The antibacterial peptide pyrrocoricin inhibits the ATPase actions of DnaK and prevents chaperone-assisted protein folding. Biochemistry 40:3016–3026

Kurata S (2014) Peptidoglycan recognition proteins in Drosophila immunity. Dev Comp Immunol 42(1):36–41. https://doi.org/10.1016/j.dci.2013.06.006

Kwon H, Bang K, Lee M, Cho S (2014) Molecular cloning and characterization of a lysozyme cDNA from the mole Gryllotalpa orientalis (Orthoptera: Gryllotalpidae). Mol Biol Rep 41:5745–5754

Ladendorff NE, Kanost MR (1991) Bacteria–induced P4 (hemolin) from Manduca sexta: a member of the immunoglodulin superfamily which can inhibit hemocyte aggregation. Arch Insect Biochem Physiol 18:285–300

Lamberty M, Ades S, Uttenweiler–Joseph S, Brookhart G, Bushey D, Hoffmann JA, Bulet P (1999) Insect immunity. Isolation from the lepidopteran Heliothis virescens of a novel insect defensin with potent antifungal activity. J Biol Chem 274(14): 9320–9326

Langen G, Imani J, Altincicek B, Kieseritzky G, Kogel KH, Vilcinskas A (2006) Transgenic expression of gallerimycin, a novel antifungal insect defensin from the greater wax moth Galleria mellonella, confers resistance to pathogenic fungi in tobacco. Biol Chem 387(5):549–557

Lanz-Mendoza H, Bettencourt R, Fabbri M, Faye I (1996) Regulation of the insect immune response: The effect of hemolin on cellular immune mechanism. Cell Immunol 169:47–54

Lee K-Y, Horodyjski FM, Valaitis AP, Denlinger DL (2002) Molecular characterization of the insect immune protein hemolin and its high induction during embryonic diapause in the gypsy moth, Lymantria dispar. Insect Biochem Mol Biol 32:1457–1467

Lee YS, Yun EK, Jang WS, Kim I, Lee JH, Park SY, Ryu KS, Seo SJ, Kim CH, Lee IH (2004) Purification, cDNA cloning and expression of an insect defensin from the great wax moth. Galleria mellonella. Insect Mol Biol 13(1):65–72

Leon LJ, Idangodage H, Wan CPL, Weers PMM (2006) Apolipophorin III: lipopolysaccharide binding requires helix bundle opening. Biochem Biophys Res Commun 348:1328–1333

Li B, Paskewitz SM (2006) A role for lysozyme in melanization of sephadex beads in Anopheles gambiae. J Insect Physiol 52:936–942

Li D, Scherfer C, Korayem AM, Zhao Z, Schmidt O, Theopold U (2002) Insect hemolymph clotting: evidence for interaction between the coagulation system and the prophenoloxidase activating cascade. Insect Biochem Mol Biol 32:919–928

Li B, Calvo E, Marinotti O, James AA, Paskewitz SM (2005) Characterization of the c–type lysozyme gene family in Anopheles gambiae. Gene 360:131–139

Li Y, Zhao P, Liu S, Dong Z, Chen J, Xiang Z, Xia Q (2012) A novel protease inhibitor in Bombyx mori is involved in defense against Beauveria bassiana. Insect Biochem Mol Biol 42(10):766–775. https://doi.org/10.1016/j.ibmb.2012.07.004

Li XJ, Yang L, Li D, Zhu YT, Wang Q, Li WW (2018) Pathogen–specific binding soluble Down syndrome cell adhesion molecule (Dscam) regulates phagocytosis via membrane–bound Dscam in crab. Front Immunol 9:801. https://doi.org/10.3389/fimmu.2018.00801

Li S, Yu X, Feng Q (2019) Fat body biology in the last decade. Ann Rev Entomol 64:315–333. https://doi.org/10.1146/annurev-ento-011118-112007

Lindström-Dinnetz I, Sun S-C, Faye I (1995) Structure and expression of Hemolin, an insect member of the immunoglobulin gene superfamily. Eur J Biochem 230:920–925

Ma G, Hay D, Li D, Asgari S, Schmidt O (2006) Recognition and inactivation of LPS by lipophorin particles. Dev Comp Immunol 30:619–626

Mak P, Zdybicka-Barabas A, Cytryńska M (2010) A different repertoire of Galleria mellonella antimicrobial peptides in larvae challenged with bacteria and fungi. Dev Comp Immunol 34(10):1129–1136. https://doi.org/10.1016/j.dci.2010.06.005

Makarova O, Rodriguez-Rojas A, Eravci M, Weise C, Dobson A, Johnston P, Rolff J (2016) Antimicrobial defence and persistent infection in insects revisited. Phil Trans R Soc B 371:20150296. https://doi.org/10.1098/rstb.2015.0296

Marmaras VJ, Lampropoulou M (2009) Regulators and signalling in insect haemocyte immunity. Cell Signal 21:186–195

Marxer M, Vollenweider V, Schmid–Hempel P (2016) Insect antimicrobial peptides act synergistically to inhibit a trypanosome parasite. Philos Trans R Soc Lond B Biol Sci 371(1695). https://doi.org/10.1098/rstb.2015.0302

Matsuzaki K (1999) Why and how are peptide–lipid interactions utilized for self–defense? Magainins and tachyplesins as archetypes. Biochim Biophys Acta 1462(15):1–10

Mattson MP, Calabrese EJ (eds) (2010) Hormessis: a revolution in biology, toxicology and medicine. Springer, Dordrecht

Mendes AM, Schlegelmilch T, Cohuet A, Awono-Ambene P, De Iorio M, Fontenille D, Morlais I, Christophides GK, Kafatos FC, Vlachou D (2008) Conserved mosquito/parasite interactions affect development of Plasmodium falciparum in Africa. PLoS Pathog 4(5):e1000069. https://doi.org/10.1371/journal.ppat.1000069

Merkling SH, Overheul GJ, van Mierlo JT, Arends D, Gilissen C, van Rij RP (2015) The heat shock response restricts virus infection in Drosophila. Sci Rep 5:12758. https://doi.org/10.1038/srep12758

Middha S, Wang X (2008) Evolution and potential function of fibrinogen–like domains across twelve Drosophila species. BMC Genom 9:260. https://doi.org/10.1186/1471-2164-9-260

Mohamed AA, Zhang I, Dorrah MA, Elmogy M, Yousef HA, Bassal TTM, Duvic B (2016) Molecular characterization of a c–type lysozyme from the desert locust, Schistocerca gregaria (Orthoptera: Acrididae). Dev Comp Immunol 61:60–69

Mohrig W, Messner B (1968) Immunoreaktionen bei Insekten. II. Lysozym als antimikrobielles Agens in Darmtrakt von Insekten. Biol Zentralbl 87:705–744

Morishima I, Horiba T, Iketani M, Nishioka E, Yamano Y (1995) Parallel induction of cecropin and lysozyme in larvae of the silkworm Bombyx mori. Dev Comp Immunol 19(5):357–363

Mukherjee K, Twyman RM, Vilcinskas A (2015) Insects as models to study the epigenetic basis of disease. Progr Biophys Mol Biol 118:69–78

Mylonakis E, Podsiadlowski L, Muhammed M, Vilcinskas A (2016) Diversity, evolution and medical applications of insect antimicrobial peptides. Philos Trans R Soc Lond B Biol Sci 371(1695). https://doi.org/10.1098/rstb.2015.0290

Narajanaswami V, Kiss RS, Weers PM (2010) The helix bundle: a reversible lipid binding motif. Comp Biochem Physiol A Mol Integr Physiol 155(2):123–133. https://doi.org/10.1016/j.cbpa.2009.09.009

Neves G, Zucker J, Daly M, Chess A (2004) Stochastic yet biased expression of multiple Dscam splice variants by individual cells. Nat Genet 36(3):240–246. https://doi.org/10.1038/ng1299

Ng TH, Chiang YA, Yeh YC, Wang HC (2014) Review of Dscam–mediated immunity in shrimp and other arthropods. Dev Comp Immunol 46(2):129–138. https://doi.org/10.1016/j.dci.2014.04.002

Nguyen LT, Haney EF, Vogel HJ (2011) The expanding scope of antimicrobial peptide structures and their modes of action. Trends Biotechnol 29:464–472

Niere M, Dettloff M, Maier T, Ziegler M, Wiesner A (2001) Insect immune activation by apolipoprotein III is correlated with the lipid–binding properties of this protein. Biochemistry 40:11502–11508

Niu G, Franc AC, Zhang G, Roobsoong W, Nguitragool W, Wang X, Prachumsri J, Butler NS, Li J (2017a) The fibrinogen–like domain of FREP1 protein is a broad–spectrum malaria transmission–blocking vaccine antigen. J Biol Chem 292(28):11960–11969. https://doi.org/10.1074/jbc.M116.773564

Niu G, Zhang G, Franca C, Cui Y, Munga S, Afrane Y, Li J (2017b) FBN30 in wild Anopheles gambiae functions as a pathogen recognition molecule against clinically circulating Plasmodium falciparum in malaria endemic areas in Kenya. Sci Rep 7(1):8577. https://doi.org/10.1038/s41598-017-09017-3

Oizumi Y, Hemmi H, Minami M, Asaoka A, Yamakawa M (2005) Isolation, gene expression and solution structure of a novel moricin analogue, antibacterial peptide from a lepidopteran insect. Spodoptera litura. Biochim Biophys Acta 1752(1):83–92

Omae Y, Hanada Y, Sekimizu K, Kaito C (2013) Silkworm apolipophorin protein inhibits hemolysin gene expression of Staphylococcus aureus via binding to cell surface lipoteichoic acids. J Biol Chem 288(35):25542–25550. https://doi.org/10.1074/jbc.M113.495051

Orozco-Flores A, Valadez-Lira JA, Oppert B, Gomez-Flores R, Tamez-Guerra R, Rodrígueez-Padilla C, Tamez-Guerra P (2017) Regulation by gut bacteria of immune response, Bacillus thuringiensis susceptibility and hemolin expression in Plodia interpunctella. J Insect Physiol 98:275–283

Otvos L Jr (2000) Antibacterial peptides isolated from insects. J Pept Sci 6(10):497–511

Oztug M, Martinon D, Weers PM (2012) Characterization of the apoLp–III/LPS complex: insight into the mode of binding interaction. Biochemistry 51(31):6220–6227

Palusińska-Szysz M, Zdybicka-Barabas A, Pawlikowska-Pawlęga B, Mak P, Cytryńska M (2012) Anti-Legionella dumoffii activity of Galleria mellonella defensin and apolipophorin III. Int J Mol Sci 13(12):17048–17064. https://doi.org/10.3390/ijms131217048

Palusińska-Szysz M, Zdybicka-Barabas A, Reszczyńska E, Luchowski R, Kania M, Gisch N, Waldow F, Mak P, Danikiewicz W, Gruszecki WI, Cytryńska M (2016) The lipid composition of Legionella dumoffii membrane modulates the interaction with Galleria mellonella apolipophorin III. Biochim Biophys Acta 7:617–629. https://doi.org/10.1016/j.bbalip.2016.04.011

Paskewitz SM, Li B, Kajla MK (2008) Cloning and molecular characterization of two invertebrate–type lysozymes from Anopheles gambiae. Insect Mol Biol 17:217–225

Patiño-Márquez IA, Patiño-González E, Hernández-Villa L, Ortíz-Reyes B, Manrique-Moreno M (2018) Identification and evaluation of Galleria mellonella peptides with antileishmanial activity. Anal Biochem 546:35–42. https://doi.org/10.1016/j.ab.2018.01.029

Pendar H, Aviles J, Adjerid K, Schoenewald C, Socha JJ (2019) Functional compartmentalization in the hemocoel of insects. Sci Rep 9(1):6075. https://doi.org/10.1038/s41598-019-42504-3

Peuß R, Wensing KU, Woestmann L, Eggert H, Milutinović B, Sroka MG, Scharsack JP, Kurtz J, Armitage SA (2016) Down syndrome cell adhesion molecule 1: testing for a role in insect immunity, behaviour and reproduction. R Soc Open Sci 3(4):160138. https://doi.org/10.1098/rsos.160138

Phillips DR, Clark KD (2017) Bombyx mori and Aedes aegypti form multi–functional immune complexes that integrate pattern recognition, melanization, coagulants, and hemocyte recruitment. PLoS ONE 12(2):e0171447. https://doi.org/10.1371/journal.pone.0171447

Pirkalla L, Nykanen P, Sistonen L (2001) Roles of the heat shock transcription factors inregulation of the heat shock response and beyond. FASEB J 15(7):1118–1131

Pohar N, Godenschwege TA, Buchner E (1999) Invertebrate tissue inhibitor of metalloproteinase: structure and nested gene organization within the synapsin locus is conserved from Drosophila to human. Genomics 57:293–296

Qian C, Wang F, Zhu B, Wang L, Wei G, Sun Y, Li S, Liu Ch (2017) Identification of a hemolin protein from Actias selene mediates response to pathogens. Int Immunopharmacol 42:74–80

Rahman MM, Ma G, Roberts HL, Schmidt O (2006) Cell–free immune reactions in insects. J Insect Physiol 52:754–762

Rasmuson T, Boman HG (1979) Insect immunity—V. Purification and same properties of immune protein P4 from haemolymph of Hyalophora cecropia pupae. Insect Biochem 17:133–140

Rawlings ND, Tolle DP, Barrett AJ (2004) Evolutionary families of peptidase inhibitors. Biochem J 378(Pt 3):705–716
Ren Q, Zhao XF, Wang JX (2009) Molecular characterization and expression analysis of a chicken–type lysozyme gene from housefly (Musca domestica). J Genet Genomics 36:7–16
Ritossa F (1962) Disocovery of the heat shock response. Cell Stress Chaperones 1(2):97–98
Rono MK, Whitten MM, Oulad-Abdelghani M, Levashina EA, Marois E (2010) The major yolk protein vitellogenin interferes with the anti–Plasmodium response in the malaria mosquito Anopheles gambiae. PLoS Biol 8(7):e1000434. https://doi.org/10.1371/journal.pbio.1000434
Ryan RO, Van der Horst DJ (2000) Lipid transport biochemistry and its role in energy production. Ann Rev Entomol 45:233–260
Satyavathi VV, Mohamed AA, Kumari S, Mamatha DM (2018) The IMD pathway regulates lysozyme–like proteins (LLP) in the silkmoth Antheraea mylitta. J Invertebr Pathol 154:102–108
Schmidt O, Faye I, Lindström-Dinnetz I, Sun S-C (1993) Specific immune recognition of insect hemolin. Dev Comp Immunol 17:195–200
Schmucker D, Clemens JC, Shu H, Worby CA, Xiao J, Muda M, Dixon JE, Zipursky SL (2000) Drosophila Dscam is an axon guidance receptor exhibiting extraordinary molecular diversity. Cell 101(6):671–684
Schuhmann B, Seitz V, Vilcinskas A, Podsiadlowski L (2003) Cloning and expression of gallerimycin, an antifungal peptide expressed in immune response of greater wax moth larvae. Galleria mellonella. Arch Insect Biochem Physiol 53(3):125–133
Scocchi M, Tossi A, Gennaro R (2011) Proline–rich antimicrobial peptides: converging to a non–lytic mechanism of action. Cell Mol Life Sci 68:2317–2330. https://doi.org/10.1007/s00018-011-0721-7
Seufi AM, Hafez EE, Galal FH (2011) Identification, phylogenetic analysis and expression profile of an anionic insect defensin gene, with antibacterial activity, from bacterial–challenged cotton leafworm Spodoptera littoralis. BMC Mol Biol 12:47. https://doi.org/10.1186/1471-2199-12-47
Shaik HA, Sehnal F (2009) Hemolin expression in the silk glands of Galleria mellonella in response to bacterial challenge and prior to cell disintegration. J Insect Physiol 55:781–787
Shen DX, Liu Y, Zhou F, Wang GR, An CJ (2014) Identification of immunityrelated genes in Ostrinia furnacalis against entomopathogenic fungi by RNA–seq analysis. PLoS ONE 9:e86436
Shin SW, Park SS, Park DS, Kim MG, Kim SC, Brey PT, Park HY (1998) Isolation and characterization of immune–related genes from the webworm, Hyphantria cunea, using PCR–based differential display and subtractive cloning. Insect Biochem Mol Biol 28:827–837
Sidén I, Boman HG (1983) Escherichia coli mutants with an altered sensitivity to cecropin D. J Bacteriol 154:170–176
Simões ML, Dong Y, Hammond A, Hall A, Crisanti A, Nolan T, Dimopoulos G (2017) The Anopheles FBN9 immune factor mediates Plasmodium species–specific defense through transgenic fat body expression. Dev Comp Immunol 67:257–265. https://doi.org/10.1016/j.dci.2016.09.012
Sinclair BJ, Ferguson LV, Salehipour–shirazi G, MacMillan HA (2013) Cross–tolerance and cross–talk in the cold: relating low temperatures to desiccation and immune stress in insects. Integ Comp Biol 53(4):545–556
Singer MA, Lindquist S (1998) Thermotolerance in Saccharomyces cerevisiae: the Yin and Yang of trehalose. Trends Biotechnol 16(11):460–468
Singh V, Aballay A (2006) Heat shock and genetic activation of HSF–1 enhance immunity to bacteria. Cell Cycle 5(21):2443–2446
Smith PH, Mwangi JM, Afrane YA, Yan G, Obbard DJ, Ranford-Cartwright LC, Little TJ (2011) Alternative splicing of the Anopheles gambiae Dscam gene in diverse Plasmodium falciparum infections. Malaria J 10:156
Sowa-Jasiłek A, Zdybicka-Barabas A, Stączek S, Wydrych J, Mak P, Jakubowicz T, Cytryńska M (2014) Studies on the role of insect hemolymph polypeptides: Galleria mellonella anionic peptide 2 and lysozyme. Peptides 53:194–201. https://doi.org/10.1016/j.peptides.2014.01.012

Stacey DA, Thomas MB, Blanford S, Pell JK, Pugh C, Fellowes MDE (2003) Genotype and temperature influence pea aphid resistance to a fungal entomopathogen. Physiol Entomol 28:75–81

Stączek S, Zdybicka-Barabas A, Mak P, Sowa-Jasiłek A, Kedracka-Krok S, Jankowska U, Suder P, Wydrych J, Grygorczuk K, Jakubowicz T, Cytryńska M (2018) Studies on localization and protein ligands of Galleria mellonella apolipophorin III during immune response against different pathogens. J Insect Physiol 105:18–27. https://doi.org/10.1016/j.jinsphys.2017.12.009

Stuart LM, Ezekowitz RA (2008) Phagocytosis and comparative innate immunity: learning on the fly. Nat Rev Immunol 8(2):131–141. https://doi.org/10.1038/nri2240

Sun Y, Dai L, Sun Y, Wang L, Qian C, Wei G, Zhu B-J, Liu CL (2016) Gene expression patterns in response to pathogen challenge and interaction with hemolin suggest that the Yippee protein of Antheraea pernyi is involved in the innate immune response. J Invertebr Phathol 138:10–17

Tanaka H, Ishibashi J, Fujitak NY, Sagisaka A, Tomimoto K, Suzuki N et al (2008) A genome–wide analysis of genes and gene families involved in innate immunity of Bombyx mori. Insect Biochem Mol Biol 38:1087–1110

Taszłow P, Wojda I (2015) Changes in the hemolymph protein profiles in Galleria mellonella infected with Bacillus thuringiensis involve apolipophorin III. The effect of heat shock. Arch Insect Biochem Physiol 88(2): 123–143. https://doi.org/10.1002/arch.21208

Taszłow P, Vertyporokh L, Wojda I (2017) Humoral immune response of Galleria mellonella after repeated infection with Bacillus thuringiensis. J Invert Pathol 149:87–96

Terenius O (2008) Hemolin—a lepidopteran anti–viral defense factor? Dev Comp Immunol 32:311–316

Terenius O, Bettencourt R, Lee SY, Li W, Söderhäll Faye I (2007) RNA interference of Hemolin causes depletion of phenoloxidase activity in Hyalophora cecropia. Dev Comp Immunol 31:571–575

Terenius O, Popham HJR, Shelby KS (2009) Bacterial, but not baculoviral infections stimulate Hemolin expression in noctuid moth. Dev Comp Immunol 33:1176–1185

Torres MDT, Sothiselvam S, Lu TK, de la Fuente-Nunez C (2019) Peptide design principles for antimicrobial applications. J Mol Biol pii S0022–2836(18):31289. https://doi.org/10.1016/j.jmb.2018.12.015

Vallet–Gely I, Lemaitre B, Boccard F (2008) Bacterial strategies to overcome insect defenses. Nat Rev 6:302–313

Van der Horst DJ, Rodenburg KW (2010) Locust flight activity as a model for hormonal regulation of lipid mobilization and transport. J Insect Physiol 56(8):844–853. https://doi.org/10.1016/j.jinsphys.2010.02.015

Vertyporokh L, Wojda I (2017) Expression of the insect metalloproteinase inhibitor IMPI in the fat body of Galleria mellonella exposed to infection with Beauveria bassiana. Acta Biochim Pol 64(2):273–278

Vertyporokh L, Taszłow P, Samorek-Pieróg M, Wojda I (2015) Short–term heat shock affects the course of immune response in Galleria mellonella naturally infected with the entomopathogenic fungus Beauveria bassiana. J Invertebr Pathol 130:42–51. https://doi.org/10.1016/j.jip.2015.07.001

Vertyporokh L, Kordaczuk J, Mak P, Hułas-Stasiak M, Wojda I (2019) Host-pathogen interactions upon the first and subsequent infection of Galleria mellonella with Candida albicans. J Insect Physiol. https://doi.org/10.1016/j.jinsphys.2019.103903

Vilcinskas A (2016) The role of epigenetics in host–parasite coevolution: lessons from the model host insects Galleria mellonella and Tribolium castaneum. Zoology (Jena) 119(4):273–280. https://doi.org/10.1016/j.zool.2016.05.004

Vilcinskas A, Wedde M (1997) Inhibition of Beauveria bassiana proteases and fungal development by inducible protease inhibitors in the haemolymph of Galleri mellonella larvae. Biocont Sci Tech 7:591–601

Vilcinskas A, Wedde M (2002) Insect inhibitors of metalloproteinases. IUBMB Life 54(6):339–343

Vogel H, Badapanda C, Vilcinskas A (2011) Identification of immunity–related genes in the burying beetle Nicrophorus vespilloides by suppression subtractive hybridization. Insect Mol Biol 20:787–800

Vogel H, Badapanda C, Knorr E, Vilcinskas A (2014) RNA–sequencing analysis reveals abundant developmental stage–specific and immunity–releted genes in pollen beetle Meligethes aeneus. Insect Mol Biol 23:98–112

Wang Y, Willott E, Kanost MR (1995) Organization and expression of the hemolin gene, a member of the immunoglobulin superfamily in an insect Manduca sexta. Insect Mol Biol 4(2):112–123

Wang X, Rocheleau TA, Fuchs JF, Hillyer JF, Chen CC, Christensen BM (2004) A novel lectin with a fibrinogen–like domain and its potential involvement in the innate immune response of Armigeres subalbatus against bacteria. Insect Mol Biol 13(3):273–282

Wang G, Li X, Wang Z (2016) APD3: the antimicrobial peptide database as a tool for research and education. Nucleic Acids Res 44:D1087–D1093

Wang Z-Z, Zhan LQ, Chen XX (2018) Two types of lysozymes from the whitefly Bemisia tabaci: Molecular characterization and functional diversification. Dev Comp Immunol 81:252–261

Watson FL, Püttmann-Holgado R, Thomas F, Lamar DL, Hughes M, Kondo M, Rebel VI, Schmucker D (2005) Extensive diversity of Ig–superfamily proteins in the immune system of insects. Science 309:1874–1878

Wedde M, Weise C, Kopacek P, Franke P, Vilcinskas A (1998) Purification and characterization of an inducible metalloprotease inhibitor from the hemolymph of greater wax moth larvae. Galleria mellonella. Eur J Biochem 255(3):535–543

Wedde M, Weise C, Nuck R, Altincicek B, Vilcinskas A (2007) The insect metalloproteinase inhibitor gene of the lepidopteran Galleria mellonella encodes two distinct inhibitors. Biol Chem 388(1):119–127

Weers PM, Ryan RO (2006) Apolipophorin III: role model apolipoprotein. Insect Biochem Mol Biol 36(4):231–240

Wen H, Lan X, Cheng T, He N, Shiomi K, Kajiura Z, Zhou Z, Xia Q, Xiang Z, Nakagaki M (2009) Sequence structure and expression pattern of a novel anionic defensin–like gene from silkworm (Bombyx mori). Mol Biol Rep 36:711–716. https://doi.org/10.1007/s11033-008-9233-4

Wen D, Wang X, Shang L, Huang Y, Li T, Wu C, Zhang R, Zhang J (2016) Involvement of a versatile pattern recognition receptor, apolipophorin–III in prophenoloxidase activation and antibacterial defense of the Chinese oak silkworm, Antheraea pernyi. Dev Comp Immunol 65:124–131. https://doi.org/10.1016/j.dci.2016.07.001

Wen D, Luo H, Li T, Wu C, Zhang J, Wang X, Zhang R (2017) Cloning and characterization of an insect apolipoprotein (apolipophorin–II/I) involved in the host immune response of Antheraea pernyi. Dev Comp Immunol 77:221–228. https://doi.org/10.1016/j.dci.2017.08.010

Westwood JT, Wu C (1993) Activation of Drosophila heat shock factor: conformational change associated with a monomer–to–trimer transition. Mol Cell Biol 13:3481–3486

Whitten MM, Tew IF, Lee BL, Ratcliffe NA (2004) A novel role for an insect apolipoprotein (apolipophorin III) in beta–1,3–glucan pattern recognition and cellular encapsulation reactions. J Immunol 172(4):2177–2185

Wiesner J, Vilcinskas A (2010) Antimicrobial peptides: The ancient arm of the human immune system. Virulence 1(5):440–464. https://doi.org/10.4161/viru.1.5.12983

Wijeratne TU, Weers PMM (2019) Lipid–bound apoLp–III is less effective in binding to lipopolysaccharides and phosphatidylglycerol vesicles compared to the lipid–free protein. Mol Cell Biochem. https://doi.org/10.1007/s11010-019-03530-x

Wojda I (2010) Szok termiczny a podatność organizmów na zakażenie i stres osomotyczny. Postępy Biochemii 56(1):86–94

Wojda I (2017a) Immunity of the greater wax moth Galleria mellonella. Insect Sci 24(3):342–357. https://doi.org/10.1111/1744-7917.12325

Wojda I (2017b) Temperature stress and insect immunity. J Therm Biol. https://doi.org/10.1016/j.jtherbio.2016.12.002

Wojda I, Jakubowicz T (2007) Humoral immune response upon mild heat-shock conditions in Galleria mellonella larvae. J Insect Physiol 53(11):1134–1144
Wojda I, Kowalski P (2013) Galleria mellonella infected with Bacillus thuringiensis involves Hsp90. Cent Eur J Biol 8(6):561–569
Wojda I, Taszłow P (2013) Heat shock affects host–pathogen interaction in Galleria mellonella infected with Bacillus thuringiensis. J Insect Physiol 59:894–905
Wojda I, Kowalski P, Jakubowicz T (2009) Humoral immune response of Galleria mellonella larvae after infection by Beauveria bassiana under optimal and heat-shock conditions. J Insect Physiol 55(6):525–531
Wu Q,Patocka J, Kuca K (2018) Insect antimicrobial peptides, a minireview. Toxins 10:461. https://doi.org/10.3390/toxins10110461, www.mdpi
Xia XF, Li Y, Yu XQ, Lin JH, Li SY, Li Q, You MS (2017) Cloning and functional identification of moricins from the diamondback moth, Plutella xylostella (L.). Insect Mol Biol 26(5):564–573. https://doi.org/10.1111/imb.12319
Xu B, Shi Y, Wu Y, Meng Y, Jin Y (2019) Role of RNA secondary structures in regulating Dscam alternative splicing. Biochim Biophys Acta Gene Regul Mech pii S1874–9399(18):30503. https://doi.org/10.1016/j.bbagrm.2019.04.008
Yamakawa K, Huot YK, Haendelt MA, Hubert R, Chen XN, Lyons GE, Korenberg JR (1998) DSCAM: a novel member of the immunoglobulin superfamily maps in a Down syndrome region and is involved in the development of the nervous system. Hum Mol Genet 7(2):227–237
Yi HY, Chowdhury M, Huang YD, Yu XQ (2014) Insect antimicrobial peptides and their applications. Appl Microbiol Biotechnol 98:5807–5822. https://doi.org/10.1007/s00253-014-5792-6
Yu X-Q, Kanost MR (2002) Binding of hemolin to bacterial lipopolysaccharide and lipoteichoic acid. An immunoglobulin superfamily member from insects as a pattern–recognition receptor. Eur J Biochem 269:1827–1834
Yu KH, Kim KN, Lee JH, Lee HS, Kim SH, Cho KY, Nam MH, Lee IH (2002) Comparative study on characteristics of lysozymes from the hemolymph of three lepidopteran larvae, Galleria mellonella, Bombyx mori. Agrius convolvuli. Dev Comp Immunol 26(8):707–713
Zakarian RJ, Dunphy GB, Albert PJ, Rau ME (2002) Apolipophorin–III affects the activity of the haemocytes of Galleria mellonella larvae. J Insect Physiol 48(7):715–723
Zdybicka-Barabas A, Cytryńska M (2011) Involvement of apolipophorin III in antibacterial defense of Galleria mellonella larvae. Comp Biochem Physiol Part B 158:90–98
Zdybicka-Barabas A, Cytryńska M (2013) Apolipophorins and insect immune response. Invert Surviv J 10:58–68
Zdybicka-Barabas A, Januszanis B, Mak P, Cytryńska M (2011) An atomic force microscopy study of Galleria mellonella apolipophorin III effect on bacteria. Biochim Biophys Acta 1808(7):1896–1906. https://doi.org/10.1016/j.bbamem.2011.03.013
Zdybicka-Barabas A, Mak P, Klys A, Skrzypiec K, Mendyk E, Fiołka MJ, Cytryńska M (2012a) Synergistic action of Galleria mellonella anionic peptide 2 and lysozyme against Gram–negative bacteria. Biochim Biophys Acta 1818:2623–2635
Zdybicka-Barabas A, Stączek S, Mak P, Piersiak T, Skrzypiec K, Cytryńska M (2012b) The effect of Galleria mellonella apolipophorin III on yeasts and filamentous fungi. J Insect Physiol 58(1): 164–177. https://doi.org/10.1016/j.jinsphys.2011.11.003
Zdybicka-Barabas A, Stączek S, Mak P, Skrzypiec K, Mendyk E, Cytryńska M (2013) Synergistic action of Galleria mellonella apolipophorin III and lysozyme against Gram–negative bacteria. Biochim Biophys Acta 1828(6):1449–1456. https://doi.org/10.1016/j.bbamem.2013.02.004
Zdybicka-Barabas A, Mak P, Jakubowicz T, Cytryńska M (2014a) Lysozyme and defense peptides as suppressors of phenoloxidase activity in Galleria mellonella. Arch Insect Biochem Pysiol 87(1):1–12
Zdybicka-Barabas A, Palusińska-Szysz M, Gruszecki WI, Mak P, Cytryńska M (2014b) Galleria mellonella apolipophorin III—an apolipoprotein with anti–Legionella pneumophila activity. Biochim Biophys Acta 1838(10):2689–2697. https://doi.org/10.1016/j.bbamem.2014.07.003

Zdybicka-Barabas A, Sowa-Jasiłek A, Stączek S, Jakubowicz T, Cytryńska M (2015) Different forms of apolipophorin III in Galleria mellonella larvae challenged with bacteria and fungi. Peptides 68:105–112. https://doi.org/10.1016/j.peptides.2014.12.013

Zdybicka-Barabas A, Stączek S, Pawlikowska-Pawlęga B, Mak P, Luchowski R, Skrzypiec K, Mendyk E, Wydrych J, Gruszecki WI, Cytryńska M (2019) Studies on the interactions of neutral Galleria mellonella cecropin D with living bacterial cells. Amino Acids 51(2):175–191. https://doi.org/10.1007/s00726-018-2641-4

Zhang ZT, Zhu SY (2009) Drosomycin, an essential component of antifungal defence in Drosophila. Insect Mol Biol 18(5):549–556. https://doi.org/10.1111/j.1365-2583.2009.00907.x

Zhang G, Niu G, Franca CM, Dong Y, Wang X, Butler NS, Dimopoulos G, Li J (2015a) Anopheles midgut FREP1 mediates Plasmodium invasion. J Biol Chem 290(27):16490–16501. https://doi.org/10.1074/jbc.M114.623165

Zhang X, He Y, Cao X, Gunaratna RT, Chen YR, Blissard G, Kanost MR, Jiang H (2015b) Phylogenetic analysis and expression profiling of the pattern recognition receptors: Insights into molecular recognition of invading pathogens in Manduca sexta. Insect Biochem Mol Biol 62:38–50. https://doi.org/10.1016/j.ibmb.2015.02.001

Zhao P, Dong Z, Duan J, Wang G, Wang L, Li Y, Xiang Z, Xia Q (2012) Genome-wide identification and immune response analysis of serine protease inhibitor genes in the silkworm Bombyx mori. PLoS One 7(2):e31168. https://doi.org/10.1371/journal.pone.0031168

Zipursky SL, Grueber WB (2013) The molecular basis of self–avoidance. Ann Rev Neurosci 36:547–568

Chapter 5
Insect Hemolymph Immune Complexes

Kevin D. Clark

Abstract Insects possess powerful immune systems that have evolved to defend against wounding and environmental pathogens such as bacteria, fungi, protozoans, and parasitoids. This surprising sophistication is accomplished through the activation of multiple immune pathways comprised of a large array of components, many of which have been identified and studied in detail using both genetic manipulations and traditional biochemical techniques. Recent advances indicate that certain pathways activate arrays of proteins that interact to form large functional complexes. Here we discuss three examples from multiple insects that exemplify such processes, including pathogen recognition, melanization, and coagulation. The functionality of each depends on integrating recognition with the recruitment of immune effectors capable of healing wounds and destroying pathogens. In both melanization and coagulation, protein interactions also appear to be essential for enzymatic activities tied to the formation of melanin and for the recruitment of hemocytes. The importance of these immune complexes is highlighted by the evolution of mechanisms in pathogens to disrupt their formation, an example of which is provided. While technically difficult to study, and not always readily amenable to dissection through genetics, modern mass spectrometry has become an indispensable tool in the study of these higher-order protein interactions. The formation of immune complexes should be viewed as an essential and emerging frontier in the study of insect immunity.

Keywords Insect immunity · Pattern recognition receptors · Pathogen-associated molecular patterns · C3 proteins · TEP1 convertase · Thioester proteins · Leucine repeat proteins · Clip-domain serine proteases · Serine protease homologs · Tyrosinases · Melanization · Phenoloxidase · Phenoloxidase cascade inhibitors · Hemocyanins · Coagulation · Coagulogen · Clotting proteins · Clot associated proteins · Hemocytes · Lipopolysaccharides · Transglutaminase · Plasmatocyte spreading peptide

---

K. D. Clark (✉)
Department of Chemistry, University of Georgia, 140 Cedar Street, 30602 Athens, GA, USA
e-mail: kclark@uga.edu

U. Hoeger and J. R. Harris (eds.), *Vertebrate and Invertebrate Respiratory Proteins, Lipoproteins and other Body Fluid Proteins*, Subcellular Biochemistry 94,
https://doi.org/10.1007/978-3-030-41769-7_5

Abbreviations

PPO	Prophenoloxidase
PO	Phenoloxidase
MC	Melanization complex
IC	Immune complex
HP	Hemolymph proteases
PAP	Phenoloxidase-activating protease
SPH	Serine protease homolog
PTU	Phenylthiourea
KD	Knockdown
CLIP	An arthropod-specific class of SPs and SPHs
LPS	Lipopolysaccharide
Tyr	Tyrosine
DOPA	Dihydroxyphenylalanine
GFP	Green fluorescent protein
B-cad	Biotin-cadaverine
PSP	Plasmatocyte spreading peptide
ROS	Reactive oxygen species
AMP	Antimicrobial peptide
SDS-PAGE	Sodium dodecyl sulfate-polyacrylamide gel electrophoresis
PAMP	Pathogen-associated molecular pattern
DA	Dopamine
SFP	Substrate-free plasma
PRR	Pattern recognition receptor
CD	N-terminal catalytic domain
RD	C-terminal repeat domain

Introduction

The study of insect immunity has a long and fascinating history that through the efforts of many researchers has led to the identification of a large array of molecular components. These interact in a variety of pathways, the activation of which can both heal wounds and defend against pathogens. Both genetic manipulation, primarily in smaller insects such as fruit flies and mosquitos, and traditional biochemical techniques, primarily in the much larger Lepidoptera larvae, have been employed to elucidate these pathways. Moreover, the ability to rapidly and inexpensively sequence DNA has provided access to the genomes of many model insects, thus allowing for further identification and study of immune-related molecules.

This body of accrued knowledge has led to a more detailed picture of the intricacies and complexity of the insect immune system, although a number of important and challenging questions remain unanswered. Many of these involve the extensive protein interactions essential to fundamental immune processes such as pathogen recognition, melanization, and coagulation. While their importance in the large-scale assembly of proteins found in coagulation may be obvious, other processes also rely on such higher order interactions, including melanin production by the enzyme phenoloxidase (PO) and pathogen recognition by pattern-recognition receptors (PRRs). Many hemolymph components that contribute to these processes have been identified, and various functions have been assigned to them; however, a holistic picture of the extensive interactions required to successfully fend off an attack or heal a wound remains mysterious in many ways.

Immune responses may be initiated by wounding, pathogen challenge or, in many cases, a combination of the two. The prevalent paradigm, that pathogen-associated molecular patterns (PAMPs) activate downstream pathways through interactions with PRRs, has been extensively documented; however, aseptic wounding also activates immune responses through a mostly unknown mechanism that likely involves the production of endogenous *danger* signals (Krautz et al. 2014). Furthermore, collection of hemolymph inadvertently results in disruption of the cuticle, and this wounding alone is often enough to induce melanization and clotting reactions (Clark 2015). Whether initiated through danger signals or the introduction of pathogens, immune induction can be rapid and irreversible.

Once activated, the insect immune system mounts many complementary responses. For example, the presence of bacteria in the hemocoel can simultaneously induce coagulation, activate the enzymes required to produce toxic reactive oxygen species (ROS) and melanin production, and initiate the expression of antimicrobial peptides (AMPs). However, each of these must function in a coordinated fashion to maximize injury to pathogens and minimize injury to the host. For example, though AMPs may harmlessly flood the host hemocoel, coagulation must specifically target wounds and trap microbes. Likewise, to avoid serious damage to the host, it is essential that melanin production and hemocyte (insect blood cells) recruitment are specifically targeted to pathogen surfaces through PRRs.

The expression levels of a large number of proteins are altered upon immune challenge. Yet, important effectors like melanization and coagulation can manifest within minutes after injury, therefore bypassing any need for protein expression, which may require up to 6 h (Levy et al. 2004). Thus, many of the components discussed in this review appear to be constitutively present. The source of these components may be soluble hemolymph proteins present as inactive zymogens or hemocytes capable of rapid exocytosis. In fact, recent studies of *D. melanogaster* have shown that two key proteins involved in melanization and coagulation, both lacking traditional signal sequences, are rapidly released from specific hemocytes by an unknown signal through alternative non-classical pathways, adding yet another level of sophistication to the insect immune system (Dziedziech et al. 2019; Schmid et al. 2019).

To add to the complexity, the type and extent of these reactions can be highly variable between species, and even within certain species the responses may depend on development. For example, plasma from *Bombyx mori* larvae rapidly melanizes at all stages of development, whereas *Manduca sexta* plasma melanizes early in the instar, but shows little to no response late in the instar when larvae are typically harvested for testing (Clark 2015). In *Chrysodeixis* (formerly *Pseudoplusia*) *includens*, a specialized subset of hemocytes called plasmatocytes exhibit significant differences in their responses to the cytokine plasmatocyte spreading peptide (PSP) during larval development (Clark et al. 2005). While there is no obvious explanation for these differences, they must be considered when designing experiments and analyzing results.

The insect immune system *interactome* contains many examples of transient and often binary interactions where an enzyme modifies a target, but there are also larger scale, non-enzymatic protein associations that serve other important functions. These include stabilization of highly reactive proteins, delivery of immune effectors to their targets, activation of enzyme activity, and regulation of immune effectors. This review will discuss several of these interactions in depth, rather than extensively cataloging every known immune-related protein interaction that occurs in the hemocoel, an impossible and overwhelming task.

To accomplish this, this review will focus on a few species that exemplify the role of protein complex formation in accomplishing the tasks outlined above. These include pathogen recognition by the TEP convertase in *Anopheles gambiae*, the formation of a melanization complex (MC) in Lepidoptera and Coleoptera, and coagulation in *Drosophila melanogaster* and *B. mori*. Each of these provides a well-developed conceptual model that can be used to query the immune systems of other species, and in addition, provide a roadmap for many of the major questions that remain in the field. The reader will likely notice a special interest in the myriad functions of serine protease homologs (SPHs), so named due to their obvious homology to serine proteases except for a lack of proteolytic activity due to substitutions at the catalytic Ser. These molecules play central roles in regulating pathogen recognition and melanization and they associate with coagulants yet lack clearly defined functions. However, they do have a propensity to form high mass multimers, the reasons for which are undoubtedly important, but elusive.

The roles of these immune complexes (ICs) have been explored using a variety of modern methods in addition to more traditional techniques, including the powerful arsenal of genetic tools available to study innate immunity in model insects like *D. melanogaster* (Troha and Buchon 2019). Further, the availability of high-resolution mass spectrometry (MS) has permitted detailed analyses of large protein ensembles involved in innate immunity in a manner previously impossible (Rebsamen et al. 2013). The use of these methods to reveal formerly unknown details of the host response to immune challenge will be highlighted throughout this review. Moreover, delving into the complexity of these reactions reveals issues that must be addressed before a full understanding of the surprising sophistication of the insect immune system can be attained.

The Mosquito TEP1 Convertase System—A Multifaceted Protein Network

The complement system, named by Paul Ehrlich in the late 1800 s due to its ability to complement cellular-based immunity, is a key component of the mammalian innate immune system (Ricklin et al. 2016). Further, it is an ancestral form of immunity that evolved long before the development of adaptive immunity (Dodds and Law 1998). One of the central components of complement is the multi-functional C3 protein, which is involved in pathogen recognition, opsonization, and blood cell recruitment. C3 is a member of a unique family of proteins that form a thioester (TE) bond between specific Cys and Gln side chains in the GCGEQ consensus sequence, thus forming an internal ring structure within proteins. TEs form a high-energy bond that is subject to attack by external nucleophiles such as serine protease active site hydroxyls (to form an ester) or primary amines (to form an amide). This results in cross-linking of nucleophiles on the pathogen surface to the TE-containing protein. Within the C3 protein the TE is normally inaccessible; however, proteolytic cleavage by a C3 convertase produces a C3b fragment, exposing this region and promoting pathogen opsonization.

The search for complement-like proteins in insects led to the discovery of numerous TE-containing proteins (TEPs). Four TEPs were discovered in *D. melanogaster*, each related to mammalian TE proteins and containing the canonical GCGEQ sequence (Lagueux et al. 2000). A TEP (TEP1) was also cloned from the *Anopheles gambiae* mosquito (Levashina et al. 2001) and was shown to opsonize both *E. coli* and *S. aureus*, which required an intact TE bond. Both wounding and bacterial challenge induce proteolytic cleavage of TEP1 to produce $TEP1_{cut}$, reminiscent of the activation of C3 (Levashina et al. 2001; Blandin and Levashina 2004). TEP1 was also shown to have a profound effect on survival of the malaria parasite, *Plasmodium berghei*, in *A. gambiae* (Blandin et al. 2004). In this study TEP1 alleles isolated from either a susceptible or a refractory strain were each shown to bind to ookinetes; however, they exhibited drastically different abilities to recruit effectors resulting in melanization and lysis. A *tep1*-resistant allele KD was also shown to completely block ookinete melanization and to increase the number of surviving parasites. These groundbreaking experiments provided the first functional evidence of a complement-like activity in *A. gambiae* and became important building blocks for subsequent discoveries. These include the protein-protein interactions required for stabilizing $TEP1_{cut}$, how it is targeted to non-self through ookinete surface nitration mediated by the midgut epithelium (Oliveira et al. 2012), and the mechanism by which it activates downstream effector pathways.

Leucine-Rich Repeat (LRR) Proteins

Concurrent with continuing work on TEPs, a protein superfamily recognized for their ability to promote protein interactions was found to be involved in immune responses to *Plasmodium* infection. Members of this family of leucine-rich repeat (LRR) proteins have been identified in a wide range of life forms and possess an equally wide range of functions, including involvement in immune responses (Bella et al. 2008). Each LRR-protein possesses various numbers of tandem LRRs containing the consensus sequence LxxLxLxxNxL (with various Leu substitutions permitted), which form solenoid folds well-suited to promote protein interactions. Gene silencing in *A. gambiae* first showed that an LRR-protein antagonizes the development of *Plasmodium* ookinetes to oocysts (Osta et al. 2004), and that specific silencing of the LRR family member LRIM1 (leucine-rich repeat immune protein 1) reduces phagocytosis during immune responses (Moita et al. 2005). Frolet and colleagues defined pre- and post-invasion phases of *Plasmodium* defenses and showed that the NF-kB transcription factors *Rel*1 and *Rel*2 decreased expression of both TEP1 and LRIM1 and abolished resistance (Frolet et al. 2006). An additional LRR-protein, APL1 (Anopheles *Plasmodium*-responsive leucine-rich repeat 1), was shown to be involved in *Plasmodium* resistance, and is comprised of three separately transcribed genes (APL1A, B, and C), of which only APL1C is responsible for resistance to *P. berghei* (Riehle et al. 2006, 2008).

In 2009, two seminal publications showed that LRIM1, APL1, and TEP1 interact to form a protein complex essential to TEP1 function and targeting (Fig. 5.1a). Fraiture and colleagues found that in the circulation $TEP1_{cut}$ is unstable unless maintained by an LRIM1/APL1 heterodimer, and that KD of either LRR protein results in a loss of $TEP1_{cut}$, likely through binding to self-tissue (Fraiture et al. 2009). In addition, LRIM1 and APL1 were found to be individually unstable, and thus require each other, but do not require TEP1 binding for their stability. Also, Povelones et al. (2009) found a circulating LRR protein complex composed of LRIM1 and APL1 and showed that it is bound together through an intermolecular disulfide bond, co-immunoprecipitates with TEP1, and is responsible for activating a form of insect complement against *P. berghei*. Intriguingly, the authors found an approximately 260 kDa LRIM1/APL1C complex, which suggests the formation of extensive higher order structures. In addition, the LRR proteins were shown to have an important role in pathogen recognition and downstream effector activation, as KD of either not only blocked TEP1 binding to parasites, but also prevented ookinete melanization, resulting in an increased number of viable oocysts.

Further structural characterization showed that LRIM1 and APL1C preferentially form a heterodimer through C-terminal coiled-coil domains in each protein and that the interaction is stabilized by a single disulfide between Cys 352 (LRIM) and Cys 551 (APL1). The authors also determined the crystal structure of the heterodimer, defined the structural and interactive roles of the LRRs, and showed specific determinants required for TEP1 binding and stabilization (Baxter et al. 2010). In a separate,

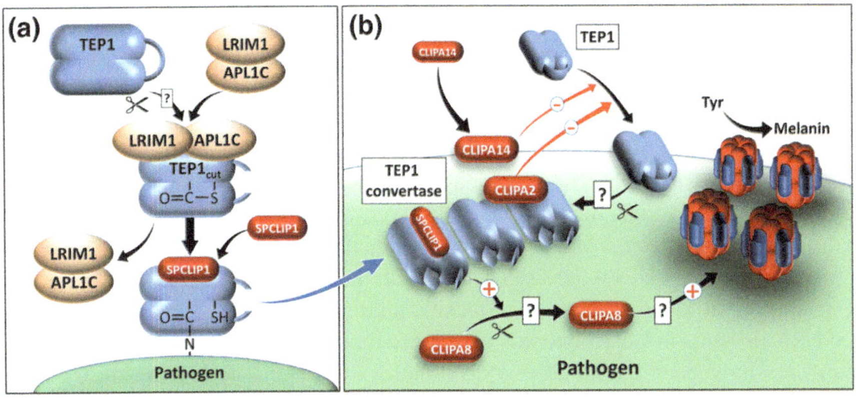

Fig. 5.1 Model of TEP1 convertase formation on a pathogen surface. **a** Covalent attachment of the thioester-containing TEP1 to the surface of pathogens, such as ookinetes or microbes, requires an ensemble of LRR proteins, a protease, and an SPH. Hemolymph TEP1 is cut by an unknown protease to expose its internal thioester, then is chaperoned by the LRIM1/APL1C complex to prevent attachment to self-tissues and ensure thioester stability. The LRIM1/APL1C complex delivers $TEP1_{cut}$ to the pathogen, which then covalently attaches to its surface and forms a complex with the SPH SPCLIP1. **b** The bound $TEP1_{cut}$/SPCLIP1 complex promotes translocation of additional TEP1s to the surface, which are then cleaved by an unknown protease. Association of these TEPs forms a TEP1 convertase capable of eliciting effectors that promote ookinete lysis and melanization. Melanization requires SPCLIP1 to induce the proteolytic activation of the SPH CLIPA8, which then promotes melanization through an unknown mechanism. Melanin formation may be catalyzed by a high mass melanization complex (MC) as described in the section on melanization. MC assembly requires proteolytically activated SPHs, suggesting the possibility that CLIPA8 may act in a similar fashion to bind and activate the phenoloxidases (POs) required for melanin production. Both CLIPA2 and CLIPA14 inhibit TEP1 convertase activity and melanization, although KDs of each indicate there are phenotypic differences. CLIPA2 colocalizes with TEP1 on the pathogen surface but does not appear to be cleaved. CLIPA14 appears to associate with the pathogen surface independent from TEP1, and is cleaved by an unknown protease that is, like CLIPA8, dependent on TEP1 and SPCLIP1

extensive analysis of *A. gambiae, Culex quinquefasciatus* and *Aedes aegypti,* numerous LRIMs found in mosquitos were classified into four subfamilies determined by the number of LRR repeats (6–14 of variable lengths), the presence or absence of a coiled-coil domain, and whether there is a transmembrane region at the C-terminus (Waterhouse et al. 2010).

The importance of the LRIM1/APLC1 complex in mosquito immunity became even more apparent when it was discovered that pull-down experiments in *A. gambiae* not only captured TEP1, but also additional TEPs, including 3, 4, and 9 (Povelones et al. 2011). This ability to carry different TEP *cargoes* indicates that the complex can pass information through different pathogen targeting pathways. For example, KD of *tep1* causes a large increase in *P. berghei* oocysts, while KD of *tep3* only results in a partial increase, and KD of *tep4* has no effect. Interestingly, TEP3 and TEP4 were previously shown to be involved in bacterial defense, suggesting there is pathogen specificity to TEP binding (Dong et al. 2006). The effects of these

TEPs on melanization were tested in a *CTL4* (C-type lectin) KD, which produces a *Plasmodium*-resistant phenotype and a high degree of ookinete melanization. KD of both *CTL4* and *TEP1* completely blocked melanization, while a *CTL4/TEP3* double knockdown (DKD) showed only partial inhibition, and KD of both *CTL4 and TEP4* had no effect. Thus, this shows that TEP1 plays a key role not only in regulating melanization but is likely also involved in the targeting of PO and cofactors. Similar connections to melanization were observed in *D. melanogaster* injected with pathogenic *Photorhabdus* bacteria, where *tep2* and *tep4* mutants increased PO activity and a *tep6* mutant decreased PO activity (Shokal and Eleftherianos 2017; Shokal et al. 2017). In contrast to these extensive studies in Diptera, the study of TEPs and LRR proteins in Lepidoptera is extremely limited. In *M. sexta*, 92 LRR-protein sequences have been identified, but no orthologs of either LRIM1 or APLC1 have been discovered. Moreover, three TEPs have been identified, but only one, TEP2, contains the canonical GCGEQ sequence required for TE formation (Zhang et al. 2015). Thus, this suggests that mosquitos have greatly expanded their TEP repertoire (the *A. gambiae* genome contains 19 TEP gene homologs) in response to a variety of pathogens. Although the functions of these proteins are essentially unknown in Lepidoptera, the discovery of TEP1 activity and its interaction with multiple LRR proteins in Diptera provided profound insights into the mechanism of pathogen recognition. As a result, a new avenue of research into the role of serine-protease homologs (SPHs) in both regulating the response to non-self and linking it to activation of the PO cascade has been opened.

Clip-Domain Serine Proteases (CLIP-SPs)

Clip-domain serine proteases (CLIP-SPs) have emerged as essential regulators of immune responses in many insects (Kanost and Jiang 2015). These modular proteins, unique to invertebrates, contain variable numbers of N-terminal CLIP domains that are separated by a connecting region from a larger chymotrypsin-like C-terminal protease domain. CLIP-SPs can be organized into five different clades, CLIPA, B, C, D or E, depending on their CLIP-type, number of CLIP-domains and whether they are functional proteases. CLIPB, CLIPC, and CLIPD proteins all possess a prototypical His-Asp-Ser triad that is catalytically active, while CLIPA and CLIPE proteins are rendered inactive by a substitution at the nucleophilic Ser residue and are referred to as serine-protease homologs (SPHs). CLIPB, C, and D are all activated by proteolytic cleavage in proximity to their protease domains, while an additional disulfide bond spanning this region keeps the fragments joined. In contrast, CLIPA is cleaved within its CLIP domain, while CLIPE (discussed below) does not appear to be cleaved (Povelones et al. 2013).

The myriad roles of the proteolytically active CLIP-SPs, including their roles in PO cascade activation, have been well-characterized in larger insects such as *M. sexta* and *Tenebrio molitor*, as well as in Diptera (Zhao et al. 2010; Veillard et al. 2016). Moreover, in *A. gambiae*, KD of *CLIPB14* and *CLIPB15* indicate that both are involved in the killing of *Plasmodium* ookinetes and Gram-negative bacteria, and that CLIPB14 plays a role in the PO cascade (Volz et al. 2006). In addition, KD of *CLIPB4* and *CLIPB8* have been shown to inhibit Sephadex bead melanization (Paskewitz et al. 2006), CLIPB8 has been implicated in the PO cascade (Zhang et al. 2016), and CLIPB9 has been shown to activate melanization and cleave prophenoloxidase (PPO) (An et al. 2011). In *A. aegypti*, separate pathways have been proposed for PO cascade activation. Immune melanization proteases IMP-1 and IMP-2 are thought to be involved in defense against parasites, and tissue melanization proteases TMP and IMP-1 involved in wound healing (Zou et al. 2010). While interesting, the proteolytically active CLIP-SPs do not appear to be involved in protein complex formation. In contrast, a series of discoveries in *A. gambiae* have shown the vital importance of protein interactions in pathogen defense and the regulation of melanization mediated by members of the non-proteolytic CLIPA and CLIPE families.

Mosquito Serine Protease Homologs (SPHs)

The *A. gambiae* CLIPA and CLIPE SPH proteins differ from the other CLIP-SP clades, as previously noted, in that they possess the hallmark domains of a CLIP protease yet lack a Ser at the active site. Thus, their clear homology to SPs contrasts with their lack of protease function and has left many unanswered questions regarding the protein-protein interactions that underlie their real function. Nonetheless, research into the *A. gambiae* CLIP SPHs has provided tantalizing details regarding the many regulatory roles of SPHs in the assembly and activity of the TEP1 convertase (Fig. 5.1b).

The earliest enquires into the role of CLIP SPHs utilized KDs of *CLIPA* genes to assess the melanization of *P. berghei* ookinetes in susceptible (S) and resistant (R) strains of *A. gambiae*. The most dramatic phenotype arose from silencing of *CLIPA8* in the melanization-competent R strain, which resulted in a drastic decrease in melanin formation (Volz et al. 2006). Despite this, oocysts did not thrive, suggesting a compensatory killing mechanism and implying that while CLIPA8 is an essential and specific link to PO activation, it is not important for other downstream effects. These findings were in agreement with previous results showing a dual killing mechanism in the R strain whereby ookinetes are killed by lysis when passing through the midgut epithelium and are then cleared through the melanization process (Blandin et al. 2004). The C-type lectin CTL4 is a potent agonist of *Plasmodium*, and KD of *CTL4* in the S strain dramatically increases melanization of transiting ookinetes in the midgut epithelium, thus rendering the S strain refractory (Osta et al. 2004). However, a *CTL4/CLIP8* DKD also abolished melanization, thus confirming its superseding role over CTL4 and its direct link to the PO cascade.

CLIPA8

Subsequent experiments showed that CLIPA8 also controls melanization of both Gram-negative and Gram-positive bacteria and demonstrated that infection induces the cleavage, and likely the activation, of CLIPA8 (Schnitger et al. 2007). Intriguingly, the authors noted depletion of truncated CLIPA8 occurred relative to the intact form, suggesting it may be sequestered in the PO complex. Thus, this is suggestive of the high mass self-organizing SPH multimers observed in the PO cascades in both *M. sexta* and *Holotrichia diomphalia* (Gupta et al. 2005; Piao et al. 2005); however, further study is necessary to determine if higher mass forms of CLIPA8 are formed. Surprisingly, wounds were still melanized following KD of *CLIPA8*, suggesting a separate mechanism for PO deposition at the damaged cuticle.

The pathogen-centric nature of CLIPA8 activity was confirmed by experiments employing the entomopathogenic fungus *Beauveria bassiana*, which, by directly breaching the adult insect cuticle and invading the hemocoel, obviates the need for manual injection of pathogens such as bacteria and bypasses any immune responses mounted by the mosquito midgut epithelium against *Plasmodium* ookinetes (Yassine et al. 2012). In this study, the authors used either direct injection of conidia or a natural infection route to demonstrate a strong melanization response by adult *A. gambiae*, mediated by an early stage hemocyte response, and a late stage humoral response that specifically targeted hyphae. Both TEP1 and CLIPA8 were required for PPO localization and melanization of the hyphal surface but were not required for the hemocyte-specific melanization response. While the presence of TEP1 on hyphae was found to be a prerequisite for PPO localization, a *TEP1* KD succumbed more quickly to infection than a *CLIPA8* KD. Thus, this suggests that TEP1 may play a central role in recruiting effectors in addition to PO through the formation of an *immune protein complex* on microbial surfaces.

While these results indicate the pathogen-centric nature of CLIPA8-mediated PO activation, unanswered questions as to how CLIPA8 activates the PO cascade and how TEP1 can be targeted to so many disparate microbial surfaces remain. Nonetheless, these investigations into pathogen recognition and effector recruitment led to the discoveries of several additional CLIP SPHs that interact with and regulate the TEP1-dependent convertase in mosquitos.

SPCLIP1

The *A. gambiae* CLIP-SPH, SPCLIP1, previously identified through its involvement in immune defense against several pathogens (Dong et al. 2006), became the focus of renewed interest due to its developmental co-regulation with the LRR-protein LRIM1 (Povelones et al. 2013). Although classified as an SPH due to a substitution of the catalytic Ser, it also contains amino acid substitutions at each residue of the catalytic triad and, being phylogenetically distinct from other SPHs, is placed in the unique CLIPE clade.

A series of KD experiments defined the SPCLIP1 interactors and again demonstrated the central role of TEP1 in the process of PO activation. For example, KD of *TEP1* does not affect SPCLIP1 hemolymph levels, but KD of *LRIM1* greatly reduces it, suggesting association with $TEP1_{cut}$ as it deposits on self tissue (Fraiture et al. 2009). However, a double KD of *TEP1/LRIM1* restored SPCLIP1 to its normal levels, again suggesting an association with $TEP1_{cut}$ (Povelones et al. 2013). In contrast, a double *LRIM1/SPCLIP1* KD did not restore $TEP1_{cut}$ levels, indicating continued $TEP1_{cut}$ binding to self-tissue without assistance from SPCLIP1. However, immunostaining experiments using *P. berghei* parasites indicated mutually dependent surface co-localization of each, which was lost upon silencing of either gene. Surprisingly, the injection of *E. coli* bioparticles resulted in the depletion of SPCLIP1 and TEP1, but not $TEP1_{cut}$, from the hemocoel, a process that was blocked on the *SPCLIP1* KD background. However, analysis of the surface-bound proteins showed the presence of SPCLIP1 and $TEP1_{cut}$, but not TEP1, in this study. Thus, this suggests the possible association of SPCLIP1 with TEP1 and their surface binding followed by an unknown protease activity that cleaves TEP1 to form $TEP1_{cut}$, action reminiscent of a convertase. Injection of bioparticles also demonstrated the time-dependent cleavage and apparent activation of CLIPA8, the SPH responsible for PO activation. In addition, KD of *SPCLIP1* blocked both CLIPA8 cleavage and melanization and led to a large increase in parasite numbers. Numerous questions remain regarding the precise interactions of SPCLIP1 with TEP1 and $TEP1_{cut}$, the identity of the proposed proteases that cleave surface-bound TEP1 and CLIPA8, and whether SPCLIP1, like other SPHs, is itself activated by cleavage. Nonetheless, the prominent role played by SPCLIP1 in regulating melanization through interactions with $TEP1_{cut}$ demonstrates the exquisite control mechanisms at play both in assembling a TEP1 convertase and in eliciting the activation and localization of PO to produce melanin, both of which occur on the pathogen surface.

CLIPA2

The importance of the CLIPA8 and SPCLIP1 SPHs in activating melanization suggests negative regulation may also be essential in managing this process. Although CLIPA2 was initially suggested to negatively regulate melanization (Volz et al. 2006), KD of *CLIPA2* was found to generate increased resistance to a variety of pathogens, including *E. coli*, *B. bassiana*, and *P. berghei* (Yassine et al. 2014). Further, injection of *E. coli* bioparticles into *A. gambiae* demonstrated that both CLIPA2 and TEP1 are depleted from the hemolymph due to their sequestration on the bioparticle surface. Interestingly, KD of *TEP1* prevented binding of CLIPA2 to the bioparticles, while KD of *CLIPA2* did not alter the amount of bound TEP1, indicating that CLIPA2 binding is dependent on TEP1. Knockdown of *LRIM1*, which was previously shown to deplete TEP1, was found to deplete both $TEP1_{cut}$ and CLIPA2 from the hemolymph, likely due to the deposition of $TEP1_{cut}$ on self tissue. Thus, this confirmed that sequestration of CLIPA2 is TEP1-dependent. As discussed above, KD of *TEP1* was found to completely block melanization and prevent the activation and cleavage of CLIPA8. In contrast, a *CLIPA2* KD exhibited exacerbated melanotic responses to both microbial and *P. berghei* infections, and increased hemolymph PO activity, indicating it is a negative regulator of the complement-like TEP1 convertase. This may occur through the inhibition of TEP1 convertase's ability to convert TEP1 to its active $TEP1_{cut}$ form on the pathogen surface by CLIPA2. Interestingly, immunoprecipitation of CLIPA2 in *B. bassiana*-infected *A. gambiae* pulled down several proteins involved in coagulation, including an ortholog of *D. melanogaster* fondue, the lipophorin ApoLp I/II, and gelsolin (Kamareddine et al. 2016). Gelsolin was also identified in the *A. aegypti* immune complex along with ApoLp I/II, PPOs and LRIM proteins, suggesting intimate ties between TEP1 convertase, melanization, and the recruitment of coagulants (Phillips and Clark 2017).

CLIPA14

The identification of the important role of CLIPA2 in negatively regulating the TEP1 convertase led to a general search for additional interacting proteins. Immunoprecipitations using a CLIPA2-specific antibody in fungal-infected *A. gambiae* resulted in the discovery of a variety of interactors, including four additional CLIPA SPHs (Kamareddine et al. 2016). Of these, CLIPA14, like CLIPA2, was also found to negatively regulate the immune response to both *Plasmodium* and bacterial infections and dramatically increases PO activity. However, CLIPA14 acts in ways quite unlike CLIPA2 (Nakhleh et al. 2017). For example, while neither SPCLIP1 nor CLIPA2 are cleaved during infection, CLIPA14 is cleaved but yields only a small band indicative of limited activation or loss of the active form through sequestering or complex formation. Through KD experiments, CLIPA14 cleavage was also shown to be dependent on TEP1 and SPCLIP1; however, unlike $TEP1_{cut}$ and SPCLIP1, CLIPA14 was not depleted from the medium in *LRIM1* KD organisms. Bioparticle infection resulted in the deposition of uncleaved CLIPA14 while KD of *TEP1* did not affect the amount of bound protein. Thus, this indicates that the localization of CLIPA14 depends on a different and unknown mechanism.

Outside of the extensive, albeit narrowly focused, research on the role of SPHs in promoting PO activity in Lepidoptera, the work described here represents a major expansion in knowledge of the capabilities of SPHs in regulating an overall immune response. This ranges from pathogen recognition and targeting to elicitation of effectors. Moreover, formation of TEP1 convertase appears to be a significant target of regulation, which is not surprising given its central role in eliciting downstream effects such as parasite lysis and melanization. The ability of SPCLIP1 to promote TEP1 convertase formation on the pathogen surface suggests it plays a key role as a cofactor in the assembly of $TEP1_{cut}$ onto the actively forming convertase. CLIPA2 and CLIPA14 both appear to negatively regulate this assembly yet act through entirely different mechanisms. In addition, the role of CLIPA2 in immunity may be far more extensive than just those functions dependent on its interaction with TEP1, and it should be noted that the above-mentioned CLIPA2 immunoprecipitations also pulled down TEP3 and TEP15. Nonetheless, CLIPA8 plays an entirely distinct role as an intermediary between TEP1 convertase and PO activation and may even have a direct role in the PO cascade. These fascinating insights provoke numerous questions including the mechanisms used by SPHs to regulate its targets, whether upstream proteolytic cleavage is required for their activation, whether they form high mass multimers as seen in Lepidoptera, and if so, whether these multimers might be formed from the assembly of different SPHs, perhaps to fine-tune their regulatory roles.

Melanization

Melanization is a wide-ranging and multi-functional phenomenon found in animals, plants, fungi, insects, and bacteria (Seo et al. 2003; Jiang et al. 2010; Vavricka et al. 2010, 2014; Sendovski et al. 2011). While melanin production lends itself to facile biochemical assays due to its easy visibility and simple spectrophotometric measurements, the underlying enzymatic mechanism(s) and the large number of reactive intermediates produced makes its study challenging (Ito 2003; Nappi et al. 2009). The synthesis of melanin is initiated by a broad class of structurally heterogeneous enzymes called tyrosinases, named for their ability to oxidize tyrosine (Tyr) to form the high molecular weight melanin polymer.

Importance of the Enzymatic Mechanism

While structurally diverse across kingdoms, all tyrosinases share homologous histidine-coordinated type 3 copper-binding sites that serve to bind the molecular O_2 needed to oxidize Tyr (Decker et al. 2007). Two oxidation steps are required, the first and rate-limiting being the conversion of the monophenol Tyr into its diphenolic derivative, DOPA, and the second being the oxidation of the ortho hydroxyl groups into dopaquinone (Vavricka et al. 2014). Subsequent enzymatic and non-enzymatic transformations convert dopaquinone into highly reactive components that ultimately cross-link to polymerize into the various forms of melanin (Nappi et al. 2009).

Complexity in the mechanism arises because efficient Tyr oxidation is itself DOPA-dependent, thus requiring a catalytic amount of the intermediate DOPA to be present, and because some DOPA-quinone is reduced back to DOPA (Peñalver et al. 2005). In addition, dopaquinone is subject to nucleophilic attack by thiol-containing peptides such as glutathione (or cysteine) found in plasma, to form the adduct glutathionyl-DOPA, which removes it from the melanin formation pathway (Ito et al. 1985; Clark et al. 2010). This inherently non-linear mechanism can manifest itself as a variable lag phase before melanin accumulation is observable. In insects, melanin production is catalyzed by POs. Each species expresses a variable number of PO isozymes that, although catalytically similar, may be expressed in different tissues and involved in different processes (Dudzic et al. 2015). While POs are functionally related to tyrosinases, they are structurally related to arthropod hemocyanins, a large family of copper-containing proteins involved in oxygen transport (Hughes 1999; Burmester 2001).

PPO Processing to PO; Is This the End of the PO Cascade?

Hemolymph POs play a key role in innate immunity and are therefore expressed and stored as inactive zymogens called prophenoloxidases (PPOs). PPO processing occurs through the PO cascade, a series of sequentially activated serine proteases (SPs) that culminates in the cleavage of PPO by phenoloxidase activating proteases (PAPs), commonly at a canonical R51, to form PO (Jiang et al. 2010). In Lepidoptera, activation of the PO cascade commences when PAMPs bind to and activate multifunctional receptor-proteases, but other interesting activation mechanisms also play a role, such as through CLIPA8 in *A. gambiae* (Volz et al. 2006). In addition, PO activation requires the presence of serine protease homologs (SPHs), inactive versions of SPs that play a key role in PO processing. Extensive research has contributed to our current detailed understanding of the PO cascade, partly through genetic dissections in model organisms such as *Drosophila melanogaster* (Li et al. 2012), and primarily through the isolation and characterization of PO cascade components from *Manduca sexta* (Jiang et al. 2010). However, there is substantial evidence that cleavage of PO at R51, the final enzymatic step of the long-standing PO cascade paradigm, does not produce an enzyme capable of metabolizing the substrate (tyrosine) for which this class of enzymes is named (Clark and Strand 2013). Clearly, questions regarding the physiological substrate used for melanin production and the PO proteoform required to metabolize it must be addressed.

Does PO, a Tyrosinase, Metabolize Tyr?

Until recently, the question of substrate use by insect POs had not been unequivocally addressed. While Tyr would be a likely physiological candidate in hemolymph, its use has been almost universally ignored in both PO characterizations and melanization assays in favor of faster-acting diphenolic substrates. While appearing to be a rather mundane research question, the answer has far reaching implications regarding post-PO cascade processing and PO complex formation in the insect hemolymph.

This problem was first addressed in *B. mori* larvae by assessing physiological substrate use in the plasma during melanization (Clark and Strand 2013). Levels of common PO substrates likely found in plasma, such as Tyr, dopamine (DA), and DOPA, were measured under different experimental conditions to determine what POs were metabolizing to synthesize melanin. A series of experiments in *B. mori* (Clark and Strand 2013; Phillips and Clark 2017), *M. sexta* (Clark 2015), and *A. aegypti* (Phillips and Clark 2017) all consistently demonstrated that Tyr is the physiological substrate for melanization by showing that: (1) only Tyr is maintained at a sufficiently high level to support melanization, (2) DA and DOPA are present at very low levels in rapidly prepared plasma, (3) increasing melanization causes a concomitant reduction in plasma Tyr levels, and (4) the PO catalysis inhibitor phenylthiourea (PTU) blocks both melanization and Tyr metabolism.

Knowing that Tyr is the physiological substrate for melanization not only informs us in our search for the PO proteoform that will metabolize it, but also illustrates a potentially serious problem regarding how melanization is commonly assessed in the insect literature. A prime example of this is in *M. sexta*, the source of so much of the biochemical knowledge regarding the PO cascade. The plasma of *M. sexta* typically does not melanize after preparation (which by necessity requires wounding), although it does produce melanin in response to the subsequent addition of PAMPs, which activate the PO cascade. However, experiments studying the developmental dependence of immune responses indicated that wounding induces spontaneous wound-induced melanization early in the 4th and 5th instars (e.g. Day 1), but not late in the instars when plasma is typically collected (e.g. Day 4 or 5) (Clark 2015). The role for this dependence is unknown, but fortuitously provided an additional way to confirm whether Tyr is the physiological substrate for POs in *M. sexta*, and what is required to metabolize it.

To demonstrate the importance of using the proper substrate, substrate-free-plasmas (SFPs) were prepared using molecular weight filters (30 kDa cutoff) from D1 and D5 5th instar larvae. Either Tyr or DA was then added to each to determine melanizing capability. Surprisingly, addition of DA to D5 SFP resulted in melanin production, even though the plasma it was prepared from was unable to melanize. In contrast, addition of Tyr at physiological levels found in hemolymph did not result in melanization. Not surprisingly, addition of either Tyr or DA to D1 SFP resulted in melanin production. Importantly, immunoblots of the samples showed the presence of POs in both, indicating PO cascade activation and providing an explanation for DA metabolism in D5 SFP. Native gels showed a high mass band trapped at the stacking/running gel interface in both D1 and D5 samples, but only the D1 sample melanized upon Tyr addition. These and other experiments showed that wounding can induce melanization in *M. sexta*, that PO is incorporated into a high mass band that can metabolize Tyr early in the instar, that Tyr is the physiological PO substrate, and, most importantly, that commonly used diphenolic substrates, such as DOPA or DA, can provide a false positive result for melanization.

What Is the Form of PO That Uses Tyr?

The substantial evidence showing that melanin production begins with Tyr oxidation begs the question as to what is required to convert PO into a metabolically active form. Experiments from several laboratories using different model insects have suggested interesting possibilities. These include additional PO processing beyond the PO cascade, the mysterious and multi-faceted involvement of SPHs during and after PO processing, and the formation of high mass PO-containing bands that appear to be cross-linked.

Some of the earliest evidence of additional PO processing and formation of higher mass forms of PO came from a series of experiments in Coleoptera from Bok Luel Lee's laboratory. In these studies, a PPO, a PAP (PPAF-I), and an SPH (PPAF-II) were each purified from *Holotrichia diomphalia* and extensively characterized. Interestingly, while the SPH appeared as a 40 kDa protein on reducing SDS-PAGE, the native form eluted at a mass of 400 kDa during gel filtration. In addition, the combination of these components caused PPO to be processed in the presence of Ca^{2+} from 79 to 76 kDa, and also to a 60 kDa form. Intriguingly, when separated by non-reducing SDS-PAGE and stained in-gel using DOPA as a substrate, these combined components developed into high mass forms indicative of active PO, one of which barely entered the gel (Lee et al. 1998; Kim et al. 2002). In the *Aedes aegypti* hemolymph, PPO bands were found from 70 to 75 kDa, although cleavage reactions produced processed forms consisting of an 18 to 20 kDa N-terminal fragment and a 50 kDa C-terminal fragment, suggesting non-canonical activation (Zou et al. 2010). *A. aegypti* PO3, which was shown to have an immune role against entomopathic fungi, was also processed in the presence of hemolymph from the 79 kDa intact PPO to two lower mass forms, a weak 60 kDa band and a strong 50 kDa band, as well as a high mass complex that contained PPO and additional proteins (Wang et al. 2017). In arthropods, an example of alternative processing was found in the crayfish *Pacifastacus leniusculus*, which processed PPO at a non-canonical downstream Arg (R176) producing a 62 kDa band (Aspán et al. 1995).

These initial suggestions of higher mass PO proteoforms in Coleoptera was followed by more experiments demonstrating that SPHs are required during PO cleavage to provide activity (Kwon et al. 2000) and the discovery of another PAP (PPAF-III) that could cleave PPOs (Kim et al. 2002). PPAF-III was shown to cleave the SPH PPAF-II and, as before, convert it to a high-mass 600 kDa form, which by EM appeared to be two stacked hexamers (Piao et al. 2005). The addition of purified PO to these large multimers increased their apparent size when viewed by EM, thus providing additional confirmation of association.

Similar, but more extensive, experiments performed in *Tenebrio molitor* showed comparable results and greatly increased our understanding of the PO activation process (Kan et al. 2008). A PAP (SPE) was purified and found to cleave PPO at two separate residues, first at the canonical R51 site, resulting in a 76 kDa band, and at R281, which converted the 76 kDa band into 27 and 46 kDa bands. However, co-incubation of SPE and PPO along with a purified SPH (SPH1) processed PPO only at R51, and also cleaved and activated the SPH resulting in strong melanin synthesis. Of greatest interest is that the reaction produced a high mass melanization-active peak during gel filtration, which when separated in a denaturing and reducing gel showed a ladder of high mass bands identified by immunoblot with both PO and SPH antibodies. This important observation led the authors to make the first reference to this phenomenon as the formation of a Melanization Complex (MC), nomenclature which we employ in this manuscript (Fig. 5.2).

Concurrent with many of these experiments, similar research into the activation of POs was being carried out in *M. sexta*. The serine protease homologs SPH1 and SPH2, which co-purified with immulectin-2, a C-type lectin found in hemolymph, were found to be homologous to the SPHs from *T. molitor* and *H. diomphalia* (Yu and Kanost 2000). Furthermore, incubation of purified PPO with its activating enzyme, PAP-1, did not efficiently cleave PPO or produce melanization activity until the purified SPHs were added.

Further experiments in *M. sexta* revealed that although PAP-1 cleaves PPO, PO production correlates poorly with PO activity. The addition of SPHs to the reaction improved the cleavage efficiency and provided an enormous increase in activity (Gupta et al. 2005). In an elegant and informative set of time-course experiments, cleavage of PPO by PAP-1 in the absence of SPHs was found to result in little activity,

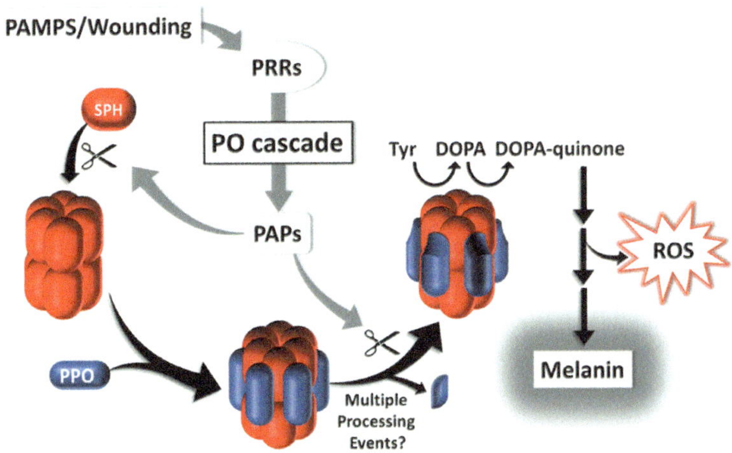

Fig. 5.2 Model of PO cascade activation and formation of the melanization complex. PO cascade activation can occur through wounding or by the interaction of PAMPs with PRRs. Either mechanism results in the proteolytic cleavage and activation of phenoloxidase activating proteases (PAPs). PAPs have been shown to cleave and activate both PPOs and SPHs, resulting in the assembly of a high mass melanization complex (MC) that is required for efficient metabolism of the physiological substrate Tyr. The conversion of Tyr into melanin involves two PO-mediated oxidations and several additional steps that produce toxic reactive oxygen species as a byproduct, making it essential that melanin production is targeted to a pathogen to avoid harm to the host. The model shown here is based on physical and biochemical evidence demonstrating that protein complex formation is a prerequisite for melanin production. In *H. diomphalia*, electron microscopy and gel filtration of activated SPHs in the absence of POs indicates self-association resulting in the formation of two stacked hexamers of approximately 600 kDa. The addition of POs results in an approximate mass increase of 400 kDa, suggesting the addition of six POs to the stacked hexamers. Results in *M. sexta* indicate association of PPO with SPHs is required for PAPs to provide efficient conversion of PPOs to POs. Extensive experiments in *B. mori*, *M. sexta*, and *A. aegypti* all indicate metabolism of the physiological PO substrate Tyr requires the formation of a high mass protein complex that includes SPHs

which could not be rescued by later addition of SPHs after PAP-1 inactivation. Thus, this indicates that *SPHs must be present during PPO cleavage to produce activity*. Like the experiments in Coleoptera, the combination of activated POs and SPHs were found to run at a high mass on a native gel, as shown by in-gel DA-mediated melanin production, and eluted at a high mass range when subjected to gel filtration. Additional evidence of the importance of SPHs was shown using expressed proSPH1 and 2 from *M. sexta* (Lu and Jiang 2008). When activated together along with PPO, the SPHs worked synergistically to produce the highest PO activity, although this did not replicate the higher level of PO activity found when SPHs isolated from hemolymph were added to the reaction. Similar to previous experiments, immunoblots using SPH antibodies showed high mass reactive species spread across the gel. As in Coleoptera, the mechanism by which these *M. sexta* proteins associate is unclear. Therefore, the nature and biological relevance of the apparent interactions between PO and the SPH need to be addressed as an important aspect of MC formation.

Although the experiments performed in *M. sexta* suggest the importance of SPHs in efficient PPO cleavage, of arguably greater importance is the finding that PPO-cleavage in the presence of SPHs results in the formation of high mass complexes with marked melanization activity. The appearance of these higher mass bands during separation by denaturing SDS-PAGE suggests a strong and apparently covalent association between PO and SPH. However, questions remain including how does this bond form and is it possible that such covalent cross-linking could be explained as experimental artifact? The only sources of potential cross-linking activity in these experiments are either an unknown SPH activity or an inherent activity of POs, which could oxidize backbone Tyr residues on adjacent SPHs, thus producing highly reactive protein-linked quinones subject to nucleophilic attack by nearby residues on POs (Heck et al. 2013). While not excluding the possibility that such a non-specific and potentially uninteresting PO-SPH interaction is responsible for high mass protein complex formation, these groundbreaking experiments provided the earliest glimpses into the actual fate of PO, the additional processing steps it may undergo, its conversion into higher mass bands, and the formation of an MC dependent on interactions with SPHs.

A series of experiments in *B. mori* built upon the previously described linkage between melanization and Tyr metabolism to address the unknowns regarding post-PO cascade PO processing (Clark and Strand 2013). First, the fate of PPO was followed immediately after bleeding to assess processing and conversion of POs into higher mass proteoforms. Hemolymph samples from a time course of melanizing plasma were separated by reducing SDS-PAGE and immunoblotted with α-PPO. Within minutes, PPO was converted to PO, and a ladder of high mass PO-containing bands appeared across the gel, likely indicating covalent linkages between PO and other hemolymph proteins. However, while plasma prepared in the presence of PTU showed the normal conversion of PPO to PO, it prevented the formation of the higher mass PO-containing bands. This suggests, as previously hypothesized, that the association of POs with other hemolymph proteins is catalyzed by PO activity.

Surprisingly, a contradictory result was obtained when these samples were tested for melanizing activity. Separation by gel filtration chromatography resulted in a high mass peak eluting at 670 kDa for both samples; however, when assayed for activity, the sample to which PTU had been added showed much more rapid Tyr-dependent melanin production than the sample without PTU (Note: gel filtration removes the PTU). To assess PPO processing in these samples they were again subjected to reducing SDS-PAGE. Whereas it was previously shown that PTU inhibited cross-linking between PO and other hemolymph proteins, the high mass peak eluted from gel filtration was found to contain a high mass smear below the well/gel interface that was far more intensely labeled than the sample prepared in the absence of PTU. While these counterintuitive results confirmed the existence of PO-induced cross-linking, importantly, they showed that such reactions do not promote, but instead inhibit, the inclusion of PO into the nascent MC through parasitic reactions. Moreover, they showed for the first time that complex formation is stabilized by a separate cross-linking activity in the plasma, likely through a transglutaminase (see section on coagulation). These experiments also demonstrated that the MC is the only entity capable of metabolizing the physiological substrate Tyr, and that the MC is the only form of PO capable of melanin formation.

Additional experimental confirmation of high mass MC formation was shown in Lepidoptera and Diptera using in-gel PO assays (Clark 2015; Phillips and Clark 2017). Plasma preparations from *B. mori*, *M. sexta*, and *A. gambiae* were separated by native-PAGE and then tested for DA- and Tyr-dependent activity. Both DA- and Tyr-dependent melanin formation was found exclusively at the well/gel interface in Lepidoptera. In *A. aegypti* pupae and adults Tyr-dependent melanization occurred primarily at the well/gel interface although there were several additional lower mass bands in the native gel that showed DA-dependent melanization, likely indicating the presence of other activated POs not incorporated into the MC that can only metabolize DA. This difference may be attributed to the fact that while *B. mori* and *M. sexta* only express two POs, *A. aegypti* may produce up to ten PO isoforms, many of which have not been functionally characterized. MS analysis of the *B. mori* and *A. aegypti* MCs also confirmed the presence of SPHs (Phillips and Clark 2017), cofactors of PO processing previously described as essential in Coleoptera and *M. sexta*. Many unknowns remain regarding PO processing and how it may enable PO incorporation into the high mass MC. These include the mechanism of apparent cross-linking within the MC, the role of SPHs in MC assembly, and how the MC enables Tyr metabolism. Thus, it is apparent that there is much more to be learned beyond our current view of the PO cascade.

Coagulation and Immune Complexes

Coagulation

The study of coagulation, traditionally viewed from the perspective of wound-healing, has revealed its intimate connections to innate immunity in both mammals (Delvaeye and Conway 2009) and arthropods (Iwanaga 2002; Iwanaga and Lee 2005). How clotting is initiated in insects is not well understood; however, as with innate immune responses, it likely relies on external PAMPS and/or endogenous wound-derived danger signals (Berisha et al. 2013; Krautz et al. 2014). These signals induce the formation of a large-scale, but localized, coagulant capable of sealing wounds and providing a toxic barrier to invading pathogens.

Coagulation is fundamentally different from the more discrete and self-limiting interactions found in the TEP convertase or the putative SPH/PPO melanization complex. A clot is composed of a large ensemble of proteins (and hemocytes) that form fibers and sheets of enormous molecular masses subject to extensive cross-linking, all of which make its study particularly challenging. Extensive studies in the horseshoe crab (*Limulus polyphemus* and *Tachypleus tridentatus*) and the freshwater crayfish (*Pacifastacus leniusculus*) provided our earliest understanding of coagulation and innate immunity in arthropods. In horseshoe crabs, LPS binding to hemocytes results in degranulation and the release of numerous clotting components and activating proteases. Subsequent binding of either LPS or β-glucan to specific proteases initiates the sequential activation of a protease cascade that ultimately cleaves the humoral protein coagulogen to from coagulin, which forms a gel that is stabilized through cross-linking by the enzyme transglutaminase (TG). In the crayfish, large clotting protein dimers polymerize in the presence of Ca^{2+}, while TG released from hemocytes results in their cross-linking and stabilization (Theopold et al. 2004; Cerenius and Söderhäll 2010).

Early studies of wound healing in insects were limited primarily to microscopic observations of morphology (Wigglesworth 1937) and the identification of abundant coagulants such as lipophorin (Lp) in the cockroach (Barwig 1985) and Lp, PPO, and a hexamer in in *Galleria mellonella* (Li et al. 2002). The experimental challenges of small organism size, as well as limited materials and methods to isolate and identify high mass protein aggregates were ultimately overcome through the power and sensitivity of modern mass spectrometry and the availability of sequenced insect genomes, assisted by many other tools first developed in *D. melanogaster* (Troha and Buchon 2019). These breakthroughs have provided a more modern and detailed molecular view of coagulation and its integration with immune complex (IC) formation (Fig. 5.3).

Early experiments in *D. melanogaster* demonstrated that hemolymph isolated from larvae rapidly develops extensive clot-associated fibers that, through peanut agglutinin (PNA) labeling, were shown to be heavily glycosylated (Scherfer et al. 2004). The fibers were also found to trap both gram-positive and negative bacteria as well as paramagnetic beads. Bead-bound proteins were magnetically isolated,

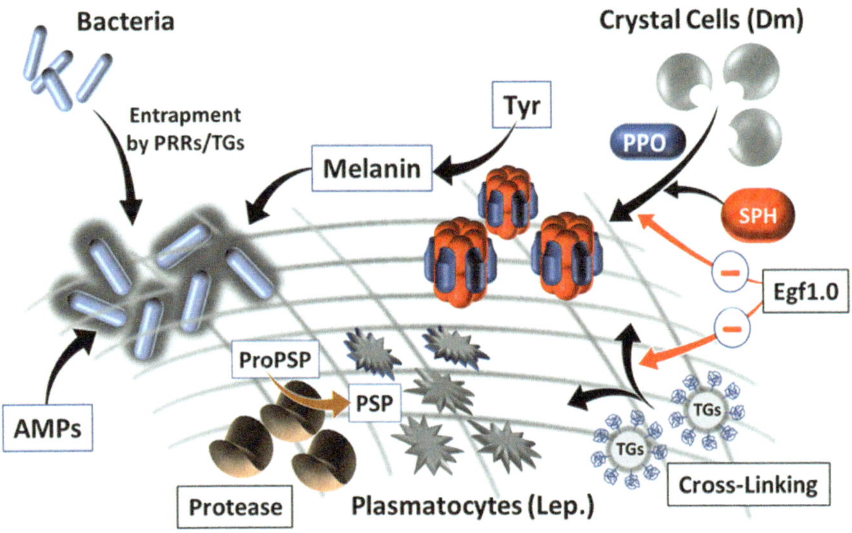

Fig. 5.3 Generalized model of Coagulation and Immune Complex formation. The physical structure of the clot in *D. melanogaster* is comprised of an extensive matrix of fibers (shown as a gray lattice). Studies in multiple insect species have shown that clotting and fiber formation involves several common components, including lipophorins, hemolectin (hemocytin), hexamerins, and gelsolin, although some may be species-specific, such as fondue in *D. melanogaster*. Other immune effectors integrated into the clot serve a variety of functions. Multiple PRRs found in the IC and in clots likely enable microbial entrapment or may induce association of the forming IC/clot with larger targets such as parasitoids in Lepidoptera. The association of PO with coagulants has been observed in *D. melanogaster*, *B. mori*, and *A. aegypti*, where it can function to kill trapped pathogens through production of cytotoxic intermediates during melanization. In *D. melanogaster*, POs are released from crystal cells (oenocytoids in Lepidoptera). Both POs and SPHs were found in the *B. mori* and *A. aegypti* ICs, likely forming the MC proposed in Fig. 5.2. Whether the MC forms in the hemolymph and then associates with the clot or whether the components assemble on the clot is unknown. Hemolymph TGs are released from hemocytes and are bound to vesicles in the hemolymph. Numerous TG targets have been identified in both *D. melanogaster* and *B. mori*, including clotting proteins, POs, and bacteria. The virally derived protein Egf1.0 inhibits PO production and appears to inhibit TG activity. Hemocytes are commonly associated with the clot and other immune processes such as encapsulation and nodulation, where they form large aggregates. The *B. mori* IC was shown to contain proteolytic activity capable of producing the cytokine PSP, which induces the localized spreading and aggregation of plasmatocytes in Lepidoptera. Interestingly, both the *B. mori* and *A. aegypti* ICs were found to contain previously uncharacterized SPs that may be responsible for PSP-like cytokine production

excised, separated by SDS-PAGE, and subjected to identification by MS. Major proteins were determined to be hemolectin, lipophorin, the extracellular matrix protein tiggrin, and, using enhanced methods, a larval serum protein (hexamerin) and a PPO, thus linking melanization to clotting. Hemolectin (Hml) was shown to be an important target of PNA, and an *hml* KD model did not form clotting fibers or effectively trap bacteria, indicating a key role of this molecule in the clotting process. In separate studies, KD of *hml* was also shown to inhibit coagulation, although it did not affect

PO activity or AMP production, indicating independent processes (Goto et al. 2003). In contrast, survival after wounding during the larval feeding stage was found to be significantly reduced in the *hml* KD (Chang et al. 2012).

Hemocytin was the first *B. mori* ortholog of Hml to be cloned and isolated due to its hemagglutinating activity against red blood cells. Further, its expression was shown to be induced by either *E. coli* or LPS (Mori et al. 1992; Kotani et al. 1995). *D. melanogaster* Hml, first isolated in 2001, is a large 400 kDa protein containing multiple functionally significant regions, including discoidin domains and several von Willebrand factor (VWF) domains important in mammalian blood clotting (Goto et al. 2001). Lesch et al. further characterized the role of Hml in *D. melanogaster* and proposed that the discoidin domains likely act as lectins, which may account for the agglutinating activity of this molecule (Lesch et al. 2007).

Lipophorins

The structure of insect lipophorin (Lp) particles and their physiological role in lipid transport have been studied extensively, primarily in Lepidoptera (Shapiro et al. 1984; Jouni et al. 2003). Lp particles, also referred to as lipid transfer particles in *B. mori*, are comprised of three subunits, Apo-I, II and III, that together produce an intact complex with an approximate mass of 620 kDa (Yokoyama et al. 2013). The dual role of Lp in lipid transport and immunity has been documented extensively. An Lp-LPS complex was shown to sequester and diminish immune activation in *B. mori* (Kato et al. 1994), and Lp particles from the desert locust *Schistocerca gregaria* have been shown to bind to dsRNA and to aggregate pathogens such as Gram-positive bacteria and fungi (Wynant et al. 2014). Lp particles isolated from the flour moth *Ephestia kuehniella* can aggregate bacteria and contain active PO that produces melanin in the presence of DOPA (Rahman et al. 2006). Likewise, an association between Lp and PO was found in the larval clot from *A. gambiae* hemolymph (Agianian et al. 2007). Lp particles may also be directly involved in clot-mediated sequestration of bacteria. Further, Schmidt and colleagues describe hypothetical scenarios whereby Lp particles bind to pathogen-derived lipopolysaccharide (LPS) or glycans and then aggregate in the presence of hemolin, a hemolymph protein that is a member of the immunoglobulin superfamily (Su et al. 1998; Schmidt et al. 2010).

PO Involvement in Clot Formation

PPOs, unlike most other hemolymph immune components, are sequestered in hemocytes prior to immune activation. Their release and recruitment to the nascent clot serves both immune and structural functions. In *D. melanogaster*, 2D PAGE and MS identification showed that clotting led to a depletion of PO in the non-clotted serum, indicating a large amount of PO association with the clotted proteins (Engström et al.

2004). Interestingly, the release and activation of POs that become associated with the clot can be induced by apoptotic hemocytes and phosphatidyl serine (PS), suggesting an alternative pathway that involves wounding rather than microbial activation (Bidla et al. 2007, 2009). The role of PO in the clotting process of *D. melanogaster* was investigated in the Bc (Black cell) mutant (Bidla et al. 2005). These flies, defective in the ability to release PO from crystal cells, still form a soft clot yet are incapable of clot-associated melanin production, showing that POs are not directly involved in clot assembly. In a later study, soft clot formation was shown to be stabilized by TG-induced cross-linking, which also covalently immobilized pathogens to the clot (Wang et al. 2010). Further hardening of the clot is subsequently catalyzed through PO-mediated secondary cross-linking.

Other Clot Proteins

In addition to Hml and Lp, other important clot components have been identified in insects, including the *D. melanogaster*-specific protein Fondue. A GFP-Fondue construct was found to label both the clot and the cuticle, and KD of *fondue* affected clot integrity (Lindgren et al. 2008). A more ubiquitous clotting component not lacking in insect orthologs is hexamerins, a large group of multifunctional proteins belonging to the arthropod hemocyanin superfamily (Telfer and Kunkel 1991; Burmester 2001; Hughes 1999). Hexamerins are non-enzymatic proteins traditionally studied for their role as storage proteins, which function as amino acid and energy repositories during periods of non-feeding. Closely related to the enzymatically active PPO family, hexamerins are named for their propensity to form hexamers (Pietrzyk et al. 2013). Further, hexamerins have appeared in multiple immune-related roles including clotting, and their levels are significantly affected by immune challenges and parasitism (Beckage and Kanost 1993; Hayakawa 1994; Shelby and Webb 1997; Morais Guedes et al. 2005). Despite this, little is known regarding any specific function they play in the immune system.

Cross-Linking by Transglutaminase (TG)

The study of coagulation is replete with an extensive list of interacting proteins, some of which are specific to a particular genus, but one clotting process that appears to be ubiquitous in both insects and mammals is that of covalent cross-linking by the Ca^{2+}-dependent enzyme transglutaminase (TG). TGs catalyze cross-linking through a two-step process that requires the formation of a thioester intermediate between a Cys thiol in TG and a Gln side chain in a target protein. The highly reactive thioester is then subject to attack from a primary amine, such as Lys, within a secondary target protein, which cross-links the two target proteins and frees the TG for more catalysis (Pedersen et al. 1994). Clotting reactions involving TGs have been intensively studied in both mammals (Muszbek et al. 2011) and in arthropods (Tokunaga et al. 1993; Iijima et al. 2005).

Early studies in *D. melanogaster* showed that major clotting proteins, including Lp and a hexamerin, are targets of TG. This was evidenced by labeling with the artificial TG substrate biotin-cadaverine (B-cad), which supplies a primary amine that attacks the thioester intermediate and a biotin moiety for labeling (Li et al. 2002). Additional experiments using B-cad confirmed Lp labeling and identified gelsolin and fondue as TG targets (Karlsson et al. 2004; Lindgren et al. 2008). Moreover, Ca^{2+} chelation and the thiol blocking agent iodoacetamide also inhibited clotting, strongly suggesting an important role for TG (Bohn and Barwig 1984; Li et al. 2002; Scherfer et al. 2004). Further expanding its role, TG was found to label microbial surfaces with B-cad, while KD of *TG* led to serious immune defects against the pathogenic nematode (*Heterorhabditis bacteriophora*) and its symbiotic bacteria (*Photorhabdus luminescens*), resulting in increased mortality (Wang et al. 2010). In *Anopheles*, KD of *TG* and the use of a TG inhibitor both resulted in increased *Plasmodium berghei* infection, indicating the importance of cross-linking in mosquito immunity (Silveira et al. 2012).

Immune Complexes in Lepidoptera and Mosquitos

The formation of high mass immune complexes (ICs) comprised of both coagulants and immune effectors occurs rapidly, leading to large changes in the plasma proteome from post-translational processing, protein interactions, and cross-linking. In *M. sexta*, He et al. used SDS-PAGE and MS analysis to observe overall changes in hemolymph protein abundance and mobility before and after bacterial immune challenge (He et al. 2016). More than 130 altered protein migrations were observed, some of which were expected protein processing events such as proteolytic activation of PO cascade components, as well as inactivation of numerous proteases by specific serpins. Of greatest interest was the large number of proteins with mobility shifts into the 80 to >400 kDa range that withstood denaturation and reduction. Many of these proteins, including SPHs and PPOs, possessed immune functions previously discussed in relationship to high mass MC formation. Surprisingly, 253 of the 654 identified proteins do not contain signal sequences for secretion, including PPO1 and PPO2. Although oenocytoids from Lepidoptera and crystal cells in *D. melanogaster* are known sources of PPOs, the mechanism of their release has been a long-standing mystery. Recent studies in *D. melanogaster* revealed that a GFP-PPO2 fusion protein is cytosolic when first expressed but ultimately forms membrane-free crystalline inclusions. Crystal cell rupture upon immune stimulation causes the rapid release and dissolution of the PPO2 crystals into the hemolymph, after which the now soluble PPO undergoes processing through the PO cascade (Schmid et al. 2019). TG also lacks a signal sequence, making its secretion equally mysterious. However, Shibata et al. demonstrated that TG-A, one of two TGs expressed as alternatively spliced isoforms in *D. melanogaster*, undergoes an unusual post-translational N-myristoylation and S-palmitoylation that targets it to multivesicular bodies for release by exosomes (Shibata et al. 2017). Moreover, Schmid and colleagues confirmed that this non-classical TG-A secretion through exosomes occurs in immune-stimulated plasmatocytes, and that TG-A remains bound to vesicular membranes or is associated with the clot.

Experiments targeting IC formation in *B. mori* specifically analyzed the high mass protein complex trapped at the well/gel interface during native-PAGE separations of hemolymph (Phillips and Clark 2017). These studies revealed that IC formation is not dependent on immune induction other than wounding and is the only entity in the gel capable of metabolizing the physiological substrate Tyr to form melanin. MS analysis of the IC demonstrated the presence of many known immune and clot-related proteins and identified several new candidates with possible roles in immune function. Some of these were previously identified in *D. melanogaster* as clot-associated, including ApoLp I and II, hemocytin (hemolectin), two hexamerins, and PPO1/2. Unlike the *D. melanogaster* clot, SPH1 and SPH2, previously discussed for their ability to form high mass multimers during PO activation, were also present in the IC, as was the Lp particle component ApoLp III, which is known to possess immune functions (Whitten et al. 2004). The IC also contained numerous PRRs, including C-type lectins and several members of the large 30 kDa lipoprotein family, most of which do not have clearly defined functions (Zhang et al. 2012). Surprisingly, the IC also contained a protease (*B. mori* SP3) that is an ortholog of the HP1 protease in *M. sexta*, although no clear functions have been ascribed to either (Yang et al. 2016). Experiments using both B-cad and the novel, high efficiency substrate KGC-B showed that TG labels a large array of IC proteins during assembly, including coagulants such as Lp, hemocytin, and hexamerins, as well as PPOs and 30 kDa lipoproteins (Phillips and Clark 2017).

Inhibition of IC Formation by EGF1.0

The analysis of *B. mori* IC proteins also assessed the effects of an important melanization inhibitor derived from the endoparasitoid wasp, *Microplitis demolitor*. *M. demolitor* propagates on its lepidopteran hosts by egg oviposition into the host hemocoel but must avoid host immune system detection to survive. It accomplishes this by co-injecting a polydnavirus that infects host tissues, resulting in the expression of numerous viral gene products that affect host functions (Strand and Burke 2012). These include two unique proteins, Egf1.0 and Egf1.5, that are potent inhibitors of PO activation.

The Egf proteins contain two functional domains: the N-terminal catalytic domain (CD), which contains a TIL-like inhibitor capable of inhibiting SPs within the PO cascade, and the C-terminal repeat domain (RD), which consists of short repeats, the number of which differs between Egf1.0 and Egf1.5 (Beck and Strand 2007; Lu et al. 2008, 2010) (Fig. 5.4a). The CD was considered to be the primary source of inhibition due to its TIL-like structure; however, numerous experiments using

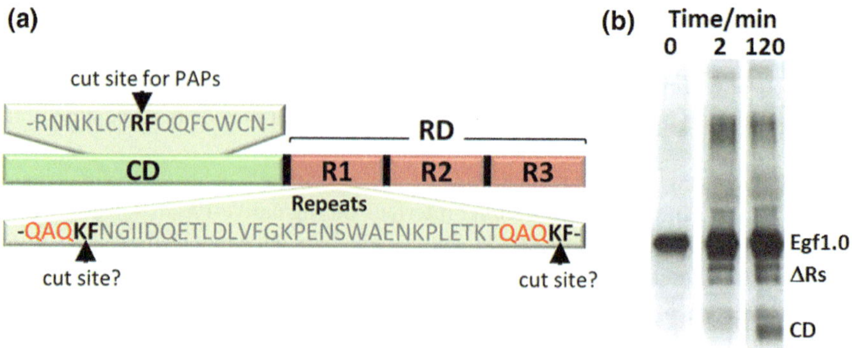

Fig. 5.4 Egf1.0 structure and processing. **a** The structure of Egf1.0 is comprised of a catalytic domain (**CD**) and a repeat domain (**RD**). The CD contains a TIL-like region capable of inhibiting activated PAPs from the PO cascade. PAPs likely target the **RF** cut site, which is equivalent to the RF cut site in POs, both of which are found at position R51. The RD contains three repeats, each defined by a potential cleavage site (**KF**) for a trypsin-like protease. The RD alone, minus the CD, is a potent inhibitor of melanization, and Egf1.0 appears to inhibit TG activity, but the mechanisms are unknown. Each repeat contains a conserved **QAQ** site that could be a target of TGs. **b** Egf1.0 processing in (**b**). *mori* hemolymph assessed by reducing SDS-PAGE and immunoblotting with α-Egf1.0. Within two minutes of Egf1.0 addition three lower mass bands form, which suggests a sequential cleavage of the repeats, and within 120 min the N-terminal region of the CD appears, likely due to cleavage at the RF cut site. In addition, several higher mass bands rapidly appear, indicating covalent attachment to hemolymph proteins

truncated constructs have demonstrated that the RD, which has no obvious homology to any other known protein, more potently inhibits melanization than the CD domain, leaving the actual inhibitory mechanism an open question.

Analysis of the *B. mori* IC in the presence and absence of Egf1.0 showed dramatic structural and functional differences. Egf1.0 inhibited the inclusion of PPO1, PPO2, and SPH1, but surprisingly, SPH2 remained in the IC. Likewise, only one of the two storage proteins (hexamerins) was missing, as were two of the 30 kDa lipoproteins. Egf1.0 also inhibited the incorporation of the SP3 protease, providing additional evidence regarding its relevance as a key, but unknown, aspect of the immune response. Further, Egf1.0 itself undergoes multiple processing events that produce both lower and higher mass bands (Fig. 5.4b), and was found in the IC, suggesting its association with one or more of the IC proteins (Phillips and Clark 2017). Additional experiments to assess the role of TG in IC formation showed that B-cad labeled many IC proteins and surprisingly, in addition to its capability as a protease inhibitor, Egf1.0 effectively inhibited B-cad labeling in both *B. mori* and *M. sexta*. Thus, this added yet more mystery to Egf1.0's mode of action.

Plasmatocyte Spreading Peptide

Although not a prime subject of this chapter, it is pertinent to mention that the prominent contributions of hemocytes to both coagulation and melanization and their involvement in other defensive reactions such as nodulation and encapsulation have been studied extensively (Strand 2008; Kim-Jo et al. 2019). In lepidoptera, plasmatocytes are a specialized hemocyte subtype that migrate toward and spread on pathogen surfaces, ultimately encapsulating them. The cytokine plasmatocyte spreading peptide (PSP), originally isolated from *Pseudoplusia includens* (now *Chrysodeixis includens*), was purified based on its potent ability to stimulate spreading, migration, and aggregation (Clark et al. 1997; Wang et al. 1999; Kamimura et al. 2001). However, the protease that cleaves the 23 amino acid peptide from its larger proprotein (Pro-PSP), and the mechanism by which its production is localized to a pathogen surface to avoid systemic plasmatocyte activation are unknown. Previous failed attempts to identify the activating protease indicated its presence in a high mass complex, suggesting that the IC itself may contain the activity. Addition of pro-PSP to the excised gel band containing the IC led to pro-PSP cleavage and production of the mass-correct PSP peptide. Thus, for the first time this linked the production of PSP to the clot- or pathogen-bound IC and provided a putative function for the elusive SP3 protease.

Discussion

Here we have described evidence demonstrating the importance of protein complexes in three key insect immune pathways. Large scale protein interactions are inherent to coagulation whereas smaller, more discrete interactions provide the functionality inherent in the mosquito TEP convertase as well as the MC found in Lepidoptera. Even though each process may be activated through separate pathways, downstream events join them into multifunctional units designed to destroy their targets. For example, the construction of the *A. aegypti* TEP convertase on the surface of an ookinete eventually recruits effectors to induce lysis and melanin production, with melanization likely carried out by a mosquito MC that resembles that found in Lepidoptera. Interestingly, the CLIPA2 SPH from *A. gambiae*, which negatively regulates TEP1 convertase-mediated ookinete killing, co-precipitates with ApoLpI/II in mosquitos infected with *Beauveria bassiana*, suggesting a link between the convertase and clot-associated proteins (Kamareddine et al. 2016). Likewise, in *D. melanogaster* the isolation of clotting proteins reveals the almost ubiquitous presence of POs, again indicating they are recruited to the forming clot. In the *B. mori* IC, all the components that appear to be necessary for clotting are present, including ApoLp proteins, hemocytin and hexamerins, but are consistently found with essential components for MC formation, including PPO1/2 and SPH1/2. In the *A. aegypti IC*, proteins involved in coagulation are present, such as ApoLp, hexamerins, and gelsolin, but are associated with components that comprise a putative mosquito MC, including PPO1/3, multiple SPHs, and several LRIM proteins (Phillips and Clark 2017). In the horseshoe crab, *Tachypleus tridentatus*, the enzyme responsible for inducing clot formation is the same as that which activates PO activity in the hemolymph, providing a link between these pathways not yet identified in insects (Nagai and Kawabata 2000). Further links between the immune response mechanisms that connect these vital pathways have yet to be elucidated.

TEP Convertase

The decision to include a discussion of the TEP1 convertase in this chapter was based on its importance in pathogen recognition, the large repertoire of protein interactions required for its function, its ability to activate downstream effectors, and the fact that no such comparably multifaceted protein network devoted to pattern recognition has previously been described in insects.

In contrast to the protein complex comprising the MC, which is presumably static once formed, the complexes involved in the formation and regulation of the TEP1 convertase are surprisingly dynamic. First, the nascent $TEP1_{cut}$, with its now-exposed thioester, must be stabilized and protected by the two LRR proteins, LRIM1 and APL1C, until delivered to the desired target. This likely depends on modification of the ookinete surface by midgut epithelial nitration (Oliveira et al. 2012). Upon arrival, the LRR proteins dissociate and $TEP1_{cut}$ undergoes nucleophilic attack by primary amines on the pathogen surface, resulting in covalent linkage. $TEP1_{cut}$ then recruits additional TEP1 that is subsequently converted to $TEP1_{cut}$ by an unknown protease to form a surface-associated protein complex comprising the mature TEP1 convertase. The amount of regulatory machinery that interacts with the convertase is exceptional, as is the fact that the regulation is apparently performed exclusively by SPHs. While the exact functional roles each play is unclear, phenotypically the SPCLIP1 acts to positively regulate the convertase, while CLIPA2 and CLIP8 act as negative regulators. Finally, the SPH CLIPA8, again through interactions with the convertase, elicits melanization on the pathogen surface through an unknown mechanism.

The protein resources devoted to assembling and regulating the TEP1 convertase are substantial, and more discoveries are likely to come, but the insights into pattern recognition they provide are profound. They also highlight how little is known about their function in model Lepidopterans such as *B. mori* and *M. sexta*, which have, in contrast, provided such insight into the PO cascade. A recent study of pattern recognition receptors (PRRs) in *M. sexta* identified 76 LRR proteins and 3 TEPs (although only one contains the GCGEQ motif) (Zhang et al. 2015). It will be interesting to see if some mimic the roles played by LRIM1/APL1C in forming the mosquito TEP1 convertase, and whether such a Lepidopteran convertase can activate melanization or other effectors.

The MC and SPHs

Substantial evidence shows that POs undergo additional processing beyond that imposed by the PO-cascade and are incorporated into a high mass structure containing SPHs and other hemolymph proteins. In addition, TG-induced cross-linking may structurally stabilize the mature MC and link it to coagulants. These results generate several important questions regarding assembly of the MC, the mechanism by which it enables Tyr metabolism, and how the MC is targeted to specific sites of wounding or pathogen invasion.

The ability to metabolize Tyr appears to be a key characteristic that distinguishes the PO-containing MC from monomeric or dimeric POs found in the hemolymph, although the mechanism remains unclear. One possibility is that association with SPHs or other components alters the PO conformation such that Tyr metabolism is enabled. However, the association of multiple POs into an organized structure may also solve a problem discussed earlier, whereby the MC structure effectively traps the catalytic amounts of DOPA needed to efficiently metabolize Tyr by preventing its diffusion into the hemolymph (Peñalver et al. 2005). It is tempting to speculate that the addition of SPHs along with PPOs during processing is sufficient to provide for Tyr metabolism. This is suggested, for example, by the massive increase in DA-dependent activity described in *M. sexta* (Lu and Jiang 2008). Unfortunately, Tyr metabolism has not been measured in experiments where PO is activated in the presence of SPHs. Thus, further experiments to assess the necessary components and structural requirements for Tyr metabolism are needed.

SPHs play a key role in forming the high mass MC. An SPH from *H. diomphalia* (PPAF-II) was shown to form a homo-oligomer upon proteolytic activation that also associates with POs to form a high mass hetero-oligomer through interactions with the central cleft of the clip domain (Piao et al. 2005). However, Wang et al. characterized multiple proteolytic processing events involved in SPH activation in *M. sex*ta and showed that the presence of both SPH1 and SPH2 are required to provide PO activation (Wang et al. 2014). The mechanism by which these closely related SPHs function synergistically is unknown, but the contrasting results in *M. sexta* and *H. diomphalia* regarding homo- versus hetero-oligomers suggests that SPH assemblies may be comprised of single or multiple SPHs, and that their composition could be used to fine tune their regulatory roles. Given the role of SPHs in efficient PO processing and assembly into higher order structures in Lepidoptera, it will be fascinating to eventually understand how CLIPA8, the *A. aegypti* SPH that links the TEP convertase to melanization, either activates POs or recruits an already formed PO-containing MC to the surface of an ookinete. However, no evidence to date suggests that mosquito SPHs form higher mass multimers.

Outside of these roles, SPHs have been shown to possess a broad range of functions. In *M. sexta*, Felföldi et al. identified a novel SPH, SPH-3, whose KD prevented the transcription of PPO and other antimicrobial effectors (Felföldi et al. 2011). In *B. mori*, SPH-1 has been shown to be involved in *E. coli*-dependent nodule formation (Sakamoto et al. 2011) and in tiger shrimp (*Penaeus monodon*) a clip-domain

SPH was revealed to promote hemocyte adhesion (Lin et al. 2006). SPHs have even been employed by parasitoids as part of their extensive arsenal used to evade various aspects of the host immune response; venom from the Braconid wasp *Cotesia rubecula* was found to contain an SPH (Vn50) that, when injected into its host, *Pieris rapae*, inhibited the activation of PPO (Zhang et al. 2004; Asgari 2006). Although the mechanism is unknown, it is tempting to hypothesize that the parasitoid SPH can displace the host SPH to prevent proper PO processing and/or MC formation. In summary, SPHs perform multiple essential regulatory roles in pathogen recognition and MC formation, yet lack a clearly defined function, making this group one of the most enigmatic in the insect immune system.

PO/Hexamerins

It is important to reiterate that POs functionally act as Tyrosinases, but structurally belong to the arthropod hexamerin family. Although there is no evidence that insect POs associate into hexamers, they may have a structural propensity to do so as seen with other members of this family (Telfer and Kunkel 1991; Pietrzyk et al. 2013). In an early study of monophenol oxidase activity in potato juice, gel chromatography indicated various association states ranging from 36 to 800 kDa (Matheis and Belitz 1979). Melanoma cells also show high mass forms of tyrosinase that associate into what the authors presciently refer to as a *Melanogenic Complex* (Orlow et al. 1994). In addition, numerous studies in Crustaceans indicate that POs form hexamers (Jaenicke and Decker 2003; Brack et al. 2008). Thus, two major components of the MC are prone to forming higher order structures that may contribute to their ability to metabolize Tyr, although some mechanism must also be present to drive targeting of the MC to the required sites. Whether POs and or SPHs alone are sufficient or whether other components present in the IC are responsible is an open question. Many of these questions, along with the role of POs in tissues other than hemolymph, have been discussed (Whitten and Coates 2017).

IC Formation and Coagulation

The related studies of coagulation in *D. melanogaster* and IC formation in Lepidoptera and *A. aegypti* have together identified the presence of numerous clotting proteins, PRRs, and immune effectors such as the PO/SPH-containing MC. However, it remains unclear whether all components are integral to clot formation, or whether some associate after clot formation. It is tempting to hypothesize that a core set of clotting proteins forms the underlying structure, which then provides a scaffold for the association of other immune factors, the composition of which may depend on the immune stimulus to which the host is subjected. For example, multiple POs, SPHs, hexamerins, 30 kDa lipoproteins (in Lepidoptera), and LRR-proteins (in mosquitos) may be present in hemolymph, but to date only certain subsets of these proteins have been found incorporated into the IC.

We must also consider the role of the hemolymph TG in the process. Results in *B. mori* show clot-associated effectors, including PO and SP3, are targets of TG-induced cross-linking, suggesting that they may require TG to provide a covalent linkage to the clot/IC (Phillips and Clark 2017). To date no function has been ascribed to SP3, but studies of its ortholog in *M. sexta* (HP1) demonstrate an unusual non-proteolytic activation mechanism that may occur through conformational change (He et al. 2017). This raises the question of whether association with the IC induces the conformational change in SP3 needed for its activation. The presence of SP3 also raises more general questions of whether functionally related proteases might be found in immune complexes in other insect orders, and whether they may similarly recruit hemocytes by cleaving a functionally related ortholog of PSP. In fact, a protease of unknown function was found in the *A. aegypti* IC (Phillips and Clark 2017). In addition, a Dipteran homolog of PSP, presumably functionally related, has been isolated from *D. melanogaster* (Tsuzuki et al. 2012) and other PSP-related peptides have been identified in mosquitos, thus providing the framework for testing whether similar cytokine activation occurs in these species (Matsumoto et al. 2012).

Whether immune challenge by various pathogens can alter the composition of the clot in a pathogen-specific manner and whether TG activity is ultimately responsible for linking effectors to the clot are questions amenable to experimentation. It is also important to note that pathogens have evolved numerous countermeasures to these responses. The *M. demolitor* Bracovirus-derived protein Egf1.0 has long been known to block melanization; however, as discussed above, it not only inhibits MC formation but also TG activity. While much remains to be learned about the mechanism of function of Egf1.0, if the importance of an immune response can be judged by whether a pathogen has evolved a response to it, MC formation and TG cross-linking activity must play key roles in the immune process.

References

Agianian B, Lesch C, Loseva O, Dushay MS (2007) Preliminary characterization of hemolymph coagulation in Anopheles gambiae larvae. Dev Comp Immunol 31(9):879–888. https://doi.org/10.1016/j.dci.2006.12.006

An C, Budd A, Kanost M, Michel K (2011) Characterization of a regulatory unit that controls melanization and affects longevity of mosquitoes. Cell Mol Life Sci 68(11):1929–1939. https://doi.org/10.1007/s00018-010-0543-z

Asgari S (2006) Venom proteins from polydnavirus-producing endoparasitoids: their role in host-parasite interactions. Arch Insect Biochem Physiol 61(3):146–156

Aspán A, Huang TS, Cerenius L, Söderhäll K (1995) cDNA cloning of prophenoloxidase from the freshwater crayfish Pacifastacus leniusculus and its activation. Proc Natl Acad Sci USA 92(4):939–943

Barwig B (1985) Isolation and characterization of plasma coagulogen (PC) of the cockroach Leucophaea maderae (Blattaria). J Comp Physiol B 155(2):135–143. https://doi.org/10.1007/bf00685206

Baxter RHG, Steinert S, Chelliah Y, Volohonsky G, Levashina EA, Deisenhofer J (2010) A heterodimeric complex of the LRR proteins LRIM1 and APL1C regulates complement-like immunity in Anopheles gambiae. Proc Natl Acad Sci USA 107(39):16817. https://doi.org/10.1073/pnas.1010575107

Beck MH, Strand MR (2007) A novel polydnavirus protein inhibits the insect prophenoloxidase activation pathway. Proc Natl Acad Sci USA 104(49):19267–19272. https://doi.org/10.1073/pnas.0708056104

Beckage NE, Kanost MR (1993) Effects of parasitism by the braconid wasp Cotesia congregata on host hemolymph proteins of the tobacco hornworm. Manduca sexta. Insect Biochem Mol Biol 23(5):643–653. https://doi.org/10.1016/0965-1748(93)90038-T

Bella J, Hindle KL, McEwan PA, Lovell SC (2008) The leucine-rich repeat structure. Cell Mol Life Sci 65(15):2307–2333. https://doi.org/10.1007/s00018-008-8019-0

Berisha A, Mukherjee K, Vilcinskas A, Spengler B, Römpp A (2013) High-resolution mass spectrometry driven discovery of peptidic danger signals in insect immunity. PLoS ONE 8(11):e80406. https://doi.org/10.1371/journal.pone.0080406

Bidla G, Dushay MS, Theopold U (2007) Crystal cell rupture after injury in Drosophila requires the JNK pathway, small GTPases and the TNF homolog Eiger. J Cell Sci 120(7):1209–1215. https://doi.org/10.1242/jcs.03420

Bidla G, Hauling T, Dushay MS, Theopold U (2009) Activation of Insect phenoloxidase after injury: endogenous versus foreign elicitors. J Innate Immun 1(4):301–308

Bidla G, Lindgren M, Theopold U, Dushay MS (2005) Hemolymph coagulation and phenoloxidase in Drosophila larvae. Dev Comp Immunol 29(8):669–679. https://doi.org/10.1016/j.dci.2004.11.007

Blandin S, Levashina EA (2004) Thioester-containing proteins and insect immunity. Mol Immunol 40(12):903–908

Blandin S, Shiao S-H, Moita LF, Janse CJ, Waters AP, Kafatos FC, Levashina EA (2004) Complement-like protein TEP1 Is a determinant of vectorial capacity in the malaria vector Anopheles gambiae. Cell 116(5):661–670. https://doi.org/10.1016/S0092-8674(04)00173-4

Bohn H, Barwig B (1984) Hemolymph clotting in the cockroach Leucophaea maderae (Blattaria). J Comp Physiol B 154(5):457–467. https://doi.org/10.1007/bf02515150

Brack A, Hellmann N, Decker H (2008) Kinetic properties of hexameric tyrosinase from the crustacean palinurus elephas. Photochem Photobiol 84(3):692–699. https://doi.org/10.1111/j.1751-1097.2008.00349.x

Burmester T (2001) Molecular evolution of the arthropod hemocyanin superfamily. Mol Biol Evol 18(2):184–195

Cerenius L, Söderhäll K (2010) Coagulation in invertebrates. J Innate Immun 3(1):3–8

Chang H-J, Dhanasingh I, Gou X, Rice AM, Dushay MS (2012) Loss of Hemolectin reduces the survival of Drosophila larvae after wounding. Dev Comp Immunol 36(2):274–278. https://doi.org/10.1016/j.dci.2011.04.009

Clark KD (2015) Altered tyrosine metabolism and melanization complex formation underlie the developmental regulation of melanization in Manduca sexta. Insect Biochem Mol Biol 58:66–75. https://doi.org/10.1016/j.ibmb.2015.01.002

Clark KD, Kim Y, Strand MR (2005) Plasmatocyte sensitivity to plasmatocyte spreading peptide (PSP) fluctuates with the larval molting cycle. J Insect Physiol 51(5):587–596. https://doi.org/10.1016/j.jinsphys.2005.03.002

Clark KD, Lu Z, Strand MR (2010) Regulation of melanization by glutathione in the moth Pseudoplusia includens. Insect Biochem Mol Biol 40(6):460–467

Clark KD, Pech LL, Strand MR (1997) Isolation and identification of a plasmatocyte-spreading peptide from the hemolymph of the lepidopteran insect pseudoplusia includens. J Biol Chem 272(37):23440–23447. https://doi.org/10.1074/jbc.272.37.23440

Clark KD, Strand MR (2013) Hemolymph melanization in the silkmoth bombyx mori involves formation of a high molecular mass complex that metabolizes tyrosine. J Biol Chem 288(20):14476–14487. https://doi.org/10.1074/jbc.M113.459222

Decker H, Schweikardt T, Nillius D, Salzbrunn U, Jaenicke E, Tuczek F (2007) Similar enzyme activation and catalysis in hemocyanins and tyrosinases. Gene 398(1):183–191. https://doi.org/10.1016/j.gene.2007.02.051

Delvaeye M, Conway EM (2009) Coagulation and innate immune responses: can we view them separately? Blood 114(12):2367–2374. https://doi.org/10.1182/blood-2009-05-199208

Dodds AW, Law SKA (1998) The phylogeny and evolution of the thioester bond-containing proteins C3, C4 and and alpha-2-macroglobulin. Immunol Rev 166(1):15–26

Dong Y, Aguilar R, Xi Z, Warr E, Mongin E, Dimopoulos G (2006) Anopheles gambiae immune responses to human and rodent plasmodium parasite species. PLoS Pathog 2(6):e52. https://doi.org/10.1371/journal.ppat.0020052

Dudzic JP, Kondo S, Ueda R, Bergman CM, Lemaitre B (2015) Drosophila innate immunity: regional and functional specialization of prophenoloxidases. BMC Biol 13(1):81. https://doi.org/10.1186/s12915-015-0193-6

Dziedziech A, Schmid M, Arefin B, Kienzle T, Krautz R, Theopold U (2019) Data on Drosophila clots and hemocyte morphologies using GFP-tagged secretory proteins: Prophenoloxidase and transglutaminase. Data in Brief 25:104229. https://doi.org/10.1016/j.dib.2019.104229

Engström Y, Loseva O, Theopold U (2004) Proteomics of the drosophila immune response. Trends Biotechnol 22(11):600–605. https://doi.org/10.1016/j.tibtech.2004.09.002

Felföldi G, Eleftherianos I, ffrench-Constant RH, Venekei I (2011) A Serine Proteinase Homologue, SPH-3, Plays a Central Role in Insect Immunity. J Immunol 186 (8):4828–4834. https://doi.org/10.4049/jimmunol.1003246

Fraiture M, Baxter RHG, Steinert S, Chelliah Y, Frolet C, Quispe-Tintaya W, Hoffmann JA, Blandin SA, Levashina EA (2009) Two mosquito LRR proteins function as complement control factors in the TEP1-mediated killing of plasmodium. Cell Host Microbe 5(3):273–284. https://doi.org/10.1016/j.chom.2009.01.005

Frolet C, Thoma M, Blandin S, Hoffmann JA, Levashina EA (2006) Boosting NF-kB-dependent basal immunity of Anopheles gambiae aborts development of plasmodium berghei. Immunity 25(4):677–685. https://doi.org/10.1016/j.immuni.2006.08.019

Goto A, Kadowaki T, Kitagawa Y (2003) Drosophila hemolectin gene is expressed in embryonic and larval hemocytes and its knock down causes bleeding defects. Dev Biol 264(2):582–591. https://doi.org/10.1016/j.ydbio.2003.06.001

Goto A, Kumagai T, Kumagai C, Hirose J, Narita H, Mori H, Kadowaki T, Beck K, Kitagawa Y (2001) A drosophila haemocyte-specific protein, hemolectin, similar to human von willebrand factor. Biochem J 359(1):99–108. https://doi.org/10.1042/bj3590099

Gupta S, Wang Y, Jiang H (2005) Manduca sexta prophenoloxidase (proPO) activation requires proPO-activating proteinase (PAP) and serine proteinase homologs (SPHs) simultaneously. Insect Biochem Mol Biol 35(3):241–248

Hayakawa Y (1994) Cellular immunosuppressive protein in the plasma of parasitized insect larvae. J Biol Chem 269(20):14536–14540

He Y, Cao X, Zhang S, Rogers J, Hartson S, Jiang H (2016) Changes in the plasma proteome of manduca sexta larvae in relation to the transcriptome variations after an immune challenge: evidence for high molecular weight immune complex formation. Mol Cell Proteomics: MCP 15(4):1176–1187. https://doi.org/10.1074/mcp.M115.054296

He Y, Wang Y, Yang F, Jiang H (2017) Manduca sexta hemolymph protease-1, activated by an unconventional non-proteolytic mechanism, mediates immune responses. Insect Biochem Mol Biol 84:23–31. https://doi.org/10.1016/j.ibmb.2017.03.008

Heck T, Faccio G, Richter M, Thöny-Meyer L (2013) Enzyme-catalyzed protein crosslinking. Appl Microbiol Biotechnol 97(2):461–475. https://doi.org/10.1007/s00253-012-4569-z

Hughes AL (1999) Evolution of the arthropod prophenoloxidase/hexamerin protein family. Immunogenetics 49(2):106–114. https://doi.org/10.1007/s002510050469

Iijima M, Hashimoto T, Matsuda Y, Nagai T, Yamano Y, Ichi T, Osaki T, Kawabata S-I (2005) Comprehensive sequence analysis of horseshoe crab cuticular proteins and their involvement in

transglutaminase-dependent cross-linking. FEBS J 272(18):4774–4786. https://doi.org/10.1111/j.1742-4658.2005.04891.x
Ito S (2003) A chemist's view of melanogenesis. Pigment Cell Res 16(3):230–236. https://doi.org/10.1034/j.1600-0749.2003.00037.x
Ito S, Palumbo A, Prota G (1985) Tyrosinase-cytalyzed conjugation of dopa with glutathione. Cell Mol Life Sci 41(7):960–961. https://doi.org/10.1007/bf01970033
Iwanaga S (2002) The molecular basis of innate immunity in the horseshoe crab. Curr Opin Immunol 14(1):87–95. https://doi.org/10.1016/s0952-7915(01)00302-8
Iwanaga S, Lee BL (2005) Recent advances in the innate immunity of invertebrate animals. J Biochem Mol Biol 38(2):128–150
Jaenicke E, Decker H (2003) Tyrosinases from crustaceans form hexamers. Biochem J 371:515–523
Jiang H, Vilcinskas A, Kanost MR (2010) Immunity in lepidopteran insects. In: Söderhäll K (ed) Invertebrate Immunity, vol 708. Advances in Experimental Medicine and Biology. Springer US, pp 181–204. https://doi.org/10.1007/978-1-4419-8059-5_10
Jouni ZE, Takada N, Gazard J, Maekawa H, Wells MA, Tsuchida K (2003) Transfer of cholesterol and diacylglycerol from lipophorin to Bombyx mori ovarioles in vitro: role of the lipid transfer particle. Insect Biochem Mol Biol 33(2):145–153. https://doi.org/10.1016/S0965-1748(02)00102-9
Kamareddine L, Nakhleh J, Osta MA (2016) Functional interaction between Apolipophorins and complement regulate the mosquito immune response to systemic infections. J Innate Immun 8(3):314–326. https://doi.org/10.1159/000443883
Kamimura M, Nakahara Y, Kanamori Y, Tsuzuki S, Hayakawa Y, Kiuchi M (2001) Molecular cloning of silkworm paralytic peptide and its developmental regulation. Biochem Biophys Res Commun 286(1):67–73
Kan H, Kim C-H, Kwon H-M, Park J-W, Roh K-B, Lee H, Park B-J, Zhang R, Zhang J, Söderhäll K, Ha N-C, Lee BL (2008) Molecular control of phenoloxidase-induced melanin synthesis in an insect. J Biol Chem 283(37):25316–25323. https://doi.org/10.1074/jbc.M804364200
Kanost MR, Jiang H (2015) Clip-domain serine proteases as immune factors in insect hemolymph. Curr Opin Insect Sci 11:47–55. https://doi.org/10.1016/j.cois.2015.09.003
Karlsson C, Korayem AM, Scherfer C, Loseva O, Dushay MS, Theopold U (2004) Proteomic analysis of the drosophila larval hemolymph clot. J Biol Chem 279(50):52033–52041. https://doi.org/10.1074/jbc.M408220200
Kato Y, Motoi Y, Taniai K, Kadono-Okuda K, Yamamoto M, Higashino Y, Shimabukuro M, Chowdhury S, Xu J, Sugiyama M, Hiramatsu M, Yamakawa M (1994) Lipopolysaccharide-lipophorin complex formation in insect hemolymph: a common pathway of lipopolysaccharide detoxification both in insects and in mammals. Insect Biochem Mol Biol 24(6):547–555. https://doi.org/10.1016/0965-1748(94)90090-6
Kim MS, Baek MJ, Lee MH, Park JW, Lee SY, Söderhäll K, Lee BL (2002) A new easter-type serine protease cleaves a masquerade-like protein during prophenoloxidase activation in holotrichia diomphalia larvae. J Biol Chem 277(42):39999–40004. https://doi.org/10.1074/jbc.M205508200
Kim-Jo C, Gatti J-L, Poirié M (2019) Drosophila cellular immunity against parasitoid wasps: a complex and time-dependent process. Front Physiol 10(603). https://doi.org/10.3389/fphys.2019.00603
Kotani E, Yamakawa M, Iwamoto S-i, Tashiro M, Mori H, Sumida M, Matsubara F, Taniai K, Kadono-Okuda K, Kato Y, Mori H (1995) Cloning and expression of the gene of hemocytin, an insect humoral lectin which is homologous with the mammalian von Willebrand factor. Biochim Biophys Acta 1260(3):245–258. https://doi.org/10.1016/0167-4781(94)00202-e
Krautz R, Arefin B, Theopold U (2014) Damage signals in the insect immune response. Front Plant Sci 5(342). https://doi.org/10.3389/fpls.2014.00342
Kwon TH, Kim MS, Choi HW, Joo CH, Cho MY, Lee BL (2000) A masquerade-like serine proteinase homologue is necessary for phenoloxidase activity in the coleopteran insect, Holotrichia diomphalia larvae. Eur J Biochem 267(20):6188–6196

Lagueux M, Perrodou E, Levashina EA, Capovilla M, Hoffmann JA (2000) Constitutive expression of a complement-like protein in Toll and JAK gain-of-function mutants of Drosophila. Proc Natl Acad Sci USA 97(21):11427. https://doi.org/10.1073/pnas.97.21.11427

Lee SY, Kwon TH, Hyun JH, Choi JS, Kawabata S-I, Iwanaga S, Lee BL (1998) In vitro activation of pro-phenol-oxidase by two kinds of pro-phenol-oxidase-activating factors isolated from hemolymph of coleopteran, Holotrichia diomphalia larvae. Eur J Biochem 254(1):50–57. https://doi.org/10.1046/j.1432-1327.1998.2540050.x

Lesch C, Goto A, Lindgren M, Bidla G, Dushay MS, Theopold U (2007) A role for Hemolectin in coagulation and immunity in Drosophila melanogaster. Dev Comp Immunol 31(12):1255–1263. https://doi.org/10.1016/j.dci.2007.03.012

Levashina EA, Moita LF, Blandin S, Vriend G, Lagueux M, Kafatos FC (2001) Conserved Role of a complement-like protein in Phagocytosis revealed by dsRNA knockout in cultured cells of the mosquito, Anopheles gambiae. Cell 104(5):709–718. https://doi.org/10.1016/s0092-8674(01)00267-7

Levy F, Bulet P, Ehret-Sabatier L (2004) Proteomic analysis of the systemic immune response of Drosophila. Mol Cell Proteomics 3(2):156. https://doi.org/10.1074/mcp.M300114-MCP200

Li D, Scherfer C, Korayem AM, Zhao Z, Schmidt O, Theopold U (2002) Insect hemolymph clotting: evidence for interaction between the coagulation system and the prophenoloxidase activating cascade. Insect Biochem Mol Biol 32(8):919–928. https://doi.org/10.1016/s0965-1748(02)00030-9

Li X, Ma M, Liu F, Chen Y, Lu A, Ling Q-Z, Li J, Beerntsen BT, Yu X-Q, Liu C, Ling E (2012) Properties of Drosophila melanogaster prophenoloxidases expressed in Escherichia coli. Dev Comp Immunol 36(4):648–656. https://doi.org/10.1016/j.dci.2011.11.005

Lin C-Y, Hu K-Y, Ho S-H, Song Y-L (2006) Cloning and characterization of a shrimp clip domain serine protease homolog (c-SPH) as a cell adhesion molecule. Dev Comp Immunol 30(12):1132–1144. https://doi.org/10.1016/j.dci.2006.03.006

Lindgren M, Riazi R, Lesch C, Wilhelmsson C, Theopold U, Dushay MS (2008) Fondue and transglutaminase in the Drosophila larval clot. J Insect Physiol 54(3):586–592. https://doi.org/10.1016/j.jinsphys.2007.12.008

Lu Z, Beck MH, Strand MR (2010) Egf1.5 is a second phenoloxidase cascade inhibitor encoded by Microplitis demolitor bracovirus. Insect Biochem Mol Biol 40(7):497–505. https://doi.org/10.1016/j.ibmb.2010.04.009

Lu Z, Beck MH, Wang Y, Jiang H, Strand MR (2008a) The viral protein Egf1.0 Is a dual activity inhibitor of prophenoloxidase-activating proteinases 1 and 3 from Manduca sexta. J Biol Chem 283(31):21325–21333. https://doi.org/10.1074/jbc.m801593200

Lu Z, Jiang H (2008b) Expression of Manduca sexta serine proteinase homolog precursors in insect cells and their proteolytic activation. Insect Biochem Mol Biol 38(1):89–98

Matheis G, Belitz H-D (1979) Multiple forms of soluble monophenol, dihydroxyphenylalanine: oxygen oxidoreductase (EC 1.14.18.1) from potato tubers (solanum tuberosum). Z Lebensmittelunters Forsch A 169(3):165–169. https://doi.org/10.1007/bf01333317

Matsumoto H, Tsuzuki S, Date-Ito A, Ohnishi A, Hayakawa Y (2012) Characteristics common to a cytokine family spanning five orders of insects. Insect Biochem Mol Biol 42(6):446–454. https://doi.org/10.1016/j.ibmb.2012.03.001

Moita LF, Wang-Sattler R, Michel K, Zimmermann T, Blandin S, Levashina EA, Kafatos FC (2005) In vivo identification of novel regulators and conserved pathways of Phagocytosis in A. gambiae. Immunity 23(1):65–73. https://doi.org/10.1016/j.immuni.2005.05.006

Morais Guedes Sd, Vitorino R, Domingues R, Tomer K, Correia AJF, Amado F, Domingues P (2005) Proteomics of immune-challenged drosophila melanogaster larvae hemolymph. Biochem Biophys Res Commun 328(1):106–115. https://doi.org/10.1016/j.bbrc.2004.12.135

Mori H, Iwamoto S-i, Kotani E, Sumida M, Matsumoto T, Matsubara F (1992) Isolation of cDNA clones coding for humoral lectin of silkworm, Bombyx mori, larvae. J Invertebr Pathol 59(1):40–45. https://doi.org/10.1016/0022-2011(92)90109-H

Muszbek L, Bereczky Z, Bagoly Z, Komáromi I, Katona É (2011) Factor XIII: a coagulation factor with multiple plasmatic and cellular functions. Physiol Rev 91(3):931–972. https://doi.org/10.1152/physrev.00016.2010

Nagai T, Kawabata S-i (2000) A link between blood coagulation and prophenol oxidase activation in arthropod host defense. J Biol Chem 275(38):29264–29267. https://doi.org/10.1074/jbc.M002556200

Nakhleh J, Christophides GK, Osta MA (2017) The serine protease homolog CLIPA14 modulates the intensity of the immune response in the mosquito Anopheles gambiae. J Biol Chem 292(44):18217–18226. https://doi.org/10.1074/jbc.M117.797787

Nappi A, Poirié M, Carton Y (2009) Chapter 4 The role of melanization and cytotoxic by-products in the cellular immune responses of Drosophila against parasitic wasps. In: Genevieve P (ed) Advances in parasitology, vol 70. Academic Press, pp 99–121

Oliveira GdA, Lieberman J, Barillas-Mury C (2012) Epithelial nitration by a peroxidase/NOX5 system mediates mosquito antiplasmodial immunity. Science (New York, NY) 335(6070):856–859. https://doi.org/10.1126/science.1209678

Orlow SJ, Zhou B-K, Chakraborty AK, Drucker M, Pifko-Hirst S, Pawelek JM (1994) High-molecular-weight forms of Tyrosinase and the Tyrosinase-related proteins: evidence for a melanogenic complex. J Investig Dermatol 103(2):196–201

Osta MA, Christophides GK, Kafatos FC (2004) Effects of mosquito genes on plasmodium development. Science 303(5666):2030–2032. https://doi.org/10.1126/science.1091789

Paskewitz SM, Andreev O, Shi L (2006) Gene silencing of serine proteases affects melanization of Sephadex beads in Anopheles gambiae. Insect Biochem Mol Biol 36(9):701–711. https://doi.org/10.1016/j.ibmb.2006.06.001

Pedersen LC, Yee VC, Bishop PD, Trong IL, Teller DC, Stenkamp RE (1994) Transglutaminase factor XIII uses proteinase-like catalytic triad to crosslink macromolecules. Protein Sci 3(7):1131–1135. https://doi.org/10.1002/pro.5560030720

Peñalver MJ, Fenoll LG, Rodríguez-López JN, García-Ruiz PA, García-Molina F, Varón R, García-Cánovas F, Tudela J (2005) Reaction mechanism to explain the high kinetic autoactivation of tyrosinase. J Mol Catal B 33(1):35–42. https://doi.org/10.1016/j.molcatb.2005.02.002

Phillips DR, Clark KD (2017) Bombyx mori and Aedes aegypti form multi-functional immune complexes that integrate pattern recognition, melanization, coagulants, and hemocyte recruitment. PLoS ONE 12(2):e0171447. https://doi.org/10.1371/journal.pone.0171447

Piao S, Song Y-L, Kim JH, Park SY, Park JW, Lee BL, Oh B-H, Ha N-C (2005) Crystal structure of a clip-domain serine protease and functional roles of the clip domains. EMBO J 24(24):4404–4414

Pietrzyk AJ, Bujacz A, Mueller-Dieckmann J, Łochynska M, Jaskolski M, Bujacz G (2013) Crystallographic identification of an unexpected protein complex in silkworm haemolymph. Acta Crystallogr Sect D: Biol Crystallogr 69(12):2353–2364. https://doi.org/10.1107/S0907444913021823

Povelones M, Bhagavatula L, Yassine H, Tan LA, Upton LM, Osta MA, Christophides GK (2013) The CLIP-domain serine protease homolog SPCLIP1 regulates complement recruitment to microbial surfaces in the malaria mosquito Anopheles gambiae. PLoS Path 9(9):e1003623. https://doi.org/10.1371/journal.ppat.1003623

Povelones M, Upton LM, Sala KA, Christophides GK (2011) Structure-function analysis of the Anopheles gambiae LRIM1/APL1C complex and its interaction with complement C3-Like protein TEP1. PLoS Path 7(4):e1002023. https://doi.org/10.1371/journal.ppat.1002023

Povelones M, Waterhouse RM, Kafatos FC, Christophides GK (2009) Leucine-rich repeat protein complex activates mosquito complement in defense against plasmodium parasites. Science 324(5924):258–261. https://doi.org/10.1126/science.1171400

Rahman MM, Ma G, Roberts HLS, Schmidt O (2006) Cell-free immune reactions in insects. J Insect Physiol 52(7):754–762. https://doi.org/10.1016/j.jinsphys.2006.04.003

Rebsamen M, Kandasamy RK, Superti-Furga G (2013) Protein interaction networks in innate immunity. Trends Immunol 34(12):610–619. https://doi.org/10.1016/j.it.2013.05.002

Ricklin D, Reis ES, Mastellos DC, Gros P, Lambris JD (2016) Complement component C3—The "Swiss Army Knife" of innate immunity and host defense. Immunol Rev 274(1):33–58. https://doi.org/10.1111/imr.12500

Riehle MM, Markianos K, Niaré O, Xu J, Li J, Touré AM, Podiougou B, Oduol F, Diawara S, Diallo M, Coulibaly B, Ouatara A, Kruglyak L, Traoré SF, Vernick KD (2006) Natural malaria infection in Anopheles gambiae is regulated by a single genomic control region. Science 312(5773):577. https://doi.org/10.1126/science.1124153

Riehle MM, Xu J, Lazzaro BP, Rottschaefer SM, Coulibaly B, Sacko M, Niare O, Morlais I, Traore SF, Vernick KD (2008) Anopheles gambiae APL1 Is a family of variable LRR proteins required for Rel1-mediated protection from the malaria parasite. Plasmodium berghei. PLOS ONE 3(11):e3672. https://doi.org/10.1371/journal.pone.0003672

Sakamoto M, Ohta M, Suzuki A, Takase H, Yoshizawa Y, Kitami M, Sato R (2011) Localization of the serine protease homolog BmSPH-1 in nodules of E. coli-injected Bombyx mori larvae and functional analysis of its role in nodule melanization. Dev Comp Immunol 35(5):611–619. https://doi.org/10.1016/j.dci.2011.01.003

Scherfer C, Karlsson C, Loseva O, Bidla G, Goto A, Havemann J, Dushay MS, Theopold U (2004) Isolation and characterization of hemolymph clotting factors in drosophila melanogaster by a pullout method. Curr Biol 14(7):625–629. https://doi.org/10.1016/j.cub.2004.03.030

Schmid MR, Dziedziech A, Arefin B, Kienzle T, Wang Z, Akhter M, Berka J, Theopold U (2019) Insect hemolymph coagulation: kinetics of classically and non-classically secreted clotting factors. Insect Biochem Mol Biol 109:63–71. https://doi.org/10.1016/j.ibmb.2019.04.007

Schmidt O, Söderhäll K, Theopold U, Faye I (2010) Role of Adhesion in arthropod immune recognition. Annu Rev Entomol 55(1):485–504. https://doi.org/10.1146/annurev.ento.54.110807.090618

Schnitger AKD, Kafatos FC, Osta MA (2007) The melanization reaction is not required for survival of Anopheles gambiae mosquitoes after bacterial infections. J Biol Chem 282(30):21884–21888. https://doi.org/10.1074/jbc.M701635200

Sendovski M, Kanteev M, Ben-Yosef VS, Adir N, Fishman A (2011) First structures of an active bacterial tyrosinase reveal copper plasticity. J Mol Biol 405(1):227–237. https://doi.org/10.1016/j.jmb.2010.10.048

Seo S-Y, Sharma VK, Sharma N (2003) Mushroom Tyrosinase: recent prospects. J Agric Food Chem 51(10):2837–2853. https://doi.org/10.1021/jf020826f

Shapiro JP, Keim PS, Law JH (1984) Structural studies on lipophorin, an insect lipoprotein. J Biol Chem 259(6):3680–3685

Shelby KS, Webb BA (1997) Polydnavirus infection inhibits translation of specific growth-associated host proteins. Insect Biochem Mol Biol 27(3):263–270. https://doi.org/10.1016/S0965-1748(96)00095-1

Shibata T, Hadano J, Kawasaki D, Dong X, Kawabata S-i (2017a) Drosophila TG-A transglutaminase is secreted via an unconventional Golgi-independent mechanism involving exosomes and two types of fatty acylations. J Biol Chem 292(25):10723–10734. https://doi.org/10.1074/jbc.M117.779710

Shokal U, Eleftherianos I (2017b) Thioester-containing protein-4 regulates the drosophila immune signaling and function against the pathogen photorhabdus. J Innate Immun 9(1):83–93. https://doi.org/10.1159/000450610

Shokal U, Kopydlowski H, Eleftherianos I (2017) The distinct function of Tep2 and Tep6 in the immune defense of Drosophila melanogaster against the pathogen Photorhabdus. Virulence 8(8):1668–1682. https://doi.org/10.1080/21505594.2017.1330240

Silveira H, Gabriel A, Ramos S, Palma J, Felix R, Custódio A, Collins LV (2012) CpG-containing oligodeoxynucleotides increases resistance of Anopheles mosquitoes to plasmodium infection. Insect Biochem Mol Biol 42(10):758–765. https://doi.org/10.1016/j.ibmb.2012.07.003

Strand MR (2008) The insect cellular immune response. Insect Sci 15(1):1–14. https://doi.org/10.1111/j.1744-7917.2008.00183.x

Strand MR, Burke GR (2012) Polydnaviruses as symbionts and gene delivery systems. PLoS Pathog 8(7):e1002757. https://doi.org/10.1371/journal.ppat.1002757
Su X-D, Gastinel LN, Vaughn DE, Faye I, Poon P, Bjorkman PJ (1998) Crystal structure of hemolin: a horseshoe shape with implications for homophilic Adhesion. Science 281(5379):991. https://doi.org/10.1126/science.281.5379.991
Telfer WH, Kunkel JG (1991) The function and evolution of insect storage hexamers. Annu Rev Entomol 36(1):205–228. https://doi.org/10.1146/annurev.en.36.010191.001225
Theopold U, Schmidt O, Söderhäll K, Dushay MS (2004) Coagulation in arthropods: defence, wound closure and healing. Trends Immunol 25(6):289–294. https://doi.org/10.1016/j.it.2004.03.004
Tokunaga F, Yamada M, Miyata T, Ding YL, Hiranaga-Kawabata M, Muta T, Iwanaga S, Ichinose A, Davie EW (1993) Limulus hemocyte transglutaminase. Its purification and characterization, and identification of the intracellular substrates. J Biol Chem 268(1):252–261
Troha K, Buchon N (2019) Methods for the study of innate immunity in Drosophila melanogaster. Developmental Biology, Wiley Interdisciplinary Reviews e344. https://doi.org/10.1002/wdev.344
Tsuzuki S, Ochiai M, Matsumoto H, Kurata S, Ohnishi A, Hayakawa Y (2012) Drosophila growth-blocking peptide-like factor mediates acute immune reactions during infectious and non-infectious stress. Scient Rep 2:210–210. https://doi.org/10.1038/srep00210
Vavricka C, Christensen B, Li J (2010) Melanization in living organisms: a perspective of species evolution. Protein Cell 1(9):830–841. https://doi.org/10.1007/s13238-010-0109-8
Vavricka CJ, Han Q, Mehere P, Ding H, Christensen BM, Li J (2014) Tyrosine metabolic enzymes from insects and mammals: a comparative perspective. Insect Sci 21(1):13–19. https://doi.org/10.1111/1744-7917.12038
Veillard F, Troxler L, Reichhart J-M (2016) Drosophila melanogaster clip-domain serine proteases: structure, function and regulation. Biochimie 122:255–269. https://doi.org/10.1016/j.biochi.2015.10.007
Volz J, Müller H-M, Zdanowicz A, Kafatos FC, Osta MA (2006) A genetic module regulates the melanization response of Anopheles to plasmodium. Cell Microbiol 8(9):1392–1405. https://doi.org/10.1111/j.1462-5822.2006.00718.x
Wang Y, Jiang H, Cheng Y, An C, Chu Y, Raikhel AS, Zou Z (2017) Activation of Aedes aegypti prophenoloxidase-3 and its role in the immune response against entomopathogenic fungi. Insect Mol Biol 26(5):552–563. https://doi.org/10.1111/imb.12318
Wang Y, Jiang H, R. Kanost M (1999) Biological activity of Manduca sexta paralytic and plasmatocyte spreading peptide and primary structure of its hemolymph precursor. Insect Biochem Mol Biol 29 (12):1075–1086. https://doi.org/10.1016/S0965-1748(99)00086-7
Wang Y, Lu Z, Jiang H (2014) Manduca sexta proprophenoloxidase activating proteinase-3 (PAP3) stimulates melanization by activating proPAP3, proSPHs, and proPOs. Insect Biochem Mol Biol 50:82–91. https://doi.org/10.1016/j.ibmb.2014.04.005
Wang Z, Wilhelmsson C, Hyrsl P, Loof TG, Dobes P, Klupp M, Loseva O, Mörgelin M, Iklé J, Cripps RM, Herwald H, Theopold U (2010) Pathogen entrapment by Transglutaminase—a conserved early innate immune mechanism. PLoS Pathog 6(2):e1000763. https://doi.org/10.1371/journal.ppat.1000763
Waterhouse R, Povelones M, Christophides G (2010) Sequence-structure-function relations of the mosquito leucine-rich repeat immune proteins. BMC Genom 11(1):531
Whitten MMA, Coates CJ (2017) Re-evaluation of insect melanogenesis research: views from the dark side. Pigment Cell Melanoma Res 30(4):386–401. https://doi.org/10.1111/pcmr.12590
Whitten MMA, Tew IF, Lee BL, Ratcliffe NA (2004) A novel role for an insect apolipoprotein (Apolipophorin III) in β-1,3-Glucan pattern recognition and cellular encapsulation reactions. J Immunol 172(4):2177–2185
Wigglesworth VB (1937) Wound healing in an insect (Rhodnius Prolixus Hemiptera). J Exp Biol 14(3):364–381
Wynant N, Duressa TF, Santos D, Van Duppen J, Proost P, Huybrechts R, Vanden Broeck J (2014) Lipophorins can adhere to dsRNA, bacteria and fungi present in the hemolymph of the desert

locust: A role as general scavenger for pathogens in the open body cavity. J Insect Physiol 64:7–13. https://doi.org/10.1016/j.jinsphys.2014.02.010
Yang F, Wang Y, He Y, Jiang H (2016) In search of a function of Manduca sexta hemolymph protease-1 in the innate immune system. Insect Biochem Mol Biol 76:1–10. https://doi.org/10.1016/j.ibmb.2016.06.009
Yassine H, Kamareddine L, Chamat S, Christophides GK, Osta MA (2014) A Serine protease homolog negatively regulates TEP1 consumption in systemic infections of the malaria vector Anopheles gambiae. J Innate Immun 6(6):806–818. https://doi.org/10.1159/000363296
Yassine H, Kamareddine L, Osta MA (2012) The mosquito melanization response is implicated in defense against the entomopathogenic fungus beauveria bassiana. PLoS Path 8(11):e1003029. https://doi.org/10.1371/journal.ppat.1003029
Yokoyama H, Yokoyama T, Yuasa M, Fujimoto H, Sakudoh T, Honda N, Fugo H, Tsuchida K (2013) Lipid transfer particle from the silkworm, Bombyx mori, is a novel member of the apoB/large lipid transfer protein family. J Lipid Res 54(9):2379–2390. https://doi.org/10.1194/jlr.M037093
Yu X-Q, Kanost MR (2000) Immulectin-2, a Lipopolysaccharide-specific Lectin from an Insect, Manduca sexta, Is induced in response to gram-negative bacteria. J Biol Chem 275(48):37373–37381. https://doi.org/10.1074/jbc.M003021200
Zhang G, Lu Z-Q, Jiang H, Asgari S (2004) Negative regulation of prophenoloxidase (proPO) activation by a clip-domain serine proteinase homolog (SPH) from endoparasitoid venom. Insect Biochem Mol Biol 34(5):477–483
Zhang X, An C, Sprigg K, Michel K (2016) CLIPB8 is part of the prophenoloxidase activation system in Anopheles gambiae mosquitoes. Insect Biochem Mol Biol 71:106–115. https://doi.org/10.1016/j.ibmb.2016.02.008
Zhang X, He Y, Cao X, Gunaratna RT, Chen Y-r, Blissard G, Kanost MR, Jiang H (2015) Phylogenetic analysis and expression profiling of the pattern recognition receptors: insights into molecular recognition of invading pathogens in Manduca sexta. Insect Biochem Mol Biol 62:38–50. https://doi.org/10.1016/j.ibmb.2015.02.001
Zhang Y, Dong Z, Liu S, Yang Q, Zhao P, Xia Q (2012) Identification of novel members reveals the structural and functional divergence of lepidopteran-specific Lipoprotein_11 family. Funct Integr Genomics 12(4):705–715. https://doi.org/10.1007/s10142-012-0281-4
Zhao P, Wang GH, Dong ZM, Duan J, Xu PZ, Cheng TC, Xiang ZH, Xia QY (2010) Genome-wide identification and expression analysis of serine proteases and homologs in the silkworm Bombyx mori. BMC Genom 11:405. https://doi.org/10.1186/1471-2164-11-405
Zou Z, Shin SW, Alvarez KS, Kokoza V, Raikhel AS (2010) Distinct melanization pathways in the mosquito aedes aegypti. Immunity 32(1):41–53. https://doi.org/10.1016/j.immuni.2009.11.011

Chapter 6
Hemoglobin in Arthropods—*Daphnia* as a Model

Bettina Zeis

Abstract Hemoglobin is the respiratory protein of many arthropods, enhancing the oxygen transport capacity of the hemolymph. One example, that has been subject of extensive studies, is the hemoglobin of the crustacean genus *Daphnia*. Here the characteristics of this oxygen binding protein are reviewed. The genetic structure is the result of repeated duplication events in the evolution, leading to a variety of di-domain isoforms. Adjustments to environmental changes thus result from differential expression of these paralogs. The biochemical properties, including spectral characteristics, concentration ranges, molecular mass of monomers and native oligomers, are compared. Structural differences between isoforms can be correlated to functional properties of oxygen binding characteristics. The mechanism of hemoglobin induction via hypoxia-inducible factor 1 allows the response to altered oxygen and temperature conditions. Changes of the hemoglobin suite in quantity and functional quality can be linked to their benefits for the animals' physiological performance. However, there is a large inter- and intra-specific variability of this induction potential. The consequences of altered hemoglobin characteristics for the animals' success within their habitat are discussed.

Keywords Hemoglobin · Oxygen transport capacity · Oxygen affinity · Hypoxia · Hypoxia-inducible factor · Temperature · Differential isoform expression · Respiration rate · Physiological limits · Clonal variation

Introduction

Hemoglobin is a ubiquitous protein. Sequences coding for this oxygen-binding pigment are present in the genomes of organisms of all kingdoms, ranging from plants to fungi and animals (for a review, see Weber and Vinogradov 2001). Their common structure includes a globular protein binding a heme goup with the ferrous iron providing the oxygen binding site. The protein conformation is arranged in eight

B. Zeis (✉)
Institut für Zoophysiologie, Westfälische Wilhelms-Universität Münster, Schlossplatz 8, 48149 Münster, Germany
e-mail: zeis@uni-muenster.de

U. Hoeger and J. R. Harris (eds.), *Vertebrate and Invertebrate Respiratory Proteins, Lipoproteins and other Body Fluid Proteins*, Subcellular Biochemistry 94,
https://doi.org/10.1007/978-3-030-41769-7_6

alpha helices in vertebrate hemoglobins, but their number may deviate in other phyla (Weber and Vinogradov 2001). Additionally, the aggregation state of subunits can range from monomeric globins to multimeres of more than hundred subunits providing a large variety of oxygen binding pigments, the different structures resulting in a broad range of functional characteristics. Finally, hemoglobin is found within cells or may be exported to the body fluids (Burmester and Hankeln 2014).

Within the arthropods, hemoglobin is the respiratory protein in the hemolymph of some insects as well as some crustacea. While many insects do not need any respiratory pigments due to their tracheal oxygen supply, there are Diptera and Hemiptera using hemoglobin during juvenile stages. Well-known examples are the aquatic larvae of midges of the genus *Chironomus* living in oxygen-poor habitats in the benthal region of lakes and ponds, which show deep red colour due to their high concentrations of extracellular respiratory hemoglobins. Cellular variants of hemoglobins (Burmester and Hankeln 2014), which are present in far lower concentrations, have also been discovered in adult *Drosophila* (Burmester et al. 2006). Within the crustacea, hemoglobin occurs in Branchiopoda, Copepoda and Thecostraca (Burmester 2015). Malacostraca, including e.g. lobsters, crabs, shrimp, crayfish, krill, amphipods and isopods usually use hemocyanin as main oxygen carriers (Burmester 2015). However, there is also evidence for co-occurrence of hemocyanin and hemoglobin, as was reported for the ectoparasitic amphipod *Cyamus scammoni* (Terwilliger 1991, 2008).

Sequence analyses allowing insights into evolution patterns suggest that in the last common ancestor of pancrustacea, including insects and crustaceans, hemocyanin was the oxygen transport protein present in the animals' hemolymph (Burmester 2015). In clades that lost hemocyanin, an extracellular respiratory hemoglobin may have evolved from cellular globins (Burmester 2015).

Hemoglobin Evolution in the *Crustacea*—A Story of Repeated Duplication Events

Gene duplication and subsequent diversification are basic events for the evolution of protein families in general, and in the case of haemoglobin the role of duplication processes for isoform multiplicity is well-documented (Storz 2018). Characteristic for invertebrates, however, incomplete duplications resulted in multi-domain subunits (Weber and Vinogradov 2001). The extracellular hemoglobin of crustacea is the result of duplication events, which added the entire globin gene or only parts of it to another globin gene during recombination processes. This way, a di-domain structure could establish on the gene level, where a stop-codon is missing after the first globin gene. In the gene product, the second globin is directly connected to the preceding one via a linker sequence (preA'). Then, the globular structure of eight alpha helices (A-H) in a 3-over-3 helix sandwich is a domain of the fusion product, and thus the result is a bar-bell-shaped protein with two heme groups. This type of hemoglobin evolved in Cladocera (e.g. genus *Daphnia*), Conchostraca and Notostraca (Weber and Vinogradov 2001). Further duplications of the entire di-domain gene led to hemoglobin clusters. In *Daphnia*, the cluster of eight hemoglobins must be the result of three more duplication events of the di-domain structured gene (Colbourne et al. 2011). Sixteen di-domains aggregate to the native molecule. Other aggregation types of the di-domain subunits are 10mers in *Czycus* (Conchostraca) with 20 oxygen binding sites (Weber and Vinogradov 2001). In Notostraca, 24mers were reported for *Lepidurus*, and 16-18mers in *Triops* (Rousselot et al. 2006).

In Anacostraca, a multitude of partial duplications must have taken place. The hemoglobin of *Artemia* is a 130 kDa subunit of 9 domains, indicating at least 4 recombination events with partial overlay of the coding regions. A subsequent duplication led to two different isoforms that form homo- or heterodimers for the native molecule (Moens and Kondo 1978; Weber and Vinogradov 2001).

A common remarkable characteristic of many of these crustaceans is the potential for hemoglobin induction under hypoxic conditions. Many of them live in habitats with considerable changes in available O_2. Additionally, the quantitative change is often accompanied by altered affinities. This physiological plasticity to adjust the functional properties of hemoglobin is subsequently reviewed in detail for the daphniid hemoglobin, giving insights in established knowledge and recent findings on *Daphnia magna* and *Daphnia pulex* hemoglobin, two well-known models representing the hemoglobins of crustacea.

Hemoglobin in *Daphnia*

The observation of red specimens of waterfleas was reported long ago (Lankester 1871). Indeed, the variations of the hemoglobin content in the hemolymph of *Daphnia magna* may be perceived with the naked eye in these transparent animals. The differences in hemoglobin concentration and oxygen affinity have gained interest of the group of H. M. Fox, in London in the 1940s (Fox 1948; Fox et al. 1949). Despite its potential for reversible oxidation, its role as respiratory pigment was questioned at first, as animals kept up their swimming activity even when the hemoglobin was inactivated by carbon monoxide (Fox 1948). Yet the induction of hemoglobin under low oxygen conditions led to enhanced survival rates, which confirmed its function in oxygen transport (Fox 1948; Fox et al. 1949, 1950). Hemoglobin levels showed fluctuations with the moulting cycle, which was associated to egg production also synchronized to the moulting state. A loss of hemoglobin in the hemolymph and a simultaneous increase of this protein in the parthenogenetic eggs led to the assumption of hemoglobin transfer to the offspring (Fox et al. 1949, 1950; Dresel 1948). In consequence, adult fertile females had lower hemoglobin concentrations than juvenile specimens (Green 1956). The highest hemoglobin concentrations were observed in males (Green 1956). Moreover, hemoglobin production was increased with the available amount of food, and in turn, the filtration rates were enhanced with higher hemoglobin levels (Fox et al. 1951). In hypoxic water, hemoglobin was beneficial for enhanced muscular activity, concerning the beating of the thoracic limbs and the antennae, the former establishing the water current providing algae as well as oxygen, the latter being responsible for the animals'swimming activity (Fox et al. 1951). As a further environmental factor inducing hemoglobin production, the ambient temperature was identified (Fox and Phear 1953). Moreover, iron supply was found to be crucial for hemoglobin production (Fox and Phear 1953). The rate of hemoglobin increase in water of low oxygen and its loss upon the return to normoxia was determined to take place in the range of few days (Fox and Phear 1953; Fox 1955). During hemoglobin degradation, iron was detected in the animals 'fat cells (Smaridge 1954; Green 1955). It was also excreted via the shell glands (Fox 1948; Smaridge 1954, 1956). In summary, these early studies already disclosed the phenotypic plasticity of hemoglobin quantity and quality in *Daphnia magna* and its beneficial role for the animals' physiological performance under oxygen deprivation.

Biochemical Properties of *Daphnia* Hemoglobin

Hemoglobin is freely dissolved in the *Daphnia* hemolymph, which amounts to 58–61% of the body volume (Kobayashi 1983). Its concentration was derived from its absorption spectrum, showing the characteristic maxima of hemoglobins (Fig. 6.1). In the oxygenated molecule, absorption peaks were observed at 414, 542 and 576 nm (Hoshi and Kobayashi 1971; Sugano and Hoshi 1971). Deoxygenation shifted the Soret peak to 423 nm. The smaller double peaks (α and β) disappeared and one peak at 561 nm characterized the hemoglobin spectrum without oxygen. For purified hemoglobin, the absorption at 414 nm can be more than four times the value at 280 nm resulting from aromatic residues as characteristic for all proteins (Kimura et al. 1999). These absorption spectra are similar to other penta-coordinated hemoglobins (Burmester and Hankeln 2014). In the deoxygenated molecule, four ligands of the heme group and the proximal histidine bind to the ferrous iron, and the sixth potential coordination position is free—until the molecule binds oxygen. Quantification can make use of extinction coefficients of $\varepsilon_{414nm} = 126.6\ mM^{-1}cm^{-1}$ and $\varepsilon_{576nm} = 14.8\ mM^{-1}cm^{-1}$ (Hoshi and Kobayashi 1971; Sugano and Hoshi 1971) using Lambert-Beer's law. However, in animals from air-saturated medium or in individuals from habitats with sufficient oxygen supply, the low hemoglobin concentration along with the overlay from other components in the animals' tissues and hemolymph, e.g. cytochromes and carotenoids, lead to difficulties in concentration measurements. A calculation of the hemoglobin concentration from its absorption directly according to Lambert-Beer's law would lead to a large overestimation of the hemoglobin content. Thus the absorption difference of oxygenated and deoxygenated hemoglobin can be used instead, evoking the latter by adding dithionite. The area between the intercepts of both curves at 421 and 452 nm (isobestic points) is correlated to the concentration and can be determined even at low concentrations in crude extracts (Cambronero et al. 2018; Schwerin et al. 2010). The concentrations can be calculated from this area according to a calibration line from different dilutions of a purified hemoglobin sample previously quantified from absorption at 576 nm (Fig. 6.1).

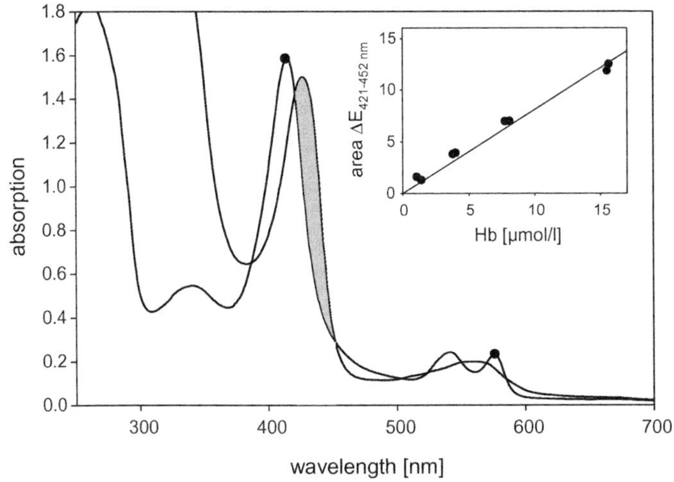

Fig. 6.1 Absorption spectra of hemoglobin from *Daphnia magna*. The oxygenated form shows the characteristic peaks at 414, 542 and 576 nm. Extinction coefficients were determined for the first and latter wavelength (circles, see text for details). Deoxygenated hemoglobin causes maximal absorption at 427 nm and 561 nm. Desoxygenation was performed with sodium dithionite, leading to the high absorption observed below 400 nm. At 421 nm and 452 nm absorption is invariant (isobestic points). In this range the area between the curves can be used for quantification. A calibration line (insert) was established from purified hemoglobin correlating concentrations derived according to Lamber Beer's law from absorbance at 576 nm to the area between both curves in the range of 421–452 nm: Hb [μmol/l] = 1.24* area Δ E

Depending on ambient oxygen concentration, animals were of pale or red colour, corresponding to hemoglobin values of 1–17 mg/ml crude extract (Kobayashi 1981), 0.5–16 mg/ml (Kobayashi and Hoshi 1982), 1.3–11.7 mg/ml (79–733 μmol/l, Lamkemeyer et al. 2003); even 0.24–241 mg/g dry mass was reported for *Daphnia magna* (Kobayashi and Nezu 1986) and values up to 257 mg/g dry mass (ca. 26% of the animals' dry mass) in *Daphnia pulex* (Landon and Stasiak 1983), demonstrating a considerable induction range and physiological plasticity. The potential to induce hemoglobin when exposed to hypoxic conditions was correlated to the animal's age. It was largest in premature specimens of 1.5 mm body size and declined in older females (Kobayashi and Hoshi 1984; Kobayashi et al. 1990c). Hemoglobin content reached a maximum just before maturation (Kobayashi 1982a; Kobayashi and Yamagata 2000). Animals larger than 3.0 mm were unable to raise their hemoglobin concentration in response to hypoxia (Kobayashi 1982a; Kobayashi and Nezu 1986). Beyond its function as a respiratory protein, a role as a protein reserve in addition to the yolk protein vitellogenin was assumed. In addition to the transfer of proteins to the offspring, the hemoglobin loss after moulting may reflect a dilution effect, as the animals' growth is restricted to that period, and a volume increase of 14% was observed subsequent to the moulting event (Fox et al. 1949; Kobayashi and Hoshi 1984).

The molecular mass of the native molecule was determined from its sedimentation in ultracentrifugation (Sugano and Hoshi 1971; Ilan et al. 1982; Lamkemeyer et al. 2006), gel electrophoresis or gel filtration (Yamauchi and Ochiai 1984; Zeis et al. 2003a) to amount to 494 kDa (Ilan et al. 1982), 590 (Lamkemeyer et al. 2006) or 670 kDa (Yamauchi and Ochiai 1984; Sugano and Hoshi 1971) in *Daphnia magna*. The hemoglobin molecule was composed of subunits of a molecular mass of 31 kDa (Ilan et al. 1982; Ilan and Daniel 1979) or 37–39 kDa (Kobayashi et al. 1990c; Lamkemeyer et al. 2006). The resolution of subunits was affected by the buffer system used in gel electrophoresis, yielding heterogenous bands in Tris-glycine systems and only one band of 39 kDa using phosphate buffer (Lamkemeyer et al. 2006). Each subunit consisted of two domains of 16–18 kDa (Ilan et al. 1982), both carrying a heme group. Glycosylation of oxygen-containing residues in the N-terminals range (Lamkemeyer et al. 2006) increased the hydrophilic nature of the protein. The native oligomers were composed of several different subunit types which, upon denaturation, could be oligomers distinguished by classical or two-dimensional gel electrophoresis, detecting 6 to 8 isoform types (Yamauchi and Ochiai 1984; Kimura et al. 1999; Zeis et al. 2004; Lamkemeyer et al. 2005) (*see* Fig. 6.2). These subunits were arranged to multimers, containing 12 to 16 subunits (Hebert et al. 1999; Kobayashi and Yamagata 2000; Lamkemeyer et al. 2003). Combinations of different isoform types led to three different types of native oligomers (Kobayashi et al. 1988). The isoelectric points of these multimers determined by isoelectric focussing were pI 4.8–5.3 (Kobayashi et al. 1988) or pI 4.4 to 5.9 (Lamkemeyer et al. 2005), those of the subunits pI 4.5–5.7 (Kobayashi et al. 1988, 1990a, 1994). Two-dimensional gels yielded larger values in the range of pI 6.5–8.1 (Zeis et al. 2003a). First determinations of the iron content of 40 iron ions per hemoglobin molecule probably over-estimated the number of hemes in the oligomer (Hoshi and Kobayashi 1971) it seems clear now that the native molecule is arranged of 16 two-domain subunits and thus comprises 32 oxygen binding sites.

The hemoglobin of *Daphnia pulex* also was subject of various studies on structural and functional characteristics. The size of the native aggregate was determined from its sedimentation coefficient (16.4–17.1 s) to 420–470 kDa (Svedberg and Eriksson-Quensel 1934; Dangott and Terwilliger 1980; Wolf et al. 1983; Peeters et al. 1990). It was reported to be composed of 12 di-domain subunits of 31–37 kDa (Dangott and Terwilliger 1980; Wolf et al. 1983; Peeters et al. 1990). The isoelectric points of the subunits were determined to be pI 7.5–7.7 (Peeters et al. 1990), those of the native aggregate pI 5.0–6.5 (Wolf et al. 1983). Comparing both *Daphnia magna* and *Daphnia pulex* hemoglobins, the size of the native oligomers can be estimated from cross-linked molecules in SDS-PAGE analyses (our unpublished results). It is obvious then, that oligomers of both species are of the same size, corresponding to 16 times the subunit molecular mass. Thus, hemoglobin of both species seems to form oligomers of 16 monomers.

Meanwhile, the information available on genomes and transcriptomes allows a closer view on the subunit characteristics (Fig. 6.3). Di-domain hemoglobin genes in *Daphnia magna* are organized in a cluster resulting from several gene duplication events. The first hemoglobin cDNA sequence of this species was published in 1997

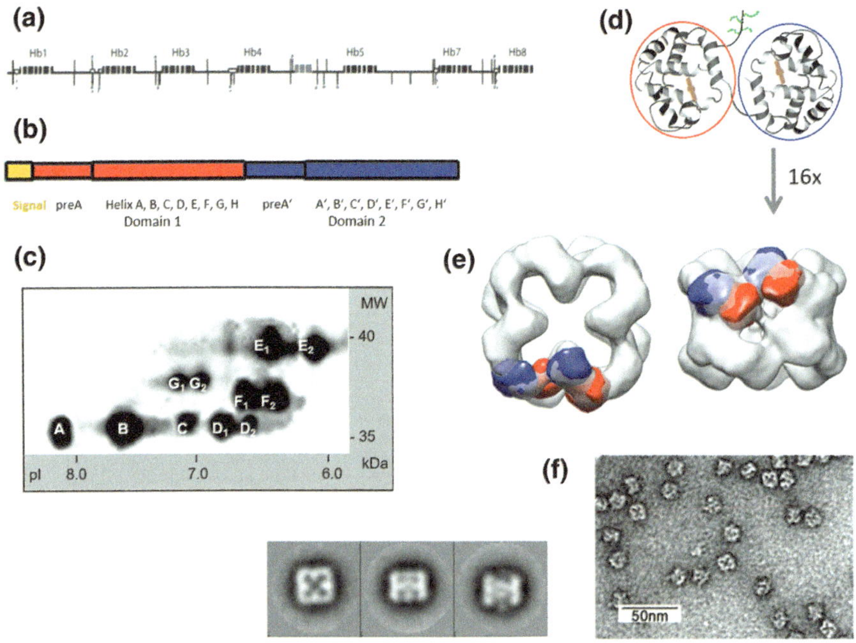

Fig. 6.2 Molecular structure of *Daphnia magna* hemoglobin. The gene cluster (**a**) resulting from repeated gene duplication events comprises seven dmhb isoforms. Each protein subunit DmHb (**b**) has a N-terminal signal sequence for export from the synthesizing cells to the hemolymph, followed by a two-domain globin with eight alpha helices in each domain. The isoforms can be separated by two-dimensional electrophoresis resulting in a characteristic pattern (**c**). The subunits may be glycosylated at N-terminal residues (**d**, green: carbohydrate chain), each domain folding into the globin structure of eight alpha-helices binding a heme group. The two linked globular domains result in a bar-belled shape of the molecule. Sixteen of these subunits are arranged in the native hemoglobin aggregate (**e**), which was modelled from electron micrographs (**f**) (Colbourne et al. 2011; Dejung 2014; Lamkemeyer et al. 2005)

(Tokishita et al. 1997). Two further complete sequences and a partial sequence were added (Kimura et al. 1999) and denominated Hb02 to Hb04. Their position in the cluster was downstream of Hb01 in the case of Hb02 while Hb03 and Hb04 were upstream of the first sequenced gene. Sequencing of the latter gene was completed (Nunes et al. 2005). Sequence information on 3 additional upstream genes and knowledge on the position of the respective genes in the cluster (Colbourne et al. 2011) led to deviating denominations (*see* Table 6.1). The hemoglobin gene sequences of *Daphnia pulex* (Dewilde et al. 1999) revealed the large homology between the corresponding genes of the sister species. Moreover, it indicated the loss of one hemoglobin isoform, Hb06, in *Daphnia magna*, thus in this species the cluster contains seven hemoglobin sequences. In *Daphnia pulex*, three di-domain hemoglobin genes were found beyond those of the cluster containing eight hemoglobin genes (Colbourne et al. 2011). A phylogenetic tree shows a complex pattern of orthologs and paralogs. Only for Hb08, the isoforms of both species clearly match (Fig. 6.4).

```
        Signal_sequence       Pre_A                           Helix_A
DpHb01  -MQFFNVALVFGVLAIA---SAYSQAPGTTTTTVTTTVTTVTAD--EGTDSGLLSAHERS     54
DpHb02  -MQFLKIALFFALVALASSSPSCSQAPGTTITSVTTTVTTVTAD--EDSDNGLLSSHDRS     57
DpHb03  ---MASFKIVFLLSVLA---FACAYKPGTTTTTVTTTVTTVSAD--EGN-EGILSSHDRS     51
DpHb04  --MAFKFALLFGLVAFA---SACSQAPGTTTTTVTTTVTTVSAD--EGD-EGILSSHDRS     52
DpHb05  --MAFKFALLFGLVAFA---SACSQAPGTTTTTVTTTVTTVSAD--EGD-EGILSSHDRS     52
DpHb06  ----------------------MATPGKTSYAIAMSIMT---TEDDEMGSGLLSTQERA     34
DpHb07  --MAFKFALLFGLVAFA---SACSQAPGTTTTTVTTTVTTVSAD--EGD-EGILSSHDRS     52
DpHb08  -MQVLSLALFIGIAAAVS-----AYAPGTKVTTVTTSVTTVTLD--EES-TGILSSHERS     51
DpHb09  MLASFKLVVLLSVVALAC---AWPQFGSSSMT-TGPTTSTVPAKENSQGPTPKLDCFDHS     56
DpHb10  --MAFKLALLFGVIAFA---SACSYAPGTTVTTVTTAVTTVSAD--EGE-EGILSSHDRS     52
DpHb11  --MAFKLVLLFGVIA-----SACSYAVSQPGTSVTTAVTTVSAD--EGE-EGILSSHDRS     50

DmHb01  -MQLFNLALVFGVVAFV---SACSQTPGTTTTTVTTTVTTVTAD--DDGEAGLLTANDRR     54
DmHb02  -MQSLKIALLFAFVALA---STCSQAPGTTTTTVTTTVTTVTAD--DDSDSGLLTANDRR     54
DmHb03  -MASFKIALLFGVIAFV---SACSQAPGTTTTTVTTTVTTVSAD--DGGEAGLLSAHERS     54
DmHb04  -MASFKIALLLGVIAFV---NACSQAPGTTTTTVTTTVTTVSAD--DGSEAGLLSAHERS     54
DmHb05  -MASFKIALLFGVVAFV---SAYSQAPGTTTTTVTTTVTTVSAD--DGSEAGILSAHERS     54
DmHb07  -MASFKIALLLGVIAFV---NACSQAPGTTTTTVTTTVTTVSAD--DGSEAGLLSAHERS     54
DmHb08  -MQVLTIALFLGIVATAS-ACASMAAPGTTVTTVTTSVTTVSAD--EES-TGILSANERS     55
                                   .        :  *      .      *    ::
        Helix_A   AB Helix_B     Helix_C CD    Helix_D Helix_E
DpHb01  VIRKTWDQAKKDGDVPPKILFRFIVANPEYQKMFKSFATVPQNELMGNGNFLAQAYTILA     114
DpHb02  VIRKTWDQAKKDGDVPPKILFRFIVANPEYQKKFKSFAAVPQNELLGNGNFLAQAYTILA     117
DpHb03  VIRKTWDQAKKDGDVPPQVLFRFVTAHPEYQKMFSKFATVPQNELLGNGNFLAQAYTILA     111
DpHb04  VIRKTWDQAKKDGDVPPQVLFRFVTAHPEYQKMFSKFATVPQNELLGNGNFLAQAYTILA     112
DpHb05  VIRKTWDQAKKDSDVAPQVLYRFVTAHPEYQKMFSKFASVPQNELMGNGNFLAQAYTILA     112
DpHb06  IIRTTWNKARKDGDVAPKLLFKFLKAYPEYQKKFSKFADVPQSNLLSNGNFLAQAYTILA     94
DpHb07  VIRKTWDQAKKDGDVAPQVLYRFVTAHPEYQKMFSKFASVPQNELLGNGNFLAQAYTILA     112
DpHb08  IIRKTWDQAKKDGDVAPQVLYRFIKAHPEYQKKFSKFADVPQSELLSNGNFLAQAYTILA     111
DpHb09  IIRNTWDQAKKNGEFAPQVLLRFIKAHPEYQQMFGKFASVPHYNLLRNGDFLAQAYTISA     116
DpHb10  VIRKTWDQAKRDGDVAPQILFRFVKAHPEYQKMFSKFANVPQNELLSNGNFLAQAYTILA     112
DpHb11  VIRITWDQAKRDGDVAPQILFRYVKAHPEYQKMFSKFANMPQNELLSNGNFLAQAYTILD     110

DmHb01  IIRKTWDQAKKDGDVPPQILFRFIKAHPEYQKMFKSFANVPQAELLSNGNFLAQAYTILA     114
DmHb02  IIRKTWDQAKKDGDVPPQILFRFIKAHPEYQKMFKSFADVPQAELLGNGNFLAQAYTILA     114
DmHb03  LIRKTWDQAKKDGDVAPQVLFRFVKAHPEYQKMFSKFANVPQAELLSNGNFLAQAYTILA     114
DmHb04  LIRKTWDQAKKDGDVAPQVLFRFVKAHPEYQKMFSKFANVPQSELLSNGNFLAQAYTILA     114
DmHb05  LIRKTWEQARKDGDVAPQILFRFVKAHPEYQKMFSKFANVPQAELLTNGNFLAQAYTILA     114
DmHb07  LIRKTWDQAKKDGDVAPQVLFRFVKAHPEYQKMFSKFANVPQSELLSNGNFLAQAYTILA     114
DmHb08  LIRKTWDQAKKDGDVAPQVLFRFIKAHPEYQKKFSKFADVPQSELLTNGNFLAQAYTILA     115
        :** **::*:::.:. *::* ::: * ****: * .** :*: :*: **:********

        Helix_E    EF     Helix_F   FG Helix_G               Helix_H
DpHb01  GLNVVIQSLSSQELLANQINALGGAHQPRGATPIMF--EQFGAITEEVLAEELGIAFNAE     172
DpHb02  GLNVVIQSLSSQELLANQLNALGGAHQPRGITPIMF--EQFGTVTEEVLAEELGNAFNAE     175
DpHb03  GLNVVVQSLSSQELLANQLNALGGAHQARGATPIMF--EQFGEILTGVLAEELGSAFNAE     169
DpHb04  GLNVVVQSLSSQELLANQLNALGGAHQARGATPIMF--EQFGEILTGVLAEELGSAFNAE     170
DpHb05  GLNVVVQSLSSQELLASQLNSLGGAHQARGATPIMF--EQFGEILTGVLAEELGSAFNAE     170
DpHb06  GLNVIVQSLSSQELMANQLNALGGAHQPRGVTTTILELKEFGVILIQVLEEEIGSAMTID     154
DpHb07  GLNVVVQSLSSQELLANQLNALGGAHQARGATPIMF--EQFGEILTGVLAEELGSAFNAE     170
DpHb08  GLNVVIQSLSSQELLANQLNALGGAHQPRGVTPAMF--EEFGVIVEQVLEEELGSTFNAE     169
DpHb09  GLNVVIQSLSSQELLAAQFNLLGSAYQPRGVTPAMF--EEFSVILEQVLEETLGSTFNVE     174
DpHb10  GLNVVIQSLSSQELMASQLNALGAAHQPRSATPIMF--EQFGAVLEEVLAEELGSTFNSE     170
DpHb11  GLNVAIQSLSSQELMAGQLNALGAAHHPRGATPIMF--EQFGTILEEGLAEELGSIFNAE     168

DmHb01  GLNVVIQSLSSQELIANQINALGGAHKPRGATPIMF--EQFGAVTEEVLAEELGSSFNAE     172
DmHb02  GLNVVIQSLSSQELIANQINALGGAHKPRGATPIMF--EQFGAVTEEVLAEELGSSFNAE     172
DmHb03  GLNVVIQSLSSQELMANQLNALGGAHKPRGATPVMF--EQFGGILEEVLAEELGSGFTAE     172
DmHb04  GLNVVIQSLFSQELMANQLNALGGAHQARGATPVMF--EQFGGILEEVLAEELGSGFTAE     172
DmHb05  GLNVVVQSLSSQELMANQLNALGGAHKPRGATPVMF--EQFGGILEEVLAEELGSSFTAE     172
DmHb07  GLNVVIQSLFSQELMANQLNALGGAHQPRGATPVMF--EQFGGILEEVLAEELGSGFTAE     172
DmHb08  GLNVVIQSLSSQELLANQLNVLGGAHQPRGVTPAMF--EEFGVIVEQVLEEELGNGFSAE     173
        **** :*** ****:* *:* **.*:: *. *  ::  ::*. :    * * :*  :. :
```

Fig. 6.3 Hemoglobin protein sequences of *Daphnia pulex* and *Daphnia magna*. The alignment was performed with sequences from Colbourne et al. 2011 (see Table 6.1). Helices are shaded, different shading intensities serve to distinguish adjacent helices. Underlined amino acids are involved in heme binding: W(A12); P(C2); F(CD1) and H(F8), the proximal histidine complexing Fe^{2+}. Further residues important for oxygen affinity are Q(E7), replacing the distal histidine common in vertebrate hemoglobins. Myoglobin B10, which is changed to L(B8) in domain 1 and F(B9) or M(B9) in domain 2 indicates different affinities of both domains. This is also true for position E11, which is isoleucine in domain 1 and valine in domain 2 (Lamkemeyer et al. 2005). Asterisks, dots and colons indicate identical amino acids and conservative replacements, respectively

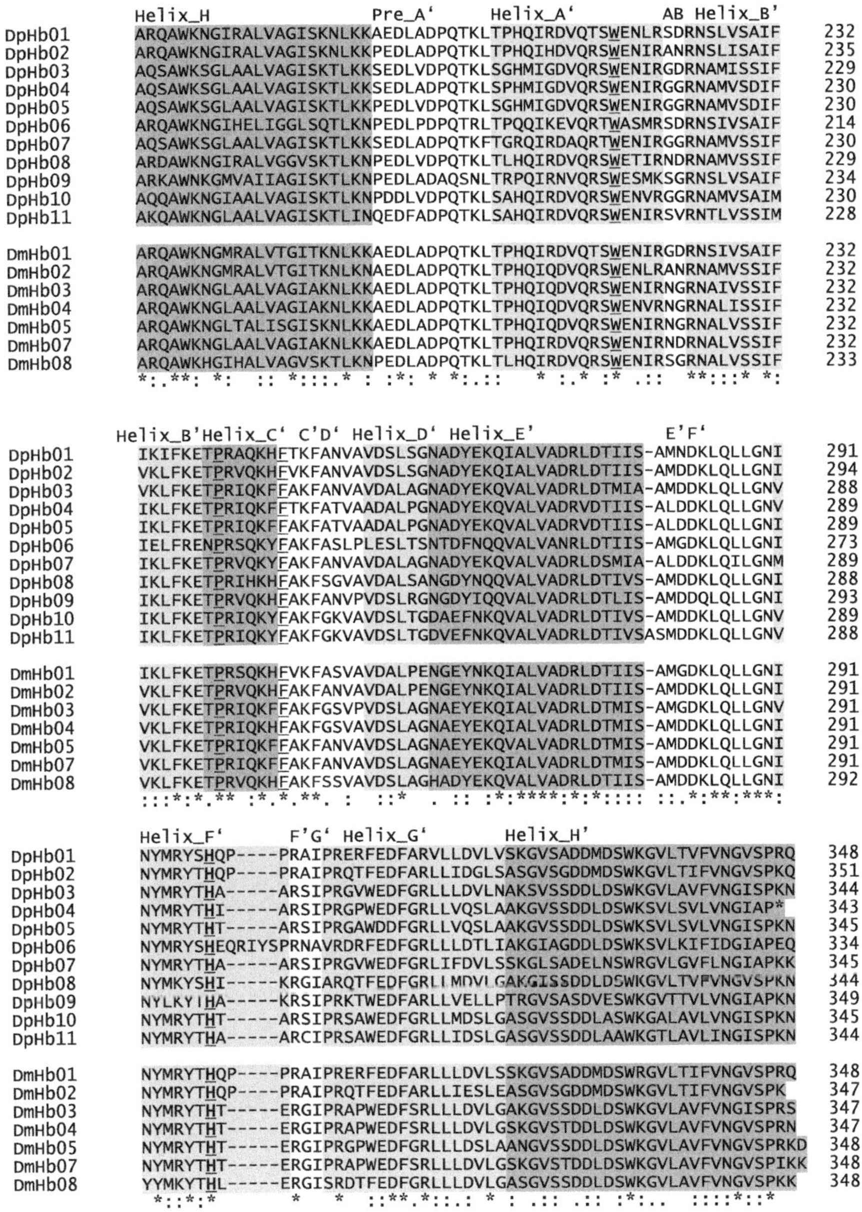

Fig. 6.3 (continued)

Table 6.1 Hemoglobins of *Daphnia pulex* and *Daphnia magna*. Sequence information form Colbourne et al. (2011) (Fig. 6.3) was used to determine number of amino acids, the molecular weight and the isoelectric point. Denominations from different references were correlated. Additionally, the induction factors between Hb concentrations in animals from normoxic and hypoxic acclimation were calculated (Zeis et al. 2003a; Gerke et al. 2011). n.f.: not found, +++ strong induction, but not detectable at normoxia, no calculation of ratio possible

Daphnia pulex				MW (Da)	pI	Amino acids		Hypoxic induction
Joint genome institute accession nr. (> jgi Dappu 1.1)			Colbourne et al. (2011)					
DpHb01	96311		DpHb01	38303	6.99	348		2.5
DpHb02	230332		DpHb02	38570	7.03	351		2.2
DpHb03	311662		DpHb03	37425	6.54	344		13.1
DpHb04	234836		DpHb04	37067	6.00	343		52.6
DpHb05	234837		DpHb05	37268	6.23	354		n.f.
DpHb06	234838		DpHb06	37362	6.37	336		n.f.
DpHb07	234839		DpHb07	37468	8.49	345		56.8
DpHb08	230333		DpHb08	37838	6.67	344		63.4
DpHb09	316317		DpHb09	38645	9.30	349		n.f.
DpHb10	92880		DpHb10	37505	6.54	345		+++
DpHb11	93831		DpHb11	37571	6.10	344		n.f.
Daphnia magna				MW (Da)	pI	Amino acids		Hypoxic induction
NCBI GenBank accession nr. AB518060.1	Tokishita et al. (1997)	Kimura et al. (1999), Nunes et al. (2005)[a]	Colbourne et al. (2011)					
DmagHb1			DmHb01	38440	7.81	348		n.f.
DmagHb2			DmHb02	38243	6.25	347	DmHbD	63.4
DmagHb3			DmHb03	37674	7.80	347	DmHbA	+++
DmagHb4		DmHb4	DmHb04	37826	6.61	347	DmHbC	5.6
DmagHb5		DmHb3	DmHb05	38071	6.55	348	DmHbG	n.f.
DmagHb6	DmHb	DmHb1	DmHb07	38089	6.98	348	DmHbB	3.2
DmagHb7		DmHb2	DmHb08	37978	6.09	348	DmHbF	11.6

[a]These protein numbers were also used in Gorr et al. (2004), Kobayashi and Yamagata (2000), Paul et al. (2004b), Zeis et al. (2003a, b, 2004)

Further di-domain hemoglobins were also detected in *Daphnia magna* (Orsini et al. 2016). Here, in the EvidentialGene database (Orsini et al. 2016; Gilbert 2014) for the *Daphnia magna* genome twelve di-domain hemoglobin genes were found and denominated DmHbX01–DmHbX12 (Orsini et al. 2016). Among these, DmHbX03–DmHbX09 are arranged in a cluster. However, the sequences show considerable

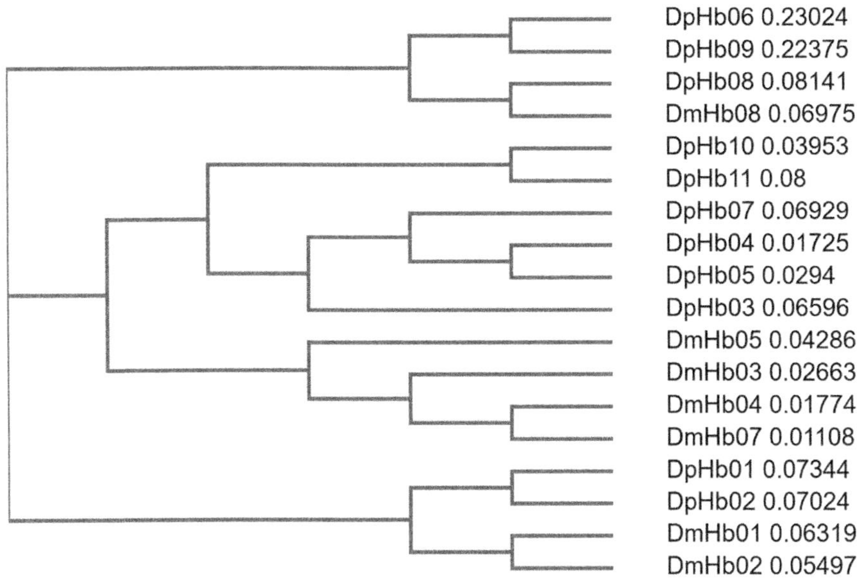

Fig. 6.4 Phylogenetic tree of hemoglobins from *Daphnia magna* and *Daphnia pulex*. Based on the protein sequences (see Fig. 6.3), the analysis was carried out with Clustal Omega (EMBL search tool, Madeira et al. 2019). The cladogram was generated as a Neighbour-joining tree without distance corrections. Numbers indicate branch lengths

deviation from those published earlier (Colbourne et al. 2011). This may result from differences in the clones used for genomic analysis and reflects the genomic variation within this fast-evolving species. A recent search in the NCBI GenBank (Geer et al. 2010) revealed more than 600 entries for *Daphnia magna* di-domain hemoglobins, the majority being the result of a large sequencing project comparing variations of a multitude of clones from two populations assembled and annotated with Evidential-Gene (Orsini et al. 2016; Gilbert 2014), analysing a variety of stressors (Orsini et al. 2016). Additionally, hemoglobin sequences of *Daphnia exilis* and *Daphnia spinulata* were studied with a focus on differences due to gene conversion events (Sutton and Hebert 2002). Sequencing projects of further species (*Daphnia similoides*, Gilbert 2017), as well as clones from the *Daphnia galeata/longispina* complex, (Huylmans et al. 2016; Herrmann et al. 2018) will further contribute information on the variability of the hemoglobin gene loci and proteins. Apart from the hemoglobin in the animals' hemolymph, hemoglobin was also detected in muscle and nerve cells (Fox 1955). These hemoglobin subtypes correspond to monomeric single-domain globins like myoglobin, neuroglobin and cytoglobin (Burmester and Hankeln 2014; Burmester 2015; Orsini et al. 2016; Pesce et al. 2002). Both, the variation in sequences of genes from the di-domain hemoglobin gene cluster and the sequences of one-domain cellular globin types need further analyses and bear high potential to shed light on the evolution of respiratory proteins in this invertebrate model organism (Fig. 6.5).

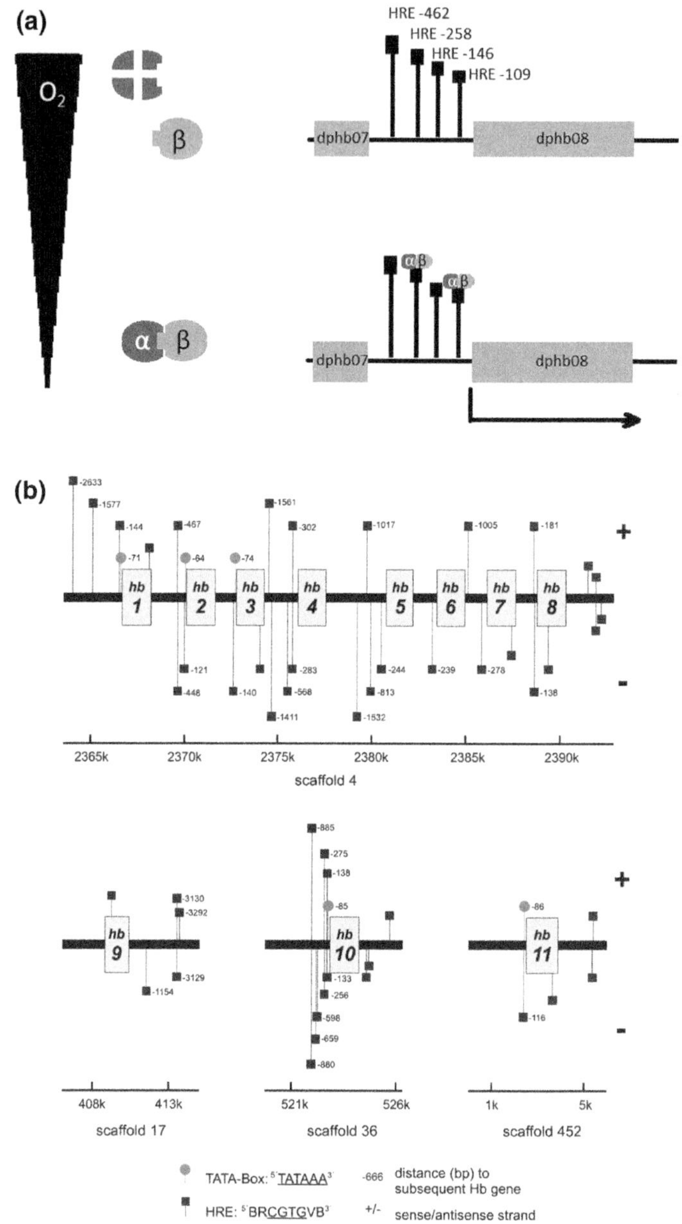

Fig. 6.5 Hemoglobin expression is regulated via hypoxia-responsive elements as binding site for the hypoxia-inducible factor (HIF). This transcription factor binds to conserved motifs upstream of hemoglobin genes in *Daphnia magna* (**a**), here two sites upstream of dmhb08 (–107 and –258 bp) were necessary for maximal induction (modified from Gorr et al. 2004). In *Daphnia pulex* (**b**), HIF-binding sites are present in all hemoglobin genes; although the strong induction of dphb10 matches a high number of hypoxia-responsive elements, a general correlation pattern of HREs and induction rate was not observed (Gerke et al. 2011)

Two different cell types contribute to the hemoglobin secretion into the hemolymph. Fat cells have been identified as sites of hemoglobin synthesis (Smaridge 1956; Kobayashi and Hoshi 1984). They are arranged laterally near the intestine, and the fat droplets are visible under the microscope in well-fed animals. Besides, expression of hemoglobin takes place in the epithelial cells of the epipodites (Goldmann et al. 1999), which was shown by in situ hybridisation with hemoglobin mRNA probes. Due to the red colour of the epipodites, they had been assigned a function as gills, however their rigid carapace structure of considerable thickness hinders diffusive processes. Instead, the oxygen uptake is conferred via the large area of the inner sides of the carapace with a very thin cuticle (Pirow et al. 1999). Ventilation with fresh water by the movements of the thorakopods sullies oxygen-rich water to this areas of diffusive exchange. The water current can be adjusted to ambient oxygen conditions. Sensing oxygen deficiency despite the increase of ventilation, the cells of epipodite epithelia and the fat cells can increase hemoglobin expression to raise the oxygen transport capacity.

In these hemoglobin synthesising cells, the secretion of the hemoglobin gene product into the hemolymph is conferred via a signal sequence of 18 amino acids. This element is missing completely in DpHb06 (Fig. 6.3). The cleavage site is absent in DpHb09, thus with SignalP 3.0 (Emanuelsson et al. 2007) no secretory signal was detected (Gerke et al. 2011). Accordingly, both isoforms may not contribute to the hemolymph hemoglobin pool, and they have not been found in 2D-gels (Gerke et al. 2011). After the signal sequence has been cleaved off in the endoplasmatic reticulum, the new N-terminus is mostly a cysteine. No further cysteines are present in the molecule, thus there are no intra-molecular disulfide bonds (Yamauchi and Ochiai 1984; Kobayashi et al. 1990b). However, a dimeric structure can be stabilized by this covalent link of two N-termini, resulting in an arrangement of two di-domain hemoglobin subunits, carrying four heme groups. With a molecular mass of 72 kDa, this complex is slightly larger than the vertebrate hemoglobin tetramer. A corresponding band in SDS-polyacrylamide electrophoresis without reducing agents can be observed, vanishing when dithionite is added to the probe (our unpublished observations). The threonine-rich region in the pre-A segment may be the site of the observed O-glycosylation (Lamkemeyer et al. 2006). Eight alpha helices are arranged in the globin fold of a 3-over-3 helix sandwich, binding the prosthetic heme group with residues in helix A, C and F. The tryptophan in position 12 of helix A (W(A12)), proline in position 2 of helix C, P(C2), and phenylalanine in the first position following helix C, F(CD1) confer hydrogen bonds to the heme and H(F8), the proximal histidine in position 8 in helix F complexes the Fe^{2+} (Tokishita et al. 1997; Kimura et al. 1999). The distal histidine present in helix E in vertebrate hemoglobins is replaced by glutamine (Q(E7)) in *Daphnia*. In vertebrate myoglobin, this exchange lowers oxygen affinity by a factor of seven (Mathews et al. 1989). In all *Daphnia* hemoglobin isoforms, this substitution of the distal histidine has taken place in both domains, suggesting it occurred before the duplication events leading to the hemoglobin cluster. Other positions have been subject to different mutations in both subunits. E11 is isoleucine in domain 1 and valine in domain 2 in all isoforms of *Daphnia magna* and *Daphnia pulex*. This residue is important for oxygen affinity. It is decisive for the competition of carbon monoxide and cyanide with oxygen, as a shift and rotation of this residue was described upon the binding of CO and CN^- ions, respectively (Osmulski and Leyko 1986). Thus, both domains of *Daphnia* hemoglobin probably differ in their susceptibility to both ligands replacing oxygen (Lamkemeyer et al. 2005). A further residue deviating in both domains is the equivalent to position B10 in the human respiratory protein. As the helix B is shorter by two amino acids in domain 1 and a single amino acid in domain 2, this residue is L(B8) and F(B'9) in *Daphnia* hemoglobin (Fig. 6.3). This substitution affects hemoglobin oxygen binding, the phenylalanine increasing affinity by a factor of 10 compared to the leucine variant, balancing the stability against auto-oxidation versus the kinetics of oxygen release (Carver et al. 1992). Thus the second domain can be assumed to bind oxygen with a considerably higher affinity in *Daphnia* (Lamkemeyer et al. 2005) giving a slower kinetic of O_2 release. While the leucine is invariant in all isoforms of both species, the phenylalanine is replaced by a methionine (M(B'9)) in the second domain of DpHb10 and DpHb11 (Fig. 6.3). The effect on oxygen affinity would be a subject worthwhile for further studies, as it is subunit type DpHb10 which is most strongly induced under hypoxic conditions, correlated with a substantial increase in oxygen affinity (Gerke et al. 2011).

Differential Isoform Expression—Response to Oxygen Changes

The hemoglobin characteristics are adjusted to changing environmental conditions (e.g. oxygen and temperature, see above) by differential isoform expression. Studies on the transcript level (Zeis et al. 2003a) using *Daphnia magna* revealed constitutive isoforms (dmhb07) and genes that were rapidly induced by a transfer of animals to severe hypoxia (dmhb08 and dmhb05) within two minutes. Deviating from the nomenclature in the original reference, the gene numbers of the sequences of the entire cluster, which were available only later (Colbourne et al. 2011), are used here (*see* Table 6.1). After 20 h, the total amount of *dmhb* transcripts had already increased by 80%. For long-term acclimated animals, a RNA level even 12 times higher in hypoxically acclimated animals has been reported (Tokishita et al. 1997).

Matching the increase on the protein level, identification of the corresponding proteins was required. The isoelectric points derived from the amino acid sequence and the position in two-dimensional gels are deviating slightly, but the relations allow to match genes and proteins in most cases. Additionally, the N-terminal sequences (Kimura et al. 1999) can now be matched to the gene bank sequences. However, the determination of tryptophan at the N-terminus of two subunits seems to be an artefact, as none of the genes code for this amino acid in the first position of the processed molecule. The subunit pattern in two-dimensional electrophoresis of hemolymph samples of *Daphnia magna* (Fig. 6.2) allows quantification of the contribution of specific isoforms to the hemoglobin pool. Subunit B is a major component of the hemoglobin pool at all oxygen conditions. This gene product of dmhb07 shows constitutive expression and is only slightly (3-fold) induced by hypoxia (Zeis et al. 2003a; Lamkemeyer et al. 2005). Subunit A, which corresponds to *dmhb03*, is absent at normoxia, therefore, an induction factor cannot be calculated. However, under hypoxic conditions, its concentration is in the range of the dominant subunits B, D and F. Subunits D and F, correlating to *dmhb02* and *dmhb08*, are induced by a factor of 63 and 11, respectively (calculated from data in (Zeis et al. 2003a). Newly synthesized proteins were identified by autoradiography 18 h after the animals' transfer to hypoxia (Zeis et al. 2003b). A significant increase on the protein level was observed after three days of oxygen deprivation, reaching a new steady state within 11 days. Hemoglobin then was the major component, amounting to 38% of all extracted proteins (Zeis et al. 2003a).

Along with the quantitative changes in the contribution of subunits to the hemoglobin pool, the oxygen affinity was increased significantly. The hemoglobin of hypoxia-acclimated animals reached half-saturation with oxygen already at half the amount necessary for the specimens from normoxic conditions (P50 of 0.5 and 1.0 kPa, respectively) (Zeis et al. 2003a). Similarly, the affinity was increased within a few days corresponding to the changes in subunit contribution (Zeis et al. 2003b).

Corresponding changes were observed in *Daphnia pulex*. In a proteomic approach using two-dimensional gel electrophoresis of crude extracts from entire animals (Zeis et al. 2009), 23 spots were identified as hemoglobin. They revealed up-regulation of hemoglobin under hypoxic conditions by induction factors of 9 (DpHb05), 11.3 (DpHb03) and 19 (DpHb04). Using only the animals' hemolymph, the subunit pattern successfully detected was extended, and high inductions of subunits DpHb04, DpHb07 and DpHb08 of 53, 57 and 63, respectively, were measured in animals from hypoxic acclimation, relative to the normoxic control (Gerke et al. 2011). For DpHb03, an induction factor of 13 was determined in this case. Like DpHb01 and DpHb02 (ratios of 2.5 and 2.3 comparing hypoxia, 2 kPa and normoxia, 21 kPa) this subunit is part of the hemoglobin pool in all oxygen conditions. DpHb03 makes up for at least 25% in all cases. The highest induction was observed for DpHb10. This subunit was not detected in hemolymph of animals raised under normoxia, but it's concentration was increased under hypoxic conditions to 23% of the hemoglobin pool. Then it was the subunit with the second highest abundance, following DpHb03, which contributed 33% to the hemoglobin pool in animals acclimate at 2 kPa Po_2 (Gerke et al. 2011). Correlated with this increase, the oxygen affinity was raised substantially (*see* below). In this context, the unusual methionine residue in position B'9 in the second domain (*see* above) might contribute to the enhanced affinity, deserving further future attention.

Differential Isoform Expression—Response to Temperature Changes

For poikilothermic aquatic animals, elevated temperatures result in oxygen deprivation. Oxygen solubility is decreased in warmer water bodies, at the same time the respiration rates of its inhabitants are considerably increased, when referring to invertebrates, fish and amphibians. As cladocerans, along with copepods, are major crustacean groups contributing to the limnic zooplankton, the response of waterfleas to ambient temperature variation has been in the focus of several studies. Long-term-incubation of *Daphnia magna* to 10, 20 and 30 °C at normoxic conditions caused differences in the hemoglobin content by a factor of 2.9 (70 μmol/l to 200 μmol/l) along with the temperature increase (Lamkemeyer et al. 2003). The quantitative changes were also brought about by differential expression of the isoforms. Comparable to hypoxic induction, subunit type A (DmHb03) was induced by a factor of 4, increasing its contribution from 9% at 10 °C to 24% at 30 °C. The constitutive subunit B (DmHb07) amounted to 23–35% at all temperatures, along with the overall increase at elevated temperatures, its concentration was raised four-fold from 17 to 73 μmol/l. Subunits C, D and F (DmHb04, DmHb02; DmHb08) were induced to a lesser extent (ratios of 2.5–3.5). Subunit G showed the opposite reaction, it disappeared at high acclimation temperatures (Lamkemeyer et al. 2003). The differences were not attributed to the changes of oxygen concentration of the medium alone, which could be estimated comparing warm-acclimated animals with those of medium of the same oxygen content at 20 °C (Lamkemeyer et al. 2003). Thus, the tissues probably experienced lower oxygen concentrations than due to the reduced solubility, alone. This further oxygen deprivation might result from the higher oxygen consumption rates, further lowering the O_2-concentrations available in the animals' tissues.

Similarly, in *Daphnia pulex* the total number and the relation of isoforms was affected by exposure to different temperatures (Gerke et al. 2011; Zeis et al. 2013). Comparing animals from acclimation at 20 °C and 24 °C, an increase in total hemoglobin content of 71 μmol/l to 136 μmol/l was evident (Zeis et al. 2013). Isoforms DpHb01, DpHb02, DpHb03 and DpHb07 were significantly increased. The range of temperature-induced changes in hemoglobin was 1.5 to threefold. The first three isoforms were constitutively expressed, while DpHb07 contributed less, but showed the largest induction (Zeis et al. 2013). As the sequences were available at that time, the induction was also studied on the transcript level. The amount of mRNA was affected by temperature to a lesser extent. Here only isoform dmhb02 was increased significantly. The total concentration of hemoglobin transcripts was only raised by 2% (Zeis et al. 2013). Thus, the raised protein expression may rely on elevated transcription rates at the warmer temperature conditions from more or less unchanged mRNA levels. Experimental series on the dynamics of differential isoform expression in response to a temperature increase were studied after a transfer of animals acclimated at 20–30 °C, a temperature which is beyond the optimum and does not allow stable cultures of *Daphnia pulex* (Zeis et al. 2013). Within 8 h after the onset of heat stress, all hemoglobin mRNAs were induced. Transcripts of *dphb01*, *dphb02* and *dphb10* showed the highest induction of 1.8, *dphb03*, *dphb04*, *dphb05* and *dphb07* increased by a factor of 1.7, indicating that transcription was almost doubled within 8 h. The contribution of RNA from *dphb06*, *dphb09* and *dphb11* was very low under all ambient conditions, suggesting that there is no noteworthy expression of these genes and accordingly no functional role of these isoforms (Zeis et al. 2013).

Mechanisms of Hemoglobin Induction—The Role of Hypoxia Inducible Factor

While the ultimate reason—adjusting oxygen transport capacity—of hemoglobin induction is quite obvious, the proximate causes require the mechanistic understanding of cellular regulation processes. The increase in hemoglobin expression intensity in response to oxygen deprivation is regulated by the hypoxia-inducible factor (HIF), a heterodimer of HIF1α and HIF1β. The former is continuously synthesized but instantaneously degraded in the presence of oxygen (Wang et al. 1995) reviewed in Semenza (2014a, b). A hydroxylation by prolyl hydroxylase is the oxygen-dependent step initiating ubiquitinylation and thus degradation in the proteasome (Semenza 2014b). If stabilized under hypoxic conditions, HIF1α will aggregate with its dimerization partner HIF1β, and the dimer can bind to hypoxia-responsive elements in the promotor-region of hemoglobin and further hypoxia-sensitive genes (Semenza 2014b). This transcription factor of the basic helix-loop-helix-PAS family is also described as ARNT (aryl hydrocarbon receptor nuclear translocator) in arthropods, including *Daphnia* (Tokishita et al. 2006). Gorr et al. (2004) not only were the first to prove HIF-conveyed induction of hemoglobin in *Daphnia magna*, but they also presented a detailed and elegant study on the relationship between the number and position of HIF-binding sites and the strength of induction. Hypoxia responsive elements were present in the upstream regions of all three hemoglobins known at that time. They performed reporter studies using luciferase activity in assays applying site-directed mutagenesis of the HREs in the promoter-region of DmHb08 (at that time denominated Hb2) (Gorr et al. 2004). Hypoxia responsive elements with the core sequence of 5'-CGTG-3' at nucleotide positions –107 and –258 were necessary for maximal hypoxic induction. A third HRE at position –147 relative to the start of transcription site interfered with an additional transcription factor (upstream stimulating factor) (Hu et al. 2011). HIF-binding was demonstrated by electromobility shifts (EMSA), detecting HIF in red animals from hypoxic acclimation, but not in the pale ones raised from normoxic cultures (Gorr et al. 2004). EMSA series with probes specific for the three HREs proved HIF binding to the positions –107 and –258 (Gorr et al. 2004). The *Daphnia magna* genes subsequently detected also carried upstream HIF binding sites (Colbourne et al. 2011). As in this species, the numbers and positions of

HIF binding sites *Daphnia pulex* showed no direct correlation to hemoglobin induction strength. The strongest induction observed for DpHb04; DpHb07 and DpHb08, increasing more than 50-fold at the protein level during hypoxia, had 5 or 2 upstream HREs, which was comparable to the constitutively expressed genes DpHb01 and DpHb02 with three HIF binding sites (Gerke et al. 2011). In these genes, HIF binding might also interfere with other constitutive transcription factors. The strongest induction was observed, however, for DpHb10, which was absent at normoxia and contributed to the hemoglobin suite to 23% at hypoxia. This gene, which is not part of the hemoglobin cluster, is preceded by 8 HREs. A detailed analysis, of which HREs are crucial for the induction of hemoglobin genes beyond DmHb08 is still missing.

The role of HIF1α and HIF1β (ARNT) in *Daphnia* was elucidated in whole-mount immuno-staining assays (Tokishita et al. 2006; Hoogewijs et al. 2007). During embryonic development, HIF1β was present in a variety of tissues. In adult animals, however, it was observed only in the epipodite epithelia, irrespective of ambient oxygen conditions (Tokishita et al. 2006). HIF1α was detected in the epipodite epithelia of embryos, which were exposed to hypoxia (Hoogewijs et al. 2007). This immuno-staining assay thus confirms that both dimerization partners may combine in this site of hemoglobin synthesis under oxygen deprivation, bind to the specific DNA motifs and initiate elevated expression of hemoglobin isoforms.

In addition to the HIF-conveyed induction, hemoglobin expression is affected by juvenile hormone (Rider et al. 2005), and a hormone-responsive element was also identified in the upstream region of *dmhb08* (formerly hb2; Gorr et al. 2006) at nucleotide position –230 in relation to the transcription start site. The exposure to pyriproxyfen, an analogue of juvenile hormone, caused an eightfold induction of dmhb08 expression, but had considerably lower effect on *dmhb05* and *dmhb07* (hb2, hb3 and hb1, respectively, Gorr et al. 2006). This interaction was inhibited in hypoxic conditions, indicating a competition of the transcription factors HIF and the juvenoid-hormone receptor for this part of the promotor. As variations of juvenoid hormone concentrations are linked to the animals' moulting process, the observed changes in hemoglobin concentration with the moulting state (Fox et al. 1949; Kobayashi 1982a) may be correlated with this cyclic changes in hormone levels affecting binding sites in the intergenic region of the hemoglobin cluster. It may further be involved in changes in hemoglobin concentration in the context of potentially hormonal disruptive toxins (Mu et al. 2005; Rider and LeBlanc 2006). Finally, the production of males includes regulation via methyl farnesoate (Rider et al. 2005). This crustacean terpenoid hormone and synthetic analogues induced both elevated hemoglobin levels and male production. The higher hemoglobin concentration in males may also be plausible in this context (Green 1956).

Functional Properties

The ability of respiratory proteins to bind oxygen is characterized by the oxygen tension, which is necessary to evoke half-maximal saturation of the molecule. Oxygen equilibrium curves revealed affinities of *Daphnia magna* hemoglobin to its ligand in the range of 0.15–0.71 kPa (1.1 Torr to 5.3 Torr, Kobayashi et al. 1988) or 0.39–1.06 kPa (Lamkemeyer et al. 2003; Zeis et al. 2003a) in dependence of the ambient oxygen and temperature. Animals acclimated to low oxygen conditions produced hemoglobin of higher affinity (Table 6.2). The hemoglobin of *Daphnia pulex* showed lower affinity in animals acclimated to normoxia (1.99 kPa at 20 °C acclimation and measuring temperature, Table 6.2), however the effect of oxygen deprivation on the characteristics of the hemoglobin suite was more pronounced, decreasing the P_{50} to 0.17 kPa at 20 °C. Thus the oxygen-induced plasticity was higher in this species. *Daphnia magna* increased the hemoglobin O_2-affinity with rising temperature, while in *Daphnia pulex*, the P_{50} of animals acclimated at 10 °C and 24 °C did not differ. The lowest affinity was observed at 20 °C (highest P_{50}, *see* Table 6.2). Accordingly, the temperature-induced plasticity was larger in *Daphnia magna*. The altered affinity was the result of the described differential isoform expression (*see* above). The hypoxia-adjusted hemoglobin of higher affinity in hemoglobin-rich animals corresponded to multimers of elevated pI values (Kobayashi et al. 1988; Paul et al. 2004b), indicating changes of subunit composition. The subunits DmHb03 (A) and DpHb08 and DpHb10, which are induced at low oxygen or high temperatures, thus must be assumed to have higher oxygen binding affinity (Lamkemeyer et al. 2003; Gerke et al. 2011; Zeis et al. 2013), although experimental confirmation by oxygen equilibrium

Table 6.2 Oxygen affinity is affected by acclimation and measuring temperature. The oxygen partial pressure of half-maximal O_2-saturation was determined in animals acclimated at normoxia (21 kPa) or hypoxia (2 kPa) at 20 °C. Additionally, the hemoglobin suite of animals from colder or warmer acclimation was used for oxygen equilibrium curves. For *Daphnia pulex* hemoglobin, P_{50} was also determined at a measuring temperature of 10 °C. Data are taken from Lamkemeyer et al. (2003) and Gerke et al. (2011). Bold numbers allow the comparison of oxygen effects at 20 °C

	Acclimation condition	Acclimation Po_2	Measuring condition	Measuring condition
	(°C)	(kPa)	10 °C	20 °C
			P_{50} (kPa)	P_{50} (kPa)
Daphnia magna	10	21		1.1
	20	**21**		**0.9**
	30	21		0.8
	20	**2**		**0.4**
Daphnia pulex	10	21	0.8	1.8
	20	**21**	**1.1**	**2.0**
	24	21	0.8	1.5
	20	**2**	**0.1**	**0.2**

curves of isolated subunit types is still lacking. The ambient temperature additionally affects the affinity of a given hemoglobin suite, showing stronger oxygen binding at cold temperatures. Affinity must always balance the potential to take up oxygen at the respiratory epithelia and to release it at the site of oxygen demand in the tissues. Accordingly, lower oxygen affinity at elevated temperatures facilitates oxygen release and thus optimal supply of the tissues in a situation of higher demand due to raised metabolic rates.

Hemoglobin can bind or release oxygen within a very small window of oxygen tension due to the molecular cooperativity. Hemoglobin oxygen equilibrium curves are of sigmoidal shape, with a very steep part around half-maximal saturation. These binding characteristics are brought about by conformational changes upon oxygenation, changing between the low-affinity T-form and the high affinity R-form, differing in salt bridges in the subunit interface regions constraining flexibility (T, tense vs. R, relaxed, Perutz 1972; Perutz et al. 1998; Storz 2018). This cooperative effect can be estimated by Hill plots and quantified by the Hill coefficient n, which is an indicator of how many subunits are affected by the conformational transition. In tetrameric vertebrate hemoglobin, this value may approach 4, indicating maximal mutual impact of the 4 subunits. For invertebrates, this value is generally lower (Weber and Vinogradov 2001). In *Daphnia magna* hemoglobin, the slope of the Hill plot was determined, resulting in a value of 2 (Kobayashi et al. 1994) and own unpublished observations. Thus, within the large molecule of 16 di-domain subunits, resulting in 32 oxygen binding sites, the cooperative effects are rather low. Accordingly, the conformational changes of subunits are largely independent from adjacent ones. Although Hill coefficients are generally low, the cooperative effects are changed along with the changes in concentration and affinity induced by oxygen deprivation. In red animals acclimated at low oxygen concentrations, the cooperative effect is higher than detected for the hemoglobin pool of pale animals from normoxia, the value of n_{50} shifting from 1.6 to 1.2, respectively (Kobayashi et al. 1988). Again, this result underlines the fact that hypoxic hemoglobin induction not only increases the concentration, but affects hemoglobin binding characteristics via differential expression of isoforms of deviating functional quality.

Oxygen affinity can also be affected by allosteric modulators, small molecules binding to the respiratory protein at sites remote from the ligand-binding heme group. Detailed analyses of the response to altered pH and the role of histidine residues in this context shed light on the evolution of haemoglobin adjusting this Bohr effect (Berenbrink 2006). Beyond protons, organic phosphates confer changes in affinity, stabilising the deoxygenated conformation of hemoglobin (Storz 2018). So far, the knowledge on potential low molecular weight effectors of *Daphnia* hemoglobin is limited. In another branchiopod crustacean, *Triops*, however, effects of Ca^{2+} and Mg^{2+} ions have been described (Horne and Beyenbach 1971; Weber and Vinogradov 2001; Pirow et al. 2007). Concerning the Bohr effect, the susceptibility to pH changes is generally low in extracellular hemoglobins (Weber and Vinogradov 2001). In *Triops*, direct variation of proton concentration as well as indirect effects via changes of CO_2 concentrations lowered the oxygen affinity of hemoglobin (Bohr coefficient of

–0.18). Concomitantly, the cooperativity declined as well (Pirow et al. 2007). *Daphnia magna* hemoglobin oxygen affinity showed only minor changes in response to altered pH (our own unpublished observations). Consequently, the regulation of affinity via allosteric modulators, though effective under the conditions of red blood cells, concentrating hemoglobin in a small volume, would be difficult to realise for extracellular hemoglobins. The large volume of hemolymph in invertebrates in general, in *Daphnia magna* making up to more than 50% (Kobayashi 1983), would require large amounts of effectors to be secreted into this fluid. Moreover, there is only restricted homeostasis of the hemolymph conditions, which are subject to changes along with the ambient amount of specific ions to a larger extent than erythrocytes of vertebrates. Similarly, ambient pH and CO_2 affect the proton concentration in the hemolymph (Weber and Pirow 2009). Thus the limited regulatory capacity and the large volume in which the hemoglobin is diluted would be disadvantageous for the regulation of oxygen binding characteristics by modulators. Instead, adjusting the protein equipment by differential isoform expression (functional isoform multiplicity) (Weber and Vinogradov 2001) seems the more effective solution favoured by the evolutionary processes.

Physiological Benefits

Respiratory measurements quantified the role of hemoglobin for oxygen uptake. The oxygen consumption rate rose exponentially with temperature, increasing from about 100 nmol/h per mg dry mass at 5 °C to 660 nmol/h per mg dry mass at 30 °C with a Q_{10} of 1.7 (Lamkemeyer et al. 2003; Paul et al. 2004b). When animals from 20 °C acclimation were exposed to different temperatures at normoxic or moderately hypoxic conditions, a high saturation of the hemolymph hemoglobin with oxygen was observed between 5 °C and 30 °C under normoxic conditions, declining beyond 30 °C. When ambient oxygen was lowered to 10 kPa, i.e. 50% of the normoxic value, the saturation status of the hemolymph hemoglobin was lowered. Only at cold temperatures (10 °C), was a full saturation reached. A reduction to 90% saturation was observed already at 20 °C and further decreased to 80% at 30 °C, despite of a compensatory increase in ventilation and perfusion (Paul et al. 2004a).

Comparing the oxygen uptake at 20 °C under normoxic and hypoxic conditions, the consumption rate was slightly reduced from 244 nmol/h per mg dry mass to 217 nmol/h per mg dry mass in the latter case. The lowered respiration might reflect a metabolic depression lowering the oxygen demand. The role of the higher hemoglobin concentration may be estimated from the hemolymph oxygen saturation. At an ambient Po_2 of 4 kPa, hemoglobin-poor animals from normoxic acclimation can only rely on about 100 pmol oxygen per μl hemolymph, while the enhanced transport capacity of hemoglobin-rich animals from hypoxic acclimation still provides about 350 pmol oxygen per μl hemolymph. Thus the critical oxygen tension that forces animals to rely on anaerobic metabolism and severely impeding ventilation and perfusion is shifted from 4.8 kPa for hemoglobin-poor animals to 1.3 kPa for hemoglobin-rich animals (Paul et al. 2004b).

Exposure to carbon monoxide, which binds with high affinity at the heme group and thus competes with oxygen for its binding site in the hemoglobin molecule, reduced the amount of oxygen taken up by the animals, affecting the loading of hemoglobin with oxygen. The effect was maximal at 14% air saturation (Hoshi et al. 1977; Hoshi and Yajima 1970). Accordingly, hemoglobin is most important for convective oxygen transport under hypoxia of this extent. Exposure to cyanide (KCN) or azide also decreased the oxygen uptake (Hoshi and Takahashi 1981). However, the effect of these two compounds is different in nature compared to CO, as revealed from in vivo oxygenation kinetics, rather affecting the unloading of oxygen from hemoglobin in the tissues. Both substances bind to cytochromes with high affinity, blocking mitochondrial respiration. Thus, the oxygen partial pressure does not decrease to low value in undisturbed tissues and in the surrounding hemolymph, and deoxygenation of hemoglobin is slowed down considerably (Hoshi and Takahashi 1981).

The transparency of waterfleas allows a further approach documenting the role of hemoglobin for the animals' respiratory physiology. Oxygen dissociation curves by spectral analysis in the living animal under the microscope show the saturation of hemoglobin with oxygen in vivo as a function of ambient oxygen tension. At 20 °C, an ambient Po_2 of 2 kPa (15 Torr) (Kobayashi and Tanaka 1991) led to a 50% saturation of hemoglobin with oxygen in hemoglobin-poor animals from normoxic acclimation in vivo (Kobayashi and Tanaka 1991). This value shifted to 0.5 kPa (3.5 Torr) (Kobayashi and Tanaka 1991) in hemoglobin-rich animals from hypoxic acclimation (Kobayashi and Tanaka 1991). Thus the increase in hemoglobin content allowed oxygen uptake in the animals at four-fold lower ambient Po_2. This in vivo P_{50} of 14% of the normoxic supply (Hoshi 1975; Kobayashi et al. 1990b) corresponds to the range of hypoxia with the largest effect of CO on respiration (Hoshi et al. 1977; *see* above). Additionally, the lowest oxygen concentration still allowing oxygen uptake of the animals was determined (Kobayashi 1982b). A value of 4% of the normoxic oxygen tension, which is 0.8 kPa (0.3 ml/l at 15 °C) (Kobayashi 1982b) was limiting the survival of smaller animals (1.2 mm), while the oxygen concentration necessary for survival increased in larger animals (3.0 mm). This shift of the limiting oxygen concentration reflects the larger amount of oxygen supply by diffusion processes in the smaller animals, while for large specimens, convective supply of the tissues is crucial. Again, the higher hemoglobin concentration in hypoxia-acclimated animals was beneficial, as the limiting oxygen concentration was lower in individuals with higher hemoglobin content (Kobayashi 1982b; Kobayashi and Hoshi 1984). In large animals from normoxic acclimation, ambient oxygen content of more than 10% (corresponding to Po_2 of 2 kPa) were limiting for oxygen uptake. All in all, the benefit of increased concentrations of hemoglobin with adjusted functional properties allows aerobic energy metabolism in conditions of lower ambient oxygen. In other words, the differential hemoglobin expression in response to changes in oxygen and temperature enlarges the window available on the scale of the respective abiotic factor.

Variability of Hemoglobin Induction Potential

The laboratory studies on the differential isoform expression of hemoglobin in *Daphnia* were carried out on one specific clone of each studied species (*see* above). This way, the changes of hemoglobin quantity and quality were the response of genetically identical animals to altered ambient conditions, excluding genetic differences contributing to the observed effects. However, the potential to induce hemoglobin shows considerable variations between the species, and even within a given species, there is a large variability of hemoglobin concentrations and changes in its expression level. The response to temperature is stronger in *Daphnia magna*, the changes induced by hypoxia are stronger in *Daphnia pulex* (Table 6.2) (Lamkemeyer et al. 2003; Gerke et al. 2011). Analysing clones from a lake inhabited by *Daphnia galeata, Daphnia longispina* (formerly *Daphnia hyalina*) and their hybrids revealed differences in the basic amount of hemoglobin present as well as in the hemoglobin increase in response to temperature changes (Fig. 6.6b; data from Pinkhaus et al. 2007). While most clones increased their hemoglobin concentration at 18 °C compared to 12 °C, a further increase upon exposition to 24 °C was only realised by one clone of *Daphnia longispina*. All studied clones of *Daphnia galeata* failed to increase hemoglobin at elevated temperatures (Pinkhaus et al. 2007). Only one hybrid was studied in which hemoglobin concentration was unchanged at all acclimation temperatures (12, 18 and 24 °C). Analyses of 39 *Daphnia magna* clones revitalised from the sediment of the Danish Lake Ring also differed in their hemoglobin content when raised at 20 °C and 30 °C, moreover the degree of hemoglobin induction varied considerably (Fig. 6.6a; data from Cambronero et al. 2018). The range of concentrations measured in this study covered 13–124 μmol/l. The highest induction potential observed revealed a doubled concentration of hemoglobin at 30 °C compared to 20 °C. In other clones, the level declined, amounting to half the value at the elevated temperature. When these clones were tested in competition experiments at both temperatures, hemoglobin-poor genotypes and hemoglobin-rich genotypes were equally successful at 20 °C, while at 30 °C, the hemoglobin-rich animals were more successful, contributing to about two thirds to the experimental population (Cambronero et al. 2018). Additionally, parameters characterising the physiological performance were measured. Here, the time of immobilisation increased more strongly in genotypes expressing high concentrations of hemoglobin (Cambronero et al. 2018). Thus, the intra-species variation of the hemoglobin pool characteristics affected the animals' physiological potentials. These results from laboratory studies raise the question on the role of variable hemoglobin concentrations and functional properties in the habitat situation.

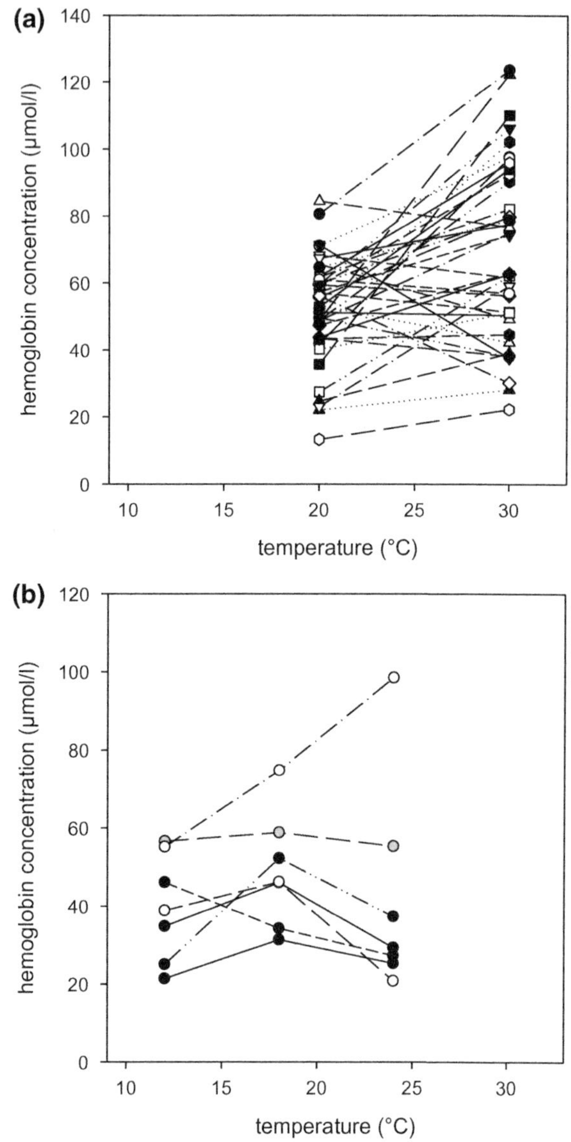

Fig. 6.6 Variation of hemoglobin induction potential in *Daphnia*. The hemoglobin concentration of 39 clonal lines of *Daphnia magna* was determined after acclimation to 20 and 30 °C (**a**). Clones of *Daphnia galeata* (black), *Daphnia longispina* (white) and their hybrid (grey) were acclimated to 12, 18 and 24 °C before their hemoglobin content was determined (**b**). The concentration range as well as the induction potential varied considerably. Data were taken from Cambronero et al. 2018 (**a**) and Pinkhaus et al. 2007 (**b**)

Consequences of Altered Hemoglobin Status for Limits of Distributions in the Animals' Habitat

The remarkable plasticity of the quantity and quality of hemoglobin in waterfleas allows these planktonic animals to inhabit water bodies of highly variable abiotic conditions. In lakes and ponds, temperature and oxygen gradients are realised on a temporal and spatial scale. Temperature varies along with seasonal processes, but in stratified water bodies, the surface water of the epilimnion can be much warmer than the lower hypolimnic layers in the summer. The opposite is true for the oxygen content, here the upper layers of high algal photosynthesis are mostly rich in oxygen, while in deeper layers, which are lacking light and are characterised by predominant degradation of organic matter, oxygen may be scarce. Thus, freshwater habitats can be considered a multidimensional space of combinations of abiotic factors affecting oxygen supply and demand. These niches would even be more complex, if the biotic parameters affecting oxygen needs were taken into consideration, as life table characteristics like size, developmental and reproductive stage, pathogen stress, parasitic burden or predation pressure must be neglected here. But there is a mutual relation between the ambient conditions experienced by the animals and their hemoglobin pools, which is adjusted by the differential isoform expression as outlined before. But vice versa, the hemoglobin status of waterflea specimens determines which conditions can be tolerated. The characteristics of oxygen transport capacity define the limiting conditions for a sufficient aerobic energy supply. Consequently, the animals' hemoglobin pool expressed affects their distribution patterns. The coexisting species *Daphnia pulex* and *Daphnia rosea* differed in their hemoglobin concentration, which was about 3 times higher in adult *D. pulex* compared to *D. rosea* of the same size (Sell 1998). The hemoglobin-poor animals were found predominatly in the upper layer of the stratified eutrophic lake Gräfenhain (Germany), where the hemoglobin-rich specimens of *D. pulex* contributed less than 10% to the waterflea assemblage. In contrast, this species made up for more than 96% in the metalimnion of low oxygen conditions (Sell 1998). Similar observations were made in a shallow pond in Melbourne, where *Daphnia carinata* specimens from the surface were poor in hemoglobin, while animals of the bottom layer had a higher hemoglobin content (Wiggins and Frappell 2002). Finally, the daphnid assemblage of lake Saidenbach (Germany) was sampled during summer stratification. In the warmer surface layers of high oxygen concentration, animals had lower hemoglobin concentrations than in the thermocline of lower temperature and oxygen content (Fig. 6.7). The determination of the animals' genotypes revealed their vertical distribution. Individuals of *Daphnia galcata* were dominant in the upper layers, while *Daphnia longispina* dominated in the thermocline. The latter is characterised by a higher basal level of hemoglobin (Pinkhaus et al. 2007). One genotype, T06, showing the highest hemoglobin induction potential, made up for 23% in the epilimnion and 39% in the deeper thermocline layer (Fig. 6.7), correlating with the higher hemoglobin concentration of animals from this depth. These examples from several daphnid species indicate that the vertical distribution of these animals in the natural habitat may be the result of the preference of ambient temperature and oxygen conditions, which match their oxygen transport capacity as determined by the hemoglobin concentration in the hemolymph. The potential to adjust hemoglobin properties to the ambient conditions thus expands the range of oxygen and temperature conditions tolerated in the animals' habitat.

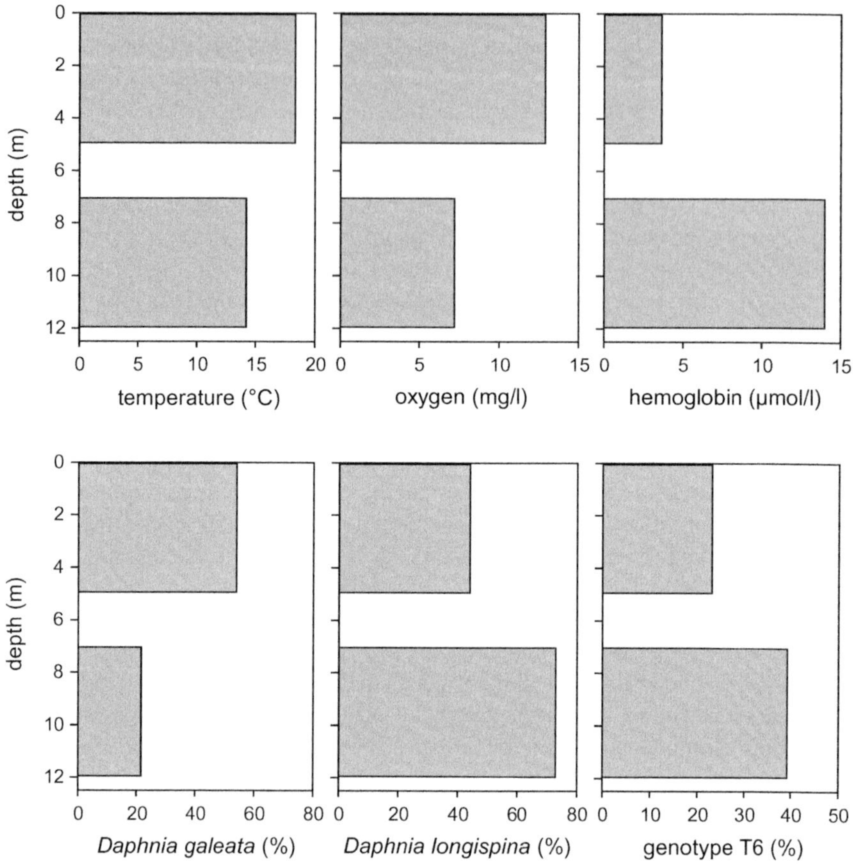

Fig. 6.7 Vertical distribution of animals varying in hemoglobin concentration. During summer stratification in lake Saidenbach, probes were taken in the epilimnion (0–5 m) and the thermocline (7–12 m). Data are given for temperature and oxygen concentration. The hemoglobin concentration of daphnids of the respective depth were analysed, and their genotypes were determined. Accordingly, the respective contribution of *Daphnia galeata*, *Daphnia longispina* and the hemoglobin-inducing genotype T06 were calculated (For details see text)

References

Berenbrink M (2006) Evolution of vertebrate haemoglobins: histidine side chains, specific buffer value and Bohr effect. Resp Physiol Neurobi 154(1–2):165–184. https://doi.org/10.1016/j.resp.2006.01.002

Burmester T (2015) Evolution of respiratory proteins across the pancrustacea. Integr Comp Biol 55(5):792–801. https://doi.org/10.1093/icb/icv079

Burmester T, Hankeln T (2014) Function and evolution of vertebrate globins. Acta Physiol 211(3):501–514. https://doi.org/10.1111/apha.12312

Burmester T, Storf J, Hasenjager A, Klawitter S, Hankeln T (2006) The hemoglobin genes of Drosophila. FEBS J 273(3):468–480. https://doi.org/10.1111/j.1742-4658.2005.05073.x

Cambronero MC, Zeis B, Orsini L (2018) Haemoglobin-mediated response to hyper-thermal stress in the keystone species Daphnia magna. Evol Appl 11(1):112–120. https://doi.org/10.1111/eva.12561

Carver TE, Brantley RE, Singleton EW, Arduini RM, Quillin ML, Phillips GN, Olson JS (1992) A novel site-directed mutant of myoglobin with an unusually high O-2 affinity and low autooxidation rate. J Biol Chem 267(20):14443–14450

Colbourne JK, Pfrender ME, Gilbert D, Thomas WK, Tucker A, Oakley TH, Tokishita S, Aerts A, Arnold GJ, Basu MK, Bauer DJ, Caceres CE, Carmel L, Casola C, Choi JH, Detter JC, Dong QF, Dusheyko S, Eads BD, Frohlich T, Geiler-Samerotte KA, Gerlach D, Hatcher P, Jogdeo S, Krijgsveld J, Kriventseva EV, Kultz D, Laforsch C, Lindquist E, Lopez J, Manak JR, Muller J, Pangilinan J, Patwardhan RP, Pitluck S, Pritham EJ, Rechtsteiner A, Rho M, Rogozin IB, Sakarya O, Salamov A, Schaack S, Shapiro H, Shiga Y, Skalitzky C, Smith Z, Souvorov A, Sung W, Tang ZJ, Tsuchiya D, Tu H, Vos H, Wang M, Wolf YI, Yamagata H, Yamada T, Ye YZ, Shaw JR, Andrews J, Crease TJ, Tang HX, Lucas SM, Robertson HM, Bork P, Koonin EV, Zdobnov EM, Grigoriev IV, Lynch M, Boore JL (2011) The ecoresponsive genome of Daphnia pulex. Science 331(6017):555–561. https://doi.org/10.1126/science.1197761

Dangott LJ, Terwilliger RC (1980) The subunit structure of Daphnia pulex hemoglobin. Comp Biochem Phys B 67(2):301–306. https://doi.org/10.1016/0305-0491(80)90148-0

Dewilde S, Van Hauwaert ML, Peeters K, Vanfleteren J, Moens L (1999) Daphnia pulex didomain hemoglobin: Structure and evolution of polymeric hemoglobins and their coding genes. Mol Biol Evol 16(9):1208–1218. https://doi.org/10.1093/oxfordjournals.molbev.a026211

Dresel EIB (1948) Passage of haemoglobin from blood into eggs of Daphnia. Nature 162(4123):736–737. https://doi.org/10.1038/162736b0

Emanuelsson O, Brunak S, von Heijne G, Nielsen H (2007) Locating proteins in the cell using TargetP. SignalP and related tools. Nat Protoc 2(4):953–971. https://doi.org/10.1038/nprot.2007.131

Fox HM (1948) The haemoglobin of Daphnia. Proc R Soc Ser B-Bio 135(879):195–212. https://doi.org/10.1098/rspb.1948.0006

Fox HM (1955) The effect of oxygen on the concentration of haem in invertebrates. Proc R Soc Ser B-Bio 143(911):203–214. https://doi.org/10.1098/rspb.1955.0005

Fox HM, Phear EA (1953) Factors Influencing Haemoglobin synthesis by Daphnia. Proc R Soc Ser B-Bio 141(903):179–189. https://doi.org/10.1098/rspb.1953.0034

Fox HM, Hardcastle SM, Dresel EIB (1949) Fluctuations in the haemoglobin content of Daphnia. Proc R Soc Ser B-Bio 136(884):388–399. https://doi.org/10.1098/rspb.1949.0032

Fox HM, Gilchrist BM, Phear EA (1950) Functions of haemoglobin in Daphnia. Nature 166(4223):609–610. https://doi.org/10.1038/166609a0

Fox HM, Gilchrist BM, Phear EA (1951) Functions of haemoglobin in Daphnia. Proc R Soc Ser B-Bio 138(893):514–527. https://doi.org/10.1098/rspb.1951.0038

Geer LY, Marchler-Bauer A, Geer RC, Han LY, He J, He SQ, Liu CL, Shi WY, Bryant SH (2010) The NCBI biosystems database. Nucleic Acids Res 38:D492–D496. https://doi.org/10.1093/nar/gkp858

Gerke P, Bording C, Zeis B, Paul RJ (2011) Adaptive haemoglobin gene control in Daphnia pulex at different oxygen and temperature conditions. Comp Biochem Phys A 159(1):56–65. https://doi.org/10.1016/j.cbpa.2011.01.017

Gilbert D (2014) Genes Daphnia Similoides Reconstr EvidentialGene. https://doi.org/10.5967/npn4-sd09

Gilbert D (2017) Genes Daphnia Similoides Reconstr EvidentialGene. https://doi.org/10.5967/f5hr-yk11

Goldmann T, Becher B, Wiedorn KH, Pirow R, Deutschbein ME, Vollmer E, Paul RJ (1999) Epipodite and fat cells as sites of hemoglobin synthesis in the branchiopod crustacean Daphnia magna. Histochem Cell Biol 112(5):335–339. https://doi.org/10.1007/Pl00007905

Gorr TA, Cahn JD, Yamagata H, Bunn HF (2004) Hypoxia-induced synthesis of hemoglobin in the crustacean Daphnia magna is hypoxia-inducible factor-dependent. J Biol Chem 279(34):36038–36047. https://doi.org/10.1074/jbc.M403981200

Gorr TA, Rider CV, Wang HY, Olmstead AW, LeBlanc GA (2006) A candidate juvenoid hormone receptor cis-element in the Daphnia magna hb2 hemoglobin gene promoter. Mol Cell Endocrinol 247(1–2):91–102. https://doi.org/10.1016/j.mce.2005.11.022

Green J (1955) Haemoglobin in the fat-cells of Daphnia. Q J Microsc Sci 96(2):173–176

Green J (1956) Variation in the haemoglobin content of Daphnia. Proc R Soc Ser B-Bio 145(919):214–232. https://doi.org/10.1098/rspb.1956.0029

Hebert PDN, Um YM, Prokopowich CD, Taylor DJ (1999) Gene conversion and evolution of daphniid hemoglobins (Crustacea, Cladocera). J Mol Evol 49(6):769–779. https://doi.org/10.1007/PI00006599

Herrmann M, Ravindran SP, Schwenk K, Cordellier M (2018) Population transcriptomics in Daphnia: the role of thermal selection. Mol Ecol 27(2):387–402. https://doi.org/10.1111/mec.14450

Hoogewijs D, Terwilliger NB, Webster KA, Powell-Coffman JA, Tokishita S, Yamagata H, Hankeln T, Burmester T, Rytkonen KT, Nikinmaa M, Abele D, Heise K, Lucassen M, Fandrey J, Maxwell RH, Pahlman S, Gorr TA (2007) From critters to cancers: bridging comparative and clinical research on oxygen sensing, HIF signaling, and adaptations towards hypoxia. Integr Comp Biol 47(4):552–577. https://doi.org/10.1093/icb/icm072

Horne FR, Beyenbach KW (1971) Physiological properties of hemoglobin in branchiopod crustacean Triops. Am J Physiol 220(6):1875–1881. https://doi.org/10.1152/ajplegacy.1971.220.6.1875

Hoshi TYN (1975) Studies on physiology and ecology of plankton. XXIX. In vivo equilibrium of oxygen with blood haemoglobin of *Daphnia magna*. Sci Rep Niigata Univ Ser D 142:1–7

Hoshi T, Kobayashi K (1971) Studies on physiology and ecology of plankton. XXV. Iron-content and milli-molar extinction coefficient of the Daphnia hemoglobin. Sci Rep Niigata Univ Ser D 7:65–68

Hoshi T, Takahashi K (1981) Studies on physiology and ecology of plankton. XXXIII. In vitro O2-response to blood haemoglobin of *Daphnia magna* with special reference to time-course in oxygenation and deoxygenation. Sci Rep Niigata Univ Ser D 18:21–28

Hoshi T, Yajima T (1970) Studies on physiology and ecology of plankton. XXIV possible role of blood haemoglobin induced by low oxygen culture in the respiration of *Daphnia magna*. Sci Rep Niigata Univ Ser D 7:107–115

Hoshi T, Yahagi N, Watanabe T (1977) Studies on physiology and ecology of plankton. XXXI. Further studies on O2-response to haemoglobin of *Daphnia magna* in vivo. Sci Rep Niigata Univ Ser D 14:7–13

Hu JM, Stiehl DP, Setzer C, Wichmann D, Shinde DA, Rehrauer H, Hradecky P, Gassmann M, Gorr TA (2011) Interaction of HIF and USF signaling pathways in human genes flanked by hypoxia-response elements and E-box palindromes. Mol Cancer Res 9(11):1520–1536. https://doi.org/10.1158/1541-7786.MCR-11-0090

Huylmans AK, Ezquerra AL, Parsch J, Cordellier M (2016) De novo transcriptome assembly and sex-biased gene expression in the cyclical parthenogenetic Daphnia galeata. Genome Biol Evol 8(10):3120–3139. https://doi.org/10.1093/gbe/evw221

Ilan E, Daniel E (1979) Structural diversity of arthropod extracellular hemoglobins. Comp Biochem Phys B 63(3):303–308. https://doi.org/10.1016/0305-0491(79)90253-0

Ilan E, Weisselberg E, Daniel E (1982) Erythrocruorin from the water-flea daphnia magna—quaternary structure and arrangement of subunits. Biochem J 207(2):297–303. https://doi.org/10.1042/Bj2070297

Kimura S, Tokishita S, Ohta T, Kobayashi M, Yamagata H (1999) Heterogeneity and differential expression under hypoxia of two-domain hemoglobin chains in the water flea. Daphnia magna. J Biol Chem 274(15):10649–10653. https://doi.org/10.1074/jbc.274.15.10649

Kobayashi M (1981) Influences of hemoglobin concentration and temperature on oxygen-uptake of Daphnia magna under low oxygen concentration. Comp Biochem Phys A 69(4):679–682
Kobayashi M (1982a) Influence of body size on hemoglobin concentration and resistance to oxygen deficiency in Daphnia magna. Comp Biochem Phys A 72(3):599–602
Kobayashi M (1982b) Relationship between oxygen-consumption and resistance to oxygen deficiency of pale Daphnia magna. Comp Biochem Phys A 73(2):239–241
Kobayashi M (1983) Estimation of the hemolymph volume in Daphnia magna by hemoglobin determination. Comp Biochem Phys A 76(4):803–805
Kobayashi M, Hoshi T (1982) Relationship between the hemoglobin concentration of Daphnia magna and the ambient oxygen concentration. Comp Biochem Phys A 72(1):247–249. https://doi.org/10.1016/0300-9629(82)90040-8
Kobayashi M, Hoshi T (1984) Analysis of respiratory role of hemoglobin in Daphnia magna. Zool Sci 1(4):523–532
Kobayashi M, Nezu T (1986) Variation of hemoglobin content in Daphnia magna. Physiol Zool 59(1):35–42. https://doi.org/10.1086/physzool.59.1.30156087
Kobayashi M, Tanaka Y (1991) Oxygen-transporting function of hemoglobin in Daphnia magna. Can J Zool 69(12):2968–2972. https://doi.org/10.1139/Z91-418
Kobayashi M, Yamagata H (2000) Structure and function of haemoglobin in Daphnia magna. Trends Comp Biol Physiol 6:163–174
Kobayashi M, Fujiki M, Suzuki T (1988) Variation in and oxygen-binding properties of Daphnia magna hemoglobin. Physiol Zool 61(5):415–419. https://doi.org/10.1086/physzool.61.5.30161263
Kobayashi M, Fujiki M, Suzuki T (1990a) Oxygen binding characteristics of Daphnia hemoglobin. J Therm Biol 15(1):57–59. https://doi.org/10.1016/0306-4565(90)90048-M
Kobayashi M, Kato S, Ninomiya M, Fujiki M, Igarashi Y, Kajita A (1990b) Hemoglobin in the parthenogenetic eggs of Daphnia magna. J Exp Zool 254(1):18–24. https://doi.org/10.1002/jez.1402540105
Kobayashi M, Nezu T, Tanaka Y (1990c) Hypoxic induction of hemoglobin-synthesis in Daphnia magna. Comp Biochem Phys A 97(4):513–517
Kobayashi M, Sato G, Ishigaki K, Igarashi Y (1994) Hill and scatchard plots of Daphnia magna hemoglobin. Comp Biochem Phys B 107(1):99–102. https://doi.org/10.1016/0305-0491(94)90230-5
Lamkemeyer T, Zeis B, Paul RJ (2003) Temperature acclimation influences temperature-related behaviour as well as oxygen-transport physiology and biochemistry in the water flea Daphnia magna. Can J Zool 81(2):237–249. https://doi.org/10.1139/Z03-001
Lamkemeyer T, Paul RJ, Stocker W, Yiallouros I, Zeis B (2005) Macromolecular isoforms of Daphnia magna haemoglobin. Biol Chem 386(11):1087–1096. https://doi.org/10.1515/Bc.2005.125
Lamkemeyer T, Zeis B, Decker H, Jaenicke E, Waschbusch D, Gebauer W, Markl J, Meissner U, Rousselot M, Zal F, Nicholson GJ, Paul RJ (2006) Molecular mass of macromolecules and subunits and the quaternary structure of hemoglobin from the microcrustacean Daphnia magna. Febs J 273(14):3393–3410. https://doi.org/10.1111/j.1742-4658.2006.05346.x
Landon MS, Stasiak RH (1983) Daphnia hemoglobin concentration as a function of depth and oxygen availability in arco lake. Minnesota. Limnol Oceanogr 28(4):731–737. https://doi.org/10.4319/lo.1983.28.4.0731
Lankester E (1871) Ueber das Vorkommen von Haemoglobin in den Muskeln der Mollusken und die Verbreitung desselben in den lebendigen Organismen. Pflüger Arch 4(1):315–320. https://doi.org/10.1007/BF01612492
Mathews AJ, Rohlfs RJ, Olson JS, Tame J, Renaud JP, Nagai K (1989) The Effects of E7 and E11 mutations on the kinetics of ligand-binding to R-state human-hemoglobin. J Biol Chem 264(28):16573–16583

Moens L, Kondo M (1978) Evidence for a dimeric form of artemia-salina extracellular hemoglobins with high-molecular-weight subunits. Eur J Biochem 82(1):65–72. https://doi.org/10.1111/j.1432-1033.1978.tb11997.x

Mu XY, Rider CV, Hwang GS, Hoy H, LeBlanc GA (2005) Covert signal disruption: anti-ecdysteroidal activity of bisphenol a involves cross talk between signaling pathways. Environ Toxicol Chem 24(1):146–152. https://doi.org/10.1897/04-063r.1

Nunes F, Spiering S, Wolf M, Wendler A, Pirow R, Paul RJ (2005) Sequencing of hemoglobin gene 4 (dmhb4) and southern blot analysis provide evidence of more than four members of the Daphnia magna globin family. Biosci Biotech Bioch 69(6):1193–1197. https://doi.org/10.1271/Bbb.69.1193

Orsini L, Gilbert D, Podicheti R, Jansen M, Brown JB, Solari OS, Spanier KI, Colbourne JK, Rush D, Decaestecker E, Asselman J, De Schamphelaere KAC, Ebert D, Haag CR, Kvist J, Laforsch C, Petrusek A, Beckerman AP, Little TJ, Chaturvedi A, Pfrender ME, De Meester L, Frilander MJ (2016) Daphnia magna transcriptome by RNA-Seq across 12 environmental stressors. Sci Data 3:160030. https://doi.org/10.1038/Sdata.2016.30

Osmulski PA, Leyko W (1986) Structure, function and physiological-role of chironomus hemoglobin. Comp Biochem Phys B 85(4):701–722. https://doi.org/10.1016/0305-0491(86)90166-5

Paul RJ, Lamkemeyer T, Maurer J, Pinkhaus O, Pirow R, Seidl M, Zeis B (2004a) Thermal acclimation in the microcrustacean Daphnia: a survey of behavioural, physiological and biochemical mechanisms. J Therm Biol 29(7–8):655–662. https://doi.org/10.1016/j.jtherbio.2004.08.035

Paul RJ, Zeis B, Lamkemeyer T, Seidl M, Pirow R (2004b) Control of oxygen transport in the microcrustacean Daphnia: regulation of haemoglobin expression as central mechanism of adaptation to different oxygen and temperature conditions. Acta Physiol Scand 182(3):259–275. https://doi.org/10.1111/j.1365-201X.2004.01362.x

Peeters K, Mertens J, Hebert P, Moens L (1990) The globin composition of Daphnia-Pulex hemoglobin and the comparison of the amino-acid-composition of invertebrate hemoglobins. Comp Biochem Phys B 97(2):369–381. https://doi.org/10.1016/0305-0491(90)90295-5

Perutz MF (1972) Nature of Heme-Heme interaction. Nature 237(5357):495–499. https://doi.org/10.1038/237495a0

Perutz MF, Wilkinson AJ, Paoli M, Dodson GG (1998) The stereochemical mechanism of the cooperative effects in hemoglobin revisited. Annu Rev Bioph Biom 27:1–34. https://doi.org/10.1146/annurev.biophys.27.1.1

Pesce A, Bolognesi M, Bocedi A, Ascenzi P, Dewilde S, Moens L, Hankeln T, Burmester T (2002) Neuroglobin and cytoglobin—Fresh blood for the vertebrate globin family. Embo Rep 3(12):1146–1151. https://doi.org/10.1093/embo-repots/kvf248

Pinkhaus O, Schwerin S, Pirow R, Zeis B, Buchen I, Gigengack U, Koch M, Horn W, Paul RJ (2007) Temporal environmental change, clonal physiology and the genetic structure of a Daphnia assemblage (D-galeata-hyalina hybrid species complex). Freshwater Biol 52(8):1537–1554. https://doi.org/10.1111/j.1365-2427.2007.01786.x

Pirow R, Wollinger F, Paul RJ (1999) The sites of respiratory gas exchange in the planktonic crustacean Daphnia magna: an in vivo study employing blood haemoglobin as an internal oxygen probe. J Exp Biol 202(22):3089–3099

Pirow R, Hellmann N, Weber RE (2007) Oxygen binding and its allosteric control in hemoglobin of the primitive branchiopod crustacean Triops cancriformis. Febs J 274(13):3374–3391. https://doi.org/10.1111/j.1742-4658.2007.05871.x

Rider CV, LeBlanc GA (2006) Atrazine stimulates hemoglobin accumulation in Daphnia magna: is it hormonal or hypoxic? Toxicol Sci 93(2):443–449. https://doi.org/10.1093/toxsci/kfl067

Rider CV, Gorr TA, Olmstead AW, Wasilak BA, Leblanc GA (2005) Stress signaling: coregulation of hemoglobin and male sex determination through a terpenoid signaling pathway in a crustacean. J Exp Biol 208(1):15–23. https://doi.org/10.1242/Jeb.01343

Rousselot M, Jaenicke E, Lamkemeyer T, Harris JR, Pirow R (2006) Native and subunit molecular mass and quarternary structure of the hemoglobin from the primitive branchiopod crustacean Triops cancriformis. Febs J 273(17):4055–4071. https://doi.org/10.1111/j.1742-4658.2006.05408.x

Schwerin S, Zeis B, Horn W, Horn H, Paul RJ (2010) Hemoglobin concentration in Daphnia (D. galeata-hyalina) from the epilimnion is related to the state of nutrition and the degree of protein homeostasis. Limnol Oceanogr 55(2):639–652. https://doi.org/10.4319/lo.2009.55.2.0639

Sell AF (1998) Adaptation to oxygen deficiency: contrasting patterns of haemoglobin synthesis in two coexisting Daphnia species. Comp Biochem Phys A 120(1):119–125. https://doi.org/10.1016/S1095-6433(98)10019-3

Semenza GL (2014a) Hypoxia-inducible factor 1 and cardiovascular disease. Annu Rev Physiol 76:39–56. https://doi.org/10.1146/annurev-physiol-021113-170322

Semenza GL (2014b) Oxygen sensing, hypoxia-inducible factors, and disease pathophysiology. Annu Rev Pathol-Mech 9:47–71. https://doi.org/10.1146/annurev-pathol-012513-104720

Smaridge MW (1954) Iron excretion by Daphnia during haemoglobin loss. Nature 173(4408):782–783. https://doi.org/10.1038/173782a0

Smaridge MW (1956) Distribution of iron in Daphnia in relation to haemoglobin synthesis and breakdown. Q J Microsc Sci 97(2):205–214

Storz JF (2018) Hemoglobin. insights into protein structure, function and evolution. Oxford University Press, London

Sugano H, Hoshi T (1971) Purification and properties of blood hemoglobin from fresh-water cladocera, Moina macrocopa and Daphnia magna. Biochim Biophys Acta 229(2):349–358. https://doi.org/10.1016/0005-2795(71)90194-2

Sutton RA, Hebert PDN (2002) Patterns of sequence divergence in Daphniid hemoglobin genes. J Mol Evol 55(4):375–385. https://doi.org/10.1007/s00239-002-0334-2

Svedberg T, Eriksson-Quensel IB (1934) The molecular weight of erythrocruorin. II. J Am Chem Soc 56856(8):1700–1706

Terwilliger NB (1991) Arthropod (Cyamus scammoni, Amphipoda) hemoglobin structure and function. In: Vinogradov SN, Kapp OH (eds) Structure and function of invertebrate oxygen carriers. Springer, New York, pp 59–64. https://doi.org/10.1007/978-1-4612-3174-5

Terwilliger NB (2008) Whale rider: the co-occurrence of haemoglobin and haemocyanin in Cyamus scammoni. In: Bolognesi M, di Prisco G, Verde C (eds) Dioxygen binding and sensing proteins, vol 9. Springer, Milan, pp 203–209. https://doi.org/10.1007/978-88-470-0807-6

Tokishita S, Shiga Y, Kimura S, Ohta T, Kobayashi M, Hanazato T, Yamagata H (1997) Cloning and analysis of a cDNA encoding a two-domain hemoglobin chain from the water flea Daphnia magna. Gene 189(1):73–78. https://doi.org/10.1016/S0378-1119(96)00836-0

Tokishita SI, Kimura S, Mandokoro Y, Kato K, Shiga Y, Takahashi Y, Ohta T, Yamagata H (2006) Tissue specific expression of a bHLH-PAS protein homologous to ARNT during the development of crustacean Daphnia magna. Gene 376(2):231–239. https://doi.org/10.1016/j.gene.2006.03.022

Wang GL, Jiang BH, Rue EA, Semenza GL (1995) Hypoxia-inducible factor-1 is a basic-helix-loop-helix-PAS heterodimer regulated by cellular O2 tension. P Natl Acad Sci USA 92(12):5510–5514. https://doi.org/10.1073/pnas.92.12.5510

Weber AK, Pirow R (2009) Physiological responses of Daphnia pulex to acid stress. BMC Physiol 9:9. https://doi.org/10.1186/1472-6793-9-9

Weber RE, Vinogradov SN (2001) Nonvertebrate hemoglobins: functions and molecular adaptations. Physiol Rev 81(2):569–628

Wiggins PR, Frappell PB (2002) Behavioural thermoregulation in Daphnia carinata from different depths of a natural water body: influence of environmental oxygen levels and temperature. Comp Biochem Phys A 133(3):771–780. https://doi.org/10.1016/S1095-6433(02)00238-6

Wolf GH, Smet J, Decleir W (1983) Oxygen binding-properties of hemoglobins from Daphnia pulex (De Geer). Comp Biochem Phys A 75(2):261–265. https://doi.org/10.1016/0300-9629(83)90080-4

Yamauchi K, Ochiai T (1984) Dissociation of the extracellular hemoglobin of Daphnia magna. Comp Biochem Phys B 79(3):465–471. https://doi.org/10.1016/0305-0491(84)90407-3
Zeis B, Becher B, Goldmann T, Clark R, Vollmer E, Bolke B, Bredebusch I, Lamkemeyer T, Pinkhaus O, Pirow R, Paul RJ (2003a) Differential haemoglobin gene expression in the crustacean Daphnia magna exposed to different oxygen partial pressures. Biol Chem 384(8):1133–1145. https://doi.org/10.1515/Bc.2003.126
Zeis B, Becher B, Lamkemeyer T, Rolf S, Pirow R, Paul RJ (2003b) The process of hypoxic induction of Daphnia magna hemoglobin: subunit composition and functional properties. Comp Biochem Phys B 134(2):243–252. https://doi.org/10.1016/S1096-4959(02)00253-1
Zeis B, Lamkemeyer T, Paul RJ (2004) Molecular adaptation of Daphnia magna hemoglobin. Micron 35(1–2):47–49. https://doi.org/10.1016/j.micron.2003.10.015
Zeis B, Buers I, Morawe T, Paul RJ (2009) The role of the lactate dehydrogenase of Daphnia magna and Daphnia pulex for the tolerance of elevated temperatures. Comp Biochem Phys A 154(1):S8–S8. https://doi.org/10.1016/j.cbpa.2009.05.035
Zeis B, Becker D, Gerke P, Koch M, Paul RJ (2013) Hypoxia-inducible haemoglobins of Daphnia pulex and their role in the response to acute and chronic temperature increase. Bba-Proteins Proteom 1834(9):1704–1710. https://doi.org/10.1016/j.bbapap.2013.01.036

Chapter 7
Molluscan Hemocyanins

Sanae Kato, Takashi Matsui and Yoshikazu Tanaka

Abstract Instead of the red blood of vertebrates, most molluscs have blue hemolymph containing hemocyanin, a type-3 copper-containing protein. The hemoglobin of vertebrate blood is replaced in most molluscs with hemocyanin, which plays the role of an oxygen transporter. Oxygen-binding in hemocyanin changes its hue from colorless deoxygenated hemocyanin into blue oxygenated hemocyanin. Molecules of molluscan hemocyanin are huge, cylindrical multimeric proteins—one of the largest protein molecules in the natural world. Their huge molecular weight (from 3.3 MDa to more than 10 MDa) are the defining characteristic of molluscan hemocyanin, a property that has complicated structural analysis of the molecules for a long time. Recently, the structural analysis of a cephalopod (squid) hemocyanin has succeeded using a hybrid method employing both X-ray crystallography and cryo-EM. In a biochemical breakthrough for molluscan hemocyanin, the first quaternary structure with atomic resolution is on the verge of solving the mystery of molluscan hemocyanin. Here we describe the latest information about the molecular structure, classification and evolution of the molecule, and the physiology of molluscan hemocyanin.

Keywords Hemocyanin · Oxygen transporter · Protein structure · Electron microscopy · X-ray crystallography · Cryo-electron microscopy · Glycoprotein · Oxygen binding protein

---

S. Kato (✉)
Faculty of Fisheries, Kagoshima University, Kagoshima 890-0056, Japan
e-mail: kato@fish.kagoshima-u.ac.jp

Course of Biological Science and Technology, The United Graduate School of Agricultural Sciences, Kagoshima University, Kagoshima 890-0056, Japan

T. Matsui
School of Science, Kitasato University, Sagamihara 252-0373, Japan
e-mail: matsui@kitasato-u.ac.jp

Y. Tanaka
Graduate School of Life Sciences, Tohoku University, Sendai 980-8577, Japan
e-mail: yoshikazu.tanaka@tohoku.ac.jp

U. Hoeger and J. R. Harris (eds.), *Vertebrate and Invertebrate Respiratory Proteins, Lipoproteins and other Body Fluid Proteins*, Subcellular Biochemistry 94,
https://doi.org/10.1007/978-3-030-41769-7_7

Abbreviations

EM	Electron microscopy
FU	Functional unit
PDB	Protein data bank
TEM	Transmission electron microscopy
SPHIRE, SPARX	For High-Resolution Electron Microscopy

Introduction

The hemolymph of molluscs is analogous to the blood of vertebrates. In contrast to vertebrate red blood containing iron-binding hemoglobin, most molluscs have blue hemolymph containing hemocyanin, a type-3 copper-binding protein, as the respiratory protein. Hemocyanin is the main protein component in the hemolymph, and functions as an oxygen transporter; that is, hemocyanin is an essential protein to support life in most molluscs. An active site of hemocyanin contains two copper atoms and six histidine residues. When an oxygen molecule (O_2) binds these copper atoms to form a Cu_2O_2 cluster, the oxygenated hemocyanins change their color, becoming blue, whilst deoxygenated hemocyanins are colorless. Unlike vertebrate hemoglobins, which are found in the blood cells, molluscan hemocyanins are huge extracellular glycoproteins, freely suspended directly in the hemolymph. These characteristics of hemocyanin are common to both molluscan and arthropod hemocyanins, but their molecular structures are entirely different. Arthropod hemocyanins are multi-hexamers of subunits containing a single active site, whereas molluscan hemocyanins are hollow cylindrical decamers or multi-decamers made up of subunits of 330–550 kDa, each of which comprises more than seven paralogous functional units (FUs), each of which includes a single active site (Markl 2013). In other words, molluscan hemocyanins are huge molecules each containing at least 70 active sites. With a molecular mass of approximately 3.3 MDa to 13.5 MDa, they are one of the largest proteins in the natural world. Why did molluscs selected this huge protein assembly as an oxygen transporter? Are the macromolecule hemocyanins containing numerous oxygen-binding sites beneficial for molluscan survival? To solve the above-mentioned mysteries, an understanding of the molecular structure of molluscan hemocyanins is necessary.

The structural biology of molluscan hemocyanin has posed difficulties for a long time, due to their huge molecular weight, unstable subunit assembly, and complicated molecular structure. Sample preparation of molluscan hemocyanins for structural biology studies would appear to be relatively straightforward because molluscan hemocyanins are freely suspended at a high concentration in the hemolymph. However, some molluscan hemocyanins easily dissociate from decamer/multi-decamer molecules into subunits in vitro. It is hard to isolate subunits dissociated from their decamer/multi-decamer molecules, because general biochemical techniques are not suitable for the treatment of molecules of great size (some are above 10 MDa), or of subunits comprising approximately 400 kDa. The first step in successfully overcoming these difficulties is the cautious treatment of the hemocyanin sample so as not to dissociate the molecule into subunits.

The second step is the structural analysis of huge molecules with mega-Da mass. Observation of molluscan hemocyanins by transmission electron microscopy (TEM) reveals that the molecules are commonly of a hollow cylindrical shape, comprising a cylindrical outer wall region, and several inner collar domains located inside the wall. Taking advantage of these molecular characteristics of molluscan hemocyanin—that is, the huge size of the entire molecule, and the paralogous structure of its component FUs—the structure has been investigated essentially by the cooperation of two different types of analyses: cryo-electron microscopy (cryo-EM) analysis for the entire molecule as a top-down study (Boisset and Mouche 2000; Gatsogiannis et al. 2007, 2015; Gatsogiannis and Markl 2009; Lamy et al. 1998; Meissner et al. 2000; Zhu et al. 2014), and high resolution X-ray crystallography for FUs as a bottom-up study (Cuff et al. 1998; Jaenicke et al. 2010, 2011; Perbandt et al. 2003). Many molluscan hemocyanins have been structurally investigated by superposing the high resolution crystal structure of some FUs on an electron density map of the cryo-EM on the assumption that all FUs have a common structure. These structures however, have not satisfied the consideration of the differences in each FU structure, and the resolutions of these whole structures were low.

The third step is high resolution structural analysis of the decamer molecules by which FUs assemble in a complicated way. Recently, there has been some progress in the structural analysis of molluscan hemocyanin (Zhang et al. 2013). These workers reported a near-atomic resolution cryo-EM structure of *Haliotis diversicolor* hemocyanin. In the study, the Cα model was constructed based solely on the density of cryo-EM. Gai et al. further reported the first X-ray crystallography of the entire squid hemocyanin (Matsuno et al. 2015; Gai et al. 2015). The atomic structure model—including side chain and carbohydrates—of the wall region was revealed at a resolution of 3.0 Å; the inner collar regions, however, were not resolved. Although the X-ray structure showed a typical hollow cylindrical wall with D_5 symmetry, the D_5 symmetry of the inner structure was apparently incorrect. The inner collar structure has subsequently been resolved using cryo-EM and an optimized refinement protocol implemented in *SPHIRE* (Tanaka et al. 2019). This latest structure is the first entire atomic molecular structure of molluscan hemocyanin.

These recent breakthroughs have given us a far deeper knowledge of the hemocyanin structure. In this chapter, we describe the molecular characteristics of molluscan hemocyanin mainly from a structural point of view, covering classification according to their molecular structures, evolutionary significance, molecular structure, glycosylation, and physiological function.

Classification of Molluscan Hemocyanins

Molecules of molluscan hemocyanin are multi-subunit assemblies, categorized into two groups by the number of compositional subunits. The molecules of one group are decamer molecules, each containing 10 subunits, and comprising the cephalopod hemocyanin group. The molecules of the other group are di-/multi-decamer molecules and comprise the gastropod hemocyanin group. Subunits of molluscan hemocyanin are composed of several paralogous FUs. FUs are classified into basal FUs (FU-a, -b, -c, -d, -e, -f, -g, and -h), and their relative FUs (FU-d*, -f1, -f2, -f3, -f4, -f5, and -f6) according to the sequence similarity. Among the basal FUs, six basal FUs—FU-a, -b, -c, -d, -e, and -f—are situated in the outer wall region of the molecule. The remaining two basal FUs—FU-g and -h—are located in inner collar domains located inside the wall. FU-h is an exceptional FU among basal FUs, with a molecular mass larger than that of other FUs, due to an extra cupredoxin-like domain at the C-terminus (Jaenicke et al. 2010). The relative FUs are located not in the wall region but rather in the inner collar domains. All subunits of both the gastropod and the cephalopod groups commonly contain basal FUs except for FU-h—namely, seven basal FUs: FU-a, -b, -c, -d, -e, -f, and -g. Most subunits possess additional FU(s): FU-h and/or relative FUs. FU-h, while not contained in subunits of the cephalopod group, is present in the subunits of the gastropod group. In other words, subunits containing FU-h comprise di-/multi-decamer molecules, while mono-decamer molecules do not have subunits containing FU-h. Both the gastropod and the cephalopod group hemocyanins are further classified into two sub-groups, respectively, by the number of composite FUs. Taken together, molluscan hemocyanins can be classified into four types according to their subunit composition and overall architecture—namely, (1) type 1, (2) type 2, (3) type 3, and (4) type 4 (Fig. 7.1).

Type 1 molluscan hemocyanin is typically that of gastropods. Most gastropods, such as keyhole limpets and abalone, have this type of hemocyanin—a hollow cylindrical structure with a wall and inner collar regions, which assembles into mostly di-decamers with a molecular size of 8 MDa. The di-decamer is stacked back to back by two decamers (Fig. 7.1a). The subunit of type 1 comprises all basal FUs (FU-a, -b, -c, -d, -e, -f, -g, and -h) (Fig. 7.1a). The wall region comprises FU-a, -b, -c, -d, -e, and -f, whereas FU-g and -h are located in the collar region (Lieb et al. 2000; Lieb and Markl 2004; Altenhein et al. 2002; Gatsogiannis and Markl 2009; Markl 2013). However, polyplacophora hemocyanin, categorized in this type, exceptionally does not form a di-decamer, forming instead a stable decamer, in spite of the identical FU composition (Harris et al. 2004; Lambert et al. 1994).

Type 2 is hemocyanin of cerithioid snails of the gastropod group. This type of hemocyanin is called mega-hemocyanin because the molecular size is approx. 13.5 MDa, comprising a tri-decamer assembly of subunits (Fig. 7.1b). Type 2 hemocyanin includes two different subunits with a molecular mass of 550 kDa (mega-subunit) and 400 kDa (typical-subunit). Ten mega-subunits and 20 typical subunits assemble into the tri-decamer with a molecular mass of 13.5 MDa. Although the typical-subunit has FU composition similar to that of the keyhole limpet type, the mega-subunit

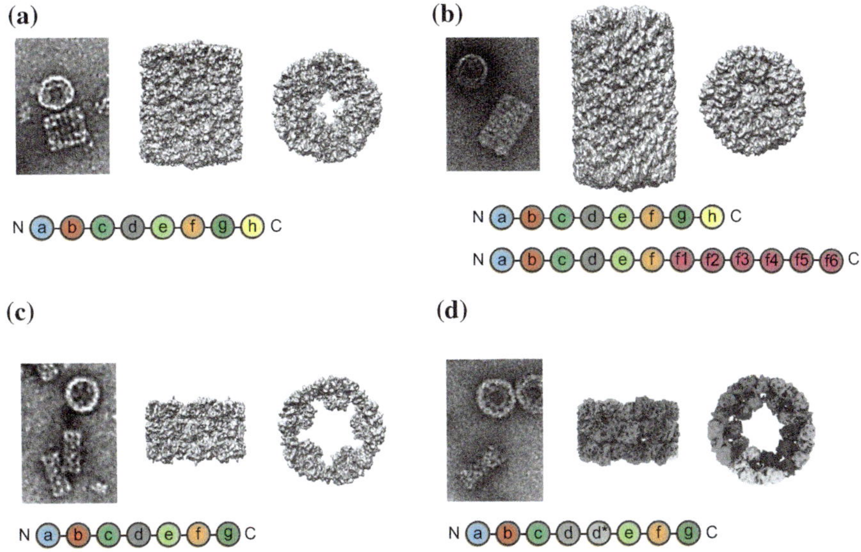

Fig. 7.1 Structure of molluscan hemocyanin and its subunit. **a–d** Representative area of negatively stained transmission electron microscopy (TEM) image (left), scheme of hemocyanin subunit (bottom), and three-dimensional (3D) model of four molluscan hemocyanin types (right) are shown. **a** Type 1 hemocyanin (Keyhole limpet-type). (TEM image: *Nordotis discus hannai* hemocyanin; 3D structure: cryo-electron microscopy (cryo-EM) structure of keyhole limpet hemocyanin.) **b** Type 2 hemocyanin (Mega-hemocyanin, *Melanoides tuberculata*). **c** Type 3 hemocyanin (Nautilus-type). (TEM image: *Enterooctopus dofleini* hemocyanin; 3D structure: *Nautilus* hemocyanin). **d** Type 4 hemocyanin (Squid-type). (*Todarodes pacificus* hemocyanin; 3D structure: cryo-EM structure, side-view and top view). *C* C-terminus, *N* N-terminus. Scheme of hemocyanin subunits: *a*, *b*, *c*, *d*, *e*, *f*, *g*, *h* basal functional units (FUs) of molluscan hemocyanins, *d**, *f1*, *f2*, *f3*, *f4*, *f5*, *f6* relative FUs. Created based upon figure from Kato et al. (2018)

comprises 12 FUs (FU-a, -b, -c, -d, -e, -f, -f1, -f2, -f3, -f4, -f5, and -f6), in which, instead of FU-g and -h, 6 additional special FUs are located at the C-terminus (Lieb et al. 2010; Gatsogiannis et al. 2015) (Fig. 7.1b). As with type 1 hemocyanins, the wall region comprises FU-a, -b, -c, -d, -e, and -f. The remaining FUs—that is, FU-g, -h, and the additional FU-f relatives (FU-f1, -f2, -f3, -f4, -f5, and -f6), are located in the collar region, making up the peculiar inner core, and filling up the central hollow.

Type 3 are nautilus and octopus hemocyanins. This type of hemocyanin is a decamer with a molecular size of approx. 3.5 MDa. Ten subunits assemble into a hollow cylindrical decamer with a wall and five inner collar regions (Miller et al. 1988; Bergmann et al. 2006; Gatsogiannis et al. 2007; Markl 2013) (Fig. 7.1c). Consequently, the height is approximately half that of type 1. The subunit comprises seven basal FUs (FU-a, -b, -c, -d, -e, -f, and -g), and lacks FU-h (Fig. 7.1c). As with type 1 and 2 hemocyanins, the wall region comprises FU-a, -b, -c, -d, -e, and -f, while the collar region contains only FU-g; therefore the central hollow is wider than that of type 1 (Miller et al. 1990).

Type 4 are decapodiformes of cephalopods—namely, squid or cuttlefish. This type of hemocyanin is a decamer with a molecular size of 3.8 MDa (Fig. 7.1d). The subunit of type 4 hemocyanin includes eight FUs, the number of FUs of the subunit being the same as that of the type 1 subunit; the composition of FUs, however, is different from that of type 1. The type 4 subunit comprises FU-a, -b, -c, -d, -d*, -e, -f, and -g (Fig. 7.1d). The subunit comprises seven basal FUs and, like type 3 subunits, lacks FU-h, and has FU-d* situated as the fifth FU. FU-d* is considered to be generated by gene duplication of FU-d (Lamy et al. 1998; Boisset and Mouche 2000; Thonig et al. 2014; Gai et al. 2015). The wall region comprises FU-a, -b, -c, -d, -e, and -f, as do other types of hemocyanin, whereas the collar domains contains FU-d* and -g. The collar domain includes center FU (FU-d*) as well as C-terminus FU (FU-g). Although the entire molecule has a hollow cylindrical structure with a size similar to that of type 3, its top view of the negative staining images shows inner collar domains more crowded than those of type 3, due to the presence of FU-d* additionally located in the collar region (Fig. 7.1d).

Molecular Structures

Structure of the FU

The crystal structure of the entire squid hemocyanin revealed that all basal FUs except FU-h have relatively similar structures regardless of their location on the wall or collar region (Gai et al. 2015; Kato et al. 2018). The similarity of these FUs indicates that the huge cylindrical molecule is built up through the assembly of similar structural building blocks. Crystallography of isolated FU-e, FU-g, and FU-h has determined their higher resolution structure (Cuff et al. 1998; Perbandt et al. 2003; Jaenicke et al. 2010, 2011). The structure of FUs commonly comprises two domains: the N-terminal core domain (NTD) having α-helices mainly, and the C-terminal β-sandwich domain (CTD) (Fig. 7.2a). It is important to note that FU-h has an extra domain—the cupredoxin domain—at the C-terminus, in addition to the common FU structure comprising the NTD and the CTD (Fig. 7.2b).

At their active sites, FUs possess a type-3 copper located in the NTD for binding to oxygen (Fig. 7.3c). Two oxygen atoms are bound between two copper atoms, Cu1 and Cu2, to form a Cu_2O_2 cluster (Fig. 7.2c). The copper ions are coordinated by six His residues: Cu1, coordinated by His41, His60, and His69, and Cu2, coordinated by His179, His183, and His210. These residues are completely conserved among all FUs. Importantly, His60 forms a thioether bridge between its Cε atom and a sulfur atom of Cys58 (Fig. 7.3c). The geometry around the Cu_2O_2-binding site is well conserved in all FUs. Each FU contains an oxygen-binding site, so that the entire hemocyanin has the same numerical oxygen-binding site as FU. That is, type 1 hemocyanin comprising 160 FUs has 160 oxygen-binding sites, and type 4 hemocyanin has 80 oxygen-binding sites.

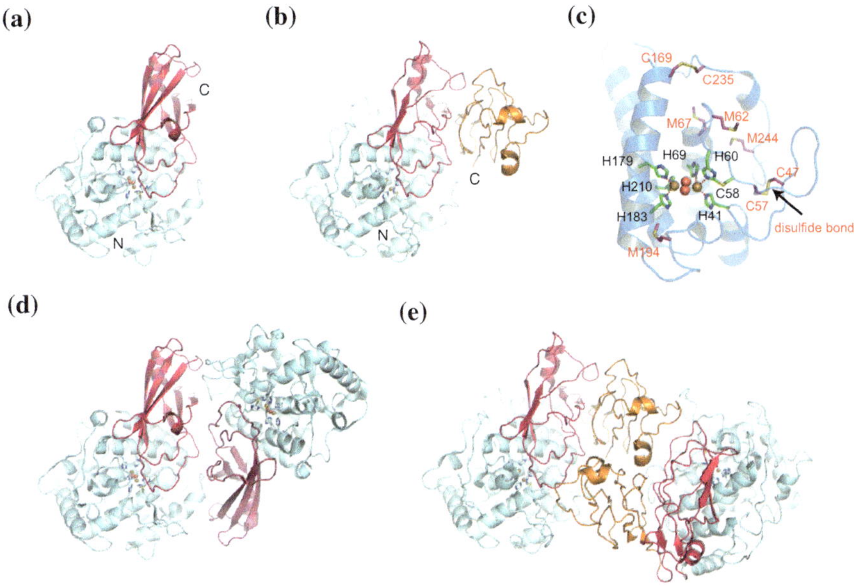

Fig. 7.2 Structures of the functional units. **a** Ribbon diagrams of FU-g. **b** Ribbon diagram of FU-h. **c** Active sites of FU. His coordinating with coppers and Cys58 forming a thioether linkage to His60 are shown as green sticks. Highly conserved sulfur-containing residues surrounding the Cu_2O_2-binding site (pink sticks). **d** Structure of FU-g dimer. **e** Structure of FU-h dimer. Cu and O atoms are shown as red and orange spheres, respectively. C C-terminus, N N-terminus. The N-terminal core domain (NTD), C-terminal β-sandwich domain (CTD), and cupredoxin domain are shown in cyan, magenta, and orange, respectively. Created based upon figures from Kato et al. (2018) and Gai et al. (2015)

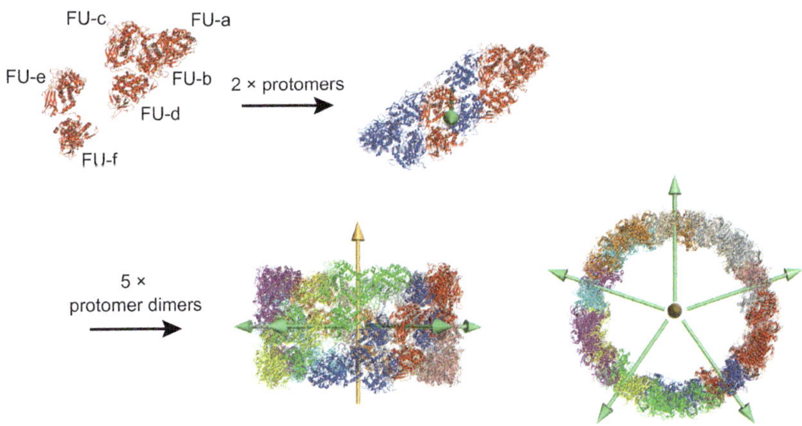

Fig. 7.3 Architecture of the wall region. Two subunits assemble into one plate-like subunit dimer with a two-fold symmetry, shown as a green arrow. Five subunit dimers assemble to form a decamer with a five-fold symmetry, shown as an orange arrow, which generates D_5 symmetry. Each subunit is represented by a separate color. Reprinted from Kato et al. (2018), with permission

Structure of Wall Region

In this section, we explain the architecture and assembly of the hemocyanin wall region. All molluscan hemocyanin molecules are constructed with a cylindrical decameric structure, with a common wall structure. In the case of gastropod hemocyanin, type 1 and type 2 hemocyanin are formed by stacking two and three identical decameric walls, respectively (Boisset and Mouche 2000; Gatsogiannis and Markl 2009; Gatsogiannis et al. 2007, 2015; Harris et al. 2004; Lamy et al. 1998; Lieb et al. 2010; Meissner et al. 2007; Zhang et al. 2013). Recently, the wall structure has solved in detail by the high resolution cryo-EM structure of *Haliotis diversicolor* hemocyanin (Zhang et al. 2013) and by crystal structure of squid hemocyanin at 3.0 Å resolution (Gai et al. 2015).

In all known hemocyanins, the architecture of the wall region, including FU composition, is common. The wall region comprises ten subunits assembled with D_5 symmetry. Namely, two subunits assemble to form a plate-like subunit dimer with 2-fold symmetry, and five of these plate-like subunit dimers are arranged with 5-fold symmetry resulting in a D_5 symmetrical cylinder (Fig. 7.3). The wall region comprises FU-a, -b, -c, -d, -e, and -f, and all the FUs form similar dimers (FU dimers) within the plate-like subunit dimer, with FU-a and -b (FU-(a-b)), FU-c and -f (FU-(c-f)), and FU-d and -e (FU-(d-e)) assembling to quite similar *pseudo* 2-fold symmetrical dimers, respectively (Fig. 7.4a–e). That is, in the plate-like subunit dimer, there are six FU dimers (Fig. 7.4b). The formation of symmetrical FU dimers in the plate-like dimer is essential for the construction of the entire wall region, such that the FU dimer is an essential structural unit. Intriguingly, using FU-e and -f, two subunits are tangled by domain swapping during the step of forming the plate-like subunit dimer (Fig. 7.3). Due to the domain swapping, two of the FU dimers—FU-(d-e) and FU-(c-f)—are formed between the subunits, meaning that four of the six FU dimers within the plate-like subunit dimer are formed by inter-subunit interaction.

The plate-like subunit dimer is further divided into three parts: the top, middle, and bottom regions (Fig. 7.4a). The top and bottom regions comprise two different FU dimers, FU-(a–b) and FU-(c–f) (Fig. 7.4f, h). The middle region comprises two identical FU dimers, FU-(d–e) (Fig. 7.4g). The relative orientation of the four FUs in the middle region is relatively similar to those in the top and bottom regions. These observations indicate that the dimer of FU dimers is a unit forming the plate-like subunit dimer, in which three units are stacked.

To summarize, in the structure of the wall region, the FU dimer plays an essential role as a building element in the architecture of the cylindrical wall. A couple of FU dimers further associate into a dimer of FU dimers, which stacks to form the plate-like subunit dimer. Five of the subunit dimers further assemble to form the wall region with 5-fold symmetry, which generates the D_5 symmetrical cylinder. This hierarchical architecture is generally observed in all known hemocyanin structures.

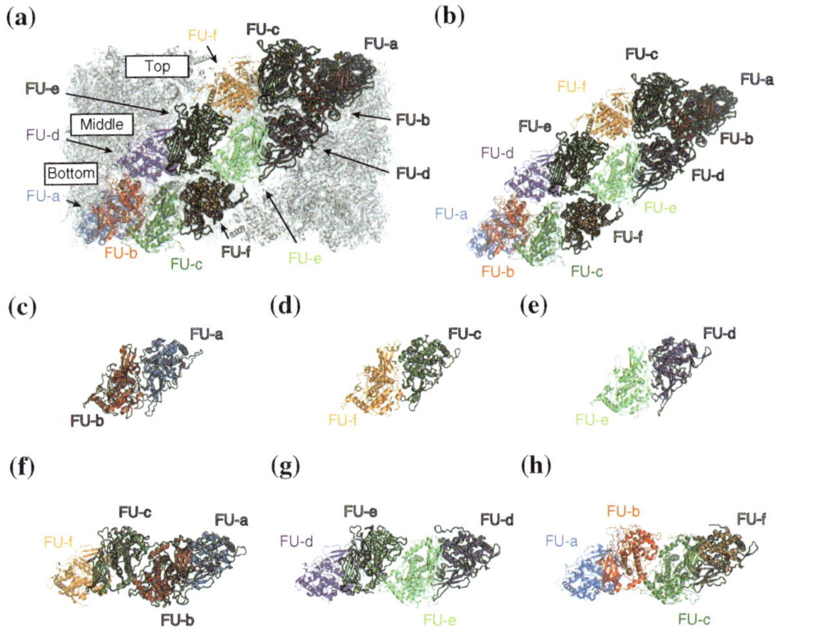

Fig. 7.4 Architecture of the plate-like dimer. **a** FUs in one plate-like dimer are highlighted in separate colors. Outline of one the subunit is highlighted. **b** Architecture of plate-like dimer. **c–e** Three types of FU dimers comprising the plate-like dimer. **f–h** Relative orientation of the four FUs in the top (**f**), middle (**g**), and bottom regions (**h**) are shown. Each subunits in one plate-like dimer are represented by light and shade colors, respectively. Reprinted from Kato et al. (2018), with permission

Type 1 (Keyhole Limpet-Type)

Type 1 hemocyanin, of which keyhole limpet hemocyanin is representative, is a typical gastropod hemocyanin. Type 1 hemocyanin has been extensively studied. The cryo-EM structure has been reported for keyhole limpet hemocyanin and *H. diversicolor* hemocyanin (PDB ID: 3J32) (Zhang et al. 2013) (Fig. 7.5). This type of hemocyanin is di-decamer and includes all basal FUs—FU-a, -b, -c, -d, -e, -f, -g and -h. The di-decamer is formed by the stacking of two identical decamers in a tail-to-tail manner. Six FUs in N-terminal segment, FU-a, -b, -c, -d, -e, and -f, form the wall region. In the wall region, two of the decameric walls are stacked with D_5 symmetry. On the other hand, FU-g and -h in C-terminal segment form the inner collar region, the structure of which is different for each type of hemocyanin; the structure of wall region, however, is common to all types of hemocyanin, as described above. In this section, we focus on the inner collar region of type 1 hemocyanin.

In the inner collar region, FU-g and -h form an FU dimer, comprising an FU-g dimer and an FU-h dimer, respectively (Fig. 7.2d, e). The FU-g dimer is formed between two subunits, each situated in different plate-like dimers in the wall region.

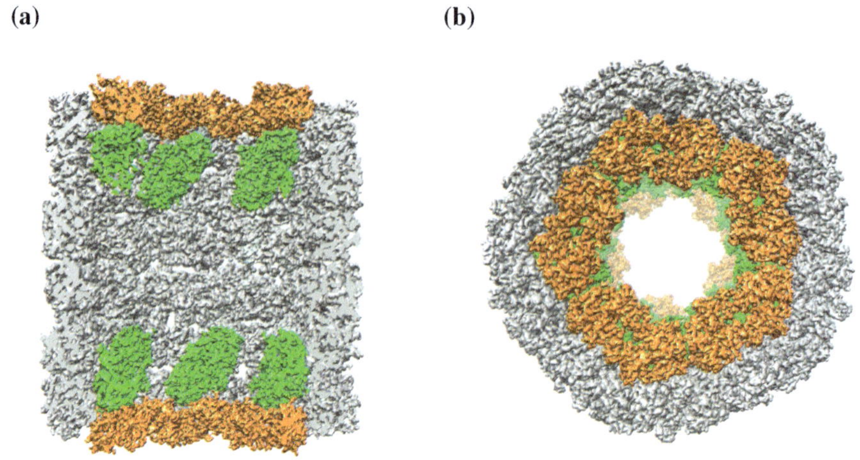

Fig. 7.5 Type 1 hemocyanin. Conformation of type 1 hemocyanin illustrated from lateral (**a**) and top views (**b**). Wall region, FU-g, and FU-h are shown as gray, green, and orange surfaces, respectively. Reprinted from Kato et al. (2018), with permission

In constructing a decameric structure, the formation of an FU-g dimer between neighboring plate-like dimers is important. The formation of the FU-g dimer is described below. Five FU-g dimers form inter-plate-like dimers arranged 5-fold symmetrically at one side of the lumen of the typical D_5 decameric wall. That is, the di-decamer molecule has a pair of the five FU-g dimers on the top and bottom (Fig. 7.5a).

FU-h covers the top of the FU-g dimer, also forming an inter-subunit dimer and five FU-h dimers arranged with 5-fold symmetry (Fig. 7.5). Since FU-h possesses an extra cuppredoxin domain at the C-terminus, the dimer assembly is different from that of other FU dimers (Fig. 7.2d). Consequently, FU-g and FU-h form homodimers with their counterparts, respectively. The combination of the subunits connected by inter-subunit FU dimers of FU-g and FU-h are identical, and the two plate-like subunit dimers are connected by both an FU-g dimer and an FU-h dimer. Accordingly, a decamer assembled in this manner has C_5 symmetry. Then, a couple of the identical C_5 decamers associate to form a di-decamer, wherein a 2-fold symmetrical association is formed through the surface with no inner collar region. The inner collar domains formed by inter-subunit FU-g and FU-h dimers are located at both outer sides of the cylinder. As a consequence of the 2-fold symmetrical association of the C_5 decamers, the entire di-decamer has D_5 symmetry.

Type 2 (Mega-Hemocyanin)

Type 2 hemocyanin is a unique hemocyanin called "mega-hemocyanin" found from cerithioid snails. This type of hemocyanin is a tri-decamer comprising two different polypeptides—one, a typical-subunit with 400 kDa having an FU composition similar to the type 2 hemocyanin comprising all basal FUs (FU-a, -b, -c, -d, -e, -f, -g, and -h), and the other, mega-subunits with 550 kDa comprising FU-a, -b, -c, -d, -e, -f, -f1, -f2, -f3, -f4, -f5, and -f6. The wall region structure is common of type 1 hemocyanin, and FU-g, -h, -f1, -f2, -f3, -f4, -f5, and -f6 form the inside structure of the cylinder of the tri-decamer. The typical-subunit forms a C_5 symmetrical decamer identical to a type 1 hemocyanin, comprising a D_5 wall, and 5-fold symmetrically arranged inter-subunit FU-g and FU-h dimers. On the other hand, the mega-subunit forms a mega-decamer, comprising a typical D_5 wall and a rhombus-shaped cylinder core (Fig. 7.6). The rhombus-shaped cylinder core comprises additional FU-f relatives, FU-f1, -f2, -f3, -f4, -f5, and -f6. Although all ten mega-subunits have identical sequences, the relative arrangement of these FUs is not identical. These FUs complicatedly fold into their own conformations to build up the rhombus-shaped cylinder core, with two-fold symmetry between the top and bottom halves. Consequently, the decamer of the mega-subunits forms a two-fold symmetrical decamer, in which a two-fold symmetrical rhombus core is enveloped by the typical D_5 wall (Lieb et al. 2010; Gatsogiannis et al. 2015). The tri-decamer is formed by stacking two copies of typical-subunit decamers onto each side of the mega-subunit decamer (Fig. 7.6d). The inner rhombus core of the mega-subunit decamer fills the inner region of the tri-decamer (Gatsogiannis et al. 2015) (Fig. 7.6b). This complicated folding manner is designated "protein origami." Due to the disappearance of the 5-fold symmetry in the mega-subunit decamer, the entire tri-decamer has two-fold symmetry.

Type 3 (Nautilus-Type)

Type 3 hemocyanin is present nautilus and octopus. This hemocyanin is a decamer, and includes seven basal FUs—FU-a, -b, -c, -d, -e, -f, and -g—and lacks FU-h. The cryo-EM structure of nautilus-type hemocyanin has been reported (EMDB-ID: 1434) (Gatsogiannis et al. 2007). The structure is a cylindrical decamer comprising a wall and five inner collar domains (Fig. 7.7).

The wall is a typical decameric wall with D_5 symmetry, as described in the previous section. The decameric wall structure of type 3 corresponds to one half of a type 1 di-decamer. Each inner collar domain comprises FU-g dimers which assemble in a manner relatively similar to that of type 1. The FU-g dimer is well superposed on the FU dimers observed in the wall region, meaning that the entire nautilus hemocyanin is therefore constructed by FU dimers. As this dimer was also observed in its crystal structure as being FU-g separated by protease treatment from an octopus hemocyanin decamer, this FU plausibly has an inherent propensity to form a homo dimer. The FU-g dimer is formed between two subunits located in neighboring plate-like subunit dimers; for example, FU-g of the subunits A and D form an inter-subunit FU-g dimer, subunit A forms a plate-like subunit dimer with subunit B, and subunit D forms a plate-like subunit dimer with subunit C (Fig. 7.7c, d). Therefore, the inner collar

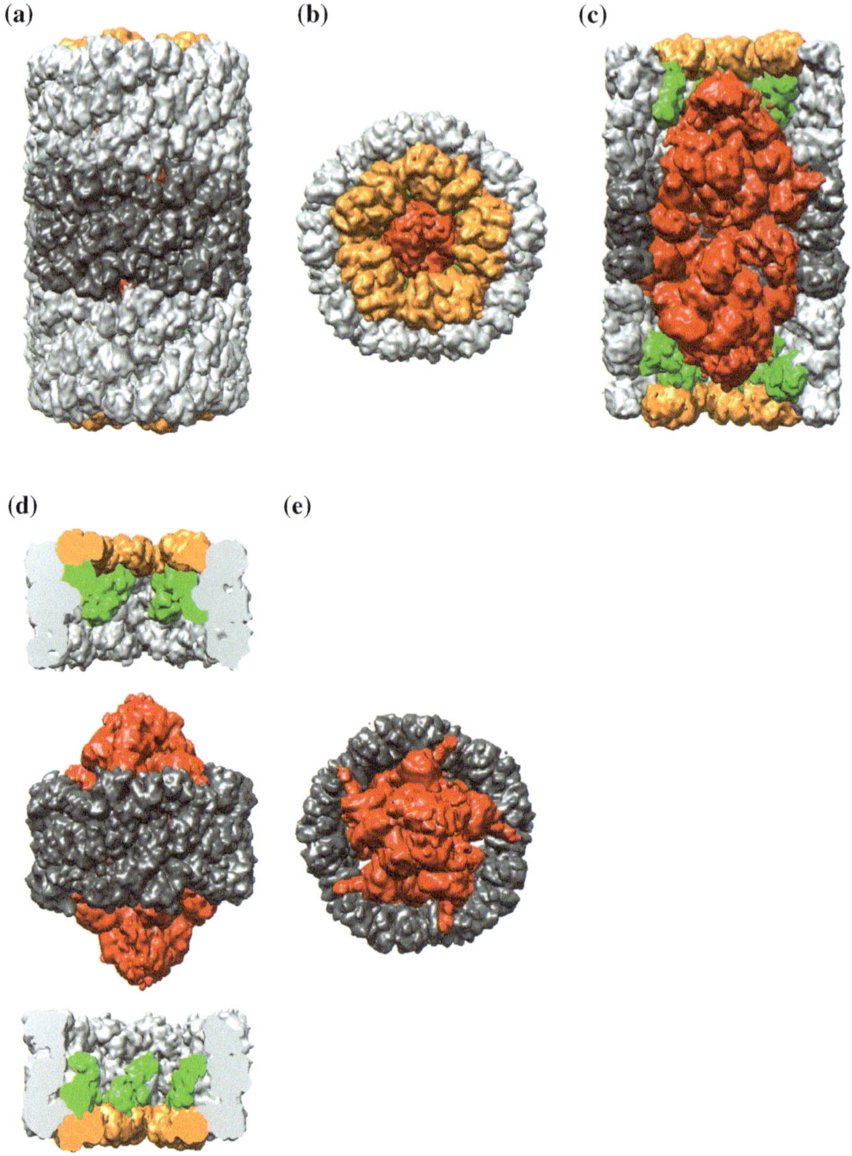

Fig. 7.6 Type 2 hemocyanin, mega-hemocyanin. **a**–**c** Architecture of mega-hemocyanin viewed from lateral (**a**) and top (**b**) and as a sectional view (**c**). **d** Composition of each decamer of mega-hemocyanin. Decamer of type 1 and rhombus-shaped cylindrical core are shown. For clarity, the sectional representation of a decamer of type 1 is also shown. **e** Top view of rhombus-shaped cylindrical core. Reprinted from Kato et al. (2018), with permission

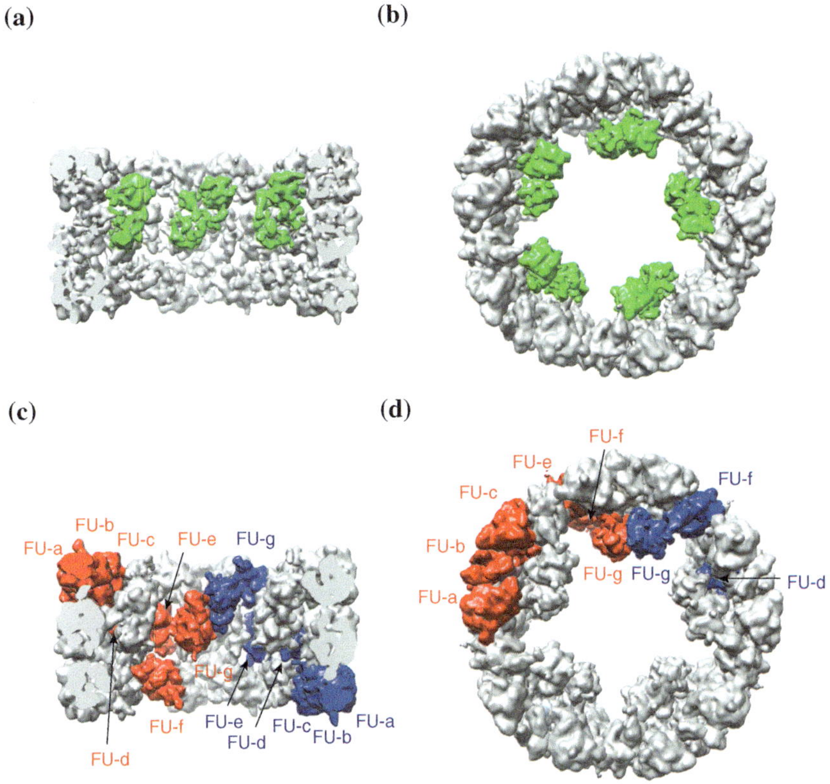

Fig. 7.7 Type 3 hemocyanin. Conformation of type 3 hemocyanin illustrated from lateral (**a**, **c**) and top views (**b**, **d**). **a**, **b** Wall region and FU-g are shown as gray and green surfaces, respectively. **c**, **d** Subunits A and D, which cooperatively form an inner collar domain, are shown in red and blue, respectively. Reprinted from Kato et al. (2018), with permission

domain is considered to reinforce the interaction between adjacent warped plate-like protomer dimers. Ten copies of FU-g form the five inter-subunit FU-g dimers as an inner collar. These collar domains are arranged in a circular manner with C5 symmetry shifted toward one cylinder opening (Gatsogiannis et al. 2007) (Fig. 7.7a). In conclusion, type 3 hemocyanin has a cylindrical wall with D_5 symmetry and five inner collar domains arranged with C_5 symmetry at one end of the cylinder.

Type 4 (Squid-Type)

Type 4 hemocyanins are decapodiformes of cephalopod, squid, and cuttlefish hemocyanin. Type 4 hemocyanin is a decamer and includes seven basal FUs—FU-a, -b, -c, -d, -e, -f, and -g, and an additional FU, FU-d*. The additional FU-d* is located as the fifth FU in the subunit, the FU order in the subunit is FU-a, -b, -c, -d, -d*, -e, -f, and -g (Fig. 7.1d). Since this FU order is the only molluscan hemocyanin type with the additional FU-d* within an N-terminal segment that usually forms the wall region, the subunit sequence is exceptional.

Structural studies of this particular hemocyanin type are crucial to unveiling the effect of this exceptional gene duplication on the overall arrangement of the decamer. The type 4 hemocyanin structure was determined by X-ray crystallography at 3.0 Å resolution (PDB ID: 4YD9) (Gai et al. 2015). The X-ray structure showed a typical decameric hollow cylindrical wall region with D_5 symmetry comprising FU-a, -b, -c, -d, -e, and -f; the inner collar structure, comprising FU-d* and -g, however, was not resolved and the D_5 symmetry of the inner collar structure was apparently incorrect. Recently, Tanaka et al. (2019) succeeded in solving the structure of type 4 hemocyanin using cryo-EM and an optimized refinement protocol for single-particle analysis of symmetry-mismatched complexes using *SPHIRE* (Moriya et al. 2017).

The so revealed squid hemocyanin structure shows a typical wall with D_5 symmetry and a characteristic asymmetric architecture of the inner collar domains (Fig. 7.8). In the cryo-EM structure of type 4, the FU-gs form typical FU dimers, but the FU-g dimers, unusually, occupy the upper and lower sides of the inner space alternately. Because of the upper and lower alternate orientation of the five FU-g dimers, the first and last dimers are necessarily located on the same side next to each other, breaking the symmetrical structure (Fig. 7.8). The duplicated FU-d*s do not form FU dimers, but are located instead in the void space opposite the typical FU-g dimers (Fig. 7.8).

In comparison with structures of ten subunits, there are two different possible conformations for each FU-g and FU-d* within a subunit, resulting in four different subunit types—namely conformers. While the four conformers possess the same primary sequence, they differ in the position of their FU-d*s and FU-gs (Fig. 7.8b, c). Figure 7.8c shows categorization of the conformers; subunits 1, 2, 5, and 6 are categorized within conformer 1; subunits 3, 4, 7, and 8 are categorized within conformer 2. Because of the distinctive conformations of subunits 9 and 10, these subunits are categorized within conformer 3 and conformer 4, respectively. The decamer contains four copies of conformer 1, four copies of conformer 2, and a single of each conformer 3 and 4. Conformer 1 and 2 subunits form homodimers exclusively, whereas conformers 3 and 4 form a heterodimer (Fig. 10.8). The conformer dimers are alternately arranged in the manner of [homodimer of conformer1]–[homodimer of conformer 2]–[homodimer of conformer 1]–[homodimer of conformer 2]–[heterodimer of conformer 3 and 4]. Type 4 hemocyanin is thereby a distinctive cylindrical decamer with a D_5 outer wall surrounding a complex asymmetric inner structure.

Glycosylation

Molluscan hemocyanin is known to possess oligosaccharides. Several FUs have consensus sequences of N-linked glycosylation—that is, Asn-X-Ser or Thr (X is not P) (Gai et al. 2015; Thonig et al. 2014; Stoeva et al. 1999, 2002; Kato et al.

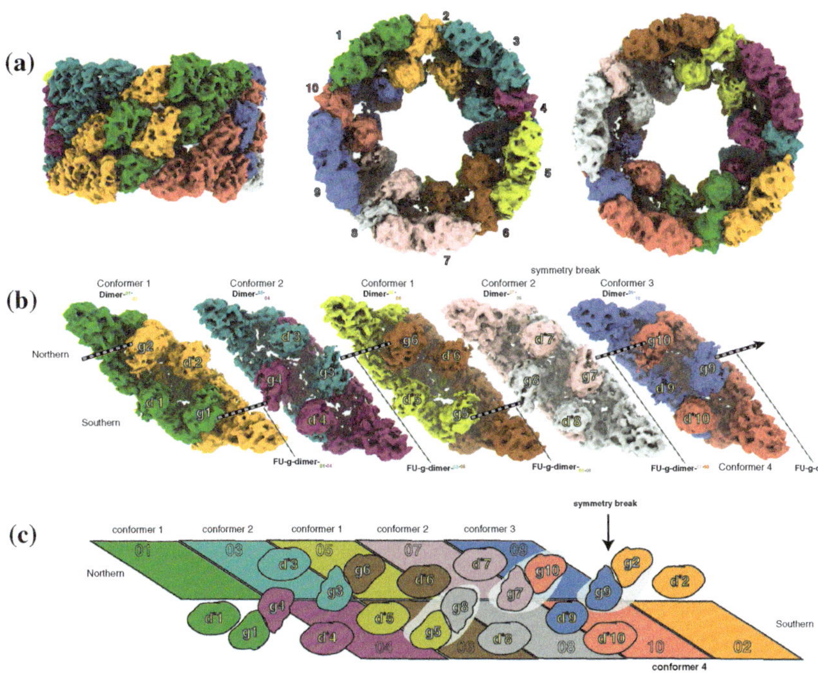

Fig. 7.8 Type 4 hemocyanin. **a** Cryo-EM structure of squid hemocyanin in lateral, top, and bottom views. The ten subunits are highlighted in color. **b** Five subunit dimers viewed from inside the cylinder. Colors correspond to those of (**a**). FUs of the inner collar (g and d*) are indicated. FU-gs forming FU-g dimers are indicated. **c** Schematic of the decameric assembly shown with the cylinder unrolled and shown from the inside. FU-g dimers are highlighted by the surrounding white background. Colors of each subunit correspond to those of (**a**) and (**b**). The arrow indicates the region of the symmetry break. Reprinted from Tanaka et al. (2019), with permission

2018). The subunit of squid *Todarodes pacificus* hemocyanin (type 4 hemocyanin) has seven of the consensus sequences, found on FU-a, -b, -d, -e, and -d*. The crystal structure of squid hemocyanin has identified respective carbohydrates, fifty of which were all located on the outer surface of the wall. These carbohydrate chains were found on five of the consensus sequences on FU-a, -b, -d, and -e. Four of these sites were located at identical positions on the CTD. FU-d possessed an additional glycan on the NTD (Fig. 7.9a). FU-d* has two consensus sequences on similar sites to that of FU-d. However, because the carbohydrate chains on the inner color region were not observed in the crystal structure, these sites are probably not glycosylated (Gielens et al. 2004; Thonig et al. 2014). All found carbohydrates were located at the interface among three subunits on the wall surface, forming carbohydrate clusters (Fig. 7.9b). Based on these observations, the carbohydrates are proposed to reinforce, through the wall regions, the interaction between plate-like subunit dimers. Most of the glycosylation sites are conserved in other cephalopod hemocyanins; cuttlefish and octopus hemocyanins possess all five glycosylated sites, and nautilus hemocyanin

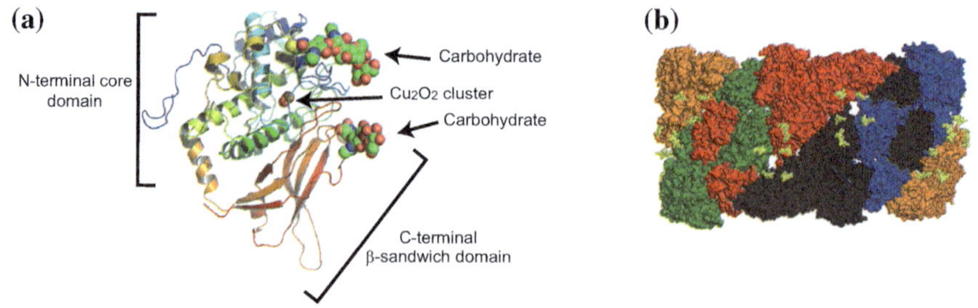

Fig. 7.9 Glycosylation of type 4 hemocyain. **a** Representative ribbon diagram of FU-d with two carbohydrate chains. Reprinted from Gai et al. (2015). **b** Carbohydrate clusters. Carbohydrate chains are represented as yellow spheres. Each protomer is illustrated as an individual color

possesses four, whereas two of the glycosylation sites are conserved in gastropod hemocyanin. In the cryo-EM structure of keyhole limpet hemocyanin, carbohydrates were bound on the rim of the cylinder wall as well as on the surface of the wall region (Gatsogiannis and Markl 2009). The additional carbohydrate might act as an inhibitor to further stacks of decamers.

Biochemical data showed the removal of the carbohydrates to induce dissociation of the decamer molecule. The subunits of squid hemocyanin could assemble to a decamer in the presence of a divalent cation such as Ca^{2+}, whereas subunits treated with glycosidase F could not reassociate into decamers (Gai et al. 2015; Kato et al. 2018). The inhibition of reassociation by deglycosylation suggests the important role played by carbohydrates in assembly. Furthermore, glucose addition caused dissociation of the decamer molecule (Gai et al. 2015). The added glucose might inhibit competitively carbohydrates to interact in the wall region. Similar phenomenon has been reported in an annelid giant hemoglobin which formed a double-layered hexagonal structure, that is to say, the giant hemoglobin was dissociated into smaller units in the presence of GalNAc, and deglycosylation of the giant hemoglobin inhibited the reassociation into the double-layered hexagonal structure (Ebina et al.1995). It's interesting that two extracellular oxygen transporters, molluscan hemocyanin and the giant hemoglobin, have such common "carbohydrate gluing".

Glycosylation is considered to contribute to the allosteric effect on oxygen-binding and release as well as interprotomer interaction. Zhang and colleagues reported that four FUs located at the interface between three protomers form a "communication cluster" involved in the allosteric effect of hemocyanin (Zhang et al. 2013). In oxygen-binding and release, in which interactions by two long loops contribute significantly, these FUs are proposed to act cooperatively. In the crystal structure of squid hemocyanin, these four FUs form a carbohydrate cluster, and two glycans are located on one of the interacting loops. This information suggests the proposition that glycosylation plays an important role in the allosteric effect of hemocyanin as well as structure reinforcement.

Evolutionary Significance

Hemocyanins exist in a wide range of species of molluscs—Gastropoda, Cephalopoda, and Polyplacophora, and so on. In phylogenetic systematics, the polyplacophora are located upstream of gastropods and cephalopods (Scheltema 1996; Salvini-Plawen and Steiner 1996; Waller 1998; Smith et al. 2011). In terms of gastropods, the first appearance of gastropoda is early Cambrian, and organisms having mega-hemocyanin—such as *Melanoides tuberculate*—are thought to have appeared later than those having keyhole limpet-type hemocyanin (Lieb et al. 2010). Cephalopods diverged into nautiloids and coleoids. The nautilus is called a living fossil, and the placement of the ancestor to the nautilus is close to that of the ammonites in the phylogram. Decapoda such as squid and cuttlefish diverged from coleioids (Kröger et al. 2011).

In this chapter, depending on the subunit composition and overall architecture, molluscan hemocyanins are classified into four types: type 1 (keyhole limpet-type); type 2 (mega-hemocyanin); type 3 (nautilus-type); and type 4 (squid-type) (see above). Taking the structure of assembly into consideration, we introduced an exceptional polyplacophora-type, which forms a decamer in spite of its FU composition with all basal FUs. Polyplacophora hemocyanin, categorized within type 1 hemocyanin on the basis of its FU composition, is known to form a decamer (Harris et al. 2004; Lambert et al. 1994), although another type 1 hemocyanin, the typical gastropod hemocyanin, assembles to a di-decamer. The structure of polyplacophora hemocyanin has not yet been determined in high resolution; its decameric assembly is considered to be identical to that of a decamer comprising the di-decamer of typical type 1 hemocyanin, since its FU composition is the same as that of type 1. The polyplacophora-type is thus an archetypal molluscan hemocyanin.

Evolution of gastropod hemocyanin, in the main, caused multimerization of subunits and FUs. During the appearance of type 1 hemocyanin from the archetypal hemocyanin, di-decameric assembly was formed with no alteration in the FU structure. Mega-hemocyanin as a type 2 hemocyanin—which appeared later—further assembled to a tri-decamer. Mega-hemocyanin developed from the typical gastropod hemocyanin by acquiring the mega-subunit, which has six relative FUs (FU-f1, -f2, -f3, -f4, -f5, and -f6) substituted for FU-g and -h. The core region comprising the obtained FU-f1-6 would reinforce the stacking of decamers by filling the lumen completely. Interestingly, the archetype, type 1, and type 2 hemocyanin all contain basal FUs in their assembly; in other words, there was no lack of basal FU during the evolution of gastropod hemocyanin.

Evolution of cephalopod hemocyanin caused the deletion of basal FU-h and the addition of relative FU-d* from the archetypal hemocyanin to type 4 (squid-type) through type 3 (nautilus-type). During its evolution, the multimerization of subunits found in gastropod evolution did not occur. Figure 7.10, based on the amino acid sequences of subunits, shows a phylogenic tree of these four types molluscan hemocyanins. The phylogenic tree indicates trichotomy and has three main nodes. The first node includes type 1 hemocyanin; the second node includes type 2 hemocyanin;

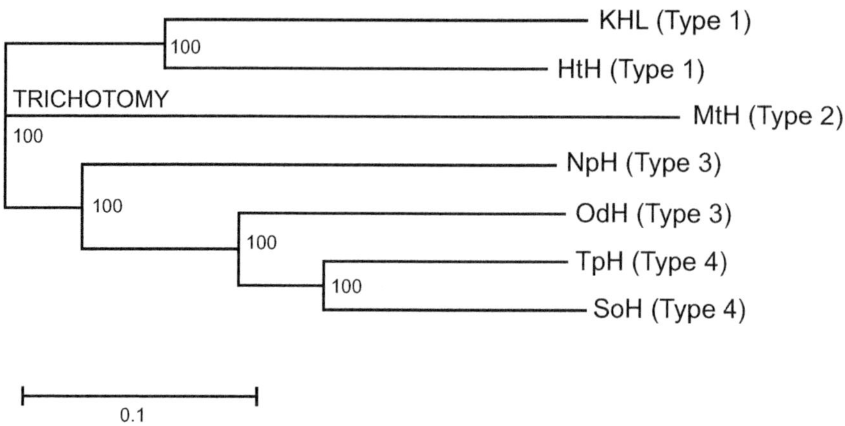

Fig. 7.10 Phylogenic tree of molluscan hemocyanins. The phylogram was constructed with ClustalW2.1 using the neighbor-joining method based on the amino acid sequences (Thompson et al. 1994). KLH (Keyhole limpet hemocyanin [DDBJ: AJ698341]) and HtH (*Haliotis tuberculate* hemocyanin [DDBJ: AJ252741]) are type 1, MtH (*Melanoides tuberculate* hemocyanin [DDBJ: KC405576]) is type 2, NpH (*Nautilus pompilius* hemocyanin [DDBJ: AJ619741]) and OdH (*Enteroctopus dofleini* hemocyanin [DDBJ: AY751301]) are type 3, and TpH (*Todarodes pacificus* hemocyanin [DDBJ: AB897790]) and SoH (*Sepia officinalis* hemocyanin [DDBJ: DQ388569]) are type 4. Bootstrap values are given at the nodes. Scale bar, 0.1 expected changes per site. Reprinted from supporting figure of Tanaka et al. (2019), with permission

and the third node includes cephalopod hemocyanin. The third node has two subnodes, second of which is divided into type 3 and type 4 hemocyanins. Type 3 has evolved from the archetype hemocyanin by losing FU-h. Due to the lack of FU-h, the protrusion of the inner collar domains were stored within the lumen of the wall. The change of these FU components caused rearrangement and optimization of the inner collar region. Evolution of type 4 hemocyanin from type 3 came about through the acquisition of FU-d*. The additional FU-d* is not elongate at the C-terminal of the subunit, but instead at the center of the FU comprising the wall region, FU-d and FU-e (Mouche et al. 1999). The additional FU-d* is ineffective in the architecture of the wall region, but has a great impact on the architecture of the inner collar.

Here, we introduce a proposed model for the evolution from C_5-symmetric type 3 (nautilus-type) to asymmetric type 4, squid hemocyanin (Fig. 7.11). In step (i) of the model, type 3 hemocyanin acquired FU-d* by gene duplication. The additional FU-d*s are entrapped on the thermodynamically stable sites, which correspond to FU-d* sites of conformers 1 and 2. In step (ii) of the model, due to steric repulsion between FU-d*s, subunit dimers dissociate into two subunits. In step (iii), each dissociated subunit spontaneously re-associates to form homodimers, both the dimer of conformer 1 and the dimer of conformer 2, wherein domain swapping occurs to avoid steric repulsion. In step (iv), the homodimers interact alternatively to assemble circularly. In step (v) the dissociated monomers bind between the terminal subunits to complete the circular association. In the final step (vi), FU-d*s rearange to avoid

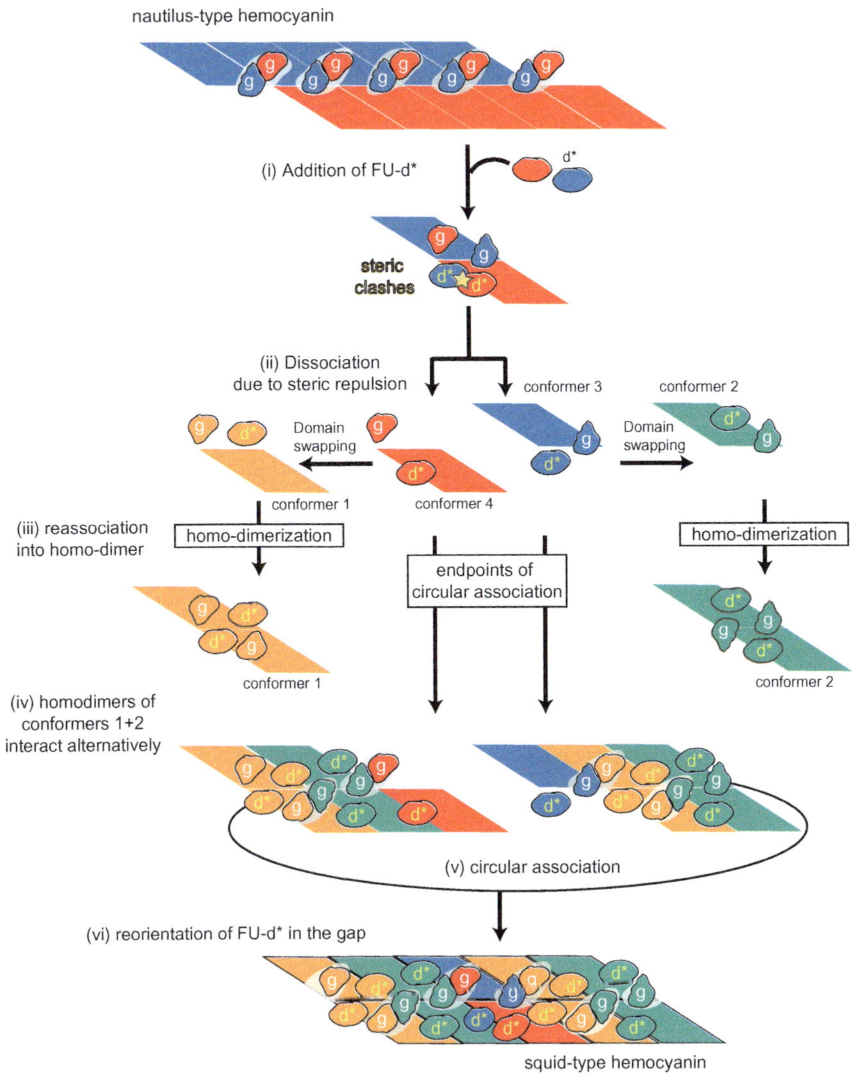

Fig. 7.11 Model for the evolution from type 3 to type 4 hemocyanin. Step (i) type 3 hemocyanin acquired FU-d* by gene duplication during evolution. FU-d* are entrapped on the sites that are thermodynamically stable, which correspond to FU-d* sites of conformers 1 and 2. Step (ii) Due to steric repulsion between FU-d*s, subunit dimers favor dissociation to monomers. Step (iii) Each dissociated monomer re-associates to form a homodimer, wherein domain swapping occurs to avoid steric repulsion. Step (iv): Homodimers assemble circularly. Step (v) The dissociated monomers bind between the terminal subunits to close the circular association. Step (vi) FU-d*s rearrange to avoid steric repulsions and a hetero-subunit dimer is formed, which closes the circle. Reprinted from Tanaka et al. (2019), with permission

steric repulsions, forming a hetero-subunit dimer with conformer 3 and 4, in this way closing the circle.

Physiology

Squid is a difficult animal to keep in captivity, and in artificial environments, is prone to die. *T. pacificus* can live for only few days under aquarium conditions, even under controlled water quality—that is, constant water temperature, saturation level dissolved oxygen, and removal of waste products. We determined the etiology of squid death under aquarium conditions, finding that in vivo dissociation of the hemocyanin decamer in the hemolymph was deeply implicated (Yoshioka et al. 2012; Kato et al. 2018). From TEM observation of hemolymph from weakened squid, we determined that the hemocyanin decamer dissociated to subunit-oligomers or subunits (Fig. 7.12b). By hematological tests, we ascertained that the hemocyanin concentration in the hemolymph was reduced in captive squid (Fig. 7.12c). The dissociation of hemocyanin decamers might trigger the degradation of hemocyanin, which would disturb the biosynthesis of hemocyanin in the mollusc. Decreases in hemocyanin concentration in the hemolymph may have resulted in circulatory failure, which is assumed to be the cause of the critical illness of captive squid. The assembled structure may prevent in vivo degradation of hemocyanin; in other words, retention of the hemocyanin molecule structure in vivo is considered to be important to health maintenance in molluscs. The reduction of hemocyanin concentration is reportedly also found in the captive Atlantic horseshoe crab *Limulus polyphemus* of Arthropoda (Coates et al. 2012).

While some studies report on the physiological function of mollusc hemocyanin, detailed information of the oxygen-binding properties remains limited. Reports have described the oxygen-binding cooperativity and affinity of *Concholepas concholepas* hemocyanin (González et al. 2017).

Closing Remarks

Molluscan hemocyanin was discovered 130 years ago, yet studies of the biochemistry and structural biology of the molecule have remained complicated over a long period. Over recent 10 years, there have been great strides in the structural biology of molluscan hemocyanin. Zhang and colleagues resolved the cryo-EM structure of type 1 hemocyanin with high resolution (Zhang et al. 2013). Gai et al. (2015) reported the first crystal structure of type 4 hemocyanin with atomic resolution. From the crystal structure, the structure and architecture of the cylindrical wall was clarified, yet the

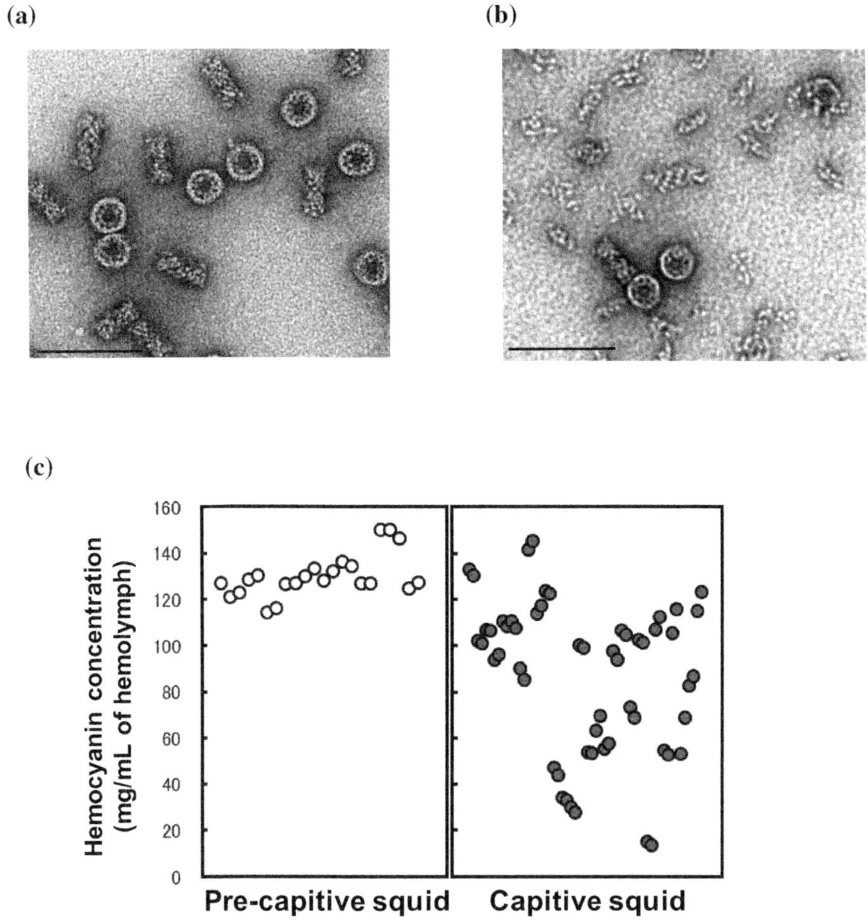

Fig. 7.12 In vivo dissociation and reduction of hemolymph hemocyanin in the squid *T. pacificus*. **a**, **b** TEM image of hemocyanin in hemolymph from (**a**) healthy animal and (**b**) weakened animal after captive test. **c** hemolymph hemocyanin concentrations collected from individual pre-captive animals (54 animals) and captive animals for three days (42 animals)

inner collar domains remained unresolved, and the topology of FU-g and FU-d* was unclear. Tanaka et al. (2019) attempted to solve the mysterious structure of the inner region using protocols designed to address the structures of a symmetry mismatch complex. Their attempts have succeeded in resolving the asymmetrical inner region and the entire decamer molecule of type 4 hemocyanin. These structural studies have made a breakthrough in the biochemistry of molluscan hemocyanin. The physiological information on molluscan hemocyanin is currently insufficient, but on the basis of recently discovered structural information, many new branches of research into this topic are anticipated.

Acknowledgements The authors would like to express thanks to the book editors Professor J. Robin Harris and Professor Ulrich Hoeger for giving this precious opportunity to review the recent progress in structural study of molluscan hemocyanin. We are grateful Dr. Christos Gatsogiannis for his collaboration work of cryo-EM structure of squid hemocyanin.

References

Altenhein B, Markl J, Lieb B (2002) Gene structure and hemocyanin isoform HtH2 from the mollusc Haliotis tuberculata indicate early and late intron hot spots. Gene 301(1–2):53–60

Bergmann S, Lieb B, Ruth P, Markl J (2006) The hemocyanin from a living fossil, the cephalopod Nautilus pompilius: protein structure, gene organization, and evolution. J Mol Evol 62(3):362–374

Boisset N, Mouche F (2000) Sepia officinalis hemocyanin: a refined 3D structure from field emission gun cryoelectron microscopy. J Mol Biol 296(2):459–472

Coates CJ, Bradford E, Krom C, Nair J (2012) Efect of temperature on biochemical and cellular properties of captive Limulus polyphemus. Aquaculture 334–337:30–38

Cuff ME, Miller KI, van Holde KE, Hendrickson WA (1998) Crystal structure of a functional unit from Octopus hemocyanin. J Mol Biol 278(4):855–870

Ebina S, Matsubara K, Nagayama K, Yamaki M, Gotoh T (1995) Carbohydrate gluing, an architectural mechanism in the supramolecular structure of an annelid giant hemoglobin. Proc Natl Acad Sci U S A 92:7367–7371

Gai Z, Matsuno A, Kato K, Kato S, Khan M, Shimizu T, Yoshioka T, Kato Y, Kishimura H, Kannno G, Miyabe Y, Terada T, Tanaka Y, Yao M (2015) Crystal structure of the 3.8 MDa respiratory supermolecule hemocyanin at 3.0 Å resolution. Structure 23:2204–2212

Gatsogiannis C, Markl J (2009) Keyhole limpet hemocyanin: 9-A CryoEM structure and molecular model of the KLH1 didecamer reveal the interfaces and intricate topology of the 160 functional units. J Mol Biol 385(3):963–983

Gatsogiannis C, Moeller A, Depoix F, Meissner U, Markl J (2007) Nautilus pompilius Hemocyanin: 9 Å Cryo-EM structure and molecular model reveal the subunit pathway and the interfaces between the 70 functional units. J Mol Biol 374(2):465–486

Gatsogiannis C, Hofnagel O, Markl J, Raunser S (2015) Structure of mega-hemocyanin reveals protein origami in snails. Structure 23:93–103

Gielens C, De Geest N, Compernolle F, Préaux G (2004) Glycosylation sites of hemocyanins of Helix pomatia and Sepia officinalis. Micron 35(1):99–100

González A, Nova E, Del Campo M, Manubens A, De Ioannes A, Ferreira J (2017) The oxygen-binding properties of hemocyanin from the mollusc Concholepas concholepas. Biochimica et Biophysica Acta (BBA)—Proteins and Proteomics 1865(12):1746–1757

Harris JR, Meissner U, Gebauer W, Markl J (2004) 3D reconstruction of the hemocyanin subunit dimer from the chiton Acanthochiton fascicularis. Micron 35(1–2):23–26

Jaenicke E, Büchler K, Markl J, Decker H, Barends TRM (2010) Cupredoxin-like domains in haemocyanins. Biochem J 426(3):373–378

Jaenicke E, Buchler K, Decker H, Markl J, Schroder GF (2011) The refined structure of functional unit h of keyhole limpet hemocyanin (KLH1-h) reveals disulfide bridges. IUBMB Life 63(3):183–187

Kato S, Matsui T, Gatsogiannis C, Tanaka Y (2018) Molluscan hemocyanin: structure, evolution, and physiology. Biophysical Reviews 10:191–202

Kröger B, Vinther J, Fuches D (2011) Cephalopod origin and evolution: a congruent picture emerging from fossils, development and molecules. BioEssays 33:602–613

Lambert O, Boisset N, Taveau JC, Lamy JN (1994) Three-dimensional reconstruction from a frozen-hydrated specimen of the chiton Lepidochiton sp. hemocyanin. J Mol Biol 244(5):640–647

Lamy J, You V, Taveau JC, Boisset N, Lamy JN (1998) Intramolecular localization of the functional units of Sepia officinalis hemocyanin by immunoelectron microscopy. J Mol Biol 284(4):1051–1074
Lieb B, Markl J (2004) Evolution of molluscan hemocyanins as deduced from DNA sequencing. Micron 35(1–2):117–119
Lieb B, Altenhein B, Markl J (2000) The sequence of a gastropod hemocyanin (HtH1 from Haliotis tuberculata). J Biol Chem 275(8):5675–5681
Lieb B, Gebauer W, Gatsogiannis C, Depoix F, Hellmann N, Harasewych MG, Strong EE, Markl J (2010) Molluscan mega-hemocyanin: an ancient oxygen carrier tuned by a ~550 kDa polypeptide. Frontiers in zoology 7:14
Markl J (2013) Evolution of molluscan hemocyanin structures. Biochem Biophys Acta 1834:1840–1852
Matsuno A, Gai Z, Tanaka M, Kato K, Kato S, Katoh T, Shimizu T, Yoshioka T, Kishimura H, Tanaka Y, Yao M (2015) Crystallization and preliminary X-ray crystallographic study of a 3.8-MDa respiratory supermolecule hemocyanin. J Struct Biol 190:379–382
Meissner U, Dube P, Harris JR, Stark H, Markl J (2000) Structure of a molluscan hemocyanin didecamer (HtH1 from Haliotis tuberculata) at 12 Å resolution by cryoelectron microscopy. J Mol Biol 298(1):21–34
Meissner U, Gatsogiannis C, Moeller A, Depoix F, Harris JR, Markl J (2007) Comparative 11 Å structure of two molluscan hemocyanins from 3D cryo-electron microscopy. Micron 38(7):754–765
Miller KI, Van Holde KE, Toumadje A, Johnson WC Jr, Lamy J (1988) Structure and function of the carboxyl-terminal oxygen-binding domain from the subunit of Octopus dofleini hemocyanin. Biochemistry 27(19):7282–7288
Miller KI, Schabtach E, van Holde KE (1990) Arrangement of subunits and domains within the Octopus dofleini hemocyanin molecule. Proc Natl Acad Sci USA 87(4):1496–1500
Moriya T, Saur M, Stabrin M, Merino F, Voicu H, Huang Z, Penczek PA, Raunser S, Gatsogiannis C (2017) High-resolution single particle analysis from electron cryo-microscopy images using SPHIRE. J Vis Exp 123:e55448
Mouche F, Boisset N, Lamy J, Zal F, Lamy JN (1999) Structural comparison of cephalopod hemocyanins: phylogenetic significance. J Struct Biol 127(3):199–212
Perbandt M, Guthöhrlein EW, Rypniewski W, Idakieva K, Stoeva S, Voelter W, Genov N, Betzel C (2003) The structure of a functional unit from the wall of a gastropod hemocyanin offers a possible mechanism for cooperativity. Biochemistry 42(21):6341–6346
Salvini-Plawen L, Steiner G (1996) Origin and evolutionary radiation of the mollusca. Oxford University Press, Oxford, pp p29–p51
Scheltema A (1996) Origin and evolutionary radiation of the mollusca. Oxford University Press, Oxford, pp p53–p85
Smith S, Wilson N, Goetz F, Feehery C, Andrade S, Rouse G, Giribet G, Dunn C (2011) Resolving the evolutionary relationships of molluscs with phylogenomic tools. Nature 480:364–367
Stoeva S, Schutz J, Gebauer W, Hundsdorfer T, Manz C, Markl J, Voelter W (1999) Primary structure and unusual carbohydrate moiety of functional unit 2-c of keyhole limpet hemocyanin (KLH). Biochem Biophys Acta 1435(1–2):94–109
Stoeva S, Idakieva K, Betzel C, Genov N, Voelter W (2002) Amino acid sequence and glycosylation of functional unit RtH2-e from Rapana thomasiana (gastropod) hemocyanin. Arch Biochem Biophys 399:149–158
Tanaka Y, Kato S, Stabrin M, Raunse S, Matsui T, Christos G (2019) Cryo-EM reveals the asymmetric assembly of squid hemocyanin. IUCrJ 6:426–437
Thompson JD, Higgins DG, Gibson TJ (1994) CLUSTAL W: improving the sensitivity of progressive multiple sequence alignment through sequence weighting, position-specific gap penalties and weight matrix choice. Nucleic Acids Research 22 (22):4673–4680
Thonig A, Oellermann M, Lieb B, Mark F (2014) A new haemocyanin in cuttlefish (Sepia officinalis) eggs: sequence analysis and relevance during ontogeny. Evodevo:6

Waller T (1998) Bivalves: an Eon of evolution. University Calgary Press, pp 1–45
Yoshioka T, Kato S, Okamoto A (2012) Short-term rearing conditions for shipping live squids. Quality improvement of coastal fish and marine invertebrates—Achievement by short-term rearing and associated systems for transportation and marketing. Kouseisyakouseikaku, Tokyo, pp 106–129
Zhang Q, Dai X, Cong Y, Zhang J, Chen DH, Dougherty MT, Wang J, Ludtke SJ, Schmid MF, Chiu W (2013) Cryo-EM structure of a molluscan hemocyanin suggests its allosteric mechanism. Structure 21(4):604–613
Zhu H, Zhuang J, Feng H, Liang R, Wang J, Xie L, Zhu P (2014) Cryo-EM structure of isomeric molluscan hemocyanin triggered by viral infection. PLoS ONE 9:e98766

Chapter 8
Arachnid Hemocyanins

Monica Cunningham, Aldana Laino, Sofia Romero and C. Fernando Garcia

Abstract Hemocyanin (Hc), a copper-containing extracellular multimeric protein, is the major protein component of hemolymph in different arachnid groups. Hc possesses 7 or 8 very well-characterized types of monomers with molecular weights ranging from 70 to 85 kDa, organized in hexamers or multiple of hexamers. The present chapter compiles the existing data with relation to the function of this protein in the arachnids. Hc has as main function the reversible transport of O_2, but it shows many secondary though not less important functions. With reference to this, it has been described that Hc can transport hydrophobic molecules (lipid-derived hormones and lipids) to the different organs, having a key role in the lipid transport system. In arachnids, like in other arthropods and invertebrates, Hc has phenoloxidase function which is related to different metabolic processes such as melanin formation and defense against pathogens. In addition, Hc has additional defensive functions since it can serve as precursor for the production of antimicrobial peptides. In short, the evolution of this protein has led to the development of multiple functions essential for organisms possessing this protein.

Keywords Arachnids · Hemocyanin · Oxygen transport · Lipid transport · Phenoloxidase · Hexamers · Antimicrobial peptides · Immune response

M. Cunningham (✉) · A. Laino · S. Romero · C. Fernando Garcia
INIBIOLP (CONICET-UNLP) - Facultad de Ciencias Médicas, UNLP, Calles 60 y 120, 1900 La Plata, Argentina
e-mail: cunninghammoni@gmail.com

A. Laino
e-mail: aldana_laino@hotmail.com

S. Romero
e-mail: sofiaromero321@msn.com

C. Fernando Garcia
e-mail: cfgarcia1123@yahoo.com.ar

U. Hoeger and J. R. Harris (eds.), *Vertebrate and Invertebrate Respiratory Proteins, Lipoproteins and other Body Fluid Proteins*, Subcellular Biochemistry 94,
https://doi.org/10.1007/978-3-030-41769-7_8

Introduction

Hemocyanin (Hc), a copper-containing extracellular multimeric protein, is the major protein component of the hemolymph in different arthropod groups (chelicerates, crustaceans and larval stages of some insects) and molluscs. Its main function is the reversible binding of O_2 mediated by two Cu^{+2} ions which are complexed by 6 histidine residues (Markl and Decker 1992; van Holde and Miller 1995; Markl 2013; Salvato and Beltramini 1990; Markl et al. 1986). The particular name of this molecule is due to the distinctive bluish colour resulting from the presence of copper in the oxygenated hemocyanin (cyan = blue). Hc concentration in the hemolymph differs considerably between species, ranging approximately between 20 and 80 mg mL^{-1}. However, in the xiphosuran *Limulus polyphemus* it was calculated in 140 mg mL^{-1} (Coates and Decker 2017). The size of this protein in arthropods can range from 500,000 to 3.5 million Daltons (van Holde and Miller 1982).

Chelicerate Hc has 7 or 8 types of monomers with molecular weights ranging from 70 to 85 kDa, organized in hexamers or multiple of hexamers, whose level of native aggregation is characteristic of each group. Each subunit takes up a different position in the native oligomer, having the same binding capacity for O_2 but dissimilar physicochemical properties and different evolutionary origin (Rehm et al. 2012). The strategies to generate these large macromolecular structures with subunits that are added and/or polymerized to produce giant molecules with many binding sites not only decrease osmotic pressure in hemolymph but also provide a solution to the problem of loading oxygen with high affinity and discharging it where needed (Linzen et al. 1985; van Holde et al. 2001).

Out of the eleven Orders comprising the Class Arachnida, in four of them, Scorpions, Amblypygi, Uropygi and Araneae, hemocyanin was found present. Some arachnids, however, have no hemocyanin or any other O_2 transport protein in their hemolymph, as occurs in the Orders Opiliones, Pseudoscorpiones, Solifugae and Acari. These arachnids have no book lungs and breathe through their trachea; this trait being evidently sufficient to perform the aerobic metabolism (Markl 1986; Markl et al. 1986; Rehm et al. 2012; Franz-Guess and Starck 2016). Surprisingly, the lack of an oxygen transport protein in spiders of the genus *Dysdera*, which are—according to the present knowledge—the only Araneae without hemocyanin, may also be explained by the possession of a highly developed tracheal system (Schmitz 2013; Rehm et al. 2012).

Substantial evidence has brought to light the multiple functionalities of Hc. Not only is it involved in the oxygen transport and storage (Foelix 2011), but also it has been demonstrated that it fulfils other functions such as lipid transport (Cunningham et al. 2007; Cunningham and Pollero 1996; Hall et al. 1995; Laino et al. 2009, 2015a), utilization as cuticle component (Paul et al. 1994), molting hormone transport (Jaenicke 1999), phenoloxidase (Baird et al. 2007; Decker and Rimke 1998; Decker et al. 2001; Jaenicke and Decker 2008; Laino et al. 2015b), defense against pathogens (Coates and Nairn 2014; Riciluca and Silva 2012) and a key role in clot formation, being the main protein component (Sanggaard et al. 2016).

Structure and Organization of Hemocyanins

Hemocyanins represent a heterogeneous protein class. The sequence homology between hemocyanins of arthropods and molluscs is low (10%) (Salvato and Beltramini 1990). They also differ in terms of structural organization, subunit size and copper content (van Holde and Miller 1995; Grossmann et al. 2000).

Arthropod hemocyanins are oligomers, forming hexamers (1 × 6 mers), where the monomeric subunits of about 75 kDa have approximately 650 amino acid residues (Martin et al. 2007). These hexamers serve as 'building blocks' and can be added to greater structures formed by 2, 4, 6 and even 8 hexamers (Markl et al. 1986). The oxygen-binding capacity is also very variable, since they possess between 6 and 160 oxygen-binding sites, some of them displaying the highest molecular cooperativity observed in nature (see below).

Most arachnids have seven primary subunits, which combine to form a 24-mer (4 × 6) quaternary structure. In the case of scorpions such as *Androctonus australis, Buthus sindicus, Buthus occitanus, Pandinus imperator* and *Scorpio maurus*, the native structure of hemocyanin is a polymer composed of 8 different polypeptide chains of ~68–75 kDa, arranged in four cubic hexamers (4 × 6-mers) (Lamy et al. 1980, 1981; Ali et al. 2007; Cong et al. 2009). The latter authors have built a pseudo atomic model based on cryo-EM density map (see https://www.rcsb.org/structure/3IXV). Within some spider lineages, however, hemocyanin evolution has been a dynamic process with extensive paralog duplication and loss (Starrett et al. 2013). The number of hexamer units is species-specific. In the Order Araneae, for example the structure and the composition of hemocyanin has been studied in over 40 spider species belonging to 25 families (Markl 1986; Markl et al. 1986; Burmester 2013). In these the quaternary structure of the hemocyanin (its native form) is organized in di- or tetrahexamers (Van Bruggen et al. 1980; Markl 1986; Markl and Decker 1992; Voit et al. 2000; Averdam et al. 2003). These studies showed that all the species belonging to the Infraorder Mygalomorphae possess tetrahexamers (4 × 6) of hemocyanin. Specifically in *Aphonopelma hentzi* (= *Eurypelma californicum*; Nentwig 2012), 7 different subunits of 624–631 amino acids and molecular weights between 71.68 and 72.18 kDa have been described. Each subunit seems to occupy a determined position in the native structure, necessary for a complete biological activity (Voit et al. 2000; Ali et al. 2007). In some species of the Infraorder Araneomorphae, deviations from the "classical" 4 × 6 hemocyanin scheme have been observed, as in the case of *Cupiennius salei* in which it was found that a mixture of 1 × 6 and 2 × 6-mer hemocyanins occurs in an approximate 1:2 ratio (Markl et al. 1976; Markl 1980; Ballweber et al. 2002). In these oligomers of hemocyanin, some types of monomers are absent (Markl 1986; Markl et al. 1986). The 2 × 6-mer structure has been observed in several families such as Agelenidae, Salticidae, Thomisidae, Clubionidae and Lycosidae (Markl 1986), with only two hemocyanin subunit types identified by immunological means, suggesting a extensive rearrangement of the subunits.

Different Functions of Hemocyanin

Transport of O_2 and Binding of Hydrophobic Molecules

Some arachnids (as detailed in the Introduction) have hemocyanin as a respiratory pigment, which is found freely dissolved in the hemolymph. The saturation curves of the oxygen binding to the hemocyanin of *A. hentzi* (Loewe 1978; Decker and Sterner 1990) and of *Pandinus imperator* (Decker 1990) have revealed the molecular mechanisms of the cooperative effect for such binding. These mechanisms, require the existence of different conformations, the interaction of 4 hexamers and an adequate regulation through allosteric effectors (van Holde and Miller 1982; Savel-Niemann et al. 1988; Decker 1990; Hartmann and Decker 2002) that result in a sigmoidal curve similar to that of haemoglobin in vertebrates. The O_2 binding capacity to the protein depends on different effectors such as pH and temperature (Paul et al. 1994; van Holde and Miller 1995). In the spiders *Lasiodora erythrocythara, A. hentzi, C. salei* and in the scorpion *P. imperator*, studies have shown that the hemocyanin affinity for O_2 expressed as $\log P_{50}$ and the Bohr effect expressed as $\Delta\log P_{50}/\Delta$Ph are pH-dependent. On the other hand, the cooperativity expressed as n_{50} is also pH-dependent increasing their values when it increases as occurred in the 3 previously cited spiders, but independent of pH changes in the scorpion (Bridges 1988; Loewe 1978; Loewe and Linzen 1975; Paul et al. 1994; Decker and Sterner 1990; Decker 1990) (Table 8.1).

As previously mentioned, apart from the widespread function as respiratory pigment, other functions have been described for the hemocyanin, one of them acting as apolipoprotein associated with circulating lipids. Although in other groups such as molluscs and crustaceans there are reports on this possible role of the hemocyanin in the lipid transport (Heras and Pollero 1990, 1992; Hall et al. 1995), this function has been reported for arachnids only in three species of spiders, *Polybetes pythagoricus*, *Latrodectus mirabilis* and *Grammostola rosea.* In these species, hemocyanin was found to be associated with lipids, forming part of the hemolymph lipoproteins (Fig. 8.1). These lipoproteins are of high and very high density in *P. pythagoricus* (HDL2 and VHDL), of high density in *L. mirabilis* (HDL2), and of very high density in *G. rosea* (Gr-VHDL). In *P. pythagoricus*, HDL_2 and VHDL transport between 25% and 49% of the total hemolymph lipids, respectively; therefore, hemocyanin is associated with the majority of circulating lipids, with free fatty acids and phospholipids being the predominant lipids (Cunningham and Pollero 1996). VHDL has a major role in lipid dynamics and is involved in lipid uptake from midgut diverticula and lipid transport. This has been demonstrated in assays performed both in vivo and in vitro Laino et al. (2009, 2011). In *L. mirabilis*, HDL2 transports 19% of total hemolymphatic lipids, major lipids being free fatty acids, phosphatidylcholine and cholesterol (Cunningham et al. 2000). Finally, *Gr-VHDL* of *G. rosea* transports 89.3% of the lipids present in the hemolymph, phospholipids being the major lipids (Laino et al. 2015a). In other species of arachnids, at present no data are available on the existence of lipoproteins whose protein component is constituted by hemocyanin

Table 8.1 Hemocyanin affinities for O_2 expressed as $\log P_{50}$, the Bohr effect expressed as $\Delta \log P_{50}/\Delta Ph$ and the cooperativity expressed as n_{50} at different pH's from the spiders *L. erythrocythara, A. hentzi, C. salei* and in the scorpion *P. imperator*

Species	Cooperativity (n_{50})	Bohr coeffiecient	Oxygen infinity (log P_{50})	References
L. erythrocythara (Theraphosidae)	2.2 at pH 7.17 3.4 at pH 7.64 3.5 at pH 8.06	−0.71 at pH from 7.3 to 8.1	1.6 at pH 7.17 1.36 at pH 7.64 1.0 at pH 8.06	Bridges (1988)
C. salei (Ctenidae)	2.8 at pH 7.5 5.2 at pH 7.5 4.6 at pH 7.5	−0.89 at pH 8–8.5	1.69 at pH 7.5 1.46 at pH 7.5	Loewe and Linzen (1975) Paul et al. (1994)
A. hentzi (sub. Eurypelma californicum) (Theraphosidae)	2.0/2.2/2.4 at pH 7.0 4.3/3.0 at pH 7.5 6.3 at pH 7.75 6.0/7.0 at pH 8	−0.01 at pH 7–7.5 −0.6 at pH 7.5–7.75 −0.7 at pH 7.75–8 −1.04 at pH 7.7–8 −12 at pH 8–8.5	1.73/1.5/1.49 at pH 7.0 1.68/1.4/1.3 at pH 7.5 1.35/1.1/0.96 at pH 8	Paul et al. (1994) Loewe (1978) Decker and Sterner (1990)
P. imperator (Scorpionidae)	4 and 6 independent of pH	−0.02 at pH 7–7.25 −0.23 at pH 7.25–7.5 −0.52 at pH 7.5–7.75 −1.3 at pH 7.8	1.65/1.1 at pH 7.0 1.59/0.98 at pH 7.5 1.46/0.78 at pH 7.75	Paul et al. (1994) Decker (1990)

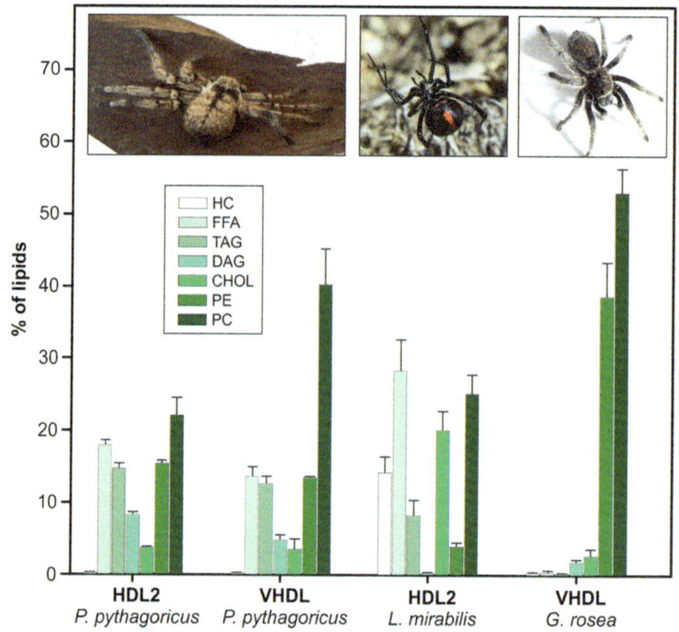

Fig. 8.1 Lipid class composition of HDL2 and VHDL isolated from hemolymphatic lipoproteins of *P. pythagoricus*, HDL2 of *L. mirabilis* and VHDL of *G. rosea*. Results are averages of 3 determinations ± SD. HC: hydrocarbons, FFA: free fatty acid, TAG: triacylglycerides, DAG: diacylglycerides, CHOL: cholesterol, PE: phosphatidylethanolamine, PC: phosphatidylcholine

(Cunningham et al. 2007). Cunningham and collaborators (1999) studied in vitro the hemocyanin capacity of the spider *P. pythagoricus* to take up different lipids, and found that it is able to bind a quantity of lipids significantly greater than that observed in vivo, suggesting a possible adaptation for the transport of different quantities of hemolymphatic lipids in response to physiological or environmental variations. With regards to additional functions of hemocyanin, it was observed that the hemocyanin of the tarantula *A. hentzi* binds ecdysone although with low affinity. The binding site for this molecule would be the domain 1 of the hemocyanin, which contains a hydrophobic pocket preserved over 450 million years in extracellular hemolymphatic proteins of arthropods (Jaenicke 1999).

Hemocyanin-Derived Phenoloxidase Activity

The family of metalloproteins with a copper-containing binuclear active site capable of reversibly binding dioxygen (binuclear site Type 3) includes the hemocyanin of molluscs and arthropods as well as phenoloxidases (tyrosinases (Tys) and catecholoxidasas (COs)). Phenoloxidases (POs) are catalytic proteins widely distributed in nature, that catalyse the initial stages of a series of enzymatic reactions of the biosynthetic route that finally leads to the formation of the melanin pigment (Cerenius et al. 2010). PO activation is modulated by the cascade of activation of proPO (Cerenius and Soderhall 2004), which is triggered by cuticular damage and/or the detection of microbes/ligands present in the hemolymph. During an immunological challenge, the recognition of pathogen-associated molecular patterns (PAMPs), for instance, lipopolysaccharides (LPSs) or β-glucans, through pathogen receptors (PRRs) present in the hemolymph unchain a cascade of serine-proteases, which convert proPO zymogen into an active PO. PO is released from arthropod hemocytes and is deposited in the infection site (Smith 2010). To date, Hc-derived PO activity (Hc-d PO) has been characterized in over 40 species of invertebrates, among them a large number of crustaceans, molluscs and only very few species of chelicerates (Coates and Nairn 2014). In this last group, no PO has been identified so far (Terwilliger and Ryan 2006), therefore, the observed activity is presumably Hc-derived. Hc can be converted not only by an in vivo but also in vitro activation to an enzyme similar to PO through two main ways of activation: either by the proteolytic treatment with proteases (for example, trypsin) or through a physical disruption of the protein conformation, utilizing for instance detergents, solvents, salts and phospholipids (Decker and Jaenicke 2004; Goldfeder et al. 2013; Decker et al. 2001; Baird et al. 2007).

Particularly in arachnids, Hc-derived PO activity has been studied only in the spiders *A. hentzi* and *P. pythagoricus* and in the scorpion *P. imperator* (Table 8.2). Following proteolysis with trypsin and chymotrypsin, the 4 × 6-meric hemocyanin of the tarantula *A. hentzi* exhibited monophenoloxidase and *o*-diphenoloxidase activities. The proteolytic cleavage removes the N-terminal fragment of the hemocyanin, thus exposing the active site for the entrance of different phenolic substances (Decker and Rimke 1998). Some years later, Decker and collaborators demonstrated that in the hemocyanin of the same tarantula, *o*-diphenoloxidase activity could be induced

Table 8.2 Characterization of the in vitro phenoloxidase activity (inductors, substrates and enzymatic parameters) of hemocyanin in different arachnids. (1) (Decker and Rimke 1998); (2) (Jaenicke and Decker 2008), (3) (Laino et al. 2015b); (4) (Schenk et al. 2015); (5) (Nillius et al. 2008)

	Inductor	Substrate	Km	Kcat	Vmax
Aphonopelma hentzi (1)	Trypsin Chymotrypsin	L-DOPA Tyrosine	–	–	–
Aphonopelma hentzi (2)	SDS	Dopamine	1.45 mM ± 0.16	0.1 ± 0.4 min^{-1}	0.091 μmoles min^{-1}
Polybetes pythagoricus (3)	SDS	Dopamine	0.407 mM ± 0.042	1.6 min^{-1}	0.081 μmoles min^{-1}
Aphonopelma hentzi (4)	Lipoprotein (HDL-2)	Dopamine	–	–	–
Pandinus imperator (5)	SDS	Dopamine Tyramine	–	–	–

in the presence of sodium dodecyl sulphate (SDS), a detergent frequently used to activate prophenoloxidases (Espin and Wichers 1999; Decker et al. 2001; Baird et al. 2007; Jaenicke and Decker 2008; Cong et al. 2009). Recently, it has been demonstrated that the interaction between high density lipoproteins and hemocyanin of *A. hentzi* induces the phenoloxidase activity in the hemocyanin. The authors propose that activation is due to the protein-protein interaction instead of the protein-lipid interaction (Schenk et al. 2015). Laino and collaborators determined that the hemocyanin of the spider *P. pythagoricus* also exhibits PO activity when treated with SDS, characterizing at the same time the kinetic parameters of this Hc-derived PO (Laino et al. 2015b). Phenoloxidase activity has also been determined for the scorpion *P. imperator*, which is also activated with SDS (Baird et al. 2007). In this same species, Nillius and collaborators have shown the change of both tyrosinase and catecholoxidase activity with the addition of Mg^{2+} ions as allosteric effectors of the SDS-activated hemocyanin (Nillius et al. 2008).

Hemocyanin as Precursor of Antimicrobial Peptides

Hemocyanin can also have other immunological properties apart from those previously mentioned due to its antimicrobial activity. One of the mechanisms of the innate immunity is the production of anti-microbial peptides (AMPs), which mostly have molecular weights below 25–30 kDa (Bulet et al. 2004). These peptides kill or inhibit the growth of microorganisms and have a key role in organisms lacking an adaptive immune system. Depending on the considered organism, AMPs can be present constitutively within secretory cells (hemocytes) or their production can be induced at the moment of infection, following their release into the hemolymph by exocytosis (Iwanaga and Kawabata 1998). The cationic nature of AMPs, associated with their tendency to adopt amphipathicity, facilitates their interaction and the entry in the anionic cellular walls and the phospholipidic membranes of microorganisms. Other AMPs are generated by partial hydrolysis of large protein precursors such as haemoglobin (Fogaca et al. 1999; Lai et al. 2004) and hemocyanin (Lee et al. 2003; Destoumieux-Garzon et al. 2001).

Within arthropods, numerous AMPs have been characterized, isolated from hemocytes or derived from the C-terminal end of hemocyanin (Iwanaga and Kawabata 1998; Schnapp et al. 1996; Khoo et al. 1999; Destoumieux et al. 1997; Lee et al. 2003).

Particularly in arachnids, different AMPs isolated from hemocytes of non-infected animals or from venom gland have been characterized (Ehret-Sabatier et al. 1996; Silva et al. 2000; Lorenzini et al. 2003; Kuhn-Nentwig et al. 2002; Corzo et al. 2002; Baumann et al. 2010). There is only one report on spiders where hemocyanin cleavage participates in the production of small peptides involved in defence reactions against an infection. This is the case of an AMP with antifungal activity called rondonine, isolated from the hemolymph of the spider *Acanthoscurria rondoniae* which inhibited the growth of *Candida albicans* MDM8 in a few minutes. It shares sequence homology with the C-terminal fragment of the subunit “d” of the Hc of *A. hentzi* and *Acanthoscurria gomesiana* and showed 90% similarity with a fragment of the subunit “f” of *A. gomesiana* (Riciluca and Silva 2012).

Conclusions

Hemocyanin, a respiratory pigment widely distributed within arachnids, shows a very conserved structure made up of hexamers or oligomers whose monomeric subunits have a molecular weight ranging from 70 to 85 kDa. The oxygen binding capacity of this protein is similar in the different species of arachnids, depending on factors such as temperature, the presence of allosteric effectors, and pH. This protein is the most abundant macromolecule in the hemolymph of arachnids, with other diverse and key roles apart from its classic function of storage, transport and release of oxygen, such as contributing to the transport of hydrophobic molecules and acting in the immune response. These findings suggest that hemocyanin has evolved to diversify its original function and has become one of the multifunctional components of greatest importance present in nature.

Acknowledgements This work was supported by grants from Consejo Nacional de Investigaciones Científicas y Técnicas (CONICET-PIP no. 515), UNLP and Agencia Nacional de Promoción Científica y Tecnológica (PICT no. 2017-0684), Argentina. A.L., F.G. and M.C. are researchers of CONICET, Argentina. S.R. is a CONICET scholarship holder. We are very grateful to Rosana del Cid for the review of the manuscript, Mario Ramos for the figure design, and Francisco Giambelluca for the photograph of the spider *Latrodectus mirabilis*.

References

Ali SA, Grossmann JG, Abbasi A, Voelter W (2007) Structural and conformational analysis of scorpion (*Buthus sindicus*) hemocyanin using low resolution techniques. Protein Pept Lett 14(5):481–488. https://doi.org/10.2174/092986607780782731

Averdam A, Markl J, Burmester T (2003) Subunit sequences of the 4 × 6-mer hemocyanin from the golden orb-web spider, *Nephila inaurata*. Eur J Biochem 270(16):3432–3439. https://doi.org/10.1046/j.1432-1033.2003.03730.x

Baird S, Kelly SM, Price NC, Jaenicke E, Meesters C, Nillius D, Decker H, Nairn J (2007) Hemocyanin conformational changes associated with SDS-induced phenol oxidase activation. Biochim Biophys Acta 1774(11):1380–1394. https://doi.org/10.1016/j.bbapap.2007.08.019

Ballweber P, Markl J, Burmester T (2002) Complete hemocyanin subunit sequences of the hunting spider *Cupiennius salei*: recent hemocyanin remodeling in entelegyne spiders. J Biol Chem 277(17):14451–14457. https://doi.org/10.1074/jbc.M111368200

Baumann T, Kuhn-Nentwig L, Largiader CR, Nentwig W (2010) Expression of defensins in non-infected araneomorph spiders. Cell Mol Life Sci 67(15):2643–2651. https://doi.org/10.1007/s00018-010-0354-2

Bridges CR (1988) The haemocyanin of the tarantula *Lasiodora erythrocythara*—the influence of C02, organic cofactors and temperature on oxygen affinity. Comp Biochem Physiol 89(4):661–667

Bulet P, Stöcklin R, Menin L (2004) Anti-microbial peptides: from invertebrates to vertebrates. Immunol Rev 198:169–184. https://doi.org/10.1111/j.0105-2896.2004.0124.x

Burmester T (2013) Evolution and adaptation of hemocyanin within spiders. In: Nentwig W (ed) Spider ecophysiology. Springer, Berlin, pp 3–14

Cerenius L, Kawabata S, Lee BL, Nonaka M, Soderhall K (2010) Proteolytic cascades and their involvement in invertebrate immunity. Trends Biochem Sci 35(10):575–583. https://doi.org/10.1016/j.tibs.2010.04.006

Cerenius L, Soderhall K (2004) The prophenoloxidase-activating system in invertebrates. Immunol Rev 198:116–126. https://doi.org/10.1111/j.0105-2896.2004.00116.x

Coates CJ, Decker H (2017) Immunological properties of oxygen-transport proteins: hemoglobin, hemocyanin and hemerythrin. Cell Mol Life Sci 74(2):293–317. https://doi.org/10.1007/s00018-016-2326-7

Coates CJ, Nairn J (2014) Diverse immune functions of hemocyanins. Dev Comp Immunol 45(1):43–55. https://doi.org/10.1016/j.dci.2014.01.021

Cong Y, Zhang Q, Woolford D, Schweikardt T, Khant H, Dougherty M, Ludtke SJ, Chiu W, Decker H (2009) Structural mechanism of SDS-induced enzyme activity of scorpion hemocyanin revealed by electron cryomicroscopy. Structure 17(5):749–758. https://doi.org/10.1016/j.str.2009.03.005

Corzo G, Villegas E, Gomez-Lagunas F, Possani LD, Belokoneva OS, Nakajima T (2002) Oxyopinins, large amphipathic peptides isolated from the venom of the wolf spider *Oxyopes kitabensis* with cytolytic properties and positive insecticidal cooperativity with spider neurotoxins. J Biol Chem 277(26):23627–23637. https://doi.org/10.1074/jbc.M200511200

Cunningham M, Garcia F, Pollero RJ (2007) Arachnid lipoproteins: comparative aspects. Comp Biochem Physiol Part C Toxicol Pharmcol 146(1–2):79–87. https://doi.org/10.1016/j.cbpc.2006.06.011

Cunningham M, Gomez C, Pollero R (1999) Lipid binding capacity of spider hemocyanin. J Exp Zool 284(4):368–373. https://doi.org/10.1002/(SICI)1097-010X(19990901)284:4<368::AID-JEZ2>3.0.CO;2-I

Cunningham M, Gonzalez A, Pollero RJ (2000) Characterization of lipoproteins isolated from the hemolymph of the spider *Latrodectus mirabilis* (Araneae, Theridiidae). J Arachnol 28:49–55

Cunningham M, Pollero RJ (1996) Characterization of lipoprotein fractions with high content of hemocyanin in the hemolymphatic plasma of *Polybetes pythagoricus*. J Exp Zool 274(5):275–280. https://doi.org/10.1002/(SICI)1097-010X(19960401)274:5<275::AID-JEZ2>3.0.CO;2-M

Decker H (1990) Nested allostery in scorpion hemocyanin (*Pandinus imperator*). Biophys Chem 37(1–3):257–263. https://doi.org/10.1016/0301-4622(90)88025-N

Decker H, Jaenicke E (2004) Recent findings on phenoloxidase activity and antimicrobial activity of hemocyanins. Dev Comp Immunol 28(7–8):673–687. https://doi.org/10.1016/j.dci.2003.11.007

Decker H, Rimke T (1998) Tarantula hemocyanin shows phenoloxidase activity. J Biol Chem 273(40):25889–25892. https://doi.org/10.1074/jbc.273.40.25889

Decker H, Ryan M, Jaenicke E, Terwilliger N (2001) SDS-induced phenoloxidase activity of hemocyanins from *Limulus polyphemus*, *Eurypelma californicum*, and *Cancer magister*. J Biol Chem 276(21):17796–17799. https://doi.org/10.1074/jbc.M010436200

Decker H, Sterner R (1990) Nested allostery of arthropodan hemocyanin (*Eurypelma californicum* and *Homarus americanus*). The role of protons. J Mol Biol 211(1):281–293. https://doi.org/10.1016/0022-2836(90)90027-j

Destoumieux D, Bulet P, Loew D, Van Dorsselaer A, Rodriguez J, Bachere E (1997) Penaeidins, a new family of antimicrobial peptides isolated from the shrimp *Penaeus vannamei* (Decapoda). J Biol Chem 272(45):28398–28406. https://doi.org/10.1074/jbc.272.45.28398

Destoumieux-Garzon D, Saulnier D, Garnier J, Jouffrey C, Bulet P, Bachere E (2001) Crustacean immunity antifungal peptides are generated from the C terminus of shrimp hemocyanin in response to microbial challenge. J Biol Chem 276(50):47070–47077. https://doi.org/10.1074/jbc.m103817200

Ehret-Sabatier L, Loew D, Goyffon M, Fehlbaum P, Hoffmann JA, van Dorsselaer A, Bulet P (1996) Characterization of novel cysteine-rich antimicrobial peptides from scorpion blood. J Biol Chem 271(47):29537–29544. https://doi.org/10.1074/jbc.271.47.29537

Espin JC, Wichers HJ (1999) Activation of a latent mushroom (*Agaricus bisporus*) tyrosinase isoform by sodium dodecyl sulfate (SDS). Kinetic properties of the SDS-activated isoform. J Agric Food Chem 47(9):3518–3525. https://doi.org/10.1021/jf981275p

Foelix RF (2011) Biology of spiders, 3rd edn. Oxford University Press, New York

Fogaca AC, da Silva PI Jr, Miranda MT, Bianchi AG, Miranda A, Ribolla PE, Daffre S (1999) Antimicrobial activity of a bovine hemoglobin fragment in the tick *Boophilus microplus*. J Biol Chem 274(36):25330–25334. https://doi.org/10.1074/jbc.274.36.25330

Franz-Guess S, Starck JM (2016) Histological and ultrastructural analysis of the respiratory tracheae of *Galeodes granti* (Chelicerata: Solifugae). Arthropod Struct Dev 45(5):452–461. https://doi.org/10.1016/j.asd.2016.08.003

Goldfeder M, Kanteev M, Adir N, Fishman A (2013) Influencing the monophenolase/diphenolase activity ratio in tyrosinase. Biochim Biophys Acta 1834(3):629–633. https://doi.org/10.1016/j.bbapap.2012.12.021

Grossmann JG, Ali SA, Abbasi A, Zaidi ZH, Stoeva S, Voelter W, Hasnain SS (2000) Low-resolution molecular structures of isolated functional units from arthropodan and molluscan hemocyanin. Biophys J 78(2):977–981. https://doi.org/10.1016/S0006-3495(00)76655-0 [pii]

Hall M, van Heusden MC, Soderhall K (1995) Identification of the major lipoproteins in crayfish hemolymph as proteins involved in immune recognition and clotting. Biochem Biophys Res Commun 216(3):939–946. https://doi.org/10.1006/bbrc.1995.2711

Hartmann H, Decker H (2002) All hierarchical levels are involved in conformational transitions of the 4 × 6-meric tarantula hemocyanin upon oxygenation. Biochim Biophys Acta 1601(2):132–137. https://doi.org/10.1016/s1570-9639(02)00459-4

Heras H, Pollero RJ (1990) Occurrence of plasma lipoproteins in octopods. Partial characterization and interorgan transport of lipids. J Exp Mar Biol Ecol 140:29–38

Heras H, Pollero RJ (1992) Hemocyanin as an apolipoprotein in the hemolymph of the cephalopod *Octopus tehuelchus*. Biochim Biophys Acta 1125:245–550

Iwanaga S, Kawabata S (1998) Evolution and phylogeny of defense molecules associated with innate immunity in horseshoe crab. Front Biosci 3:D973–D984. https://doi.org/10.2741/a337

Jaenicke E, Decker H (2008) Kinetic properties of catecholoxidase activity of tarantula hemocyanin. FEBS J 275(7):1518–1528. https://doi.org/10.1111/j.1742-4658.2008.06311.x

Jaenicke R (1999) Stability and folding of domain proteins. Prog Biophys Mol Biol 71(2):155–241. https://doi.org/10.1016/s0079-6107(98)00032-7

Khoo L, Robinette DW, Noga EJ (1999) callinectin, an antibacterial peptide from blue crab, *Callinectes sapidus*, hemocytes. Mar Biotechnol (NY) 1(1):44–51. https://doi.org/10.1007/pl00011750

Kuhn-Nentwig L, Muller J, Schaller J, Walz A, Dathe M, Nentwig W (2002) Cupiennin 1, a new family of highly basic antimicrobial peptides in the venom of the spider *Cupiennius salei* (Ctenidae). J Biol Chem 277(13):11208–11216. https://doi.org/10.1074/jbc.M111099200

Lai R, Lomas LO, Jonczy J, Turner PC, Rees HH (2004) Two novel non-cationic defensin-like antimicrobial peptides from haemolymph of the female tick, amblyomma hebraeum. Biochem J 379(Pt 3):681–685. https://doi.org/10.1042/BJ20031429

Laino A, Cunningham ML, Garcia F, Heras H (2009) First insight into the lipid uptake, storage and mobilization in arachnids: role of midgut diverticula and lipoproteins. J Insect Physiol 55(12):1118–1124. https://doi.org/10.1016/j.jinsphys.2009.08.005

Laino A, Cunningham ML, Heras H, Garcia F (2011) *In vitro* lipid transfer between lipoproteins and midgut-diverticula in the spider *Polybetes pythagoricus*. Comp Biochem Physiol 160(4):181–186. https://doi.org/10.1016/j.cbpb.2011.08.003

Laino A, Cunningham M, Suarez G, Garcia CF (2015a) Identification and characterization of the lipid transport system in the tarantula *Grammostola rosea*. OJAS 5(1):9–20. https://doi.org/10.4236/ojas.2015.51002

Laino A, Lavarias S, Suarez G, Lino A, Cunningham M (2015b) Characterization of phenoloxidase activity from spider *Polybetes pythagoricus* hemocyanin. J Exp Zool 323(8):547–555. https://doi.org/10.1002/jez.1947

Lamy J, Bijlholt MC, Sizaret PY, van Bruggen EF (1981) Quaternary structure of scorpion (*Androctonus australis*) hemocyanin. Localization of subunits with immunological methods and electron microscopy. Biochem 20(7):1849–1856. https://doi.org/10.1021/bi00510a021

Lamy J, Bonaventura J, Bonaventura C (1980) Structure, function, and assembly in the hemocyanin system of the scorpion Androctonus australis. Biochem 19(13):3033–3039. https://doi.org/10.1021/bi00554a031

Lee SY, Lee BL, Soderhall K (2003) Processing of an antibacterial peptide from hemocyanin of the freshwater crayfish *Pacifastacus leniusculus*. J Biol Chem 278(10):7927–7933. https://doi.org/10.1074/jbc.M209239200

Linzen B, Soeter NM, Riggs AF, Schneider HJ, Schartau W, Moore MD, Yokota E, Behrens PQ, Nakashima H, Takagi T et al (1985) The structure of arthropod hemocyanins. Science 229(4713):519–524. https://doi.org/10.1126/science.4023698

Loewe R (1978) Hemocyanin in spiders. V. Fluorimetric recording of oxygen binding curves, and its application to the analysis of allosteric interactions in *Eurypelma californicum* hemocyanin. J Comp Physiol 128:161–168

Loewe R, Linzen B (1975) Haemocyanins in spiders. II. Automatic recording of oxygen binding curves, and the effect of Mg ++ on oxygen affinity, cooperativity, and subunit association of *Cupiennius salei* haemocyanin. J Comp Physiol 98:147–156

Lorenzini DM, da Silva PI Jr, Fogaca AC, Bulet P, Daffre S (2003) Acanthoscurrin: a novel glycine-rich antimicrobial peptide constitutively expressed in the hemocytes of the spider *Acanthoscurria gomesiana*. Dev Comp Immunol 27(9):781–791. doi:S0145305X03000582 [pii]

Markl J (1980) Hemocyanins in spiders. XI. The quaternary structure of Cupiennius hemocyanin. J Comp Physiol B 140:199–207

Markl J (1986) Evolution and function of structurally diverse subunits in the respiratory protein hemocyanin from arthropods. Biol Bull 171:90–115

Markl J (2013) Evolution of molluscan hemocyanin structures. Biochim Biophys Acta 1834(9):1840–1852. https://doi.org/10.1016/j.bbapap.2013.02.020

Markl J, Decker H (1992) Molecular structure of the arthropod hemocyanins, vol 13. Springer, Berlin, Heidelberg. https://doi.org/10.1007/978-3-642-76418-9_12

Markl J, Schmid R, Czichos-Tiedt S, Linzen B (1976) Haemocyanins in spiders, III. Chemical and physical properties of the proteins in *Dugesiella* and *Cupiennius* blood. Hoppe Seylers Z Physiol Chem 357(12):1713–1725

Markl J, Stöcker W, Runzler R, Precht E (1986) Immunological correspondences between the hemocyanin subunits of 86 arthropods: evolution of a multigene protein family. In: Linzen B (ed) Invertebrate oxygen carriers. Springer, Berlin, Heidelberg, pp 281–292. https://doi.org/10.1007/978-3-642-71481-8_50

Martin AG, Depoix F, Stohr M, Meissner U, Hagner-Holler S, Hammouti K, Burmester T, Heyd J, Wriggers W, Markl J (2007) *Limulus polyphemus* hemocyanin: 10 A cryo-EM structure, sequence analysis, molecular modelling and rigid-body fitting reveal the interfaces between the eight hexamers. J Mol Biol 366(4):1332–1350. https://doi.org/10.1016/j.jmb.2006.11.075

Nentwig W (2012) The species referred to as *Eurypelma californicum* (Theraphosidae) in more than 100 publications is likely to be *Aphonopelma hentzi*. J Arachnol 40:128–130. https://doi.org/10.2307/41804580

Nillius D, Jaenicke E, Decker H (2008) Switch between tyrosinase and catecholoxidase activity of scorpion hemocyanin by allosteric effectors. FEBS Lett 582(5):749–754. https://doi.org/10.1016/j.febslet.2008.01.056

Paul R, Bergner B, Pfeffer-Seidl A, Decker H, Efinger R, Storz H (1994) Gas transport in the Haemolymph of Arachnids-Oxygen transport and the physiological role of Haemocyanin. J Exp Biol 188(1):25–46

Rehm P, Pick C, Borner J, Markl J, Burmester T (2012) The diversity and evolution of chelicerate hemocyanins. BMC Evol Biol 12:19. https://doi.org/10.1186/1471-2148-12-19

Riciluca KC, Silva PI (2012) Rondonin an antifungal peptide from spider (*Acanthoscurria rondoniae*) haemolymph. Results Immunol 2:66–71. https://doi.org/10.1016/j.rinim.2012.03.001

Salvato B, Beltramini M (1990) Hemocyanins: molecular architecture, structure and reactivity of binuclear copper active site. Life Chem Rep 8:1–47

Sanggaard KW, Dyrlund TF, Bechsgaard JS, Scavenius C, Wang T, Bilde T, Enghild JJ (2016) The spider hemolymph clot proteome reveals high concentrations of hemocyanin and von Willebrand factor-like proteins. Biochim Biophys Acta 1864(2):233–241. https://doi.org/10.1016/j.bbapap.2015.11.004

Savel-Niemann A, Markl J, Linzen B (1988) Hemocyanins in spiders. XXII. Range of allosteric interaction in a four-hexamer hemocyanin. Co-operativity and Bohr effect in dissociation intermediates. J Mol Biol 204(2):385–395. https://doi.org/10.1016/0022-2836(88)90583-9

Schenk S, Schmidt J, Hoeger U, Decker H (2015) Lipoprotein-induced phenoloxidase-activity in tarantula hemocyanin. Biochim Biophys Acta 1854(8):939–949. https://doi.org/10.1016/j.bbapap.2015.03.006

Schnapp D, Kemp GD, Smith VJ (1996) Purification and characterization of a proline-rich antibacterial peptide, with sequence similarity to bactenecin-7, from the haemocytes of the shore crab Carcinus maenas. Eur J Biochem 240(3):532–539. https://doi.org/10.1111/j.1432-1033.1996.0532h.x

Silva PI Jr, Daffre S, Bulet P (2000) Isolation and characterization of gomesin, an 18-residue cysteine-rich defense peptide from the spider *Acanthoscurria gomesiana* hemocytes with sequence similarities to horseshoe crab antimicrobial peptides of the tachyplesin family. J Biol Chem 275(43):33464–33470. https://doi.org/10.1074/jbc.M001491200

Smith VJ (2010) Immunology of Invertebrates: Cellular. Encyclopedia of life sciences (ELS), Chichester. https://doi.org/10.1002/9780470015902.a0002344.pub3

Schmitz A (2013) Tracheae in spiders—respiratory organs for special functions. In: Nentwig W (ed) Spider ecophysiology. Springer, Heidelberg

Starrett J, Hedin M, Ayoub N, Hayashi CY (2013) Hemocyanin gene family evolution in spiders (Araneae), with implications for phylogenetic relationships and divergence times in the infraorder Mygalomorphae. Gene 524(2):175–186. https://doi.org/10.1016/j.gene.2013.04.037

Terwilliger NB, Ryan MC (2006) Functional and phylogenetic analyses of phenoloxidases from brachyuran (*Cancer magister*) and branchiopod (*Artemia franciscana*, *Triops longicaudatus*) crustaceans. Biol Bull 210(1):38–50. https://doi.org/10.2307/4134535

Van Bruggen EFJ, Bijlholt MMC, Schutter WG, Wichertjes T, Bonaventura J, Bonaventura C, Lamy J, Lamy J, Leclerc M, Schneider H-J, Markl J, Linzen B (1980) The role of structurally diverse subunits in the assembly of three cheliceratan hemocyanins. FEBS Lett 116:207–210

van Holde KE, Miller KI (1982) Haemocyanins. Q Rev Biophys 15(1):1–129

van Holde KE, Miller KI (1995) Hemocyanins. Adv Protein Chem 47:1–81. https://doi.org/10.1016/S0065-3233(08)60545-8

van Holde KE, Miller KI, Decker H (2001) Hemocyanins and invertebrate evolution. J Biol Chem 276(19):15563–15566. https://doi.org/10.1074/jbc.R100010200

Voit R, Feldmaier-Fuchs G, Schweikardt T, Decker H, Burmester T (2000) Complete sequence of the 24-mer hemocyanin of the tarantula *Eurypelma californicum*. Structure and intramolecular evolution of the subunits. J Biol Chem 275(50):39339–39344. https://doi.org/10.1074/jbc.m005442200

Chapter 9
Multifunctional Roles of Hemocyanins

Christopher J. Coates and Elisa M. Costa-Paiva

Abstract The copper-containing hemocyanins are proteins responsible for the binding, transportation and storage of dioxygen within the blood (hemolymph) of many invertebrates. Several additional functions have been attributed to both arthropod and molluscan hemocyanins, including (but not limited to) enzymatic activity (namely phenoloxidase), hormone transport, homeostasis (ecdysis) and hemostasis (clot formation). An important secondary function of hemocyanin involves aspects of innate immunity—such as acting as a precursor of broad-spectrum antimicrobial peptides and microbial/viral agglutination. In this chapter, we present the reader with an up-to-date synthesis of the known functions of hemocyanins and the structural features that facilitate such activities.

Keywords Innate immunity · Phenoloxidase · Invertebrate · Antimicrobial · Oxygen transport protein · Cryptides

Introduction

Hemocyanins are large, extracellular glycoproteins tasked with increasing the oxygen carrying capacity of the invertebrate hemolymph (blood). The taxonomic distribution of hemocyanins was once considered restricted to arthropods and molluscs, but recent evidence demonstrates the presence of hemocyanin genes from a much broader range of animals, including annelids (Aguilera et al. 2013; Martín-Durán et al. 2013; Costa-Paiva et al. 2018). Although phylogenetically and structurally

This chapter is dedicated to the work and fond memory of the late Professor Heinz Decker (1950–2018).

C. J. Coates (✉)
Department of Biosciences, College of Science,
Swansea University, Swansea, Wales SA2 8PP, UK
e-mail: c.j.coates@swansea.ac.uk

E. M. Costa-Paiva
Departamento de Zoologia, Instituto Biociências, Universidade de São Paulo, São Paulo, Brazil
e-mail: elisam.costapaiva@gmail.com

U. Hoeger and J. R. Harris (eds.), *Vertebrate and Invertebrate Respiratory Proteins, Lipoproteins and other Body Fluid Proteins*, Subcellular Biochemistry 94,
https://doi.org/10.1007/978-3-030-41769-7_9

distinct, hemocyanins from molluscs and arthropods are both members of the type 3 copper protein family, which also counts insect hexamerins, crustacean cryptocyanins (pseudo-hemocyanins) and phenoloxidases (tyrosinase and catecholoxidase) as members (Burmester 1999, 2002, 2015). The active site of hemocyanin subunits consist of two copper atoms (CuA and CuB) held in position by six conserved histidine residues (Markl and Decker 1992; Fig. 9.1). This structural arrangement within each active site permits the reversible binding of dioxygen in a side-on bridging coordination. Arthropod hemocyanin subunits (circa 72 kDa, 650 amino acids) consist of three structural domains: an *N*-terminus formed by 5 or 6 α-helices (I), a central bundle of 4 α-helices that house the dicopper active site (II), and a *C*-terminus formed by a 7-stranded anti-parallel β-barrel (III) (Rehm et al. 2012; Fig. 9.1). These subunits combine to form the typical arthropod hemocyanin hexamers (1 × 6 mer; 2 × 6 mer; 4 × 6 mer; 8 × 6 mer). Conversely, mollusc hemocyanin subunits (350–550 kDa) consist of 7 or 8 paralogous functional units (FUa–FUh; Markl 2013; Gatsogiannis et al. 2015; Kato et al. 2018). These large polypeptide chains form cylindrical decamers (1 × 10 mer; 2 × 10 mer; 3 × 10 mer; Fig. 9.1)—offering in excess of 240 oxygen binding sites in some species. The functional unit consists of two domains,

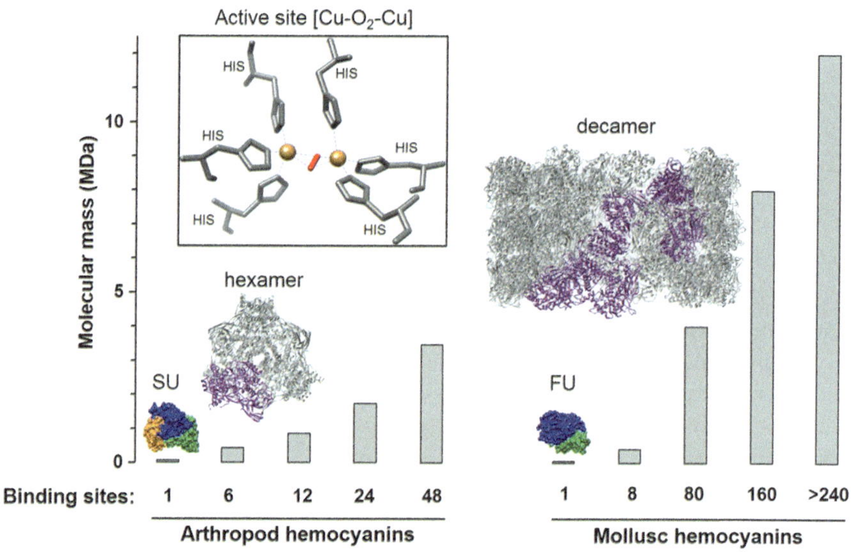

Fig. 9.1 Structural hierarchy of hemocyanins. Arthropod hemocyanin subunit (SU) domains I, II and III are coloured orange, blue and green, respectively. Molluscan hemocyanin functional unit (FU) domains α and β are coloured blue and green, respectively. A single subunit (~72 kDa) is coloured purple on the arthropod hemocyanin hexamer (~432 kDa). A single subunit (~400 kDa; up to 8 FUs) is coloured purple on the mollusc decamer (~4 MDa). Mollusc tri-decamers (i.e., mega hemocyanins) are in excess of 12 MDa. The number of oxygen binding sites are provided. Inset: conserved type-3 copper active site consisting of the six invariable histidine residues, two copper atoms (orange spheres) and bound dioxygen (red). Hartmann and Decker (2004) inspired this figure

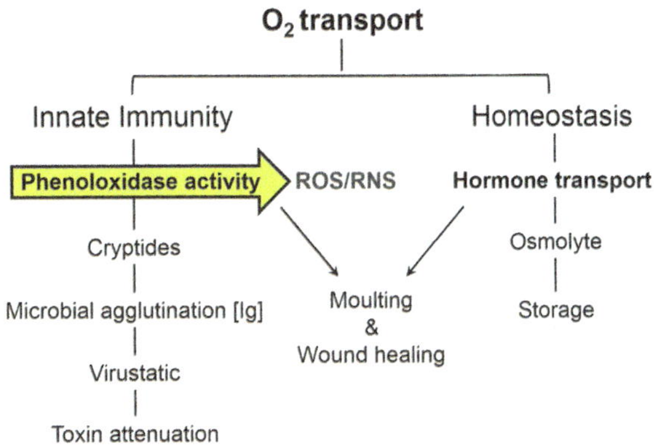

Fig. 9.2 Roles of hemocyanin beyond oxygen transport—innate immunity and homeostasis. The scheme represents all confirmed functions of hemocyanins from the available literature. ROS/RNS: reaction oxygen/nitrogen species; Ig: immunoglobulin-like domain

α and β, which are equivalent to domains II and III of the arthropod hemocyanin subunits, respectively (see Chap. 10 for further details).

Over 70 years ago, Klotz et al. (1948) suggested that horseshoe crab hemocyanin likely undertakes an additional role within the hemolymph, such as the transport of small ions. Research undertaken within the past two decades has confirmed this early observation of hemocyanin multi-functionality. In the following chapter, we examine invertebrate hemocyanins across several broad themes, innate immunity, wound healing and homeostasis (Fig. 9.2).

Hemocyanin and Invertebrate Innate Immunity

Phenoloxidase Activities

The conversion of hemocyanin from an oxygen carrier into a phenoloxidase-like enzyme has received much attention (reviewed by Decker and Tuczek 2000; Coates and Nairn 2014). Phenoloxidases (POs) are widely distributed in nature—found in animals, plants and microbes. PO enzymes (catecholoxidase [EC1.10.3.1] and tyrosinase [EC 1.14.18.1]) catalyse the hydroxylation of mono-phenols into diphenols and subsequently oxidise those diphenols into quinones and semi-quinone intermediates. After several further enzymatic/non-enzymatic steps, the quinone derivatives form the antimicrobial pigment melanin (Fig. 9.3). Invertebrates weaponize melanic polymers to help agglutinate microbes, seal wounds, entrap parasitoids, and sclerotize the cuticle post ecdysis. The proPhenoloxidase activation cascade modulates phenoloxidase-mediated immunity in crustaceans and insects (Cerenius and Soderhall 2004). Briefly, pathogen-associated molecular patterns (PAMPs) are identified by soluble and hemocyte-bound pathogen recognition proteins/receptors (PRPs) in the hemolymph, which in turn activate the expression of immune factors (including antimicrobial peptides) and the release of proPhenoloxidase (proPO) via cellular

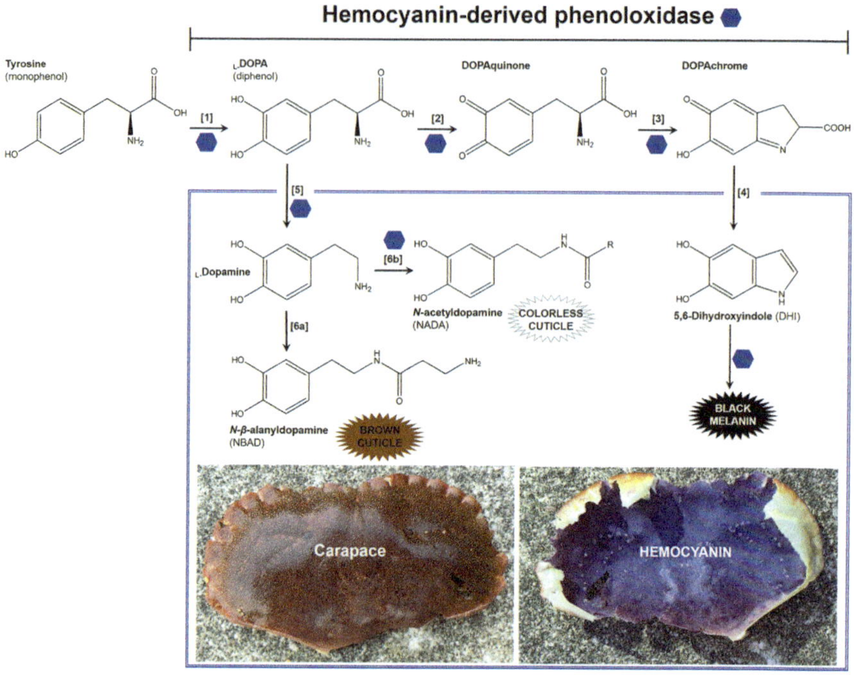

Fig. 9.3 Hemocyanin-derived phenoloxidase activities associated with melanogenic and sclerotinogenic pathways. The blue hexagons represent hemocyanin, and their positions indicate confirmed activities. Inset: dorsal (left) and ventral (right) views of the carapace of an edible crab, *Cancer pagurus*, collected from the shore of Limeslade Bay (Wales, UK). The underside of the carapace is blue-ish purple in colour due to hemocyanin in the cuticle underlying the shell [1]. Hydroxylation of tyrosine into L-DOPA [2]. Oxidation of L-DOPA into DOPAquinone [3]. Autocatalysis of DOPAquinone into DOPAchrome [4]. DOPAchrome is converted to DHI by an isomerase. DHI will eventually go on to form black/brown eumelanin [5]. L-DOPA is converted to L-Dopamine by a decarboxylase [6a and 6b]. Dopamine is converted into *N*-acetyldopamine and/or *N*-β-alanyldopamine by *N*-acetyl transferase and NBAD synthetase, respectively. Dopamine derivatives, NADA and NBAD, will eventually go on to form colourless- and brown-pigmented cuticle components. Coates and Nairn (2014) and Whitten and Coates (2017) inspired this figure

degranulation. Serine-proteases cleave the inactive proPO into an active PO that can accept phenolic substrates (Cerenius et al. 2008).

POs and hemocyanins (HCs) share a high degree of sequence homology and display almost identical structural architecture with regards the copper-based active site. Conformational changes in HC are essential for its activation into HC-derived PO, which can be induced either through proteolytic cleavage of the *N*-terminal domain (I) or through interactions with a number of allosteric factors (e.g., urea, detergents; Coates and Nairn 2014). Upon incubating both arthropod and mollusc HCs with the anionic detergent, sodium dodecyl sulphate (SDS), the structural features in and around the dicopper site loosen, thereby creating a passage for bulky phenolic substrates to gain access to the oxygen stored within (Zlateva et al. 1996; Salvato et al. 1998; Decker and Rimke 1998; Decker et al. 2001; Siddiqui et al. 2006; Baird et al. 2007; Naresh et al. 2015). Using cryoelectron microscopy, Cong et al. (2009) observed the flexibility of HC subunits; domain I twists away from domain II (location of Cu-active site) when exposed to SDS micelles and this leads to cooperative activation of the entire tetra-hexamer of scorpion HC (*Pandinus imperator*). However, SDS is an exogenous activator and concentrations exceeding micelle formation eventually cause the protein to denature. In the search for endogenous activators of HC-derived PO, antimicrobial peptides (Nagai et al. 2001), clotting factors (Nagai and Kawabata 2000), hemocyte lysates (Adachi et al. 2003), phospholipids (Coates et al. 2011, 2013) and lipoproteins (Schenk et al. 2015) have all been able to induce catalytic activity.

Chelicerates such as spiders, scorpions and horseshoe crabs lack a hemocyte-associated PO (Jaenicke et al. 2009)—making them ideal models to study HC-derived PO. In our previous work, we exposed horseshoe crab (*Limulus polyphemus*) HC to inner membrane phospholipids as they mimic several properties of SDS, such as their net negative charge and ability to form micelles/liposomes spontaneously in vitro (Coates et al. 2011, 2013). The conversion of HC into a PO by phosphatidylserine was accompanied by subtle secondary and tertiary structural changes in a manner similar to SDS. Crucially, the phospholipid-hemocyanin complex appeared more stable when compared to the SDS-hemocyanin complex. Prolonged incubation (2 days) of HC with SDS led to complete protein denaturation, whereas HC remained structurally and functionally intact with phosphatidylserine (Coates et al. 2011; Coates 2012). In chelicerate hemolymph, access to phosphatidylserine is restricted as they are located within the cytoplasmic (inner) layer of cellular membranes. However, when horseshoe crab immune cells, i.e., amebocytes, were exposed to fungal spores in vitro they ingested the microbial targets and subsequently underwent apoptosis. Apoptotic amebocytes displayed phosphatidylserine on their exoplasmic (outer) membranes, which is a universally conserved feature of cell death. The redistribution of phosphatidylserine led to a concomitant increase in HC-derived PO activity (Coates et al. 2013), which could be disrupted by the phospholipid-binding protein Annexin V.

Unlike chelicerates, many crustaceans contain both HC and hemocyte-associated PO in their hemolymph. The relative contribution of HC-derived PO to biological defences is difficult to determine in this scenario. Crustacean PO is a far superior enzyme, demonstrating catalytic power up to 100-fold greater than HC (Lee et al. 2004; Terwilliger and Ryan 2006; Coates and Nairn 2013), yet HC is present in concentrations several hundred times higher than PO. Nonetheless, due to the conserved active sites of HC-derived POs and POs, the mechanism of substrate catalysis is the same. Diphenols enter the active site and are orientated toward the copper-bound dioxygen where they undergo a two-electron oxidation into quinone(s) (Decker et al. 2018). A by-product of this process is the release of reactive oxygen and nitrogen radicals, which have been confirmed as potent antimicrobials for both PO (Cerenius et al. 2010a) and HC-derived PO (Coates and Talbot 2018). In fact, Jiang et al. (2007) provided strong evidence that ROS/RNS generated from hemoglobin and HC represent an ancient antimicrobial strategy. Hemocyanins can also be deactivated by common PO-specific inhibitors such as tropolone and kojic acid (Jaenicke and Decker 2008; Wright et al. 2012). Collectively, these data suggest that HC can function as a PO in vivo. In the absence of a pro-HC activation cascade (unlike PO), there remains a lack of knowledge on the regulation of HC between an oxygen-carrier state and an enzymatic state in vivo, which should be addressed in future research.

While the *ortho*-diphenoloxidase activity of HC is well characterised (see above), the apparent *para*-diphenoloxidase (laccase-like) activity has been reported in two studies (Le Bris et al. 2016; Besser et al. 2018). In the shrimp *Penaeus monodon*, PO-activities were assessed in different body parts (abdomen, cephalothorax, pereiopod, tail) using two distinct substrates, L-DOPA (*ortho*-isomer) and hydroquinone (*para*-isomer). The authors concluded that the thermostable PO-like activity in the cephalothorax was 'probably' due to HC. It is possible that a laccase (EC 1.10.3.2.) enzyme rather than HC was responsible for oxidising hydroquinone as these enzymes are found throughout the exoskeletal layers of many arthropods. Besser et al. (2018) found HC to be abundant in the hindgut contents of wood boring marine crustaceans (genus *Limnoria*), and that HC likely assists in lignocellulose destruction. In the presence of SDS, HC converted pyrogallol into purpurogallin—an oxidase reaction commonly performed by laccases.

In 2012, Šobotník et al. described a striking behaviour of so-called 'blue workers' of the termite *Neocapritermes taracua* when they are fighting. These older termites mix copper-containing protein crystals with a salivary gland product in order to 'burst' a toxic droplet onto their conspecific—thereby sacrificing themselves in a bid to kill the enemy. HC-like proteins were identified as the major component of the blue crystals. Interestingly, the authors noted a change in the toxic droplet from blue to clear as it became sticky. As HC goes from an oxygenated state (blue) to a deoxygenated state, it will lose its pigment properties (Coates and Nairn 2014). Therefore, we speculate that HC-derived PO activity acts as a catalyst for the contents of the salivary gland—oxygen is used up in such a reaction and likely accounts for the change in colour.

Cryptide Formation

A cryptide is a peptide or polypeptide derived from a much larger protein that demonstrates alternative or exaggerated activity. To date, >10 antimicrobial peptides (AMPs) originating from the *C*-terminus of HC subunits have been recovered from invertebrates (Coates and Decker 2017), including shrimp (*Penaeus vannamei*; Destoumieux-Garzon et al. 2001), crayfish (*Pacifastacus leniusculus*; Lee et al. 2003), a spider (*Acanthoscurria rondoniae*; Riciluca et al. 2012) and abalone (*Haliotis tuberculata*; Zhuang et al. 2015). The best characterised is PvHCt—an amphipathic, anionic, linear α-helix extracted from the hemolymph of *P. vannamei*. This peptide is a potent antifungal agent at concentrations between 3 and 50 μM but does not have any measurable negative impact on bacterial growth (>12 species tested; Destoumieux-Garzon et al. 2001). Petit et al. (2016) confirmed that PvHCt is a pore forming antimicrobial peptide—binding to the surface of fungal cells, disrupting the lipid bilayer and causing membrane leakiness. A HC-derived antibacterial peptide, Astacidin 1, was isolated from the hemolymph of freshwater crayfish (Lee et al. 2003). Unlike PvHCt, Astacidin 1 forms a cationic β-sheet structure in citric acid buffer (pH 4). Using a Wenxiang diagram to visualise Astacidin 1 in α-helical form (Fig. 9.4), we note an amphipathic distribution of amino acids that is reminiscent of pore-forming peptides. Most antimicrobial peptides (AMPs) adopt their final structures when in contact with a membrane, and therefore it is possible that

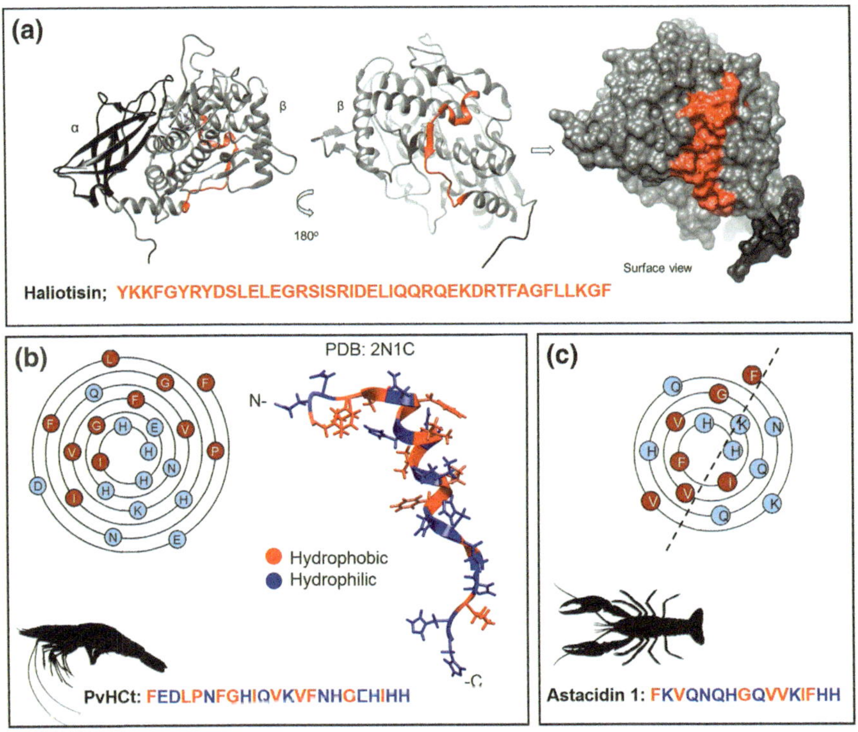

Fig. 9.4 Antimicrobial peptides generated from hemocyanins. a The crystal structure of giant keyhole limpet (*Megathura crenulata*; PDB-4BED) hemocyanin FU-E was used to illustrate the location of haliotisin at the surface of abalone hemocyanin (coloured red). The peptide remains accessible within the hemocyanin decamer (see Zhuang et al. 2015). **b** The structure and amino acid distribution of PvHCt—an antifungal peptide derived from the *C-terminus* of *P. vannamei* hemocyanin. The Wenxiang diagram and crystal structure (PDB-2N1C) confirm that PvHCt is a linear, amphipathic a-helix similar to most pore forming antimicrobial peptides (Petit et al. 2016). **c** When depicting the crayfish (*Pacifastacus leniusculus*) hemocyanin-derived peptide, Astacidin 1, using a Wenxiang diagram it seems to have distinct hydrophobic and hydrophilic sides. Lee et al. (2003) determined Astacidin 1 to be a β-sheet structure under acidic conditions. Protein structures were visualised in UCSF Chimera (Pettersen et al. 2004) and Wenxiang diagrams (Chou et al. 1997) were produced using the online server (http://www.jci-bioinfo.cn/wenxiang2)

Astacidin 1 will function in a similar manner to PvHCt. In agreement with our observations, Choi and Lee (2014) exposed yeast cells (*Candida albicans*) to Astacidin 1 and recorded membrane damage in the form of pores up to 4.3 nm in diameter. Immune-stimulation of shrimp and crayfish leads to increased concentrations of HC-derived AMPs in the respective hemolymph (Destoumieux-Garzon et al. 2001; Lee et al. 2003); with additional evidence suggesting that HC is targeted by trypsin-like proteases to release the bioactive fragments (Li et al. 2018).

To date, HC-derived AMPs have been observed in a single mollusc species. An in silico survey of abalone HC led to the identification of a looped region within the β-domain of functional unit E, which contained several putative AMPs (Zhuang et al. 2015) By synthesising these peptides and exposing them to Gram-positive and Gram-negative bacteria their antibiotic properties were confirmed. Bacteria incubated with the most potent peptide, termed haliotisin, showed membrane irregularities and significantly fewer flagella under transmission electron microscopy. Like all known AMPs from arthropod hemocyanins, haliotisin is located at the surface of the protein oligomer where it is easily accessible for proteases to cut out during immune challenge (Fig. 9.4). A degenerated fragment (ELVRKNVDHLSTPDVLELV) of whelk (*Rapana venosa*) hemocyanin FU-E was recovered from the hemolymph but did not show any measurable antimicrobial properties in vitro—its function remains unknown (Dolashka et al. 2011).

Microbial Agglutination and Immunoglobulin-Like Domains

Broadly, invertebrates lack adaptive immune mechanisms such as the capacity to produce clonal immunoglobulins (Rowley and Powell 2007), yet they do contain immune effectors that are members of the immunoglobulin (Ig) superfamily. In a series of articles, Zhang and co-workers found Ig-like domains within the *C*-terminus of *P. vannamei* HC using affinity chromatography and immunoblotting (Zhang et al. 2004b, 2006, 2017a). When the shrimp HC was incubated with a range of bacterial targets and mammalian erythrocytes, cellular agglutination was observed. The agglutination properties of HC were attributed to the dodecameric form, which offers two-carbohydrate binding sites as opposed to a single site present on the hexameric form (Pan et al. 2008). The addition of excess carbohydrates such as glucose and *N*-acetylneuraminic acid to solutions containing shrimp and mud crab (*Scylla serrata*) HCs inhibited microbial agglutination (Yan et al. 2011a, b). Furthermore, *Escherichia coli* strains containing mutations to outer membrane proteins (e.g., OmpX and OmpA) were not susceptible to the HC-induced aggregation, which suggests a degree of specificity toward Gram-negative bacteria. Zhang et al. (2017a) surveyed the sequence variability of the putative Ig-like domain (C-terminus) of *P. vannamei* HC. The authors describe this region as rich in single nucleotide polymorphisms (SNPs) and likely acts as a pathogen recognition receptor in shrimp hemolymph. Previously, SNPs were identified in the same region of shrimp HC and linked to enhanced resistance to *Vibrio parahaemolyticus* infection (Zhao et al. 2012;

Guo et al. 2013). Several HC isoforms have been identified in shrimp—those that contain an IgM-like domain and those that do not. Interestingly, these HC isoforms differ in the extent of glycosylation but display similar levels of HC-derived PO activity (Zhang et al. 2017a). Removal of sugars using O-glycosidase can reduce the agglutination properties of these IgM-positive HCs. Moreover, after challenging shrimp with either *Vibrio alginolyticus* or *V. fluvialis* the HCs purified at 72 h post-inoculation contained substantially higher contents of glycans (Zhang et al. 2017b). To the best of our knowledge, no study has focussed on the potential agglutinative properties of mollusc HC, however, a lectin-like agglutinin from octopus (*Octopus maya*) hemolymph showed partial identity to HC (Alpuche et al. 2010).

Hemolytic activity of HC was measured routinely alongside agglutination activity in the studies mentioned above. HC oligomers and individual subunits purified from shrimp (*P. vannamei*; Zhang et al. 2009) and mud crabs (*S. serrata*; Yan et al. 2011a, b) cause vertebrate erythrocytes (human, chicken, cow) to rupture in vitro, but do not affect the crustacean hemocytes. HC-induced hemolysis is influenced by environmental factors such as divalent cations and pH, and can be inhibited by the addition of osmoprotectants (e.g., polyethylene glycol). Despite such evidence, it remains unclear what biological role such hemolytic properties of HC would play in vivo.

Toxin Attenuation

The Gram-negative bacterium, *V. parahaemolyticus,* is a major cause of disease in commercial shrimp, notably acute hepatopancreatic necrosis disease. While studying shrimp-*Vibrio* antibiosis, Boonchuen et al. (2018) injected *Vibrio* toxins directly into the shrimp and measured differentially expressed genes—HC subunits 3 and 4 were upregulated significantly within 3 h. Pre-incubation of HC with the toxin prior to injection into the shrimp led to increased survival rates. Further still, the authors demonstrated that HC could agglutinate *V. parahaemolyticus*, and neutralise the toxin by binding to the PirA domain but not PirB.

Antiviral Properties (Virustatic)

In a similar manner to microbial agglutination, HC isolated from shrimp *Penaeus monodon* was capable of binding non-specifically to a range of fish DNA and RNA viruses (Zhang et al. 2004a). When investigating white spot syndrome virus (WSSV)-infected shrimp, the authors recovered two HC subunits interacting with viral particles. In a follow-on study, Lei et al (2008) pre-incubated WSSV particles with purified shrimp HC prior to infection. Those animals having received a dose of HC-WSSV contained 10-fold fewer viruses in the gills compared to the WSSV only controls. Dolashka and colleagues recorded virustatic properties of mollusc

(*Rapana venosa*) HC toward respiratory syncytial (RSV) virus and herpes simplex virus (HSV) (Dolashka et al. 2009, 2010). Glycosylated functional units (FUc and FUe) were effective against RSV and HSV, respectively, whereas non-glycosylated FUb or native HC decamers did not show any anti-viral properties. Transcriptomic surveys of many crustaceans—e.g., *Penaeus japonicus, Macrobrachium rosenbergii, Procambarus clarkii*—detail substantial increases in HC mRNA levels and corresponding proteins during viremia (Pan et al. 2005; Chongsatja et al. 2007; Bourchookarn et al. 2008; Rattanarojpong et al. 2007; Ravi et al. 2010; Shi et al. 2010; Wang et al. 2013). Interestingly, abalone (*Haliotis diversicolor*) suffering from shrivelling syndrome-associated virus (AbSV) generate a novel isoform of HC1 (Zhu et al. 2014). When compared to the native HC from uninfected abalone, HC1 presented a more 'relaxed' structural state (visualised using cryo-electron microscopy), which corresponded to a 5-fold increase in PO activity.

Havanapan et al. (2009) documented a substantial increase in HC *C*-terminal fragments in taura syndrome virus-infected shrimp, and, located a putative extracellular regulated kinase (ERK) 1/2 domain spanning arginine520 to serine532 on HC also. These data suggest that HC-derived cryptides (see section above) may act as immune signalling molecules when confronted with pathogens. Most recently, five putative anti-viral peptides derived from HC were analysed from WSSV-infected *P. vannamei* (Zhan et al. 2019). One of these peptides, referred to as LvHcL48, was capable of interfering with WSSV transcriptional activities and could bind to a key viral envelope protein (VP28). Sequencing of LvHcL48 using Edman degradation confirmed its origin as the C-terminus of HC subunits.

Hemocyanin and Invertebrate Homeostasis

Considering oxygen transport/storage is the primary function of HC, it is no surprise that hypoxia, salinity, thermal and osmotic stresses (which all affect oxygen saturation) lead to increased levels of HC in the hemolymph (Boone and Schoffeniels 1979; Paul and Pirow 1998; Decker and Föll 2000; Tanner et al. 2006). In fact, such environmental parameters influence HC subunit expression—as each subunit has unique oxygen-binding capacity (Spicer and Hodgson 2003; Brown-Peterson et al. 2005). HC is present in large amounts in invertebrate hemolymph, which is linked to buffering capacity and nonspecific (small) molecule transport (Paul and Pirow 1998; Terwilliger 2015). HC can be present at concentrations up to 50 mg/ml in tarantula and 160 mg/ml in horseshoe crabs (Coates et al. 2012), yet some authors have suggested that as little as 5 mg/ml is sufficient to meet metabolic demands (Paul et al. 1994; Decker et al. 2007). If this is true, then the 'excess' HC is free to participate in the various immune activities described above or co-opted into the exoskeleton for PO-associated hardening of the cuticle (Table 9.1). Indeed, the development of *post-mortem* hyperpigmentation in commercial crustaceans, notably shrimp, has been attributed to HC-derived PO activity due to its thermostability (resistance to

Table 9.1 Hemocyanin contributes to hemostasis and sclerotisation

Organism	Description	Reference
Acanthoscurria geniculata [spider] *and Penaeus vannamei* [shrimp]	The 'clot proteomes' of spider and shrimp hemolymph were found to contain high levels of HC. HC (mRNA) expression in infected/wounded shrimp followed similar patterns to transglutaminase, and Histidine-pull down experiments suggest that HC and transglutaminase interact	Sanggaard et al. (2016), Yao et al. (2019)
Cherax quadricarinatus [crayfish]	HC with PO activity was located in the chitin matrix of the gastrolith—a transient calcium deposit similar to exoskeleton—during pre-molt preparations	Glazer et al. (2013)
Eurypelma californicum [spider]	HC binds ecdysone—a moulting hormone—via a conserved hydrophobic pocket in domain I and carries it around the hemolymph	Jaenicke et al. (1999)
Eurypelma californicum [spider], *Limulus polyphemus* [horseshoe crab] *and Penaeus japonicus* [shrimp]	HCs are capable of catalysing late steps in the melanogenic and sclerotinogenic pathways (Fig. 9.3). HC-derived POs can convert dihydroxyindoles directly into eumelanin, and oxidise a diverse range of diphenols/catecholamines (dopamine, epinephrine, norepinephrine, *N*-acetyldopamine) into cuticle hardening agents. These enzymatically active HCs have been detected throughout the exo- and endocuticle layers of shrimp exoskeleton	Adachi et al. (2005a, b), Jaenicke and Decker (2008), Coates and Talbot (2018)

(continued)

Table 9.1 (continued)

Organism	Description	Reference
Limulus polyphemus [horseshoe crab]	HC is converted into a PO-like enzyme upon binding to soluble phosphatidylserine, or, apoptotic cells displaying phosphatidylserine on their outer membranes. This mode of activation is reminiscent of the conversion of prothrombin (clotting zymogen) into thrombin—via platelet membranes with high phosphatidylserine content	Coates et al. (2011, 2013), Vance and Steenbergen (2005)
Portunus pelagicus [crab]	Transcriptomic studies on crustaceans (e.g., *Portunus pelagicus*) post-moult, intermoult and pre-moult demonstrated that HC expression patterns are consistent with known PO activators	Kuballa and Elizur (2008), Kuballa et al. (2011)
Tachypleus tridentatus [horseshoe crab]	HC interacts with both clotting factors and chitin-binding antimicrobial peptides, and in doing so, switches into HC-derived PO. The authors postulated the HC binds to the AMP, tachyplesin, which anchors to exposed chitin during wound healing. HC is thus activated into a PO in close proximity to a developing clot	Nagai and Kawabata (2000), Nagai et al. (2001)

freeze-thawing) and high concentration in contrast to hemocyte-derived PO (García-Carreño et al. 2008; Martínez-Alvarez et al. 2008; Coates and Nairn 2013; Coates and Albalat 2014). Hyperpigmentation (or melanosis) is a non-infectious condition that causes cuticular darkening, resulting in reduced market value of shellfish and large accumulation of waste.

Preventing the loss of vital body fluid (hemolymph) is essential to an invertebrate's survival. Damage to the exoskeleton, whether a pathogen, parasite or predator causes it, will initiate a series of reactions collectively known as the hemostatic system. This will function alongside the immune response in order to repair the wound (reviewed by Theopold et al. 2004). The polymerisation of lipophorins and vitellogenin-like proteins is achieved via the action of a calcium-dependent cross-linking enzyme, namely transglutaminase (E.C. 2.3.2.13 (Cerenius et al. 2010b). These clotting proteins contain a cysteine rich domain homologous to the mammalian blood clotting proteins of the Von Willebrand factor. Moreover, the complete replacement of the crustacean exoskeleton and the hardening of the new cuticle layers post moulting is a major physiological task. Terwilliger (2007, 2015) has argued that it is under such extreme conditions that HC is recruited to assist PO and other proteins. Several key pieces of evidence indicate HC is involved in wound healing and sclerotisation (Fig. 9.3, Table 9.1).

Concluding Remarks

In 2014, we concluded that the *C*-terminal domain of both arthropod and mollusc HCs is key to their expanded role(s) in innate immunity (Coates and Nairn 2014). Many subsequent studies (including our own) have provided compelling evidence in support of this. Examples include the characterisation of the first antibacterial peptide derived from the C-terminal region of abalone (*H. diversicolor*) HC (Zhuang et al. 2015), and the immunoglobulin-like features of shrimp HC domain III (Zhang et al. 2017a). Now, we are certain that HC is an essential component of innate immunity and homeostasis. What remains unclear is how all of these activities are regulated. How is the labour of each activity distributed amongst a seemingly heterogeneous population of HC subunits, or is each HC oligomer made-up of subunit combinations capable of performing all activities? This question is even more intriguing when we consider the independent evolution of mollusc and arthropod HCs.

Acknowledgements We should like to thank Prof Andrew Rowley (Swansea University) for providing comments on the text. No direct funding was provided to undertake this review; however, the content was presented and discussed at the SafeAqua workshop on Invertebrate Immunology held in Thailand, July 2019. The SafeAqua project has received funding from the European Union's Horizon 2020 research and innovation programme under the grant agreement No. 734486 (CJC is the PI for Swansea University).

References

Adachi K, Hirata T, Nishioka T et al (2003) Hemocyte components in crustaceans convert hemocyanin into a phenoloxidase-like enzyme. Comp Biochem Phys B 134(1):135–141

Adachi K, Endo H, Watanabe T et al (2005a) Hemocyanin in the exoskeleton of crustaceans: enzymatic properties and immunolocalization. Pigment Cell Res 18(2):136–143

Adachi K, Wakamatsu K, Ito S et al (2005b) An oxygen transporter hemocyanin can act on the late pathway of melanin synthesis. Pigment Cell Res 18(3):214–219

Aguilera F, McDougall C, Degnan BM (2013) Origin, evolution and classification of type-3 copper proteins: lineage-specific gene expansions and losses across the Metazoa. BMC Evol Biol 13(1):96

Alpuche J, Pereyra A, Mendoza-Hernández G et al (2010) Purification and partial cnharacterization of an agglutinin from *Octopus maya* serum. Comp Biochem Phys B 156(1):1–5

Baird S, Kelly SM, Price NC et al (2007) Hemocyanin conformational changes associated with SDS-induced phenol oxidase activation. BBA Proteins Proteom 1774(11):1380–1394

Besser K, Malyon GP, Eborall WS et al (2018) Hemocyanin facilitates lignocellulose digestion by wood-boring marine crustaceans. Nat Commun 9(1):5125

Boonchuen P, Jaree P, Tassanakajon A et al (2018) Hemocyanin of Litopenaeus vannamei agglutinates Vibrio parahaemolyticus AHPND (VPAHPND) and neutralizes its toxin. Dev Comp Immunol 84:371–381

Boone WR, Schoffeniels E (1979) Hemocyanin synthesis during hypo-osmotic stress in the shore crab Carcinus maenas (L.). Comp Biochem Physiol B 63(2):207–214

Bourchookarn A, Havanapan PO, Thongboonkerd V et al (2008) Proteomic analysis of altered proteins in lymphoid organ of yellow head virus infected Penaeus monodon. BBA Proteins Proteom 1784(3):504–511

Brown-Peterson NJ, Larkin P, Denslow N et al (2005) Molecular indicators of hypoxia in the blue crab Callinectes sapidus. Mar Ecol Prog Ser 286:203–215

Burmester T (1999) Evolution and function of the insect hexamerins. Eur J Entomol 96:213–226

Burmester T (2002) Origin and evolution of arthropod hemocyanins and related proteins. J Comp Physiol B 172(2):95–107

Burmester T (2015) Evolution of respiratory proteins across the Pancrustacea. Integr Comp Biol 55(5):792–801

Cerenius L, Söderhäll K (2004) The prophenoloxidase-activating system in invertebrates. Immunol Rev 198(1):116–126

Cerenius L, Lee BL, Söderhäll K (2008) The proPO-system: pros and cons for its role in invertebrate immunity. Trends Immunol 29(6):263–271

Cerenius L, Babu R, Söderhäll K et al (2010a) In vitro effects on bacterial growth of phenoloxidase reaction products. J Invert Pathol 103(1):21–23

Cerenius L, Kawabata SI, Lee BL et al (2010b) Proteolytic cascades and their involvement in invertebrate immunity. Trends Biochem Sci 35(10):575–583

Choi H, Lee DG (2014) Antifungal activity and pore-forming mechanism of astacidin 1 against Candida albicans. Biochimie 105:58–63

Chongsatja PO, Bourchookarn A, Lo CF et al (2007) Proteomic analysis of differentially expressed proteins in Penaeus vannamei hemocytes upon Taura syndrome virus infection. Proteomics 7(19):3592–3601

Chou KC, Zhang CT, Maggiora GM (1997) Disposition of amphiphilic helices in heteropolar environments. Proteins: Struct Funct Bioinform 28(1):99–108

Coates CJ (2012) Hemocyanin-derived phenoloxidase; biochemical and cellular investigations of innate immunity. Ph.D. thesis. http://hdl.handle.net/1893/12228

Coates CJ, Albalat (2014) Engaging with strategies to impede post-mortem hyperpigmentation in commercial crustaceans. In: Hay RM (ed) Shellfish: human consumption, health implications and conservation concerns. Nova Science Publishers, pp 169–194. ISBN: 978-163321196-4, 978-163321195-7

Coates CJ, Decker H (2017) Immunological properties of oxygen-transport proteins: hemoglobin, hemocyanin and hemerythrin. Cell Mol Life Sci 74(2):293–317
Coates CJ, Nairn J (2013) Hemocyanin-derived phenoloxidase activity: a contributing factor to hyperpigmentation in Nephrops norvegicus. Food Chem 140(1–2):361–369
Coates CJ, Nairn J (2014) Diverse immune functions of hemocyanins. Dev Comp Immunol 45(1):43–55
Coates CJ, Talbot J (2018) Hemocyanin-derived phenoloxidase reaction products display anti-infective properties. Dev Comp Immunol 86:47–51
Coates CJ, Kelly SM, Nairn J (2011) Possible role of phosphatidylserine–hemocyanin interaction in the innate immune response of Limulus polyphemus. Dev Comp Immunol 35(2):155–163
Coates CJ, Bradford EL, Krome CA et al (2012) Effect of temperature on biochemical and cellular properties of captive Limulus polyphemus. Aquaculture 334:30–38
Coates CJ, Whalley T, Wyman M et al (2013) A putative link between phagocytosis-induced apoptosis and hemocyanin-derived phenoloxidase activation. Apoptosis 18(11):1319–1331
Cong Y, Zhang Q, Woolford D et al (2009) Structural mechanism of SDS-induced enzyme activity of scorpion hemocyanin revealed by electron cryomicroscopy. Structure 17(5):749–758
Costa-Paiva EM, Schrago CG, Coates CJ et al (2018) Discovery of novel hemocyanin-like genes in metazoans. Biol Bull 235(3):134–151
Decker H, Föll R (2000) Temperature adaptation influences the aggregation state of hemocyanin from Astacus leptodactylus. Comp Biochem Physiol A 127(2):147–154
Decker H, Rimke T (1998) Tarantula hemocyanin shows phenoloxidase activity. J Biol Chem 273(40):25889–25892
Decker H, Tuczek F (2000) Tyrosinase/catecholoxidase activity of hemocyanins: structural basis and molecular mechanism. Trends Biochem Sci 25(8):392–397
Decker H, Ryan M, Jaenicke E et al (2001) SDS-induced phenoloxidase activity of hemocyanins from limulus polyphemus, eurypelma californicum, and cancer magister. J Biol Chem 276(21):17796–17799
Decker H, Hellmann N, Jaenicke E et al (2007) Minireview: recent progress in hemocyanin research. Integr Comp Biol 47(4):631–644
Decker H, Solem E, Tuczek F (2018) Are glutamate and asparagine necessary for tyrosinase activity of type-3 copper proteins? Inorg Chim Acta 481:32–37
Destoumieux-Garzón D, Saulnier D, Garnier J et al (2001) Crustacean immunity antifungal peptides are generated from the c terminus of shrimp hemocyanin in response to microbial challenge. J Biol Chem 276(50):47070–47077
Dolashka P, Velkova L, Shishkov S et al (2010) Glycan structures and antiviral effect of the structural subunit RvH2 of Rapana hemocyanin. Carbohydr Res 345(16):2361–2367
Dolashka P, Moshtanska V, Borisova V et al (2011) Antimicrobial proline rich peptides from the hemolymph of marine snail Rapana venosa. Peptides 32(7):1477–1483
Dolashka-Angelova P, Lieb B, Velkova L et al (2009) Identification of glycosylated sites in Rapana hemocyanin by mass spectrometry and gene sequence, and their antiviral effect. Bioconjug Chem 20(7):1315–1322
García-Carreño FL, Cota K, Navarrete del Toro MA (2008) Phenoloxidase activity of hemocyanin in whiteleg shrimp Penaeus vannamei: conversion, characterization of catalytic properties, and role in postmortem melanosis. J Agricul Food Chem 56(15):6454–6459
Gatsogiannis C, Hofnagel O, Markl J et al (2015) Structure of mega-hemocyanin reveals protein origami in snails. Structure 23(1):93–103
Glazer L, Tom M, Weil S et al (2013) Hemocyanin with phenoloxidase activity in the chitin matrix of the crayfish gastrolith. J Exp Biol 216(10):1898–1904
Guo L, Zhao X, Zhang Y, Wang Z, Zhong M, Li S, Lun J, (2013) Evidences of SNPs in the variable region of hemocyanin Ig-like domain in shrimp litopenaeus vannamei. Fish Shellfish Immunol 35(5):1532–1538
Hartmann H, Decker H (2004) Small-angle scattering techniques for analyzing conformational transitions in hemocyanins. In: Methods in enzymology, vol 379. Academic Press, pp 81–106

Havanapan PO, Kanlaya R, Bourchookarn A et al (2009) C-terminal hemocyanin from hemocytes of Penaeus vannamei interacts with ERK1/2 and undergoes serine phosphorylation. J Proteome Res 8(5):2476–2483

Jaenicke E, Decker H (2008) Kinetic properties of catecholoxidase activity of tarantula hemocyanin. FEBS J 275(7):1518–1528

Jaenicke E, Föll R, Decker H (1999) Spider hemocyanin binds ecdysone and 20-OH-ecdysone. J Biol Chem 274(48):34267–34271

Jaenicke E, Fraune S, May S et al (2009) Is activated hemocyanin instead of phenoloxidase involved in immune response in woodlice? Dev Comp Immunol 33(10):1055–1063

Jiang N, Tan NS, Ho B et al (2007) Respiratory protein–generated reactive oxygen species as an antimicrobial strategy. Nat Immunol 8(10):1114

Kato S, Matsui T, Gatsogiannis C et al (2018) Molluscan hemocyanin: structure, evolution, and physiology. Biophys Rev 10(2):191–202

Klotz IM, Schlesinger AH, Tietze F (1948) Comparison of the binding ability of hemocyanin and serum albumin for organic ions. Biol Bull 94(1):40–44

Kuballa AV, Elizur A (2008) Differential expression profiling of components associated with exoskeletal hardening in crustaceans. BMC Genom 9(1):575

Kuballa AV, Holton TA, Paterson B et al (2011) Moult cycle specific differential gene expression profiling of the crab Portunus pelagicus. BMC Genom 12(1):147

Le Bris C, Cudennec B, Dhulster P et al (2016) Melanosis in Penaeus monodon: involvement of the laccase-like activity of hemocyanin. J Agricul Food Chem 64(3):663–670

Lee SY, Lee BL, Söderhäll K (2003) Processing of an antibacterial peptide from hemocyanin of the freshwater crayfish Pacifastacus leniusculus. J Biol Chem 278(10):7927–7933

Lee SY, Lee BL, Söderhäll K (2004) Processing of crayfish hemocyanin subunits into phenoloxidase. Biochem Biophys Res Commun 322(2):490–496

Lei K, Li F, Zhang M et al (2008) Difference between hemocyanin subunits from shrimp Penaeus japonicus in anti-WSSV defense. Dev Comp Immunol 32(7):808–813

Li C, Wang F, Aweya JJ et al (2018) Trypsin of Litopenaeus vannamei is required for the generation of hemocyanin-derived peptides. Dev Comp Immunol 79:95–104

Markl J (2013) Evolution of molluscan hemocyanin structures. BBA Proteins Proteom 1834(9):1840–1852

Markl J, Decker H (1992) Molecular structure of the arthropod hemocyanins. In: Blood and tissue oxygen carriers. Springer, Berlin, pp 325–376

Martín-Durán JM, de Mendoza A, Sebé-Pedrós A et al (2013) A broad genomic survey reveals multiple origins and frequent losses in the evolution of respiratory hemerythrins and hemocyanins. Genome Biol Evol 5(7):1435–1442

Martínez-Alvarez O, Gómez-Guillén C, Montero P (2008) Presence of hemocyanin with diphenoloxidase activity in deepwater pink shrimp (Parapenaeus longirostris) post mortem. Food Chem 107(4):1450–1460

Nagai T, Kawabata SI (2000) A link between blood coagulation and prophenol oxidase activation in arthropod host defense. J Biol Chem 275(38):29264–29267

Nagai T, Osaki T, Kawabata SI (2001) Functional conversion of hemocyanin to phenoloxidase by horseshoe crab antimicrobial peptides. J Biol Chem 276(29):27166–27170

Naresh KN, Sreekumar A, Rajan SS (2015) Structural insights into the interaction between molluscan hemocyanins and phenolic substrates: an in silico study using docking and molecular dynamics. J Mol Graph Model 61:272–280

Pan D, He N, Yang Z et al (2005) Differential gene expression profile in hepatopancreas of WSSV-resistant shrimp (Penaeus japonicus) by suppression subtractive hybridization. Dev Comp Immunol 29(2):103–112

Pan JY, Zhang YL, Wang SY et al (2008) Dodecamer is required for agglutination of Litopenaeus vannamei hemocyanin with bacterial cells and red blood cells. Mar Biotechnol 10(6):645–652

Paul RJ, Pirow R (1998) The physiological significance of respiratory proteins in invertebrates. Zoology 100(4):298–306

Paul R, Bergner B, Pfeffer-Seidl A et al (1994) Gas transport in the haemolymph of arachnids-oxygen transport and the physiological role of haemocyanin. J Exp Biol 188(1):25–46
Petit VW, Rolland JL, Blond A et al (2016) A hemocyanin-derived antimicrobial peptide from the penaeid shrimp adopts an alpha-helical structure that specifically permeabilizes fungal membranes. BBA Gen Subj 1860(3):557–568
Pettersen EF, Goddard TD, Huang CC et al (2004) UCSF Chimera—a visualization system for exploratory research and analysis. J Comput Chem 25(13):1605–1612
Rattanarojpong T, Wang HC, Lo CF et al (2007) Analysis of differently expressed proteins and transcripts in gills of Penaeus vannamei after yellow head virus infection. Proteomics 7(20):3809–3814
Ravi M, Basha AN, Taju G et al (2010) Clearance of Macrobrachium rosenbergii nodavirus (MrNV) and extra small virus (XSV) and immunological changes in experimentally injected Macrobrachium rosenbergii. Fish Shellfish Immunol 28(3):428–433
Rehm P, Pick C, Borner J et al (2012) The diversity and evolution of chelicerate hemocyanins. BMC Evol Biol 12(1):19
Riciluca KCT, Sayegh RSR, Melo RL et al (2012) Rondonin an antifungal peptide from spider (Acanthoscurria rondoniae) haemolymph. Results Immunol 2:66–71
Rowley AF, Powell A (2007) Invertebrate immune systems–specific, quasi-specific, or nonspecific? J Immunol 179(11):7209–7214
Salvato B, Santamaria M, Beltramini M et al (1998) The enzymatic properties of Octopus vulgaris hemocyanin: o-diphenol oxidase activity. Biochem 37(40):14065–14077
Sanggaard KW, Dyrlund TF, Bechsgaard JS et al (2016) The spider hemolymph clot proteome reveals high concentrations of hemocyanin and von Willebrand factor-like proteins. BBA Proteins Proteom 1864(2):233–241
Schenk S, Schmidt J, Hoeger U, Decker H (2015) Lipoprotein-induced phenoloxidase-activity in tarantula hemocyanin. BBA Proteins Proteom 1854(8):939–949
Shi XZ, Li XC, Wang S et al (2010) Transcriptome analysis of hemocytes and hepatopancreas in red swamp crayfish, Procambarus clarkii, challenged with white spot syndrome virus. Invertebrate Surviv J 7(1):119–131
Siddiqui NI, Akosung RF, Gielens C (2006) Location of intrinsic and inducible phenoloxidase activity in molluscan hemocyanin. Biochem Biophys Res Commun 348(3):1138–1144
Šobotník J, Bourguignon T, Hanus R et al (2012) Explosive backpacks in old termite workers. Science 337(6093):436
Spicer JI, Hodgson E (2003) Structural basis for salinity-induced alteration in oxygen binding by haemocyanin from the estuarine amphipod Chaetogammarus marinus (L.). Physiol Biochem Zool 76(6):843–849
Tanner CA, Burnett LE, Burnett KG (2006) The effects of hypoxia and pH on phenoloxidase activity in the Atlantic blue crab, Callinectes sapidus. Comp Biochem Physiol A 144(2):218–223
Terwilliger NB (2007) Hemocyanins and the immune response: defense against the dark arts. Integr Comp Biol 47(4):662–665
Terwilliger NB (2015) Oxygen transport proteins in crustacean: hemocyanin and hemoglobin. In: Chang ES, Thiel M (eds) The natural history of the crustacea: physiology. Oxford University Press. ISBN: 9780199832415
Terwilliger NB, Ryan MC (2006) Functional and phylogenetic analyses of phenoloxidases from brachyuran (Cancer magister) and branchiopod (Artemia franciscana, Triops longicaudatus) crustaceans. Biol Bull 210(1):38–50
Theopold U, Schmidt O, Söderhäll K et al (2004) Coagulation in arthropods: defence, wound closure and healing. Trends Immunol 25(6):289–294
Vance JE, Steenbergen R (2005) Metabolism and functions of phosphatidylserine. Prog Lipid Res 44(4):207–234
Wang DL, Sun T, Zuo D et al (2013) Cloning and tissue expression of hemocyanin gene in Cherax quadricarinatus during white spot syndrome virus infection. Aquaculture 410:216–224

Whitten MM, Coates CJ (2017) Re-evaluation of insect melanogenesis research: views from the dark side. Pigment Cell Melanoma Res 30(4):386–401
Wright J, Clark WM, Cain JA et al (2012) Effects of known phenoloxidase inhibitors on hemocyanin-derived phenoloxidase from Limulus polyphemus. Comp Biochem Physiol B 163(3–4):303–308
Yan F, Qiao J, Zhang Y et al (2011a) Hemolytic properties of hemocyanin from mud crab Scylla serrata. J Shellfish Res 30(3):957–963
Yan F, Zhang Y, Jiang R et al (2011b) Identification and agglutination properties of hemocyanin from the mud crab (Scylla serrata). Fish Shellfish Immunol 30(1):354–360
Yao D, Wang Z, Wei M et al (2019) Analysis of Litopenaeus vannamei hemocyanin interacting proteins reveals its role in hemolymph clotting. J Proteomics 201:57–64
Zhan S, Aweya JJ, Wang F et al (2019) Litopenaeus vannamei attenuates white spot syndrome virus replication by specific antiviral peptides generated from hemocyanin. Dev Comp Immunol 91:50–61
Zhang X, Huang C, Qin Q (2004a) Antiviral properties of hemocyanin isolated from shrimp Penaeus monodon. Antiviral Res 61(2):93–99
Zhang Y, Wang S, Peng X (2004b) Identification of a type of human IgG-like protein in shrimp Penaeus vannamei by mass spectrometry. J Exp Mar Biol Ecol 301(1):39–54
Zhang Y, Wang S, Xu A et al (2006) Affinity proteomic Approach for identification of an IgA-like protein in Litopenaeus vannamei and study on its Agglutination characterization. J Proteome Res 5(4):815–821
Zhang Y, Yan F, Hu Z et al (2009) Hemocyanin from shrimp Litopenaeus vannamei shows hemolytic activity. Fish Shellfish Immunol 27(2):330–335
Zhang YL, Peng B, Li H et al (2017a) C-terminal domain of hemocyanin, a major antimicrobial protein from Litopenaeus vannamei: structural homology with immunoglobulins and molecular diversity. Front Immunol 8:611
Zhang Z, Wang F, Chen C et al (2017b) Glycosylation of hemocyanin in Litopenaeus vannamei is an antibacterial response feature. Immunol Lett 192:42–47
Zhao X, Guo L, Zhang Y et al (2012) SNPs of hemocyanin C-terminal fragment in shrimp Litopenaeus vannamei. FEBS Lett 586(4):403–410
Zhu H, Zhuang J, Feng H et al (2014) Cryo-EM structure of isomeric molluscan hemocyanin triggered by viral infection. PLoS ONE 9(6):e98766
Zhuang J, Coates CJ, Zhu H et al (2015) Identification of candidate antimicrobial peptides derived from abalone hemocyanin. Dev Comp Immunol 49(1):96–102
Zlateva T, Di Muro P, Salvato B et al (1996) The o-diphenol oxidase activity of arthropod hemocyanin. FEBS Lett 384(3):251–254

Chapter 10
Recent Insights into the Diversity and Evolution of Invertebrate Hemerythrins and Extracellular Globins

Elisa M. Costa-Paiva and Christopher J. Coates

Abstract There are three broad groups of oxygen-transport proteins found in the haemolymph (blood) of invertebrates, namely the hemocyanins, the hemerythrins and the globins. Both hemerythrins and extracellular globins are iron-based proteins that are understudied when compared to the copper-containing hemocyanins. Recent evidence suggests that hemerythrins and (giant) extracellular globins (and their linker chains) are more widely distributed than previously thought and may have biological functions beyond oxygen transport and storage. Herein, we review contemporary literature of these often-neglected proteins with respect to their structural configurations on formation and ancestral states.

Keywords Hemerythrin · Giant hexagonal bilayer hemoglobin · Extracellular hemoglobin · Evolutionary history · Bioinformatics

Introduction

The vast majority of extant animals use aerobic metabolism for their maintenance and growth, thereby making the transport of respiratory gases such as oxygen (O_2) a vital physiological task (Hill et al. 2016). The need for oxygen in an animal is mistakenly associated with respiratory function exclusively—oxygen is needed for other purposes such as collagen synthesis and wound healing (Schreml et al. 2010; Mills and Canfield 2014). Collagen is the most abundant protein in extant animals, serving as the primary component of the metazoan extracellular matrix (Shoulders and Raines 2009) and even in sponges, collagen forms spongin fibres to aid cell adhesion and mechanical support (Garrone 1999).

E. M. Costa-Paiva (✉) · C. J. Coates
Department of Zoology, Biociences Institute, University of Sao Paulo, Sao Paulo, Brazil
e-mail: elisam.costapaiva@gmail.com

C. J. Coates
e-mail: c.j.coates@swansea.ac.uk

Department of Biosciences, College of Science, Swansea University, Swansea SA2 8PP, UK

U. Hoeger and J. R. Harris (eds.), *Vertebrate and Invertebrate Respiratory Proteins, Lipoproteins and other Body Fluid Proteins*, Subcellular Biochemistry 94,
https://doi.org/10.1007/978-3-030-41769-7_10

Considering the close relationship between animals and oxygen, for the past half-century, it has been reasoned that low oxygen availability prevented the origin of animals until the late Neoproterozoic Era (1.000–541 million years ago, Ma) (Nursall 1959). This "oxygen control hypothesis" described by Knoll (1992), stated that rising oxygen levels created the permissive environment critical for animal evolution to occur. This is why increased dioxygen availability is frequently referred to as a trigger for animal evolution coupled with aerobic metabolism (Mills and Canfield 2014). However, recent studies challenge this canonical view that animal origin was primarily controlled by atmospheric oxygen levels (Mills et al. 2014; 2018).

Early animals were likely small, soft-bodied, and collagen-poor (although not necessarily collagen-free) and lived under low oxygen levels, restricting their oxygen use to high-priority physiological functions (Mills and Canfield 2014). Nevertheless, the diversification of lineages enabled an increase of morphological, physiological, and ecological complexity. In 1970, Raff and Raff, calculated that oxygen tensions of ~1.25 to 6.25% present atmospheric levels constrain diffusion-based animals to thicknesses of 0.1–0.5 mm. This constraint implies that any increase in body size would demand a circulatory system, involving the operation of respiratory pigments for internal oxygen transport (Raff and Raff 1970; Schmidt-Rhaesa 2007).

The emergence of respiratory pigments may represent a polyphyletic response to a strong selective pressure imposed by the accumulation of free oxygen during Precambrian (Alvarez-Carreño et al. 2016). A strong natural selection, therefore, benefited the emergence and improvement of circulatory systems and the presence of oxygen binding proteins that would increase their transport capacity by up to 20 times in relation to direct dissolution in liquids (Schmidt-Rhaesa 2007). These carrier molecules are proteins that likely originated from enzymes whose primary function would be to protect the organism from dioxygen toxicity, having acquired the potential for molecule transport later (Terwilliger 1998). The evolutionary history of metazoans, with an increase of body sizes and homeostasis dependent on the separation between the internal and external parts of the body was only possible with the evolution of oxygen-binding proteins (Manwell 1960). These proteins play an essential role in organismal homeostasis, and hence have been extensively investigated (Burmester 2015). The evolutionary history of animals and oxygen binding proteins are intertwined for over millions of years. Thus, in order to better understand the origin and diversification of animals, we must study the diversity and evolutionary history of oxygen binding proteins.

Oxygen carrier proteins are biological macromolecules that can reversibly bind molecular dioxygen and are often referred to as respiratory or blood pigments, since they tend to exhibit colour when bound to the oxygen (Terwilliger 1998). They are divided into three chemical categories: hemoglobins (Hbs), hemerythrins (Hrs) and hemocyanins (Hcs) (Terwilliger et al. 1976). Although these proteins can bind O_2 reversibly in a highly specific manner by noncovalent bonds, their binding affinities and evolutionary origins differ greatly. Furthermore, recent studies have shown that within these large categories there has been an underestimation of the diversity of these protein families in metazoans (Martín-Durán et al. 2013; Koch et al. 2016; Costa-Paiva et al. 2017a, b, 2018; Belato et al. 2019) (Fig. 10.1). In addition

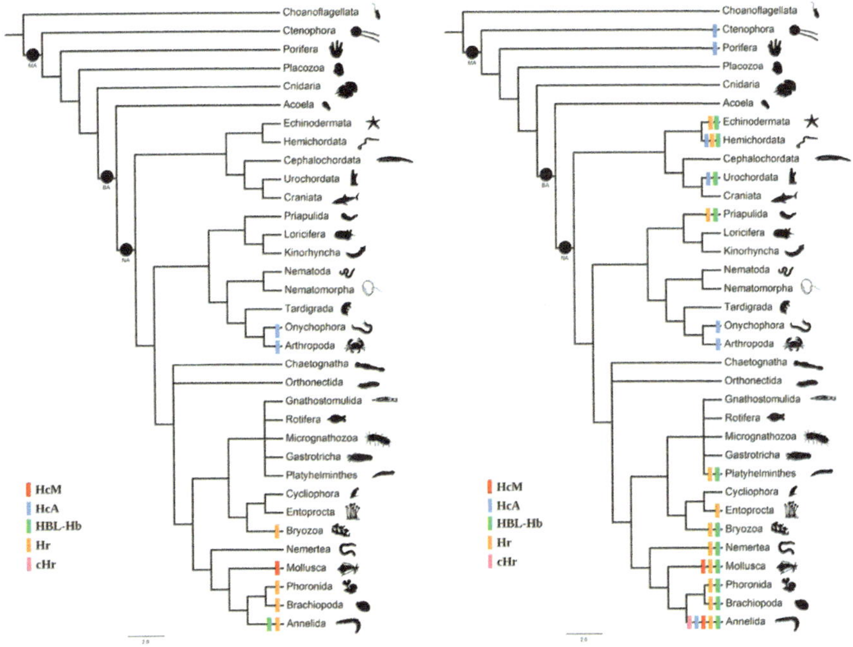

Fig. 10.1 Hypothesized relationships among metazoan phyla derived from recent phylogenomic studies (Whelan et al. 2015; Halanych 2016; Cannon et al. 2016; Kocot et al. 2017). Red rectangles represent Hcs from molluscs, blue rectangles represent Hcs from arthropods and onychophorans; green rectangles represent giant hexagonal bilayer hemoglobins (HBL-Hbs), also known as erythrocruorins and chlorocruorins, an extracellular hemoglobin; orange rectangles represent Hrs; and pink rectangles represent an exclusively annelid Hr subtype (cHr). MA is metazoan ancestor, BA is bilaterian ancestor, and NA is nephrozoan ancestor. Left panel–records of oxygen binding proteins prior to 2013. Right panel–updated records as of 2019

to academic studies, the pharmaceutical industry has concentrated great efforts in investigating the application of these proteins in a series of marketable products, e.g., the treatment of degenerative diseases and as blood substitutes for transfusions (Rousselot et al. 2006; Deng et al. 2015; Jones 2015; Mandler et al. 2015a, b; Zal 2015).

The diversity and evolutionary history of Hcs and intracellular Hbs have been covered extensively (Lecomte et al. 2005; Vinogradov et al. 2006; Decker et al. 2007; Burmester 2001, 2002, 2015), however, Hrs and extracellular Hbs, although predating Hc studies, have received far less attention (usually a restricted taxonomic context; (Weber and Vinogradov 2001; Chabasse et al. 2006a, b; Vanin et al. 2006; Bailly et al. 2008). Nevertheless, with the recent popularization of high-throughput sequencing methods and the advancement of bioinformatics to handle massive amounts of genomic data, much more has been discovered about these proteins (Martín-Durán et al. 2013; Belato et al. 2019). In the following chapter, we focus on the expanded biological insights into Hrs and giant hexagonal bilayer hemoglobins (HBL-Hbs), an extracellular globin.

Hemerythrins

The Molecule

The hemerythrins (Hrs) belong to a family of proteins whose evolutionary history seems to be quite complex. Biochemically, they belong to the class of alpha-proteins (SCOP, Andreeva et al. 2004) and are relatively small, with about a hundred amino acids organized into a single domain (Mangum 1992). Despite its name, Hrs contains no heme group but rather a nonheme diiron site that reversibly binds two molecules of oxygen in an 'end-on' coordination (Coates and Decker 2017).

These proteins appear to have a relationship between their quaternary structure and localization within the organism (Kurtz 1992). The largest quantity of Hr in an organism is contained in hemerythrocytes, which freely float in the coelomic fluid or in the circulatory system—designated circulating hemerythrin (cHr) (Costa-Paiva et al. 2017b). However, there is also a non-circulating Hr found in muscle tissues, i.e., myohemerythrin (myoHr) (Manwell 1960). Hr from coelomic hemerythrocytes are the best characterised, due to the convenience of isolating large quantities and its relevant stability after purification (Kurtz1992). The coelomic Hr from most species consists of an octamer of molecular weight ~108,000 Da (Ward et al. 1975), while myoHr is always found in monomeric form (Fig. 10.2). Other Hr oligomers include

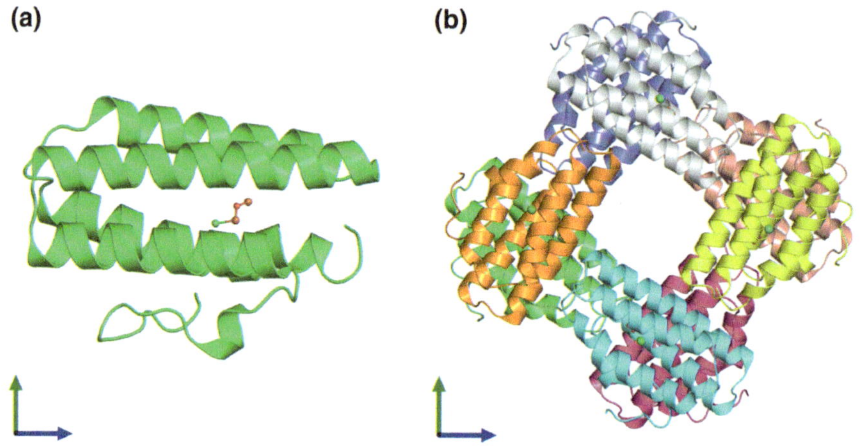

Fig. 10.2 **a** Tertiary structure of a Hr monomer from *Themiste zostericola* at a resolution of 1.8 Å by X-ray diffraction, modified from PDB (1A7D) (Martins et al. 1997); **b** Quaternary structure of a Hr octamer from *Phacolopsis gouldii* at a resolution of 1.8 Å by X-ray diffraction, modified from PDB (1I4Y) (Farmer et al. 2001)

dimers, trimers and tetramers (Kurtz 1992). Strong inter-subunit interactions are responsible for the remarkable stability of these oligomers (Holmes and Stenkamp 1991). Hr tertiary structures consist of four nearly parallel α-helices in a left-twisted bundle named A, B, C, and D representing 70% of the secondary structure (Klippenstein et al. 1972; Kurtz 1992). Several aromatic side chains on the interior surfaces of the four helical regions are interlaced with each other, and these hydrophobic interactions presumably contribute to the stabilization of the helical bundle (Sieker et al. 1982).

Until 1991, just three Hr sequences from sipunculids were available: two octameric coelomic Hrs from *Phascolosoma gouldii* and *Themiste dyscritum*, and a monomeric Hr from the retractor muscles of *Themiste zostericola* (Sanders-Loehr and Loehr 1979). For these sequences, the cHrs consisted of 113 residues each and the monomer myoHrs consisted of 118 residues, presenting about 40% similarity between the two (Kurtz 1992). The conserved residues include those that furnish ligands to iron (5 histidines) and several residues in the immediate vicinity of the diiron site. By exploiting genomic methodologies, we now have several hundred Hr sequences to interrogate (Martín-Durán et al. 2013; Costa-Paiva et al. 2017a, b).

Each subunit of Hr contains two iron ions, Fe1 and Fe2, which bind directly to the polypeptide chain forming a diiron site where the dioxygen binds (Fig. 10.3a). The iron atoms are coordinated by seven amino acid side chains of the protein being: five terminal ligands that are exclusive imidazoles from histidine residues, three to Fe1 and two for Fe2 (Stenkamp et al. 1984). Moreover, three further ligands bridge Fe1 and Fe2: two carboxylate moieties from aspartate and glutamate residues and a single oxygen molecule (Kurtz 1992; Fig. 10.3b).

Concerning the oxygen binding process, O_2 binds to the open coordination site of Fe2 and starts to oxidize this atom. The incipient negative charge that is formed on the coordinated O_2 attracts the proton of the hydroxo bridge causing the shortening of the Fe-O_2 bonds and thereby facilitates the oxidation of Fe1 (Kurtz 1992). One advantage of Hr in relation to simple heme-based proteins is their oxygen-binding capacity is about 15–28% greater (Mangum 1992).

Hemerythrin Gene Family

Four Hr subtypes have already been reported in the literature: (a) polymeric circulating Hrs (cHr); (b) myohemerythrins (myoHrs); (c) ovohemerythrins (ovoHrs); and (d) neurohemerythrins (nHrs) (Mangum 1992; Baert et al. 1992; Vergote et al. 2004; Vanin et al. 2006). The hemerythrocytes (pink blood cells) that circulate in the coelomic fluid and in the circulatory system of sipunculans contain an intracellular polymeric Hr (cHr) that exists in trimeric or octameric quaternary structure (Robitaille and Kurtz 1988; Vanin et al. 2006). Both structural conformations were previously reported for some sipuculid species (Ferrell and Kitto 1971; Addison and Bruce 1977; Smith et al. 1983; Wilkins and Harrington 1983; Uchida et al. 1990; Satake et al. 1990). Conversely, myoHr is a cytoplasmic monomeric protein present

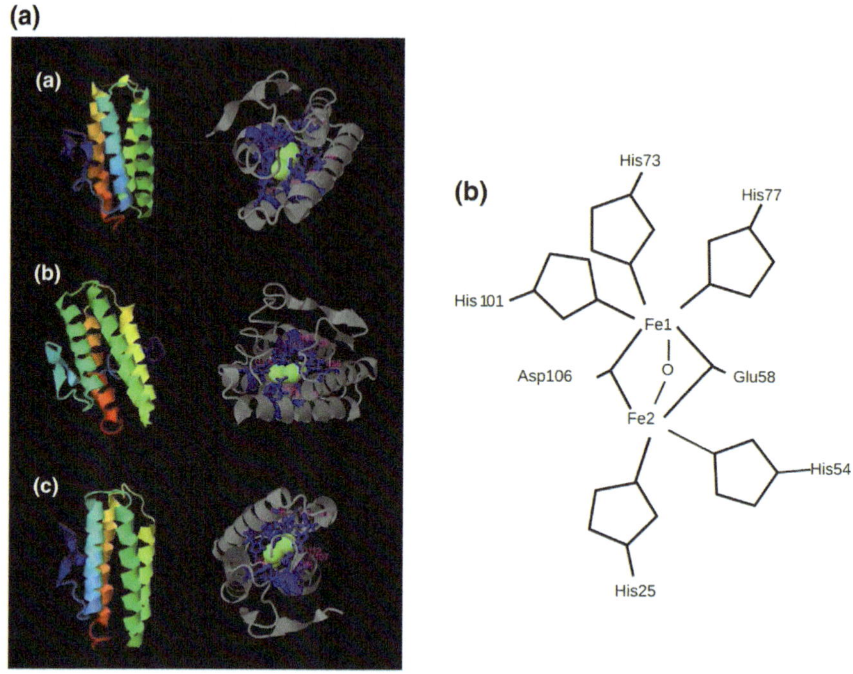

Fig. 10.3 **a** Tertiary structure of Hrs from **a** Echinodermata; *Leptosynapta clarki*; **b** Mollusca; *Nuculana pernula*; **c** Phoronida; *Phoronis psammophila.* All structures were inferred using bioinformatic methods and are modified from Costa-Paiva et al. (2017b). **b** Cartoon representation of the diiron site in oxygenated Hr from the sipunculid Themiste dyscritum

in the muscles of annelids, including sipuculans (Hendrickson et al. 1975; Ward et al. 1975; Tagagi and Cox 1991).

Although the number of residues in Hr sequences varies moderately, there is a conspicuous difference between the primary structure of cHrs and myoHrs: the sequence of myoHrs always contains five additional residues between the last two α-helices when compared to cHrs. This difference was assumed to be a five-residue insertion in the myoHrs, leading to a non-helical turn between helixes C and D (Kurtz 1992). Recent phylogenomic analysis of the evolutionary history of animal Hr sequences demonstrated that it is in fact a loss of five residues for cHrs (Costa-Paiva et al. 2017b) (Fig. 10.4).

In 1992, Baert and collaborators discovered a new Hr subtype, a 14-kDa Hr identified as a major component of mature oocytes of the leech *Theromyzon tessulutum*. They observed a strong resemblance between this new protein and sipunculid Hrs based on physiochemical properties, including: molecular mass, iron content, UV/visible spectra, amino acid composition, and *N*-terminal sequence. These similarities with Hrs and the protein in the oocyte justify the chosen name, ovohemerythrin (ovoHr). This ovoHr can be found in large amounts in the coelomic fluid during

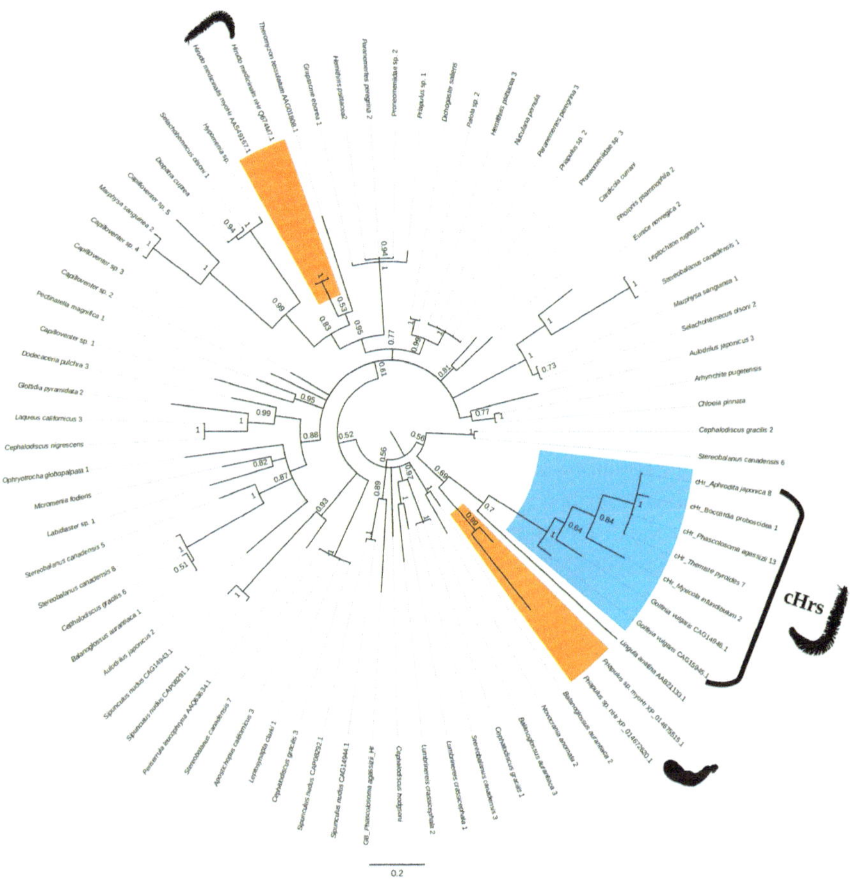

Fig. 10.4 Gene genealogy of animal Hrs. This tree is midpointed rooted and the blue clade represents cHrs with the five residue deletion between C and D α-helices; orange clades represent sequences of myoHr and 'nHr' from a leech and a priapulid. Inspired by Costa-Paiva et al. (2017b)

gametogenesis. However, at the time of oocyte enlargement, the ovoHr concentration decreases dramatically in contrast to vitellogenin—a precursor protein of egg yolk—which is detected specifically in the coelomic fluid at that stage. The authors suggested that a more complex function for ovoHr in oxygen transport is likely, rather than being merely a yolk nutrient for the embryo.

Almost ten years later, Coutte et al. (2001) reported a cDNA sequence that encoded a leech Hr. The mRNA was expressed widely in leech tissues and is stage specific, appearing during the period preceding oogenesis and disappearing after egg laying. These findings could validate the presence of ovoHr described almost a decade before however, the authors were unconvinced that the cDNA encoded ovoHr. The highest identity to the sequence was to myoHr sequences, suggesting that the expression observed during the period of oocyte enlargement indicates an important role of the leech protein in oogenesis and reproduction that may be related to the need of increased oxygen levels in relation to metabolic demands.

Subsequently, Vergote and collaborators (2004) discovered an additional subtype of the Hr family, namely neurohemerythrin (nHr), from proteomic analysis of changes in protein expression in the leech *Hirudo medicinalis* in response to septic injury. In situ hybridization and immunocytochemistry approaches showed that the new leech protein (nHr) was present in the leech central nervous system. The authors postulated that nHr could be directly involved in the innate immune response of the leech to bacterial invasion.

Although being mentioned just a few times in the literature (Baert et al. 1992; Coutte et al. 2001; Vergote et al. 2004), both ovoHr and nHr represent a substantial part of Hr knowledge and should be further investigated. Unfortunately, the complete amino acid sequence of ovoHrs remains unavailable. In addition, it is important to highlight here that annelids represent the only animal phylum with records of the presence of all four Hr subtypes. This makes annelids a key lineage for evaluating whether the subtypes described reflect true genomic clades in gene trees or whether they are in fact just the differential expression products (isoforms) of the same gene across different tissue types.

Hemerythrins in Annelida

Annelids are the metazoan lineage that presents the highest diversity of oxygen transport proteins reported to date, presenting all three proteins (Fig. 10.1b). This may be due to the ancient origin of annelids, the great diversification of life forms and habits they represent, and in addition to the great variety of oxygen absorption and transport strategies inside the body (Schumway 1979; Rouse and Pleijel 2001). Aside from oxygen binding protein diversity within annelids, it is also possible to observe distinct protein diversity in individuals, e.g., species of Terebellidae and Opheliidae, and in some species of Sipuncula (Bailly et al. 2008; Liu et al. 2013). Such organisms may simultaneously express more than one protein or may have different protein expression in different parts of the body (Bailly et al. 2008). However, in annelids whose records indicate the expression of only one type of oxygen carrying protein, it is not clear whether the occurrence of only one type of molecule is due to the silencing/inactivation of the gene expressing of others or if the occurrence reflects the absence of the associated gene.

Until recently, annelid Hrs were considered to be restricted to a sipunculid family only, the Magelonidae (Kurtz 1992). However, more recent genomic studies, such as Vanin et al. (2006), Bailly et al. (2008) and Costa-Paiva et al. (2017a), expanded significantly the records of Hrs in annelids. Bailly et al. (2008) discovered Hrs genes in seven annelid species, showing that Hrs are broadly distributed and suggesting that such genes could be present in all annelids. Unfortunately, this work was based on genomic data, thus not evidencing whether or not Hr genes would be expressed in these species. Costa-Paiva et al. (2017a) examined a wide diversity of annelid taxa to further understand how different forms of Hrs are evolutionarily related to each other by employing bioinformatic approaches to survey Hrs from annelid transcriptomes. They identified Hrs in 44 taxa and further describe the molecular diversity and evolution of Hrs in the light of annelid phylogeny. Their findings demonstrated that sequence diversity among Hr-bearing annelid species is much greater than expected and that many of these Hrs are actively expressed.

Classification of tissue specific Hr subtypes (Baert et al. 1992; Coutte et al. 2001; Vergote et al. 2004) was not validated by Costa-Paiva et al. (2017a) gene genealogy. Although their analyses used whole organisms (including reproductive and nerve tissues), their results failed to recover Hr proteins that corresponded to ovoHrs or nHrs. They argued that these categories were described based on limited differences in amino acid sequence and do not reflect distinct monophyletic subgroups within the Hr gene family. Given this, they suggested the recognition of only two primary types of Hrs, circulating Hrs (cHr) and non-circulating myoHrs (myoHrs), instead of the four subtypes previously reported in the literature.

Incongruence in Hr gene genealogy obtained by Costa-Paiva et al. (2017a) relative to the current knowledge of annelid evolutionary history indicated that Hrs have a complex history within this lineage. The evolution of annelid Hrs possibly involved several events of gene loss and duplication, paralogous substitutions, and even lateral gene transfer (Alvarez-Carreño et al. 2016). Earlier work suggested that all Hrs arose from a single ancestral monomeric protein (myoHr) in the annelid ancestor (Mangum 1992; Vanin et al. 2006). However, the presence of cHrs in Amphinomidae, Oweniidae, and Sipuncula, members of lineages close to the root of the Annelida tree (Weigert et al. 2014; Weigert and Bleidorn 2016), contradicts this hypothesis. Costa-Paiva et al. (2017a) suggested that both subtypes (cHrs and myoHrs) would already be present in an ancestral annelid in the Cambrian (Liu et al. 2015) (Fig. 10.5).

Distribution of Hemerythrin Genes in Metazoa

Regarding the presence of Hr in metazoans, until 2013, homologues were recorded for few marine invertebrate lineages belonging to four phyla: Annelida (including sipunculids, Weigert et al. 2014), Brachiopoda, Priapulida, and Bryozoa (Kurtz Jr 1992; Terwilliger 1998; Bailly et al. 2008). In 2013, Martín-Durán and collaborators carried out a large genomic survey and reported the presence of Hr genes for two other metazoan species, one cnidarian (*Nematostella vectensis*) and one arthropod (*Calanus finmarchicus*). However, this work was based on gDNA only, which would not show if such genes are actively expressed. In addition, their work had a low phylogenetic coverage within each metazoan phylum and was performed using rather inclusive e-values for bioinformatic inference of homologies.

In 2017, Costa-Paiva and collaborators investigated metazoan Hr diversity by employing in silico approaches to survey Hrs from 120 metazoan transcriptomes and genomes. They found 58 new Hr genes actively transcribed in 36 species distributed across 11 animal phyla, with new records in Echinodermata, Hemichordata, Mollusca, Nemertea, Phoronida, and Platyhelminthes. These results showed evidence of the presence of Hrs in just nephrozoan lineages, even though searches were carried

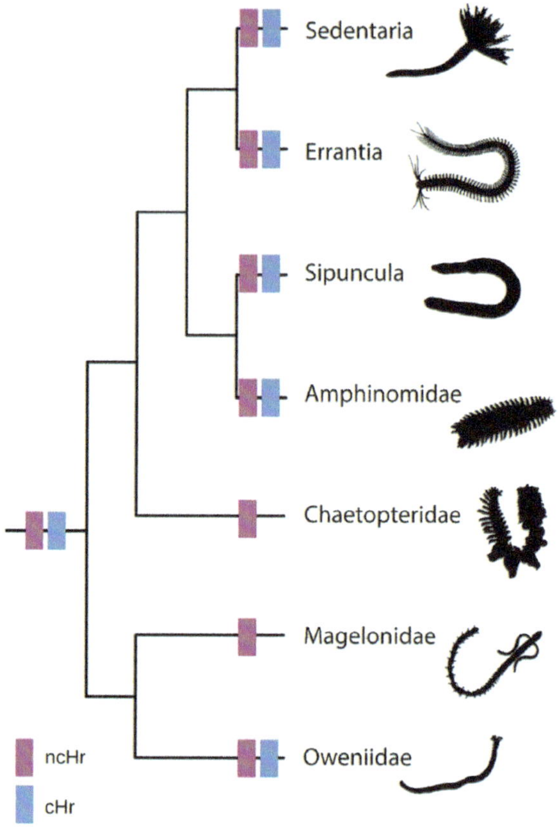

Fig. 10.5 Phylogenetic hypothesis of the main Annelida lineages derived from recent phylogenomic studies showing the lineages that have records of the presence of cHr (blue rectangles) and non-circulating (ncHr) or myoHr (pink rectangles). Modified from Weigert and Bleidorn (2016)

out for bilaterian genomes (Fig. 10.6). The records of Hr in the cnidarian *N. vectensis* and the arthropod *C. finmarchicus* attributed by Martín-Durán and collaborators (2013) were not homologous to the Hrs observed in the other Metazoa.

Traditional classification of specific Hr subtypes, cHrs, myoHrs, ovoHrs, and nHrs, (Baert et al. 1992; Coutte et al. 2001; Vergote et al. 2004) as discussed above was not validated by the gene genealogy observed in Costa-Paiva et al. (2017b), but further corroborated previous findings of Costa-Paiva et al. (2017a). Again, classification of myoHrs and cHrs had traditionally relied on differences regarding the monomeric or polymeric form, respectively, and the presence or absence of a five-amino-acid insertion/deletion between the C and D α-helices (Sanders-Loehr and Loehr 1979; Kurtz 1992; Vanin et al. 2006; Costa-Paiva et al. 2017a). The cHr subtype of Hrs, which lacks those five residues, is present exclusively in Annelida and represents a novelty in bilaterian Hr evolution. Mangum (1992) suggested that it was difficult to eliminate the possibility that annelid Hr had a separate origin and she was entirely correct about cHrs.

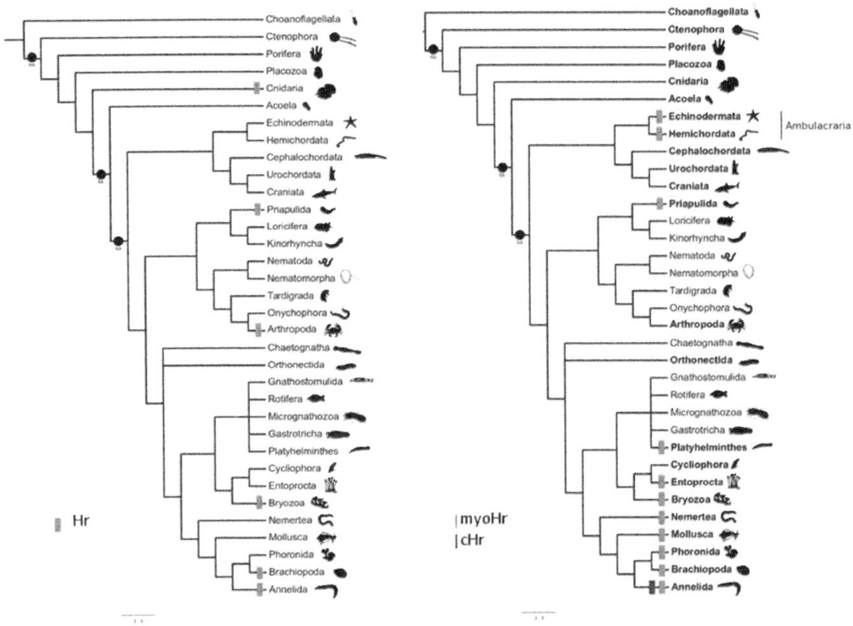

Fig. 10.6 Hypothesized relationships among metazoan phyla derived from recent phylogenomic studies (Whelan et al. 2015; Halanych 2016; Cannon et al. 2016; Kocot et al. 2017). Pink rectangles represent myoHrs and purple rectangles represent cHrs. MA is metazoan ancestor, BA is bilaterian ancestor, and NA is nephrozoan ancestor. Left panel—records of oxygen binding proteins until 2013. Right panel –updated records as of 2017

Evolutionary History

Although Hrs have a wide biological distribution and are present in Archaea, Bacteria, and Eukarya (Bailly et al. 2008; French et al. 2008; Li et al. 2015), the specific taxon that Hr would have arisen from remains unclear. The phylogenetic distribution currently known for this protein suggests that the ability to bind oxygen reversibly must first have been explored by early prokaryotes to serve different functions, e.g., protection against oxidative stress and supply oxygen to certain metabolic pathways (Alvarez-Carreño et al. 2016). Subsequently, this functional domain would have co-opted oxygen binding, which would be the framework of Hr itself, in pre-existing proteins.

Concerning metazoans, Bailly et al. (2008) suggested that the Hr gene would have arisen after the separation of the protostome and deuterostome lineages, which would explain its presence in only a few protostomes and complete absence in the deuterostomes. At that time, the confirmed presence of Hr was limited to Annelida, Bryozoa, Priapulida, and Brachiopoda, which justifies the scenario proposed by Bailly et al. (2008). Moreover, their results suggested that animal Hr genes seemed to have evolved from a common ancestral gene since all of them share the same pattern of distribution of exons (Bailly et al. 2008). This view is in contrast to a homoplastic appearance of animal Hrs as suggested by Kurtz (1992). In 2013, Martín-Durán and collaborators suggested, unlike Bailly et al. (2008), that Hrs had multiple origins and frequent losses along the evolution of animals. However, this study also included a limited number of metazoans, only 13 species distributed in five phyla, which could cause some bias in their results.

Contrary to previous suggestions that Hr genes were absent in deuterostomes (Bailly et al. 2008), Costa-Paiva et al. (2017b) found Hrs in deuterostome and were likely present in nephozoans, but not in nonbilaterian animal lineages. Additionally, their results also contradicted Martín-Durán et al. (2013) who suggested that the ancestral metazoan would already have a functional Hr gene whose evolutionary history would then be followed by frequent loss of these genes in several lineages. Costa-Paiva et al. (2017b) proposed a different evolutionary scenario, with the origin of the Hr gene in the nephrozoan ancestor, since no evidence of the presence of Hrs in non-bilateral metazoans was found (Fig. 10.6b). As expected, their Hr gene tree did not mirror metazoan phylogeny, suggesting that Hrs evolutionary history was complex and besides the oxygen carrying capacity, the drivers of Hr evolution may also consist of secondary functional specializations of the proteins.

Although these insights about the evolutionary history of Hrs showed some plausible scenarios for the origin of Hrs, the selection pressure for Hr remains a point of interest. Most investigators of molecular structure have concluded that the Hrs are so unlike the heme proteins that they must have originated entirely independently (Klippenstein 1980). Nonetheless, Sippl (1984) and Volbeda and Hol (1989) have pointed out structural similarities between the metal-binding helices of both heme and Hr proteins. Volbeda and Hol (1989) even suggested that Hbs, Hrs, and Hcs must have a single, common origin.

Secondary Functional Specializations

Although the distribution of Hrs in animals is likely tied to the need to deliver oxygen to tissues, many animal species can express more than one oxygen binding protein, certain annelid species possess HBL-Hbs and Hrs simultaneously (Costa-Paiva et al. 2017a; Belato et al. 2019). Costa-Paiva et al. (2017b) suggested that although the observed pattern could be explained by the need to deliver oxygen, secondary functional specializations could also be important for driving diversification (Coates and Decker 2017). In 1993, Florkin suggested that Hrs could serve as oxygen stores for transient periods of hypoxia. However, at the time, it was considered that none of the oxygen carriers found in worm-like animals, which included HBL-Hbs and Hrs, could function in gas transport.

Secondary functional specializations in Hrs may also have been an important trigger for the diversification of these molecules throughout their evolutionary history. There are records of Hrs molecules involved in other functions besides oxygen loading, for example, in the storage of iron atoms, in the detoxification of heavy metals and even involved in metabolic pathways related to the innate immunity

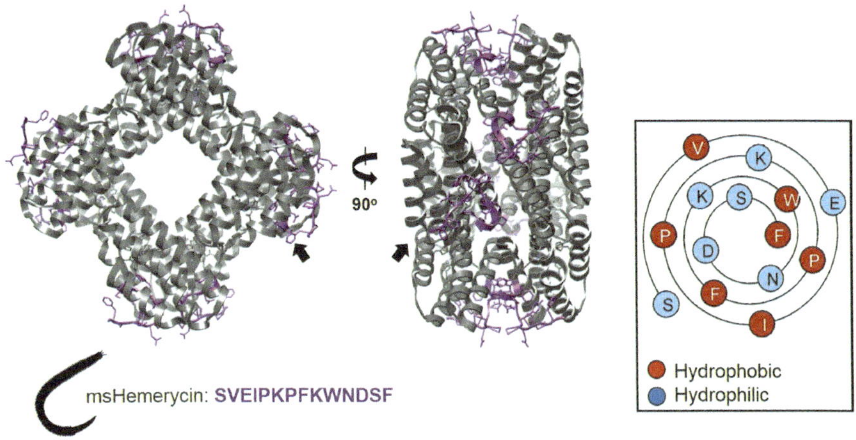

Fig. 10.7 Hemerythrin-derived antibacterial peptide. The crystal structure of hemerythrin from *Phascolopsis gouldii* (PDB–1I4Y) was used to illustrate the location of msHemerycin (coloured purple) from *Marphysa sanguinea*. This peptide is present at the *N*-terminus of each protein monomer, making it highly accessible for proteases to cut out. Inset: *ms*Hemerycin is presented as a Wenxiang diagram (NB. No distinct amphipathicity can be seen)

of some annelids (Baert et al. 1992; Demuynck et al. 1993; Vergote et al. 2004; Coates and Decker, 2017). For example, Coutte et al. (2001) suggested that the leech *T. tessulatum* expresses an Hr-related protein facilitates detoxification of free iron after a blood meal. Most recently, Seo et al. (2016) isolated a hemerythrin-derived peptide from an acidified body extract of Lugworm (*Marphysa sanguinea*). This 1.7 kDa peptide derived from the N-terminal of hemerythrin monomers (Fig. 10.7), referred to as msHemerycin, was effective at inhibiting the growth of bacteria (Gram-positive/negative) in vitro but had no negative impacts on yeast (*C. albicans*). Preliminary structural models revealed an unordered state in solution, and our visualisation of the peptide using a Wenxiang diagram does not support an amphipathic distribution of amino acids (Fig. 10.7), indicating that pore formation is not the mechanism of action. This finding is very interesting, as many hemoglobin- and hemocyanin-derived bioactive peptides have been identified already (Coates and Decker 2017; see Chapter 9 Coates and Costa-Paiva). Additional studies of the gene structure of Hr proteins and physiological aspects of organisms are the next important steps toward a better understanding of the evolutionary patterns involved in this dynamic family of oxygen carrier proteins.

Hexagonal Bilayer Hemoglobins

The (Mega-)Molecule

In 1872, Lankester described a large extracellular red and green respiratory protein of some annelid species. Although they resembled red blood cell Hbs, Lankester concluded that these proteins were distinctly different from vertebrate Hbs and named them 'erythrocruorin (red) and chlorocruorin (green)'. Many years later, several researches have indicated remarkable similarities between annelid and vertebrate heme-containing respiratory proteins. For example, Jhiang and Riggs (1988) showed that the earthworm *Lumbricus terrestris* globin gene has two introns in the exactly same location as vertebrate globin genes. In 1951, Keilin and Hartree even suggested that all of them should be referred to as hemoglobins (Hbs). Until recently, these giant extracellular Hbs were still referred to as erythrocruorins or chlorocruorins, even with a lack of uniformity in usage of these terms. Several papers used the term erythrocruorin incorrectly to describe different types of invertebrate cell-associated Hbs, such as those found in molluscs, nematodes and crustaceans, (Terwilliger et al. 1976; Ilan et al. 1981, 1982; Darawshe et al. 1987). The term chlorocruorin refers to the distinct green colour of some polychaete blood caused by extracellular hemoglobin. The green coloration, however, is only a result of a modified heme in which a formyl group replaces the vinyl group (Vinogradov et al. 1993; Weber and Vinogradov 2001), and should not be considered a different protein to erythrocruorin. Therefore, those terms are obsolete and the most accurate name is hexagonal bilayer hemoglobin (HBL-Hb).

HBL-Hb represents one of the four groups of invertebrate extracellular Hbs, which is comprised of a multi-subunit complex that forms a unique quaternary structure with high molecular mass in the range of 3,000–4,000 kDa (Vinogradov 1985; Weber and Vinogradov 2001). This protein complex comprises 180 polypeptides chains, which are grouped into two categories: globins and non-globins (Lamy et al. 1996). The structure of earthworm (*L. terrestris)* HBL-Hb is the best studied and contains 144 globin chains arranged into 12 dodecamers that assemble around a central core of 36 linker chains organized into 12 linker trimers (Fig. 10.8) (Lamy et al. 2000; Royer et al. 2005, 2006). Each globin chain binds reversibly to oxygen, which represents an extraordinary oxygen transport capacity.

Although being known as extracellular proteins, HBL-Hbs are synthesized intracellularly and then secreted into the circulatory system or coelomic fluid (Terwilliger 1992). Their extracellular location seems to be due to their large molecular size, and also prevents their excretory loss through tissue membranes (Weber and Vinogradov 2001). The protein complex is extremely stable, being resistant to autoxidation and is capable of transporting oxygen to tissues when transfused into mammals without producing side effects (Harrington et al. 2007). Such features make HBL-Hb a promising candidate to act as a blood substitute in human transfusions/therapeutics and for the preservation of organs, tissues and cells ex vivo (Zal and Rousselot 2014; Zal 2015, 2017).

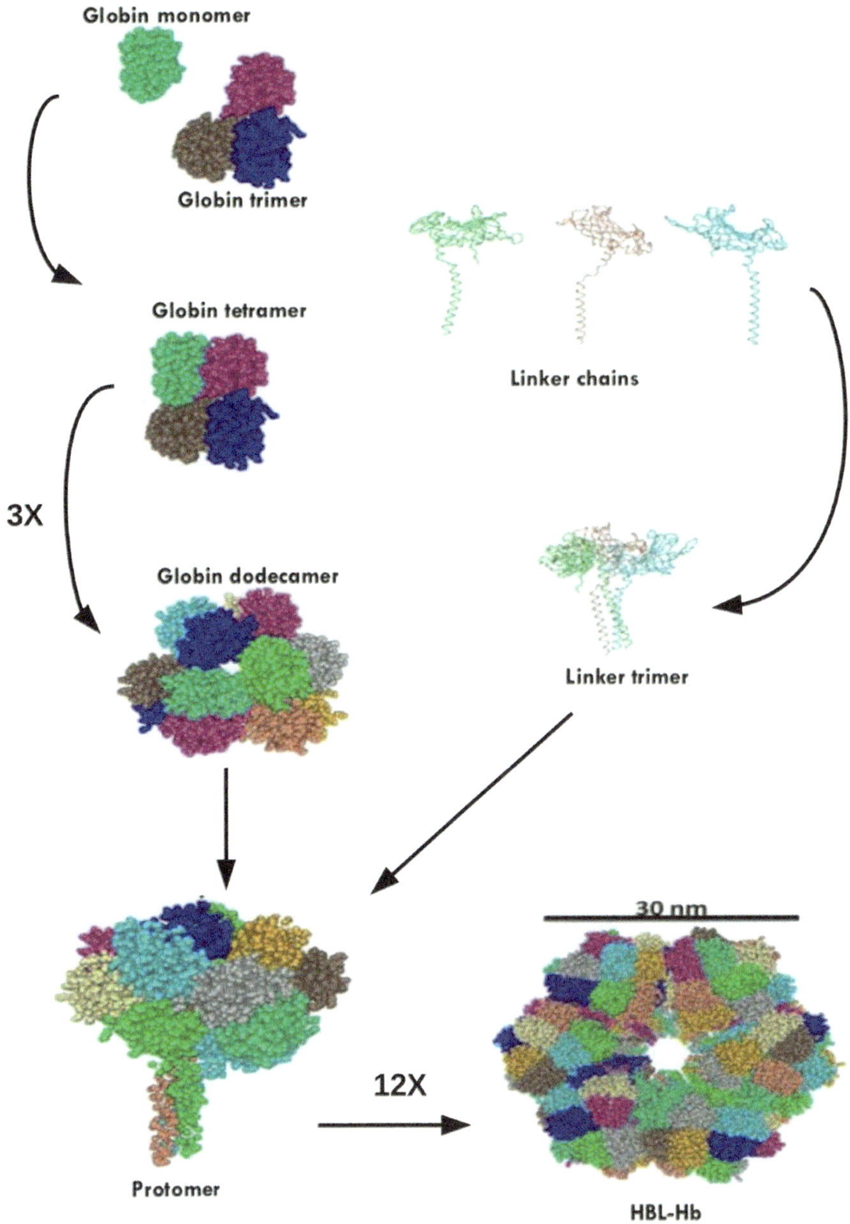

Fig. 10.8 The assembled structure of the earthworm *Lumbricus terrestris* HBL-Hb. The subunits assemble into a globin trimer that pairs with a globin monomer to form a globin tetramer which then associates with two more tetramers to form a globin dodecamer. Three linker subunits form a linker trimer which binds the dodecamer to form a protomer. Finally, 12 protomers assemble into the hexagonal bilayer structure. Modified from Elmer and Palmer (2012)

Distribution

Hemoglobins (Hbs) occur in all domains of life displaying an extraordinary diversity of form and function, among which vertebrate Hbs stand out as the most studied blood pigments of all (Royer 1992; Vinogradov et al. 2006, 2007; Gell 2018). Even though invertebrate Hbs present a wider variation of structures, much less is known about their diversity, distribution, and evolutionary history. Until recently, the known taxonomic distribution of HBL-Hbs was restricted only to annelids (Terwilliger 1992; Weber and Vinogradov 2001). However, our recent work demonstrated that within respiratory pigments there has been a gross underestimation regarding the diversity and taxonomic distribution of HBL-Hbs (Belato et al. 2019).

To better understand the diversity and expression of extracellular globins and linkers that are used in making HBL-Hbs, we interrogated transcriptomic data from across bilaterians, including heavy samplings of annelids. We found actively transcribed globin and linker genes in 16 animal species from nine phyla (other than Annelida), including the first record of extracellular globin genes in deuterostomes (Fig. 10.9).

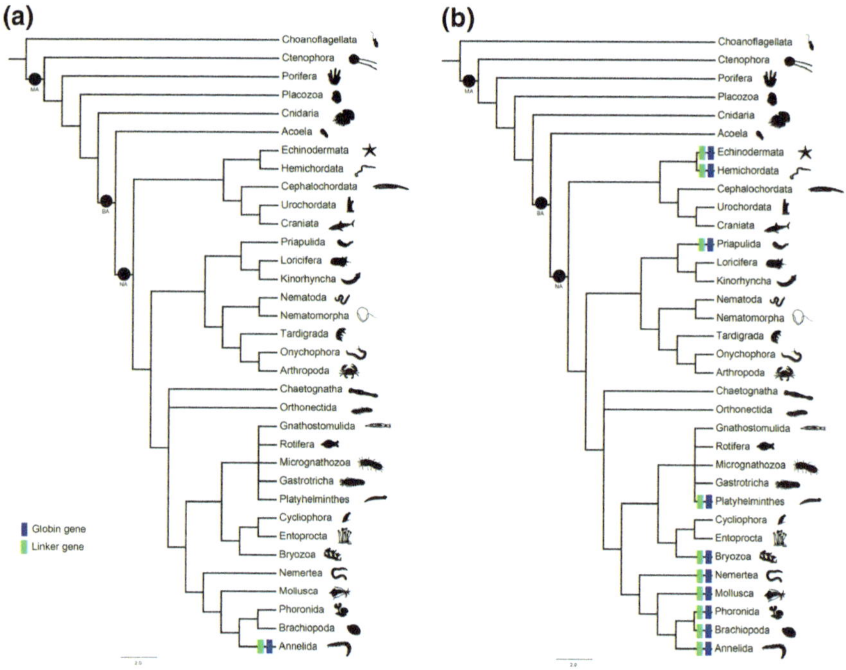

Fig. 10.9 Hypothesized relationships among metazoan phyla derived from recent phylogenomic studies (Whelan et al. 2015; Halanych 2016; Cannon et al. 2016; Kocot et al. 2017). Blue rectangles represent records of globin genes and green rectangles represent records of linker genes. MA is metazoan ancestor, BA is bilaterian ancestor, and NA is nephrozoan ancestor. Left panel—records of oxygen binding proteins until 2018. Right panel –updated records as of 2019

Evolutionary History

The first study on the molecular evolution of globin genes was the pioneering work performed by Gotoh et al. (1987) who divided HBL-Hb globins into two main chains, A and B, based on differences in their primary structures and with a common origin that would already be present in the ancestor of Annelida. This division would be analogous to the division of the α and β chains of vertebrate monomers (Gotoh et al. 1987; Shishikura et al. 1997). Subsequently, Yuasa et al. (1996) constructed a phylogenetic tree with 15 extracellular globin sequences from annelids that corroborated the existence of the main clades A and B and suggested a further division into four paralogous globin types: A1/A2 and B1/B2. According to Yuasa et al. (1996), the gene duplication events that led to the four subtypes occurred before the Annelida diversification since, at that time, HBL-Hb proteins were only known in annelids. In the work of Negrisolo et al. (2001), 26 extracellular globin sequences were partitioned into clades A and B, with high statistical support values. However, the subdivisions A1, A2, B1 and B2 could not be recovered as monophyletic groups. These results indicated that only the division in the two main clades A and B would be phylogenetically representative.

Bailly and collaborators (2002), using 15 sequences of globin genes, recovered both clades A and B, as well as subgroups A1, A2, B1 and B2, but these subdivisions were based on low values of statistical support. Finally, Chabasse et al. (2006a) corroborated the results of Bailly et al. (2002). Using 52 annelid globin sequences, they succeeded in recovering the main groups A and B, as well as A1, A2, B1 and B2, but again all subgroups presented low values of statistical support. In 2019, we used a much larger number of both annelid and other metazoan globin sequences and confirmed the early hypothesis of two main paralogous globin chains, A and B (Gotoh et al. 1987; Belato et al. 2019), both in annelids and other metazoans. However, further subdivisions (A1/A2 and B1/B2) could not be recovered, either in annelids or in other metazoans, suggesting that there are only two monophyletic globin chains, A and B (Belato et al. 2019).

Concerning the linker gene evolutionary history, Suzuki et al. (1990) were the first to describe two distinct linker chains, L1 and L2, based on dissimilarities in their sequences. In 1993, Suzuki and Riggs described one additional linker chain (L3) and suggested it was the result of an ancestral globin gene duplication event and sets of paralogous genes. They also demonstrated that linkers have a conserved cysteine-rich domain that is not present in globin sequences, but homologous to the low-density lipoprotein receptor class A (LDL-A) found in many metazoans, e.g., humans and frogs (Sudhof et al. 1985; Mehta et al. 1991). The LDL-A module enables linkers to combine the globin dodecamers to form the hexagonal bilayer structure of the HBL-Hb complex (Suzuki and Riggs, 1993). In 1996, Fushitani and collaborators also described a fourth linker chain (L4), however this chain appears to be a minor component, most likely a variation of the L3 chain (Vinogradov 2004; Royer et al. 2006).

In 2001, Negrisolo and collaborators analyzed seven linker sequences and were unable to recover the three linker subtypes as monophyletic groups. Using 22 linker sequences, Chabasse et al. (2006b) clustered linkers in two main clades but with low statistical support values. We surveyed more than 400 linker sequences and generated results that supported the existence of three linker types L1, L2 and L3 among Annelida in agreement with Suzuki et al. (1990) and Suzuki and Riggs (1993) (Belato et al. 2019). However, when other metazoan taxa were included in their analyses, none of the three linker types could be recovered as monophyletic groups. Thus, there are three linker units for annelids, but this does not appear to be the case for higher taxonomic levels. Lamy and collaborators (2000) suggested that the presence of the three linker types is not required for the assembly of the HBL-Hb structure. Nevertheless, they showed that every linker chain can replace each other in the HBL-Hb assembly, and the molecule can assemble with only one or two types of linker chains. Their results also suggested that the globin dodecamer is unstable without these linkers (Lamy et al. 2000).

As HBL-Hbs distribution was thought to be exclusive to Annelida, all works on evolutionary hypotheses of the emergence of HBL-Hbs suggested that the molecule was already present in, but limited to, the common annelid ancestor (Yuasa et al. 1996; Negrisolo et al. 2001; Bailly et al. 2002). Considering that HBL-Hb oligomers are formed by 144 extracellular globins and 36 linkers chains, the independent evolution of this protein complex is unlikely. Although searching in numerous genomes and transcriptomes of metazoans, we did not find any HBL-Hb sequence in non-nephrozoan species, such as sponges and cnidarians (Belato et al. 2019). Consequently, we suggested that HBL-Hbs must have arisen in the nephrozoan ancestor (Fig. 10.9), just prior to the deuterostome-protostome split.

Acknowledgements We should like to thank Fernando Avila Queiroz (Federal University of Rio de Janeiro) for valuable help with the figures and Flavia A. Belato (University of Sao Paulo) for critically reading the text. A Fellowship to EMC-P was provided by FAPESP (Sao Paulo Research Foundation, Brazil).

References

Addison AW, Bruce RE (1977) Chemistry of Phascolosoma lurco hemerythrin. Arch Biochem Biophys 183:328–332

Alvarez-Carreño C, Becerra A, Lazcano A (2016) Molecular evolution of the oxygen-binding hemerythrin domain. PLoS ONE 11(6):e0157904

Andreeva A, Howorth D, Brenner SE et al (2004) SCOP database in 2004: refinements integrate structure and sequence family data. Nucleic Acids Res 32:226–229

Baert JL, Britel M, Sautière P et al (1992) Ovohemerythrin, a major 14-kDa yolk protein distinct from vitellogenin in leech. Eur J Biochem 209:563–569

Bailly X, Jollivet D, Vanin S et al (2002) Evolution of the sulfide-binding function within the globin multigenic family of the deep-sea hydrothermal vent tubeworm Riftia pachyptila. Mol Biol Evol 19(9):1421–1433

Bailly X, Vanin S, Chabasse C et al (2008) A phylogenomic profile of hemerythrins, the nonheme diiron binding respiratory proteins. BMC Evol Biol 8:244

Belato FA, Schrago CG, Coates CJ et al (2019) Newly discovered occurrences and gene tree of the extracellular globins and linker chains from the giant hexagonal bilayer hemoglobin in metazoans. Genome Biol Evol 11(3):597–612
Burmester T (2001) Molecular evolution of the arthropod hemocyanin superfamily. Mol Biol Evol 18(2):184–195
Burmester T (2002) Origin and evolution of arthropod hemocyanins and related proteins. J Comp Physiol B 172(2):95–107
Burmester T (2015) Evolution of respiratory proteins across the Pancrustacea. Integr Comp Biol 55(5):765–770
Cannon JT, Vellutini BC, Smith J et al (2016) Xenacoelomorpha is the sister group to Nephrozoa. Nature 530(7588):89–93
Chabasse C, Bailly X, Rousselot M et al (2006a) The multigenic family of the extracellular hemoglobin from the annelid polychaete *Arenicola marina*. Comp Biochem Physiol B: Biochem Mol Biol 144(3):319–325
Chabasse C, Bailly X, Sanchez S et al (2006b) Gene structure and molecular phylogeny of the linker chains from the giant annelid hexagonal bilayer hemoglobins. J Mol Evol 63(3):365–374
Coates CJ, Decker H (2017) Immunological properties of oxygen-transport proteins: hemoglobin, hemocyanin and hemerythrin. Cell Mol Life Sci 74(2):293–317
Costa-Paiva EM, Whelan NV, Waits DS et al (2017a) Discovery and evolution of novel hemerythrin genes in annelid worms. BMC Evol Biol 17(1):85–96
Costa-Paiva EM, Schrago CG, Halanych KM (2017b) Broad phylogenetic occurrence of the oxygen-binding hemerythrins in bilaterians. Genome Biol Evol 9(10):2580–2591
Costa-Paiva EM, Schrago CG, Coates CJ et al (2018) Discovery of novel hemocyanin-like genes in Metazoans. Biol Bull 235(3):134–151
Coutte L, Slomianny MC, Malecha J et al (2001) Cloning and expression analysis of a cDNA that encodes a leech hemerythrin. Biochim Biophys Acta 1518(3):282–286
Darawshe S, Tsafadyah Y, Daniel E (1987) Quaternary structure of erythrocruorin from the nematode *Ascaris suum*. Evidence for unsaturated haem-binding sites. Biochem J 242(3): 689–694
Decker H, Hellmann N, Jaenicke E et al (2007) Recent progress in hemocyanin research. Integr Comp Biol 47(4):631–644
Deng SX, Landry DW, Kalantarov G et al (2015) Anti-drug vaccines. U.S. Patent 0343042A1, December 3, 2015
Demuynck S, Li KW, van der Schors R et al (1993) Amino acid sequence of the small cadmium-binding protein (MP-II) from *Nereis diversicolor* (Annelida, Polychaeta)—evidence for a myohemerythrin structure. Eur J Biochem 217:151–156
Elmer J, Palmer A (2012) Biophysical properties of *Lumbricus terrestris* erythrocruorin and its potential use as a red blood cell substitute. J Funct Biomater 3(4):49–60
Farmer CS, Kurtz DM, Liu ZJ et al (2001) The crystal structures of *Phascolopsis gouldii* wild type and L98Y methemerythrins: structural and functional alterations of the O2 binding pocket. J Biol Inorg Chem 6(4):418–429
Ferrell RF, Kitto GB (1971) Structural studies on *Dendrostomum pyroides* hemerythrin. Biochemistry 10:2923–2929
Florkin M (1933) Recherches sur les hémérythrines. Arch Int Physiol 36(2–3):247–282
French CE, Bell JML, Ward FB (2008) Diversity and distribution of hemerythrin-like proteins in prokaryotes. FEMS Microbiol Lett 279(2):131–145
Fushitani K, Higashiyama K, Asao M et al (1996) Characterization of the constituent polypeptides of the extracellular hemoglobin from *Lumbricus terrestris:* heterogeneity and discovery of a new linker chain L4. Biochim Biophys Acta (BBA) 1292(2):273–280
Garrone R (1999) Collagen, a common thread in extracellular matrix evolution. Proc Indian Acad Sci - Chem Sci 111(1):51–56
Gell D (2018) Structure and function of haemoglobins. Blood Cells Mol Dis 70:13–42

Gotoh T, Shishikura F, Snow J et al (1987) Two globin strains in the giant annelid extracellular haemoglobins. Biochem J 241(2):441–445

Halanych KM (2016) How our view of animal phylogeny was reshaped by molecular approaches: lessons learned. Org Divers Evol 16(2):319–328

Harrington J, Kobayashi S, Dorman S et al (2007) Acellular invertebrate hemoglobins as model therapeutic oxygen carriers: unique redox potentials. Artif Cells Blood Substit Biotechnol 35(1):53–67

Hendrickson WA, Klippenstein GL, Ward KB (1975) Tertiary structure of myohemerythrin at low resolution. Proc Natl Acad Sci USA 72:2160–2164

Hill RW, Wyse GA, Anderson M (2016) Animal Physiology. Sinauer Associates, Inc. Publishers. Sunderland

Holmes MA, Stenkamp RE (1991) Structures of met and azidomet hemerythrin at 1.66 A resolution. J Mol Biol 220:723–737

Ilan E, David M, Daniel E (1981) Erythrocruorin from the crustacean *Caenestheria inopinata*. Quaternary structure and arrangement of subunits. Biochem 20(21): 6190–6194

Ilan E, Weisselberg E, Daniel E (1982) Erythrocruorin from the water-flea *Daphnia magna*. Quaternary structure and arrangement of subunits. Biochem J 207(2): 297–303

Jhiang SM, Riggs AF (1988) Exon-intron organization in genes of earthworm and vertebrate globins. Science 240(4850):334–336

Jones L (2015) Recent advances in the molecular design of synthetic vaccines. Nature Chem 7(12):952–960

Keilin D, Hartree EF (1951) Relationship between haemoglobin and erythrocruorin. Nature 168(4268):266

Klippenstein GL (1980) Structural aspects of hemerythrin and myohemerythrin. Am Zool 20(1):39–51

Klippenstein GL, Van Riper DA, Oosterom EA (1972) A comparative study of the oxygen transport proteins of *Dendrostomum pyroides*. Isolation and characterization of hemerythrins from the muscle, the vascular system, and the coelom. J Biol Chem 247(18): 5959–5963

Knoll AH (1992) Biological and biogeochemical preludes to the ediacaran radiation. In: Lipps JH, Signor PW (eds) Origin and Early Evolution of the Metazoa. Plenum Press, New York

Koch J, Lüdemann J, Spies R et al (2016) Unusual diversity of myoglobins genes in the lungfish. Mol Biol Evol 33(12):3033–3041

Kocot KM, Struck TH, Merkel J et al (2017) Phylogenomics of Lophotrochozoa with consideration of systematic error. Syst Biol 66(2):256–282

Kurtz Jr DM (1992) Molecular structure/function relationships of hemerythrins. In: Mangum ChP (ed) Advances in Comparative & Environmental Physiology Vol 13—Blood and Tissue Oxygen Carriers. Springer-Verlag, Berlin, pp 151–171

Lamy J, Green B, Toulmond A et al (1996) Giant hexagonal bilayer hemoglobins. Chem Rev 96(8):3113–3124

Lamy J, Kuchumov A, Taveau J et al (2000) Reassembly of *Lumbricus terrestris* hemoglobin: A study by matrix-assisted laser desorption/ionization mass spectrometry and 3D reconstruction from frozen-hydrated specimens. J Mol Biol 298(4):633–647

Lankester E (1868) Preliminary notice of some observations with the spectroscope on animal substances. J Anat Physiol 2(1):114–116

Lecomte JT, Vuletich DA, Lesk AM (2005) Structural divergence and distant relationships in proteins: evolution of the globins. Curr Opin Struct Biol 15(3):290–301

Li X, Tao J, Hu X et al (2015) A bacterial hemerythrin-like protein MsmHr inhibits the SigF-dependent hydrogen peroxide response in mycobacteria. Front Microbiol 5:1–11

Liu J, Ou Q, Han J et al (2015) Lower Cambrian polychaete from China sheds light on early annelid evolution. Naturwissenschaften 102(5):1–7

Liu Y, Li C, Su X et al (2013) Cloning and characterization of hemerythrin gene from sipuncula *Phascolosoma esculenta*. Genes Genomics 35(1):95–100

Mandler M, Gruber P, Mattner F et al (2015a). Method for treating a synucleiopathy. U.S. Patent 0093431 A1, April 2, 2015
Mandler M, Zauner W, Mattner F et al (2015b). Method for treating a beta-Amyloidosis. U.S. Patent 0093432 A1, April 2, 2015
Mangum CP (1992) Physiological function of the hemerythrins. In: Mangum CP (ed) Advances in Comparative & Environmental Physiology Vol 13—Blood and Tissue Oxygen Carriers. Springer-Verlag, Berlin, pp 173–192
Manwell C (1960) Comparative physiology: blood pigments. Annu Rev Physiol 22(1):191–244
Martín-Durán JM, De Mendoza A, Sebé-Pedrós A et al (2013) A broad genomic survey reveals multiple origins and frequent losses in the evolution of respiratory hemerythrins and hemocyanins. Genome Biol Evol 5:1435–1442
Martins LJ, Hill CP, Ellis WR (1997) Structures of wild-type chloromet and L103N hydroxomet *Themiste zostericola* myohemerythrins at 1.8 Å resolution. Biochem 36(23): 7044–7049
Mehta K, Chen W, Goldstein J et al (1991). The low density lipoprotein receptor in *Xenopus laevis*. I. Five domains that resemble the human receptor. J Biol Chem 266(16): 10406–10414
Mills DB, Canfield DE (2014) Oxygen and animal evolution: Did a rise of atmospheric oxygen "trigger" the origin of animals? BioEssays 36(12):1145–1155
Mills DB, Francis WR, Vargas S et al (2018) The last common ancestor of animals lacked the HIF pathway and respired in low-oxygen environments. Elife 7:e31176
Mills DB, Ward LM, Jones C et al (2014) Oxygen requirements of the earliest animals. PNAS 111(11):4168–4172
Negrisolo E, Pallavicini A, Barbato R et al (2001) The evolution of extracellular hemoglobins of annelids, vestimentiferans, and pogonophorans. J Biol Chem 276(28):26391–26397
Nursall JR (1959) Oxygen as a prerequisite to the origin of the Metazoa. Nature 183(4669):1170
Raff RA, Raff EC (1970) Respiratory mechanisms and the metazoan fossil record. Nature 228:1003–5
Robitaille PML, Kurtz DM Jr (1988) Phosphorus-31 NMR probes of sipunculan erythrocytes containing the oxygen-carrying protein hemerythrin. Biochemistry 27(12):4458–4465
Rouse G, Pleijel F (2001) Polychaetes. Oxford University Press, New York
Rousselot M, Delpy E, Rochelle CD et al (2006) *Arenicola marina* extracellular hemoglobin: A new promising blood substitute. Biotechnol J 1(3):333–345
Royer W (1992) Structures of red blood cell hemoglobins. In: Mangum CP (ed) Advances in Comparative & Environmental Physiology Vol 13—Blood and Tissue Oxygen Carriers. Springer-Verlag, Berlin, pp 87–116
Royer W, Sharma H, Strand K et al (2006) *Lumbricus* erythrocruorin at 3.5 å resolution: architecture of a megadalton respiratory complex. Structure 14(7): 1167–1177
Royer W, Zhu H, Gorr T et al (2005) Allosteric hemoglobin assembly: Diversity and similarity. J Biol Chem 280(30):27477–27480
Sanders-Loehr J, Loehr TM (1979) Hemerythrin: a review of structural and spectroscopic properties. Adv Inorg Biochem 1:235–252
Satake K, Yugi M, Kamo M et al (1990) Hemerythrin from *Lingula unguis* consists of two different subunits, alpha and beta. Protein Seq Data Anal 3:1–5
Schmidt-Rhaesa A (2007) The Evolution of Organs Systems. Oxford University Press, New York
Schumway SE (1979) The effects of body size, oxygen tension, and mode of life on the oxygen uptake rates of polychaetes. Comp Biochem Physiol A Physiol 62:273–278
Seo JK, Nam BH, Go HJ, Jeong M, Lee KY, Cho SM et al (2016) Hemerythrin-related antimicrobial peptide, msHemerycin, purified from the body of the Lugworm, *Marphysa sanguinea*. Fish Shellfish Immunol 57:49–59
Shishikura F, Ochiai T, Yamanaka I (1997) Leech extracellular hemoglobin: two globin strains that are akin to vertebrate hemoglobin α and β chains. Zool Sci 14(6):923–930
Shoulders MD, Raines RT (2009) Collagen structure and stability. Annu Rev Biochem 78:929–958
Schreml S, Szeimies RM, Prantl L et al (2010) Oxygen in acute and chronic wound healing. Bri J Dermatol 163(2):257–268

Sieker LC, Stenkamp RE, Jensen LH (1982) The environment of the binuclear iron coordination complex in methemerythrin. In: Dunford HB, Dolphin D, Raymond KS, Sieker LC (eds) The Biological Chemistry of Iron. Springer, Dordrecht

Sippl MJ (1984) *On the origin of globins: structural relations between the globin exon products.* Life Chem Rep Suppl 1:223–224

Smith JL, Hendrickson WA, Addison AW (1983) Structure of trimeric hemerythrin. Nature 303:86–88

Stenkamp RE, Sieker LC, Jensen LH (1984). Binuclear iron complexes in methemerythrin and azidomethemerythrin at 2.0-. ANG. Resolution. J Am Chem Soc 106(3): 618–622

Sudhof T, Goldstein J, Brown M et al (1985) The LDL receptor gene: a mosaic of exons shared with different proteins. Science 228(4701):815–822

Suzuki T, Riggs A (1993) Linker chain L1 of earthworm hemoglobin. Structure of gene and protein: homology with low density lipoprotein receptor. J Biol Chem 268(18): 13548–13555

Suzuki T, Takagi T, Gotoh T (1990) Primary structure of two linker chains of the extracellular hemoglobin from the polychaete *Tylorrhynchus heterochaetus*. J Biol Chem 265(21):12168–12177

Tagagi T, Cox JA (1991) Primary structure of myohemerythrin fron the annelid *Nereis diversicolor*. FEBS Lett 285:25–27

Terwilliger NB (1992) Molecular structure of the extracellular heme proteins In: Mangum CP (ed) Advances in Comparative & Environmental Physiology Vol 13—Blood and Tissue Oxygen Carriers. Springer-Verlag, Berlin, pp 193–229

Terwilliger NB (1998) Functional adaptations of oxygen-transport proteins. J Exp Biol 201:1085–1098

Terwilliger RC, Terwilliger NB, Schabtach E (1976) Comparison of chlorocruorin and annelid hemoglobin quaternary structures. Comp Biochem Physiol A 55(1):51–55

Uchida T, Yano H, Satake K et al (1990) The amino acid sequence of hemerythrin from *Siphonosoma cumanense*. Protein Seq Data Anal 3:141–147

Vanin S, Negrisolo E, Bailly X et al (2006) Molecular evolution and phylogeny of sipunculan hemerythrins. J Mol Evol 62:32–41

Vergote D, Sautière PE, Vandenbulcke F et al (2004) Up-regulation of neurohemerythrin expression in the central nervous system of the medicinal leech, *Hirudo medicinalis*, following septic injury. J Biol Chem 279(42):43828–37

Vinogradov S (1985) The structure of invertebrate extracellular hemoglobins (erythrocruorins and chlorocruorins). Comp Biochem Physiol B 82(1):1–15

Vinogradov S (2004) The stoichiometry of the four linker subunits of *Lumbricus terrestris* hemoglobin suggests an asymmetric distribution. Micron 35(1–2):127–129

Vinogradov S, Hoogewijs D, Bailly X et al (2006) A phylogenomic profile of globins. BMC Evol Biol 6(1):31

Vinogradov S, Hoogewijs D, Bailly X et al (2007) A model of globin evolution. Gene 398(1):132–142

Vinogradov S, Walz D, Pohajdak B et al (1993) Adventitious variability – the amino-acid sequences of nonvertebrate globins. Comp Biochem Physiol B 106(1):1–26

Volbeda A, Hol WG (1989) Pseudo 2-fold symmetry in the copper-binding domain of arthropodan haemocyanins: possible implications for the evolution of oxygen transport proteins. J Mol Biol 206(3):531–546

Ward KB, Hendrickson WA, Klippenstein GL (1975) Quaternary and tertiary structure of hemerythrin. Nature 257:818–821

Weber R, Vinogradov S (2001) Nonvertebrate hemoglobins: functions and molecular adaptations. Physiol Rev 81(2):569–628

Weigert A, Bleidorn C (2016) Current status of annelid phylogeny. Org Divers Evol 16(2):345–362

Weigert A, Helm C, Meyer M et al (2014) Illuminating the base of the annelid tree using transcriptomics. Mol Biol Evol 31(6):1391–1401

Whelan NV, Kocot KM, Moroz LL et al (2015) Error, signal, and the placement of Ctenophora sister to all other animals. PNAS 112(18):5773–5778

Wilkins RG, Harrington PC (1983) The chemistry of hemerythrin. Adv Inorg Biochem 5:51–85

Yuasa H, Green B, Takagi T et al (1996) Electrospray ionization mass spectrometric composition of the 400 kDa hemoglobin from the pogonophoran *Oligobrachia mashikoi* and the primary structures of three major globin chains. Biochim Biophys Acta Prot Struct Mol Enzym 1296(2):235–244

Zal F (2015) Use of haemoglobin of annelids for treating cancer. U.S. Patent Application No. 14/768,303. Washington, DC: U.S. Patent and Trademark Office

Zal F (2017) Use of annelid haemoglobin for maintaining stem cells in the undifferentiated state. U.S. Patent No. 9,663,761. Washington, DC: U.S. Patent and Trademark Office

Zal F, Rousselot M (2014) Extracellular hemoglobins from annelids, and their potential use in biotechnology. In: La Barre S, Kornprobst J-M (eds) Outstanding Marine Molecules: Chemistry, Biology, Analysis. Wiley-VCH, Weinheim, pp 361–376

Chapter 11
Embryonic and Fetal Human Hemoglobins: Structures, Oxygen Binding, and Physiological Roles

James M. Manning, Lois R. Manning, Antoine Dumoulin, Julio C. Padovan and Brian Chait

Abstract During the past two decades, significant advances have been made in our understanding of the human fetal and embryonic hemoglobins made possible by the availability of pure, highly characterized materials and novel methods, e.g., nano gel filtration, to study their properties and to correct some misconceptions. For example, whereas the structures of the human adult, fetal, and embryonic hemoglobins are very similar, it has generally been assumed that functional differences between them are due to primary sequence effects. However, more recent studies indicate that the *strengths* of the interactions between their subunits are very different leading to changes in their oxygen binding properties compared to adult hemoglobin. Fetal hemoglobin in the oxy conformation is a much stronger tetramer than adult hemoglobin and dissociates to dimers 70-times less than adult hemoglobin. This property may form the basis for its protective effect against malaria. A major source of the increased strength of fetal hemoglobin resides within the A-helix of its gamma subunit as demonstrated in studies with the hybrid hemoglobin Felix and related hybrids. Re-activating fetal hemoglobin synthesis in vivo is currently a major focus of clinical efforts designed to treat sickle cell anemia since it inhibits the aggregation of sickle hemoglobin. The mechanisms for both the increased oxygen affinity of fetal hemoglobin and its decreased response to DPG have been clarified. Acetylated fetal hemoglobin, which makes up 10–20% of total fetal hemoglobin, has a significantly weakened tetramer structure suggesting a similar role for other kinds of protein

J. M. Manning (✉) · L. R. Manning
Department of Biology, Northeastern University, Boston, MA 02115, USA
e-mail: j.manning@northeastern.edu

L. R. Manning
e-mail: l.manning@northeastern.edu

A. Dumoulin
Department of Developability, Pierre Fabre Research Centre, Castres 81106, France
e-mail: antoine.dumoulin@pierre-fabre.com

J. C. Padovan · B. Chait
Laboratory of Gaseous Ion Chemistry, Rockefeller University, New York, NY 10065, USA
e-mail: padovan@rockefeller.edu

B. Chait
e-mail: chait@rockefeller.edu

U. Hoeger and J. R. Harris (eds.), *Vertebrate and Invertebrate Respiratory Proteins, Lipoproteins and other Body Fluid Proteins*, Subcellular Biochemistry 94,
https://doi.org/10.1007/978-3-030-41769-7_11

acetylation. Embryonic hemoglobins have the weakest tetramer and dimer structures. In general, the progressively increasing strength of the subunit interfaces of the hemoglobin family during development from the embryonic to the fetal and ultimately to the adult types correlates with their temporal appearance and disappearance in vivo, i.e., ontogeny.

Keywords Hemoglobin · Sickle cell · Malaria · Acetylation · Oxygen affinity · Nano gel filtration · Ontogeny

Introduction

Adult hemoglobin has provided us with an appreciation of how a protein accomplishes its physiological goals of oxygen transport from lungs to tissues with such high efficiency (Perutz 1989). Additional insights into protein versatility can be realized from comparisons of the adult type with the fetal and embryonic hemoglobins to ascertain how changes in amino acid sequence at strategic locations can enable a particular hemoglobin type to maximize binding of oxygen when its supply is limiting. The family of human hemoglobin tetramers is an excellent system in which to study this variable extent of oxygen binding; they all possess a very similar structural architecture as a platform on which to confer type-specific properties (Fig. 11.1a). The human hemoglobins number eight in all; many, but not all, are present at various stages of development. That they bind different amounts of oxygen (Fig. 11.2) has been known for many years but the mechanism whereby this is achieved is conjectural. This chapter describes subtle differences at the subunit interfaces of normal adult, fetal, and embryonic human hemoglobins that impart important physiological properties optimized for one particular type of hemoglobin. Independent of its increased oxygen binding properties, fetal hemoglobin provides a therapeutic effect in sickle cell anemia and possibly in malaria, as also described in this chapter. The early literature describing the embryonic hemoglobins is relatively sparse with some conclusions questionable.

In order to obtain reliable data on the fetal and embryonic hemoglobins, sufficient quantitites of native and well characterized materials as well as sensitive, accurate methods to measure their properties must be available. Fetal hemoglobin purified from umbilical cord blood (Bookchin and Nagel 1971; Bookchin et al. 1975) and characterized by mass spectrometry has made it possible to study this hemoglobin in great detail. Embryonic hemoglobins are the least understood type due to difficulty in procuring them, a problem that has now been resolved since they are available in their native states from transgenic mice (He and Russell 2001) making it possible to study and compare them with the well known adult hemoglobin.

The methodology used in these studies is based on gel filtration chromatography but using very low hemoglobin concentrations to study differences in the basic foundations of the subunit assemblies of the three hemoglobin types that are responsible

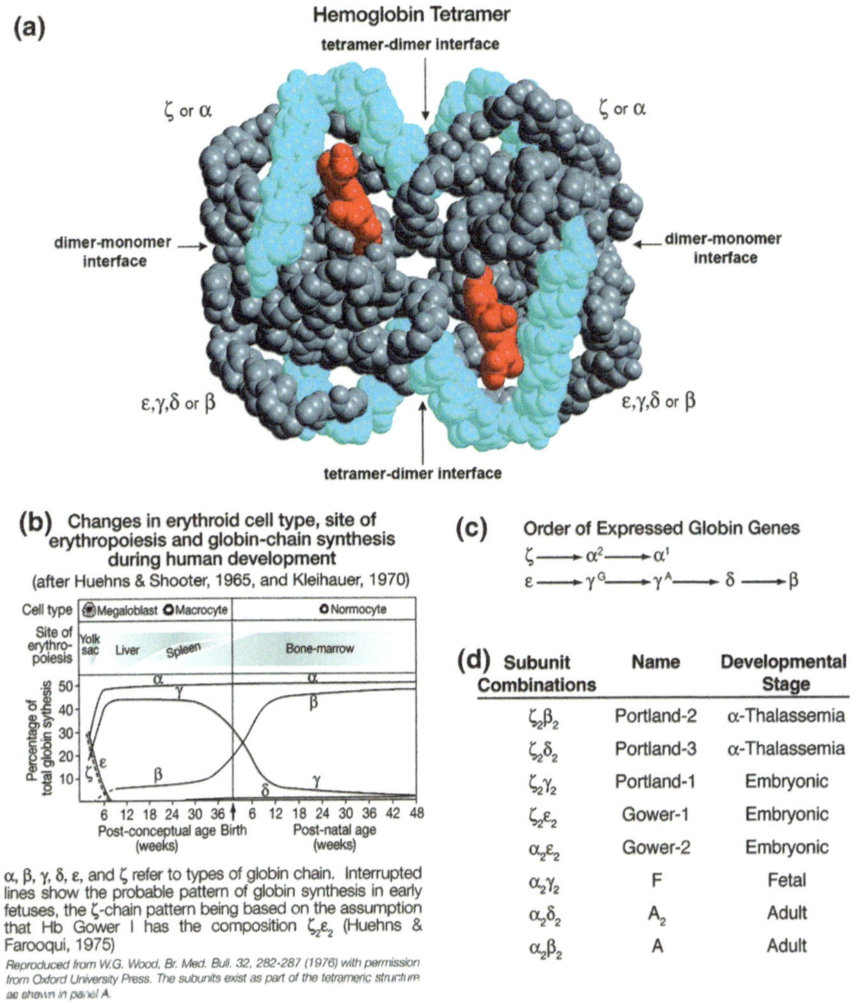

Fig. 11.1 Part A-Hemoglobin tetramer with the different types of globin subunits indicated by Greek letters and the locations of the two types of subunit interfaces designated with arrows. Part B-Changes in the expression of subunits as a function of time. Part C-Order of globin subunit expression. Part D-Names of the eight different types of human hemoglobins formed by combinations of various subunits as a function of time. Reproduced from Manning et al. (2009) with permission

for their specific oxygen bonding characteristics. Avoiding the high hemoglobin concentrations usually employed, which greatly exceed their equilibrium dissociation constants, permits one to observe all stable assembly intermediates since it shifts the equilibrium away from the predominant tetramer species and towards its constituent dimer and monomer components, as described below. We refer to this experimental approach as nano gel filtration, whose basis is reminiscent of the principle used a

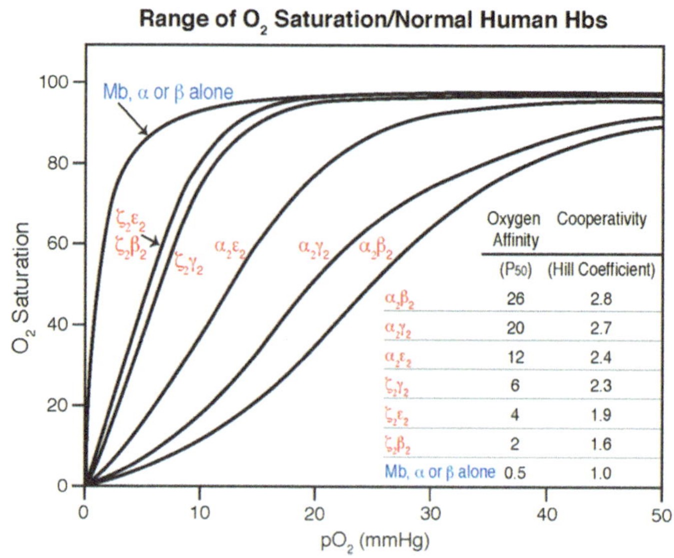

Fig. 11.2 Oxygen binding curves of human embryonic, fetal, and adult hemoglobin tetramers (red labels). The oxygen binding curves of the individual alpha and beta subunits have blue labels. Reproduced from Manning et al. (2017) with permission

century ago to observe the fundamental property of the bending of light whose magnitude was accurately measured during a solar eclipse when the intensity of the sun's intense glare, which would ordinarily obscure such observations, was minimized (Dyson et al. 1920); this experiment was an important proof of Einstein's theory of general relativity.

Hemoglobin Ontogeny

The individual globin subunits are expressed at precise times (Fig. 11.1b) and in a particular order (Fig. 11.1c) in a process known as ontogeny. The depiction in Fig. 11.1c represents the basic elements of the system that expresses the human globin subunits but it is an oversimplification since the human alpha cluster and beta cluster globin genes, which are present on chromosomes 16 and 11, respectively, are embedded in nucleosome arrays that are subject to opening by promotors or closing by repressors; these bind to specific gene regions distant from the expressed loci (sometimes referred to as the Locus Control Region, LCR) (Stamatoyannopoulos and Grosveld 2001). There are three examples of a reciprocal pattern of gene expression in Fig. 11.1b (epsilon – > gamma, gamma –> beta, and zeta –> alpha) where expression of one gene is decreasing and, concurrently, expression of another gene is increasing. Such behavior is referred to as gene switching (Ikuta et al. 1996, Baron 1996) and is a very active area of research especially as it relates to fetal hemoglobin expression in sickle cell anemia; this will be discussed later in this chapter.

Other aspects in the transition from Fig. 11.1b, c involve globin gene clusters, consisting of expressed (exons) and unexpressed (introns) regions that are subject to a process known as splicing to produce the functional intact genes shown in Fig. 11.1c. Neither this topic nor post-translational modifications (referred to as epigenetics), nor RNA polymerase with associated transcription factors for messenger RNA synthesis will be covered in this chapter. Instead, we will evaluate the final expressed products of the system shown in Fig, 1D and, specifically, how they differ from one another regarding oxygen binding and subunit interfaces.

Subunit Assembly

The expressed globin subunits bind heme and then pair together to form dimers held together by a dimer-monomer interface (Fig. 11.1a) which, in the case of adult Hb A, has an equilibrium strongly in favor of dimer formation (Perutz 1989). Pairs of the same dimers combine to form tetramers having a flexible tetramer-dimer interface that can rearrange into two conformations in equilibrium, oxy (R) and deoxy (T), which either bind or release oxygen, respectively (Perutz 1989). This chapter will cover the assembly process for the entire human hemoglobin family and describe the physiological consequences of the differences between them.

Even though there are eight possible combinations of hemoglobin subunits, only six have been found during normal development (Fig. 11.1d). The remaining two, Hb Portland-2 and Hb Portland-3, are only present in patients with alpha-thalassemia where the alpha gene has suffered deletions so the zeta gene continues to be expressed. Why these two hemoglobins are not normally expressed is unclear; a suggestion is offered here.

Some individual hemoglobin subunits are unstable and prone to aggregation, unfolding, or oxidation at the heme moiety leading to experimental difficulties in studying their assembly to tetramers directly. For example, individual beta and gamma subunits self-assemble (Adachi et al. 2000). Through the careful studies of many investigators (Antonini and Brunori 1971; Edelstein et al. 1970), such problems for individual subunits for adult Hb A have largely been solved but subunits comprising the other hemoglobins could still pose problems. Once formed, however, tetramers are generally quite stable. Hence, our experiments begin with intact tetramers and we study the dissociation process by dilutions in order to shift the tetramer <--> dimer <--> monomer equilibrium to the right. Such an experimental approach, however, requires a highly sensitive detection system in which the quality of the data and the ensuing conclusions depend on the resolution and reproducibility of the methods employed.

In this chapter a simple high resolution gel filtration method coupled with software for elution curve analysis (Manning et al. 1996) has been invaluable in yielding new insights into the fetal and embryonic types of human hemoglobin and how they differ from adult Hb A. In spite of earlier claims based on simulations that such gel filtration methods could not yield accurate results (Zimmerman and Ackers 1971), our data (Manning et al. 1996) for adult Hb A agreed well with previous results obtained by other methods. It has been generally assumed that all human hemoglobins have not only the same overall structure shown in Fig. 11.1a but also similar strengths at both types of interfaces, but in this chapter we demonstrate otherwise using this methodology.

Oxygen Binding Comparisons

The most important physiological properties of the three types of human hemoglobin- embryonic, fetal, and adult- are their oxygen binding capacities (Fig. 11.2), which differ considerably and have evolved so that each can optimally bind the oxygen to which it is exposed at a given time during development (Wood 1976; Bunn and Forget 1986). For example, adult Hb A (alpha2beta2,) (curve farthest to the right) has a midpoint value (50% oxygen saturation defined as P_{50}) of 26 mm Hg (Fig. 11.2 inset), which is close to the average oxygen tension of venous blood. Fetal and embryonic hemoglobins bind oxygen more avidly than adult Hb. In order to achieve maximum binding of available oxygen, their oxygen binding curves are shifted to the left (Fig. 11.2).

This is advantageous since the amount of oxygen in the environments to which they are exposed are lower than that used by air-breathing adults. In order for a hemoglobin to be functional, the middle part of the oxygen binding curve where the oxygen binding (y-axis) is most sensitive to changes in oxygen tension (x-axis), i.e. having the highest degree of cooperativity, should be near the available oxygen tension. Thus, if an embryonic hemoglobin was present in an adult, it would not function as a useful oxygen carrier because at the high ambient oxygen tension that adults use, an embryonic hemoglobin would be fully saturated with oxygen and unable to release it. Also, preterm neonates with greater than normal amounts of fetal hemoglobin (Fig. 11.1b) and lower 2,3-DPG may also have difficulty releasing oxygen.

The mechanisms by which these variable oxygen saturations are achieved are unclear and inspection of the structure in Fig. 11.1a offers no insight. Speculation has focused on primary sequence differences of fetal and embryonic hemoglobins compared to adult hemoglobin and how those changes could influence the heme environment where oxygen is bound. However, definitive experimental evidence is lacking for this perspective. In this chapter we describe the likely sources of their increased oxygen affinities as arising from how their subunits interact.

Fetal Hemoglobin–Oxygen Binding Without DPG

The moderately increased oxygen affinity of Hb F (Fig. 11.2) (Poyart et al. 1978; Chen et al. 2000) enables the transfer of oxygen from maternal hemoglobin to the fetus. How this important left shift in the oxygen binding curve of Hb F is imparted has long been the subject of speculation, but its mechanism was only clarified when recombinant DNA technology became available. To understand this mechanism, we start with the equilibrium between the R (oxy) and T (deoxy) tetrameric states, whose ratio is defined as the allosteric constant, L. When Hb either binds or releases oxygen, the tetramer-dimer interface (Fig. 11.1a) rearranges its subunit contacts and the position of the R/T equilibrium changes (Perutz 1989). The single amino acid difference between fetal and adult hemoglobins at this interface is the substitution of an Asp in the gamma subunit of Hb F for a Glu in the beta subunit of Hb A at position 43. A recombinant Hb A with this single substitution tightens this interface in the R-state (Chen et al. 2000) thereby shifting the R/T equilibrium so that there is less of the T-state and more of the R-state and the affinity for oxygen increases (Poyart et al. 1978). The other four sequence differences at the subunit interfaces are at the dimer-monomer interface. A recombinant hemoglobin with these four additional substitutions in addition to the Glu–> Asp substitution did not have a further increased oxygen affinity (Chen et al. 2000) indicating that the single Asp for Glu substitution at the tetramer-dimer interface was sufficient to account for the increased oxygen binding of Hb F.

Different Responses of Fetal and Adult Hemoglobins to DPG

Since it is the oxygen binding properties of adult and fetal hemoglobins within the red cell that is physiologically important, the effect of the allosteric regulator 2,3-diphosphoglyceric acid (2,3-DPG), which is normally present in red cells at equimolar concentrations with hemoglobin (Benesch and Benesch 1967), requires consideration. It binds with perfect geometry in a 1:1 ratio with the deoxy form of Hb A in a cleft between the two beta chains (Arnone 1972) and moves the oxygen binding curve further to the right. DPG has a diminished effect on fetal Hb F compared with that of adult Hb A, which further increases the difference between the two regarding oxygen binding. Since the DPG binding pocket in fetal Hb contains a Ser instead of a His at position 143, it was conjectured (Bunn and Forget 1986) that this substitution was responsible for the reduced DPG effect on Hb F. When this hypothesis was tested directly by two independent labs using recombinant mutants (Adachi et al. 1997; Fang et al. 1999), the role of the Ser—> His substitution at position at 143 was excluded as contributing to the reduced effect of DPG on Hb F. Instead, the E43D substitution at the tetramer-dimer interface, mentioned above as the main contributor to the increased oxygen binding of Hb F (Chen et al. 2000), is responsible for the lowered DPG response.

Effect of Fetal Hemoglobin in Sickle Cell Anemia

The structure of sickle hemoglobin (Hb S) has been solved (Padlan and Love 1985; Harrington et al. 1997); tetramers in the deoxy state initially adhere to one another through hydrophobic bonding involving the V6 beta mutation sites as donors and F85-L88 beta acceptor sites on adjacent tetramers. The aggregate is subsequently stabilized by a number of other inter-tetrameric interactions both laterally and longitudinally. In the oxy state, this aggregate does not form due to unfavorable contacts. As a result, oxygen binding to hemoglobin S aggregates in sickle cells compared to normal cells is decreased, i.e., the oxygen binding curve shifts to the right compared to that for normal adult Hb A (Fig. 11.2) The presence of such aggregates in red cells causes them to assume a crescent sickle shape in the venous circulation thereby occluding capillaries resulting in oxygen deprivation. One approach, which has been reviewed recently (Eaton and Bunn 2017), to treating this disease is to prevent this aggregation (polymerization) directly. Another approach is based on the observation that fetal hemoglobin impedes this polymerization process because tetramers of Hb F with gamma instead of beta subunit have unfavorable contacts for aggregation (Bookchin and Nagel 1971; Poillon et al. 1993). Hence, increasing fetal Hb F production in vivo is currently the focus of several major efforts to treat sickle cell anemia. The model on which this approach is based is an example of an "experiment of nature" described next.

Hereditary Persistence of Fetal Hemoglobin (HPFH)

As depicted in Fig. 11.1b, fetal Hb expression persists until several months after birth when gamma subunit production nearly ceases except for a small amount (~1%) which continues to form even in normal individuals (Bunn and Forget 1986; Akinsheye et al. 2011). It is not known how this residual expression continues rather than being completely suppressed. A small percentage of patients with sickle cell anemia have elevated levels of Hb F (up to 30%) and these individuals have clinically fewer episodes of sickle cell crises. Even though the mechanism is not understood, this observation forms the basis of research efforts into treating sickle cell anemia by re-starting gamma subunit synthesis, as described next.

Manipulating Fetal Hemoglobin Gene in Sickle Cell Anemia

The first generation of inducers of fetal Hb synthesis built upon the clinical observations that those sickle cell patients who also had increased amounts of Hb F (HPFH) experienced decreased severity of the disease, as described above. Since it had been reported that gamma subunit production was associated with decreased levels of

DNA methylation (van der Ploeg and Flavell 1980; Groudine et al. 1981), the strategy of using DNA hypomethylating agents to induce Hb F synthesis was devised; studies with 5-azacytidine (DeSimone et al. 1982; Lavelle et al. 2008) ensued and this goal was realized but toxicity issues led to termination of its usage. Subsequent studies focused on this approach by testing analogs of azacytidine and, subsequently, similar agents such as hydroxyurea; this pharmaceutical agent, Hydrea, is currently in use as a treatment for sickle cell anemia and can lead to a significant elevation in Hb F but is associated with a higher frequency of stroke. Based on another aspect of gene expression, i.e., the acetylation of histones within active genes, butyrates and related compounds, which inhibit histone deacetylation, were also tested. These studies have been reviewed recently (Musallam et al. 2013).

Another effort to express higher levels of Hb F and thereby lower severity in sickle cell patients involves the gamma gene repressor named BCL11A (Sankaran et al. 2008, 2010) whose interaction with a specific DNA region has been identified. This strategy is to prevent binding of BCL11A thus increasing gamma subunit expression. Patients with HPFH have mutations in one of the two regions that bind this repressor thus leading to incomplete termination of HbF expression (Brendel et al. 2016; Liu et al. 2018). This is a new and currently active avenue of study.

Other Genetic Approaches in Sickle Cell Anemia Treatment

Rapid progress in genetic techniques recently has opened new approaches to correct the effects of the sickle gene mutation. Among these are the use of corrective genes whose expressed globins do not participate in Hb S aggregation (normal beta, the T87Q gene (described below) or the gamma gene), gene corrections e.g. by CRISPR-Cas9 or direct gamma gene induction. Recent reviews on these approaches are in (Demirci et al. 2018; Lettre and Bauer 2016; Telen et al. 2019).

One of the main contributors to the inhibition of Hb S aggregation by Hb F is the T87Q gamma substitution of Hb F (Nagel et al. 1979; Adachi et al. 1996). This is part of the Val-6 binding site on adjacent tetramers, which initiates Hb S aggregation. A recent genetic approach directed at treating sickle cell anemia has been the use of lentiglobin (Leonard and Tisdale 2018), where this substitution is fused with the remainder of the normal Hb A beta subunit, so both donor and acceptor sites that have been altered. This approach has shown encouraging initial results (Ribeil et al. 2017).

Another natural hemoglobin that inhibits Hb S aggregation just as effectively as Hb F is Hb A2, a minor hemoglobin occurring normally (Nagel et al. 1979; Adachi et al. 1996). Hb A2 also contains the Glu-6 site as well as the T87Q substitution so both donor and acceptor sites present in Hb S are replaced. Thus, genetic manipulations to increase the expression of Hb A2 as a treatment for sickle cell anemia are also worthy of consideration.

Fetal Hemoglobin, Malaria, and Thalassemia

A general understanding of thalassemia is helpful prior to a discussion of fetal hemoglobin and its relationship to malaria. Deletions at the alpha loci result in the disease alpha thalassemia (Higgs et al. 2005) and deletions at the beta loci lead to beta thalassemia. Of particular interest is the experiment of nature, e.g., in beta thalassemia, the *preceding* gene continues to be expressed (Fig. 11.1c), so fetal hemoglobin is produced in moderately high amounts. In alpha thalassemia, the zeta gene is expressed so the embryonic hemoglobin, Portland-2 (zeta2beta2) (as well as beta4 tetramers (HbH) and gamma4 tetramers (Hb Barts), are produced (Randhawa et al. 1984). Why hemoglobin Portland-2 is not found in normal embryos has not been explained but below we offer a suggestion.

The two alpha genes normally expressed in humans are identical but the two gamma genes encode for two different gamma subunits with either Gly or Ala at position 136. The Gly/Ala ratio changes from 75/25 before birth to 40/60 in adult red cells (Schroeder et al. 1968; Steinberg et al. 2014), another example of gene switching. It is not known whether these two types of Hb F have any different properties since neither has been studied separately.

In 1976 a link was proposed between decreased malaria occurrences and increased fetal hemoglobin concentrations in red cells both in newborns and in those afflicted with beta-thalassemia compared (Pasvol et al. 1976; Mosca et al. 2009). As shown in Fig. 11.1b, normal newborns still retain a significant amount of Hb F in their red cells. Hence, the decreased incidence of malaria in populations with high levels of beta-thalassemia is thought to be due to prevention of the normal switch from gamma to beta subunit production due to beta gene deletions. Thus, gamma subunit synthesis continues so Hb F concentration in such red cells is comparatively high and growth of the malaria parasite is impeded thereby conferring resistance. A rationale for this observation was provided recently (Shear et al. 1998) which stated that the protease of the malaria parasite digests Hb F very poorly because, as we had reported earlier (Dumoulin et al. 1997), Hb F is a tetramer with very strong subunit contacts and therefore not readily prone to dissociation (see below). Hence, the presence of protease-susceptible subunits formed by dissociation of Hb A in normal cells permits growth of the malaria parasite in adult cells. When cells contain high amounts of Hb F, subunits are present in much lower concentrations because Hb F dissociates 70-fold less than Hb A (Dumoulin et al. 1997) and a nutrient source for the parasite is depleted. This source of the increased strength of the fetal Hb tetramer is described below.

It has been known for many years that individuals who are heterozygous for sickle cell anemia, i.e., one sickle gene from one parent and one normal gene from the other parent are less likely to develop malaria (Allison 1954). This observation has been studied by many investigators; the mechanism has been reviewed recently (Luzzatto 2012).

Fetal Hemoglobin, a Strong Tetramer

All human hemoglobin tetramers have a very similar architecture (Fig. 11.1a) even though their oxygen-carrying capacities are significantly different (Fig. 11.2). A study of the assembly properties of each tetramer using a common laboratory instruments and a high resolution gel support with very high sensitivity and resolving capability coupled with software for elution curve analysis has provided evidence how this increased strength is achieved (Manning et al. 1996). The first proof of concept that this procedure was capable of detecting very small changes in subunit contact strengths was with the liganded form of fetal hemoglobin (Dumoulin et al. 1997). In spite of an earlier erroneous report (Bunn 1969) that Hb F and Hb A had the same dissociation properties, Fig. 11.3 shows that their tetramer-dimer dissociation properties are vastly different (~70-fold). Perutz had reported earlier that the deoxy

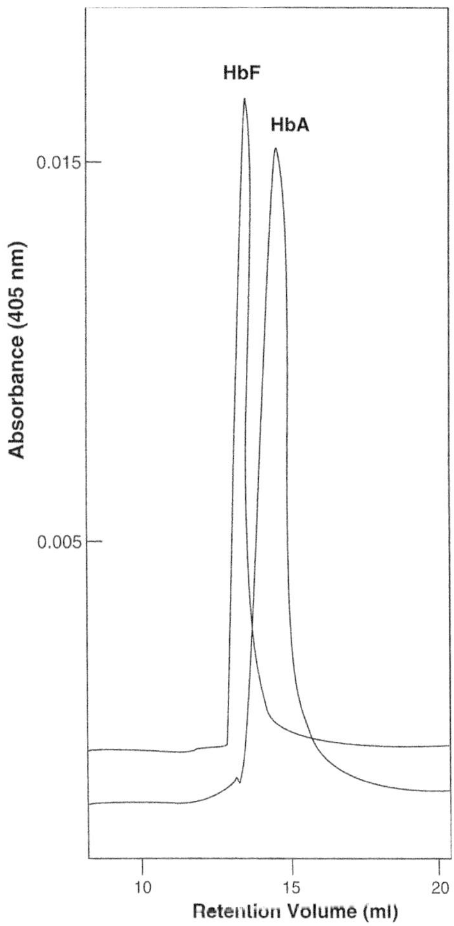

Fig. 11.3 Nano Gel Filtration of Human Fetal (Hb F) and Adult (Hb A) Hemoglobins. At the very dilute concentration used (500 nM), Hb A is predominantly dimeric and Hb F is tetrameric, which forms the basis for their separation. Reproduced from Dumoulin et al. (1997) with permission

form of fetal hemoglobin had a more compact overall structure (Frier and Perutz 1977) but the corresponding structure of the oxy conformation of Hb F was never solved. The molecular basis for this increased strength is described next.

Sources of Fetal Hemoglobin Enhanced Strength

In order to ascertain which segments of the gamma subunit sequence of Hb F endowed it with the 70-fold enhanced strength of its tetramer-dimer subunit contacts (Fig. 11.3), we substituted the 5 amino acids that were different in both types of interfaces between Hb F and Hb A; hence, this recombinant Hb A/F had the single difference at the tetramer-dimer interface and the four differences at the dimer-monomer interfaces (Dumoulin et al. 1997). Its oxygen-binding properties in response to DPG was the same as that for Hb F and not Hb A, despite the fact that it possessed all of the DPG binding residues of Hb A. Its tetramer-dimer dissociation was increased only 5-fold towards that of Hb F instead of 70-fold if all of the increased tetramer strength were due to the five interface differences. This observation clearly indicated that segments other than the subunit interfaces themselves contributed to the increased stability of the Hb F tetramer. This source of the other site(s) responsible for increased tetramer strength was investigated with a recombinant Hb called Hb Felix (Dumoulin et al. 1998), as described next.

Hemoglobin Felix

The recombinant Hb Felix has the A-helix (amino acids 1–18) of the gamma subunit of Hb F (comprising eight amino acid differences compared with the beta subunit of Hb A) joined to helices B-H (the remaining sequence of Hb A) (Dumoulin et al. 1998). Studies on this hybrid hemoglobin have provided many new insights into hemoglobin function. This recombinant Hb is completely stable and has the native conformation of Hb. A completely unexpected finding was that its tetramer strength is close to that of Hb F despite the fact that the A-helix is not part of the tetramer-dimer interface. This suggests that the A-helix influences the tetramer-dimer interface due to a long-distance effect perhaps by protruding deeper into the central cavity than the A-helix of Hb A (Dumoulin et al. 1998; Manning et al. 1998, 1999). In dissecting which amino acids of the A-helix were responsible for this enhanced effect, individual and combination substitutions were performed (Yagami et al. 2002). The major contributors were the V1G, P5E, and E7D at sites 1, 5, and 7 respectively. The DPG response of Hb Felix is unaffected by the A-helix substitution.

Acetylated Fetal Hemoglobin

A fraction (~15%) of Hb F normally present in red cells and referred to as Hb F1 is acetylated at the N-terminal Gly of its gamma subunit; adult Hb A has Val at this position; Val is not a favorable substrate for the enzyme that adds an acetyl group to N-terminal sites. It is not clear whether acetylated Hb F1 has a function although it does have some very interesting properties. Hence, Hb F1 has a reduced binding of DPG compared to Hb F (Bunn and Briehl 1970); thus, its oxygen affinity is higher than that of HbF in the presence of DPG. Another important feature of Hb F1 is its significantly reduced tetramer strength, about 30-fold increased dissociation constant compared to that of unacetylated Hb F rendering it similar to adult Hb A in this respect (Manning and Manning 2001). An interpretation of this post-translational modification is that it represents an unfavorable energetic event for the Hb F type of hemoglobin since it weakens major subunit interactions. In terms of influencing gene expression (Fig. 11.1c), such a property could disfavor gamma subunit expression thereby lowering Hb F formation in favor of beta subunit expression and Hb A formation, i.e. "switching" (Manning et al. 2009), as described more fully below.

Embryonic Hemoglobins Are Weak Tetramers with Increased Oxygen Affinity

Most of the early information on embryonic hemoglobins is derived from studies conducted about 50 years ago (Huehns et al. 1964; Huehns and Shooter 1965; Huehns and Faroqui 1975) when purification and characterization methods were incompletely developed; hence not all of the embryonic hemoglobins were identified correctly. In addition, difficulties in procuring them due to their rarity and ethical considerations hampered investigations on them. Later studies by Brittain and colleagues using improved methodology made a significant advance (Brittain 2002). Our studies on the embryonic hemoglobins (Manning et al. 2007) were made possible through the work of He and Russell (2001) who transfected the human globin genes into mice permitting the isolation of adequate amounts of the embryonic hemoglobin in the native state.

There are three embryonic human hemoglobins—Portland-1 (zeta2gamma2), Gower-1 (zeta2epsilon2), and Gower-2 (alpha2epsilon2) normally present during the first few months of life but two others—Portland-2 (zeta2beta2) and Portland-3 (zeta2delta2) have not been found during normal development, but reported only in cases of severe alpha-thalassemia when alpha subunits are not synthesized (Fig. 11.1d). Gene switching facilitated by promotors and repressors to change subunit types is usually invoked to explain the normal appearance and disappearance of the first three hemoglobins (Fig. 11.1b) but there has been no adequate explanation for the absence of the other two hemoglobins during normal development. In collaboration with Dr. J. E. Russell, we isolated, characterized, and studied all the embryonic hemoglobins except for Hb Portland-3 and compared them with fetal and adult hemoglobins. Each embryonic hemoglobin purified from transgenic mouse blood showed a single band upon electrophoresis, a single elution peak during chromatography, and had the correct mass and circular dichroism pattern (Manning et al. 2007). An unexpected observation during purification was that some of them, in contrast to adult hemoglobin, did not remain tetrameric due to their ease of dissociation, as described below.

The embryonic hemoglobins bind oxygen extremely avidly (Hofmann et al. 1995; Hofmann and Brittain 1996; Brittain 2002) (Fig. 11.2), thus providing them with the ability to capture the very small amounts of oxygen to which they are exposed in embryonic tissue. However, it has never been established on a molecular basis how this property is attained. It would be potentially useful to have this information as a basis for further investigations in situations where the oxygen supply is low.

We compared purified embryonic hemoglobins with fetal and adult hemoglobin using a number of proven methods and made the following conclusions:

(1) From a structural standpoint, Hb Gower-2, alpha2epsilon2 (Sutherland-Smith et al. 1998), which occurs during normal development, is the most thoroughly studied embryonic Hb. Its structure is remarkably similar to that of adult and fetal hemoglobins with minor differences but there is nothing obvious from the structure how its very high oxygen affinity is bestowed. Models attempting to interpret its primary sequence differences from adult hemoglobin have been suggested to explain its increased oxygen binding capacity but these are not conclusive.
(2) Attempts to solve the structure of Hb Portland-1, alpha2gamma2, were hampered by its dissociation to free gamma subunits which then re-associated to gamma4 (Hb Bart's) (Kidd et al. 2001). Gamma subunits undergo self-assembly to form stable dimers and tetramers. This finding is consistent with the weakness of the subunit interfaces of the embryonic hemoglobins that we reported.
(3) Generally, both tetramer-dimer and dimer-monomer subunit interfaces are weak in the embryonic hemoglobins compared to corresponding interfaces in fetal and adult hemoglobin. This was evident when we reported a systematic study of these properties for all the embryonic hemoglobins (Fig. 11.4) (Manning et al. 2012) by high resolution gel filtration.
(4) When embryonic hemoglobins are mixed, they slowly dissociate to individual subunits which compete with each other to recombine and form the energetically most favorable tetramer. For example, when three embryonic hemoglobins containing alpha, gamma, or beta subunits are mixed, Hb A is formed but Hb F is not, suggesting that Hb A is the more stable tetramer (Manning et al. 2009). If normal adult and fetal hemoglobins are mixed in the same manner, they do not exhibit this behavior.
(5) The tetramer-dimer subunit interfaces dissociation constants of adult, fetal, and embryonic hemoglobins do not show a regular pattern but the values for the embryonic hemoglobins are about 10-30 times weaker than those for fetal and

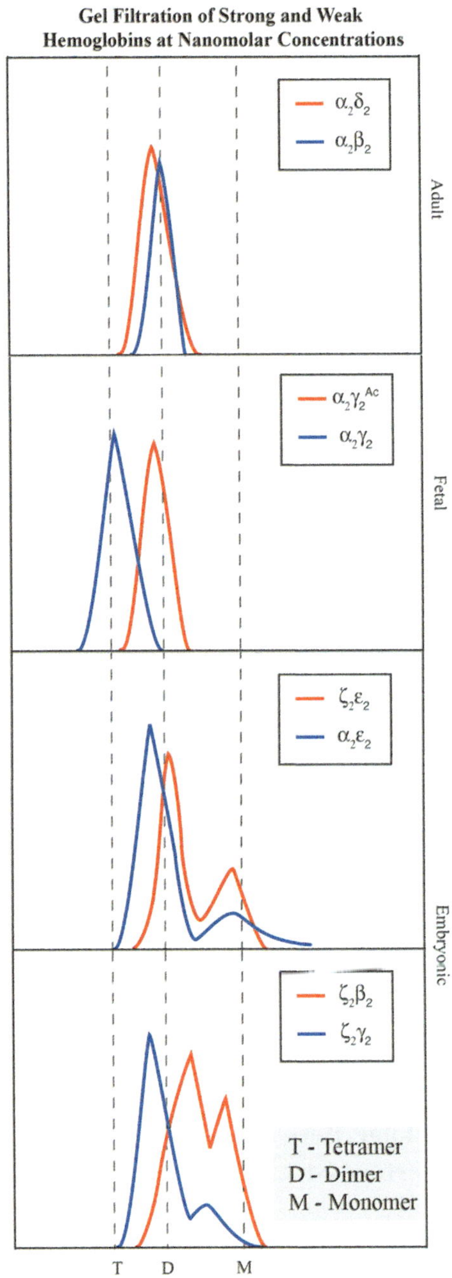

Fig. 11.4 Nano Gel Filtration of Strong and Weak Hemoglobins. At the concentrations used (10–100 nM), the human adult and fetal hemoglobins in the upper two panels have strong tetrameric and dimeric structures. The embryonic hemoglobins in the lower two panels generally have weak structures shown by a significant monomer population. Reproduced from Manning et al. (2017) with permission

adult tetramers (Manning et al. 2007) thus correcting some erroneous reports (Brittain 2002).

(6) Dissociation of the dimer-monomer interfaces have an ordered degree of strength corresponding to embryonic < fetal < adult (Fig. 11.4) (Manning et al. 2007). The alpha-beta and alpha-gamma dimers are very stable but the corresponding embryonic zeta-containing dimers are weak. In fact, the dimer-monomer dissociation equilibrium constant of Hb Portland (zeta-beta) was accurately determined to be 1.4 nM. That such a measurement can be made indicates that this interface is so weak that the equilibrium between dimers and monomers is freely reversible; in contrast, the adult Hb dimer-monomer dissociation constant (alpha-beta) is beyond the detection limits of current instruments. It is possible that the acetylation of the N-terminus of the zeta subunit contributes significantly to the weakness of the embryonic hemoglobin structures by analogy to the acetylation of fetal hemoglobin (Hb F1) described above.

(7) There is a correlation between dimer dissociation rates and oxygen binding. Thus, in addition to showing decreased oxygen affinity from left to right, Fig. 11.2 also depicts the increasing strengths of the interface between monomers in the same direction. The individual subunits and myoglobin represent the completely dissociated state whereas adult Hb has the strongest dimer interface; the embryonic and fetal hemoglobins have intermediate strengths. The alpha-epsilon dimer behaves like a transition Hb meaning that it shows an *intermediate* dimer strength and also has an *intermediate* degree of oxygen affinity (Fig. 11.2) clearly indicating a correlation between dimer strength and oxygen binding (Manning et al. 2010, 2017).

(8) A comparison of tetramer and dimer interface strengths of all the human hemoglobins by gel filtration at nanomolar concentrations dramatically demonstrate the order of their stability (Fig. 11.4) to be embryonic < fetal < adult (Manning et al. 2012).

(9) The embryonic hemoglobins have moderately reduced Hill coefficients (n values of 1.6–2.4 compared to n value of 2.8 for adult Hb, (see inset to Fig. 11.2) (Brittain 2002; Manning et al. 2017). This feature further enhances the ability of the embryonic hemoglobins to bind oxygen efficiently.

Linkage Between Oxygen Binding Affinity and Dimer Interface Strength

The degree of dimer interface strength has a significant influence on oxygen binding potential of the *normal* human hemoglobins. The wide range of oxygen binding potential shown in Fig. 11.2 starts from the left with the very high affinity of the individual subunits (and myoglobin) and proceeds to the far right with the low oxygen affinity of the adult hemoglobin tetramer; the embryonic and fetal hemoglobin tetramers are intermediate between these two extremes that parallels the strength between different monomer types in the heterodimer of each hemoglobin (Manning et al. 2017). Hence, whereas the embryonic dimers with high oxygen affinity are quite weak and dissociate fairly readily, the fetal and adult dimers with lower affinities dissociate very little and have a low affinity. Hence, the variations in dimer strengths have a corresponding relationship to oxygen binding potential (Manning et al. 2007). Therefore, an unexplored area for future research is whether oxygen itself (Genbacev et al. 1997; Simon and Keith 2008) might have some role in promoting expression of a hemoglobin type.

The very tight binding of oxygen to the embryonic hemoglobins cannot readily be explained as arising from local effects of individual amino acid substitutions, e.g. on the heme moiety, but rather due to more global effects on the bonds that hold the subunits together. The well known oxygen binding curves of the human hemoglobin family (Fig. 11.2) are better envisaged as a spectrum of tetramers with different binding potential that depends on the degree to which the individual subunits bond together. Moderate weakening of the adult Hb A tetramer with its relatively low oxygen affinity will increase the population of dimers and monomers in the overall equilibrium not readily discernable by inspection of the structure but demonstrable by gel filtration at extremely dilute concentrations (Fig. 11.4). This equilibrium shift leads to an increase in oxygen binding. A major effect of amino acid differences between different subunit types is to determine how tightly they bind to one another, which in turn affects how readily they bind oxygen, an allosteric effect. It is likely that different properties of other hemoglobin types remain to be discovered when they are studied from a different perspectives.

Cooperativity and the Bohr Effect

Cooperativity can be envisaged as arising like a compressed spring representing the tightness of the interactions between subunits to confer low oxygen binding; release of this tension facilitates the binding of oxygen (Perutz 1989). The assembly process from individual subunits, which do not display cooperativity, to tetramers with positive cooperativity is an important feature of the oxygen binding curve and achieves its maximum in the midrange of the sigmoidal curves (Fig. 11.2); it is expressed as the Hill Coefficient (n). The curves from right to left (adult–> fetal–> embryonic) have decreasing cooperativity (Fig. 11.2 inset), which is consistent with a decrease in their overall tetramer strength since their subunit contacts are weaker in that same direction. The equilibrium between the deoxy and the oxy states, T/R, is defined as the allosteric constant, L. For adult Hb A this value is 5×10^5 whereas for the embryonic hemoglobins, it is 1–2 orders of magnitude lower (Brittain 2002). This decrease is consistent with a loss in tetramer strength for the embryonic hemoglobins. The Bohr Effect of Hb F is about 20% greater than that of Hb A at physiological chloride concentrations (Poyart et al. 1978) and likely enhances oxygen supply to the fetal circulation.

Relationship Between Dimer Interface Strength and Ontogeny

The stability of the dimer-monomer interface between the various hemoglobin subunits is clearly demonstrated by gel filtration at very low concentrations (Fig. 11.4). This pattern also resembles their chronological appearances (Fig. 11.1b), i.e., there is a correlation between the stability of this type of interface in terms of how long the dimer interface remains intact and the physiological duration of that particular hemoglobin in vivo. Stronger dimer pairs supplant weaker ones in the order embryonic – > fetal –> adult. Thus, the strength and ease of formation of dimers appears to be related to ontogeny, which would represent a thermodynamic contribution to gene expression. We refer to this as an "intrinsic" contribution to gene expression in contrast to "extrinsic" contributions, such as with promotors and repressors (Manning et al. 2012). Even though the term "switching" strictly refers to individual gene expression, it is exemplified as the production of a hemoglobin type when two subunits combine, i.e., embryonic to fetal type or fetal to adult type. The enhanced ability of fetal Hb to form tetramers as demonstrated by its very low tetramer-dimer dissociation constant may contribute to the switch from embryonic to fetal type of hemoglobin during development. The slow acetylation of fetal Hb weakens its tetrameric structure considerably (similar to that of Hb A, discussed above) and could subsequently contribute to the switch to the adult type of Hb, which is not acetylated. In addition, whether the change from Gly136 to Ala-136 in the two types of gamma genes has a role in switching remains to be studied. However, it is clear that considering both acetylation and the presence of two types of gamma subunits, fetal hemoglobin represents a heterogenous system whereas adult hemoglobin is not; i.e. an energetically more favorable state likely to favor its formation.

Conclusions

Hemoglobin has proven to be a continuing source of new and important information when viewed from different perspectives with novel approaches. Although the oxygen transport mechanism is well known for the adult type of hemoglobin, increased oxygen binding by the embryonic hemoglobins involves an additional parameter; a looser tetrameric structure that leads to higher oxygen affinity. Hence, the concept put forth here is that the hemoglobin structure is a platform capable of variable oxygen carrying abilities as a result of subtle alterations holding it together, i.e., the bonding between the subunits that comprise the tetramer. Thus, not all natural hemoglobin tetramers are alike in terms of their subunit strengths although their overall structural architectures may be practically identical and this flexibility is the source of distinct properties for a given hemoglobin type, which can provide it with advantages in a particular physiological circumstance, such as sickle cell anemia, malaria, and ontogeny.

References

Adachi K, Konitzer P, Pang J, Reddy KS, Surrey S (1997) Amino acids responsible for decreased 2,3-biphosphoglycerate binding to fetal hemoglobin. Blood 90(8):2916–2920

Adachi K, Pang J, Konitzer P, Surrey S (1996) Polymerization of recombinant hemoglobin F gamma E6V and hemoglobin F gamma E6V, gamma Q87T alone, and in mixtures with hemoglobin S. Blood 87(4):1617–1624

Adachi K, Zhao Y, Yamaguchi T, Surrey S (2000) Assembly of gamma- with alpha-globin chains to form human fetal hemoglobin in vitro and in vivo. J Biol Chem 275(17):12424–12429. https://doi.org/10.1074/jbc.C000137200

Akinsheye I, Alsultan A, Solovieff N, Ngo D, Baldwin CT, Sebastiani P, Chui DH, Steinberg MH (2011) Fetal hemoglobin in sickle cell anemia. Blood 118(1):19–27. https://doi.org/10.1182/blood-2011-03-325258

Allison AC (1954) Protection afforded by sickle-cell trait against subtertian malareal infection. Br Med J 1(4857):290–294. https://doi.org/10.1136/bmj.1.4857.290

Antonini E, Brunori M (1971) Hemoglobin and myoglobin in their reactions with ligands. Elsevier Science Publishing Co., New York

Arnone A (1972) X-ray diffraction study of binding of 2,3-diphosphoglycerate to human deoxyhaemoglobin. Nature 237(5351):146–149

Baron MH (1996) Developmental regulation of the vertebrate globin multigene family. Gene Expr 6(3):129–137

Benesch R, Benesch RE (1967) The effect of organic phosphates from the human erythrocyte on the allosteric properties of hemoglobin. Biochem Biophys Res Commun 26(2):162–167

Bookchin RM, Nagel RL (1971) Ligand-induced conformational dependence of hemoglobin in sickling interactions. J Mol Biol 60(2):263–270. https://doi.org/10.1016/0022-2836(71)90292-0

Bookchin RM, Nagel RL, Balazs T (1975) Role of hybrid tetramer formation in gelation of haemoglobin S. Nature 256(5519):667–668. https://doi.org/10.1038/256667a0

Brendel C, Guda S, Renella R, Bauer DE, Canver MC, Kim YJ, Heeney MM, Klatt D, Fogel J, Milsom MD, Orkin SH, Gregory RI, Williams DA (2016) Lineage-specific BCL11A knockdown circumvents toxicities and reverses sickle phenotype. J Clin Invest 126(10):3868–3878. https://doi.org/10.1172/jci87885

Brittain T (2002) Molecular aspects of embryonic hemoglobin function. Mol Aspects Med 23(4):293–342. https://doi.org/10.1016/S0098-2997(02)00004-3

Bunn HF (1969) Subunit dissociation of certain abnormal human hemoglobins. J Clin Invest 48(1):126–138. https://doi.org/10.1172/jci105961

Bunn HF, Briehl RW (1970) The interaction of 2,3-diphosphoglycerate with various human hemoglobins. J Clin Invest 49(6):1088–1095. https://doi.org/10.1172/jci106324

Bunn HF, Forget BG (1986) Hemoglobin: Molecular, genetic, and clinical aspects. W.B Saunders, Philadelphia, PA

Chen W, Dumoulin A, Li X, Padovan JC, Chait BT, Buonopane R, Platt OS, Manning LR, Manning JM (2000) Transposing sequences between fetal and adult hemoglobins indicates which subunits and regulatory molecule interfaces are functionally related. Biochemistry 39(13):3774–3781. https://doi.org/10.1021/bi992691l

Demirci S, Uchida N, Tisdale JF (2018) Gene therapy for sickle cell disease: An update. Cytotherapy 20(7):899–910. https://doi.org/10.1016/j.jcyt.2018.04.003

DeSimone J, Heller P, Hall L, Zwiers D (1982) 5-Azacytidine stimulates fetal hemoglobin synthesis in anemic baboons. Proc Natl Acad Sci U S A 79(14):4428–4431

Dumoulin A, Manning LR, Jenkins WT, Winslow RM, Manning JM (1997) Exchange of subunit interfaces between recombinant adult and fetal hemoglobins. Evidence for a functional interrelationship among regions of the tetramer. J Biol Chem 272 (50):31326–31332

Dumoulin A, Padovan JC, Manning LR, Popowicz A, Winslow RM, Chait BT, Manning JM (1998) The N-terminal sequence affects distant helix interactions in hemoglobin. Implications for mutant proteins from studies on recombinant hemoglobin felix. J Biol Chem 273 (52):35032–35038. https://doi.org/10.1074/jbc.273.52.35032

Dyson FW, Eddington AS (1919) Davidson C (1920) A determination of the deflection of light by the sun's gravitational field, from observations made at the total eclipse of May 29. Philos Trans R Soc Lond 220(571–581):291–333. https://doi.org/10.1098/rsta.1920.0009

Eaton WA, Bunn HF (2017) Treating sickle cell disease by targeting HbS polymerization. Blood 129(20):2719–2726. https://doi.org/10.1182/blood-2017-02-765891

Edelstein SJ, Rehmar MJ, Olson JS, Gibson QH (1970) Functional aspects of the subunit association-dissociation equilibria of hemoglobin. J Biol Chem 245(17):4372–4381

Fang TY, M. Z, Simplaceanu V, Ho NT, Ho C (1999) Assessment of roles of surface histidyl residues in the molecular basis of the Bohr effect and of beta 143 histidine in the binding of 2,3-bisphosphoglycerate in human normal adult hemoglobin. Biochemistry 38 (40):13423–13432. https://doi.org/10.1021/bi9911379

Frier JA, Perutz MF (1977) Structure of human foetal deoxyhaemoglobin. J Mol Biol 112(1):97–112

Genbacev O, Zhou Y, Ludlow JW, Fisher SJ (1997) Regulation of human placental development by oxygen tension. Science 277(5332):1669–1672. https://doi.org/10.1126/science.277.5332.1669

Groudine M, Eisenman R, Weintraub H (1981) Chromatin structure of endogenous retroviral genes and activation by an inhibitor of DNA methylation. Nature 292(5821):311–317

Harrington DJ, Adachi K, Royer WE Jr (1997) The high resolution crystal structure of deoxyhemoglobin S. J Mol Biol 272(3):398–407. https://doi.org/10.1006/jmbi.1997.1253

He Z, Russell JE (2001) Expression, purification, and characterization of human hemoglobins Gower-1 (zeta(2)epsilon(2)), Gower-2 (alpha(2)epsilon(2)), and Portland-2 (zeta(2)beta(2)) assembled in complex transgenic-knockout mice. Blood 97(4):1099–1105

Higgs DR, Garrick D, Anguita E, De Gobbi M, Hughes J, Muers M, Vernimmen D, Lower K, Law M, Argentaro A, Deville MA, Gibbons R (2005) Understanding alpha-globin gene regulation: Aiming to improve the management of thalassemia. Ann N Y Acad Sci 1054:92–102. https://doi.org/10.1196/annals.1345.012

Hofmann O, Brittain T (1996) Ligand binding kinetics and dissociation of the human embryonic haemoglobins. Biochem J 315(Pt 1):65–70

Hofmann O, Mould R, Brittain T (1995) Allosteric modulation of oxygen binding to the three human embryonic haemoglobins. Biochem J 306(Pt 2):367–370

Huehns ER, Beaven GH, Stevens BL (1964) Recombination studies on haemoglobins at neutral pH. Biochem J 92(2):440–444

Huehns ER, Faroqui AM (1975) Oxygen dissociation properties of human embryonic red cells. Nature 254(5498):335–337

Huehns ER, Shooter EM (1965) Human haemoglobins. J Med Genet 2(1):48–90

Ikuta T, Papayannopoulou T, Stamatoyannopoulos G, Kan YW (1996) Globin gene switching. In vivo protein-DNA interactions of the human beta-globin locus in erythroid cells expressing the fetal or the adult globin gene program. J Biol Chem 271 (24):14082–14091. https://doi.org/10.1074/jbc.271.24.14082

Kidd RD, Mathews A, Baker HM, Brittain T, Baker EN (2001) Subunit dissociation and reassociation leads to preferential crystallization of haemoglobin Bart's (gamma4) from solutions of human embryonic haemoglobin Portland (zeta2gamma2) at low pH. Acta Crystallogr D Biol Crystallogr 57(Pt 6):921–924

Lavelle D, Saunthararajah Y, Desimone J (2008) DNA methylation and mechanism of action of 5-azacytidine. Blood 111 (4):2485; author reply 2486. https://doi.org/10.1182/blood-2007-10-119867

Leonard A, Tisdale JF (2018) Stem cell transplantation in sickle cell disease: therapeutic potential and challenges faced. Expert Rev Hematol 11(7):547–565. https://doi.org/10.1080/17474086.2018.1486703

Liu N, Hargreaves VV, Zhu Q, Kurland JV, Hong J, Kim W, Sher F, Macias-Trevino C, Rogers JM, Kurita R, Nakamura Y, Yuan GC, Bauer DE, Xu J, Bulyk ML, Orkin SH (2018) Direct Promoter

Repression by BCL11A Controls the Fetal to Adult Hemoglobin Switch. Cell 173 (2):430–442 e417. https://doi.org/10.1016/j.cell.2018.03.016

Lettre G, Bauer DE (2016) Fetal haemoglobin in sickle-cell disease: from genetic epidemiology to new therapeutic strategies. Lancet 387(10037):2554–2564. https://doi.org/10.1016/s0140-6736(15)01341-0

Luzzatto L (2012) Sickle cell anaemia and malaria. Mediterr J Hematol Infect Dis 4(1):e2012065. https://doi.org/10.4084/mjhid.2012.065

Manning JM, Dumoulin A, Li X, Manning LR (1998) Normal and abnormal protein subunit interactions in hemoglobins. J Biol Chem 273(31):19359–19362. https://doi.org/10.1074/jbc.273.31.19359

Manning JM, Dumoulin A, Manning LR, Chen W, Padovan JC, Chait BT, Popowicz A (1999) Remote contributions to subunit interactions: lessons from adult and fetal hemoglobins. Trends Biochem Sci 24(6):211–212. https://doi.org/10.1016/S0968-0004(99)01395-X

Manning JM, Popowicz AM, Padovan JC, Chait BT, Manning LR (2012) Intrinsic regulation of hemoglobin expression by variable subunit interface strengths. FEBS J 279(3):361–369. https://doi.org/10.1111/j.1742-4658.2011.08437.x

Manning LR, Jenkins WT, Hess JR, Vandegriff K, Winslow RM, Manning JM (1996) Subunit dissociations in natural and recombinant hemoglobins. Protein Sci 5(4):775–781. https://doi.org/10.1002/pro.5560050423

Manning LR, Manning JM (2001) The acetylation state of human fetal hemoglobin modulates the strength of its subunit interactions: long-range effects and implications for histone interactions in the nucleosome. Biochemistry 40(6):1635–1639. https://doi.org/10.1021/bi002157+

Manning LR, Popowicz AM, Padovan J, Chait BT, Russell JE, Manning JM (2010) Developmental expression of human hemoglobins mediated by maturation of their subunit interfaces. Protein Sci 19(8):1595–1599. https://doi.org/10.1002/pro.441

Manning LR, Popowicz AM, Padovan JC, Chait BT, Manning JM (2017) Gel filtration of dilute human embryonic hemoglobins reveals basis for their increased oxygen binding. Anal Biochem 519:38–41. https://doi.org/10.1016/j.ab.2016.12.008

Manning LR, Russell JE, Padovan JC, Chait BT, Popowicz A, Manning RS, Manning JM (2007) Human embryonic, fetal, and adult hemoglobins have different subunit interface strengths. Correlation with lifespan in the red cell. Protein Sci 16 (8):1641–1658. https://doi.org/10.1110/ps.072891007

Manning LR, Russell JE, Popowicz AM, Manning RS, Padovan JC, Manning JM (2009) Energetic differences at the subunit interfaces of normal human hemoglobins correlate with their developmental profile. Biochemistry 48(32):7568–7574. https://doi.org/10.1021/bi900857r

Mosca A, Paleari R, Leone D, Ivaldi G (2009) The relevance of hemoglobin F measurement in the diagnosis of thalassemias and related hemoglobinopathies. Clin Biochem 42(18):1797–1801. https://doi.org/10.1016/j.clinbiochem.2009.06.023

Musallam KM, Taher AT, Cappellini MD, Sankaran VG (2013) Clinical experience with fetal hemoglobin induction therapy in patients with beta-thalassemia. Blood 121 (12):2199–2212; quiz 2372. https://doi.org/10.1182/blood-2012-10-408021

Nagel RL, Bookchin RM, Johnson J, Labie D, Wajcman H, Isaac-Sodeye WA, Honig GR, Schiliro G, Crookston JH, Matsutomo K (1979) Structural bases of the inhibitory effects of hemoglobin F and hemoglobin A2 on the polymerization of hemoglobin S. Proc Natl Acad Sci U S A 76(2):670–672

Padlan EA, Love WE (1985) Refined crystal structure of deoxyhemoglobin S. I. Restrained least-squares refinement at 3.0-A resolution. J Biol Chem 260 (14):8272–8279

Pasvol G, Weatherall DJ, Wilson RJ, Smith DH, Gilles HM (1976) Fetal haemoglobin and malaria. Lancet 1(7972):1269–1272

Perutz MF (1989) Mechanisms of cooperativity and allosteric regulation in proteins. Q Rev Biophys 22(2):139–237

Poillon WN, Kim BC, Rodgers GP, Noguchi CT, Schechter AN (1993) Sparing effect of hemoglobin F and hemoglobin A2 on the polymerization of hemoglobin S at physiologic ligand saturations. Proc Natl Acad Sci U S A 90(11):5039–5043. https://doi.org/10.1073/pnas.90.11.5039

Poyart C, Bursaux E, Guesnon P, Teisseire B (1978) Chloride binding and the Bohr effect of human fetal erythrocytes and HbFII solutions. Pflugers Arch 376(2):169–175

Randhawa ZI, Jones RT, Lie-Injo LE (1984) Human hemoglobin Portland II (zeta 2 beta 2). Isolation and characterization of Portland hemoglobin components and their constituent globin chains. J Biol Chem 259 (11):7325–7330

Ribeil JA, Hacein-Bey-Abina S, Payen E, Magnani A, Semeraro M, Magrin E, Caccavelli L, Neven B, Bourget P, El Nemer W, Bartolucci P, Weber L, Puy H, Meritet JF, Grevent D, Beuzard Y, Chretien S, Lefebvre T, Ross RW, Negre O, Veres G, Sandler L, Soni S, de Montalembert M, Blanche S, Leboulch P, Cavazzana M (2017) Gene Therapy in a Patient with Sickle Cell Disease. N Engl J Med 376(9):848–855. https://doi.org/10.1056/NEJMoa1609677

Sankaran VG, Menne TF, Xu J, Akie TE, Lettre G, Van Handel B, Mikkola HK, Hirschhorn JN, Cantor AB, Orkin SH (2008) Human fetal hemoglobin expression is regulated by the developmental stage-specific repressor BCL11A. Science 322(5909):1839–1842. https://doi.org/10.1126/science.1165409

Sankaran VG, Xu J, Orkin SH (2010) Advances in the understanding of haemoglobin switching. Br J Haematol 149(2):181–194. https://doi.org/10.1111/j.1365-2141.2010.08105.x

Scheepens A, Mould R, Hofmann O, Brittain T (1995) Some effects of post-translational N-terminal acetylation of the human embryonic zeta globin protein. Biochem J 310(Pt 2):597–600

Schroeder WA, Huisman TH, Shelton JR, Shelton JB, Kleihauer EF, Dozy AM, Robberson B (1968) Evidence for multiple structural genes for the gamma chain of human fetal hemoglobin. Proc Natl Acad Sci U S A 60(2):537–544

Shear HL, Grinberg L, Gilman J, Fabry ME, Stamatoyannopoulos G, Goldberg DE, Nagel RL (1998) Transgenic mice expressing human fetal globin are protected from malaria by a novel mechanism. Blood 92(7):2520–2526

Simon MC, Keith B (2008) The role of oxygen availability in embryonic development and stem cell function. Nat Rev Mol Cell Biol 9(4):285–296. https://doi.org/10.1038/nrm2354

Stamatoyannopoulos G, Grosveld F (2001) Hemoglobin switching. In: Majerus PW, Perlmutter RM, Varmus H (eds) Stamatoyannopoulos G. The molecular basis of blood diseases. W.B. Saunders Co., Philadelphia, PA

Steinberg MH, Chui DH, Dover GJ, Sebastiani P, Alsultan A (2014) Fetal hemoglobin in sickle cell anemia: a glass half full? Blood 123(4):481–485. https://doi.org/10.1182/blood-2013-09-528067

Sutherland-Smith AJ, Baker HM, Hofmann OM, Brittain T, Baker EN (1998) Crystal structure of a human embryonic haemoglobin: the carbonmonoxy form of Gower II (alpha2 epsilon2) haemoglobin at 2.9 A resolution. J Mol Biol 280 (3):475–484

Telen MJ, Malik P, Vercellotti GM (2019) Therapeutic strategies for sickle cell disease: towards a multi-agent approach. Nat Rev Drug Discov 18(2):139–158. https://doi.org/10.1038/s41573-018-0003-2

van der Ploeg LH, Flavell RA (1980) DNA methylation in the human gamma delta beta-globin locus in erythroid and nonerythroid tissues. Cell 19(4):947–958. https://doi.org/10.1016/0092-8674(80)90086-0

Wood WG (1976) Haemoglobin synthesis during human fetal development. Br Med Bull 32(3):282–287

Yagami T, Ballard BT, Padovan JC, Chait BT, Popowicz AM, Manning JM (2002) N-terminal contributions of the gamma-subunit of fetal hemoglobin to its tetramer strength: remote effects at subunit contacts. Protein Sci 11(1):27–35. https://doi.org/10.1110/ps.30602

Zimmerman JK, Ackers GK (1971) Molecular sieve studies of interacting protein systems. X. Behavior of small zone profiles for reversibly self-associating solutes. J Biol Chem 246 (23):7289–7292

Chapter 12
Sickle Cell Hemoglobin

Amit Kumar Mandal, Amrita Mitra and Rajdeep Das

Abstract Sickle cell hemoglobin (HbS) is an example of a genetic variant of human hemoglobin where a point mutation in the β globin gene results in substitution of glutamic acid to valine at sixth position of the β globin chain. Association between tetrameric hemoglobin molecules through noncovalent interactions between side chain residue of βVal6 and hydrophobic grooves formed by βAla70, βPhe85 and βLeu88 amino acid residues of another tetramer followed by the precipitation of the elongated polymer leads to the formation of sickle-shaped RBCs in the deoxygenated state of HbS. There are multiple non-covalent interactions between residues across intra- and inter-strands that stabilize the polymer. The clinical phenotype of sickling of RBCs manifests as sickle cell anemia, which was first documented in the year 1910 in an African patient. Although the molecular reason of the disease has been understood well over the decades of research and several treatment procedures have been explored to date, an effective therapeutic strategy for sickle cell anemia has not been discovered yet. Surprisingly, it has been observed that the oxy form of HbS and glutathionylated form of deoxy HbS inhibits polymerization. In addition to describe the residue level interactions in the HbS polymer that provides its stability, here we explain the mechanism of inhibition in the polymerization of HbS in its oxy state. Additionally, we reported the molecular insights of inhibition in the polymerization for glutathionyl HbS, a posttranslational modification of hemoglobin, even in its deoxy state. In this chapter we briefly consider the available treatment procedures of sickle cell anemia and propose that the elevation of glutathionylation of HbS within RBCs, without inducing oxidative stress, might be an effective therapeutic strategy for sickle cell anemia.

A. K. Mandal (✉)
Department of Biological Sciences, Indian Institute of Science Education and Research Kolkata, Mohanpur 741246, Nadia, West Bengal, India
e-mail: amitkm@iiserkol.ac.in

A. Mitra · R. Das
Clinical Proteomics Unit, Division of Molecular Medicine, St. John's Research Institute, St. John's National Academy of Health Sciences, 100 ft road, Koramangala, Bangalore 560034, India
e-mail: amrita@sjri.res.in

R. Das
e-mail: rajdeep@sjri.res.in

U. Hoeger and J. R. Harris (eds.), *Vertebrate and Invertebrate Respiratory Proteins, Lipoproteins and other Body Fluid Proteins*, Subcellular Biochemistry 94,
https://doi.org/10.1007/978-3-030-41769-7_12

Keywords Sickle cell anemia · Sickle cell hemoglobin · Glutathionyl sickle hemoglobin · Hydrogen/deuterium exchange mass spectrometry · Oxygen dissociation curve · Polymerization

Introduction

Human hemoglobin is a tetrameric protein consisting of α and β globin chains in duplicate. Each of the globin chain is noncovalently bound to a porphyrin nucleus, where Fe (II) is coordinated at the centre of the ring. Systematic association and dissociation of oxygen with Fe (II) tuned to the partial pressure of dissolved oxygen in blood enables the hemoglobin molecule to be a carrier of oxygen from lungs to tissues. Hemoglobinopathies are examples of human genetic disorders where mutations in the globin genes lead to substitution, deletion or insertion of amino acids in the polypeptide chains, thereby resulting in hemoglobin variants. There are 1346 hemoglobin variants discovered to date, where 315 variants are reported with functional disorders such as differential affinity for oxygen binding, unstable tetramers of hemoglobin or formation of methemoglobin (http://globin.cse.psu.edu/hbvar/menu.html).

Sickle cell hemoglobin (HbS) is an example of a genetic variant of human hemoglobin, where glutamic acid at the sixth position of β globin chain is substituted with valine, a hydrophobic amino acid residue. This is an example of missense mutation, where a single nucleotide change from A to U in either GAA or GAG, codons of glutamic acid, to GUA or GUG, codons of valine, results in a replacement of glutamic acid at sixth position of the β globin chain to valine. Sickle cell anemia, also known as sickle cell disease, is caused by the polymerization of HbS followed by its precipitation inside red blood cells (RBCs) resulting in alteration in the morphology of RBCs under low partial pressures of oxygen. Several treatment strategies have been developed over decades; however, no effective therapy for the sickle cell disease has been discovered yet.

Sickle Cell Anemia

The clinical phenotype of sickle cell anemia was first observed in 1670 in an African family. However, the abnormalities in the in vivo synthesis of hemoglobin associated with sickle cell anemia were documented in 1910 in a West Indian student in Chicago. In the year 1949, Linus Pauling first demonstrated that sickle cell anemia is a molecular disease that might be caused by the abnormality in hemoglobin present in patient's blood (Pauling et al. 1949). In 1957, Vernon Ingram reported that in the N-terminal fragment of β globin chain of HbS, glutamic acid, a negatively charged residue, at sixth position is replaced by valine, a neutral residue (Ingram 1956). The polymerization of deoxy HbS results in the formation of fibrils. Subsequently, the bending of the fibril produces sickle-shaped RBCs in patients with sickle cell anemia. Sickling of RBCs leads to a high rate of hemolysis and loss in their elasticity during circulation across fine capillaries which eventually blocks the capillaries and occludes veins that manifests as severe pain, stroke and even organ damages. It has been reported that in general, survival of patients in homozygous sickle cell disease is less than 50 years (Platt et al. 1994).

In general, hemoglobin variant genes produced by the mutations are known as alleles. The genetic heterogeneity of different alleles within a population is known as polymorphism. Most of the hemoglobin variants are rare and without any clinical manifestation. However, some are responsible for genetic diseases and sickle cell anemia is one such example. In the heterozygous state of sickle cell disease, an individual carries one copy each of a normal β globin allele and sickle β allele. RBCs of those individuals contain both the normal adult human hemoglobin (HbA) and HbS. These subjects do not suffer from sickle cell anemia and are resistant to malaria. However, in the homozygous state there are two copies of sickle β alleles, thereby producing only HbS, consisting of both β globin chains as mutant chains, β^S, and HbA is completely absent. They are also resistant to malaria but suffer from sickle cell anemia. Thus, heterozygotes have advantage over homozygotes in malarial infection. This is perhaps the reason for the persistence of the sickle cell gene within the population over centuries, especially in the continents exposed to malarial infection. In the double heterozygous states, a second genetic variant of β globin chain is found to be present along with β^S in the tetrameric hemoglobin molecule.

Characterization of HbS

In a diagnostic laboratory the identification and characterization of HbS in patients with sickle cell anemia is routinely performed using gel electrophoresis and/or an automated cation exchange chromatographic method. Based on the net reduction in negative charge on the HbS molecules compared to HbA tetramers, caused by the replacement of a negatively charged glutamic acid residue by a neutral valine, HbA migrates faster and ahead of HbS in alkaline gel electrophoresis (Ou and Rognerud 2001). However, co-migration of HbS along with other hemoglobin variants such as HbD, HbQ, Hb Lepore etc. often leads to ambiguity in its identification using this method (Mathew et al. 2011).

In the cation exchange chromatographic method, varying concentrations of phosphate buffers are used as mobile phase and are allowed to pass under pressure through a cation exchange resin that acts as the stationary phase. Here, tetrameric hemoglobin variant molecules are separated according to their interaction with the stationary phase. HbA, having a less cationic nature compared to HbS, elutes faster than HbS in the cation exchange chromatographic separation technique (Joutovsky et al. 2004). In the commercially available BioRad D10 instrument that works on the principle of cation exchange chromatography, HbA was observed to elute in the retention time window of 1.9–3.1 min, whereas HbS eluted between 4.3 and 4.7 min (Joutovsky et al. 2004). However, co-elution of different variants in cation exchange chromatographic platforms, such as HbA_2 along with glycosylated HbS, results in ambiguity in the analysis (da Fonseca et al. 2015).

Reverse phase chromatography based separation of proteolytic peptides in nano liquid chromatographic system followed by mass analysis in the proteomics platform provides unambiguous characterisation of hemoglobin variants at the amino acid residue level. However, the high sequence homology between normal and hemoglobin variants might lead to challenges in the analysis of mass spectrometric data. To overcome this hurdle, we developed a customized database consisting of only the signature peptides consisting the mutated amino acid residues of all hemoglobin variants discovered to date. We created individual databases of the signature peptides of hemoglobin variants corresponding to different proteolytic enzymes such as trypsin, chymotrypsin and GluC. The customized databases are compatible with proteomics search engine softwares and might be used for the routine diagnosis of hemoglobin variants. Using the customized databases, which are the first of their kind, we have successfully characterized HbS and several other hemoglobin variants that remained ambiguously defined by the conventional methods (Das et al. 2013, 2015, 2016).

Oxygen Binding Equilibrium of HbS

Functional activity of hemoglobin variants is assessed through oxygen binding equilibrium, where the fraction of hemoglobin saturated with oxygen is measured against varying partial pressure of dissolved oxygen and subsequently compared with that of HbA. Oxygen affinity of hemoglobin is characterised by p_{50}, partial pressure of oxygen required to saturate hemoglobin by 50%. In general, p_{50} depends on the quaternary structure of the hemoglobin molecule, pH, temperature and concentration of 2,3-diphosphoglycerate (2,3-DPG), an allosteric regulator. p_{50} of HbA in presence of equimolar amount of 2,3-DPG at 37 °C is 26.6 mmHg with Hill coefficient $n = 2.5$, that measures the heme-heme interactions between tetramers. For hemoglobin variants, a decrease in p_{50} indicates an increase in the oxygen affinity while an increase in p_{50} implies a decrease in the oxygen affinity. p_{50} and n for HbS were observed to be 27.3 mmHg and 3.0, respectively (Das et al. 2018). It is well known that compared to healthy volunteers, RBCs in sickle cell disease have reduced oxygen affinity (Eaton and Hofrichter 1987). This decrease in oxygen affinity is contributed, at least partly, by the increase in intracellular concentration of 2,3-DPG, which is a compensatory mechanism commonly observed in all forms of anemia to facilitate the release of oxygen in tissues. Both p_{50} and the concentration of 2,3-DPG vary widely in patients with sickle cell anemia. This in turn contributes in the decrease in the solubility of HbS and subsequently in the sickling of RBCs under hypoxic condition (Safo and Kato 2014).

Polymerization of HbS

The polymerization of HbS is a dynamic event. In the deoxy state of HbS, polymerization initiates through the binding of hydrophobic patch formed by the isopropyl group of the side chain of βVal6 in one of the β^S globin chains of a HbS tetramer with the hydrophobic pocket formed by βAla70, βPhe85 and βLeu88 residues of another hemoglobin tetramer, whose hydrophobic patch binds to the hydrophobic groove of a third HbS molecule (Branden and Tooze 1998; Ghatge et al. 2016). Consecutive additions of HbS tetramers in the aggregate results in a double-stranded polymer chain (Wishner et al. 1975). Subsequently, the combination of seven such half-staggered paired strands results in a long helical rope-like fourteen stranded insoluble fiber (Noguchi and Schechter 1985). The intermolecular interaction between different residues across inter-strands (lateral) and intra-strands (axial) of HbS tetramers stabilize the fiber (Wishner et al. 1975; Padlan and Love 1985; Harrington et al. 1997; Ghatge et al. 2016). The axial and lateral contacts of the HbS crystal structure have been reported (Harrington et al. 1997). The axial contacts are located between tetramers vertically along the axis of the crystal, whereas two closely packed strands are stabilized by the lateral contacts. β subunits are assigned according to their contacts, i.e. β^S subunits acting as acceptor pockets for mutant valine residues are designated as β_1^S and those acting as valine donors are termed β_2^S. α chains are named according to their positions relative to the β^S chains. The residue pairs involved in axial and lateral contacts with the nature of interactions are shown in Table 12.1.

Hydrophobic interactions across residues in the polymer make the process an endothermic. Thus, the polymerization is driven by the entropy change where release of solvated water molecules from the surface of hemoglobin tetramers to the bulk solution contributes in the decrease in free energy of the system. The kinetics of polymerization has been found to have a delay time (t_d), the lag period required for the aggregation of about 15–30 tetramers of hemoglobin molecule to form a single nucleus in the polymerization step of individual fibers. This mechanism of polymerization is defined as homogeneous nucleation. On the other hand, aggregation of tetramers on the surface of existing fiber provides more nuclei for the formation of polymer and the mechanism of polymerization is known as heterogeneous nucleation (Hoffman et al. 2017). After the delay time, polymer formation proceeds exponentially and the fiber growth continues approximately up to 250 tetramers. During polymerization, homogeneous nucleation predominates in a highly concentrated solution of deoxy forms of HbS whereas heterogeneous nucleation is favoured in a relatively less concentrated solution. Slow deoxygenation followed by the precipitation of the fiber within the cell results in a classic elongated sickle-shaped RBCs whereas rapid deoxygenation leads to granular or cobblestone texture of the RBCs with no major distortion in the shape of the cell (Lauer et al. 1980; Noguchi and Schechter 1985; Nagel 2003). In the nucleation processes studied in vitro, 't_d' depends on the protein concentration, temperature and ionic strength of the solution. (Hofrichter et al. 1974; Ferrone et al. 1985b; Christoph et al. 2005). In general, 't_d' is inversely proportional to the intracellular concentration of hemoglobin (Bunn 1997). In our study, t_d of

Table 12.1 Axial and lateral contacts in the crystal structure of deoxy sickle hemoglobin

Axial contacts	Nature of interaction	Lateral contacts	Nature of interaction
α_1 Pro114–α_2 Lys16	H-bond	β_1^S Lys66–β_2^S Pro5	van der Waals interaction
α_1 Ala115–α_2 Lys16	van der Waals interaction	β_1^S Gly69–β_2^S Pro5	van der Waals interaction
α_1 Pro114–α_2 Glu116	van der Waals interaction	β_1^S Ala70–β_2^S Pro5	van der Waals interaction
β_1^S Gly16–β_2^S Lys120	van der Waals interaction	β_1^S Asp73–β_2^S Thr4	H-bond
β_1^S Lys17–β_2^S Phe118	H-bond	β_1^S Asp73–β_2^S Val6	H-bond
β_1^S Lys17–β_2^S His117	H-bond	β_1^S Asp79–α_2 Ser49	van der Waals interaction
β_1^S Lys17–β_2^S His116	van der Waals interaction	β_1^S Asp79–α_2 His50	H-bond
β_1^S Val18–β_2^S Lys120	van der Waals interaction	β_1^S Asn80–α_2 His50	H-bond
β_1^S Glu22–α_2 His20	H-bond	β_1^S Gly83–β_2^S Pro125	van der Waals interaction
β_1^S His117–α_2 Pro114	H-bond	β_1^S Thr84–β_2^S Val6	H-bond
β_1^S His117–α_2 Ala115	van der Waals interaction	β_1^S Phe85–β_2^S Val6	H-bond
β_1^S Phe118–α_2 Ala115	van der Waals interaction	β_1^S Thr87–β_2^S Ala13	H-bond
β_1^S Phe118–α_2 Pro114	van der Waals interaction	β_1^S Thr87–β_2^S Ala10	H-bond
β_1^S Gly119–α_2 Ala115	van der Waals interaction	β_1^S Thr87–β_2^S Val126	H-bond
β_1^S Glu121–α_2 Pro114	van der Waals interaction	β_1^S Leu88–β_2^S Ser9	H-bond
		β_1^S Glu90–β_2^S Lys17	Ion pair interaction
		β_1^S Lys95–β_2^S Lys17	van der Waals interaction

deoxy HbS was found to be 47.76 s and 0.11 s for 10 and 20 μM sets of HbS, respectively (Das et al. 2018). On the other hand, the rate of polymer formation in vivo is dependent on the extent of deoxygenation, intracellular concentration of HbS, and the concentration of fetal hemoglobin (HbF) within the RBCs (Bunn 1997). The pathophysiology of sickle cell anemia depends on the polymerization kinetics and the perturbation in the duration of blood flow that causes cellular transit time to exceed the delay time of polymerization (Hoffman et al. 2017). Eventually, sickle RBCs become rigid and block the capillaries or occlude veins. As a result, sickle cell patients suffer from severe pain, stroke and chronic organ damages (Hebbel 1994; Kim-Shapiro et al. 1999). It is interesting to emphasise that the polymerization does not occur in the oxy state of HbS. Due to unavailability of the crystal structure of oxy form of HbS, the molecular insights of the inhibition of polymerization in the oxy state remained unknown until we showed recently that both the destruction of the groove structure and the increase in the dynamics of conformational flexibility of the hydrophobic patch of HbS do not favour the association of tetramers of HbS in its oxy state (Das et al. 2017). Additionally, covalent binding of a glutathione moiety to βCys93 of HbS molecule prolongs the delay time of the polymerization (Craescu et al. 1986; Embury 1994; Ghatge et al. 2016). Similar to the oxy form of HbS, we have also shown the molecular basis of inhibition caused by the glutathionylation of HbS, even in the deoxy state of hemoglobin (Das et al. 2018).

Inhibition in HbS Polymerization

Polymerization of HbS has been found to have a direct impact on the pathophysiology of sickle cell anemia (Adachi 1979a; Ferrone et al. 1985a, b). It has been reported that the polymerization becomes inhibited if 't_d' associated with the nucleation of tetramers is longer than the transit time of RBCs required in the deoxygenation of HbS through circulation across peripheral capillaries to alveoli in the lungs, where reoxygenation takes place. The stabilization of the oxy form of HbS is a challenging problem in the drug discovery for sickle cell anemia as the oxy form does not participate in the polymerization. Thus, it is crucial to understand the molecular insights of the oxy forms of HbS in terms of its conformational dynamics in the solution phase. Hydrogen/deuterium exchange (H/DX) based mass spectrometry is a potential tool in the analysis of conformation of large molecules, where the exchange kinetics of the backbone amide hydrogen of a polypeptide chain with its isotope, deuterium from solvent (D_2O) provides the information on the spatial orientation of those groups and their accessibility to the solvent. Thus, H/DX kinetics data enable us to assess the conformational flexibility/rigidity of the polypeptide backbone of different proteolytic peptides originated from the experimental protein molecule of interest. Although H/DX kinetics is monitored at the proteolytic peptide level of a protein, the isotope exchange occurs within the intact protein molecule in solution phase. Thus, exchange kinetics data can be translated in terms of the conformational dynamics of the functionally active form of the experimental protein molecule. In general, an increase in the isotope exchange rate indicates an increase in conformational flexibility whereas the decrease in the same implies an increase in the rigidity across the respective region of the molecule.

The key event in the polymerization of HbS is the association between the hydrophobic patch contributed by the mutated residue, βVal6, of one tetramer with the hydrophobic pocket formed by β globin residues, Ala70, Phe85 and Leu88, of another tetramer of hemoglobin. Thus, the appropriate stereospecific orientation of the isopropyl head group of βVal6 is crucial in the polymerization. H/DX kinetics of HbS showed that the dissociation of 2,3-DPG upon oxygenation of the tetrameric hemoglobin results in a large increase in the conformational flexibility of N-terminal region of β^S globin chain containing two of the interacting residues, βVal1 and βHis2 with the alloesteric regulator, 2,3-DPG (Craescu et al. 1986; Das et al. 2017). Thus, there might be a loss in the required stereospecific orientation of the side chain groups of βVal6 upon oxygenation of HbS. In addition, the H/DX kinetics of a peptic peptide containing residues β86-102 showed a large increase in the conformational flexibility of amino acid residues across the hydrophobic pocket in the oxy forms of HbS, thereby indicating that the constituent residues of the hydrophobic pocket might have been exposed to the solvent resulting in disappearance of the pocket in the functionally active quaternary structure of oxy form of HbS (Table 12.2) (Das et al. 2017). Therefore, the loss in the required spatial orientation of the hydrophobic patch and the disappearance of the hydrophobic pocket are crucial factors behind non-participation of oxy form of HbS in the polymerization.

Table 12.2 Analysis of H/DX kinetic parameters of peptic peptides of oxy and deoxy sickle hemoglobin

Peptide molecular ion m/z (charge state) Mass in Da	Residues	Change in conformational dynamics $\sum\{(k_iP_i)_{oxyHbS} - (k_iP_i)_{doxyHbS}\}$	Inference (Deoxy HbS → OxyHbS)
732.9 m/z (+2) 1463.8 Da	β1–14 VHLTPVEKSAVTAL	+201.78	Flexible
640.9 m/z (+3) 1920.0 Da	β86–102 ATLSELHCDKLHVDPEN	+233.29	Flexible
529.3 m/z (+3) 1584.9 Da	α34–46 LSFPTTKTYFPHF	+204.67	Flexible
728.3 m/z (+4) 2909.6 Da	α1–29 VLSPADKTNVKAAWGKVGAHAGEYGAEAL	+218.37	Flexible
600.3 m/z (+3) 1797.9 Da	β15–31 WGKVNVDEVGGEALGRL	+17.24	Flexible

HbS polymer was reported to be stabilized by both axial and lateral contacts between residues across tetramers (Harrington et al. 1997). Wishner and colleagues reported that αHis45, αPhe46 and αAsp47 residues are involved in the lateral contacts between two adjacent strands within the double strands in the crystal structure of deoxy HbS (Wishner et al. 1975). Although interactions between tetramers in 14 stranded fibrous polymer might not be identical to the linearly packed double strands of HbS, according to Edelstein's prediction that the residues αLys40, αHis45 and αAsp47 might be involved in the contacts between tetramers of deoxy HbS (Edelstein 1981). Additionally, in a genetic variant of hemoglobin, Hb Sealy, the substitution of αAsp47 with His results in a significant inhibition in the polymerization of HbS (Garel et al. 1986). In the H/DX kinetics, an α globin fragment spanning residues 34–46 was observed to show a large increase in the conformational flexibility upon oxygenation of HbS (Table 12.2) (Das et al. 2017). Thus, the large conformational flexibility in aforementioned region of α globin chain in the oxy form of HbS indicates that it might be difficult for these residues to be involved in the interaction across tetramers in the oxy state, thereby contributing towards inhibition in the polymerization upon oxygenation of HbS.

It was reported that the salt bridge between αVal1 and αArg141 of deoxy HbA disappear on oxygenation (Perutz 1970a, b). As a result pK_a of the N-terminal αNH_2 is reduced, that results in the release of a Bohr proton (Perutz 1970a, b) with concomitant increase in the conformational flexibility across residues in the oxy state (Narayanan et al. 2015). Compared to normal hemoglobin, HbS was observed to have larger Bohr effect (Ueda et al. 1979). The axial contacts between residues αLys16-αPro114 and αLys16-αAla115 were observed in the crystal structure of deoxy HbS (Harrington et al. 1997). The amino terminal of α globin chain spanning residues from 1 to 29 was reported to gain conformational flexibility upon oxygenation of HbS. Thus, the observed increase in the conformational flexibility upon deoxy to oxy transition across this region of HbS indicated that the aforementioned axial contacts might be difficult to form in the oxy state of HbS. Therefore, the absence of necessary interactions that are important to stabilize the polymer, contributes in the inhibition of polymerization of oxy HbS.

The following amino acid residues βGly16, βLys17, βVal18 and βGlu22 were found to be involved in the axial contacts in a single strand of deoxy HbS fiber (Harrington et al. 1997). Additionally, amino acid substitutions βGlu22Ala and βGly16Asp were observed to inhibit HbS polymerization (McCune et al. 1994). H/DX kinetics of a fragment of β^S globin chain containing residues 15–31 exhibited an increase in the conformational flexibility on deoxy to oxy transition. This indicated that the required conformational rigidity that are necessary for the aforementioned axial interactions between HbS tetramers to occur might be perturbed in the oxy state of HbS. Therefore, the increased conformational flexibility in this region of oxy HbS molecule might also contribute to the inhibition in polymerization.

It can be inferred from the above cited information that several parts of both α and β globin chains in the oxy state of HbS exhibited a large increase in the conformational flexibility. Hence, the fundamental molecular basis of the inhibition in the polymerization upon oxygenation of HbS appeared to be due to perturbation in

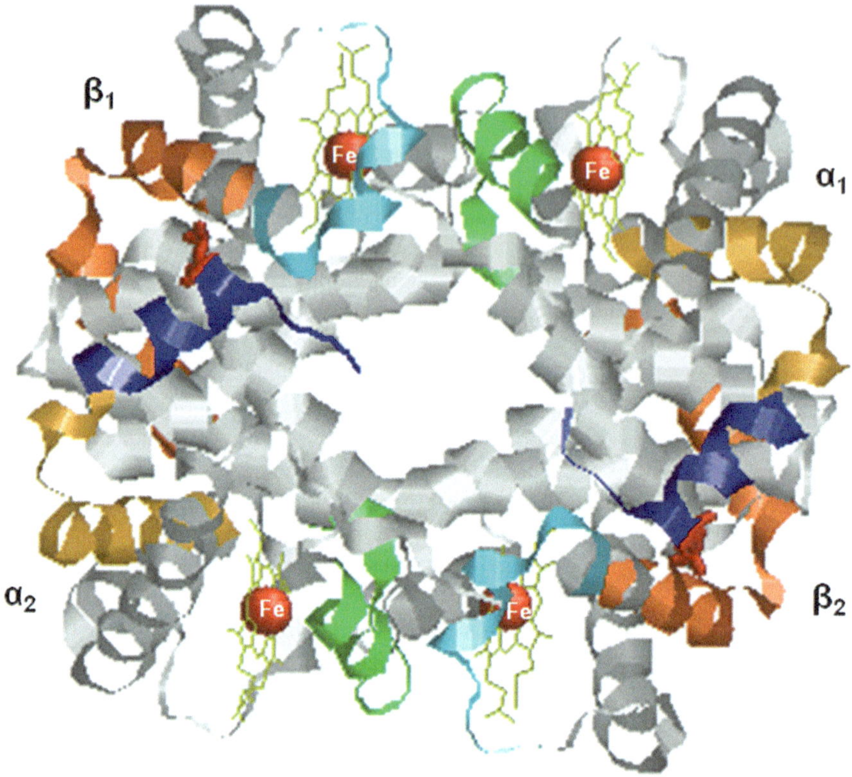

Fig. 12.1 Five peptic peptides that play significant role in the inhibition of polymerization in oxy HbS: Blue-β(1–14), 1463.8 Da; Green-β(86–102), 1920 Da; Red-orange-α(1–29), 2909.6 Da; Cyan-α(34–46), 1584.9 Da; Orange-β(15–31), 1797.9 Da

the required orientation of the hydrophobic patch, unstructured groove conformation and the destabilization of important axial and lateral contacts between residues across strands in the oxy state of HbS. All the above peptic peptides of oxy HbS that were studied in the isotope exchange kinetics are mapped in the crystal structure of deoxy HbS and represented with different colour codes in Fig. 12.1.

Inhibition in Polymerization of Glutathionyl Sickle Hemoglobin

Glutathionyl hemoglobin (GSHbS) is a posttranslational modification of human hemoglobin. Reduced Glutathione (GSH, a tripeptide), is present in the millimolar concentration range within RBCs and acts as a natural antioxidant. Under the condition of oxidative stress, GSH destroys oxidants and is simultaneously oxidised to GSSG. Besides recycling of GSSG back to GSH, elevated levels of GSSG undergoes thiol disulfide exchange with βCys93, a freely accessible cysteine residue of hemoglobin, thereby forming GSHbS within RBCs. Garel and colleagues reported that the glutathionylation of HbS inhibits its polymerization. However, the molecular basis of inhibition was not fully understood (Garel et al. 1986). We investigated the molecular mechanism of inhibition of polymerization upon glutathionylation of βCys93 residue of HbS in vitro. The difference in the conformational dynamics of the deoxy states of both HbS and GSHbS were compared using H/DX kinetics data of peptic peptides originated from different regions of both the molecules as shown in Table 12.3.

N-terminal fragment of β^S globin chain spanning residues 1–14, consisting of Val6, was reported to gain significant conformational rigidity upon glutathionylation of deoxy HbS. Thus, the surface exposure of βVal6, which is crucial for the polymerization to occur, was decreased in deoxy GSHbS. Additionally, it was reported that in the HbS fiber βSer9, βAla10 and βAla13 residues are involved in the lateral contacts with βThr87, a residue around the perimeter of the acceptor groove of HbS (Reddy et al. 1997). Therefore, gain in the conformational rigidity across N-terminal part of deoxy GSHbS indicated that the hydrophobic group of βVal6 along with other residues in this region becomes solvent shielded or buried inside the structure (Table 12.3). This might reduce the availability of hydrophobic patch for nucleation, thereby inhibiting the polymerization.

A β^S globin fragment comprising of residues 15–31, revealed a significant increase in the conformational rigidity upon glutathionylation of HbS in its deoxy form (Table 12.3). As mentioned above, the constituent residues of this peptide such as βGly16, βLys17, βVal18 and βGlu22 are involved in the important axial contacts between adjacent tetramers in a single stranded polymer of HbS (Harrington et al. 1997). Thus, H/DX data indicated that glutathionylation might lead to localized structural perturbations across residues, which results in reduced solvent exposure of this segment of the polypeptide. Therefore, the aforementioned residues in this region of deoxy GSHbS might have limited accessibility to be available for the interactions that stabilizes the polymer, thereby contributing to the inhibition of polymerization.

Another β^S globin fragment with spanning residues 32–41 also showed an increase in conformational rigidity upon glutathionylation of deoxy HbS (Table 12.3). However, the degree of acquired rigidity was less compared to the two above peptides. The peptide fragments β1–14, β15–31 and β32–41 are in a continuous stretch of amino-acid sequence in the amino terminal of β^S globin chain. Thus, it might be approximated that the conformational rigidity induced upon glutathionylation of amino terminal region of β^S globin chain contributes significantly in the inhibition of polymerization of HbS.

In the H/D exchange experiment, one of the α globin fragment with residues 34–46 showed an increase in the conformational flexibility upon glutathionylation of deoxy HbS (Table 12.3). As explained above, αLys40, αPhe46 and αAsp47 residues are expected to be involved in the contacts between double strands in the fourteen stranded fibers of HbS (Edelstein 1981; Garel et al. 1986). Thus, the observed

Table 12.3 Analysis of H/DX kinetic parameters of peptic peptides of deoxy sickle and deoxy glutathionyl sickle hemoglobin

Peptide molecular ion m/z(charge state) Mass in Da	Residues	Change in conformational dynamics $\sum\{(k_iP_i)_{deoxyGSHbS} - (k_iP_i)_{doxyHbS}\}$	Inference (Deoxy HbS → Deoxy GSHbS)
732.9 m/z (+2) 1463.8 Da	β1–14 VHLTPVEKSAVTAL	−171.62	Rigid
600.3 m/z (+3) 1797.9 Da	β15–31 WGKVNVDEVGGEALGRL	−140.15	Rigid
654.8 m/z (+2) 1307.6 Da	β 32–41 LVVYPWTQRF	−12.83	Rigid
529.3 m/z (+3) 1584.9 Da	α34–46 LSFPTTKTYFPHF	90.69	Flexible
728.3 m/z (+4) 2909.6 Da	α1–29 VLSPADKTNVKAAWGKVGAHAGEYGAEAL	67.57	Flexible

increase in conformational flexibility for this peptide upon glutathionylation indicated that the destabilization in the aforementioned contacts between double strands of deoxy GSHbS might contribute in the inhibition in polymerization.

An α globin fragment α1–29 showed an increase in conformational flexibility upon glutathionylation (Table 12.3). A similar trend was also observed in terms of conformational flexibility of HbS on its oxygenation and was expected to contribute in the inhibition of polymerization (Das et al. 2017).

In HbS, βCys93 belongs to the same helix where βPhe85 and βLeu88 reside in the quaternary structure (Garel et al. 1986). βPhe85 and βLeu88 residues are involved in the formation of hydrophobic groove where βVal6 of another tetramer binds during polymerization of HbS (Garel et al. 1986). It might be hypothesized that the covalent modification with glutathione, a tripeptide consisting of polar residues, might lead to perturbation in the conformation of deoxy HbS in the vicinity of βCys93. Thus, destabilization of the conformation near acceptor groove might be expected upon glutathionylation of deoxy HbS, which might contribute in the inhibition of HbS polymerization.

Based on the oxygen binding equilibrium curve to measure the oxygen affinity of both HbS and GSHbS, the partial pressure of oxygen at half saturation of hemoglobin (p_{50}) and Hill coefficient (n) were calculated. p_{50} and n for HbS were 27.3 mmHg and 3 respectively, whereas those values for GSHbS were 23.1 mmHg and 2.3 respectively (Das et al. 2018). Thus, glutathionylation of HbS led to increase in the oxygen affinity. A similar effect was reported in the functional assay of glutathionyl normal human hemoglobin (GSHbA) (Garel et al. 1986). The structural analysis of GSHbA showed that glutathionylation of HbA led to destabilization of the deoxy conformer with a concomitant shift towards oxy-like conformation of HbA (Garel et al. 1986; Mitra et al. 2011, 2012). The similar nature of shift in p_{50} and n values on glutathionylation of HbS might indicate similar type of conformational shift as observed in GSHbA that contributes in the inhibition of polymerization.

It can be concluded that inhibition of HbS polymerization upon glutathionylation of βCys93 might be attributed to the shielding of Val6 residue of β^S globin chain and the probable perturbation of hydrophobic acceptor groove of the tetramer. Additionally, glutathionylation of HbS also induces localized structural perturbations across residues that are involved in the axial contacts during polymerization.

Treatment of Sickle Cell Anemia

Several therapeutic strategies for sickle cell anemia have been proposed, based on the biochemical characteristics of the disease, such as polymerization of HbS in deoxy conditions, reduction in the cellular content of ions and dehydration of RBCs leading to increase in cell density (Matte et al. 2019).

One of the strategies to prevent sickling of RBCs is by stabilizing the oxy form of HbS in vivo using synthetic molecules such as tucaresol, valeresol, 5-hydroxymethyl 2-furfural, etc. which are competitive binders of 2,3-DPG in deoxy hemoglobin

(Safo and Kato 2014). However, the non-specific binding and poor selectivity of binding sites of these molecules have been a concern for the treatment of sickle cell anemia (Safo and Kato 2014). Another potential confounder has been the adverse side effects reported during clinical trials of these compounds due to the requirement for administering high dosages and their consequent short half-lives in blood circulation (Rolan et al. 1995; Oder et al. 2016).

In the treatment of sickle cell anemia, another therapeutic strategy has been developed based on the observation that the pathophysiological consequences of the expression of β^S gene is potentially ameliorated by HbF synthesis (Hinkle and Wolf 1949; Ginder 2015). 5-Azacytidine, a blocker of DNA methylation and suppressor of gene expression, had been used initially to increase the synthesis of γ-globin chain of hemoglobin (Charache et al. 1983). However, due to its carcinogenic potential it could not be administered directly to patients with sickle cell anemia (Cavaliere and Bufalari 1987). As an alternative, hydroxyurea, a key agent in cancer chemotherapy as an inhibitor of DNA synthesis (Singh and Xu 2016) and having no effect on DNA methylation, has also been observed to be beneficial to treat patients with sickle cell anemia (Halsey and Roberts 2003; Platt 2008). Although the clinical benefits of hydroxyurea in the treatment for sickle cell anemia was discovered a few decades back, the precise mechanisms by which hydroxyurea induces HbF synthesis remain incompletely understood. One of the most accepted mechanisms is of the reversible inhibition of ribonucleotide reductase by hydroxyurea in vivo. Ribonucleotide reductase is a critical enzyme that converts ribonucleosides into deoxyribonucleosides, that are required for the synthesis and repair of DNA (Elford 1968). Inhibition of ribonucleotide reductase results in decreased intracellular population of deoxyribonucleotide triphosphates and slows down the progression of cellular division by arresting the cell cycle at the S phase (Koç et al. 2004; Alvino et al. 2007). Subsequently, it is thought to recruit the earlier erythroid progenitors of the intracellular population that have a greater capacity for HbF synthesis. The elevation of HbF might increase the molecular crowding within RBCs and subsequently reduce the rate of polymerization of HbS tetramers (Das et al. 2017). However, clinical investigations following administration of hydroxyurea to patients with sickle cell disease revealed several side effects in children and adults including hematopoietic toxicity and subsequent suppression in reticulocyte count, thereby limiting the use of hydroxyurea in the treatment of sickle cell anemia (Das et al. 2017).

Recent studies have shown that N-acetyl cysteine (NAC), an antioxidant, has been effective in the treatment of patients with sickle cell anemia (Nur et al. 2012), where the levels of total glutathione in plasma and erythrocytes is reduced and also the ratio of reduced glutathione (GSH), the major intracellular antioxidant, to its oxidized form glutathione disulfide (GSSG) is reduced resulting in high oxidative stress, one of the hallmarks in the pathophysiology of sickle cell anemia (Nur et al. 2012). Studies have shown that treatment of sickle RBCs with NAC, which can readily enter the RBCs, results in an increase in the level of cellular GSH, thereby reducing the oxidative stress (Özpolat et al. 2014). A few clinical trials are being undertaken to evaluate the efficiency of NAC in treatment of patients with sickle cell anemia (Özpolat et al. 2014).

It is known that patients with sickle cell anemia produce more reactive oxygen species (ROS) than healthy individuals that causes further damage to red cell membrane (Egini ans Guillaume 2017). The major causative factor is the higher level of circulating cell-free hemoglobin in patients with sickle cell anemia. Such patients undergo continuous intravascular hemolysis caused by the circulating sickled RBCs. This results in a high concentration of cell-free ferrous hemoglobin in plasma which reduces nitric oxide, an important vasodilator and anti-inflammatory molecule, thereby inactivating it (Belhassen 2001; Reiter et al. 2002; Guzik et al. 2003). The heme unit being hydrophobic in nature inserts itself into the plasma membrane of endothelial cells and releases iron, inducing cellular oxidative damage by catalyzing the nonenzymatic generation of ROS (Balla et al. 1993; Jeney 2002). It has been reported that vitamin C, an antioxidant, is capable of inhibiting ROS formation by serving as an electron donor, thereby reducing molecular oxygen (Egini and Guillaume 2017). This prevents the formation of dense cells, a clinical manifestation of sickle cell anemia that leads to increased vaso-occlusion (Chan 2000).

Another clinical phenomenon associated with sickle cell anemia is the dehydration of RBCs resulting in the formation of dense cells (McGoron et al. 2000). It has been reported that oral administration of magnesium pidolate and clotrimazole helps to prevent dehydration of sickled RBCs (Brugnara 2018). Sickle cell dehydration is mainly caused by the loss of K^+, Cl^- ions and water. Two major ion transport pathways are involved in sickle cell dehydration: KCl cotransport (KCC) and the Ca^{2+}-activated K^+ channel (Gardos channel) (Brugnara 2001). The KCC activity is inversely dependent on intracellular magnesium, which is abnormally reduced in sickled RBCs (Brugnara 2018). Mg^{2+} pidolate administration restores the levels of magnesium in sickled RBCs and inhibits the activity of the KCC ion transport channel (Brugnara 2001). On the other hand, entry of calcium through this pathway triggers the activation of the Gardos channel, thereby resulting in rapid K^+ efflux along with water loss. Clotrimazole and derivatives of clotrimazole metabolites have been reported to specifically block the Gardos channel thereby preventing dehydration of RBCs in patients with sickle cell anemia (Joiner 2008).

Transfusion of normal RBCs (blood transfusion) is considered as one of the effective therapeutic measures for treatment of patients with sickle cell anemia (Wintrobe 2014). Transfusion therapy with normal adult blood is known to improve oxygenation of tissues compared to HbS, which has low affinity for oxygen (Marouf 2011). Transfusion of normal RBCs also reduces the events of vaso-occlusion by diluting the population of host sickled RBCs (Marouf 2011). However, the complexities involved in the procedure restricts its application to selected patients with high severity of the diseased state (Wintrobe 2014). Additionally, alloimmunization, a phenomemon of developing antibodies to multiple types of blood antigens, is a major concern with chronic blood transfusion (Chou et al. 2013). Iron overload is also a concern associated with chronic blood transfusion. Iron entering the body with transfused blood tends to accumulate in several organs such as liver, kidneys, heart, pancreas etc., resulting in severe toxicity in these organs (Wood et al. 2005).

In patients with sickle cell anemia with increased severity and the associated multiple complications such as stroke, acute chest syndrome, recurrent pain crisis and exchange transfusions, nephropathy, retinopathy, osteonecrosis of multiple joints etc., hematopoietic stem cell transplantation in bone marrow is used as a therapeutic strategy for sickle cell anemia, where human leukocyte antigen (HLA)-identical siblings are recruited as donors (Wintrobe 2014). Although the application of bone marrow transplantation as a curative treatment in sickle cell anemia is widespread broad, the common limitations of this strategy arise from the associated increased disease severity such as events of stroke, red cell alloimmunization, etc. Brain parenchymal damage has been reported in children who had undergone transplantation (Woodard et al. 2005). The possibility of infertility in patients with sickle cell anemia receiving bone marrow transplants have also been reported (Kassim and Sharma 2017). Additionally, concerns over high procedural costs, lack of expertise and lack of available HLA-matched donors along with certain ethical issues about bone marrow transplantation have restricted its application as a therapeutic strategy for sickle cell anemia (Wintrobe 2014).

Sickle cell anemia has been considered for a long time as a candidate for which gene therapy could be applicable. Several clinical trials are under progress to evaluate the efficiency of this strategy as a curative measure for the disease. The trials have been planned based on the natural regulation process of the globin genes and genetic manipulation of the hematopoietic system (Hoban et al. 2016). Several techniques such as the use of γ-retroviral vectors, novel genome engineering and gene regulation approaches are under scrutiny to claim therapeutic benefits in the treatment of sickle cell anemia (Hoban et al. 2016). Use of lentiviral vector–mediated insertion of a normal β-globin gene into hematopoietic stem cells has also been successfully tested (Ribeil et al. 2017). However, certain concerns resting with the random genomic insertion, low efficiency of gene transfer and safety issues have limited the progress of developing this technique as a therapeutic strategy (Wintrobe 2014).

A century after the discovery of sickle cell anemia, the effective medical management of the disease still remains elusive. Therefore, it is crucial to develop treatment procedures that might be considered as gold-standard therapeutic strategies for sickle cell anemia. This might be multi-agent strategies based on the pathophysiology of sickle cell anemia in the respective patients that might potentially improve the quality of life and the survival of patients.

Reduced glutathione (GSH) is a potential antioxidant, present in 2–2.5 mM concentration inside normal human RBCs, whereas in erythrocytes of individuals with sickle cell anemia, it was found to be 32–36% lower (Colombo et al. 2010; Gizi et al. 2011). Under normal physiological conditions GSH does not glutathionylate proteins (Birchmeier et al. 1973; Mitra et al. 2011). However, under the condition of elevated oxidative stress, the redox equilibrium between reduced and oxidized glutathione (GSSG) inside RBCs shifts towards its oxidized state. A part of the accumulated GSSG might participate in the glutathionylation of protein via thiol exchange with free and accessible cysteine residues of proteins such as hemoglobin, to form GSHbS. Based on the previous report published by our group on the mechanism of inhibition in polymerization of HbS by glutathionylation of βCys93, we proposed that the elevation in the levels of GSHbS might be beneficial to the patient with sickle cell anemia (Das et al. 2018). However, an increase in the level of glutathionyl hemoglobin in blood has been reported in several medical conditions that are associated with elevated oxidative stress (Lou 2000, 2003; Mandal et al. 2007; Nonaka et al. 2007; Newman et al. 2007; Mitra et al. 2011). Oxidative stress induced glutathionylation of human hemoglobin in RBCs had been reported with the exposure to tertiary butyl hydrogen peroxide (Colombo et al. 2010). Thus, glutathionylation of HbS might be achieved in patients with sickle cell anemia under the condition of oxidative stress. Although the formation of GSHbS might be beneficial to the patient in terms of inhibition in polymerization of HbS, the elevation of oxidative stress itself is not welcomed in any individual. Thus, it might be important to investigate the mechanisms by which the level of GSHbS can be elevated in vivo without inducing the oxidative stress.

Methods Used

Purification and Characterization of HbS

HbS was purified from the hemolysate of the blood sample from a patient with sickle cell anemia. In cation exchange chromatography, exploiting the surface charge of tetrameric hemoglobin, HbS was separated and purified from HbA. HbS was allowed to bind to the carboxymethyl cellulose resin (CM52), a weak cation exchanger, and was eluted from the column using pH and salt gradient of 10 mM potassium phosphate buffer, pH 6.7–20 mM potassium phosphate, pH 7.5 (Waterman et al. 1974; Das et al. 2017). Fractions were collected and analysed using liquid chromatography coupled to an electrospray ionisation mass spectrometry (LC/ESI-MS) platform, where the masses of intact globin chains of purified HbS were monitored (Mandal et al. 2008). During mass analysis, α and β globin chains were separated in reverse phase LC using a linear gradient of 2% increase in acetonitrile per minute containing 0.1% acetic acid with a flow rate of 0.2 mL/min.

Oxygen Binding

The oxygen binding assay was performed with 25 mM purified HbS dissolved in 100 mM potassium phosphate buffer, pH 7.4 in the presence of 50 mM NaCl (aqueous) and 6.25 mM of 2,3-DPG, an isotonic buffer solution with respect to whole blood (Prange et al. 2001). 3 mL of 25 mM of hemoglobin solution was completely deoxygenated by the addition of 2 mM of sodium dithionite and bubbling N_2 gas at 40 mL/min containing 5.4% CO_2. The sample was gradually re-oxygenated by bubbling O_2 gas at 40 ml/min containing 5.4% CO_2 and the absorbance of the solution was measured at 548 and 577 nm to calculate the saturation (SO_2). The partial pressure of the dissolved O_2 gas was measured using oxygen electrode inbuilt in

Blood Gas Analyser (Radiometer, Copenhagen, Denmark) (Das et al. 2018). The saturation of HbS was calculated using Eq. 12.1 and the obtained data were fitted into Hill's equation to get a fitted sigmoidal curve using OriginLab 8.1 software. The partial pressure of oxygen at 50% saturation (p_{50}) of HbS and Hill coefficient (n) were calculated from this fitted oxygen equilibrium curve.

$$SO_2(\%) = \left(\frac{(\lambda_{577}/\lambda_{548}) - 0.76}{0.424} \right) \times 100 \tag{12.1}$$

Glutathionylation of HbS

Purified HbS was glutathionylated in two steps according to Das et al. (2018). In the first step, 2,2′-dithiodipyridine (2-PDS) solution was incubated with HbS in the molar ratio of 2-PDS:HbS = 10:1 in 50 mM potassium phosphate buffer pH 7.4 for 1 h at 25 °C. The unreacted PDS was removed through a Sephadex G-50 spin column. In the second step, synthesized thiopyridinyl HbS was subjected to react with reduced GSH in the molar ratio of HbS:GSH = 1:40 at 25 °C for 1 h. Subsequently, excess GSH was removed by using a Sephadex G-50 spin column. The extent of glutathionylation of HbS was monitored by analyzing the synthesized molecules using LC/ESI-MS platform (Das et al. 2018).

Conformational Dynamics of HbS and GSHbS Using H/D Exchange Kinetics

For H/D exchange kinetics, 60 μM stock solution of both deoxy and oxy forms of HbS and deoxy form of GSHbS were diluted by 15 folds in 100 mM ammonium bicarbonate/D_2O buffer, pD 7.4 at 25 °C containing 50 mM sodium dithionite in three different sets. Prior to the dilution, the deoxy reaction sets were subjected to continuous bubbling of N_2 gas containing 5% CO_2 at a rate of 40 mL/min whereas the oxy set was bubbled with oxygen gas containing 5% CO_2. H/DX kinetics was followed for 6 h. At different time intervals, 10 μL of the reaction mixture was taken out and the isotope exchange reaction was quenched by adding 90 μL ice cold aqueous solution of 0.1% formic acid pH 2.5. 2 μL of aqueous pepsin solution was added to the aliquot maintaining enzyme: substrate molar ratio 1:10. Pepsin digestion of the isotope exchanged protein was carried out in situ for 5 min at 0 °C. 20 pmol of proteolytic peptide pool was injected in ACQUITY UPLC™ system with H/DX technology (Wales et al. 2008). The exchanged peptides were eluted from ACQUITY UPLC BEH C18 column (100 mm × 1 mm, 1.7 μm) (Waters, UK) at 0 °C using a mixture of solvents A and B, with a gradient of 3–40% of solvent B over 7 min using

40 μL/min flow rate. Solvents A and B consisted of water and acetonitrile containing 0.1% formic acid, respectively.

Analysis of H/D Exchange Data

Peptides were identified from the analysis of tandem mass spectra of undeuterated control set using the ProteinLynx Global Server (PLGS) 2.5 software (Waters, UK). MassLynx 4.1 software was used to generate the mass spectra of different peptic peptides from the respective extracted ion chromatograms. The obtained spectra were baseline corrected and the isotope averaged centroid mass of each molecular ion was measured using HX-Express 2 2007 v27 (Weis et al. 2006). The number of deuterium incorporated in place of backbone amide hydrogens within the experimental peptide ion at a given time t, D(t) was calculated following Eq. 12.2 (Zhang and Smith 1993):

$$D(t) = \frac{(M_t - M_0)}{(M_\infty - M_0)} \times N \tag{12.2}$$

where, M_t is the observed isotope average centroid mass of the peptide molecular ion at time t and M_0 and M_∞ are the isotope average centroid masses at zero and infinite times respectively (Mitra et al. 2011). N represents the total number of exchangeable backbone amide hydrogens present in the peptide molecular ion. At a fixed temperature, pH and in large excess of D_2O, the amide hydrogen exchange follows pseudo first order kinetics. In addition, at a given time t, for a peptide consisting of N exchangeable amide hydrogens, the deuterium content can be described as in Eq. 12.3.

$$D(t) = N - \sum_{i=1}^{N} \exp^{-k_i^t} \tag{12.3}$$

where, k_i is the H/DX rate constant of *i*th amide hydrogen (Hoofnagle et al. 2003). In general, each backbone amide hydrogen has a characteristic exchange rate constant, however in practice all exchangeable backbone amide hydrogens of a peptide might be categorized into fast, intermediate, and slow exchanging groups, Thus, D(t) can be expressed as Eq. 12.4:

$$D(t) = N - P_A e^{-k_1^t} - P_B e^{-k_2^t} - P_C e^{-k_3^t} \tag{12.4}$$

where P_A, P_B, and P_C are the number of fast, intermediate and slow exchanging amide hydrogens with average H/DX rate constants k_1, k_2, and k_3, respectively. Both P and k can be calculated from the best fit curve of D(t) versus t. To obtain the best fit data for deoxy HbS, oxy HbS and deoxy GSHbS, the rate constants and the populations in different groups were varied to minimize SSR (sum of squared residuals) (Das et al.

2018). The best-fit curves were obtained using MS Excel solver. SSR was calculated using the following Eq. 12.5:

$$SSR = \sum_i \left[\frac{\{(y_i)_{obs} - (y_i)_{calc}\}}{(y_i)_{obs}} \right]^2 \quad (12.5)$$

$(y_i)_{obs}$ $[(D_t)_{obs}]$ and $(y_i)_{calc}$ $[(D_t)_{cal}]$ were calculated from Eqs. 12.2 to 12.4 respectively (Kemmer and Keller 2010). Using six populations (P_A, P_B, and P_C) and six rate constants (k_1, k_2 and k_3) of both initial and final states of the process, it is possible to investigate the change in conformational dynamics associated with oxygenation and also glutathionylation of HbS using the method of initial rates (Narayanan et al. 2015). The overall H/DX rate of a peptide was calculated from the summation of initial exchange rates of three different groups of exchangeable amide hydrogens in each set. Subsequently, the above calculated rates were subtracted between two different states: oxy HbS & deoxy HbS and deoxy HbS & deoxy GSHbS, to understand both the direction and the degree of change in conformational dynamics associated with oxygenation and glutathionylation of HbS.

Turbidity Experiments to Monitor Polymerization Kinetics

Solubility of HbS was reported to be reduced by three orders of magnitude at 1.8 M phosphate buffer pH 7.4, thereby facilitating the polymerization kinetics of HbS to be monitored by absorption spectrophotometry. To deoxygenate HbS, 3 mM of sodium dithionite was maintained in 1.8 M potassium phosphate buffer, pH 7.4 at 0 °C and the solution was quickly transferred into a spectrophotometer cuvette at 37 °C. Polymerization of deoxy HbS is an as endothermic process. Thus, a temperature jump method was used to initiate the polymerization by a rapid temperature increase from 0 to 37 °C using 10 and 20 mM HbS in two different sets. The optical density (turbidity) of the solution was monitored at 700 nm (Adachi 1978, 1979b). Similarly, polymerization of GSHbS was performed following the aforementioned method. The delay time of polymerization was calculated by fitting the average optical densities [A(t)] versus time (x) into Eq. 12.6 (Yohe et al. 2000; Knee et al. 2007).

$$A(t) = \frac{(A_1 - A_2)}{1 + e^{\{x - x_0/dx\}}} + A_2 \quad (12.6)$$

where A_1 and A_2 are the initial and final absorbances respectively, and x_0 is the time corresponding to 50% change in the absorbance. x_0 denotes the delay time (t_d) of polymerization (Knee et al. 2007). The slope of the transition in absorbance profile is defined by 'dx' during the polymerization event. A small dx indicates that the transition reaches the inflection point. The delay time, t_d, was calculated from the point of intersection of a tangent to the linear portion of the rapid polymerization

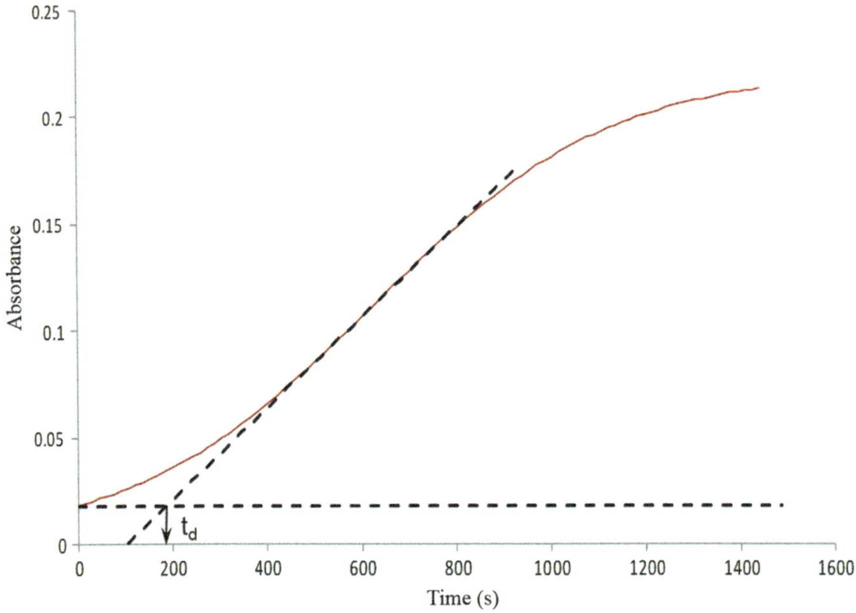

Fig. 12.2 Polymerization kinetics of 10 μM GSHbS. The time corresponding to the point of intersection of the tangent and the horizontal line represents the delay time (t_d) of polymerization

phase of HbS and the horizontal line (Fasanmade 1996). The time corresponding to the point of intersection of the tangent and the horizontal line represents the t_d of polymerization as shown in Fig. 12.2.

References

Adachi KAT (1978) Demonstration of a delay time during aggregation of diluted solutions of deoxyhemoglobin S and hemoglobin CHarlem in concentrated phosphate buffer. J Biol Chem 253:6641–6643

Adachi KAT (1979) Nucleation-controlled aggregation of deoxyhemoglobin S. Possible difference in the size of nuclei in different phosphate concentrations. J Biol Chem 254:7765–7771

Alvino GM, Collingwood D, Murphy JM et al (2007) Replication in hydroxyurea: it's a matter of time. Mol Cell Biol 27:6396–6406. https://doi.org/10.1128/mcb.00719-07

Balla J, Jacob HS, Balla G et al (1993) Endothelial-cell heme uptake from heme proteins: induction of sensitization and desensitization to oxidant damage. Proc Natl Acad Sci 90:9285–9289. https://doi.org/10.1073/pnas.90.20.9285

Belhassen L (2001) Endothelial dysfunction in patients with sickle cell disease is related to selective impairment of shear stress-mediated vasodilation. Blood 97:1584–1589. https://doi.org/10.1182/blood.V97.6.1584

Birchmeier W, Tuchschmid PE, Winterhalter KH (1973) Comparison of human hemoglobin a carrying glutathione and a mixed disulfide with the naturally occurring human hemoglobin A3. Biochemistry 12:3667–3672. https://doi.org/10.1021/bi00743a015

Branden C, Tooze J (1998) Introduction of protein structure, p 43

Brugnara C (2001) Therapeutic strategies for prevention of sickle cell dehydration. Blood Cells Mol Dis 27:71–80. https://doi.org/10.1006/bcmd.2000.0366

Brugnara C (2018) Sickle cell dehydration: pathophysiology and therapeutic applications. Clin Hemorheol Microcirc 68:187–204. https://doi.org/10.3233/CH-189007

Bunn HF (1997) Pathogenesis and treatment of sickle cell disease. N Engl J Med 337:762–769. https://doi.org/10.1146/annurev.nutr.23.011702.073027

Cavaliere A, Bufalari AVR (1987) 5-azacytidine carcinogenesis in BALB/c mice. Cancer Lett 37:51–58. https://doi.org/10.1016/0304-3835(87)90145-5

Chan AC (2000) A cocktail approach to antioxidant therapy. Nutrition 16:1098–1100. https://doi.org/10.1016/S0899-9007(00)00445-7

Charache S, Dover G, Smith K et al (1983) Treatment of sickle cell anemia with 5-azacytidine results in increased fetal hemoglobin production and is associated with nonrandom hypomethylation of DNA around the γ-δ-β-globin gene complex. Proc Natl Acad Sci USA 80:4842–4846. https://doi.org/10.1073/pnas.80.15.4842

Chou ST, Jackson T, Vege S et al (2013) High prevalence of red blood cell alloimmunization in sickle cell disease despite transfusion from Rh-matched minority donors. Blood. https://doi.org/10.1182/blood-2013-03-490623

Christoph GW, Hofrichter J, Eaton WA (2005) Understanding the shape of sickled red cells. Biophys J 88:1371–1376. https://doi.org/10.1529/biophysj.104.051250

Colombo G, Dalle-Donne I, Giustarini D et al (2010) Cellular redox potential and hemoglobin S-glutathionylation in human and rat erythrocytes: a comparative study. Blood Cells Mol Dis 44:133–139. https://doi.org/10.1016/j.bcmd.2009.11.005

Craescu CT, Poyart C, Schaeffer C et al (1986) Covalent binding of glutathione to hemoglobin. II. functional consequences and structural changes reflected in NMR spectra. J Biol Chem 261:14710–14716

da Fonseca SF, Amorim T, Purificação A et al (2015) Hemoglobin A2 values in sickle cell disease patients quantified by high performance liquid chromatography and the influence of alpha thalassemia. Rev Bras Hematol Hemoter 37:296–301. https://doi.org/10.1016/j.bjhh.2015.05.005

Das R, Mitra G, Mathew B et al (2013) Automated analysis of hemoglobin variants using nanoLC-MS and customized databases. J Proteome Res 12:3215–3222. https://doi.org/10.1021/pr4000625

Das R, Muralidharan M, Mitra G et al (2015) Mass spectrometry based characterization of Hb Beckman variant in a falsely elevated HbA1c sample. Anal Biochem 489:53–58. https://doi.org/10.1016/j.ab.2015.07.010

Das R, Mitra G, Mathew B et al (2016) Mass spectrometry-based diagnosis of hemoglobinopathies: a potential tool for the screening of genetic disorder. Biochem Genet 54:816–825. https://doi.org/10.1007/s10528-016-9758-5

Das R, Mitra A, Bhat V, Mandal AK (2017) Application of isotope exchange based mass spectrometry to understand the mechanism of inhibition of sickle hemoglobin polymerization upon oxygenation. J Struct Biol 199:76–83. https://doi.org/10.1016/j.jsb.2017.04.009

Das R, Mitra A, Mitra G et al (2018) Molecular insights of inhibition in sickle hemoglobin polmerization upon glutathionylation: hydrogen/deuterium exchange mass spectrometry and molecular dynamics simulation-based approach. Biochem J 475:2153–2166. https://doi.org/10.1042/BCJ20180306

Eaton WA, Hofrichter J (1987) Hemoglobin S gelation and sickle cell disease. Blood 70:1245–1266

Edelstein SJ (1981) Molecular topology in crystals and fibers of hemoglobin S. J Mol Biol 150:557–575. https://doi.org/10.1016/0022-2836(81)90381-8

Egini O, Guillaume EJA (2017) Effect of Vitamin C on reactive oxygen species formation in erythrocytes of sickle cell anemia patients. Blood 130:4778

Elford HL (1968) Effect of hydroxyurea on ribonucleotide reductase. Biochem Biophys Res Commun 33:129–135. https://doi.org/10.1016/0006-291X(68)90266-0

Embury S et al (1994) Sickle cell disease. Basic principles and clinical practice. Raven Press

Fasanmade AA (1996) Gelation kinetics of dilute hemoglobin from sickle cell anemia patients. Hemoglobin. https://doi.org/10.3109/03630269609005845

Ferrone FA, Hofrichter J, Eaton WA (1985a) Kinetics of sickle hemoglobin polymerization. I. studies using temperature-jump and laser photolysis techniques. J Mol Biol 183:591–610. https://doi.org/10.1016/0022-2836(85)90174-3

Ferrone FA, Hofrichter J, Eaton WA (1985b) Kinetics of sickle hemoglobin polymerization. II. A double nucleation mechanism. J Mol Biol 183:611–631. https://doi.org/10.1016/0022-2836(85)90175-5

Garel MC, Domenget C, Caburi-Martin J et al (1986) Covalent binding of glutathione of hemoglobin. I. inhibition of hemoglobin S polymerization. J Biol Chem 261:14704–14709

Ghatge MS, Ahmed MH, Omar ASM et al (2016) Crystal structure of carbonmonoxy sickle hemoglobin in R-state conformation. J Struct Biol 194:446–450. https://doi.org/10.1016/j.jsb.2016.04.003

Ginder GD (2015) Epigenetic regulation of fetal globin gene expression in adult erythroid cells. Transl Res 165:115–125. https://doi.org/10.1016/j.trsl.2014.05.002

Gizi A, Papassotiriou I, Apostolakou F et al (2011) Assessment of oxidative stress in patients with sickle cell disease: the glutathione system and the oxidant-antioxidant status. Blood Cells Mol Dis 46:220–225. https://doi.org/10.1016/j.bcmd.2011.01.002

Guzik TJ, Korbut R, Adamek-Guzik T (2003) Nitric oxide and superoxide in inflammation and immune regulation. J Physiol Pharmacol

Halsey C, Roberts IAG (2003) The role of hydroxyurea in sickle cell disease. Br J Haematol 120:177–186. https://doi.org/10.1046/j.1365-2141.2003.03849.x

Harrington DJ, Adachi K, Royer WE (1997) The high resolution crystal structure of deoxyhemoglobin S. J Mol Biol 272:398–407. https://doi.org/10.1006/jmbi.1997.1253

HbVar Menu. http://globin.cse.psu.edu/hbvar/menu.html. Accessed 12 Oct 2019

Hebbel R (1994) Sickle cell disease: basic principles and clinical practice

Hinkle LE, Wolf S (1949) Experimental study of life situations, emotions, and the occurrence of acidosis in a juvenile diabetic. Am J Med Sci 5:159–160. https://doi.org/10.1097/00000441-194902000-00002

Hoban MD, Orkin SH, Bauer DE (2016) Genetic treatment of a molecular disorder: gene therapy approaches to sickle cell disease. Blood 127:839–848. https://doi.org/10.1182/blood-2015-09-618587

Hoffman R, Benz EJ, Silberstein LE et al (2017) Hematology: basic principles and practice

Hofrichter J, Ross PD, Eaton WA (1974) Kinetics and mechanism of deoxyhemoglobin S gelation: a new approach to understanding sickle cell disease. Proc Natl Acad Sci USA 71:4864–4868. https://doi.org/10.1073/pnas.71.12.4864

Hoofnagle AN, Resing KA, Ahn NG (2003) Protein analysis by hydrogen exchange mass spectrometry. Annu Rev Biophys Biomol Struct 32:1–25. https://doi.org/10.1146/annurev.biophys.32.110601.142417

Ingram VM (1956) A specific chemical difference between the globins of normal human and sickle-cell anæmia hæmoglobin. Nature 178:792–794. https://doi.org/10.1038/178792a0

Jeney V (2002) Pro-oxidant and cytotoxic effects of circulating heme. Blood 100:879–887. https://doi.org/10.1182/blood.V100.3.879

Joiner CH (2008) Gardos pathway to sickle cell therapies? Blood

Joutovsky A, Hadzi-Nesic J, Nardi MA (2004) HPLC retention time as a diagnostic tool for hemoglobin variants and hemoglobinopathies: a study of 60000 samples in a clinical diagnostic laboratory. Clin Chem 50:1736–1747. https://doi.org/10.1373/clinchem.2004.034991

Kassim AA, Sharma D (2017) Hematopoietic stem cell transplantation for sickle cell disease: the changing landscape. Hematol Oncol Stem Cell Ther 10:259–266. https://doi.org/10.1016/j.hemonc.2017.05.008

Kemmer G, Keller S (2010) Nonlinear least-squares data fitting in excel spreadsheets. Nat Protoc 5:267–281. https://doi.org/10.1038/nprot.2009.182

Kim-Shapiro DB, King SB, Shields H et al (1999) The reaction of deoxy-sickle cell hemoglobin with hydroxyurea. Biochim Biophys Acta Gen Subj 1428:381–387. https://doi.org/10.1016/S0304-4165(99)00071-9

Knee KM, Roden CK, Flory MR, Mukerji I (2007) The role of β93 Cys in the inhibition of Hb S fiber formation. Biophys Chem 127:181–193. https://doi.org/10.1016/j.bpc.2007.02.002

Koç A, Wheeler LJ, Mathews CK, Merrill GF (2004) Hydroxyurea Arrests DNA replication by a mechanism that preserves basal dNTP pools. J Biol Chem 279:223–230. https://doi.org/10.1074/jbc.M303952200

Lauer J, Shen C-KJ, Maniatis T (1980) The chromosomal arrangement of human α-like globin genes: sequence homology and α-globin gene deletions. Cell 20:119–130. https://doi.org/10.1016/0092-8674(80)90240-8

Lou MF (2000) Thiol regulation in the lens. J Ocul Pharmacol Ther 16:137–148. https://doi.org/10.1089/jop.2000.16.137

Lou MF (2003) Redox regulation in the lens. Prog Retin Eye Res 22:657–682. https://doi.org/10.1016/S1350-9462(03)00050-8

Mandal AK, Woodi M, Sood V et al (2007) Quantitation and characterization of glutathionyl haemoglobin as an oxidative stress marker in chronic renal failure by mass spectrometry. Clin Biochem. https://doi.org/10.1016/j.clinbiochem.2007.05.006

Mandal AK, Bisht S, Bhat VS et al (2008) Electrospray mass spectrometric characterization of hemoglobin Q (Hb Q-India) and a double mutant hemoglobin S/D in clinical samples. Clin Biochem 41:75–81. https://doi.org/10.1016/j.clinbiochem.2007.09.006

Marouf R (2011) Blood transfusion in sickle cell disease. Hemoglobin. https://doi.org/10.3109/03630269.2011.596984

Mathew B, Bhat V, Mandal AK (2011) Analysis of hemoglobin variants using nondenaturing gel electrophoresis and matrix-assisted laser desorption ionization mass spectrometry. Anal Biochem 416:135–137. https://doi.org/10.1016/j.ab.2011.04.029

Matte A, Mazzi F, Federti E et al (2019) New therapeutic options for the treatment of sickle cell disease. Mediterr J Hematol Infect Dis 11:e2019002. https://doi.org/10.4084/mjhid.2019.002

McCune SL, Reilly MP, Chomo MJ et al (1994) Recombinant human hemoglobins designed for gene therapy of sickle cell disease. Proc Natl Acad Sci USA 91:9852–9856. https://doi.org/10.1073/pnas.91.21.9852

McGoron AJ, Joiner CH, Palascak MB, Claussen WJFR (2000) Dehydration of mature and immature sickle red blood cells during fast oxygenation/deoxygenation cycles: role of KCl cotransport and extracellular calcium. Blood 95:2164–2168

Mitra G, Muralidharan M, Pinto J et al (2011) Structural perturbation of human hemoglobin on glutathionylation probed by hydrogen–deuterium exchange and MALDI mass spectrometry. Bioconjug Chem 22:785–793. https://doi.org/10.1021/bc100602f

Mitra G, Muralidharan M, Narayanan S et al (2012) Glutathionylation induced structural changes in oxy human hemoglobin analyzed by backbone amide hydrogen/deuterium exchange and MALDI-mass spectrometry. Bioconjug Chem 23:2344–2353. https://doi.org/10.1021/bc300291u

Nagel RL (2003) Hemoglobin disorders. Humana Press, New Jersey

Narayanan S, Mitra G, Muralidharan M et al (2015) Protein structure-function correlation in living human red blood cells probed by isotope exchange-based mass spectrometry. Anal Chem 87:11812–11818. https://doi.org/10.1021/acs.analchem.5b03217

Newman SF, Sultana R, Perluigi M et al (2007) An increase in S-glutathionylated proteins in the Alzheimer's disease inferior parietal lobule, a proteomics approach. J Neurosci Res 85:1506–1514. https://doi.org/10.1002/jnr.21275

Noguchi CT, Schechter AN (1985) Sickle hemoglobin polymerization in solution and in cells. Annu Rev Biophys Biophys Chem 14:239–263. https://doi.org/10.1146/annurev.biophys.14.1.239

Nonaka K, Kume N, Urata Y et al (2007) Serum levels of S-glutathionylated proteins as a risk-marker for arteriosclerosis obliterans. Circ J 71:100–105. https://doi.org/10.1253/circj.71.100

Nur E, Brandjes DP, Teerlink T et al (2012) N-acetylcysteine reduces oxidative stress in sickle cell patients. Ann Hematol 91:1097–1105. https://doi.org/10.1007/s00277-011-1404-z

Oder E, Safo MK, Abdulmalik O, Kato GJ (2016) New developments in anti-sickling agents: can drugs directly prevent the polymerization of sickle haemoglobin in vivo? Br J Haematol 175:24–30. https://doi.org/10.1111/bjh.14264
Ou CN, Rognerud CL (2001) Diagnosis of hemoglobinopathies: electrophoresis versus. HPLC Clin Chim Acta 313:187–194. https://doi.org/10.1016/S0009-8981(01)00672-6
Özpolat T, Chen J, Fu X et al (2014) Effects of N-acetylcysteine in patients with sickle cell disease. Blood 124:4173. https://doi.org/10.1182/blood.V124.21.4173.4173
Padlan EA, Love WE (1985) Refined crystal structure of deoxyhemoglobin S. II. molecular interactions in the crystal. J Biol Chem 260:8280–8291
Pauling L, Itano HA, Singer SJ, Wells IC (1949) Sickle cell anemia, a molecular disease. Science 110(80):543–548. https://doi.org/10.1126/science.110.2865.543
Perutz MF (1970a) Stereochemistry of cooperative effects in haemoglobin. Nature 228:726–734. https://doi.org/10.1038/228726a0
Perutz MF (1970b) The Bohr effect and combination with organic phosphates. Nature 228:734–739. https://doi.org/10.1038/228734a0
Platt O (2008) Hydroxyurea for the treatment of sickle cell anemia. N Engl J Med 358:1362–1369. https://doi.org/10.1056/NEJMct0708272
Platt OS, Brambilla DJ, Rosse WF et al (1994) Mortality in sickle cell disease—life expectancy and risk factors for early death. N Engl J Med 330:1639–1644. https://doi.org/10.1056/NEJM199406093302303
Prange HD, Shoemaker JL, Westen EA et al (2001) Physiological consequences of oxygen-dependent chloride binding to hemoglobin. J Appl Physiol 91:33–38. https://doi.org/10.1152/jappl.2001.91.1.33
Reddy LR, Reddy KS, Surrey S, Adachi K (1997) Role of β87 Thr in the β6 Val acceptor site during deoxy Hb S polymerization. Biochemistry 36:15992–15998. https://doi.org/10.1021/bi9717439
Reiter CD, Wang X, Tanus-Santos JE et al (2002) Cell-free hemoglobin limits nitric oxide bioavailability in sickle-cell disease. Nat Med 8:1383–1389. https://doi.org/10.1038/nm1202-799
Ribeil J-A, Hacein-Bey-Abina S, Payen E et al (2017) Gene therapy in a patient with sickle cell disease. N Engl J Med 376:848–855. https://doi.org/10.1056/NEJMoa1609677
Rolan P, Mercer A, Wootton R, Posner J (1995) Pharmacokinetics and pharmacodynamics of tucaresol, an antisickling agent, in healthy volunteers. Br J Clin Pharmacol 39:375–380. https://doi.org/10.1111/j.1365-2125.1995.tb04465.x
Safo MK, Kato GJ (2014) Therapeutic strategies to alter the oxygen affinity of sickle hemoglobin. Hematol Oncol Clin North Am 28:217–231. https://doi.org/10.1016/j.hoc.2013.11.001
Singh A, Xu YJ (2016) The cell killing mechanisms of hydroxyurea. Genes (Basel). 7:99
Ueda Y, Nagel RL, Bookchin RM (1979) An increased Bohr effect in sickle cell anemia. Blood 53:472–480
Wales TE, Fadgen KE, Gerhardt GC, Engen JR (2008) High-speed and high-resolution UPLC separation at zero degrees celsius. Anal Chem 80:6815–6820. https://doi.org/10.1021/ac8008862
Waterman MR, Yamaoka K, Dahm L et al (1974) Noncovalent modification of deoxyhemoglobin S solubility and erythrocyte sickling. Proc Natl Acad Sci USA 71:2222–2225. https://doi.org/10.1073/pnas.71.6.2222
Weis DD, Engen JR, Kass IJ (2006) Semi-automated data processing of hydrogen exchange mass spectra using HX-Express. J Am Soc Mass Spectrom 17:1700–1703. https://doi.org/10.1016/j.jasms.2006.07.025
Wintrobe MM (2014) Wintrobe's clinical hematology. In: Wintrobe's clinical hematology, 13th edn, Part IV, Sect 4, Chap 33
Wishner BC, Ward KB, Lattman EE, Love WE (1975) Crystal structure of sickle-cell deoxyhemoglobin at 5 Å resolution. J Mol Biol 98:179–194. https://doi.org/10.1016/S0022-2836(75)80108-2
Wood JC, Enriquez C, Ghugre N et al (2005) MRI R2 and R2* mapping accurately estimates hepatic iron concentration in transfusion-dependent thalassemia and sickle cell disease patients. Blood 106:1460–1465. https://doi.org/10.1182/blood-2004-10-3982

Woodard P, Helton KJ, Khan RB et al (2005) Brain parenchymal damage after haematopoietic stem cell transplantation for severe sickle cell disease. Br J Haematol. https://doi.org/10.1111/j.1365-2141.2005.05491.x

Yohe ME, Sheffield KM, Mukerji I (2000) Solubility of fluoromethemoglobin S: effect of phosphate and temperature on polymerization. Biophys J 78:3218–3226. https://doi.org/10.1016/S0006-3495(00)76858-5

Zhang Z, Smith DL (1993) Determination of amide hydrogen exchange by mass spectrometry: a new tool for protein structure elucidation. Protein Sci 2:522–531

Chapter 13
Multiplicity and Polymorphism of Fish Hemoglobins

Øivind Andersen

Abstract The diversity of fish hemoglobins and the association with oxygen availability and physiological requirements during the life cycle has attracted scientists since the first report on multiple hemoglobin in fishes (Buhler and Shanks 1959). The functional heterogeneity of the fish hemoglobins enables many species to tolerate hypoxic conditions and exhausting swimming, but also to maintain the gas pressure in the swim bladder at large depths. The hemoglobin repertoire has further increased in various species displaying polymorphic hemoglobin variants differing in oxygen binding properties. The multiplicity of fish hemoglobins as particularly found in the tetraploid salmonids strongly contrasts with the complete loss of hemoglobins in Antarctic icefishes and illustrates the adaptive radiation in the oxygen transport of this successful vertebrate group.

Keywords Fish hemoglobins · Polymorphism · Isoforms · Functional heterogeneity · Oxygen affinity · Oxygen transport · Bohr effect · Atlantic cod · Salmonids · Eels · Flatfish

Diversity of Vertebrate Hemoglobins

The uptake and transport of oxygen by hemoglobin has fascinated scientists since the oxygen-carrying property of the heme protein was *discovered* in 1840 (Hünefeld 1840). The tetrameric hemoglobin molecule in jawed vertebrates is composed of two α and two β globin subunits, which interact to facilitate the reversible binding of a single O_2 molecule to each of the four heme groups (Fig. 13.1a) (Perutz 1970). The hemoglobin-oxygen affinity is modulated by pH, temperature and allosteric phosphate effectors to meet short-term variations in oxygen availability and physiological requirements (Clementi et al. 1994; Giardina et al. 2004). Furthermore, amino acid changes at key positions in the hemoglobin molecule have modified the oxygen binding properties and resulted in the adaptation of various vertebrate groups to a wide

Ø. Andersen (✉)
Norwegian Institute of Food, Fisheries and Aquaculture Research (NOFIMA), PO BOX 210,1431 Ås, Norway
e-mail: Oivind.Andersen@Nofima.no

U. Hoeger and J. R. Harris (eds.), *Vertebrate and Invertebrate Respiratory Proteins, Lipoproteins and other Body Fluid Proteins*, Subcellular Biochemistry 94,
https://doi.org/10.1007/978-3-030-41769-7_13

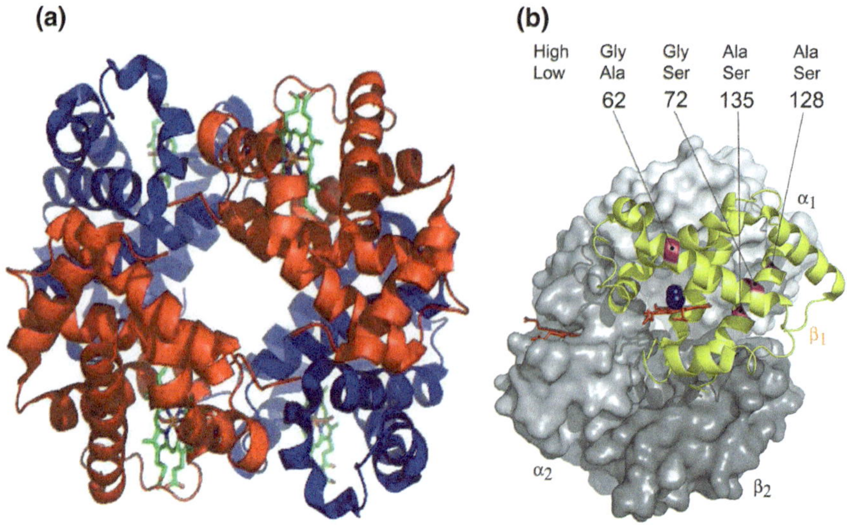

Fig. 13.1 The oxygen-binding hemoglobin molecule. **a** Vertebrate hemoglobin consists of two α and two β globin subunits (blue and red), which cooperatively bind oxygen to the four heme groups (green). **b** Four amino acid substitutions in the β globin of the deer mouse contribute to the divergence in the hemoglobin-oxygen affinity between populations adapted to high-and low-land altitudes (Storz et al. 2009)

range of habitats. Birds flying at high-altitude exhibit several amino acid substitutions in the α and β subunits that give higher oxygen affinity compared to related lowland species (Hiebl et al. 1988a, b; Scott 2011; McCracken et al. 2010). In the cold-adapted woolly mammoth (*Mammuthus primigenius*) a Glu-Gln replacement at the interface between the $\alpha_1\beta_1$ dimer subunits lowered the effect of temperature on the oxygen affinity and facilitated the release of oxygen at low temperatures (Campbell et al. 2010). Hemoglobin polymorphisms in the deer mouse (*Peromyscus maniculatus*) underly the difference in oxygen affinity and the divergence in hypoxia tolerance between the highland and lowland phenotypes (Fig. 13.1b) (Storz et al. 2007, 2009). Similarly, in toadfish (*Thalassophryne maculaso*) the polymorphic Hb-II-b variant has higher oxygen affinity than the Hb-II-a type that might be associated with the adaptation to poorly oxygenated water (Pérez 1986). Poecilid fish species colonizing sulfide springs displayed significant shifts in globin gene expression and amino acid changes in several globins indicating relaxed selection compared to sulfide-intolerant species (Barts et al. 2018). The examples illustrate the variety of adaptative changes that underly the tremendous diversity of vertebrate hemoglobins, which have evolved to optimize the oxygen transport under vastly different environmental and physiological conditions. Structural and functional variations are extensively found in fish wherein the hemoglobins in the gill capillaries make up the interface between the exothermic organism and the ambient water varying greatly in temperature and oxygen availability.

Multiplicity of Fish Hemoglobins

Fish typically exhibit multiple hemoglobin components or isoforms, and the functional heterogeneity found in many species seems to be associated with the fluctuating physicochemical environment and the varying oxygen needs during the life cycle. Anadromous salmonids and catadromous eels perform exhausting migrations between freshwater and sea water at variable oxygen levels, temperature and hydrostatic pressure, and the oxygen requirements are met by the production of functionally distinct hemoglobin isoforms (Weber 2000; De Souza and Bonilla-Rodriuez 2007). Fish hemoglobin are broadly classified as cathodic and anodic isoforms, which differ in intrinsic oxygen affinities and sensitivity to allosteric regulators (Powers 1972; Giles and Randall 1979; Weber and Jensen 1988; Weber 1990). The cathodic isoforms have high oxygen affinity and show low sensitivity to changes in pH that is important for securing oxygen transport when blood pH drops during exercise and hypoxic conditions. In comparison, the anodic isoforms bind oxygen with relative low affinity and respond to reduced pH by decreased binding affinities. In addition to this Bohr effect, nearly all anodic fish isoforms also exhibit the Root effect whereby the oxygen carrying capacity is reduced at low pH (Fig. 13.2). The inability of hemoglobin to become fully oxygen saturated even at high oxygen tension is crucial for the secretion of oxygen into the swim bladder for buoyancy adjustment and for suppling oxygen to the avascular retina needed for visual acuity (Root 1931; Pelster and Randall 1998; Waser and Heisler 2005; Berenbrink et al. 2005).

The functionally heterogeneous hemoglobins in salmonids and eels contrast with the hemoglobins in the sedentary flatfishes and carp, which possess only anodic isoforms with similar oxygenation properties (Gillen and Riggs 1972; Weber and de Wilde 1975). Carps live in habitats with large diurnal and seasonal variations in oxygen tensions and show very high tolerance to low hypoxic conditions. Common carp (*Cyprinus carpio*) exhibits three major and one minor hemoglobin components, which show strongly increased oxygen affinity during hypoxia as the result of reduced

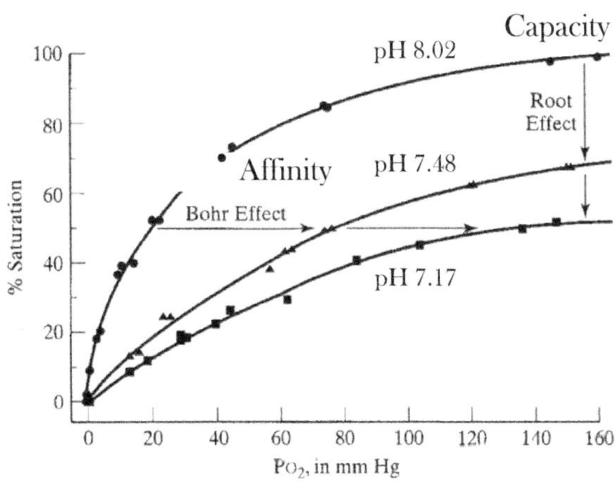

Fig. 13.2 Effects of blood pH on the oxygen binding affinity (Bohr effect) and capacity (Root effect) of winter flounder hemoglobin measured at 15 °C Modified from Hayden et al. (1975)

GTP and ATP levels (Gillen and Riggs 1972; Weber and Lykkeboe 1978; Ohkubo et al. 1993). The common carp genome suggests redundancy in the hemoglobin system by containing totally 35 globin genes that includes eight pseudogenes coding for functionally inactive globins (Table 13.1). The single cathodic β globin might be part of the larval hemoglobins, which in fish tend to display little or no Bohr effect of beneficial importance for early development (Rombough 1997; Rombough and Drader 2009). In comparison, the tetraploid genome of the related goldfish (*Carassius auratus*) harbors totally 27 globin genes, including three pseudogenes, and also novel genes encoding ethanol-producing enzymes of crucial importance for surviving anoxic conditions (Fagernes et al. 2017). The extremely hypoxia-tolerant goldfish and the closely related crucian carp (*C. carassius*) exhibit up to four hemoglobin isoforms displaying very high oxygen affinity with strong sensitivity to pH and temperature (Houston and Cyr 1974; Kamshilov and Kamshilova 2007: Sollid et al. 2005).

A variety of fish species have been shown to acclimate to hypoxia or altered temperature by differentially regulating hemoglobin isoforms with specific properties. The Tibetan schizothoracine fish (*Schizopygopsis pylzovi*) was found to respond to hypoxia by altered globin gene expression and protein synthesis (Xia et al. 2016), and a hemoglobin switch in red drum (*Sciaenops ocellatus*) acclimated to hypoxia increased the aerobic performance (Pan et al. 2017). Lake Victoria cichlids showed higher oxygen affinity after exposure to chronic hypoxia that was associated with the production of functionally distinct hemoglobin isoforms in the African cichlid *Haplochromis ishmaeli* (Rutjes et al. 2007). However, the other cichlids examined displayed higher affinity mediated by decreased allosteric interactions that seems to

Table 13.1 Diversity in number of hemoglobin genes and components in various teleost species. The genes are found in the sequenced genomes available at www.ncbi.nlm.gov. ND, no data

Species	Hb genes (pseudogenes)	Hb components	Comments
Rainbow trout	42 (4)	9+	Tetraploid
Arctic charr	36 (5)	10	Tetraploid
Northern pike	27 (3)	13+	Diploid
Common carp	35 (8)	4	Diploid
Goldfish	27 (3)	4	Tetraploid
Japanese eel	10	6	Catadromous
Swamp eel	11 (1)	6	Air-breather
Electric eel	11	1	Air-breather
Tongue sole	8	ND	Flatfish
Japanese flounder	ND	6	Flatfish
Polar cod	8	3	Arctic
Antarctic bullhead	5	2	Antarctic
Icefishes	(1)	0	Antarctic

be the general mechanism in fish for rapid adjustments in oxygenation properties in response to fluctuating oxygen availability (Wood and Johansen 1972; Weber and Lykkeboe 1978; Weber and Jensen 1988; Weber 2000).

Hemoglobins in Air-Breathing Fish

Air-breathing has evolved in many fish lineages by the development of various respiratory tissues, which differ in efficiency of gas exchange that seems to be related to the hemoglobin characteristics (Dutta and Munshi 1985; Tate et al. 2017). The facultative air-breathing Asian swamp eel (*Monopterus albus*) has reduced gills and is forced to air breathe during periods of aquatic hypoxia (Fig. 13.3a) (Liam 1967). The six distinct hemoglobin isoforms, which are coded by ten globin genes, have high oxygen carrying capacity and unusual high affinity that secure adequate oxygen uptake through the thick respiratory epithelium in the buccopharyngeal cavity (Damsgaard et al. 2014). The abantid fishes gulp air-bubbles to ventilate the labyrinth organ for gas exchange with the circulating blood (Fig. 13.3b) (Burggren 1979; Munshi et al. 1986). The obligate air-breathing climbing perch (*Anabas testudineus*) is unable to obtain enough oxygen in water of high CO_2 and low pH (Hughes and Singh 1970) suggesting hemoglobins with strong Bohr effect. Consistently, the genome contains 18 functional genes, but only a single non-Bohr β globin gene. In comparison, the blood of the obligate air-breathing electric eel (*Electrophorus electricus*) was reported to contain only one hemoglobin component (Huber and Braunitzer 1989), while the genome harbours 11 globin genes. Differences in hemoglobin properties between closely related abantid species are demonstrated by the very high oxygen affinity and large Bohr shift in the hypoxia-tolerant siamese fighting fish (*Betta splendnes*), in

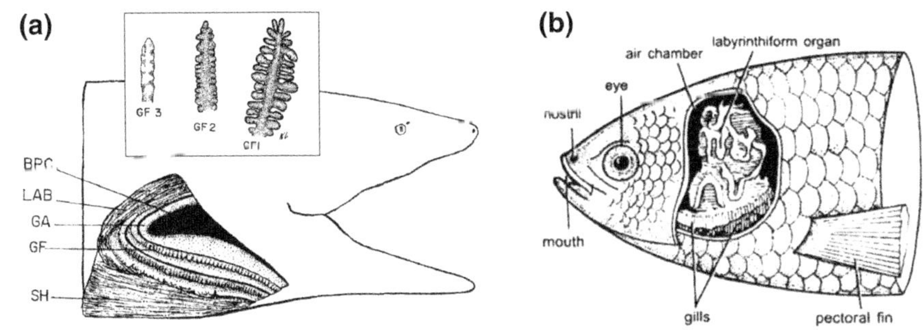

Fig. 13.3 Accessory respiratory organs in air-breathing fish. **a** The Asian swamp eel has reduced gills (inserted) and instead utilizes the buccopharyngeal cavity for gas exchange (Liam 1967). BPC; buccopharyngeal cavity, LAB; levator arcus branchialis musculature, GA; gill arch, GF; gill filaments, SH; sternohyoideus muscle. **b** The labyrinth organ of abantid fishes is covered with a vascular mucous membrane that brings about respiration

contrast to the blue gourami (*Trichopodus trichopterus*) inhabiting well-oxygenated waters (Mendez-Sanchez and Burggren 2017).

Loss of Hemoglobins in Antarctic Fish

The multiplicity of hemoglobins in most teleosts contrasts with the low number or complete loss of hemoglobins in Antarctic notothenioid fishes, which have adapted to the constant sub-zero water temperatures and high dissolved oxygen levels (Ruud 1954; Barber et al. 1981; Cocca et al. 1995; Clementi et al. 1994; Near et al. 2006). The sluggish icefishes possess only remnants of an α globin gene (Fig. 13.4a), and the oxygen demand is largely achieved by the enormous heart pumping large blood volumes through the wide vessels at extremely low pressure (Hemmingsen et al. 1972; Hemmingsen 1991; Harrison et al. 1991). The red-blooded Antarctic notothenioids typically have a single major hemoglobin component accounting for more than 90% of total hemoglobin and a minor component with similar properties (di Prisco and D'Avino 1989; di Prisco et al. 1991; D'Avino et al. 1990). Consistently, the genome of the Antarctic bullhead (*Notothenia coriiceps*) contains only three α and two β globin genes. Exceptions are found in more active species exhibiting several functionally distinct hemoglobins, and the three major components of the migratory Antarctic

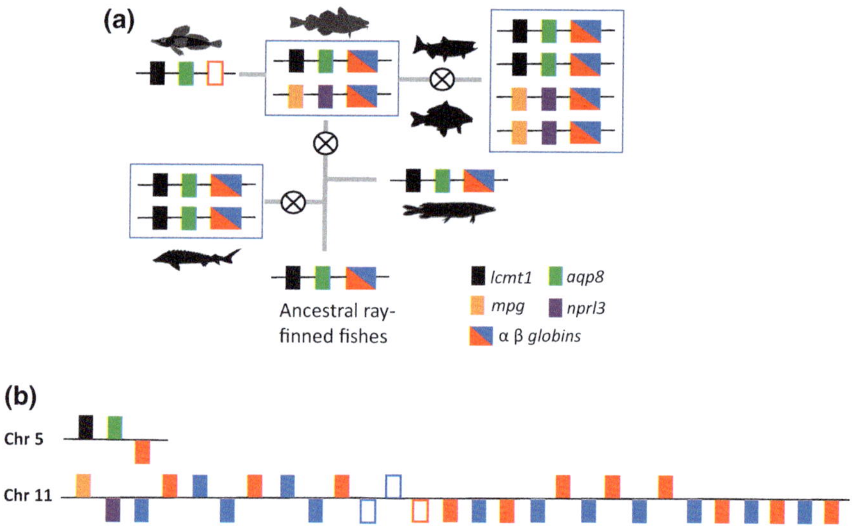

Fig. 13.4 Genomic organization of the α and β globin genes in fish. **a** Dublications (⊗) and loss of the globin clusters in different fish lineages represented by sterlet sturgeon, spotted gar, Atlantic cod, icefishes, salmonids and goldfish. **b** All globin genes in the diploid Northern pike are located in the MN cluster on chromosome 11, except for a single α gene in the LA cluster on chromosome 5. Genes transcribed in opposite directions are depicted above or below the solid line. Pseudogenes are shown in white

silverfish (*Pleurogramma antarcticum*) have higher oxygen affinity than in other notothenioids (D'Avino et al. 1994; Tamburrini et al. 1996; Riccio et al. 2002). Low hemoglobin numbers are also found in the sedentary Arctic kelp snailfish (*Liparis tunicatus*) possessing one major and one minor components, while three functionally distinct isoforms are found in the Arctic wolffish (*Anarhichas minor*), Arctic cod (*Arctogadus glacialis*) and polar cod (*Boreogadus saida*) (Verde et al. 2002, 2006; Giordano et al. 2007). In contrast, the highly adaptive Arctic charr (*Salvelinus fontinalis*) has multiple hemoglobins and the tetraploid genome contains 31 functionally globin genes of which 14 showed downregulated expression associated with heat tolerance (Quinn et al. 2011).

Genomic Organization of Fish Hemoglobins

The hemoglobin gene arose before the separation of ray-finned fish (Actinopterygians) and lobe-finned fish (Sarcopterygians) about 500 million years ago and was followed by a gene duplication in the ancestor of jawed vertebrates that yielded the α and β globin subunits (Goodman et al. 1987; Hoffmann et al. 2012; Storz et al. 2011, 2013). Subsequent duplications resulted in multiple α and β genes, which were likely organized in a single gene cluster as found in spotted gar (*Lepisosteus oculatus*) representing an unduplicated sister group of teleost fishes (Fig. 13.4a). The teleost-specific whole genome duplication event about 320 million years ago gave rise to two globin clusters designated LA and MN named after the conserved flanking gene pairs *Lcmt1-Aqp8* and *Mpg-Nprl3*, respectively (Opazo et al. 2013). The teleost globin clusters contain both α and β genes, in contrast to their location on different chromosomes in amniote vertebrates. The variable number of globin genes in different species resulted from lineage-specific gene duplications and deletions. The repertoire of α and β genes has mainly expanded in the MN cluster, as particularly demonstrated in the Northern pike (*Esox lucius*) exhibiting only a single α gene in the LA cluster, while the MN cluster harbors 26 α and β globin genes, including three pseudogenes (Fig. 13.4b). Both globin clusters have been duplicated in the tetraploid genomes of goldfish and salmonids possessing four clusters located on separate chromosomes. The primitive sturgeons separated from the other ray-finned fishes about 360 million years ago before the teleost-specific genome duplication event, but the sturgeon genome has undergone up to three separate ploidization events (Rajkov et al. 2014). Accordingly, the tetraploid genome of the sterlet sturgeon (*Acipenser ruthenus*) (Andreyushkova et al. 2017) contains two MN globin clusters that likely resulted from the duplication of the single ancestral globin cluster, while no globin genes are found adjacent to the *Lcmt1-Aqp8* gene pair.

Hemoglobin Polymorphisms in Atlantic Cod

The diversity of fish hemoglobins is further increased in various species by the expression of polymorphic globin genes coding for allelic variants. Although polymorphic globins have been reported in many teleosts, the structural and functional properties have been examined in only a few cases. Polymorphism in fish hemoglobin was first identified in Atlantic cod and whiting by electrophoretic separation of three hemoglobin variants designated HbI-11, HbI-22 and heterozygous HbI-12 (Sick 1961) (Fig. 13.5a). The molecular background of the cod hemoglobin polymorphism was revealed by the identification of two amino acid substitutions at positions 55 and 62 in the β1 globin that gave rise to two variants Met55-Lys62 and Val55-Ala62 corresponding to the HbI-1 and HbI-2 phenotypes, respectively (Andersen et al. 2009; Borza et al. 2009) (Fig. 13.5b).

The two hemoglobin variants in Atlantic cod display a north-south distribution along the Norwegian coast with the almost complete dominance of HbI-2 fish in the northern region, while the HbI-1 type mainly occurs in the mid- and southern coastal waters (Fig. 13.6). Accordingly, HbI-2 cod was shown to select lower temperature (8.2 °C) than HbI-1 fish (15.4 °C), while only the latter phenotype preferred a lower temperature (9.8 °C) during hypoxia (Petersen and Steffensen 2003). The replacement of the Met residue with the smaller Val in HbI-2 fish was predicted to increase the oxygen affinity by increasing the gap at the α1-β1 interface (Andersen et al. 2009). The importance of this subunit contact for the stability of the dimer and the oxygen affinity is demonstrated in the high-altitude tolerant Andean goose and bar-headed goose possessing the Leu55β→Ser and Pro119α→Ala changes, respectively, when compared to lowland geese with normal oxygen affinity (Jessen et al. 1991; McCracken et al. 2010; Natarajan et al. 2018). Effects of acclimation

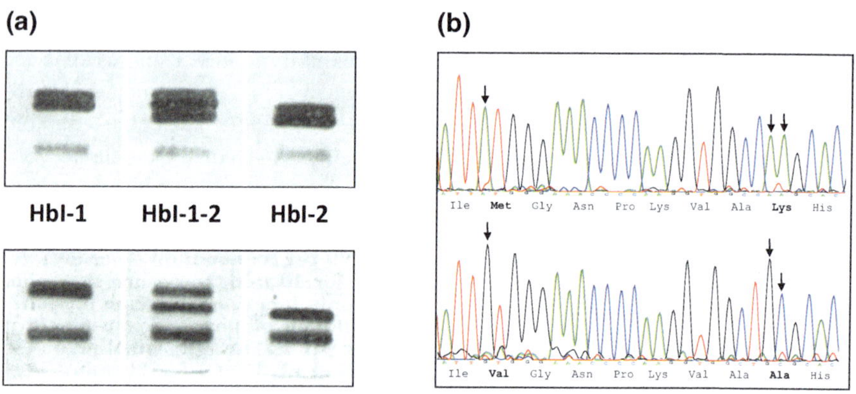

Fig. 13.5 Hemoglobin polymorphisms in codfishes. **a** Polymorphic hemoglobins in Atlantic cod (upper panel) and whiting (lower panel) separated by gel electrophoresis (Sick 1961). **b** The homozygous HbI-1 and HbI-2 types in Atlantic cod correspond to the Met55-Lys62 and Val55-Ala62 variants of the β1 globin, respectively (Andersen et al. 2009)

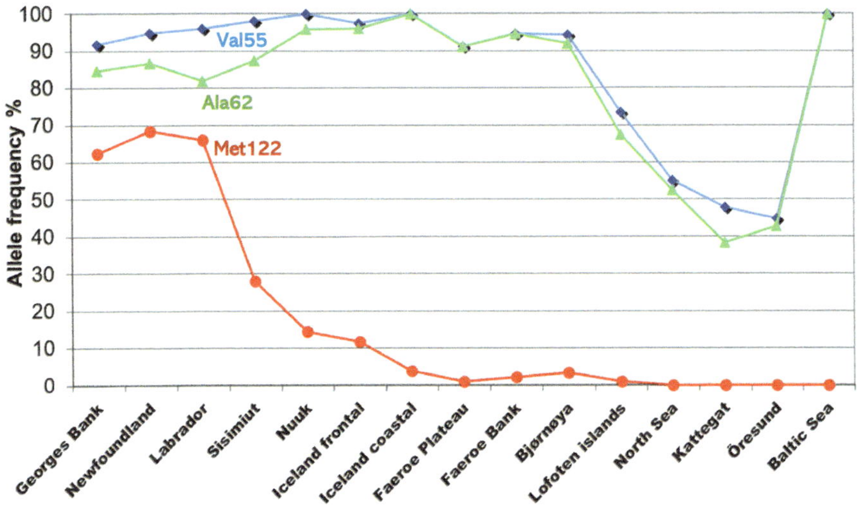

Fig. 13.6 Geographic distribution of the β1 globin polymorphisms in Atlantic cod. Allele frequencies of Val55, Ala62 and Met122 are presented. A north-south gradient of the HbI-1 (Met55-Lys62) and HbI-2 (Ala55-Val62) variants are shown in Northeast Atlantic populations (Andersen et al. 2014)

temperature on the oxygen binding properties in Atlantic cod was suggested by the significantly higher oxygen affinity in HbI-2 than in HbI-1 at both 4 °C and 12 °C when acclimatized to 4 °C, while no difference between the two types was shown after acclimation to 12 °C (Brix et al. 2004). However, the results were not confirmed in a recent study, which showed no difference in thermal sensitivity of the cod HbI genotypes when acclimatized to either 5, 12.5 or 20 °C (Barlow et al. 2017).

The functional significance of the cod hemoglobin polymorphism is further complicated by the presence of two promoter variants, which differ in length due to an insertion/deletion polymorphism (Fig. 13.7) (Star et al. 2011). The resulting short and long promoter variants are strongly associated with the HbI-1 and HbI-2 types, respectively, and the higher activity of the long promoter than the short variant at high temperatures suggests higher hemoglobin levels and higher oxygen capacity in HbI-2 fish at elevated temperatures (Star et al. 2011). In addition to the Met55Val and Lys62Ala polymorphisms, a Leu-Met substitution at position 122 of the cod β1 globin are found in Northwest Atlantic populations (Fig. 13.6) (Borza et al. 2009). The Met122 variant is almost lacking in the Northeast Atlantic, while intermediate frequencies are found in cod populations at Greenland, Iceland and Bjørnøya suggesting that this polymorphism originated in Northwest Atlantic populations and has later expanded in the eastern direction. Structural modelling predicted that the Leu122Met replacement may offer an adaptive advantage by modifying the interaction with the highly conserved Arg31α at the α1-β1 subunit interface (Andersen et al. 2014).

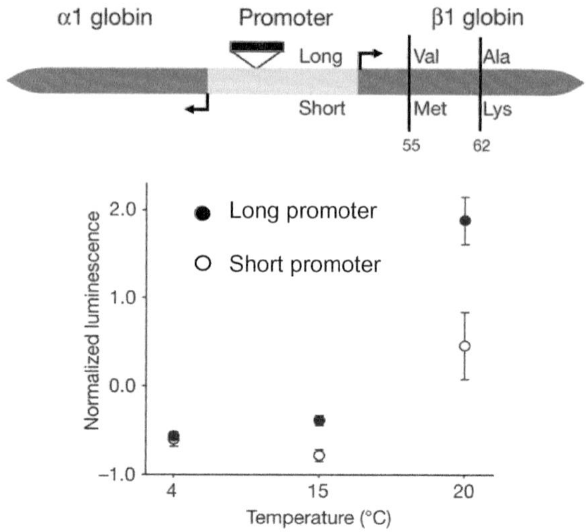

Fig. 13.7 The bi-directional promoter of the α1-β1 globin gene pair in Atlantic cod displays a polymorphic indel. The long promoter variant is associated with the HbI-2 type and showed higher activity than the short HbI-1 promoter at elevated temperatures in a luciferase assay (Star et al. 2011)

Large Repertoire of Salmonid Hemoglobins

All salmonids begin their life in freshwater and anadromous species migrate to the sea for feeding before returning to the natal river for spawning. During their homing migration salmonids undertake active vertical migrations down to about 100 m depths to obtain directional cues and for behavioral thermoregulation (Døving et al. 1985; Tanaka et al. 2000). Rather than modulating buoyancy for changing depth, the function of the salmonid swim bladder is probably to obtain neutral buoyancy in the surface water (Tanaka et al. 2001). The exhausting upward swimming against strong currents and rapids requires sufficient oxygen uptake and delivery to the skeletal and cardiac muscles. Salmonids possess multiple hemoglobin with distinct functional properties and several isoforms are produced at specific developmental stages (Tsuyuki and Ronald 1971; Giles and Vanstone 1976; Giles and Randall 1980; Houde et al. 2019). Anodic isoforms were found throughout the life cycle in king salmon (*Oncorhynchus tshawytscha*), and the additional cathodic isoforms at the adult stage showed higher oxygen affinity than the anodic components (Harrington 1986). This is consistent with the different properties of the cathodic HbIV and the anodic HbI components in rainbow trout, wherein the HbI isoform is sensitive to changes in pH, temperature and organic phosphates (Binotti et al. 1971; Weber et al. 1976b).

The molecular mechanisms underlying the pH sensitivity of the anodic hemoglobins are not clearly understood, but the C-terminal histidine residue in the anodic β globins is thought to account for about 50% of the Bohr effect (Perutz et al. 1980; Shih et al. 1993). The imidazole ring of His147 in the Bohr β globin forms a salt-bridge with the negative charged Asp, or Glu in non-salmonids, on position 94 (Fig. 13.8). This bond is broken upon oxygen binding and results in the release of so-called Bohr protons at physiological conditions (Tame et al. 1996). In comparison, the 3D structure of the non-Bohr β globin shows no interaction between

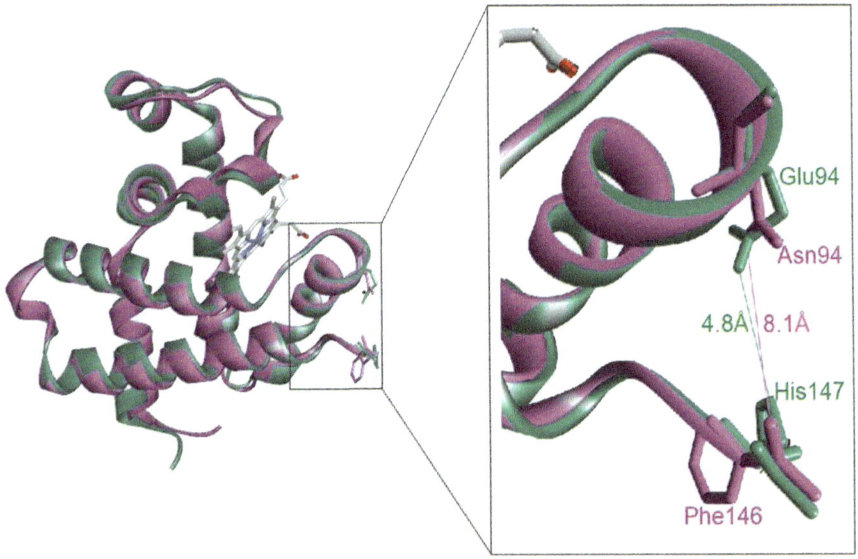

Fig. 13.8 Superposed 3D structures of the Bohr effect HbIV (green) and the non-Bohr HbI (magenta) β subunits of rainbow trout. HbIV shows a salt bridge between the C-terminal His147 and the charged Glu94, while no binding is formed in HbI between the C-terminal Phe146 and the neutral Asn94 at a distance of 8.1Å Courtesy of Dr. Maria Cristina De Rosa

the C-terminal Phe residue and the neutral Asn94. The strong pH sensitivity of the anodic isoforms can be critical for the oxygen uptake when the blood pH drops during exercise and other stressful situations. Salmonids and probably other teleosts with Root-effect hemoglobins have developed a mechanism to maintain oxygen loading under such conditions by regulating pH in the red blood cells at the gills via a β-adrenergic sodium-proton exchanger (βNHE) (Rummer et al. 2013; Randall et al. 2014; Rummer and Brauner 2015; Shu et al. 2017). In contrast, the oxygen delivery in the red muscle and possibly other tissues is facilitated by the low pH as the result of the local carbonic anhydrase activity that short-circuits the extrusion of protons by the membrane pump.

The tetraploidization of the salmonid genome about 90 million years ago resulted in the duplication of the teleost MN and LA globin clusters, which comprise about 40 α and β globins, including several pseudogenes, in salmonids. Rainbow trout exhibits 38 functional globin genes of which seven code for non-Bohr globins, and the cathodic components consistently make up about 20% of the total hemoglobin content in adult rainbow trout (Fago et al. 2002). The cathodic non-Bohr globins are not confined to active anadromous salmonids, but are also found in the freshwater whitefish (*Coregonus clupeaformis*) possessing four anodic and four cathodic hemoglobin components (Evans et al. 2012). The greater proportion of cathodic components in the dwarf ecotype than in normal whitefish was suggestive of divergent abilities to utilize different trophic niches. The lack of cathodic globin genes in the

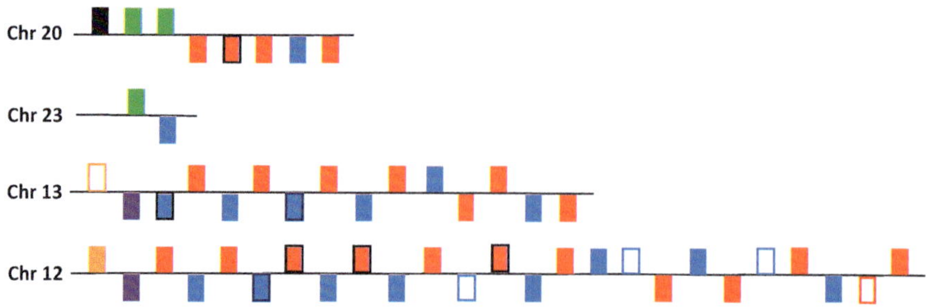

Fig. 13.9 The tetraploid genome of rainbow trout contains four globin clusters comprising totally 22 α (red) and 19 β (blue) globin genes. Four α1 and three β1 non-Bohr globins are framed, and the pseudogenes are shown in white. Color codes for the flanking genes are shown in Fig. 13.4. Genes transcribed in opposite directions are depicted above or below the solid line

closely related Northern pike suggests that the cathodic salmonid globins evolved after the tetraploidization event. Accordingly, convergent evolution of functionally similar hemoglobins was proposed in the phylogenetic analysis of multiple anodic and cathodic fish globins (Opazo et al. 2012).

Hemoglobin polymorphism has been reported in several salmonid species, but the variation in electrophoretic patterns might be due to non-genetic factors, such as environmental differences, ontogenic hemoglobin switching or individual differences in the asymmetric subunit composition of the hemoglobin tetramer. Genetic polymorphism in cutthroat trout hemoglobin was suggested by the different numbers of cathodic components in adult fish sampled in the Yellowstone Lake (Braman et al. 1980). Similarly, Arctic charr displayed three and five cathodic components in stocks from the Frasier River, Labrador, and the Storvatn Lake, Norway, respectively, while both phenotypes were found in fish from the Canadian Nauyuk Lake (Fig. 13.10)

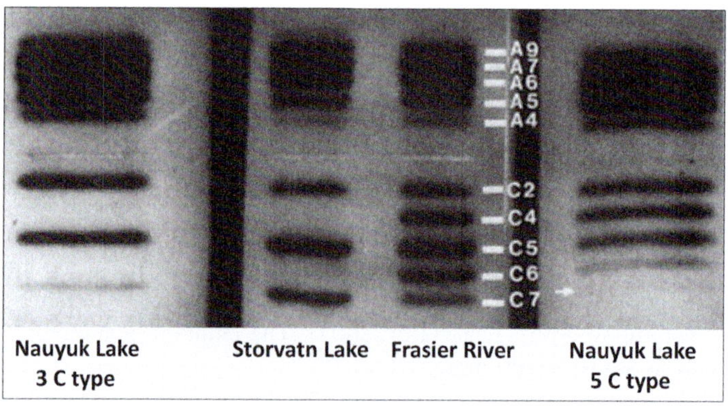

Fig. 13.10 Polymorphism in the number of cathodic hemoglobin components in adult Arctic charr originating from Nauyuk Lake, Storvatn Lake and Frasier River. White arrow indicates the C7 component (Giles 1991)

(Giles 1991). The latter stock displayed higher standard oxygen consumption rates and higher oxygen binding affinity with larger Bohr shift than the other stocks. Furthermore, hemoglobin polymorphism in an anodic component of chinook salmon (*Oncorhynchus tshawytscha*) showed different allele frequencies in Canadian stocks (Fyhn and Withler 1991). The recent sequencing of salmonid genomes will contribute to identify genetic variation in the hemoglobins and the adaptive value of functional differences in diverged populations.

Hemoglobins and the Extreme Lifestyle of Eels

Eels are extremely tolerant to varying oxygen levels and survive in hypoxic and even brief anoxic conditions, but also prolonged periods of air breathing on land (Weber et al. 1975). The slow undulations of the tail make the eels very efficient swimmers, and catadromous species swim continuously thousands of kilometers for up to six months to reach the spawning site in the ocean (Aoyama 2009; van Ginneken et al. 2005; Palstra and van den Thillart 2010). The sustained swimming activity is mainly powered by the aerobic red muscle using lipid stores as energy substrate. Migratory eels perform diurnal vertical movements down to 800 m depths (Aarestrup et al. 2009), and the high hydrostatic pressure has been shown to reduce the oxygen consumption by increasing the mitochondrial membrane fluidity (Sébert et al. 2009). The swimming costs are further reduced by the swim bladder, although the fish is probably negative buoyant at large depths due to the slow gas secretion into the swim bladder (Pelster 2015). In addition, more oxygen would be needed to fill the swim bladder than consumed by the aerobic swimming muscle.

The eel hemoglobins consist of both anodic and cathodic isoforms, which differ in oxygen affinities and sensitivity to allosteric regulators of importance for the oxygen uptake and delivery during air-breathing on land and at varying depths (Weber et al. 1976a; Pelster and Weber 1990; Fago et al. 1995). Prolonged air-exposure causes acid-base disturbances and reduced blood pH similar to the situations at exercise and hypoxia. Without sufficient regulation of erythrocytic pH at the gills as found in salmonids (Rummer et al. 2013), the oxygen transport is secured by the pH insensitive cathodic isoforms binding oxygen at higher affinity than the anodic isoforms. Contrasting with salmonids, the cathodic eel hemoglobins respond to GTP, which is the major allosteric phosphate effector in eel, and the reduced GTP levels under hypoxia increase the binding affinity that safeguards oxygen loading (Fago et al. 1995; Weber et al. 1976a; Pelster and Weber 1990). The anodic isoform is highly pH sensitive with a strong Root effect of importance for the secretion of oxygen into the swimbladder during diving as the result of the blood acidification in the gas gland (Steen 1963; Pelster 2015).

Hemoglobin polymorphism was documented in American eel (*Anguilla rostrata*) by electrophoretic separation of two homozygous variants of which one allelic variant was identified in a single individual out of 666 fish examined (Fig. 13.11a) (Sick et al. 1967). The rare type was apparently not present in the twelve eels examined

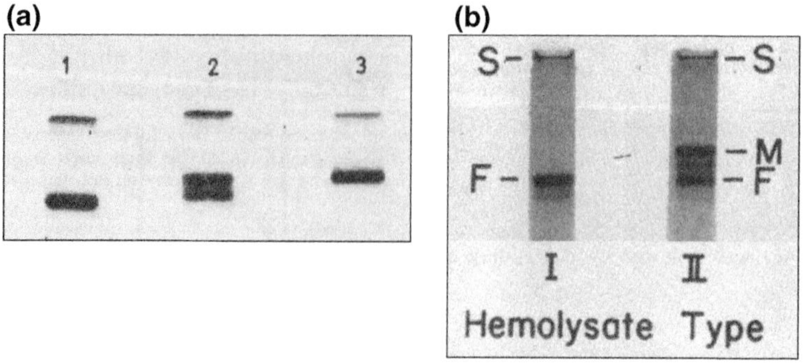

Fig. 13.11 Polymorphism in the anodic hemoglobin component of American eel. **a** Electrophoretic patterns of the common homozygous form (1), the heterozygous form (2) and the rare homozygous form (3) (Sick et al. 1967). **b** Two hemoglobin types of American eel possessing the F (type I) or the F and M (type II) major anodic components (Gillen and Riggs 1973)

by Gillen and Riggs (1973) reporting two types displaying either a single (F) or two (F + M) major cathodic components, respectively (Fig. 13.11b) of which the latter type II seems to identical to the heterozygous variant identified by Sick et al. (1967). The highly similar oxygenation properties and large Bohr effects of the F and M components suggest minor functional differences between the putative homozygous I and heterozygous II types, while the properties of the rare variant is unknown (Gillen and Riggs 1973).

Hemoglobins of the Bottom-Dwelling Flatfishes

During the larval metamorphosis flatfishes become highly specialized for a bottom-living life by the dorso-ventral flattening of the body and the movement of the eyes to the upper side. Flatfishes are all predators lying buried in the sand for long periods interrupted by short burst swimming for hunting and escape. The gill surface area is relatively small and cutaneous respiration may represent as much as 30% of the total oxygen consumption, which is significantly lower in flatfishes than in most roundfishes (Nonnotte and Kirsch 1978; Duthie 1982). The swim bladder is lost at the larval stage, but the negative buoyant flatfish takes advantage of the increased lift, or ground effect, when swimming near the bottom (Webb 2002). Many flatfish species also cover large distances by actively swimming upwards followed by powerless gliding to recover the oxygen dept in the anaerobic white muscle. The lactate produced in the muscle seems to be largely retained and utilized for glycogen resynthesis (Turner et al. 1983; Maxime et al. 2000).

Consistent with the rather inactive lifestyle, flatfishes seem to be lacking cathodic hemoglobins and the multiple anodic components display highly similar functional properties as found in flounder and plaice (Weber and de Wilde 1976). However, the genome sequences of Japanese flounder (*Paralichthyes olivaceus*) and tongue sole (*Cynoglossus semilaevis*) revealed a globin gene coding for a β globin lacking the C-terminal His residue of importance for pH sensitivity as shown in Fig. 13.8. Furthermore, many flatfish species have adapted to specific habitats and differ in tolerance to temperature and oxygen availability that seem to be associated with inter-species differences in hemoglobin functions. The European flounder (*Platichthys flesus*) inhabits very shallow waters with large fluctuations in dissolved oxygen, while the related plaice (*Pleuronectes platessa*) is common in well-aerated water at depths from 10 to 50 meters. The blood volume in the flounder is about twice the volume in plaice and has much higher hematocrit and hemoglobin levels (Weber and de Wilde 1975). In addition, the higher intrinsic oxygen affinity and less pH-sensitivity of the flounder hemoglobin isoforms, together with the higher gill ventilation during hypoxia, result in significant higher oxygen uptake in flounder than in plaice when subjected to low ambient oxygen levels (Fig. 13.12) (Weber and de Wilde 1975, 1976; Steffensen et al. 1982).

Although the flounder is well adapted to a life in shallow waters, southern populations seems to have developed higher capacity than northern populations to cope with elevated temperatures and hypoxia as documented by the contrasting patterns in energy metabolism (Pédron et al. 2017). The divergence between the populations might involve differences in functional characteristics of the hemoglobins, which showed polymorphism in flounders from Holland and Kattegat (Weber and de Wilde 1976). Consistently, genetic variants of the flounder α2 globin were identified by alignments of multiple cDNA sequences available from the species (Fig. 13.13a).

Turbot is widely distributed from the Northern Africa and the Mediterranean in the south to the Arctic circle. This flatfish tolerates low oxygen levels for long periods, and severe hypoxia induced increased gill ventilation together with increased hematocrit and hemoglobin levels (Maxime et al. 2000; Wu et al. 2016). Hemoglobin polymorphism has been reported in turbot (Manwell and Baker 1967), and a recent genetic

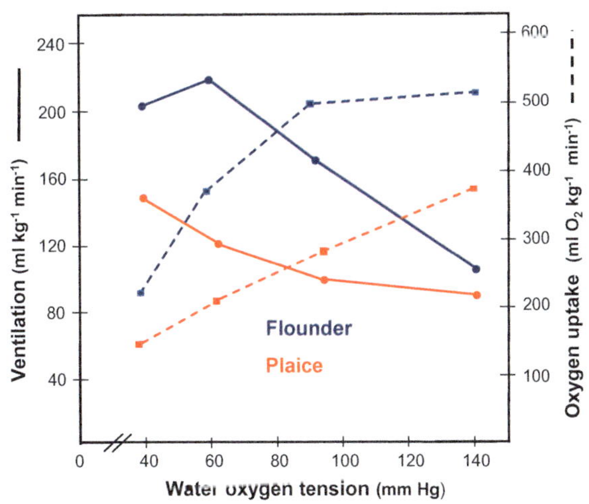

Fig. 13.12 Effects of decreasing oxygen tension on ventilation volume (solid lines) and oxygen uptake (stippled lines) in European flounder (blue) and plaice (red) Modified from Steffensen et al. (1982)

(a)

```
Flounder α2a  1 MSLSAKDKSLVRGLWAKAEGRVLDIGGEALGRMLVSYPQTKTYFAEWGTDLTPQSQKVGHHGGVIMGAVGMG  72
Flounder α2b  1 MSLSARDKSLVRGLWAKAEGRVLDIGGEALGRMLVSYPQTKTYFAEWGTDLTPQSQKVRHHGGVIMGAVGMG  72
Flounder α2c  1 MSLSAKDKSLVRGLWAKAEGRVLDIGGEALGRMLVSYPQTKAYFAEWGTNLTPQSQKVRHHGGVIMGAVGMG  72

Flounder α2a 73 VKYIVTLTEAMSSLSELHAFTLRVDPSNFKILAHSIILVMAMYYPKEFTAEVHVSFDKFLSCLAWALSEKYR 144
Flounder α2b 73 VKYIDTLTEAMSSLSELHAFTLRVDPSNFKILAHSIILVMAMYYPKEFTAEVHVSFDKFLSCLAWALSEKYR 144
Flounder α2c 73 VKYIDTLTEAMSSLSELHAFTLRVDPSNFKILAHGIILVMAMYYPKEFTAEVHVSFDKFLSCLAWALSEKYR 144
```

(b)

```
Turbot β2a  1 MVEWTDFERATIQNLFSKMDYELVGPAALSRCLVVYPWTQRYFGNFGNLYNAEAITSNENVINHGKVVLHGLDR 74
Turbot β2b  1 MVEWTDFERATIQNLFSKMDYELVGPAALSRCLVVYPWTQRYFGNFGNLYNAEAITSNENVINHGKVVLHGLDR 74

Turbot β2a 75 AVKNMDNIKDTYAELSLLHSEKLHVDPDNFRLLADCLTIVVASRMGLDFTGDVQAAFQKFLAVVVSSLGRQYH 147
Turbot β2b 75 AVKNMDNIKETYAELSLLHSEKLHVDPDNFRLLADCLTIVVASRMGSDFTGDVQAAFQKFLAVVVSSLGRQYH 147
```

Fig. 13.13 Polymorphism in flatfish hemoglobins. Sequence alignments of (**a**) three genetic variants of European flounder α2 globin and (**b**) two variants of turbot β2 globin

analysis of multiple populations revealed polymorphisms in several globins, including the Asp84Glu and Leu121Ser substitutions in the turbot β2 globin (Fig. 13.13b). The oxygen binding affinities of three phenotypes designated Hb-1/1, Hb-1/2 and Hb-2/2 were shown to be associated with differences in body growth of turbot reared at four experimental temperatures (Fig. 13.14) (Samuelsen et al. 1999; Imsland et al. 2000). The affinity was lowest in the most common Hb-1/1 fish and highest in the Hb-2/2 fish, which also showed highest body growth. Growth differences between the genotypes were largest at the optimal temperature, which was 23 °C for the Hb-2/2 type and 19 °C for Hb-1/1 (Imsland et al. 2000). Differences in distribution of the genotypes in northern turbot populations suggest adaptive changes of potential importance for the breeding of turbot in aquaculture (Imsland et al. 2003).

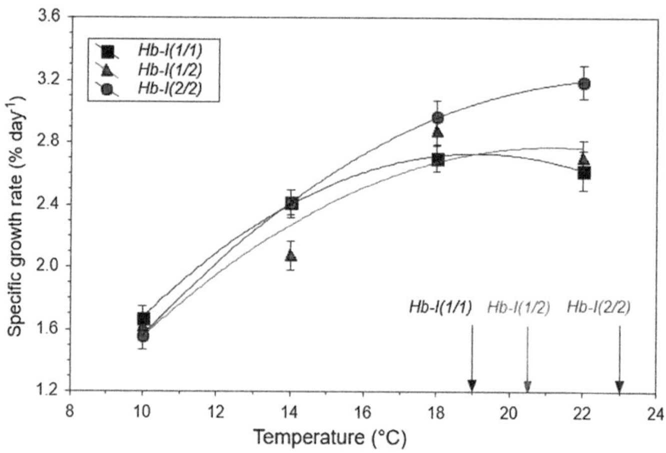

Fig. 13.14 Body growth in three hemoglobin genotypes of turbot reared at 10 °C, 14 °C, 18 °C or 22 °C for 10 weeks. Optimal temperatures for growth are indicated (Imsland et al. 2000)

References

Aarestrup K, Okland F, Hansen MM, Righton D, Gargan P, Castonguay M, Bernatchez L, Howey P, Sparholt H, Pedersen MI, McKinley RS (2009) Oceanic spawning migration of the European eel (*Anguilla anguilla*). Science 325:1660

Andersen Ø, De Rosa MC, Yadav P, Pirolli D, Fernandes JMO, Berg PR, Jentoft S, Andrè C (2014) The conserved Phe GH5 of importance for hemoglobin intersubunit contact is mutated in gadoid fish. BMC Evol Biol 14:54

Andersen Ø, Wetten OF, De Rosa MC, Andre C, Alinovi CC, Colafranceschi M, Brix O, Colosimo A (2009) Haemoglobin polymorphisms affect the oxygen binding properties in Atlantic cod populations. Proc R Soc B 276:833–841

Andreyushkova DA, Makunin AI, Beklemisheva VR, Romanenko SA, Druzhkova AS, Biltueva LB, Serdyukova NA, Graphodatsky AS, Trifonov VA (2017) Next generation sequencing of chromosome-specific libraries sheds light on genome evolution in paleotetraploid sterlet (*Acipenser ruthenus*). Genes 8:318

Aoyama J (2009) Life history and evolution of migration in catadromous eels (Genus *Anguilla*). Aqua-BioSci Monogr 2:1–42

Barber DL, Mills Westermann JE, White MG (1981) The blood cells of the Antarctic icefish *Chaenocephalus aceratus* Lönnberg: light and electron microscopic observations. J Fish Biol 19:11–28

Barlow SL, Metcalfe J, Righton DA, Berenbrink M (2017) Life on the edge: O_2 binding in Atlantic cod red blood cells near their southern distribution limit is not sensitive to temperature or haemoglobin genotype. J Exp Biol 220:414–424

Barts N, Greenway R, Passow CN, Arias-Rodriguez L, Kelley JL, Tobler M (2018) Molecular evolution and expression of oxygen transport genes in livebearing fishes (*Poeciliidae*) from hydrogen sulfide rich springs. Genome 61:273–286

Berenbrink M, Koldkjær P, Keep O, Cossins AR (2005) Evolution of oxygen secretion in fishes and the emergence of a complex physiological system. Science 307:1752–1757

Binotti I, Giovenco S, Giardina B, Antonini E, Brunori M, Wyman J (1971) Studies on the functional properties of fish hemoglobin. II. The oxygen equilibrium of the isolated hemoglobin components from trout blood. Arch Biochem Biophys 142:274–280

Borza T, Stone C, Gamperl AK, Bowman S (2009) Atlantic cod (*Gadus morhua*) hemoglobin genes: multiplicity and polymorphism. BMC Genet 10:51

Braman JC, Stalnaker CB, Klar GT, Farley TM (1980) Hemoglobin polymorphism in adult cutthroat trout (*Salmo clarkii*), hemoglobins. J Exp Zool 211:411–413

Brix O, Thorkildsen S, Colosimo A (2004) Temperature acclimation modulates the oxygen binding properties of the Atlantic cod (*Gadus morhua* L.) genotypes-HbI*1/1, HbI*1/2, and HbI*2/2 by changing the concentrations of their major hemoglobin components (results from growth studies at different temperatures). Comp Biochem Physiol A 138:241–251

Buhler DR, Shanks WE (1959) Multiple hemoglobins in fishes. Science 129:899–900

Burggren WW (1979) Bimodal gas exchange during variation in environmental oxygen and carbon dioxide in the air breathing fish *Trichogaster trichopterus*. J Exp Biol 82:197–213

Campbell KL, Roberts JEE, Watson LN, Stetefeld J, Sloan AM, Signore AV, Howatt JW, Tame JRH, Rohland N, Shen TJ, Austin JJ, Hofreiter M, Ho C, Weber RE, Cooper A (2010) Substitutions in woolly mammoth hemoglobin confer biochemical properties adaptive for cold tolerance. Nat Genet 42:536–540

Clementi MA, Condo SG, Castagnola M, Giardina B (1994) Hemoglobin function under extreme life conditions. Eur J Biochem 223:309–317

Cocca E, Ratnayake-Lecamwasam M, Parker S, Camardella L, Ciaramella M, di Prisco G, Detrich HW (1995) Genomic remnants of alpha-globin genes in the hemoglobinless antarctic icefishes. Proc Natl Acad Sci USA 92:1817–1821

Damsgaard C, Findorf I, Helbo S, Kocagoz Y, Buchanan R, Huong do TT, Weber RE, Fago A, Bayley M, Wang T (2014) High blood oxygen affinity in the air-breathing swamp eel *Monopterus albus*. Comp Biochem Physiol A 178:102–108

D'Avino R, Caruso C, Schinina ME, Rutigliano B, Romano M, Camardella L, Bossa F, Barra D, di Prisco G (1990) Hemoglobin from the antarctic fish *Notothenia coriiceps neglecta*. Amino acid sequence of the beta chain. Comp Biochem Physiol B 96:367–373

D'Avino R, Caruso C, Tamburrini M, Romano M, Rutigliano B, Polverino de Laureto P, Camardella L, Carratore V, di Prisco G (1994) Molecular characterization of the functionally distinct hemoglobins of the Antarctic fish *Trematomus newnesi*. J Biol Chem 269:9675–9681

De Souza PC, Bonilla-Rodriuez GO (2007) Fish hemoglobins. Braz J Med Biol Res 40:769–778

di Prisco G, D'Avino R (1989) Molecular adaptation of the blood of Antarctic teleosts to environmental conditions. Antarct Sci 1:119–124

di Prisco G, D'Avino R, Caruso C, Tamburrini M, Camardella L, Rutigliano B, Carratore V, Romano M (1991) The biochemistry of oxygen transport in red-blooded Antarctic fish. In: di Prisco G, Maresca B, Tota B (eds) Biology of Antarctic Fish. Springer, Berlin, pp 269–274

Duthie GG (1982) The respiratory metabolism of temperature-adapted flatfish at rest and during swimming activity and the use of anaerobic metabolism at moderate swimming speeds. J Exp Biol 97:359–373

Dutta HM, Munshi JSD (1985) Functional morphology of air-breathing fishes: a review. Proc Animal Sci 94:359–375

Døving KB, Westerberg H, Johnsen PB (1985) Role of olfaction in the behavioural and neuronal responses of Atlantic salmon, *Salmo salar*, to hydrographic stratification. Can J Fish Aquat Sci 42:1658–1667

Evans ML, Praebel K, Peruzzi S, Bernatchez L (2012) Parallelism in the oxygen transport system of the lake whitefish: the role of physiological divergence in ecological speciation. Mol Ecol 21:4038–4050

Fagernes CE, Stensløkken KO, Røhr ÅK, Berenbrink M, Ellefsen S, Nilsson GE (2017) Extreme anoxia tolerance in crucian carp and goldfish through neofunctionalization of duplicated genes creating a new ethanol-producing pyruvate decarboxylase pathway. Sci Reports 7:7884

Fago A, Bendixen E, Malte H, Weber RE (1995) The anodic hemoglobin of *Anguilla anguilla*. Molecular basis for allosteric effects in a root-effect hemoglobin. J Biol Chem 272:15628–15635

Fago A, Forest E, Weber RE (2002) Hemoglobin and subunit multiplicity in the rainbow trout (*Oncorhynchus mykiss*) hemoglobin system. Fish Physiol Biochem 24:335–342

Fyhn UEH, Withler RE (1991) A genetic polymorphism in hemoglobins of chinook salmon, (*Oncorhynchus tshawytscha*). Can J Zool 69:1904–1910

Giardina B, Mosca D de Rosa MC (2004) The Bohr effect of haemoglobin in vertebrates: an example of molecular adaption to different physiological requirements. Acta Physiol Scand 182:229–244

Giordano D1, Vergara A, Lee HC, Peisach J, Balestrieri M, Mazzarella L, Parisi E, di Prisco G, Verde C (2007) Hemoglobin structure/function and globin-gene evolution in the Arctic fish *Liparis tunicatus*. Gene 406:58–68

Goodman M, Czelusniak J, Koop BF, Tagle DA, Slightom JL (1987) Globins: a case-study in molecular phylogeny. Cold Spring Harb Sym 52:875–890

Giles MA (1991) Strain differences in hemoglobin polymorphism, oxygen consumption, and blood oxygen equilibria in three hatchery broodstocks of Arctic charr, *Salvelinus alpinus*. Fish Physiol Biochem 9:291–301

Giles MA, Randall DJ (1979) Oxygenation characteristics of the polymorphic hemoglobins of coho salmon (*Oncorhynchus kisutch*) at different developmental stages. Comp Biochem Physiol A 65:265–271

Giles MA, Vanstone WE (1976) Ontogenetic variation in the multiple hemoglobins of coho salmon (*Oncorhynchus kisutch*) and effect of environmental factors on their expression. J Fish Res Board Can 33:1144–1149

Gillen RG, Riggs A (1972) Structure and function of the hemoglobins of the carp, *Cyprinus carpio*. J Biol Chem 247:6039–6046

Gillen RG, Riggs A (1973) Structure and function of the isolated hemoglobins of the American eel, *Anguilla rostrata*. J Biol Chem 248:1961–1969
Hayden JB, Cech JJ Jr, Birdges DW (1975) Blood oxygen dissociation characteristics of the winter flounder *Psedopleuronectes americanus*. J Fish Res Board Can 32:1539–1544
Harrington JP (1986) Structural and functional studies of the king salmon, *Oncorhynchus tshawytscha*, hemoglobins. Comp Biochem Physiol B 84:111–116
Harrison P, Zummo G, Farina F, Tota B, Johnston IA (1991) Gross anatomy, myoarchitecture, and ultrastructure of the heart ventricle in the haemoglobinless icefish *Chaenocephalus aceratus*. Can J Zool 69:1339–1347
Hemmingsen EA (1991) Respiratory and cardiovascular adaptations in hemoglobin-free fish: Resolved and unresolved problems. In: di Prisco G, Maresca B, Tota B (eds) Biology of Antarctic fish. Springer, New York, pp 191–203
Hemmingsen EA, Douglas EL, Johansen K, Millard RW (1972) Aortic blood flow and cardiac output in the hemoglobin free fish *Chaenocephalus aceratus*. Comp Biochem Physiol A 43:1045–1051
Hiebl I, Braunitzer G, Schneeganss D (1988a) High-altitude respiration of geese. The primary structures of the major and minor hemoglobin-components of adult Andean goose (*Chloephaga melanoptera, Anatidae*): the mutation Leu → Ser in position 55 of the ß-chains. Biol Chem Hoppe-Seyler 368:1559–1569
Hiebl I, Weber RE, Schneeganss D, Kösters J, Braunitzer G (1988b) High-altitude respiration of birds. Structural adaptations in the major and minor hemoglobin components of adult Rüppell's griffon (*Gyps rueppellii, Aegypiinae*): a new molecular pattern for hypoxic tolerance. Biol Chem Hoppe-Seyler 369:217–232
Hoffmann FG, Opazo JC, Storz JF (2012) Whole-genome duplications spurred the functional diversification of the globin gene superfamily in vertebrates. Mol Biol Evol 29:303–312
Houde ALS, Günther OP, Strohm J, Ming TJ, Li S, Patterson DA, Farrell AP, Hinch SG, Miller KM (2019) Discovery and validation of candidate smoltification gene expression biomarkers across multiple species and ecotypes of Pacific salmonids. J Exp Biol 222:jeb198036
Houston AH, Cyr D (1974) Thermoacclimatory variation in the haemoglobin systems of goldfish (*Carassius auratus*) and rainbow trout (*Salmo gairdneri*). J Exp Biol 61:455–461
Huber F, Braunitzer G (1989) The primary structure of the hemoglobin of the electric eel (*Electrophorus electricus*). Biol Chem Hoppe Seyler 370:245–250
Hughes GM, Singh BN (1970) Respiration in an air-breathing fish, the climbing perch, *Anabas testudineus*. J Exp Biol 53:281–298
Hünefeld FL (1840) Die Chemismus in der thierischen organization. Leipzig, Brockhaus, p 160
Imsland AK, Foss A, Stefansson SO, Nævdal G (2000) Hemoglobin genotypes of turbot (*Scophthalmus maximus*): consequences for growth and variations in optimal temperature for growth. Fish Physiol Biochem 23:75–81
Imsland AK, Scanu G, Nævdal G (2003) New variants of the haemoglobins of turbot (*Scophthalmus maximus*): possible use in population genetics studies and aquaculture. Sarsia 88:55–64
Jessen TH, Weber RE, Fermi G, Tame J, Braunitzer G (1991) Adaptation of bird hemoglobins to high altitudes: demonstration of molecular mechanism by protein engineering. Proc Natl Acad Sci USA 88:6519–6522
Kamshilov IM, Kamshilova TB (2007) Effect of temperature on functional properties of hemoglobin of crucian carp (*Carassius carassius*). J Ichthyol 47:469–472
Liam KT (1967) Functional morphology of the integumentary, respiratory, and digestive systems of the synbranchoid fish *Monopterus albus*. Copeia 2:375–388
Manwell C, Baker MA (1967) Polymorphism of turbot hemoglobin: a "hybrid" hemoglobin with three kinds of polypeptide chains. Amer Zool 7:214
Maxime V, Pichavant K, Boeuf G, Nonnotte G (2000) Effects of hypoxia on respiratory physiology of tubot, *Scophthtalmus maximus*. Fish Physiol Biochem 22:51–59
McCracken KG, Barger CP, Sorenson MD (2010) Phylogenetic and structural analysis of the HbA (α(A)/β(A)) and HbD (α(D)/β(A)) hemoglobin genes in two high-altitude waterfowl from the

Himalayas and the Andes: bar-headed goose (*Anser indicus*) and Andean goose (*Chloephaga melanoptera*). Mol Phylogenetics Evol 56:649–658

Mendez-Sanchez JF, Burggren WW (2017) Cardiorespiratory physiological phenotypic plasticity in developing air-breathing anabantid fishes (Betta splendens and Trichopodus trichopterus). Physiol Rep 5:e13359

Munshi JSD, Olson KR, Ojha J, Ghosh TK (1986) Morphology and vascular anatomy of the accessory respiratory organs of the air-breathing climbing perch, *Anabas testudineus* (Bloch). Am J Anat 176:321–331

Natarajan C, Jendroszek A, Kumar A, Weber RE, Tame JRH, Fago A, Storz JF (2018) Molecular basis of hemoglobin adaptation in the high-flying bar-headed goose. PLoS Genet 14:e1007331

Near TJ, Parker SK, Detrich 3rd, HW (2006) A genomic fossil reveals key steps in hemoglobin loss by the Antarctic icefishes. Mol Biol Evol 23:2008–2016

Nonnotte G, Kirsch R (1978) Cutaneous respiration in seven sea-water teleosts. Resp Physiol 35:111–118

Ohkubo N, Watabe S, Oshiro T, Takashima F, Nakajima H (1993) Subunit structures of multiple hemoglobins in carp. J Comp Physiol B 163:445–451

Opazo JC, Butts GT, Nery MF, Storz JF, Hoffmann FG (2013) Whole-genome duplication and the functional diversification of teleost fish hemoglobins. Mol Biol Evol 30:140–153

Palstra AP, van den Thillart GEEJM (2010) Swimming physiology of European silver eels (*Anguilla anguilla* L.): energetic costs and effects on sexual maturation and reproduction. Fish Physiol Biochem 36:297–322

Pan YK, Ern R, Morrison PR, Brauner CJ, Esbaugh AJ (2017) Acclimation to prolonged hypoxia alters hemoglobin isoform expression and increases hemoglobin oxygen affinity and aerobic performance in a marine fish. Sci Reports 7:7834

Pédron N, Le Du J, Charrier G, Zambonino-Infante JL, Le Bayon N, Vasconcelos RP, Fonseca VF, Le Grand F, Laroche J (2017) Contrasting patterns of energy metabolism in northern vs southern peripheral European flounder populations exposed to temperature rising and hypoxia. Mar Environ Res 129:258–267

Pelster B (2015) Swimbladder function and the spawning migration of the European eel *Anguilla anguilla*. Front Physiol 5:486

Pelster B, Randall DJ (1998) The physiology of the root effect. In: Perry SF, Tufts BL (eds) Fish Respiration. Academic Press, San Diego, CA, pp 113–139

Pelster B, Weber RE (1990) Influence of organic phosphates on the Root effect of multiple fish haemoglobins. J Exp Biol 149:425–437

Pérez JE (1986) Hemoglobin polymorphism in the toadfish *Thalassophryne maculaso Gunther*. J Exp Mar Biol Ecol 10:287–294

Perutz MF (1970) Stereochemistry of cooperative effects in haemoglobin. Nature 228:726–739

Perutz MF, Kilmartin JV, Nishikura K, Fogg JH, Butler PJG, Rollema HS (1980) Identification of residues contributing to the Bohr effect of human haemoglobin. J Mol Biol 138:649–668

Petersen MF, Steffensen JF (2003) Preferred temperature of juvenile Atlantic cod *Gadus morhua* with different haemoglobin genotypes at normoxia and moderate hypoxia. J Exp Biol 206:359–364

Powers DA (1972) Hemoglobin adaptation for fast and slow water habitats in sympatric catostomid fishes. Science 177:360–362

Quinn NL, McGowan CR, Cooper GA, Koop BF, Davidson WS (2011) Identification of genes associated with heat tolerance in Arctic charr exposed to acute thermal stress. Physiol Genomics 43:685–696

Rajkov J, Shao Z, Berrebi O (2014) Evolution of polyploidy and functional diploidization in sturgeons: microsatellite analysis in 10 sturgeon species. J Heredity 105:521–531

Randall DJ, Rummer JL, Wilson JM, Wang S, Brauner CJ (2014) A unique mode of tissue oxygenation and the adaptive radiation of teleost fishes. J Exp Biol 217:1205–1214

Riccio A, Tamburrini M, Carratore V, di Prisco G (2002) Functionally distinct hemoglobins of the cryopelagic Antarctic teleost *Pagothenia borchgrevinki*. J Fish Biol 57:20–32

Rombough PJ (1997) Piscine cardiovascular development. In: Burggren W, Keller B (eds) Development of cardiovascular systems: molecules to organisms. Cambridge University Press, Cambridge, pp 145–165
Rombough P, Drader H (2009) Hemoglobin enhances oxygen uptake in larval zebrafish (Danio rerio) but only under conditions of extreme hypoxia. J Exp Biol 212:778–784
Root RW (1931) The respiratory function of the blood of marine fishes. Biol Bull 61:427–456
Rummer JL, Brauner CJ (2015) Root effect haemoglobins in fish may greatly enhance general oxygen delivery relative to other vertebrates. PLoS ONE 10:e0139477
Rummer JL, McKenzie DJ, Innocenti A, Supuran CT, Brauner CJ (2013) Root effect hemoglobin may have evolved to enhance general tissue oxygen delivery. Science 340:1327–1329
Rutjes HA, Nieveen MC, Weber RE, Witte F, Van den Thillart GEEJM (2007) Multiple strategies of Lake Victoria cichlids to cope with lifelong hypoxia include hemoglobin switching. Am J Physiol Regul Integr Comp Physiol 293:R1376–R1383
Ruud JT (1954) Vertebrates without erythrocytes and blood pigment. Nature 173:848–850
Samuelsen EN, Imsland AK, Brix O (1999) Oxygen binding properties of three different hemoglobin genotypes in turbot, *Scophthalmus maximus* Rafinesque. Fish Physiol Biochem 20:135–141
Scott GR (2011) Elevated performance: the unique physiology of birds that fly at high altitudes. J Exp Biol 214:2455–2462
Sébert P, Scaion D, Belhomme M (2009) High hydrostatic pressure improves the swimming efficiency of European migrating silver eel. Respir Physiol Neurobiol 165:112–114
Shih DT, Luisi BF, Myazaki G, Perutz MF, Nagai K (1993) A mutagenetic study of the allosteric linkage of His(HC3)146β in haemoglobin. J Mol Biol 230:1291–1296
Shu JJ, Harter TS, Morrison P, Brauner CJ (2017) Enhanced hemoglobin–oxygen unloading in migratory salmonids. J Comp Physiol B 188:409–419
Sick K (1961) Haemoglobin polymorphism in fishes. Nature 192:894–896
Sick K, Bahn E, Frydenberg O, Nielsen JT, von Wettstein D (1967) Hemoglobin polymorphism of the American fresh water eel *Anguilla*. Nature 214:1141–1442
Sollid J, Weber RE, Nilsson GE (2005) Temperature alters the respiratory surface area of crucian carp *Carassius carassius* and goldfish *Carassius auratus*. J Exp Biol 220:564–572
Star et al (2011) The genome sequence of Atlantic cod reveals a unique immune system. Nature 477:207–210
Steen JB (1963) The physiology of the swimbladder in the eel *Anguilla vulgaris*. III. The mechanism of gas secretion. Acta Physiol Scand 59:221–241
Steffensen JF, Lomholt JP, Johansen K (1982) Gill ventilation and O_2 extraction during graded hypoxia in two ecologically distinct species of flatfish, the flounder (*Platichthys flesus*) and the plaice (*Pleuronectes platessa*). Environ Biol Fish 7:157–163
Storz JF, Opazo JC, Hoffmann FG (2011) Phylogenetic diversification of the globin gene superfamily in chordates. IUBMB Life 63:313–322
Storz JF, Opazo JC, Hoffmann FG (2013) Gene duplication, genome duplication, and the functional diversification of vertebrate globins. Mol Phylogenet Evol 66:469–478
Storz JF, Sabatino SJ, Hoffman FG, Gering EJ, Moriyama H, Ferrand N, Monteiro B, Nachman MW (2007) The molecular basis of high-altitude adaptation in deer mice. PLoS Genet 3:448–459
Storz JF, Runck AM, Sabatino SJ, Kelly JK, Ferrand N, Moriyama H, Weber RE, Fago A (2009) Evolutionary and functional insights into the mechanism underlying high-altitude adaptation of deer mouse hemoglobin. Proc Natl Acad Sci USA 106:14450–14455
Tamburrini M, D'Avino R, Fago A, Carratore V, Kunzmann A, di Prisco G (1996) The unique hemoglobin system of *Pleuragramma antarcticum*, an antarctic migratory teleost: structure and function of the three components. J Biol Chem 271:23780–23785
Tame JRH, Wilson JC, Weber RE (1996) The crystal structure of trout HbI in the deoxy and carbonmonoxy form. J Mol Biol 259:749–760
Tanaka H, Takagi Y, Naito Y (2000) Behavioural thermoregulation of chum salmon during homing migration in coastal waters. J Exp Biol 203:1825–1833

Tanaka H, Takagi Y, Naito Y (2001) Swimming speeds and buoyancy compensation of migrating adult chum salmon *Oncorhynchus keta* revealed by speed/depth/acceleration data logger. J Exp Biol 204:3895–3904

Tate M, McGoran RE, White CR, Portugal SJ (2017) Life in a bubble: the role of the labyrinth organ in determining territory, mating and aggressive behaviours in anabanoids. J Fish Biol 91:723–749

Tsuyuki H, Ronald AP (1971) Molecular basis for multiplicity of pacific salmon hemoglobins: evidence for in vivo existence of molecular species with up to four polypeptides. Comp Biochem Physiol B 39:503–510

Turner JD, Wood CM, Hobe H (1983) Physiological consequences of severe exercise in the inactive benthic flathead sole (*Hippoglossoides elassodon*): a comparison with the active pelagic rainbow trout (*Salmo gairdneri*). J Exp Biol 104:269–288

van Ginneken V, Antonissen E, Müller UK, Booms R, Eding E, Verreth J, van den Thillart G (2005) Eel migration to the Sargasso: remarkably high swimming efficiency and low energy costs. J Exp Biol 208:1329–1335

Verde C, Balestrieri M, de Pascale D, Pagnozzi D, Lecointre G, di Prisco G (2006) The oxygen transport system in three species of the boreal fish family Gadidae. Molecular phylogeny of hemoglobin. J Biol Chem 281:22073–22084

Verde C, Carratore V, Riccio A, Tamburrini M, Parisi E, di Prisco G (2002) The functionally distinct hemoglobins of the Arctic spotted wolffish *Anarhichas minor*. J Biol Chem 277:36312–36320

Waser W, Heisler N (2005) Oxygen delivery to the fish eye: root effect as crucial factor for elevated retinal PO_2. J Exp Biol 208:4035–4047

Webb PW (2002) Kinematics of plaice, *Pleuronectes platessa*, and cod, *Gadus morhua*, swimming near the bottom. J Exp Biol 205:2125–2134

Weber RE (1990) Functional significance and structural basis of multiple hemoglobins with special reference to ectothermic vertebrates. In: Truchot JP, Lahlou B (eds), Animal nutrition and transport processes. 2. Transport, respiration and excretion: comparative and environmental aspects. Karger, Basel, pp 58–75

Weber RE (2000) Adaptation for oxygen transport: Lessons from fish hemoglobins. In: di Prisco G, Giardina B, Weber RE (eds), Hemoglobin function in vertebrates: molecular adaptation in extreme and temperate environments. Springer, Italia

Weber RE, de Wilde JAM (1975) Oxygenation properties of haemoglobins from the flatfish plaice (*Pleuronectes platessa*) and flounder (*Platichthys flesus*). J Comp Physiol B 101:99–110

Weber RE, de Wilde JAM (1976) Multiple haemoglobins in plaice and flounder and their functional properties. Comp Biochem Physiol B 54:433–437

Weber RE, Jensen FB (1988) Functional adaptations in hemoglobins from ectothermic vertebrates. Ann Rev Physiol 50:161–179

Weber RE, Lykkeboe G (1978) Respiratory adaptations in carp blood influences of hypoxia, red cell organic phosphates, divalent cations and CO_2 on hemoglobin-oxygen affinity. J Comp Physiol 128:127–137

Weber RE, Lykkeboe G, Johansen K (1975) Biochemical aspects of the adaptation of hemoglobin-oxygen affinity in eels to hypoxia. Life Sci 17:1345–1350

Weber RE, Lykkeboe G, Johansen K (1976a) Physiological properties of eel haemoglobin: hypoxic acclimation, phosphate effects and multiplicity. J Exp Biol 64:75–88

Weber RE, Wood SC, Lomholt JP (1976b) Temperature acclimation and oxygen-binding properties of blood and multiple haemoglobin from rainbow trout. J Exp Biol 65:333–345

Wood SC, Johansen K (1972) Adaptation to hypoxia by increased HbO_2 affinity and decreased red cell ATP concentration. Nature 237:278–279

Wu Z, you F, Wen A, Ma D, Zhang P (2016) Physiological and morphological effects of severe hypoxia, hypoxia and hyperoxia in juvenile turbot (*Scophthalmus maximus* L.). Aquaculture Res 47:2019–2027

Xia M, Chao Y, Jia J, Li C, Kong Q, Zhao Y, Guo S, Qi D (2016) Changes of hemoglobin expression in response to hypoxia in a Tibetan schizothoracine fish, *Schizopygopsis pylzovi*. J Comp Physiol B 186:1033–1043

Chapter 14
Hemoglobin: Structure, Function and Allostery

Mostafa H. Ahmed, Mohini S. Ghatge and Martin K. Safo

Abstract This chapter reviews how allosteric (heterotrophic) effectors and natural mutations impact hemoglobin (Hb) primary physiological function of oxygen binding and transport. First, an introduction about the structure of Hb is provided, including the ensemble of tense and relaxed Hb states and the dynamic equilibrium of Hb multistate. This is followed by a brief review of Hb variants with altered Hb structure and oxygen binding properties. Finally, a review of different endogenous and exogenous allosteric effectors of Hb is presented with particular emphasis on the atomic interactions of synthetic ligands with altered allosteric function of Hb that could potentially be harnessed for the treatment of diseases.

Keywords Hemoglobin · Allostery · Allosteric effectors · T state · Relaxed state · Oxygen affinty · X-ray crystallography · Hemoglobin variants

Introduction

Hemoglobin (Hb) is the most studied of the heme containing globulin proteins and yet is not fully understood. It was one of the first proteins to be studied by X-ray crystallography, and earned Max Perutz the Nobel Prize in Chemistry in 1962. The structural studies provided a plethora of data that offered glimpses of the magnificent molecular mechanisms behind Hb physiological functions. Hemoglobin is a polyfunctional molecule that is involved in several functions, such as catalytic (nitrite reductase, NO dioxygenase, monooxygenase, alkylhydroperoxidase, esterase, lipoxygenase); nitric

M. H. Ahmed · M. S. Ghatge · M. K. Safo
Department of Medicinal Chemistry, School of Pharmacy,
Virginia Commonwealth University, Richmond, VA 23219, USA
e-mail: ahmedmh@vcu.edu

M. S. Ghatge
e-mail: msghatge@vcu.edu

M. S. Ghatge · M. K. Safo (✉)
Institute for Structural Biology, Drug Discovery and Development,
Virginia Commonwealth University, Richmond, VA 23219, USA
e-mail: msafo@vcu.edu

U. Hoeger and J. R. Harris (eds.), *Vertebrate and Invertebrate Respiratory Proteins, Lipoproteins and other Body Fluid Proteins*, Subcellular Biochemistry 94,
https://doi.org/10.1007/978-3-030-41769-7_14

oxide metabolism; metabolic reprogramming; pH regulation and maintaining redox balance (Kosmachevskaya and Topunov 2018). This chapter however, focuses on Hb primary function of oxygen transport and how mutations, endogenous or exogenous ligands or effectors affect Hb allostery.

Structure and Function of Hemoglobin

The primary function of Hb is to transport oxygen (O_2) from the lung to tissues, binding and releasing O_2 in a cooperative manner, as demonstrated by the oxygen equilibrium curve (OEC), which represents O_2 saturation of Hb (SO_2) at varying partial pressures of O_2 (pO_2) (Fig. 14.1). The pO_2 at 50% SO_2 (expressed as P_{50}) measures the O_2-affinity for Hb, which is about 26 mmHg for normal adult human Hb (HbA) (Fig. 14.1). Historically, Hb function has been explained in terms of equilibrium between two classical states: the tense (T) state (unliganded Hb) which exhibits low affinity for O_2, and the relaxed (R) state (liganded Hb) which exhibits high affinity for O_2, providing a structural basis for cooperative effects that facilitate the efficient uptake and release of O_2 (Perutz 1972a, b; Perutz et al. 1998; Safo and Bruno 2011; Safo et al. 2011). The equilibrium between the T and R states is affected by endogenous heterotropic ligands, such as 2,3-bisphosphoglycerate (2,3-BGP), protons (H^+), carbon dioxide (CO_2), chloride (Cl^-) or synthetic allosteric effectors that modulate Hb-O_2 affinity, either by stabilizing the R state Hb (left-shift the OEC) (Fig. 14.1 colored red) or stabilizing the T state Hb (right-shift the OEC) (Fig. 14.1; cyan) (Perutz 1972a, b; Perutz et al. 1998; Safo and Bruno 2011; Safo et al. 2011).

In most vertebrates, Hb is a tetramer, consisting of two α-subunits (α_1 and α_2) and two β-subunits (β_1 and β_2) that are structurally similar and about the same size (Fig. 14.2a). The two αβ dimers (named $\alpha_1\beta_1$ and $\alpha_2\beta_2$) are arranged around a 2-fold axis of symmetry resulting in a large central water cavity in the T or unliganded or deoxygenated structure and a narrower cavity in the R or liganded or oxygenated

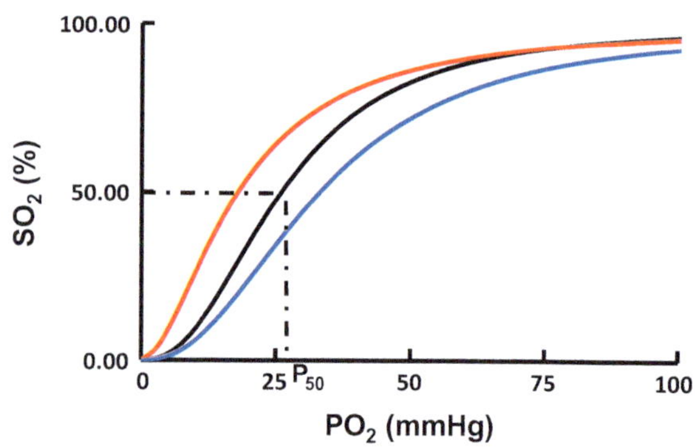

Fig. 14.1 Oxygen equilibrium curve of Hb. The normal P_{50} value is indicated by dashed lines. The left-shift and right-shift in the curves are colored red and blue respectively

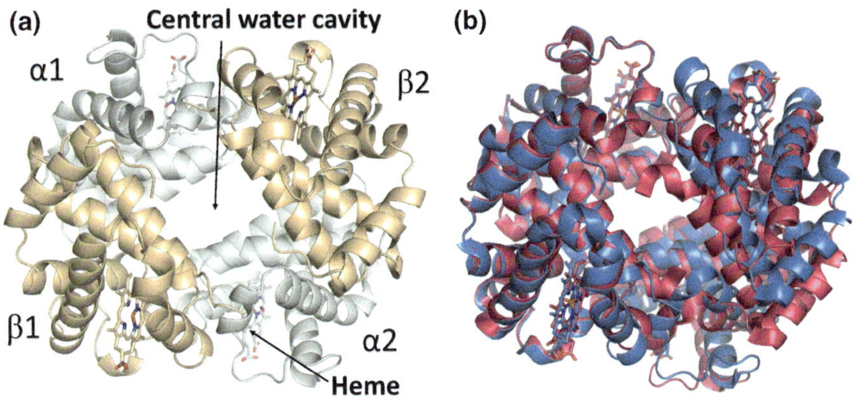

Fig. 14.2 Crystal structure of hemoglobin. **a** Overall quaternary structure of Hb with the two α chains and β chains colored grey and tan, respectively. **b** Structure of oxygenated (R state) Hb (magenta) superimposed on the structure of deoxygenated (T state) Hb (blue). Note the larger central water cavity in the T structure

structure (Fig. 14.2b) (Fermi 1975; Safo and Bruno 2011; Safo et al. 2011). The α- and β-clefts are the two entry points into the central water cavity that are larger in T state structure than R state structure. The interdimer interface ($\alpha_1\beta_1$–$\alpha_2\beta_2$) of T state structure is also characterized by more salt-bridge/hydrogen-bond interactions than R state structure. (Fermi 1975; Safo and Bruno 2011; Safo et al. 2011).

The α-subunits and β-subunits are formed of 7 and 8 helices, respectively named A–H that are joined by non-helical segments (referred to as corners). Each subunit has a binding pocket for heme formed by the E and F helices. The heme consists of a ferrous ion held in the center of a porphyrin and coordinated by the four nitrogen atoms of the porphyrin ring. The Fe is also covalently anchored to Hb at the heme proximal pocket by an imidazole of a histidine residue located on the F helix (known as the proximal histidine or His (F8)). This setup allows the Fe to bind O_2 or other gasses at the distal pocket of the heme by a covalent bond to fulfill the octahedral coordination of six ligands. The O_2 molecule binds in an "end-on bent" geometry where one oxygen atom binds to Fe, and the other protrudes at an angle. The imidazole of a histidine residue at the distal pocket (His E7) stabilizes the bound O_2 through hydrogen-bond interaction. In the absence of oxygen (in deoxygenated Hb) at the α-cleft, a very weakly bonded water molecule fills the site, forming a distorted octahedron.

Other ligands, such as nitrite ion (NO^{2-}), nitric oxide (NO), carbon monoxide (CO), cyanide (CN^-), sulfur monoxide (SO), sulfide (S^{2-}), may bind to the distal side of the heme and act as competitive inhibitors, affecting O_2 binding. Some of these compounds bind with significantly higher affinity than oxygen, making these compounds highly toxic to humans. For example, Hb binding affinity for CO is as much as 240 times greater than O_2, and a concentration of 0.1% CO in the air could lead to unconsciousness and eventually death (Winter and Miller 1976).

Ligand binding and unbinding events have far more ramifications, not just on the heme iron, but rather on the tertiary and the whole quaternary structure of Hb by inducing conformational changes to the globin E helix, CD and FG corners that extend to the heme environment (affecting the size of the distal pocket), central water cavity, α- and β-clefts, and salt-bridge/hydrogen-bond interactions across the $\alpha_1\beta_1$–$\alpha_2\beta_2$ ($\alpha_1\beta_2$ or $\alpha_2\beta_1$ or $\alpha_1\alpha_2$ or $\beta_1\beta_2$) dimer interface, subsequently, triggering cooperativity events that affect the T → R transition and giving rise to the phenomenon of allostery (Perutz 1972a, b; Baldwin and Chothia 1979; Paoli et al. 1996; Perutz et al. 1998; Safo and Bruno 2011; Safo et al. 2011).

Allosteric Models of Hemoglobin

Various *allosteric models* have been proposed to explain cooperative oxygen binding to *hemoglobin*. The earliest are the two-state Monod–Wyman–Changeux (MWC) and the Koshland–Némethy–Filmer (KNF) models. The MWC model assumes that, upon ligand binding, the T state switches to the R state without intermediate states (Monod et al. 1965), while the KNF model assumes one Hb conformation in the absence of ligand, which changes with each binding of ligand to the subunit that are sequentially transmitted to the rest of the subunits (Koshland et al. 1966). Following, Max Perutz proposed a stereochemical mechanism incorporating aspects of both the MWC and the KNF models to explain the cooperativity effect in Hb (Perutz 1972a, b; Perutz et al. 1998). According to this combined model, ligand binding to each subunit of the tetramer induces tertiary conformational changes that are transmitted to other subunits through direct communication between the α_1 and β_2 subunits; ultimately leading to a sequential increase in the affinity for the ligand at other heme sites and shifting the allosteric equilibrium from the T state towards the R state. The two-state model was based on a plethora of liganded and unliganded Hb crystal structures from different species, such as horse, human, and bovine (Muirhead and Perutz 1963; Perutz et al. 1968; Fermi 1975; Ladner et al. 1977). Liganded Hb structures are mostly co-crystallized with CO as the heme ligand since CO-liganded Hb is chemically stable, rendering it easier to manipulate and crystallize compared to the O_2-liganded Hb. Nevertheless, the structures of O_2- or CO-liganded Hb are very similar. Based on these crystal structures, Perutz partly attributed the low oxygen affinity in the T state to tension in the Fe–His (F8) bond, which restrains the Fe from moving into the porphyrin plane on ligand binding (Perutz 1972a, b; Perutz et al. 1998). This proposition was further supported by Paoli et al. (1996) who published for the first time the crystal structure of fully liganded Hb in T state (by introducing ligands to deoxygenated Hb crystals in the presence of T state stabilizing allosteric effector) demonstrating the rupture of the Fe–His(F8) bond in the α-subunits, which was cited as the reason behind uncoupling of the structural changes at the α-subunits from those at the β subunits. Barrick et al. (1997) also through site-directed mutagenesis studies noted that disrupting the Fe–His(F8) bond leads to a significant increase in ligand affinity, reduction in cooperativity, as well as slow down of the quaternary switch.

However, the group also suggested that additional communication pathways exist (other than the Fe–His(F8) bond) that is responsible for the residual cooperativity observed (Barrick et al. 1997).

Perutz's stereochemical model failed to account for the critical contributions of heterotropic ligands that can modulate the structure and function of Hb, and moreover, represented Hb as having only a binary state of either T or R without the existence of intermediary states. Several variations of Hb allosteric models were consequently put forward. Examples are the modified MWC Cooperon model of Brunori et al. (1986); the SK model of Szabo and Karplus (Szabo and Karplus 1972); the tertiary two-state (TTS) model of Henry and colleagues that describes distinct "tertiary t and r" within the T and R states conformations (Henry et al. 2002); and the global allostery model by Yonetani that also relates conformational changes to effector binding in both the T and R states and proposes that O_2 affinity is dependent on heterotropic effector-induced tertiary structural changes (Yonetani et al. 2002; Yonetani and Tsuneshige 2003; Yonetani and Kanaori 2013). There are several excellent reviews of this topic, and the reader is referred to two such publications (Yonetani and Kanaori 2013; Gell 2018).

Multi States that Lie Along the T to R Transition

The early structural studies by Max Perutz and others (Muirhead and Perutz 1963; Perutz et al. 1968; Fermi 1975; Ladner et al. 1977) provided two classical end-state conformations of Hb during the T to R transition, but several lines of evidence drawn from functional, computational, structural, thermodynamic and spectroscopic experiments suggested discrete multi states along the T $\rightarrow$ R transition (Sawicki and Gibson 1976, 1978; Samaja et al. 1987; Schumacher et al. 1995; Jayaraman et al. 1995; Wilson et al. 1996; Perrella and Cera 1999; Yonetani and Tsuneshige 2003; Samuni et al. 2004; Song et al. 2008), which is expected to be due to the fact that proteins are flexible entities, and crystal structures can only provide a subset of the conformational ensembles present under physiological conditions. NMR and wide-angle X-ray scattering (WAXS), studies showed that the time-averaged solution structure of liganded Hb is unlikely to be identical to crystallized structures (Lukin et al. 2004; Gong et al. 2006; Sahu et al. 2007; Song et al. 2008; Makowski et al. 2011; Fan et al. 2013). Using laser photolysis and in the presence of the potent allosteric effector of Hb, inositol hexaphosphate (IHP), Sawicki and Gibson observed quaternary conformational changes in CO-liganded human Hb (Sawicki and Gibson 1976, 1978). Perrella and Cera (1999) also showed, using rapid quenching of the reaction between human Hb and CO, that different ligation intermediates exist with distinct conformations and oxygen affinities. Mozzarelli and collegues reported two distinct human Hb populations that had different oxygen-binding affinities (1000 and 100 times lower than the high-O_2 affinity R state) and were non-cooperative (Mozzarelli et al. 1991). The former species was proposed to have T state like conformation, while the latter showed relaxed features that placed it closer to the R state.

Several crystal structures have also provided evidence of multi states along the T $\rightarrow$ R transition. A horse deoxygenated Hb trapped in the high-affinity relaxed state was reported to be a ligation intermediate of the R state (Wilson et al. 1996). Cross-linked human Hb, with trimesic acid exhibiting low oxygen affinity, was reported by Schumacher et al., in which the crystal structures have not yet reached the R state conformation, displaying several intermediate T and R features (Schumacher et al. 1995). Safo et al. (2002b) also showed that R state structure of CO-liganded human Hb with a phosphate molecule bound at the β-cleft exhibited subtle but significant tertiary T state features at the $\alpha_1\beta_2$ interface. In the presence of heterotropic effectors, Yonetani and Tsuneshinge (2003) observed that liganded Hb exhibits T state constraint within an R-like quaternary structure.

Other studies have also shown the ligated T state structure with significant changes in the heme pockets, as well as changes at the $\alpha_1\beta_2$ interface consistent with the presence of intermediate states along the T $\rightarrow$ R transition (Abraham et al. 1992a; Paoli et al. 1996; Song et al. 2008). An ensemble of related T-like quaternary structures induced by mutations in a cluster of residues at the αβ interface and centered at βTrp37 has been described by Arnone's group (Kavanaugh et al. 2005). This cluster of residues was previously reported by Mozzarelli's group to be the major region of the quaternary constraint (Noble et al. 2001).

Fully Liganded Hb Structure Trapped in a Tense Conformation

Several of the proposed allosteric models maintain the original MWC tenet that cooperative oxygen binding cannot occur in the absence of quaternary transition, consistent with the fact that a fully liganded T state hemoglobin heterotetramer, obtained from crystallizing ligated hemoglobin in solution, has never been reported. In a recent study, we reported the structure of such a uniquely fully liganded variant mammalian Hb $\zeta_2\beta_2^s$ (formed from Hb S ($\alpha_2\beta_2^s$) by replacing adult α-globin with embryonic ζ-globin subunits), crystallized from CO-liganded Hb solution that remains trapped in a quaternary T state-like conformation (Safo et al. 2013, 2015). Hb $\zeta_2\beta_2^s$ inhibits polymerization of deoxygenated Hb S in vitro and ameliorates pathogenic features of sickle cell disease (SCD) in mouse models (He and Russell 2004a, b). The structure displayed a central water cavity, dimer interface and salt-bridge/hydrogen-bond interactions, β-cleft, etc. that are more typical for a tense conformation (Safo et al. 2013, 2015). It is clear that ligand binding to Hb $\zeta_2\beta_2^s$ is effected mostly by tertiary structural changes within the larger T- or R state structures, providing insights into the contributions of tertiary and quaternary structures to cooperative Hb-O_2 ligand binding, as well as validating the hypothesis that Hb ligand affinity can be decoupled from overall quaternary structure (Henry et al. 2002; Yonetani et al. 2002; Yonetani and Tsuneshige 2003; Yonetani and Kanaori 2013). Moreover, the structure explains Hb

$\zeta_2\beta_2^s$ antipolymer activities by favoring an alternate T state structure that is excluded from pathological deoxygenated Hb S polymers (Safo et al. 2015).

Multi Relaxed States Beyond the T → R Transition

Several studies by the Safo group and others have unveiled different relaxed Hb states that appear to lie beyond the classical T → R transition, suggesting that the relaxed state is not unique to the classical R state but is instead an ensemble of fully-liganded states exhibiting distinct quaternary conformations (Smith et al. 1991; Doyle et al. 1992; Silva et al. 1992; Janin and Wodak 1993; Smith and Simmons 1994; Srinivasan and Rose 1994; Schumacher et al. 1997; Fernandez et al. 2000; Mueser et al. 2000; Lukin et al. 2003; Safo and Abraham 2005; Jenkins et al. 2009; Safo et al. 2011; Abdulmalik et al. 2011). Examples of these states are the R2, RR2, R3, and RR3. Assigning these states structurally depends on different key parameters, such as rigid-body screw rotation (defined in terms of screw rotation angle, screw rotation translation, the direction of the screw rotation axis, and a point on the rotation axis) of the $\alpha_1\beta_1$ dimer relative to the $\alpha_2\beta_2$ dimer, interdimer salt-bridge/hydrogen-bond interactions, heme–heme distance, size of the distal heme pocket, size of the central water cavity, and size of the α-cleft and β-cleft (Baldwin and Chothia 1979; Safo and Abraham 2005; Jenkins et al. 2009; Safo et al. 2011).

Rigid-body screw rotation to quantify the allosteric movement between the T structure and the classical R structure was first reported by Baldwin and Chothia who found an ~14° rotation and ~1 Å translation of the $\alpha_1\beta_1$ dimer relative to the $\alpha_2\beta_2$ dimer (Baldwin and Chothia 1979). It was later used by several investigators to analyze the quaternary differences between several Hb states that were, in some instances, as significant or even larger than those between the T structure and the classical R structure (Silva et al. 1992; Mueser et al. 2000; Safo and Abraham 2005; Jenkins et al. 2009; Safo et al. 2011). Here, we describe some of these structural differences with Figures using more recent high-resolution structures of the classical R (PDB ID: 2DN1), T (PDB ID: 2DN2), R2 (PDB ID: 1QXD or 1QXE) and R3 (PDB ID: 1YZI). Upon the transition from the T state to the classical R state, a sliding motion occurs between the β_2-subunit and the opposite α_1-subunit at a so called "switch region," with a fulcrum, also at a so called "hinge region" (Baldwin and Chothia 1979; Silva et al. 1992; Lukin et al. 2003; Safo and Abraham 2005; Jenkins et al. 2009; Safo et al. 2011). This motion places the β_2FG corner residue β_2His97 between α_1Thr41 and α_1Thr38 in the R structure (from its location between α_1Pro44 and α_1Thr41 in the T structure), where it forms a hydrogen-bond interaction with α_1Thr38 (Fig. 14.3a). Moreover, the T → R transition leads to the narrowing of the central water cavity and the α- and β-clefts (Fig. 14.3b), as well as an increase in $\alpha_1\beta_2$ iron–iron distance and a decrease in the $\beta_1\beta_2$ iron–iron-distance (Baldwin and Chothia 1979; Silva et al. 1992; Lukin et al. 2003; Safo and Abraham 2005; Jenkins et al. 2009; Safo et al. 2011).

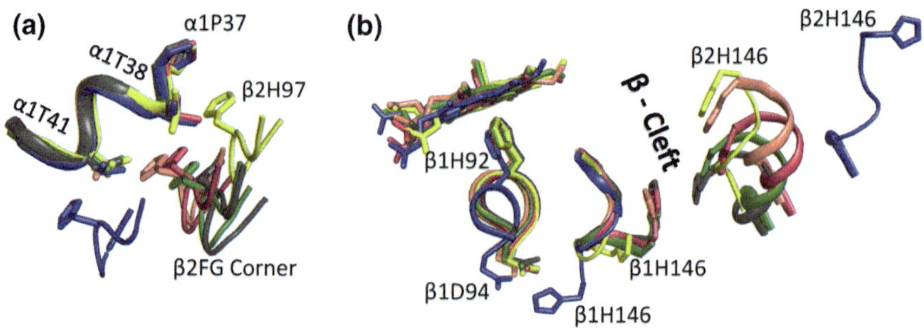

Fig. 14.3 Superposed structures of T (blue), R (magenta), R3 (yellow), RR2 (green), R2 (black), and RR3 (salmon) on their $\alpha_1\beta_1$ dimers. **a** Transitions between the different states lead to significant changes (sliding motion) at the $\alpha_1\beta_2$ dimer interface switch regions. **b** Transitions from the T state to the relaxed states breaks a T state stabilizing salt-bridge interaction between βAsp94 and βHis146. In the R2 and RR2 structures β1His146 makes close contact with β2His146, while in the other relaxed structures, βHis146 becomes highly disordered. There is also a significant size decrease in the β-cleft of the relaxed structures compared to the T structure

Interestingly, the T → R2 transition (rotation/translation of ~23°/3.1 Å), and T → RR2 transition (~17°/2.6 Å) occur approximately in the same direction of the screw rotation axis of the T → R transition, in the order of T → R → RR2 → R2 (Safo and Abraham 2005; Jenkins et al. 2009; Safo et al. 2011). It is interesting to note that the T → R2 transition was first proposed to lie along the T → R transition, with R as the end-state (Silva et al. 1992). However, as stated above, further analysis showed that R2 is not an intermediate but rather an end-state structure with the T → R2 transition first passing through the T → R transition and then the R → RR2 transition (Janin and Wodak 1993; Schumacher et al. 1997; Safo and Abraham 2005; Jenkins et al. 2009; Safo et al. 2011). The RR2 → R2 transition is characterized by a rotation/translation of ~6°/0.5 Å (Safo and Abraham 2005; Jenkins et al. 2009; Safo et al. 2011). In both R2 and RR2 structures, the β_2FG has further rotated perpendicularly from the R structure position, disengaging the hydrogen-bond interaction between β_2His97 and α_1Thr38 (Fig. 14.3a). This has led to widening of the central water cavity in these two structures, as well as bringing the two C-termini residues of βHis146 to engage in closer interactions, resulting in well-defined βHis146 positions (Fig. 14.3b) compared to the highly disordered βHis146 in the R structure (Silva et al. 1992; Lukin et al. 2003; Safo and Abraham 2005; Jenkins et al. 2009; Safo et al. 2011). This observation, as will be discussed later, is significant due to the critical contribution of βHis146 to the Bohr effect (Silva et al. 1992).

Unlike the R, RR2 and R2 structures where β_2His97 is located between α_1Thr41 and α_1Thr38, a rotation of β_2FG corner in the R3 structure has further moved β_2His97 away from the T state position, placing it between α_1Thr38 and α_1Pro37 (Fig. 14.3a), resulting in the smallest central water cavity and α-cleft (Safo and Abraham 2005; Jenkins et al. 2009; Safo et al. 2011). Interestingly a complete closure of the β-cleft is observed in the R3 structure due to extensive hydrogen-bond interactions between residues from the opposite C-terminals of the HC segments (Safo and Abraham

2005; Jenkins et al. 2009; Safo et al. 2011). The T → R3 and R → R3 transitions involve rotation/translation of ~22°/1.7 Å and ~10°/1.1 Å, respectively, and occur approximately in the same direction, with the R state mediating the T → R3 transition (Safo and Abraham 2005; Jenkins et al. 2009; Safo et al. 2011).

Another uniquely liganded structure, RR3 (PDB ID: 3D17) appears to lie in-between the R → R3 transition and in the order T → R → RR3 → R3 (Safo and Abraham 2005; Jenkins et al. 2009; Safo et al. 2011). The transition from the RR3 to either R or R3 is ~6.2°/1.0 Å. Interestingly, the RR3 structure shows significant rotation of the distal βHis63(E7) out of the distal pocket, forming what is termed a His(E7) ligand channel to the bulk solvent (Fig. 14.4). A smaller rotation of His(E7) is also observed in the R3 structure (Fig. 14.4). The rotated position of the βHis63 in both R3 or RR3 is stabilized by a salt-bridge interaction between the imidazole side chain and the β-heme propionate (Safo and Abraham 2005; Jenkins et al. 2009). Of interest is that the closest distance between the βHis63 imidazole and the bound heme ligand are 6.6 Å and 4 Å in RR3 and R3 structures, respectively, compared to the ~3 Å observed in the classical R, R2, and RR2 structures, which allows stabilization of the bound ligand (Safo and Abraham 2005; Jenkins et al. 2009; Safo et al. 2011).

Overall, two trajectories have been proposed for the transition between the T and the relaxed states (Safo and Abraham 2005; Jenkins et al. 2009; Safo et al. 2011). One trajectory suggests R2 to be an end-state with both R and RR2 lying on the T →

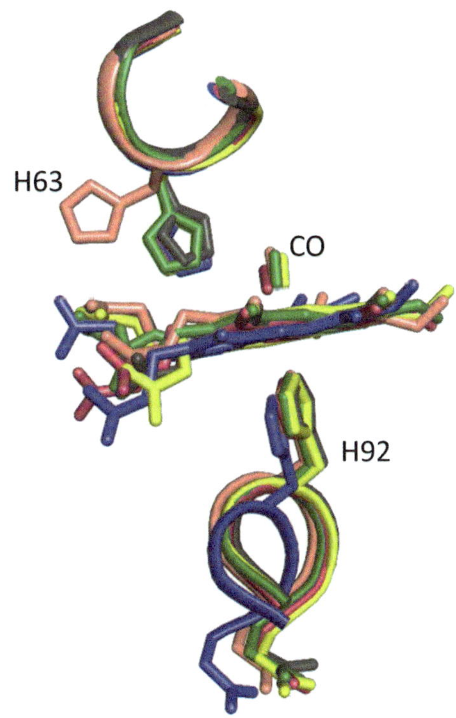

Fig. 14.4 Superposed β heme structures of R (magenta), R3 (yellow), RR2 (green), R2 (black), and RR3 (salmon) showing the positions of βHis63. Note the rotation of βHis63 out of the distal pocket in the RR3 structure, creating a ligand channel to the bulk solvent, while in R, RR2, and R2 structures, βHis63 is still located in the pocket making hydrogen-bond interaction with the ligand. The R3 structure shows a partially opened ligand channel

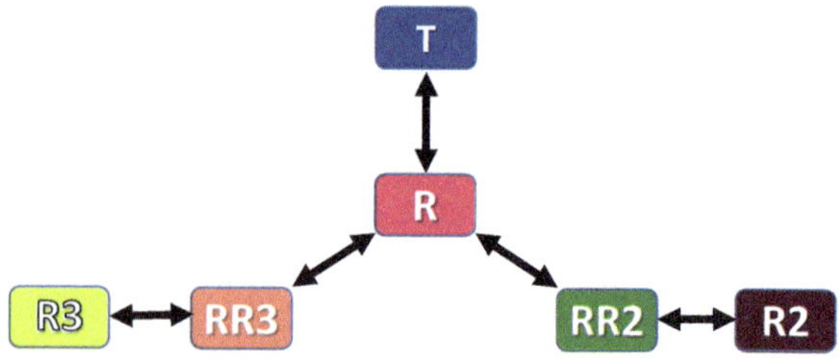

Fig. 14.5 Schematic representation of the proposed allosteric pathway between the different Hb states

R2 transition. The second shows R3 as another end-state with R and RR3 lying along the T → R3 transition (Fig. 14.5). Several other liganded Hb structures have also been reported, such as Gower II COHb (PDB ID: 1AJ9), bovine COHb (PDB ID: 1FSX, 1G08, and 1G09), and cross-linked forms of human COHb (PDB ID: 1HAB and 1HAC), that have distinct relaxed state quaternary structures with intermediate features between the R → RR2 and RR2 → R2 transitions (Schumacher et al. 1997; Fernandez et al. 2000; Mueser et al. 2000; Safo and Abraham 2005; Jenkins et al. 2009; Safo et al. 2011).

The R and R2 State Controversy

The discovery of the R2 structure in the early 1990s (Smith et al. 1991; Silva et al. 1992; Smith and Simmons 1994) initiated a controversy regarding the physiological relevance of the R2 or classical R structure, which took several years to resolve. The R structure was suggested to be an artifact and/or intermediate trapped between the T → R2 transition, and the R2 as the physiologically relaxed end-state (Srinivasan and Rose 1994; Schumacher et al. 1997). Other investigators, however suggested the R2 structure to be an intermediate between the T → R transition (Smith et al. 1991; Silva et al. 1992; Smith and Simmons 1994). Later studies and comprehensive analysis of the R and R2 structures, however, suggested that the R2 is not an intermediate, but rather an end-state relaxed structure (Doyle et al. 1992; Janin and Wodak 1993; Safo and Abraham 2005; Jenkins et al. 2009; Safo et al. 2011).

The argument for assigning the R2 structure as the physiologically relevant relaxed state, and R structure as an artifact was partly due to the former being crystallized with low-salt that mimics the in vivo environment, while the latter was crystallized with non-physiological high-salt condition. However, Safo and co-workers later crystallized the R2 structure using high-salt condition (Safo et al. 2004; Abdulmalik et al. 2011). Of interest is that the R2 crystals only form in high-salt when co-crystallized with antisickling aromatic aldehydes (Safo et al. 2004; Abdulmalik et al. 2011) usually at pH between 6.4 and 6.8. In the absence of the aldehyde, only R, RR2, RR3 or R3 or mixture of these crystals form with high-salt, and their ratio/formation appears to be a function of pH (Safo and Abraham 2005; Jenkins et al. 2009; Safo et al. 2011). The R and R3 crystals predominate at high pH (>6.5) and low pH (<6.5), respectively, while the RR2 crystals typically appear at pH around 7 and not as abundant as the R and R3 crystals (Safo and Abraham 2005; Jenkins et al. 2009; Safo et al. 2011). The RR3 crystal has only been observed once and appeared with R and R3 crystals at pH 6.4 (Safo and Abraham 2005; Jenkins et al. 2009; Safo et al. 2011). It seems that various relaxed states do exist in dynamic equilibrium and their fraction and/or distribution appears to be pH-dependent with low energy barrier in changing one state to another. Consistently, an NMR study by Lukin et al. showed that liganded Hb exists as a mixture of R and R2, and a structure intermediate between the R and R2 (Lukin et al. 2003).

The existence of a multi-relaxed Hb states is appreciated even more when one considers crystal structures of liganded Hb in the presence of allosteric effectors. Aromatic aldehydes bind to the α-cleft of liganded Hb in the R2 form in a symmetry-related fashion that ties together the two α-subunits and stabilizes the relaxed state relative to the T state (Safo et al. 2004, 2011; Abdulmalik et al. 2011), which as will be discussed later is a potential pharmacologic strategy to treat sickle cell disease. As noted above, in high-salt liganded Hb without aromatic aldehyde crystallizes in R or RR2 or RR3 form depending on the pH, but in the presence of aromatic aldehyde appears to shift the equilibrium to the R2 form that crystallizes out. The binding pockets of the R, R3, and RR3 structures are sterically crowded, explaining the preferential binding of aromatic aldehyde to the R2 Hb (Safo et al. 2004; Abdulmalik et al. 2011). Gong et al. (2006) using NMR study reported that liganded Hb in the R2 form moves toward the R state in the presence of IHP. Unlike the aldehydes which bind at the α-cleft of R2 Hb, IHP binds at the β-cleft of classical R Hb (Song et al. 2008). It is interesting to note that the size of the β-cleft in the R structure is significantly larger than the other relaxed structures (Safo and Abraham 2005; Jenkins et al. 2009; Safo et al. 2011), suggesting preferential binding of IHP to the R structure, consistent with observed binding of phosphate at the β-cleft of R structure (Safo et al. 2002b). Thiols and imidazolylacryloyls seem to effect their antisickling activities in part by preferential binding to classical R and/or R3 structures and forming covalent interaction with βCys93 that lead to increase in Hb affinity for oxygen (Nakagawa et al. 2014, 2018; Omar et al. 2015).

Not clear is how the different liganded relaxed states fit in the overall scheme of Hb oxygen transport function. Based on the fact that the His(E7) ligand channel in the R, RR2, and R2 structures are closed, while R3 and RR3 are partially or fully opened, respectively, prompted Safo and colleagues to suggest that the R, R2, and RR2 conformations are involved in heme ligand transport, while the R3 and RR3 may be involved in ligand release (Safo and Abraham 2005; Jenkins et al. 2009; Safo et al. 2011). Perutz has previously predicted such a rotation of His(E7) for access of ligand (Perutz 1989). Birukou et al. (2011) replaced the distal histidines, αHis58 and βHis63 with Trp in human Hb, which significantly reduced ligand access to the hemes, prompting the investigators to propose that over 90% of the ligand entry in human Hb is due to the His(E7) ligand channel. The dynamics of the His(E7) rotation is likely pH-dependent as relatively low pH appears to stabilize the open ligand channel conformation as observed in the RR3 structure, while high pH may favor closure of the channel (Safo and Abraham 2005; Jenkins et al. 2009; Safo et al.

2011). Consistently, the rate constant for ligand binding to myoglobin increases with a decrease in pH (Traylor et al. 1983). Not obvious is whether His(E7) rotation out of the distal pocket in the T state is also pH-dependent since the pH in the lungs is not expected to be as low as in the tissues. Nevertheless, it is suggested that ligand may diffuse into the heme cavity, especially the α-heme through several hydrophobic pathways between the B, G, and H helices (Czerminski and Elber 1991), although not in an agreement by all investigators (Elber 2010).

Hemoglobin Variants with Altered Oxygen Affinity

Over 1000 naturally occurring hemoglobin variants have been identified, and although most are yet to be associated with any disease state, significant numbers are implicated in pathologies ranging from mild to severe, such as polycythemia or anemia, methemoglobinemia, cyanosis, tissue hypoxia and respiratory distress (Reissmann et al. 1961; Bonaventura and Riggs 1968; Schneider et al. 1975; Winslow and Charache 1975; Schoenborn 1976; Efremov et al. 1978; Arous et al. 1981; Dinçol et al. 1994; Kister et al. 1995; Borg et al. 1997; Marengo-Rowe 2006; Thom et al. 2013). We describe few examples of these variants that have mutations located at the surface of the protein, heme pocket, α1β2 or α1β1 interface, or the hydrophobic interior that alter hemoglobin structure and affect its oxygen binding properties. The most well-known variant is sickle Hb, which results from a single point mutation (βGlu6Val) in the β-globin chain of normal Hb to give sickle Hb (Bunn 1997). The mutation is located at the surface of the protein. Under hypoxia or when HbS is deoxygenated, the pathogenic βVal6 from one deoxygenated HbS molecule binds to a hydrophobic pocket on an adjacent deoxygenated HbS molecule, joining the molecules together to form insoluble polymers and giving rise to the classical sickle red blood cell (RBC) morphology (Eaton and Hofrichter 1990; Bunn 1997; Harrington et al. 1997; Ghatge et al. 2016). Sickled RBCs ultimately aggregate causing microvascular obstruction leading to SCD crisis (Belcher et al. 2003; Aliyu et al. 2008; De Franceschi 2009; Akinsheye and Klings 2010). Although the mutation does not directly affect the protein oxygen-binding property, a presumed elevation of 2,3-BPG in sickle red blood cells has been proposed to worsen the disease progression (Torrance et al. 1970; Grasso et al. 1975). Under physiological conditions, Hb releases 25–40% of O_2 to tissues, aided by binding of 2,3-BPG to deoxygenated Hb. Unfortunately, in SCD, the repetitive deoxygenation-reoxygenation cycles with cell sickling lead to RBC membrane damage and hemolysis, reducing the life span to 10–20 days (from 90 to 120 days for normal RBCs), which manifests in chronic anemia (Zago and Bottura 1983; Connor et al. 1994), prompting compensatory elevation of 2,3-BPG in sickle RBC that decreases HbS affinity for O_2 (P_{50} of 34 vs 26 mmHg in normal subjects) to unload more O_2 to tissue (Torrance et al. 1970; Grasso et al. 1975). This counterproductive response, increases deoxygenated HbS concentration, worsening the HbS polymerization and thus the concomitant RBC sickling process.

In vitro manipulation to reduce 2,3-BPG content has been suggested as a means to reduce the hypoxia-induced sickling (Poillon et al. 1986).

As noted above, the $\alpha_1\beta_1$–$\alpha_2\beta_2$ dimer interface (or $\alpha_1\beta_2$ or $\alpha_2\beta_1$ interface) is one of the most important structural determinant of the T → R transition (Baldwin and Chothia 1979; Safo and Abraham 2005; Jenkins et al. 2009; Safo et al. 2011). It is, therefore, no surprise that several mutations affecting this interface have been reported to affect the oxygen affinity of Hb (Reissmann et al. 1961; Charache et al. 1966; Botha et al. 1966; Jones et al. 1967; Bunn et al. 1972; Lokich et al. 1973; Jensen et al. 1975). Hb Bassett (αAsp94Ala), with the αAla94 mutation located at the $\alpha_1\beta_2$ interface is characterized by a markedly reduced oxygen affinity (P_{50} of cell free Hb at pH 7.0 = 22.0 mmHg compared with 10.5 mmHg in HbA) and low subunit cooperativity (n = 1.4 vs. 2.6 in HbA) (Abdulmalik et al. 2004; Safo et al. 2005). It is one of the few mutations where both the T and R structures of the mutant have been elucidated (Safo et al. 2005). The deoxygenated structure is characterized by two unique inter-dimer hydrogen-bond interactions (α1Tyr42–β2Asp99 and α1Asn97–β2Asp99) that are known to contribute to stabilization of T state Hb. Interestingly, the liganded R structure maintains these two hydrogen-bond interactions, in addition to losing a native R state stabilizing hydrogen-bond interaction between α1Asp94 and β2Asn102. These unique T state features in the R structure explain Hb Bassett low affinity for oxygen (Safo et al. 2005). Other low-O_2 affinity Hb variants with mutation at the $\alpha_1\beta_2$ interface include Hb Kansas (βAsn102Thr), Hb Saint Mande Â (βAsn102Tyr), Hb Setif (αAsp94Tyr), Hb Capa (αAsp94Gly) and Hb Titusville (αAsp94Asn) (Bonaventura and Riggs 1968; Schneider et al. 1975; Arous et al. 1981; Dinçol et al. 1994; Huisman et al. 1996; Borg et al. 1997).

Hb Thionville, a variant with a αVal1Glu mutation has been reported to exhibit an increased stabilization relative to normal Hb (Vasseur et al. 1992). Structurally, this mutation results in the inhibition of the cleavage of the initiator methionine, which becomes acetylated. This results in several tertiary structural changes in the form of an extension of the α-chain N-terminus, the formation of new intra- and inter-subunit contacts in the $\alpha_1\alpha_2$ interface that stabilize the T state and ultimately result in decreased Hb–O_2 binding affinity. However, the reduced oxygen affinity was not severe enough to produce any hematological abnormalities in the male carrier (Vasseur et al. 1992). In contrast, a similar mutation in the β chains of Hb (βVal1Glu; Hb Doha) resulted in significant anemia in a female carrier upon giving birth, who was otherwise healthy (Kamel et al. 1985). The introduction of a glutamic acid residue in Hb Doha also prevented the removal of the initiator methionine extending the N-terminus by one residue (Kamel et al. 1985). Unfortunately, we are not aware of any structural data for Hb Doha.

Several βAsp99 Hb variants, such as Hb Ypsilanti (βAsp99Tyr), Hb Radcliffe (βAsp99Ala), Hb Coimbra (βAsp99Glu) are known to exhibit increased affinity for O_2 and decreased cooperativity (Jorge et al. 2018). These individuals are clinically characterized by erythrocytosis (Hardison et al. 1994, 2001; Riemer et al. 1998). The βAsp99 residue is located at the switch region of α1β2 dimer interface and forms unique hydrogen-bond interactions with α1Tyr42 and α1Asn97, stabilizing the T structure. Mutations of βAsp99 abrogate these hydrogen-bond interactions, shifting the allosteric equilibrium to the R state, and increasing Hb affinity for oxygen (Jorge et al. 2018).

Hb Rothschild is characterized by a mutation (βTrp37Arg) at the hinge region of α1β2 dimer interface, leading to significant structural perturbation at the $\alpha_1\beta_2$ interface. Interestingly, this variant exhibits variable oxygen affinities, including both low and high affinity (Sharma et al. 1980; Nienhuis 1987). The T state structure showed a chloride ion bound at the $\alpha_1\beta_2$ interface as a counterion to βArg37 (Kavanaugh et al. 1992, 2001). The chloride ion was suggested to affect the solution properties of this mutant, especially at varying chloride concentration that may explain the variations in oxygen binding properties (Kavanaugh et al. 1992, 2001).

Mutations in the heme pocket are also known to affect Hb oxygen binding properties (Thom et al. 2013). An example is Hb Kirklareli (αHis58Leu), which is associated with iron deficiency and increased CO binding (Bissé et al. 2017). The imidazole sidechain of αHis58 provides stability to the heme bound O_2 through hydrogen-bond interactions, which is lost with αLeu58. Indeed, the crystal structure of Hb Kirklareli shows that the bound O_2 is no longer stabilized, consistent with the observed increased autoxidation and loss of hemin ~200 times more rapidly than native α subunits (Bissé et al. 2017). Interestingly, Hb Kirklareli α subunit has an ~80,000-fold higher affinity for CO than O_2, causing it to rapidly take up and retain carbon monoxide (Bissé et al. 2017).

Several Hb variants with mutations at the $\alpha_1\beta_1$ ($\alpha_2\beta_2$) interface have been identified, with some of these mutations causing significant alterations at the $\alpha_1\beta_1$ interface that manifest in changes in Hb oxygen binding properties. Examples are Hb Philly (βTyr35Phe), which is characterized by increased oxygen affinity and decreased cooperativity, Hb Peterborough (βVal111Phe) and Hb Stanmore (βVal111Ala) also characterized by decreased oxygen affinities (Rieder et al. 1969; King et al. 1972; Como et al. 1991; Thom et al. 2013). These mutations cause hemolytic anemia and/or reticulocytosis (Thom et al. 2013).

Finally, mutations, deletions or insertions in the core nonpolar regions of Hb have been reported to produce unstable hemoglobins (Carrell et al. 1966; Dacie et al. 1967; Murari et al. 1977; Sakuragawa et al. 1984; Plaseska et al. 1991) that tend to undergo spontaneous oxidation causing precipitation and the formation of insoluble inclusions called Heinz bodies leading to hemolytic anemia.

Hemoglobin Modulators

Endogenous Modulators

2,3-Bisphosphoglycerate (2,3-BPG)

2,3-Bisphosphoglycerate (Fig. 14.6) plays a central role in hemoglobin allostery by reducing Hb–O_2 affinity and allowing efficient tissue oxygenation (Arnone 1972;

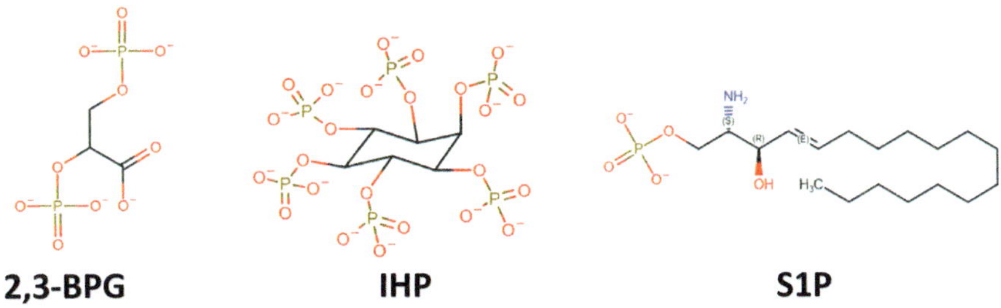

Fig. 14.6 Chemical structures of IHP, 2,3-BPG and S1P

Perutz et al. 1986; Marden et al. 1990; Richard et al. 1993). Its effect on Hb oxygen affinity was first reported by Benesch and Benesch (1967). The atomic basis of its mechanism of action was elucidated with the crystal structure of Hb that showed 2,3-BPG preferentially bind to the β-cleft of deoxygenated Hb making interactions with βHis2, βLys82, βHis143, and βHis146, and linking together the two β-subunits to stabilize the T state (Fig. 14.7) (Arnone 1972; Richard et al. 1993). In contrast to the T structure, the ensemble relaxed structures have relatively smaller β-clefts explaining their lower affinity for 2,3-BPG, although liganded R state Hb with the largest β-cleft among the relaxed structures is known to bind 2,3-BPG, albeit at a lower affinity than deoxygenated Hb (Gupta et al. 1979). Crystallographic and NMR studies with phosphate and IHP (Fig. 14.6) also suggest that the R and RR2 structures are capable of binding 2,3-BPG at the β-cleft (Gupta et al. 1979; Safo et al. 2002b; Safo and Abraham 2005). It is quite obvious that the R3 structure, with its completely closed β-cleft due to elaborate β_1–β_2 inter-subunit hydrogen-bond interactions should preclude 2,3-BPG binding (Safo and Abraham 2005). As noted above 2,3-BPG has been proposed to be involved in the pathogenesis of sickle cell disease by its elevation

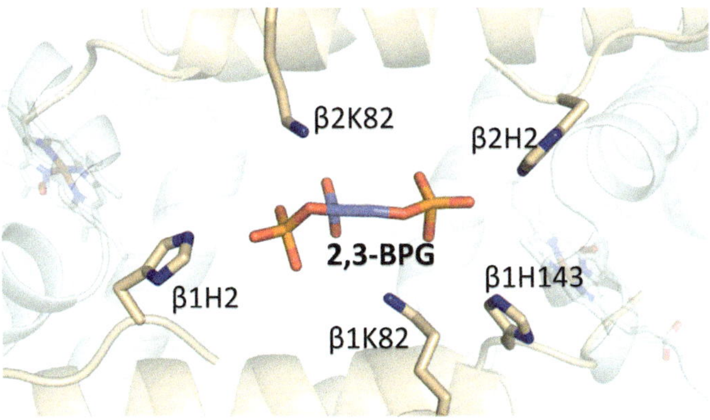

Fig. 14.7 Binding of 2,3-BPG (purple) at the β-cleft of Hb. The α-subunits are colored in gray and the β-subunits are colored in tan

in sickle RBC, which leads to further decrease in Hb affinity for oxygen to increase the concentration of the polymer forming deoxygenated HbS.

Sphingosine-1-Phosphate (S1P)

S1P (Fig. 14.6) is an important signaling molecule that is enriched in erythrocytes and is known to regulate diverse biological processes through activation of cell surface S1P receptors and/or by interaction with critical regulatory proteins within cells (Spiegel and Milstien 2003; Hänel et al. 2007; Ito et al. 2007). Interaction between S1P and hemoglobin was first uncovered by the Xia group after a metabolomic screen identified the pathological role of elevated S1P in SCD (Zhang et al. 2014). S1P at μM concentrations was found to regulate binding of deoxygenated Hb to the membrane protein Band 3 (cdB3) and ultimately resulting in metabolic reprogramming in SCD (Sun et al. 2017). The crystal structure of deoxygenated Hb in complex with S1P showed S1P to bind to the surface of the protein close to the heme pocket (making hydrophilic/hydrophobic interactions at the switch interface) (Fig. 14.8) (Sun et al. 2017). S1P only binds to the surface of Hb upon binding of 2,3-BPG at the β-cleft. The ternary complex leads to a significant conformational change that not only increase the T state character of the T structure but also sterically impede diffusion of diatomic ligands (O_2) into the heme pocket (Sun et al. 2017). These structural changes have been proposed to in part explain the observed decrease in Hb–O_2 affinity of HbS and the concomitant sickling of RBC (Sun et al. 2017). An interesting structural observation is that the last 3–4 carbon atoms of the bound S1P do not make any interaction with the Hb residue but hang out in the bulk solvent. These carbon atoms are hypothesized to mediate the hydrophobic interactions with cdB3, to also promote SCD pathogenesis (Sun et al. 2017). These findings add significant new insight to erythrocyte pathology and physiology, which paves the way for novel therapeutic interventions in SCD.

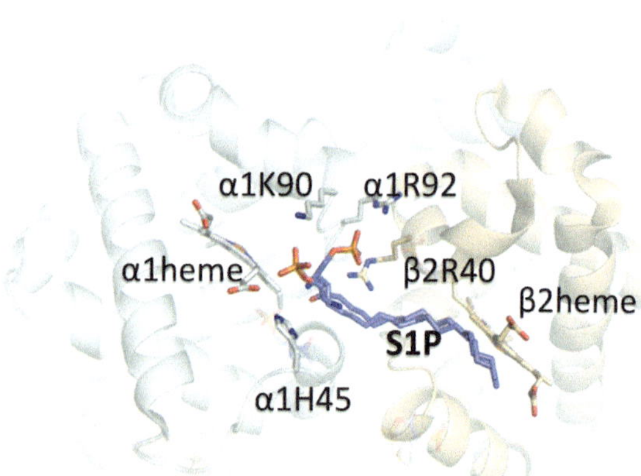

Fig. 14.8 Binding of S1P (purple) on the surface of deoxygenated Hb. The α-subunits are colored in gray and the β-subunits are colored in tan. Note that S1P only binds to the surface of the protein when 2,3-BPG binds to the β-cleft

The Bohr Effect—Carbon Dioxide, Proton, and Chloride

For most mammalian hemoglobins, the heme affinity for ligands is dependent upon ambient pH (Bohr effect), due to tertiary structural perturbations (Shibayama and Saigo 2001; Yonetani et al. 2002; Yonetani and Tsuneshige 2003; Yonetani and Kanaori 2013), as well as the equilibrium between the quaternary T and R structures (Perutz 1972a, b; Perutz et al. 1998). In 1904, CO_2 was discovered by Christian Bohr to lower the oxygen affinity of Hb, allowing efficient delivery of oxygen to tissues (Bohr et al. 1904). In the tissue, where CO_2 concentration is high, its conversion to carbamate and/or bicarbonate releases hydrogen ions which in turn lowers the pH causing the oxygen affinity of Hb to decrease. Max Perutz and others have proposed a molecular basis for this proton-dependent Bohr effect that involves the C-terminal residues of both the α- and β-chains (αArg141 and βHis146, respectively) (Perutz 1976; O'Donnell et al. 1979; Mozzarelli et al. 1991; Kavanaugh et al. 1992; Perutz et al. 1993, 1994, 1998; Bettati and Mozzarelli 1997; Bettati et al. 1998; Safo and Abraham 2005; Jenkins et al. 2009; Safo et al. 2011). At high pH in deoxygenated Hb, βHis146 makes intra-subunit and inter-subunit salt-bridge interactions with βAsp94 (Fig. 14.3b) and αLys40, while αArg141 also participates in a separate inter-subunit salt-bridge interactions with αLys127 and αAsp126 (Fig. 14.9). These interactions stabilize the low-affinity T structure, consequently, facilitating O_2 release. At low pH especially in the tissues, these salt-bridge interactions are broken (Figs. 14.3b and 14.9), which increases the mobility of both αArg141 and βHis146, facilitating the T → R transition, and consequently increasing Hb oxygen affinity. Other residues, such as αVal1, αHis122, βHis2, βLys82, βHi143, are also known to contribute to the Bohr effect through deoxygenation-linked proton binding (Perutz et al. 1969; Kilmartin et al. 1978; Perutz 1983; Lukin and Ho 2004; Berenbrink 2006).

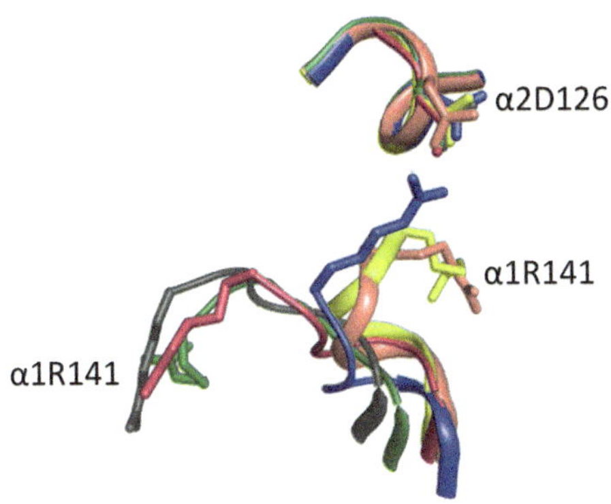

Fig. 14.9 Superposed structures of T (blue), R (magenta), R3 (yellow), RR2 (green), R2 (black), and RR3 (salmon) on their $\alpha_1\beta_1$ dimers. αArg141 participates in inter-subunit salt-bridge interaction with αAsp126 (as well as αLys127—not shown) in deoxygenated Hb stabilizing the low-affinity T state and facilitating O_2 release. At higher pH, this interaction is broken increasing the mobility of αArg141, which facilitates the T → R transition, and increase Hb oxygen affinity

Excess positively charged residues, such as αVal1 and βLys82 in the central water cavity mutually repel each other, increasing the free energy of Hb (Bonaventura et al. 1994; Perutz et al. 1998). Chloride ions contribute to the Bohr effect by neutralizing these positive charges to stabilize Hb (Perutz et al. 1993, 1994; Fronticelli et al. 1994). More chloride ions are found in the larger T structure central water cavity when compared to the smaller R structure central water cavity, leading to greater stabilization of the T state with concomitant lowering of Hb oxygen affinity. Several Hb surface located histidines have also been suggested to contribute to the Bohr effect (Busch and Ho 1990; Sun et al. 1997). For a more detailed discussion of the Bohr effect, the reader is referred to a review article by Mairbäurl & Weber (2012).

Contribution to the Bohr effect by the various relaxed Hb structures may be different due to obvious tertiary and quaternary differences. In the R, R3 and RR3 structures, βHis146 is highly disordered, while in the RR2 and R2 structures, this residue is resolved due to close proximity of the two symmetry-related βHis146 residues that make interaction with each other (Fig. 14.3b) (Silva et al. 1992; Safo and Abraham 2005; Jenkins et al. 2009; Safo et al. 2011), prompting Arnone and co-workers to suggest that the contribution of βHis146 to the Bohr effect are different in the R and R2 states (Silva et al. 1992).

Exogenous Modulators

Modulators Shifting the Allosteric Equilibrium to the Low-O_2 Affinity

A fundamental importance of Hb allostery is taking advantage of it to develop therapeutics for diseases. The discovery of 2,3-BPG prompted the search for synthetic Hb effectors for the treatment of ischemic-related diseases. One of the earliest compounds to be studied was inositol hexaphosphate (IHP) (Fig. 14.6), that binds similarly as 2,3-BPG to the β-cleft of deoxygenated Hb and was found to be 1000 × more potent than 2,3-BPG in decreasing Hb affinity for oxygen (Yonetani et al. 2002). IHP has limited absorption profile for a useful therapeutic application (Stucker et al. 1985; Biolo et al. 2009). Nonetheless, it has been beneficial in investigating the allosteric properties of Hb. Analogs of IHP, such as myoinositol trispyrophosphate that are capable of crossing the membrane of erythrocytes, have been shown to enhance the exercise capacity in mice with severe heart failure (Biolo et al. 2009).

Propionates are a class of synthetic effectors that also induce low-O_2 affinity by binding to Hb and stabilizing the T state. Examples of these compounds include the antilipidemic drug bezafibrate (BZF) and several of its urea derivatives (L35, L345, LR16) (Abraham et al. 1983a; Perutz and Poyart 1983; Lalezari et al. 1988, 1990; Shibayama et al. 2002; Yokoyama et al. 2006), RSR-13 and several of its derivatives (KDD3-138, RSR-40, RSR-4, TB-27, etc.) (Fig. 14.10); the latter group of compounds was discovered and studied by Abraham and co-workers (Randad et al. 1991; Wireko et al. 1991; Abraham et al. 1992b; Khandelwal et al. 1993; Phelps Grella et al. 2000; Safo et al. 2001, 2002a; Youssef et al. 2002). These compounds,

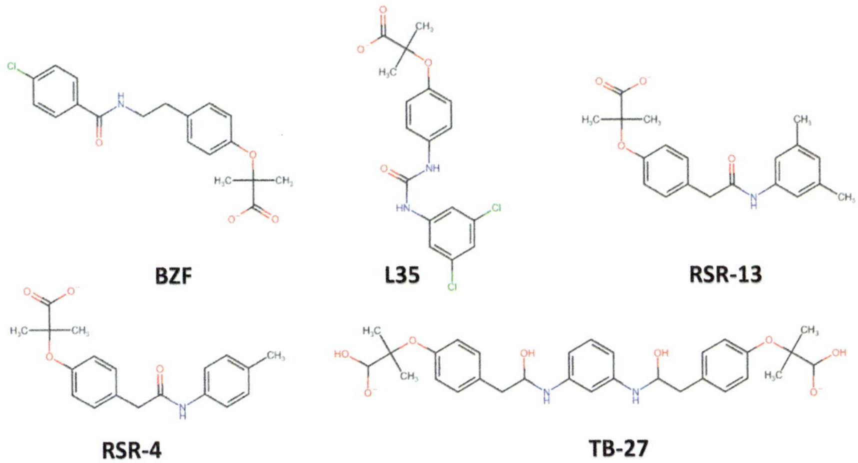

Fig. 14.10 Chemical structures of BZF (and derivative L35) and RSR-13 and derivatives (RSR-4 and TB-27)

unlike IHP or 2,3-BPG effect their allosteric activity in part by binding to the middle of the central water cavity of deoxygenated Hb. RSR-13 (aka Efaproxiral), the most well-known propionate was originally synthesized to mimic allosteric effects first seen with BZF, but with less protein binding in serum (Randad et al. 1991; Abraham et al. 1992b). The mode of atomic interactions of the propionates with deoxygenated Hb are very similar. X-ray crystal structures indicated that a pair of RSR-13 form noncovalent interactions with three subunits of the deoxygenated Hb tetramer within the central water cavity in a symmetry-related fashion (Fig. 14.11a), effectively stabilizing the T state from transitioning to the R state and thus reducing the affinity of Hb for O_2 (Wireko et al. 1991; Abraham et al. 1992a; Safo et al. 2001). Each

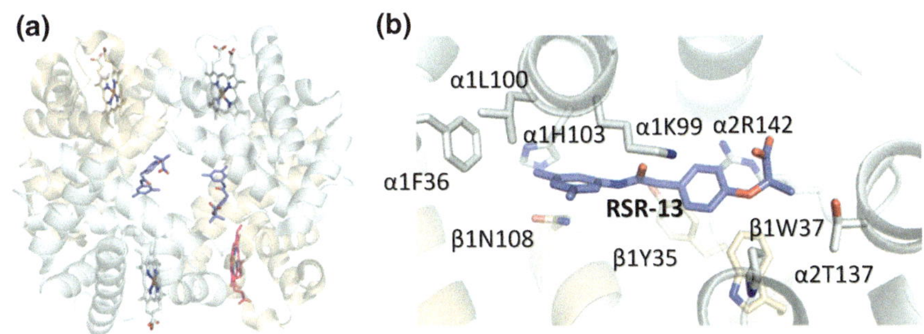

Fig. 14.11 **a** Binding of a pair of RSR-13 (purple) at the central water cavity of deoxygenated (T state) Hb. **b** Detailed interactions between one of the RSR-13 molecules and the protein. The other molecule makes similar symmetry-related interactions. The two α-subunits are colored grey, while the β-subunit is colored tan

molecule of RSR-13 makes hydrogen-bond and hydrophobic interactions with two α-subunits and one β-subunit of the protein. Its specific interactions to the protein are illustrated in Fig. 14.11b (Safo et al. 2001). The allosteric effects of RSR-13 are additive to that of 2,3-BPG, since the compounds have distinct Hb binding sites (Laberge et al. 2005). RSR-13 has undergone preclinical testing in hypoxia- and ischemia-induced conditions, as well as late-stage human testing as an adjunct to enhance the O_2-induced effects during whole-brain radiation (Abraham et al. 1992b; Khandelwal et al. 1993; Pagel et al. 1998; Woods et al. 1998; Kleinberg et al. 1999; Shaw et al. 2003).

In addition to their interactions with deoxygenated Hb as described above for RSR-13, BZF and L35 have also been shown to bind to the central water cavity and/or surface of liganded Hb close to the α-heme (in the form of the classical R structure), which sterically impede ligand access to the heme (Shibayama et al. 2002; Chen et al. 2005; Yokoyama et al. 2006). Clearly, these effectors are capable of binding to both liganded and deoxygenated Hb, and their regulatory effect on Hb allostery appears to result from their interactions with both forms of Hb.

Safo's group recently developed novel nitric oxide (NO)-releasing prodrugs of RSR-13 derivatives by attaching the NO-releasing moieties nitrooxyethyl, nitrooxypropyl, and 1-(pyrrolidin-1-yl)diazen-1-ium-1,2-diolate, respectively, to the carboxylate of RSR-13 (Xu et al. 2015). Crystallographic studies with the prodrugs showed RSR-13, the hydrolysis product of the prodrugs bound to the central water cavity of deoxygenated Hb (as described above) that explained these compounds ability to decrease Hb affinity for oxygen (Xu et al. 2015). Moreover, the released NO, was observed exclusively bound to the two α hemes, which was suggested to be due to RSR-13 decreasing the protein's affinity for the ligand at the β hemes (Xu et al. 2015).

Aryloxyalkanoic Acids are another class of right-shifting compounds that bind non-covalently to Hb (Omar et al. 2016). The effect of these compounds on Hb's oxygen affinity depends on the site of interaction. Aryloxyalkanoic acids that bind to the αTrp14 hydrophobic pocket of Hb have been shown to increase oxygen affinity, while those that bind to the central water cavity of the protein to stabilize the T state in most instances showed the opposite effect of decreasing the protein affinity for oxygen (Abraham et al. 1982, 1983b, 1984; Patwa et al. 1987; Mehanna and Abraham 1990).

Recently, a high-throughput screening campaign identified IRL-2500 (Fig. 14.12a), a synthetic peptide, that decreased the affinity of Hb for O_2 (Goldstein et al. 2018). Structural studies with deoxygenated Hb showed this compound to overlap the 2,3-BPG binding site at the β-cleft, engaging in water-mediated and direct hydrogen-bond, as well as hydrophobic interactions with the β-cleft residues of βHis2, βLys82, βAsn139, and βHis143 (Fig. 14.12b), which provide additional interactions across the two β-subunit interface of deoxygenated Hb leading to further stabilization of the T state (Goldstein et al. 2018).

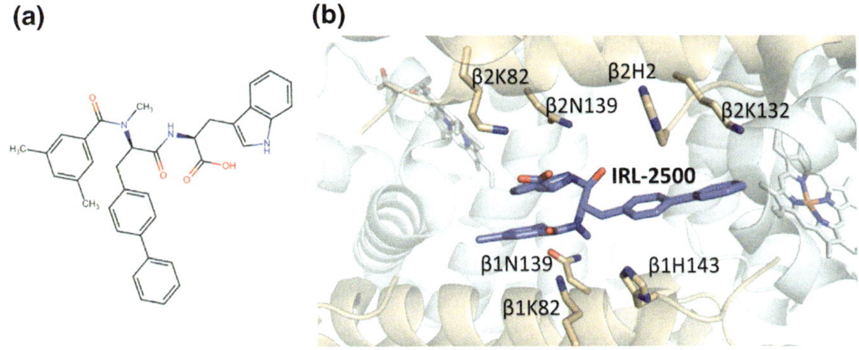

Fig. 14.12 **a** Chemical structure of IRL-2500. **b** Binding of IRL-2500 (purple) at the β-cleft of Hb. The α-subunits are colored in gray and the β-subunits are colored in tan

Modulators Shifting the Allosteric Equilibrium to the High-O_2 Affinity

Several synthetic and natural compounds have also been shown, in most instances to bind covalently to Hb to shift the allosteric equilibrium to the R state and increase the oxygen affinity of the protein. Some of these compounds have been clinically evaluated as antisickling agents for the treatment of sickle cell disease, as high-O_2 affinity sickle Hb molecules are resistant to polymer formation.

Aromatic aldehydes (Fig. 14.13), the most well-studied effectors interact with Hb and increase the protein affinity for oxygen. The food flavoring agent vanillin was one of the earliest to be studied (Fig. 14.13) (Zaugg et al. 1977; Abraham et al. 1991). Although relatively non-toxic, its limited bioavailability and the large dose needed

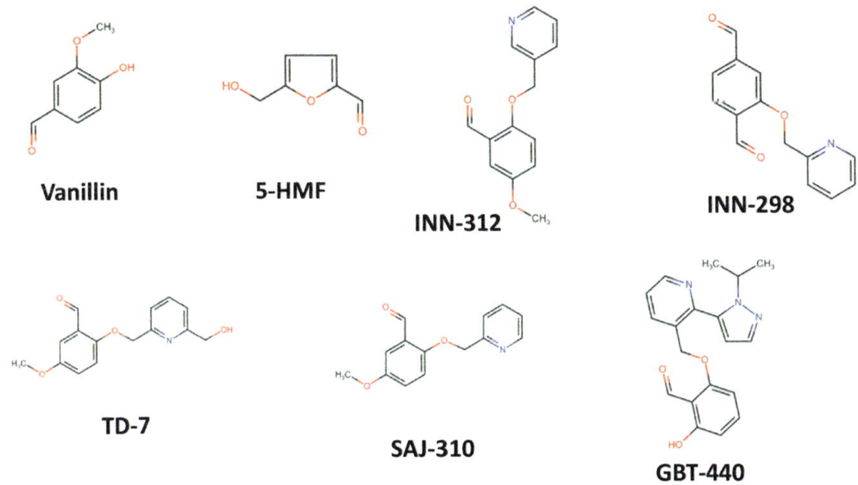

Fig. 14.13 Chemical structures of high-O_2 affinity antisickling aromatic aldehydes

to increase Hb affinity for oxygen and elicit in vivo therapeutic antisickling effects were not clinically acceptable.

Furfural and several of its analogs, such as 5-hydroxymethyl-2-furfural (5-HMF) (Fig. 14.13) (Safo et al. 2004; Abdulmalik et al. 2005; Xu et al. 2017) have also been shown to interact with Hb and increase its oxygen affinity. 5-HMF showed significant pharmacologic improvement over vanillin both in vitro and in vivo and was subsequently studied in phase II clinical trials for the treatment of SCD (Abdulmalik et al. 2005; Stern et al. 2012; Kato et al. 2013). The study was terminated due in part to in vivo bioavailability issue. Nonetheless, the study of 5-HMF helped trigger a dramatic increase in commercial interest in developing drugs for SCD that show a similar mechanism of action. It is important to note that prior to 5-HMF studies, the antisickling mechanism of action of aromatic aldehydes was suggested to be due to binding to deoxygenated Hb and destabilizing the T state and/or binding to liganded Hb and stabilizing the classical R state (Beddell et al. 1984; Abraham et al. 1991; Wireko and Abraham 1991). Safo and co-workers, however showed that mechanistically, 5-HMF, and for that matter, other antisickling aromatic aldehydes increase Hb affinity for O_2 by binding to the R2 structure and not the T or classical R as proposed (Safo et al. 2004; Abdulmalik et al. 2005, 2011; Xu et al. 2017; Pagare et al. 2018; Deshpande et al. 2018). Two molecules of the aromatic aldehydes bind in a symmetry-related fashion at the α-cleft of the R2 structure, each molecule forming Schiff-base interaction between its aldehyde moiety and the N-terminal αVal1 nitrogen of the Hb, and through additional hydrogen-bond and/or hydrophobic interactions tie the two α-subunits together and restrict the transition to the T state (Fig. 14.14) (Safo et al. 2004; Abdulmalik et al. 2005, 2011; Xu et al. 2017; Pagare et al. 2018; Deshpande et al. 2018).

To improve on vanillin and 5-HMF left-shifting potency and thus their antisickling activities, several derivatives of these lead compounds were developed. 5-HMF derivatives incorporate different substituents at the alcohol moiety of 5-HMF

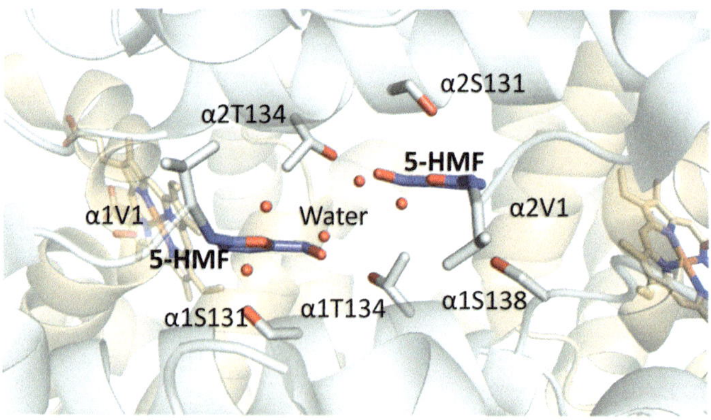

Fig. 14.14 Binding of 5-HMF (purple) in a symmetry-related fashion at the α-cleft Hb. The α-subunits are colored in gray and the β-subunits are colored in tan. Water molecules are red spheres

for additional interactions with the protein, which led to a significant increase in their allosteric activity, translating into 1.5–4.0-fold higher antisickling effects than 5-HMF (Xu et al. 2017).

Several groups have also embarked on structural modification of vanillin for more potent high-O_2 affinity allosteric effectors. Some of these novel compounds include pyridyl derivatives of benzaldehydes that have been studied by Safo and colleagues, and shown to exhibit several-fold left-shifting potency over vanillin (Nnamani et al. 2008; Abdulmalik et al. 2011; Pagare et al. 2018; Deshpande et al. 2018). The crystal structures of some of these compounds, such as SAJ-310, INN-312, INN-298, and TD-7 (Fig. 14.13) complexed to Hb have been elucidated (Abdulmalik et al. 2011; Pagare et al. 2018; Deshpande et al. 2018) and show the compounds to expectedly bind similarly to the R2 structure as observed with 5-HMF. Of note is that the covalent bond between the aldehyde and the αVal1 N forced the molecules with *meta*-positioned methoxy-pyridine group, such as in INN-298 to direct down the central water cavity (Fig. 14.15), while those with *ortho*-positioned methoxy-pyridine, such as INN-312 or TD-7 is disposed toward the mouth of the α-cavity (Fig. 14.16) (Abdulmalik et al. 2011; Pagare et al. 2018; Deshpande et al. 2018). The latter group of compounds make interactions with the solvent-exposed αF-helix to exhibit a second antisickling mechanism of direct polymer destabilization that is independent of the primary mechanism of increasing HbS-O_2 affinity (Pagare et al. 2018; Deshpande et al. 2018). The αF-helix has been shown to play an important role in polymer stabilization, and its perturbation is thought to destabilize the polymer (Safo et al. 2004, 2011; Abdulmalik et al. 2011). In line with this hypothesis, the Hb variant Stanleyville (αAsn78 → αLsy78) inhibits HbS polymerization (Benesch et al. 1979; Nagel et al. 1980; Rhoda et al. 1983). Four select representatives of

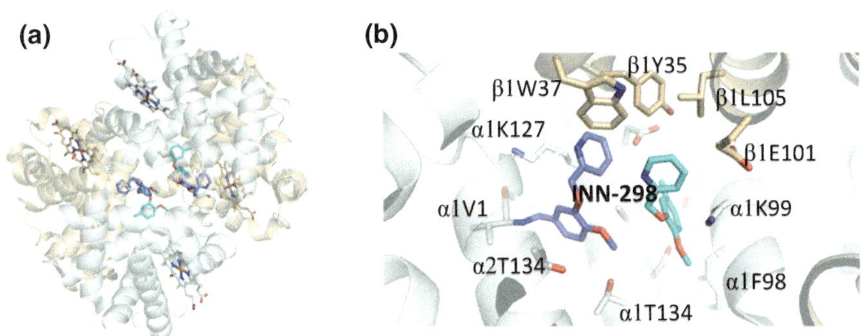

Fig. 14.15 **a** Binding of INN-298 (purple and cyan) in a symmetry-related fashion at the α-cleft Hb. Unlike most aromatic aldehydes, four molecules of INN-298 bind per one Hb molecule. The molecules in purple make Schiff-base interactions with αVal1 nitrogen (primary), while the molecules in cyan (secondary) bind non-covalently and significantly weaker. **b** Detailed interactions of two of the INN-298 molecules (purple and cyan) with Hb. The *meta*-positioned methoxy-pyridine group of the primary bound INN-298 disposes further down the central water-cavity

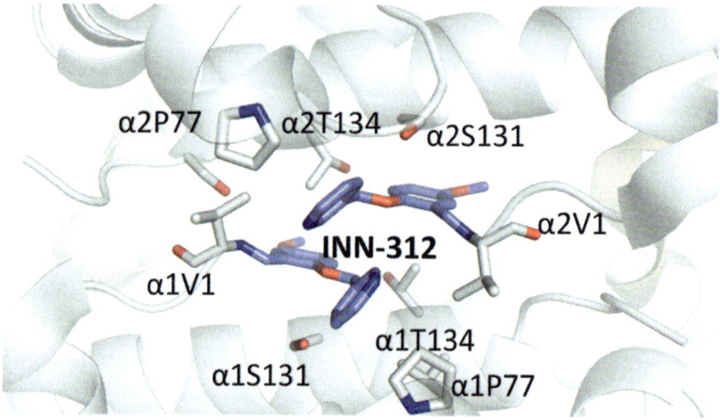

Fig. 14.16 Binding of INN-312 (purple) in a symmetry-related fashion at the α-cleft of Hb. The α-subunits are colored in gray. The *ortho*-positioned methoxy-pyridine group disposes toward the surface of the Hb to make interaction with the αF-helix residue of Pro77

these compounds, VZHE-039, PP-6, PP-10, and PP-14, are currently undergoing preclinical in vivo studies by Safo and co-workers (Safo, unreported studies).

GBT-440 (Voxelotor; Fig. 14.13) is another synthetic aldehyde analog of vanillin developed by Global Blood Therapeutics, Inc. that also increases the oxygen affinity of Hb by stabilizing the relaxed state (Oksenberg et al. 2016; Metcalf et al. 2017). GBT-440 similarly binds to the N-terminal αVal1 of one of the α-chains of the R2 structure but because of the bulky pyrazole substituent, a second molecule is precluded from binding to the opposite α-chain resulting in a single GBT-440 molecule bound per Hb tetramer, which is in contrast with the observed 2:1 stoichiometry of other antisickling aromatic aldehydes (Fig. 14.17). GBT-440 is currently being stud-

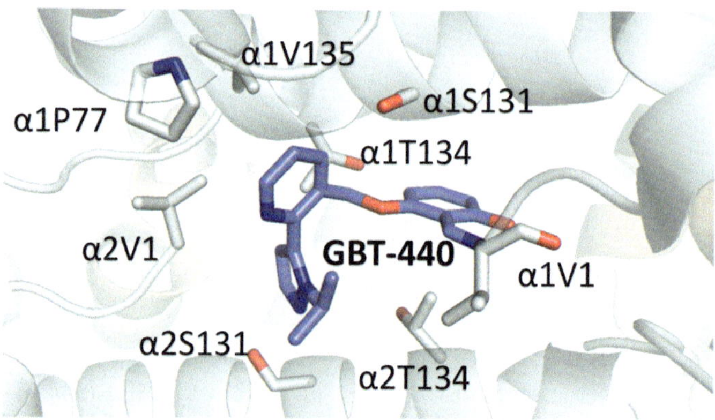

Fig. 14.17 Binding of GBT-440 (purple) at the α-cleft Hb. The α-subunits are colored in gray and the β-subunits are colored in tan. Unlike other aldehyde effectors only one molecule of GBT-440 binds per one Hb molecule

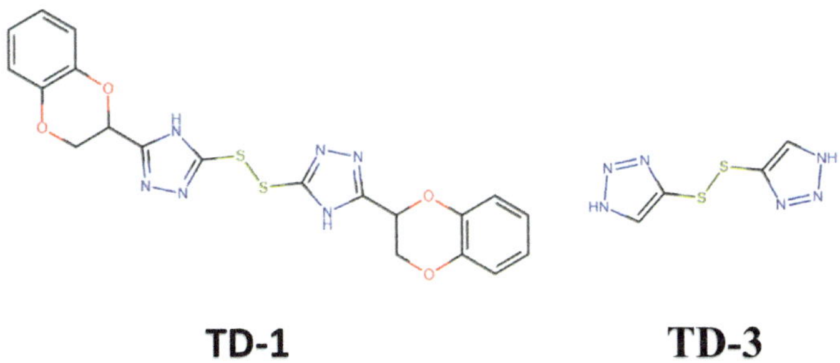

Fig. 14.18 Chemical structures of antisickling thiol-containing agents

ied in phase III clinical trials for the treatment of SCD (NCT03036813) (Oksenberg et al. 2016; Metcalf et al. 2017, p. 440; Vichinsky et al. 2019).

High-throughput screening of a small molecule library has identified thiol-containing effectors of hemoglobin, TD-1, and TD-3 (Fig. 14.18) (Nakagawa et al. 2014, 2018). These compounds, unlike the aromatic aldehydes, act via covalent disulfide bond formation with βCys93 of deoxygenated Hb that inhibits a T state salt-bridge formation between βAsp94 and βHis146 shifting the equilibrium to the relaxed state and increasing Hb affinity for oxygen (Fig. 14.19a) (Nakagawa et al. 2014, 2018). The compounds also bind to the R and/or R3 structures that stabilize the relaxed state, contributing to the increase in Hb oxygen affinity (Fig. 14.19b) (Nakagawa et al. 2014, 2018).

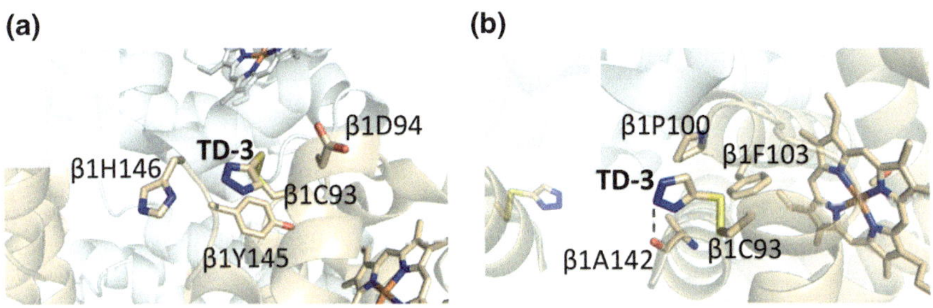

Fig. 14.19 Crystal structure of TD-3 in complex with T and R structures, where TD-3 forms disulfide bond with βCys93 sulfur atom. Hb α and β subunits are shown as grey and tan, respectively. **a** The binding of TD-3 with βCys93 in deoxygenated Hb leads to disruption of the T state stabilizing salt-bridge interaction between βAsp94 and βHis146. **b** The binding of TD-3 in CO-liganded Hb (R structure) prevents possible interaction between βAsp94 and βHis146 that is required to shift the allosteric transition to the T state

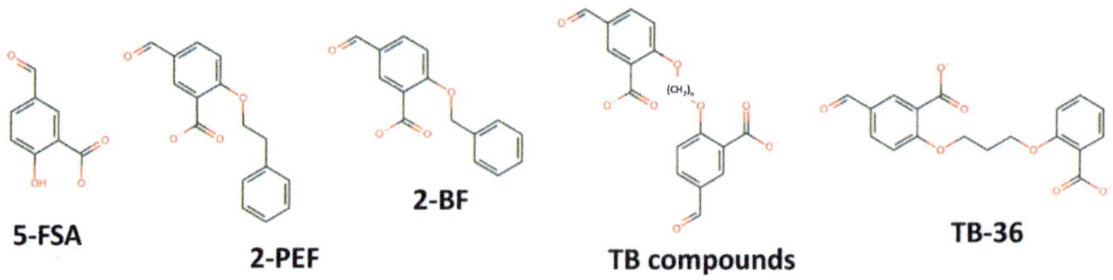

Fig. 14.20 Chemical structures of right-shifting (low-O_2 affinity) aromatic aldehydes

Allosteric Effectors that Target the Same Site but Induce Opposite Equilibrium Shifts

Above, we described several aromatic aldehydes that increase the oxygen affinity of Hb by forming Schiff-base interaction with Hb αVal1 nitrogen at the α-cleft. However, not all αVal1 Schiff-base adducts with aromatic aldehydes result in increasing Hb affinity for oxygen. Abraham's group tested several mono-aldehyde-acid effectors of Hb, such as 5-FSA, 2-BF, and 2-PEF (Fig. 14.20), that were hypothesized to produce a high-O_2 affinity Hb (Abraham et al. 1995). Unexpectedly, these compounds showed the opposite effect by reducing Hb affinity for oxygen, despite forming Schiff-base interaction with the Hb αVal1 nitrogen. Structural analysis showed that because of the carboxylate substituent on the benzene ring (relative to the aldehyde), these compounds preferentially bind to the α-cleft of deoxygenated Hb, where they form Schiff-base interaction with αVal1 nitrogen, as well as an inter-subunit salt-bridge interaction between the carboxylate and the guanidinium group of the opposite αArg141 of Hb, providing additional constraints to the T state (Fig. 14.21) (Abraham et al. 1995). The R2 structure does not have αArg141 at the correct position to make such salt-bridge interaction with the carboxylate of the compounds, while R, RR2 and R3 structures sterically preclude binding at their respective α-clefts. Removal of the carboxylate from the compounds abrogates the αArg141 interaction in deoxygenated T structure, resulting in these molecules preferentially binding to liganded Hb in the form of R2 structure as observed with vanillin and several of

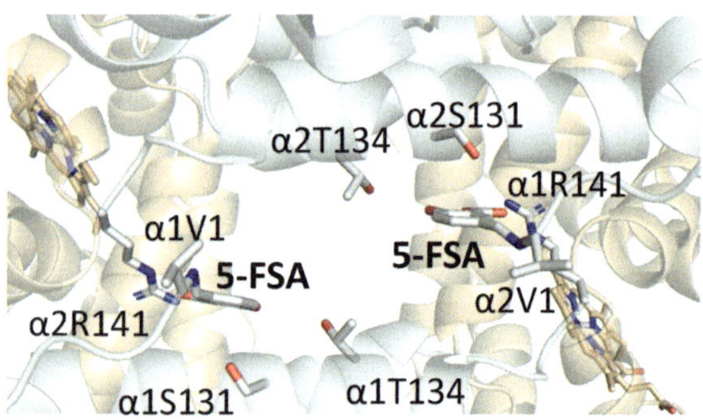

Fig. 14.21 (A) Crystal structures of deoxygenated Hb in complex with the mono-aldehyde-acid molecule of 5-FSA

its analogs (Abdulmalik et al. 2011; Pagare et al. 2018; Deshpande et al. 2018). Nonetheless, these non-carboxylate aromatic aldehydes still bind to deoxygenated Hb, albeit significantly weaker because of lack of strong inter-subunit interactions (Safo et al. 2004; Abdulmalik et al. 2011). Moreover, binding to deoxygenated Hb does not appear to contribute to the stability of the T state but rather further destabilizes it by replacing a T state stabilizing chloride ion in the central water cavity (Safo et al. 2004).

The crystallographic result with the mono-aldehyde-acid compounds prompted Abraham et al. to design di-aldehyde-acid derivatives by adding a second aromatic aldehyde at the para position of the first aldehyde, such as TB compounds (Fig. 14.20), which was expected to form a second Schiff-base adduct with αLys99 of the opposite α-subunit to further constrain the T structure (Boyiri et al. 1995). Different classes of these compounds were synthesized that showed several-fold increase in reducing Hb affinity for oxygen compared to the mono-aldehyde analogs (Boyiri et al. 1995). Crystallographic studies with the di-aldehyde compounds showed a pair of compounds bound to deoxygenated Hb as designed (Boyiri et al. 1995). The first aldehyde made a Schiff-base interaction with α1Val1 nitrogen, while the second aldehyde made an inter-subunit Schiff-base interaction with the amine of the α2Lys99. The meta-carboxylate moiety engaged in an inter-subunit salt-bridge interaction with α2Arg141 similar to the parent mono-aldehdye-acid compounds. Overall, these interactions provided additional stabilization to the T state, further reducing the oxygen affinity of Hb. It is worth noting that effectors with the shortest bridges exhibited the most potent effect, suggesting that, the tighter the two α-subunits are held together, the higher the degree of constraint on the T-structure. Such covalent cross-linkers of Hb subunits, with further optimization of cell permeability, have the potential for being used as cell-free Hb-based blood substitutes.

Concluding Remarks

Hb is a fascinating molecule that never ceases to amaze scientists. It helped to shape our understanding of one of the most complex theories in molecular biology, which is allostery, although it is yet to be fully unravelled. The remarkable advances in structural biology techniques, such as cryo-electron microscopy and time-resolved spectroscopic methods will, hopefully, provide more glimpses into the phenomenon of Hb allostery. In this chapter, we provided an overview of the molecular structure of Hb, relating to its primary function of oxygen transport. We discussed the concept of allostery and the different models that were proposed over the years to explain this phenomenon. It is clear that the oxygen carrying function of Hb involves an ensemble of states, although their specific functions, especially the relaxed states are yet to be fully understood. Finally, we provided an overview of the molecular mechanisms of how different Hb variants, as well as effectors modulate Hb oxygen binding property, and the effort being made to harness the latter for therapeutics. These allosteric effectors, depending on their mode of interactions with T or relaxed

Hb, can decrease or increase Hb affinity for oxygen, the later useful for treating sickle cell disease and the former for hypoxia- and ischemia-induced conditions. Of interest is that the α-cleft of T and R2 structures of Hb act as sinks for aromatic aldehydes, and preferential binding to the T or R2 α-cleft can have profound effect on Hb oxygen affinity, given rise to a new understanding that an allosteric effector can bind to the same site but produce opposite allosteric effects. These studies have uncovered crucial roles played by several Hb residues, such as αVal1, αLys99, αArg141, G helix, F helix residues in modulating Hb allostery.

Conflict-of-Interest Disclosure Virginia Commonwealth University and Martin K. Safo have patents related to several aromatic aldehydes mentioned in the chapter.

References

Abdulmalik O, Safo MK, Lerner NB et al (2004) Characterization of hemoglobin bassett (alpha94Asp → Ala), a variant with very low oxygen affinity. Am J Hematol 77:268–276. https://doi.org/10.1002/ajh.20184

Abdulmalik O, Safo MK, Chen Q et al (2005) 5-hydroxymethyl-2-furfural modifies intracellular sickle haemoglobin and inhibits sickling of red blood cells. Br J Haematol 128:552–561. https://doi.org/10.1111/j.1365-2141.2004.05332.x

Abdulmalik O, Ghatge MS, Musayev FN et al (2011) Crystallographic analysis of human hemoglobin elucidates the structural basis of the potent and dual antisickling activity of pyridyl derivatives of vanillin Corrigendum. Acta Crystallogr D Biol Crystallogr 67:1076. https://doi.org/10.1107/S0907444911045860

Abraham DJ, Mehanna AS, Williams FL (1982) Design, synthesis, and testing of potential antisickling agents. 1. halogenated benzyloxy and phenoxy acids. J Med Chem 25:1015–1017. https://doi.org/10.1021/jm00351a002

Abraham DJ, Perutz MF, Phillips SE (1983a) Physiological and x-ray studies of potential antisickling agents. Proc Natl Acad Sci USA 80:324–328. https://doi.org/10.1073/pnas.80.2.324

Abraham DJ, Perutz MF, Phillips SE (1983b) Physiological and x-ray studies of potential antisickling agents. PNAS 80:324–328. https://doi.org/10.1073/pnas.80.2.324

Abraham DJ, Kennedy PE, Mehanna AS et al (1984) Design, synthesis, and testing of potential antisickling agents. 4. structure-activity relationships of benzyloxy and phenoxy acids. J Med Chem 27:967–978. https://doi.org/10.1021/jm00374a006

Abraham DJ, Mehanna AS, Wireko FC et al (1991) Vanillin, a potential agent for the treatment of sickle cell anemia. Blood 77:1334–1341

Abraham DJ, Peascoe RA, Randad RS, Panikker J (1992a) X-ray diffraction study of di and tetra-ligated T-state hemoglobin from high salt crystals. J Mol Biol 227:480–492. https://doi.org/10.1016/0022-2836(92)90902-V

Abraham DJ, Wireko FC, Randad RS et al (1992b) Allosteric modifiers of hemoglobin: 2-[4-[[(3,5-disubstituted anilino)carbonyl]methyl]phenoxy]-2-methylpropionic acid derivatives that lower the oxygen affinity of hemoglobin in red cell suspensions, in whole blood, and in vivo in rats. Biochemistry 31:9141–9149. https://doi.org/10.1021/bi00153a005

Abraham DJ, Safo MK, Boyiri T et al (1995) How allosteric effectors can bind to the same protein residue and produce opposite shifts in the allosteric equilibrium. Biochemistry 34:15006–15020. https://doi.org/10.1021/bi00046a007

Akinsheye I, Klings ES (2010) Sickle cell anemia and vascular dysfunction: the nitric oxide connection. J Cell Physiol 224:620–625. https://doi.org/10.1002/jcp.22195

Aliyu ZY, Gordeuk V, Sachdev V et al (2008) Prevalence and risk factors for pulmonary artery systolic hypertension among sickle cell disease patients in Nigeria. Am J Hematol 83:485–490. https://doi.org/10.1002/ajh.21162

Arnone A (1972) X-ray diffraction study of binding of 2,3-diphosphoglycerate to human deoxyhaemoglobin. Nature 237:146–149. https://doi.org/10.1038/237146a0

Arous N, Braconnier F, Thillet J et al (1981) Hemoglobin Saint Mandé beta 102 (G4) asn replaced by tyr: a new low oxygen affinity variant. FEBS Lett 126:114–116. https://doi.org/10.1016/0014-5793(81)81046-0

Baldwin J, Chothia C (1979) Haemoglobin: The structural changes related to ligand binding and its allosteric mechanism. J Mol Biol 129:175–220. https://doi.org/10.1016/0022-2836(79)90277-8

Barrick D, Ho NT, Simplaceanu V et al (1997) A test of the role of the proximal histidines in the Perutz model for cooperativity in haemoglobin. Nat Struct Biol 4:78–83

Beddell CR, Goodford PJ, Kneen G et al (1984) Substituted benzaldehydes designed to increase the oxygen affinity of human haemoglobin and inhibit the sickling of sickle erythrocytes. Br J Pharmacol 82:397–407. https://doi.org/10.1111/j.1476-5381.1984.tb10775.x

Belcher JD, Bryant CJ, Nguyen J et al (2003) Transgenic sickle mice have vascular inflammation. Blood 101:3953–3959. https://doi.org/10.1182/blood-2002-10-3313

Benesch R, Benesch RE (1967) The effect of organic phosphates from the human erythrocyte on the allosteric properties of hemoglobin. Biochem Biophys Res Commun 26:162–167. https://doi.org/10.1016/0006-291X(67)90228-8

Benesch RE, Kwong S, Edalji R, Benesch R (1979) Alpha chain mutations with opposite effects on the gelation of hemoglobin S. J Biol Chem 254:8169–8172

Berenbrink M (2006) Evolution of vertebrate haemoglobins: Histidine side chains, specific buffer value and Bohr effect. Respir Physiol Neurobiol 154:165–184. https://doi.org/10.1016/j.resp.2006.01.002

Bettati S, Mozzarelli A (1997) T state hemoglobin binds oxygen noncooperatively with allosteric effects of protons, inositol hexaphosphate, and chloride. J Biol Chem 272:32050–32055. https://doi.org/10.1074/jbc.272.51.32050

Bettati S, Mozzarelli A, Perutz MF (1998) Allosteric mechanism of haemoglobin: rupture of salt-bridges raises the oxygen affinity of the T-structure. J Mol Biol 281:581–585. https://doi.org/10.1006/jmbi.1998.1983

Biolo A, Greferath R, Siwik DA et al (2009) Enhanced exercise capacity in mice with severe heart failure treated with an allosteric effector of hemoglobin, myo-inositol trispyrophosphate. PNAS 106:1926–1929. https://doi.org/10.1073/pnas.0812381106

Birukou I, Soman J, Olson JS (2011) Blocking the gate to ligand entry in human hemoglobin. J Biol Chem 286:10515–10529. https://doi.org/10.1074/jbc.M110.176271

Bissé E, Schaeffer-Reiss C, Van Dorsselaer A et al (2017) Hemoglobin Kirklareli (α H58L), a new variant associated with iron deficiency and increased CO binding. J Biol Chem 292:2542–2555. https://doi.org/10.1074/jbc.M116.764274

Bohr C, Hasselbalch K, Krogh A (1904) Ueber einen in biologischer Beziehung wichtigen Einfluss, den die Kohlensäurespannung des Blutes auf dessen Sauerstoffbindung übt1. Skandinavisches Archiv Für Physiologie 16:402–412. https://doi.org/10.1111/j.1748-1716.1904.tb01382.x

Bonaventura J, Riggs A (1968) Hemoglobin Kansas, a human hemoglobin with a neutral amino acid substitution and an abnormal oxygen equilibrium. J Biol Chem 243:980–991

Bonaventura C, Arumugam M, Cashon R et al (1994) Chloride masks effects of opposing positive charges in Hb A and Hb Hinsdale (β139 Asn → Lys) that can modulate cooperativity as well as oxygen affinity. J Mol Biol 239:561–568. https://doi.org/10.1006/jmbi.1994.1395

Borg I, Valentino M, Fiorini A, Felice AE (1997) Hb Setif [alpha 94(G1)Asp → Tyr] in Malta. Hemoglobin 21:91–96

Botha MC, Beale D, Isaacs WA, Lehmann H (1966) Hemoglobin J Cape Town-alpha-2 92 arginine replaced by glutamine beta-2. Nature 212:792–795. https://doi.org/10.1038/212792a0

Boyiri T, Safo MK, Danso-Danquah RE et al (1995) Bisaldehyde allosteric effectors as molecular ratchets and probes. Biochemistry 34:15021–15036. https://doi.org/10.1021/bi00046a008

Brunori M, Coletta M, Di Cera E (1986) A cooperative model for ligand binding to biological macromolecules as applied to oxygen carriers. Biophys Chem 23:215–222. https://doi.org/10.1016/0301-4622(86)85006-2

Bunn HF (1997) Pathogenesis and treatment of sickle cell disease. N Engl J Med 337:762–769. https://doi.org/10.1056/NEJM199709113371107

Bunn HF, Bradley TB, Davis WE et al (1972) Structural and functional studies on hemoglobin Bethesda (alpha2beta2 145His), a varient associated with compensatory erythrocytosis. J Clin Invest 51:2299–2309. https://doi.org/10.1172/JCI107040

Busch MR, Ho C (1990) Effects of anions on the molecular basis of the Bohr effect of hemoglobin. Biophys Chem 37:313–322. https://doi.org/10.1016/0301-4622(90)88031-M

Carrell RW, Lehmann H, Hutchison HE (1966) Haemoglobin Köln (beta-98 valine–methionine): an unstable protein causing inclusion-body anaemia. Nature 210:915–916. https://doi.org/10.1038/210915a0

Charache S, Weatherall DJ, Clegg JB (1966) Polycythemia associated with a hemoglobinopathy. J Clin Invest 45:813–822. https://doi.org/10.1172/JCI105397

Chen Q, Lalezari I, Nagel RL, Hirsch RE (2005) Liganded hemoglobin structural perturbations by the allosteric effector L35. Biophys J 88:2057–2067. https://doi.org/10.1529/biophysj.104.046136

Como PF, Wylie BR, Trent RJ et al (1991) A new unstable and low oxygen affinity hemoglobin variant: Hb Stanmore [beta 111(G13)Val–Ala]. Hemoglobin 15:53–65

Connor J, Pak CC, Schroit AJ (1994) Exposure of phosphatidylserine in the outer leaflet of human red blood cells. relationship to cell density, cell age, and clearance by mononuclear cells. J Biol Chem 269:2399–2404

Czerminski R, Elber R (1991) Computational studies of ligand diffusion in globins: I. Leghemoglobin. Proteins: Struct Funct Bioinform 10:70–80. https://doi.org/10.1002/prot.340100107

Dacie JV, Shinton NK, Gaffney PJ, Lehmann H (1967) Haemoglobin Hammersmith (beta-42 (CDI) Phe replaced by ser). Nature 216:663–665. https://doi.org/10.1038/216663a0

De Franceschi L (2009) Pathophisiology of sickle cell disease and new drugs for the treatment. Mediterr J Hematol Infect Dis 1:e2009024. https://doi.org/10.4084/MJHID.2009.024

Deshpande TM, Pagare PP, Ghatge MS et al (2018) Rational modification of vanillin derivatives to stereospecifically destabilize sickle hemoglobin polymer formation. Acta Crystallogr D Struct Biol 74:956–964. https://doi.org/10.1107/S2059798318009919

Dinçol G, Dinçol K, Erdem S et al (1994) Hb Capa or alpha (2)94(G1)Asp → Gly beta 2, a mildly unstable variant with an A → G (GAC → GGC) mutation in codon 94 of the alpha 1-globin gene. Hemoglobin 18:57–60

Doyle ML, Lew G, Turner GJ et al (1992) Regulation of oxygen affinity by quaternary enhancement: does hemoglobin ypsilanti represent an allosteric intermediate? Proteins: Struct Funct Bioinform 14:351–362. https://doi.org/10.1002/prot.340140304

Eaton WA, Hofrichter J (1990) Sickle cell hemoglobin polymerization. Adv Protein Chem 40:63–279

Efremov GD, Stojmirovic E, Lam HL et al (1978) HB Beth Israel (beta 102 [G4] Asn replaced by Ser) observed in a Yugoslavian teenager. Hemoglobin 2:75–77. https://doi.org/10.3109/03630267808999192

Elber R (2010) Ligand diffusion in globins: simulations versus experiment. Curr Opin Struct Biol 20:162–167. https://doi.org/10.1016/j.sbi.2010.01.002

Fan J-S, Zheng Y, Choy W-Y et al (2013) Solution structure and dynamics of human hemoglobin in the carbonmonoxy form. Biochemistry 52:5809–5820. https://doi.org/10.1021/bi4005683

Fermi G (1975) Three-dimensional fourier synthesis of human deoxyhaemoglobin at 2–5 a resolution: refinement of the atomic model. J Mol Biol 97:237–256. https://doi.org/10.1016/s0022-2836(75)80037-4

Fernandez EJ, Abad-Zapatero C, Olsen KW (2000) Crystal structure of Lysβ182-Lysβ282 crosslinked hemoglobin: A possible allosteric intermediate 11 Edited by K. Nagai. J Mol Biol 296:1245–1256. https://doi.org/10.1006/jmbi.2000.3525

Fronticelli C, Pechik I, Brinigar WS et al (1994) Chloride ion independence of the Bohr effect in a mutant human hemoglobin beta (V1M + H2deleted). J Biol Chem 269:23965–23969

Gell DA (2018) Structure and function of haemoglobins. Blood Cells Mol Dis 70:13–42. https://doi.org/10.1016/j.bcmd.2017.10.006

Ghatge MS, Ahmed MH, Omar ASM et al (2016) Crystal structure of carbonmonoxy sickle hemoglobin in R-state conformation. J Struct Biol 194:446–450. https://doi.org/10.1016/j.jsb.2016.04.003

Goldstein SR, Liu C, Safo MK et al (2018) Design, synthesis, and biological evaluation of allosteric effectors that enhance CO release from carboxyhemoglobin. ACS Med Chem Lett 9:714–718. https://doi.org/10.1021/acsmedchemlett.8b00166

Gong Q, Simplaceanu V, Lukin JA et al (2006) Quaternary structure of carbonmonoxyhemoglobins in solution: structural changes induced by the allosteric effector inositol hexaphosphate. Biochemistry 45:5140–5148. https://doi.org/10.1021/bi052424h

Grasso JA, Sullivan AL, Sullivan LW (1975) Ultrastructural studies of the bone marrow in sickle cell anaemia. II. the morphology of erythropoietic cells and their response to deoxygenation in vitro. Br J Haematol 31:381–389. https://doi.org/10.1111/j.1365-2141.1975.tb00869.x

Gupta RK, Benovic JL, Rose ZB (1979) Location of the allosteric site for 2,3-bisphosphoglycerate on human oxy- and deoxyhemoglobin as observed by magnetic resonance spectroscopy. J Biol Chem 254:8250–8255

Hänel P, Andréani P, Gräler MH (2007) Erythrocytes store and release sphingosine 1-phosphate in blood. FASEB J 21:1202–1209. https://doi.org/10.1096/fj.06-7433com

Hardison R, Chao KM, Schwartz S et al (1994) Globin gene server: a prototype E-mail database server featuring extensive multiple alignments and data compilation for electronic genetic analysis. Genomics 21:344–353. https://doi.org/10.1006/geno.1994.1275

Hardison RC, Chui DH, Riemer C et al (2001) Databases of human hemoglobin variants and other resources at the globin gene server. Hemoglobin 25:183–193

Harrington DJ, Adachi K, Royer WE (1997) The high resolution crystal structure of deoxyhemoglobin S. J Mol Biol 272:398–407. https://doi.org/10.1006/jmbi.1997.1253

He Z, Russell JE (2004a) Effect of zeta-globin substitution on the O_2-transport properties of Hb S in vitro and in vivo. Biochem Biophys Res Commun 325:1376–1382. https://doi.org/10.1016/j.bbrc.2004.10.180

He Z, Russell JE (2004b) Antisickling effects of an endogenous human alpha-like globin. Nat Med 10:365–367. https://doi.org/10.1038/nm1022

Henry ER, Bettati S, Hofrichter J, Eaton WA (2002) A tertiary two-state allosteric model for hemoglobin. Biophys Chem 98:149–164. https://doi.org/10.1016/S0301-4622(02)00091-1

Huisman THJ, Carver MFH, Efremov G (1996) A syllabus of human hemoglobin variants. The Sickle Cell Anemia Foundation, Augusta, GA, USA

Ito K, Anada Y, Tani M et al (2007) Lack of sphingosine 1-phosphate-degrading enzymes in erythrocytes. Biochem Biophys Res Commun 357:212–217. https://doi.org/10.1016/j.bbrc.2007.03.123

Janin J, Wodak SJ (1993) The quaternary structure of carbonmonoxy hemoglobin ypsilanti. Proteins: Struct Funct Bioinform 15:1–4. https://doi.org/10.1002/prot.340150102

Jayaraman V, Rodgers KR, Mukerji I, Spiro TG (1995) Hemoglobin allostery: resonance Raman spectroscopy of kinetic intermediates. Science 269:1843–1848. https://doi.org/10.1126/science.7569921

Jenkins JD, Musayev FN, Danso-Danquah R et al (2009) Structure of relaxed-state human hemoglobin: insight into ligand uptake, transport and release. Acta Cryst D 65:41–48. https://doi.org/10.1107/S0907444908037256

Jensen M, Oski FA, Nathan DG, Bunn HF (1975) Hemoglobin Syracuse (alpha2beta2-143(H21)His leads to Pro), a new high-affinity variant detected by special electrophoretic methods. observations on the auto-oxidation of normal and variant hemoglobins. J Clin Invest 55:469–477. https://doi.org/10.1172/JCI107953

Jones RT, Osgood EE, Brimhall B, Koler RD (1967) Hemoglobin Yakina. I. clinical and biochemical studies. J Clin Invest 46:1840–1847. https://doi.org/10.1172/JCI105674

Jorge SE, Bringas M, Petruk AA et al (2018) Understanding the molecular basis of the high oxygen affinity variant human hemoglobin Coimbra. Arch Biochem Biophys 637:73–78. https://doi.org/10.1016/j.abb.2017.11.010

Kamel K, el-Najjar A, Webber BB et al (1985) Hb Doha or alpha 2 beta 2[X-N-Met-1(NA1)Val–Glu]; a new beta-chain abnormal hemoglobin observed in a Qatari female. Biochim Biophys Acta 831:257–260. https://doi.org/10.1016/0167-4838(85)90043-3

Kato GJ, Lawrence MP, Mendelsohn LG et al (2013) Phase 1 clinical trial of the candidate anti-sickling agent Aes-103 In adults with sickle cell anemia. Blood 122:1009–1009

Kavanaugh JS, Rogers PH, Case DA, Arnone A (1992) High-resolution x-ray study of deoxy-hemoglobin Rothschild 37.beta. Trp.fwdarw. Arg: a mutation that creates an intersubunit chloride-binding site. Biochemistry 31:4111–4121. https://doi.org/10.1021/bi00131a030

Kavanaugh JS, Weydert JA, Rogers PH et al (2001) Site-directed mutations of human hemoglobin at residue 35beta: a residue at the intersection of the alpha1beta1, alpha1beta2, and alpha1alpha2 interfaces. Protein Sci 10:1847–1855. https://doi.org/10.1110/ps.16401

Kavanaugh JS, Rogers PH, Arnone A (2005) Crystallographic evidence for a new ensemble of ligand-induced allosteric transitions in hemoglobin: the T-to-T(high) quaternary transitions. Biochemistry 44:6101–6121. https://doi.org/10.1021/bi047813a

Khandelwal SR, Randad RS, Lin PS et al (1993) Enhanced oxygenation in vivo by allosteric inhibitors of hemoglobin saturation. Am J Physiol 265:H1450–H1453. https://doi.org/10.1152/ajpheart.1993.265.4.H1450

Kilmartin JV, Imai K, Jones RT et al (1978) Role of Bohr group salt bridges in cooperativity in hemoglobin. Biochim Biophys Acta 534:15–25. https://doi.org/10.1016/0005-2795(78)90471-3

King MA, Wiltshire BG, Lehmann H, Morimoto H (1972) An unstable haemoglobin with reduced oxygen affinity: haemoglobin Peterborough, 3 (GI3) Valine lead to Phenylalanine, its interaction with normal haemoglobin and with haemoglobin Lepore. Br J Haematol 22:125–134. https://doi.org/10.1111/j.1365-2141.1972.tb08794.x

Kister J, Kiger L, Francina A et al (1995) Hemoglobin Roanne [alpha 94(G1) Asp → Glu]: a variant of the alpha 1 beta 2 interface with an unexpected high oxygen affinity. Biochim Biophys Acta 1246:34–38. https://doi.org/10.1016/0167-4838(94)00190-r

Kleinberg L, Grossman SA, Piantadosi S et al (1999) Phase I trial to determine the safety, pharmacodynamics, and pharmacokinetics of RSR13, a novel radioenhancer, in newly diagnosed glioblastoma multiforme. J Clin Oncol 17:2593–2603. https://doi.org/10.1200/JCO.1999.17.8.2593

Koshland DE, Némethy G, Filmer D (1966) Comparison of experimental binding data and theoretical models in proteins containing subunits. Biochemistry 5:365–385. https://doi.org/10.1021/bi00865a047

Kosmachevskaya OV, Topunov AF (2018) Alternate and additional functions of erythrocyte hemoglobin. Biochem Mosc 83:1575–1593. https://doi.org/10.1134/S0006297918120155

Laberge M, Kövesi I, Yonetani T, Fidy J (2005) R-state hemoglobin bound to heterotropic effectors: models of the DPG, IHP and RSR13 binding sites. FEBS Lett 579:627–632. https://doi.org/10.1016/j.febslet.2004.12.033

Ladner RC, Heidner EJ, Perutz MF (1977) The structure of horse methaemoglobin at 2-0 A resolution. J Mol Biol 114:385–414. https://doi.org/10.1016/0022-2836(77)90256-x

Lalezari I, Rahbar S, Lalezari P et al (1988) LR16, a compound with potent effects on the oxygen affinity of hemoglobin, on blood cholesterol, and on low density lipoprotein. PNAS 85:6117–6121. https://doi.org/10.1073/pnas.85.16.6117

Lalezari I, Lalezari P, Poyart C et al (1990) New effectors of human hemoglobin: structure and function. Biochemistry 29:1515–1523. https://doi.org/10.1021/bi00458a024

Lokich JJ, Moloney WC, Bunn HF et al (1973) Hemoglobin brigham (alpha2Abeta2100 Pro–Leu). Hemoglobin variant associated with familial erythrocytosis. J Clin Invest 52:2060–2067. https://doi.org/10.1172/JCI107390

Lukin JA, Ho C (2004) The structure–function relationship of hemoglobin in solution at atomic resolution. Chem Rev 104:1219–1230. https://doi.org/10.1021/cr940325w
Lukin JA, Kontaxis G, Simplaceanu V et al (2003) Quaternary structure of hemoglobin in solution. PNAS 100:517–520. https://doi.org/10.1073/pnas.232715799
Lukin JA, Kontaxis G, Simplaceanu V et al (2004) Backbone resonance assignments of human adult hemoglobin in the carbonmonoxy form. J Biomol NMR 28:203–204. https://doi.org/10.1023/B:JNMR.0000013816.64039.6f
Mairbäurl H, Weber RE (2012) Oxygen transport by hemoglobin. Compr Physiol 2:1463–1489. https://doi.org/10.1002/cphy.c080113
Makowski L, Bardhan J, Gore D et al (2011) WAXS studies of the structural diversity of hemoglobin in solution. J Mol Biol 408:909–921. https://doi.org/10.1016/j.jmb.2011.02.062
Marden MC, Bohn B, Kister J, Poyart C (1990) Effectors of hemoglobin. separation of allosteric and affinity factors. Biophys J 57:397–403. https://doi.org/10.1016/S0006-3495(90)82556-X
Marengo-Rowe AJ (2006) Structure-function relations of human hemoglobins. Proc (Bayl Univ Med Cent) 19:239–245. https://doi.org/10.1080/08998280.2006.11928171
Mehanna AS, Abraham DJ (1990) Comparison of crystal and solution hemoglobin binding of selected antigelling agents and allosteric modifiers. Biochemistry 29:3944–3952. https://doi.org/10.1021/bi00468a022
Metcalf B, Chuang C, Dufu K et al (2017) Discovery of GBT440, an orally bioavailable R-state stabilizer of sickle cell hemoglobin. ACS Med Chem Lett 8:321–326. https://doi.org/10.1021/acsmedchemlett.6b00491
Monod J, Wyman J, Changeux JP (1965) On the nature of allosteric transitions: a plausible model. J Mol Biol 12:88–118. https://doi.org/10.1016/s0022-2836(65)80285-6
Mozzarelli A, Rivetti C, Rossi GL et al (1991) Crystals of haemoglobin with the T quaternary structure bind oxygen noncooperatively with no Bohr effect. Nature 351:416–419. https://doi.org/10.1038/351416a0
Mueser TC, Rogers PH, Arnone A (2000) Interface sliding as illustrated by the multiple quaternary structures of liganded hemoglobin. Biochemistry 39:15353–15364. https://doi.org/10.1021/bi0012944
Muirhead H, Perutz MF (1963) Structure of hæemoglobin: a three-dimensional Fourier synthesis of reduced human haemoglobin at 5.5 Å resolution. Nature 199:633–638. https://doi.org/10.1038/199633a0
Murari J, Smith LL, Wilson JB et al (1977) Some properties of hemoglobin Gun Hill. Hemoglobin 1:267–282
Nagel RL, Johnson J, Bookchin RM et al (1980) Beta-chain contact sites in the haemoglobin S polymer. Nature 283:832–834. https://doi.org/10.1038/283832a0
Nakagawa A, Lui FE, Wassaf D et al (2014) Identification of a small molecule that increases hemoglobin oxygen affinity and reduces SS erythrocyte sickling. ACS Chem Biol 9:2318–2325. https://doi.org/10.1021/cb500230b
Nakagawa A, Ferrari M, Schleifer G et al (2018) A triazole disulfide compound increases the affinity of hemoglobin for oxygen and reduces the sickling of human sickle cells. Mol Pharm 15:1954–1963. https://doi.org/10.1021/acs.molpharmaceut.8b00108
Nienhuis AW (1987) Hemoglobin: Molecular, genetic and clinical aspects: By H. F. Bunn and B. G. Forget. Philadelphia: W. B. Saunders Company. (1986). 690 pp. $99.00. Cell 48:731. https://doi.org/10.1016/0092-8674(87)90069-9
Nnamani IN, Joshi GS, Danso-Danquah R et al (2008) Pyridyl derivatives of benzaldehyde as potential antisickling agents. Chem Biodivers 5:1762–1769. https://doi.org/10.1002/cbdv.200890165
Noble RW, Hui HL, Kwiatkowski LD et al (2001) Mutational effects at the subunit interfaces of human hemoglobin: evidence for a unique sensitivity of the T quaternary state to changes in the hinge region of the alpha 1 beta 2 interface. Biochemistry 40:12357–12368. https://doi.org/10.1021/bi010988p

O'Donnell S, Mandaro R, Schuster TM, Arnone A (1979) X-ray diffraction and solution studies of specifically carbamylated human hemoglobin A. evidence for the location of a proton- and oxygen-linked chloride binding site at valine 1 alpha. J Biol Chem 254:12204–12208
Oksenberg D, Dufu K, Patel MP et al (2016) GBT440 increases haemoglobin oxygen affinity, reduces sickling and prolongs RBC half-life in a murine model of sickle cell disease. Br J Haematol 175:141–153. https://doi.org/10.1111/bjh.14214
Omar AM, Mahran MA, Ghatge MS et al (2015) Identification of a novel class of covalent modifiers of hemoglobin as potential antisickling agents. Org Biomol Chem 13:6353–6370. https://doi.org/10.1039/c5ob00367a
Omar AM, Mahran MA, Ghatge MS et al (2016) Aryloxyalkanoic acids as non-covalent modifiers of the allosteric properties of hemoglobin. Molecules 21. https://doi.org/10.3390/molecules21081057
Pagare PP, Ghatge MS, Musayev FN et al (2018) Rational design of pyridyl derivatives of vanillin for the treatment of sickle cell disease. Bioorg Med Chem 26:2530–2538. https://doi.org/10.1016/j.bmc.2018.04.015
Pagel PS, Hettrick DA, Montgomery MW et al (1998) RSR13, a synthetic allosteric modifier of hemoglobin, enhances recovery of stunned myocardium in dogs. Adv Exp Med Biol 454:527–531. https://doi.org/10.1007/978-1-4615-4863-8_63
Paoli M, Liddington R, Tame J et al (1996) Crystal structure of T state haemoglobin with oxygen bound at all four haems. J Mol Biol 256:775–792. https://doi.org/10.1006/jmbi.1996.0124
Patwa DC, Abraham DJ, Hung TC (1987) Design, synthesis, and testing of potential antisickling agents. 6. Rheologic studies with active phenoxy and benzyloxy acids. Blood Cells 12:589–601
Perrella M, Cera ED (1999) CO ligation intermediates and the mechanism of hemoglobin cooperativity. J Biol Chem 274:2605–2608. https://doi.org/10.1074/jbc.274.5.2605
Perutz MF (1972a) Stereochemical mechanism of cooperative effects in haemoglobin. Biochimie 54:587–588. https://doi.org/10.1016/s0300-9084(72)80142-1
Perutz MF (1972b) Nature of haem-haem interaction. Nature 237:495–499. https://doi.org/10.1038/237495a0
Perutz MF (1976) Structure and mechanism of haemoglobin. Br Med Bull 32:195–208. https://doi.org/10.1093/oxfordjournals.bmb.a071363
Perutz MF (1983) Species adaptation in a protein molecule. Mol Biol Evol 1:1–28. https://doi.org/10.1093/oxfordjournals.molbev.a040299
Perutz MF (1989) Myoglobin and haemoglobin: role of distal residues in reactions with haem ligands. Trends Biochem Sci 14:42–44. https://doi.org/10.1016/0968-0004(89)90039-X
Perutz MF, Poyart C (1983) Bezafibrate lowers oxygen affinity of haemoglobin. The Lancet 322:881–882. https://doi.org/10.1016/S0140-6736(83)90870-X
Perutz MF, Muirhead H, Cox JM, Goaman LC (1968) Three-dimensional Fourier synthesis of horse oxyhaemoglobin at 2.8 A resolution: the atomic model. Nature 219:131–139. https://doi.org/10.1038/219131a0
Perutz MF, Muirhead H, Mazzarella L et al (1969) Identification of residues responsible for the alkaline Bohr effect in haemoglobin. Nature 222:1240–1243. https://doi.org/10.1038/2221240a0
Perutz MF, Giulio Fermi, Abraham DJ et al (1986) Hemoglobin as a receptor of drugs and peptides: x-ray studies of the stereochemistry of binding. J Am Chem Soc 108:1064–1078. https://doi.org/10.1021/ja00265a036
Perutz MF, Fermi G, Poyart C et al (1993) A novel allosteric mechanism in haemoglobin: structure of bovine deoxyhaemoglobin, absence of specific chloride-binding sites and origin of the chloride-linked Bohr effect in bovine and human haemoglobin. J Mol Biol 233:536–545. https://doi.org/10.1006/jmbi.1993.1530
Perutz MF, Shih DT -b., Williamson D (1994) The chloride effect in human haemoglobin. a new kind of allosteric mechanism. J Mol Biol 239:555–560. https://doi.org/10.1006/jmbi.1994.1394
Perutz MF, Wilkinson AJ, Paoli M, Dodson GG (1998) The stereochemical mechanism of the cooperative effects in hemoglobin revisited. Annu Rev Biophys Biomol Struct 27:1–34. https://doi.org/10.1146/annurev.biophys.27.1.1

Phelps Grella M, Danso-Danquah R, Safo MK et al (2000) Synthesis and structure-activity relationships of chiral allosteric modifiers of hemoglobin. J Med Chem 43:4726–4737. https://doi.org/10.1021/jm000199q

Plaseska D, Dimovski AJ, Wilson JB et al (1991) Hemoglobin Montreal: a new variant with an extended beta chain due to a deletion of Asp, Gly, Leu at positions 73, 74, and 75, and an insertion of Ala, Arg, Cys, Gln at the same location. Blood 77:178–181

Poillon WN, Kim BC, Welty EV, Walder JA (1986) The effect of 2,3-diphosphoglycerate on the solubility of deoxyhemoglobin S. Arch Biochem Biophys 249:301–305. https://doi.org/10.1016/0003-9861(86)90006-8

Randad RS, Mahran MA, Mehanna AS, Abraham DJ (1991) Allosteric modifiers of hemoglobin. 1. Design, synthesis, testing, and structure-allosteric activity relationship of novel hemoglobin oxygen affinity decreasing agents. J Med Chem 34:752–757. https://doi.org/10.1021/jm00106a041

Reissmann KR, Ruth WE, Nomura T (1961) A human hemoglobin with lowered oxygen affinity and impaired heme-heme interactions. J Clin Invest 40:1826–1833. https://doi.org/10.1172/JCI104406

Rhoda MD, Martin J, Blouquit Y et al (1983) Sickle cell hemoglobin fiber formation strongly inhibited by the Stanleyville II mutation (alpha 78 Asn leads to Lys). Biochem Biophys Res Commun 111:8–13. https://doi.org/10.1016/s0006-291x(83)80109-0

Richard V, Dodson GG, Mauguen Y (1993) Human deoxyhaemoglobin-2,3-diphosphoglycerate complex low-salt structure at 2.5 A resolution. J Mol Biol 233:270–274. https://doi.org/10.1006/jmbi.1993.1505

Rieder RF, Oski FA, Clegg JB (1969) Hemoglobin Philly (beta 35 tyrosine phenylalanine): studies in the molecular pathology of hemoglobin. J Clin Invest 48:1627–1642. https://doi.org/10.1172/JCI106128

Riemer C, ElSherbini A, Stojanovic N et al (1998) A database of experimental results on globin gene expression. Genomics 53:325–337. https://doi.org/10.1006/geno.1998.5524

Safo MK, Abraham DJ (2005) The enigma of the liganded hemoglobin end state: a novel quaternary structure of human carbonmonoxy hemoglobin. Biochemistry 44:8347–8359. https://doi.org/10.1021/bi050412q

Safo MK, Bruno S (2011) Allosteric effectors of hemoglobin: past, present and future. In: Chemistry and biochemistry of oxygen therapeutics. Wiley, Ltd., pp 285–300

Safo MK, Moure CM, Burnett JC et al (2001) High–resolution crystal structure of deoxy hemoglobin complexed with a potent allosteric effector. Protein Sci 10:951–957

Safo MK, Boyiri T, Burnett JC et al (2002a) X-ray crystallographic analyses of symmetrical allosteric effectors of hemoglobin: compounds designed to link primary and secondary binding sites. Acta Crystallogr D Biol Crystallogr 58:634–644. https://doi.org/10.1107/s0907444902002627

Safo MK, Burnett JC, Musayev FN et al (2002b) Structure of human carbonmonoxyhemoglobin at 2.16 Å: a snapshot of the allosteric transition. Acta Crystallographica Section D 58:2031–2037. https://doi.org/10.1107/S0907444902015809

Safo MK, Abdulmalik O, Danso-Danquah R et al (2004) Structural basis for the potent antisickling effect of a novel class of five-membered heterocyclic aldehydic compounds. J Med Chem 47:4665–4676. https://doi.org/10.1021/jm0498001

Safo MK, Abdulmalik O, Lin HR et al (2005) Structures of R- and T-state hemoglobin Bassett: elucidating the structural basis for the low oxygen affinity of a mutant hemoglobin. Acta Crystallogr D Biol Crystallogr 61:156–162. https://doi.org/10.1107/S0907444904030501

Safo MK, Ahmed MH, Ghatge MS, Boyiri T (2011) Hemoglobin-ligand binding: understanding Hb function and allostery on atomic level. Biochim Biophys Acta 1814:797–809. https://doi.org/10.1016/j.bbapap.2011.02.013

Safo MK, Ko T-P, Abdulmalik O et al (2013) Structure of fully liganded Hb $\zeta 2\beta 2s$ trapped in a tense conformation. Acta Crystallogr D Biol Crystallogr 69:2061–2071. https://doi.org/10.1107/S0907444913019197

Safo MK, Ko T-P, Schreiter ER, Russell JE (2015) Structural basis for the antipolymer activity of Hb ζ2β2s trapped in a tense conformation. J Mol Struct 1099:99–107. https://doi.org/10.1016/j.molstruc.2015.06.047
Sahu SC, Simplaceanu V, Gong Q et al (2007) Insights into the solution structure of human deoxyhemoglobin in the absence and presence of an allosteric effector. Biochemistry 46:9973–9980. https://doi.org/10.1021/bi700935z
Sakuragawa M, Ohba Y, Miyaji T et al (1984) A Japanese boy with hemolytic anemia due to an unstable hemoglobin (Hb Bristol). Nippon Ketsueki Gakkai Zasshi 47:896–902
Samaja M, Rovida E, Niggeler M et al (1987) The dissociation of carbon monoxide from hemoglobin intermediate. J Biol Chem 262:4528–4533
Samuni U, Dantsker D, Juszczak LJ et al (2004) Spectroscopic and functional characterization of T state hemoglobin conformations encapsulated in silica gels. Biochemistry 43:13674–13682. https://doi.org/10.1021/bi048531d
Sawicki CA, Gibson QH (1976) Quaternary conformational changes in human hemoglobin studied by laser photolysis of carboxyhemoglobin. J Biol Chem 251:1533–1542
Sawicki CA, Gibson QH (1978) The relation between carbon monoxide binding and the conformational change of hemoglobin. Biophys J 24:21–33. https://doi.org/10.1016/S0006-3495(78)85328-4
Schneider RG, Atkins RJ, Hosty TS et al (1975) Haemoglobin Titusville: alpha94 Asp replaced by Asn. a new haemoglobin with a lowered affinity for oxygen. Biochim Biophys Acta 400:365–373
Schoenborn BP (1976) Dichloromethane as an antisickling agent in sickle cell hemoglobin. Proc Natl Acad Sci USA 73:4195–4199. https://doi.org/10.1073/pnas.73.11.4195
Schumacher MA, Dixon MM, Kluger R et al (1995) Allosteric transition intermediates modelled by crosslinked haemoglobins. Nature 375:84–87. https://doi.org/10.1038/375084a0
Schumacher MA, Zheleznova EE, Poundstone KS et al (1997) Allosteric intermediates indicate R2 is the liganded hemoglobin end state. Proc Natl Acad Sci USA 94:7841–7844. https://doi.org/10.1073/pnas.94.15.7841
Sharma VS, Newton GL, Ranney HM et al (1980) Hemoglobin Rothschild (beta 37(C3)Trp replaced by Arg): a high/low affinity hemoglobin mutant. J Mol Biol 144:267–280. https://doi.org/10.1016/0022-2836(80)90090-x
Shaw E, Scott C, Suh J et al (2003) RSR13 plus cranial radiation therapy in patients with brain metastases: comparison with the radiation therapy oncology group recursive partitioning analysis brain metastases database. J Clin Oncol 21:2364–2371. https://doi.org/10.1200/JCO.2003.08.116
Shibayama N, Saigo S (2001) Direct observation of two distinct affinity conformations in the T state human deoxyhemoglobin. FEBS Lett 492:50–53. https://doi.org/10.1016/s0014-5793(01)02225-6
Shibayama N, Miura S, Tame JRH et al (2002) Crystal structure of horse carbonmonoxyhemoglobin-bezafibrate complex at 1.55-A resolution. a novel allosteric binding site in R-state hemoglobin. J Biol Chem 277:38791–38796. https://doi.org/10.1074/jbc.M205461200
Silva MM, Rogers PH, Arnone A (1992) A third quaternary structure of human hemoglobin A at 1.7-A resolution. J Biol Chem 267:17248–17256
Smith FR, Simmons KC (1994) Cyanomet human hemoglobin crystallized under physiological conditions exhibits the Y quaternary structure. Proteins 18:295–300. https://doi.org/10.1002/prot.340180310
Smith FR, Lattman EE, Carter CW (1991) The mutation beta 99 Asp-Tyr stabilizes Y-a new, composite quaternary state of human hemoglobin. Proteins 10:81–91. https://doi.org/10.1002/prot.340100202
Song X, Simplaceanu V, Ho NT, Ho C (2008) Effector-induced structural fluctuation regulates the ligand affinity of an allosteric protein: binding of inositol hexaphosphate has distinct dynamic consequences for the T and R states of hemoglobin. Biochemistry 47:4907–4915. https://doi.org/10.1021/bi7023699
Spiegel S, Milstien S (2003) Sphingosine-1-phosphate: an enigmatic signalling lipid. Nat Rev Mol Cell Biol 4:397–407. https://doi.org/10.1038/nrm1103

Srinivasan R, Rose GD (1994) The T-to-R transformation in hemoglobin: a reevaluation. PNAS 91:11113–11117. https://doi.org/10.1073/pnas.91.23.11113

Stern W, Mathews D, McKew J et al (2012) A phase 1, first-in-man, dose-response study of Aes-103 (5-HMF), an anti-sickling, allosteric modifier of hemoglobin oxygen affinity in healthy norman volunteers. Blood 120:3210–3210

Stucker O, Laurent D, Duvelleroy M et al (1985) Incorporation of inositol hexaphosphate in stored erythrocytes: effect on tissue oxygenation. Life Support Syst 3(Suppl 1):458–461

Sun DP, Zou M, Ho NT, Ho C (1997) Contribution of surface histidyl residues in the alpha-chain to the Bohr effect of human normal adult hemoglobin: roles of global electrostatic effects. Biochemistry 36:6663–6673. https://doi.org/10.1021/bi963121d

Sun K, D'Alessandro A, Ahmed MH et al (2017) Structural and functional insight of sphingosine 1-phosphate-mediated pathogenic metabolic reprogramming in sickle cell disease. Sci Rep 7:15281. https://doi.org/10.1038/s41598-017-13667-8

Szabo A, Karplus M (1972) A mathematical model for structure-function relations in hemoglobin. J Mol Biol 72:163–197. https://doi.org/10.1016/0022-2836(72)90077-0

Thom CS, Dickson CF, Gell DA, Weiss MJ (2013) Hemoglobin variants: biochemical properties and clinical correlates. Cold Spring Harb Perspect Med 3:a011858. https://doi.org/10.1101/cshperspect.a011858

Torrance J, Jacobs P, Restrepo A et al (1970) Intraerythrocytic adaptation to anemia. N Engl J Med 283:165–169. https://doi.org/10.1056/NEJM197007232830402

Traylor TG, Deardurff LA, Coletta M et al (1983) Reactivity of ferrous heme proteins at low pH. J Biol Chem 258:12147–12148

Vasseur C, Blouquit Y, Kister J et al (1992) Hemoglobin Thionville. An alpha-chain variant with a substitution of a glutamate for valine at NA-1 and having an acetylated methionine NH_2 terminus. J Biol Chem 267:12682–12691

Vichinsky E, Hoppe CC, Ataga KI et al (2019) A phase 3 randomized trial of Voxelotor in sickle cell disease. N Engl J Med 381:509–519. https://doi.org/10.1056/NEJMoa1903212

Wilson J, Phillips K, Luisi B (1996) The crystal structure of horse deoxyhaemoglobin trapped in the high-affinity (R) state. J Mol Biol 264:743–756. https://doi.org/10.1006/jmbi.1996.0674

Winslow RM, Charache S (1975) Hemoglobin Richmond. Subunit dissociation and oxygen equilibrium properties. J Biol Chem 250:6939–6942

Winter PM, Miller JN (1976) Carbon monoxide poisoning. JAMA 236:1502–1504. https://doi.org/10.1001/jama.1976.03270140054029

Wireko FC, Abraham DJ (1991) X-ray diffraction study of the binding of the antisickling agent 12C79 to human hemoglobin. Proc Natl Acad Sci USA 88:2209–2211. https://doi.org/10.1073/pnas.88.6.2209

Wireko FC, Kellogg GE, Abraham DJ (1991) Allosteric modifiers of hemoglobin. 2. Crystallographically determined binding sites and hydrophobic binding/interaction analysis of novel hemoglobin oxygen effectors. J Med Chem 34:758–767. https://doi.org/10.1021/jm00106a042

Woods JA, Storey CJ, Babcock EE, Malloy CR (1998) Right-shifting the oxyhemoglobin dissociation curve with RSR13: effects on high-energy phosphates and myocardial recovery after low-flow ischemia. J Cardiovasc Pharmacol 31:359–363. https://doi.org/10.1097/00005344-199803000-00005

Xu GG, Deshpande TM, Ghatge MS et al (2015) Design, synthesis, and investigation of novel nitric oxide (NO)-releasing prodrugs as drug candidates for the treatment of ischemic disorders: insights into NO-releasing prodrug biotransformation and hemoglobin-NO biochemistry. Biochemistry 54:7178–7192. https://doi.org/10.1021/acs.biochem.5b01074

Xu GG, Pagare PP, Ghatge MS et al (2017) Design, synthesis, and biological evaluation of ester and ether derivatives of antisickling agent 5-HMF for the treatment of sickle cell disease. Mol Pharm 14:3499–3511. https://doi.org/10.1021/acs.molpharmaceut.7b00553

Yokoyama T, Neya S, Tsuneshige A et al (2006) R-state haemoglobin with low oxygen affinity: crystal structures of deoxy human and carbonmonoxy horse haemoglobin bound to the effector molecule L35. J Mol Biol 356:790–801. https://doi.org/10.1016/j.jmb.2005.12.018

Yonetani T, Kanaori K (2013) How does hemoglobin generate such diverse functionality of physiological relevance? Biochim Biophys Acta 1834:1873–1884. https://doi.org/10.1016/j.bbapap.2013.04.026

Yonetani T, Tsuneshige A (2003) The global allostery model of hemoglobin: an allosteric mechanism involving homotropic and heterotropic interactions. CR Biol 326:523–532. https://doi.org/10.1016/S1631-0691(03)00150-1

Yonetani T, Park S-I, Tsuneshige A et al (2002) Global allostery model of hemoglobin. Modulation of $O_{(2)}$ affinity, cooperativity, and Bohr effect by heterotropic allosteric effectors. J Biol Chem 277:34508–34520. https://doi.org/10.1074/jbc.M203135200

Youssef AM, Safo MK, Danso-Danquah R et al (2002) Synthesis and X-ray studies of chiral allosteric modifiers of hemoglobin. J Med Chem 45:1184–1195. https://doi.org/10.1021/jm010358l

Zago MA, Bottura C (1983) Splenic function in sickle-cell diseases. Clin Sci 65:297–302. https://doi.org/10.1042/cs0650297

Zaugg RH, Walder JA, Klotz IM (1977) Schiff base adducts of hemoglobin. Modifications that inhibit erythrocyte sickling. J Biol Chem 252:8542–8548

Zhang Y, Berka V, Song A et al (2014) Elevated sphingosine-1-phosphate promotes sickling and sickle cell disease progression. J Clin Invest 124:2750–2761. https://doi.org/10.1172/JCI74604

Chapter 15
Serum Albumin, Lipid and Drug Binding

Koji Nishi, Keishi Yamasaki and Masaki Otagiri

Abstract Albumin is widely conserved from vertebrates to invertebrates, and nature of mammalian albumins permit them to bind various endogenous ligands and drugs in the blood. It is known that at least two major ligand binding sites are present on the albumin molecule, which are referred to as Site I and Site II. These binding sites are thought to be almost completely conserved among mammals, even though the degree of binding to these sites are different depending on the physical and chemical properties of drugs and differences in the microenvironment in the binding pockets. In addition, the binding sites for medium and long-chain fatty acids are also well conserved among mammals, and it is considered that there are at least seven binding sites, including Site I and Site II. These bindings properties of albumin in the blood are also widely known to be important for transporting drugs and fatty acids to various tissues. It can therefore be concluded that albumin is one of the most important serum proteins for various ligands, and information on human albumin can be very useful in predicting the ligand binding properties of the albumin of other vertebrates.

Keywords Albumin · Fatty acid · Drug binding · Binding site · Ligand transport

K. Nishi · K. Yamasaki · M. Otagiri (✉)
Faculty of Pharmaceutical Sciences, Sojo University, Ikeda 4-22-1, Nishi-Ku, Kumamoto 860-0082, Japan
e-mail: otagirim@ph.sojo-u.ac.jp

K. Nishi
e-mail: knishi@ph.sojo-u.ac.jp

K. Yamasaki
e-mail: kcyama@ph.sojo-u.ac.jp

K. Yamasaki · M. Otagiri
DDS Research Institute, Sojo University, Ikeda 4-22-1, Nishi-Ku, Kumamoto 860-0082, Japan

U. Hoeger and J. R. Harris (eds.), *Vertebrate and Invertebrate Respiratory Proteins, Lipoproteins and other Body Fluid Proteins*, Subcellular Biochemistry 94,
https://doi.org/10.1007/978-3-030-41769-7_15

Introduction

Albumin is one of most widely studied proteins in body fluids. In 400 A.D., Hippocrates described the presence of foam in urine, which was subsequently proved to be albumin. After more than a thousand years, albumins from human to various mammals have now been identified. As of this writing, serum albumins from various vertebrate and invertebrate species have been identified and cloned. Vertebrate albumins are known to be the most abundant soluble protein in plasma. Albumins in mammals including human, bovine, dog, sheep, goat, cat and rabbit have been researched and their amino acid sequences (Allerton et al. 1962; Brown et al. 1989; Dixon and Sarkar 1974; Dugaiczyk et al. 1982; Hilger et al. 1996; Ho et al. 1993; Jacobs and Koj 1969; Lawn et al. 1981; Sala-Trepat et al. 1979a, b; Sargent et al. 1979; Weijers 1977) have been revealed and three-dimensional structures by X-ray crystallographic studies (Bujacz 2012; He and Carter 1992; Majorek et al. 2012; Yamada et al. 2016).

The bindings of various endogenous and exogeneous substrates in the blood control their pharmacokinetics (Fanali et al. 2012; Kragh-Hansen 1981, 1990, 2013; Peters 1970, 1985). Clinically, human albumin has been used for patients who are bleeding based on its function to maintain the colloid osmotic pressure of serum (Kobayashi 2006; Matejtschuk et al. 2000; Ohnishi et al. 2008). Since the 1970s, not only in humans, but also other mammalians such as bovine, horse and rabbit, the binding of fatty acids and various drugs to serum albumins has been investigated. These studies clearly indicate that these albumins have important roles in the delivery of endogenous and exogenous ligands in blood. In this chapter, the ligand binding properties, including fatty acid binding properties, are reviewed, focusing on albumin from mammalian species, especially, human albumin.

Human Albumin

Among the vertebrate and invertebrate albumin species, human albumin has been the most extensively studied for its function and structure (Chuang and Otagiri 2006; Fanali et al. 2012; Kragh-Hansen 1981, 1990, 2013; Kragh-Hansen et al. 2002; Otagiri 2005). Human albumin consists of 585 amino acid residues and there are no oligosaccharide chains attached to the molecule. The molecule has a heart-shaped structure, containing α-helices (67%) and no β-sheets. In addition, Human albumin is composed of three highly homologous domains, which are referred to as domain I, II and III, respectively. The physiological functions of the molecule other than ligand binding, include the maintenance of colloid osmotic pressure (Losowsky et al. 1962). Albumin has been administered to patients with hypoalbuminemia, hemorrhagic shock due to a loss of albumin (burn victims, the nephrotic syndrome, etc.) and decreased albumin biosynthesis (cirrhosis, etc.) (Reviewers 1998). Human albumin also has antioxidant properties (Anraku et al. 2001, 2003; Watanabe et al. 2017). This is due to the fact that, in addition to albumin being the most abundant molecule in blood, it contains a free cysteine residue at position 34 in the molecule. There is some evidence to show that this cysteine residue functions to scavenge reactive oxygen species that are generated during various inflammatory disease such as, for example, chronic renal and hepatic disease and diabetes (Anraku et al. 2011; Matsuyama et al. 2009; Mera et al. 2005; Nagumo et al. 2014; Oettl and Marsche 2010; Terawaki et al. 2004). It has been reported that the high glycemic state in blood during diabetes leads to the glycation of albumin molecules. This glycated albumin finally becomes an advanced glycation end product known as an AGE and this structure can exert various adverse effects (Baret et al. 2017; de Souza Pinto et al. 2012; Yamagishi et al. 2003). It has been also reported that such oxidation and glycation of human albumin also affects the drug binding capacity of the molecule (Bourdon et al. 1999; Collison et al. 2002; Keita et al. 1993; Murtiashaw and Winterhalter 1986; van Boekel et al. 1992; Yamazaki et al. 2005). Thus, human albumin acts as, not only an important ligand carrier, but also as a serum protein that can be used as a biomarker for certain disease states. Other functions, human albumin include enzymatic properties (esterase and thioesterase activities) (Diaz et al. 2001; Kurono et al. 1992; Watanabe et al. 2000). In fact, the angiotensin II receptor blocker, olmesartan medoxomil, was developed as prodrug that can be hydrolyzed by human serum albumin (Ma et al. 2005).

There are essentially two high-affinity drug binding sites on the HSA molecule, referred to as Site I and II, respectively. Most endogenous and exogenous ligands, which bind to human albumin, bind to either site in the range of 10^4–10^6 M^{-1}. Until recently it has not been possible to predict which site a ligand binds to, based on the structural characteristics of the drug. In previous reports, Site I was reported to be formed as a pocket in subdomain IIA and drugs with multiple rings and bulky structures such as warfarin, phenylbutazone, acenocoumarol and indomethacin tend to bind to this site (Kragh-Hansen 1988; Yamasaki et al. 1996). In addition, the finding that ligands as large as bilirubin bind to Site I suggests that this Site is large (Brodersen and Stern 1980). This site is considered to be flexible and to have less structural specificity because Site I-binding drugs exhibit diverse structural characteristics. Furthermore, Site I is comprised of a broad area on the molecule, called Site Ia, Ib and Ic. These subsite regions are known to be overlapped (Petersen et al. 2000). A single mutation in amino acid residues in Site I (K199, W214, R218 and H242) affects the overall structure and thermal stability of Site I (Watanabe et al. 2001). Site I is located at the boundary between domain I and domain II. In a study using recombinants of each domain, it was found that the ligand binding ability of subsites Ia and Ib could not be detected in recombinant domain II, but the binding of ligands to subsite Ic in domain II was preserved (Matsushita et al. 2004).

An X-ray crystallographic study showed that Site II is located in subdomain IIIA (He and Carter 1992; Sudlow et al. 1975). NSAIDs containing aromatic carboxylic acids such as ketoprofen and ibuprofen bind to Site II (Chuang et al. 1999; Wanwimolruk et al. 1982). Thus, unlike Site I, Site II is considered to have a narrow binding pocket, and to be slightly rigid compared to Site I. In fact, a single mutation of Arg110 or Tyr411, both of which are involved in ligand binding in Site II, significantly induced a decrease in binding capacity (Watanabe et al. 2000).

Although there are several types of ligands, binding sites other than Site I and Site II have also been reported. Human albumin contains three metal binding sites (Peters 1966). The N-terminal binding site is one such site, where the binding of Cu (II), Co (II), Ni (II) can occur, and the binding involves Asp1, Ala2, His3 (Sadler et al. 1994). In particular, the binding of Cu (II) is very strong, and its dissociation constant is as low as 6.7×10^{-17} M. The second Site is located at Cys34. As described above, Cys34 is located in subdomain IA, and approximately 40% of this residue in circulation is present as the free form without binding of a low molecular thiol (cysteine and glutathione). Therefore, free ligands can bind covalently to Cys34 which contains not only endogenous substances, nitric oxide (Ishima et al. 2009), Ag (I), Cu (II), Au (I) and Pt (II) (Coffer et al. 1986; Peters 1966), but also thiol-containing drugs, such as N-acetyl-L-cysteine (Sengupta et al. 2001), D-penicillamine and captopril (Coffer et al. 1986; Narazaki et al. 1997). The binding capacity of these ligands has been reported to fluctuate during a disease state, suggesting that the redox state of Cys34 is susceptible during a disease state. The third metal binding site, referred to as a Co (II) binding site A and B respectively, binds Zn (II) and Cd (II) other than Co (II). X-ray absorption fine structure spectroscopy and modeling studies suggest that this binding involves His67, Asn99 (domain I), His247, Asp249 (domain II) (Blindauer et al. 2009).

The saturated long-chain fatty acids, lauric acid (C12), myristic acid (C14), palmitic acid (C16), and stearic acid (C18), have all been reported to form a three-dimensional complex structure with human albumin (Fig. 15.1) (Bhattacharya et al. 2000). Among them, myristic acid and palmitic acid have been extensively studied. Previous studies indicate that the albumin molecule contains five binding sites for myristic acid, one in subdomain IB, one at the interface between subdomain IA and subdomain IIA, two in IIIA and one in IIIB respectively (Peters 1966). Furthermore, it has been reported that the sixth Site is located at the boundary between subdomains IIA and IIB. These six fatty acid binding sites have also been identified in other long chain fatty acids (C12: 0, C16: 0 and C18: 0). In addition, it has also been reported that a seventh binding Site is present in subdomain IIA.

The binding site of medium-chain fatty acids such as capric acid (C10) is also very similar to long-chain fatty acids. According to a report by Curry and colleagues there are an additional two binding sites with slightly lower affinity for the seven long-chain fatty acid binding sites (Bhattacharya et al. 2000; Petitpas et al. 2001). One is in the crevice between domain I-II and domain II-III and the other is in between domain I and III.

Although the binding site of the short-chain fatty acid octanoic acid and albumin has not been clarified, inhibition experiments using the previous Site II binding ligand suggest that it is Site II (Kragh-Hansen 1991). Furthermore, we recently reported that phenylbutyrate, a fatty acid derivative, binds strongly to Site II of the albumin molecule, and the short medium-chain fatty acid in the side chain is likely to bind to Site II (Sakurama et al. 2018, 2019). On the other hand, 2D NMR spectroscopy findings suggest that there are nine binding sites for fatty acids (Hamilton 2013), but

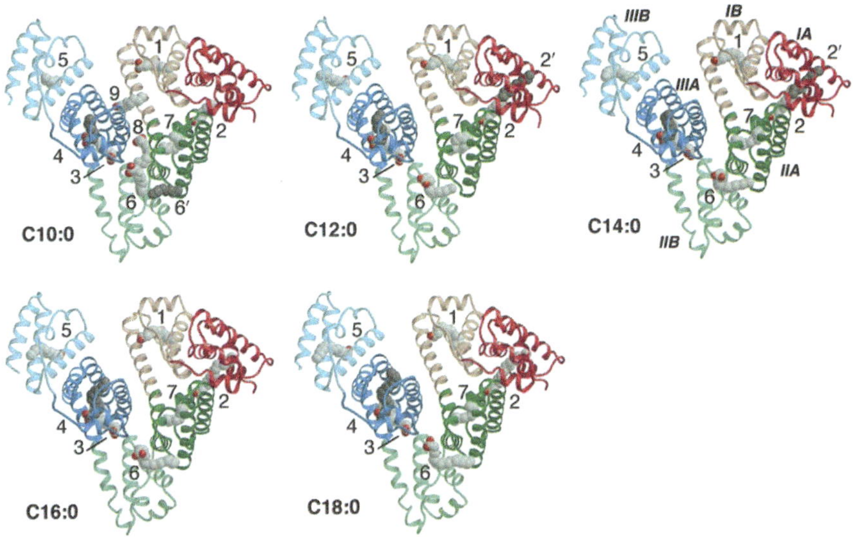

Fig. 15.1 Structures of human albumin complexed with five different fatty acids. Numbers in molecules (1–9) represent the biding site of fatty acids respectively (Bhattacharya et al. J Mol Biol. 2000; 303: 721–732.)

the number of fatty acid binding sites of human albumin is still controversial. Similar to saturated fatty acids, unsaturated fatty acids, oleic acid (cis-form) and arachidonic acid have been also reported to bind to these seven binding sites, and the binding modes appear to be very similar to those of saturated fatty acids (Petitpas et al. 2001).

Regarding other endogenous ligands, it is well known that bilirubin, various metals, and uremic toxins (indoxyl sulfate; IS, 3-carboxy-4-methyl-5-propyl-2-furanpropanoate; CMPF, which are known to be elevated during renal dysfunction and are known to be associated with renal dysfunction) bind to human albumin (Sakai et al. 1995). Dicarboxylic acids such as bilirubin or CMPF bind to Site I (Brodersen and Stern 1980; Petersen et al. 2000; Sakai et al. 1995). It is known that ligands that contain an aromatic carboxylic acid such as L-tryptophan, indole acetic acid and indoxyl sulfuric acid bind to Site II (Irikura et al. 1991; Kragh-Hansen 1990).

Other Mammalian Albumins

The three-dimensional structures of albumins have been clarified by X-ray crystal structure analysis for bovine, dog, horse, sheep, goat, cat and rabbit albumins, and they show that amino acids sequences and structures of the these molecules are extremely homologous (71–76%) to human albumin. This suggests that albumins from these species also have drug and fatty acid binding properties that are equivalent to human albumin. In fact, there are reports that the binding properties of ligands, including fatty acids, are similar to those from humans and that binding pockets to similar to those for human albumins are also present in other mammalian albumins. Superposing the X-ray crystal structures of human and other mammalian albumins suggest that other animal albumins also have quite similar binding pockets in Site I and II. The positions of amino acid residues that are considered to be important for drug binding in human albumin are also very similar in other mammalian albumins (Fig. 15.2). The width and shape of the Site I entrance appears to be different between these albumins (Fig. 15.3). However, there is a little difference in the binding characteristics in site I among human and other mammalian albumins (Kosa et al. 1997). This may be due to the fact that the entrance of Site I is sufficiently large to bind bilirubin. On the other hand, the entrance to Site II appears to be much narrower than site I, and the charge state around the entrance of this site in human albumin appears

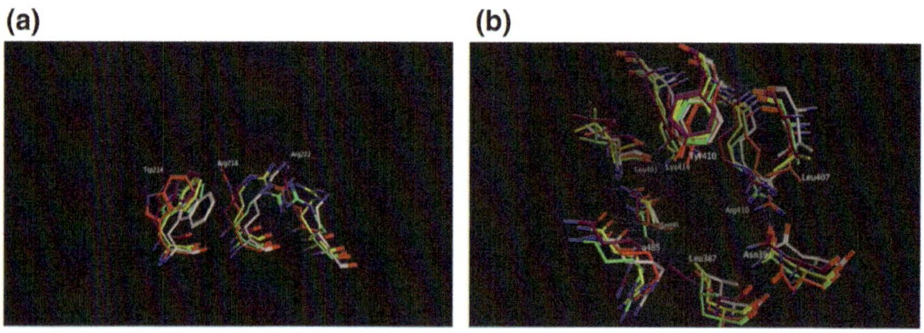

Fig. 15.2 Amino acid residues in the vicinity of site I (**a**) and II (**b**) of human (red), bovine (pink), dog (yellow), horse (purple) and rabbit (green) albumins. PDB codes for human bovine, dog, horse and rabbit albumins are 1UOR, 4F5S, 5GHK, 4F5U and 3V09 respectively

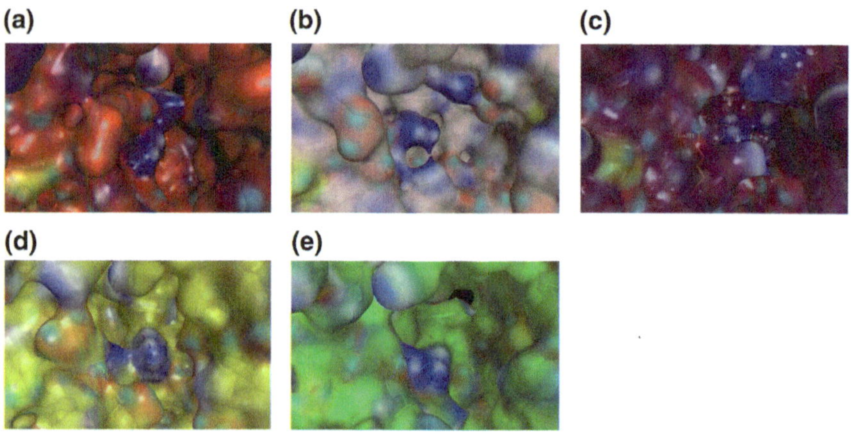

Fig. 15.3 Protein surfaces of site I of human (**a**), bovine (**b**), horse (**c**), dog (**d**) and rabbit (**e**) albumins. PDB codes for human bovine, dog, horse and rabbit albumins are 1UOR, 4F5S, 5GHK, 4F5U and 3V09 respectively

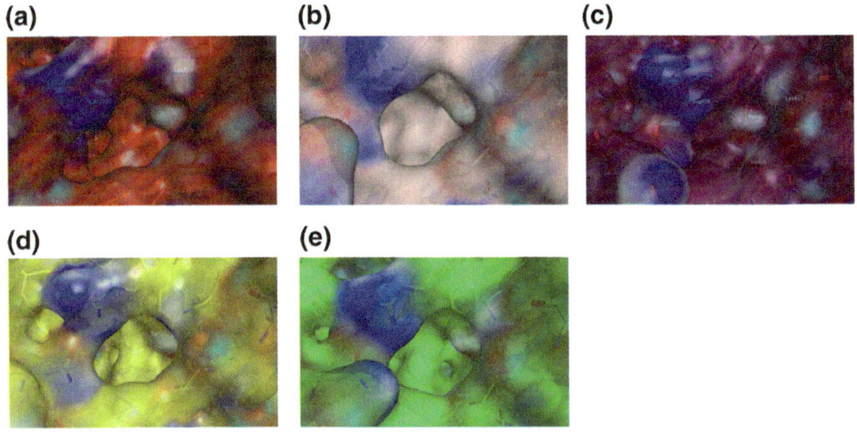

Fig. 15.4 Protein surfaces at site II of human (**a**), bovine (**b**), horse (**c**), dog (**d**) and rabbit (**e**) albumins. PDB codes for human bovine, dog, horse and rabbit albumins are 1UOR, 4F5S, 5GHK, 4F5U and 3V09 respectively

to be similar to that of dogs but different from other animal species (Fig. 15.4). These slight differences in the microenvironment may have resulted in minor differences in drug binding between animal species. This section discusses the ligand binding properties of other mammalian albumins compared to that from humans.

Bovine

Bovine albumin is also one of the most widely studied albumin from a mammal species. X-ray crystallographic study data show that three-dimensional structure of this molecule is quite similar to that of humans. From substitution experiments using fluorescent probes that bind to Site I or Site II of human albumin, it has been reported that Site I are retained in bovine albumin (Kosa et al. 1997). On the other hand, it has been reported that Site II on bovine albumin is slightly different from that of human. The results of binding experiments suggest that this difference is due to changes in the microenvironment and is caused by differences in the size and/or hydrophobicity of the binding site, rather than differences in the amino acid sequences. Isothermal titration calorimetry assays (ITC) and various spectroscopic methods show, as in human albumin, there are two binding sites for metal cations such as Cu (II), Zn (II) and Ni (II), where one is at the N-terminal moiety, and another is cysteine residue at position 34 (Kolthoff and Willeford 1958; Peters and Blumenstock 1967).

Bovine albumin is also reported to have three primary binding sites for long chain fatty acids (Hamilton et al. 1991). NMR measurements performed using a pepsin albumin digest (1–306, 307–582) and a tryptic digest (377–582) suggest that three primary sites are present on the albumin molecule, among which two are located in residues 307–582 or residues 377–582, and one is in the N-terminal fragment (1–306). Furthermore, based on studies involving the chemical modification of bovine albumin, the modification of Lys 222 in subdomain IIA was inhibited by a C8 fatty acid (but not by medium chain, long chain fatty acids (C16, C18, C18:1) (Walker 1976). A possible binding site for short-chain fatty acids may be present near Lys222. In addition, photoaffinity labeling experiment showed Lys116, Lys349 and Lys473 were labeled with palmitate (Reed 1986).

Horse (Equine)

Similar to human albumin, equine albumin is also known to bind NSAID drugs. Detailed structure of complexes of diclofenac and naproxen with equine albumin have been revealed based on X-ray crystallography and differential scanning calorimetry data (Sekula and Bujacz 2015). Albumin is known to contain 2 binding sites for both diclofenac and naproxen respectively; sites for diclofenac are located in domain III, sites for naproxen are located in domain II and III, and the binding sites in domain III for each drug overlap. This site is considered to correspond to the Site II of human albumin. Therefore, similar to human albumin, with regard to Site II, equine albumin is also considered to contain a Site II for binding drugs.

Dog (Canine)

Representative drugs that bind to Site II of human albumin, warfarin and phenylbutazone, also bind to human, bovine, rabbit and rat albumin, but not dog albumin. However, drug binding experiments using Site II binding drugs, such as ibuprofen and diazepam, indicate that ibuprofen binds to mammalian species albumins (human, bovine, dog, rabbit and rat) to the same degree ($1.0–3.0 \times 10^6$ M^{-1}), but diazepam appears to have a high affinity only to human and dog albumin (Kosa et al. 1997). Furthermore, we recently reported that the antipsychotic drug, aripiprazole, which, similar to diazepam, contains a chlorine atom binds strongly only to human and dog albumins among mammalian species (Sakurama et al. 2018). This finding suggests that the microenvironment of the Site II binding pocket of human and dog albumin are similar. On the other hand, copper binding sites observed in human and bovine albumin are lacking (Appleton and Sarkar 1971).

Role of Albumin in the Transport of Fatty Acids

Albumin is known to be one of major carriers of fatty acids in the blood. Very low density lipoprotein (VLDL) and chylomicrons, fatty acid binding proteins (FABP) and fatty acid translocase (FAT/CD36), albumin binding protein (ABP) are also involved in the transport of long chain fatty acids from the blood to the intracellular cytosol and mitochondria (Fig. 15.5) (van der Vusse et al. 2002). The first step starts when fatty acids that are bound to albumin in blood pass to the apical membrane of vascular endothelial cells. Current knowledge of this indicates that the albumin-fatty acid complex interacts directly with the lipid bilayer membrane and during the transcytosis of this complex, fatty acids are thought to be released intracellularly (Kosa et al. 2007; Horie et al. 1988; Reed and Burrington 1989), whereas albumin binds to ABP on the membrane surface and the released fatty acids are then taken up by vascular endothelial cells (Luiken et al. 1999). On other hand, triglycerides (TG) and cholesterol ester (CE) in VLDL and chylomicrons are hydrolyzed to glycerol, cholesterol and fatty acids by lipoprotein lipases that are produced by endothelial cells (van der Vusse et al. 1992). The neonatal Fc receptor (FcRn) on the vascular endothelial cell membrane is considered to be a receptor for albumin and to be important for albumin recycling (Kim et al. 2006). Therefore, FcRn may be involved in the uptake of fatty acids that are bound to albumin. The fatty acids which pass through the vascular endothelial cells again bind to albumin in the intercellular space and are then transferred to the cell membrane of tissue, but the mechanism responsible for this is not well understood. Thereafter, the fatty acids transported into tissue cells bind to FABP followed by forming Acyl-CoA, which is used for the acylation of diacylglycerol, protein and signal transduction. In addition, acyl-CoA is transported into mitochondria and undergoes β-oxidation to become acetyl-CoA, which is a fuel for the citric acid cycle (van der Vusse et al. 2002). Thus, albumin plays a very

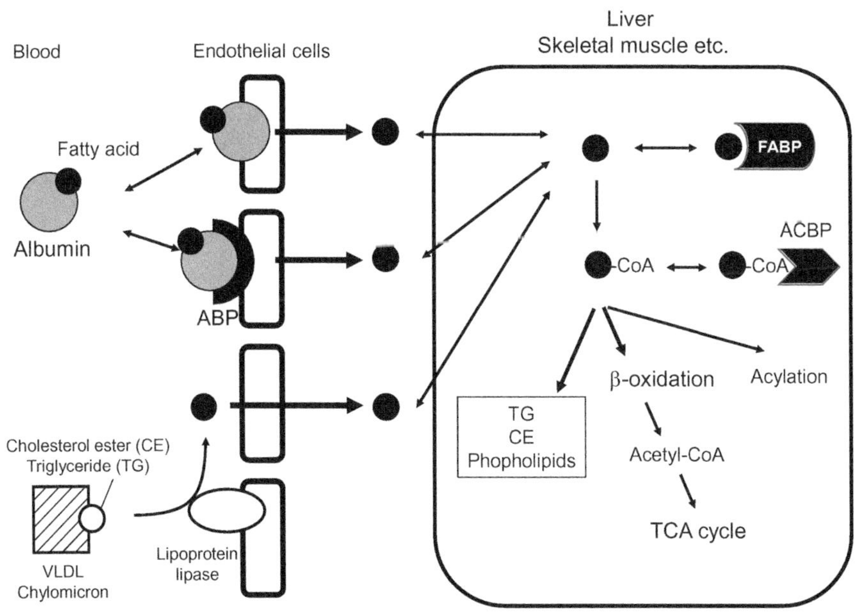

Fig. 15.5 Fatty acid transport from blood to tissue

important role in the transport of fatty acids in the blood, for not only the synthesis of various lipids, but also energy production. Furthermore, albumin not only plays an important role in fatty acid transport, it is also reported that fatty acids contribute to the structural stability and thermal stability of human albumin. Therefore, fatty acids may contribute to the structural stability of albumin, resulting in extending its retention time in the blood.

Conclusion

Albumin is one of the most extensively studied proteins, and past studies indicate that it is widely conserved from vertebrates to invertebrates. It is well known that the binding of drugs and fatty acids are a main functions of albumin, and that the amino acid sequences, three-dimensional structures and the drug binding properties (affinity and binding site) of mammalian albumins are all quite similar. In particular, the transport of fatty acids that are bound to albumin to tissues, where triglycerides and phospholipids are synthesized, is a very important process in terms of maintaining biological homeostasis. Therefore, albumin is a protein that plays an important role in a wide variety of species ranging from vertebrates to invertebrates, and data on human albumin is likely to be the useful information for many researchers.

References

Allerton SE, Elwyn D, Edsalljt SP (1962) Isolation and amino acid composition of dog plasma albumin. J Biol Chem 237:85–88

Anraku M, Yamasaki K, Maruyama T, Kragh-Hansen U, Otagiri M (2001) Effect of oxidative stress on the structure and function of human serum albumin. Pharm Res 18(5):632–639

Anraku M, Kragh-Hansen U, Kawai K, Maruyama T, Yamasaki Y, Takakura Y, Otagiri M (2003) Validation of the chloramine-T induced oxidation of human serum albumin as a model for oxidative damage in vivo. Pharm Res 20(4):684–692

Anraku M, Takeuchi K, Watanabe H, Kadowaki D, Kitamura K, Tomita K, Kuniyasu A, Suenaga A, Maruyama T, Otagiri M (2011) Quantitative analysis of cysteine-34 on the anitioxidative properties of human serum albumin in hemodialysis patients. J Pharm Sci 100(9):3968–3976. https://doi.org/10.1002/jps.22571

Appleton DW, Sarkar B (1971) The absence of specific copper (II)-binding site in dog albumin a comparative study of human and dog albumins. J Biol Chem 246(16):5040–5046

Baret P, Le Sage F, Planesse C, Meilhac O, Devin A, Bourdon E, Rondeau P (2017) Glycated human albumin alters mitochondrial respiration in preadipocyte 3T3-L1 cells. BioFactors (Oxford, England) 43(4):577–592. https://doi.org/10.1002/biof.1367

Bhattacharya AA, Grune T, Curry S (2000) Crystallographic analysis reveals common modes of binding of medium and long-chain fatty acids to human serum albumin. J Mol Biol 303(5):721–732. https://doi.org/10.1006/jmbi.2000.4158

Blindauer CA, Harvey I, Bunyan KE, Stewart AJ, Sleep D, Harrison DJ, Berezenko S, Sadler PJ (2009) Structure, properties, and engineering of the major zinc binding site on human albumin. J Biol Chem 284(34):23116–23124

Bourdon E, Loreau N, Blache D (1999) Glucose and free radicals impair the antioxidant properties of serum albumin. FASEB J: Off Publ Fed Am Soc Exp Biol 13(2):233–244. https://doi.org/10.1096/fasebj.13.2.233
Brodersen R, Stern L (1980) Binding of bilirubin to albumin. CRC Crit Rev Clin Lab Sci 11(4):307–399
Brown WM, Dziegielewska KM, Foreman RC, Saunders NR (1989) Nucleotide and deduced amino acid sequence of sheep serum albumin. Nucleic Acids Res 17(24):10495. https://doi.org/10.1093/nar/17.24.10495
Bujacz A (2012) Structures of bovine, equine and leporine serum albumin. Acta Crystallogr D Biol Crystallogr 68(Pt 10):1278–1289. https://doi.org/10.1107/S0907444912027047
Chuang VT, Otagiri M (2006) Stereoselective binding of human serum albumin. Chirality 18(3):159–166. https://doi.org/10.1002/chir.20237
Chuang VT, Kuniyasu A, Nakayama H, Matsushita Y, Hirono S, Otagiri M (1999) Helix 6 of subdomain III A of human serum albumin is the region primarily photolabeled by ketoprofen, an arylpropionic acid NSAID containing a benzophenone moiety. Biochem Biophys Acta 1434(1):18–30. https://doi.org/10.1016/s0167-4838(99)00174-0
Coffer MT, Shaw CF III, Eidsness M, Watkins J, Elder R (1986) Reactions of auranofin and chloro (triethylphosphine) gold with bovine serum albumin. Inorg Chem 25(3):333–339
Collison KS, Parhar RS, Saleh SS, Meyer BF, Kwaasi AA, Hammami MM, Schmidt AM, Stern DM, Al-Mohanna FA (2002) RAGE-mediated neutrophil dysfunction is evoked by advanced glycation end products (AGEs). J Leukoc Biol 71(3):433–444
de Souza Pinto R, Castilho G, Paim BA, Machado-Lima A, Inada NM, Nakandakare ER, Vercesi AE, Passarelli M (2012) Inhibition of macrophage oxidative stress prevents the reduction of ABCA-1 transporter induced by advanced glycated albumin. Lipids 47(5):443–450. https://doi.org/10.1007/s11745-011-3647-9
Diaz N, Suarez D, Sordo TL, Merz KM Jr (2001) Molecular dynamics study of the IIA binding site in human serum albumin: influence of the protonation state of Lys195 and Lys199. J Med Chem 44(2):250–260
Dixon JW, Sarkar B (1974) Isolation, amino acid sequence and copper(II)-binding properties of peptide (1-24) of dog serum albumin. J Biol Chem 249(18):5872–5877
Dugaiczyk A, Law SW, Dennison OE (1982) Nucleotide sequence and the encoded amino acids of human serum albumin mRNA. Proc Natl Acad Sci USA 79(1):71–75. https://doi.org/10.1073/pnas.79.1.71
Fanali G, di Masi A, Trezza V, Marino M, Fasano M, Ascenzi P (2012) Human serum albumin: from bench to bedside. Mol Aspects Med 33(3):209–290. https://doi.org/10.1016/j.mam.2011.12.002
Hamilton JA (2013) NMR reveals molecular interactions and dynamics of fatty acid binding to albumin. Biochim Biophys Acta (BBA)-Gen Subj 1830(12):5418–5426
Hamilton JA, Era S, Bhamidipati SP, Reed RG (1991) Locations of the three primary binding sites for long-chain fatty acids on bovine serum albumin. Proc Natl Acad Sci USA 88(6):2051–2054. https://doi.org/10.1073/pnas.88.6.2051
He XM, Carter DC (1992) Atomic structure and chemistry of human serum albumin. Nature 358(6383):209–215. https://doi.org/10.1038/358209a0
Hilger C, Grigioni F, Hentges F (1996) Sequence of the gene encoding cat (Felis domesticus) serum albumin. Gene 169(2):295–296
Ho JX, Holowachuk EW, Norton EJ, Twigg PD, Carter DC (1993) X-ray and primary structure of horse serum albumin (Equus caballus) at 0.27-nm resolution. European journal of biochemistry 215(1):205–212. https://doi.org/10.1111/j.1432-1033.1993.tb18024.x
Horie T, Mizuma T, Kasai S, Awazu S (1988) Conformational change in plasma albumin due to interaction with isolated rat hepatocyte. Am J Physiol 254(4 Pt 1):G465–G470. https://doi.org/10.1152/ajpgi.1988.254.4.G465

Irikura M, Takadate A, Goya S, Otagiri M (1991) 7-Alkylaminocoumarin-4-acetic acids as fluorescent probe for studies of drug-binding sites on human serum albumin. Chem Pharm Bull 39(3):724–728

Ishima Y, Kragh-Hansen U, Maruyama T, Otagiri M (2009) Albumin as a nitric oxide-traffic protein: characterization, biochemistry and possible future therapeutic applications. Drug Metab Pharmacokinet 24(4):308–317

Jacobs S, Koj A (1969) Amino acid composition of rabbit plasma albumin and fibrin. Anal Biochem 27(1):178–182

Keita Y, Kratzer W, Worner W, Rietbrock N (1993) Effect of free fatty acids on the binding kinetics at the benzodiazepine binding site of glycated human serum albumin. Int J Clin Pharmacol Ther Toxicol 31(7):337–342

Kim J, Bronson CL, Hayton WL, Radmacher MD, Roopenian DC, Robinson JM, Anderson CL (2006) Albumin turnover: FcRn-mediated recycling saves as much albumin from degradation as the liver produces. Am J Physiol Gastrointest Liver Physiol 290(2):G352–G360. https://doi.org/10.1152/ajpgi.00286.2005

Kobayashi K (2006) Summary of recombinant human serum albumin development. Biol: J Int Assoc Biol Stand 34(1):55–59. https://doi.org/10.1016/j.biologicals.2005.08.021

Kolthoff I, Willeford B Jr (1958) The interaction of copper(II) with bovine serum albumin1. J Am Chem Soc 80(21):5673–5678

Kosa T, Maruyama T, Otagiri M (1997) Species differences of serum albumins: I. Drug binding sites. Pharm Res 14(11):1607–1612

Kosa T, Nishi K, Maruyama T, Sakai N, Yonemura N, Watanabe H, Suenaga A, Otagiri M (2007) Structural and ligand-binding properties of serum albumin species interacting with a biomembrane interface. J Pharm Sci 96(11):3117–3124. https://doi.org/10.1002/jps.20887

Kragh-Hansen U (1981) Molecular aspects of ligand binding to serum albumin. Pharmacol Rev 33(1):17–53

Kragh-Hansen U (1988) Evidence for a large and flexible region of human serum albumin possessing high affinity binding sites for salicylate, warfarin, and other ligands. Mol Pharmacol 34(2):160–171

Kragh-Hansen U (1990) Structure and ligand binding properties of human serum albumin. Dan Med Bull 37(1):57–84

Kragh-Hansen U (1991) Octanoate binding to the indole- and benzodiazepine-binding region of human serum albumin. Biochem J 273(Pt 3):641–644. https://doi.org/10.1042/bj2730641

Kragh-Hansen U (2013) Molecular and practical aspects of the enzymatic properties of human serum albumin and of albumin-ligand complexes. Biochem Biophys Acta 1830(12):5535–5544. https://doi.org/10.1016/j.bbagen.2013.03.015

Kragh-Hansen U, Chuang VT, Otagiri M (2002) Practical aspects of the ligand-binding and enzymatic properties of human serum albumin. Biol Pharm Bull 25(6):695–704

Kurono Y, Kushida I, Tanaka H, Ikeda K (1992) Esterase-like activity of human serum albumin. VIII. Reaction with amino acid p-nitrophenyl esters. Chem Pharm Bull 40(8):2169–2172

Lawn RM, Adelman J, Bock SC, Franke AE, Houck CM, Najarian RC, Seeburg PH, Wion KL (1981) The sequence of human serum albumin cDNA and its expression in E. coli. Nucleic Acids Res 9(22):6103–6114. https://doi.org/10.1093/nar/9.22.6103

Losowsky M, Alltree E, Atkinson M (1962) Plasma colloid osmotic pressure and its relation to protein fractions. Clin Sci 22:249

Luiken JJ, Schaap FG, van Nieuwenhoven FA, van der Vusse GJ, Bonen A, Glatz JF (1999) Cellular fatty acid transport in heart and skeletal muscle as facilitated by proteins. Lipids 34(Suppl):S169–S175

Ma SF, Anraku M, Iwao Y, Yamasaki K, Kragh-Hansen U, Yamaotsu N, Hirono S, Ikeda T, Otagiri M (2005) Hydrolysis of angiotensin II receptor blocker prodrug olmesartan medoxomil by human serum albumin and identification of its catalytic active sites. Drug Metab Dispos: Biol Fate Chem 33(12):1911–1919. https://doi.org/10.1124/dmd.105.006163

Majorek KA, Porebski PJ, Dayal A, Zimmerman MD, Jablonska K, Stewart AJ, Chruszcz M, Minor W (2012) Structural and immunologic characterization of bovine, horse, and rabbit serum albumins. Mol Immunol 52(3–4):174–182. https://doi.org/10.1016/j.molimm.2012.05.011

Matejtschuk P, Dash CH, Gascoigne EW (2000) Production of human albumin solution: a continually developing colloid. Br J Anaesth 85(6):887–895. https://doi.org/10.1093/bja/85.6.887

Matsushita S, Isima Y, Chuang VT, Watanabe H, Tanase S, Maruyama T, Otagiri M (2004) Functional analysis of recombinant human serum albumin domains for pharmaceutical applications. Pharm Res 21(10):1924–1932

Matsuyama Y, Terawaki H, Terada T, Era S (2009) Albumin thiol oxidation and serum protein carbonyl formation are progressively enhanced with advancing stages of chronic kidney disease. Clin Exp Nephrol 13(4):308–315. https://doi.org/10.1007/s10157-009-0161-y

Mera K, Anraku M, Kitamura K, Nakajou K, Maruyama T, Otagiri M (2005) The structure and function of oxidized albumin in hemodialysis patients: Its role in elevated oxidative stress via neutrophil burst. Biochem Biophys Res Commun 334(4):1322–1328. https://doi.org/10.1016/j.bbrc.2005.07.035

Murtiashaw MH, Winterhalter KH (1986) Non-enzymatic glycation of human albumin does not alter its palmitate binding. Diabetologia 29(6):366–370

Nagumo K, Tanaka M, Chuang VT, Setoyama H, Watanabe H, Yamada N, Kubota K, Tanaka M, Matsushita K, Yoshida A, Jinnouchi H, Anraku M, Kadowaki D, Ishima Y, Sasaki Y, Otagiri M, Maruyama T (2014) Cys34-cysteinylated human serum albumin is a sensitive plasma marker in oxidative stress-related chronic diseases. PLoS ONE 9(1):e85216. https://doi.org/10.1371/journal.pone.0085216

Narazaki R, Harada K, Sugii A, Otagiri M (1997) Kinetic analysis of the covalent binding of captopril to human serum albumin. J Pharm Sci 86(2):215–219. https://doi.org/10.1021/js960234+

Oettl K, Marsche G (2010) Redox state of human serum albumin in terms of cysteine-34 in health and disease. Methods Enzymol 474:181–195. https://doi.org/10.1016/s0076-6879(10)74011-8

Ohnishi K, Kawaguchi A, Nakajima S, Mori H, Ueshima T (2008) A comparative pharmacokinetic study of recombinant human serum albumin with plasma-derived human serum albumin in patients with liver cirrhosis. J Clin Pharmacol 48(2):203–208. https://doi.org/10.1177/0091270007310549

Otagiri M (2005) A molecular functional study on the interactions of drugs with plasma proteins. Drug Metab Pharmacokinet 20(5):309–323

Peters T (1966) All about Albumin: Biochemistry, Genetics, and Medical Application, Acad. Press, Orlando, FL:42

Peters T Jr (1970) Serum albumin. Adv Clin Chem 13:37–111

Peters T Jr (1985) Serum albumin. Adv Protein Chem 37:161–245

Peters T Jr, Blumenstock FA (1967) Copper-binding properties of bovine serum albumin and its amino-terminal peptide fragment. J Biol Chem 242(7):1574–1578

Petersen CE, Ha CE, Harohalli K, Feix JB, Bhagavan NV (2000) A dynamic model for bilirubin binding to human serum albumin. J Biol Chem 275(28):20985–20995. https://doi.org/10.1074/jbc.M001038200

Petitpas I, Grune T, Bhattacharya AA, Curry S (2001) Crystal structures of human serum albumin complexed with monounsaturated and polyunsaturated fatty acids. J Mol Biol 314(5):955–960. https://doi.org/10.1006/jmbi.2000.5208

Reed RG (1986) Location of long chain fatty acid-binding sites of bovine serum albumin by affinity labeling. J Biol Chem 261(33):15619–15624

Reed RG, Burrington CM (1989) The albumin receptor effect may be due to a surface-induced conformational change in albumin. J Biol Chem 264(17):9867–9872

Reviewers CIGA (1998) Human albumin administration in critically ill patients: systematic review of randomised controlled trials. BMJ (Clin Res Ed) 317(7153):235–240. https://doi.org/10.1136/bmj.317.7153.235

Sadler PJ, Tucker A, Viles JH (1994) Involvement of a lysine residue in the N-terminal Ni2+ and Cu2+ binding site of serum albumins. Comparison with Co2+, Cd2+ and Al3+. Eur J Biochem 220(1):193–200. https://doi.org/10.1111/j.1432-1033.1994.tb18614.x
Sakai T, Takadate A, Otagiri M (1995) Characterization of binding site of uremic toxins on human serum albumin. Biol Pharm Bull 18(12):1755–1761
Sakurama K, Kawai A, Tuan Giam Chuang V, Kanamori Y, Osa M, Taguchi K, Seo H, Maruyama T, Imoto S, Yamasaki K, Otagiri M (2018) Analysis of the binding of aripiprazole to human serum albumin: the importance of a chloro-group in the chemical structure. ACS Omega 3(10):13790–13797. https://doi.org/10.1021/acsomega.8b02057
Sakurama K, Nishi K, Imoto S, Hashimoto M, Komatsu T, Morita Y, Taguchi K, Otagiri M, Yamasaki K (2019) Further evidence regarding the important role of chlorine atoms of aripiprazole on binding to the site II area of human albumin. J Pharm Sci 108(5):1890–1895. https://doi.org/10.1016/j.xphs.2018.11.045
Sala-Trepat JM, Dever J, Sargent TD, Thomas K, Sell S, Bonner J (1979a) Changes in expression of albumin and alpha-fetoprotein genes during rat liver development and neoplasia. Biochemistry 18(11):2167–2178. https://doi.org/10.1021/bi00578a006
Sala-Trepat JM, Sargent TD, Sell S, Bonner J (1979b) alpha-Fetoprotein and albumin genes of rats: no evidence for amplification-deletion or rearrangement in rat liver carcinogenesis. Proc Natl Acad Sci USA 76(2):695–699. https://doi.org/10.1073/pnas.76.2.695
Sargent TD, Wu JR, Sala-Trepat JM, Wallace RB, Reyes AA, Bonner J (1979) The rat serum albumin gene: analysis of cloned sequences. Proc Natl Acad Sci USA 76(7):3256–3260. https://doi.org/10.1073/pnas.76.7.3256
Sekula B, Bujacz A (2015) Structural insights into the competitive binding of diclofenac and naproxen by equine serum albumin. J Med Chem 59(1):82–89
Sengupta S, Chen H, Togawa T, DiBello PM, Majors AK, Budy B, Ketterer ME, Jacobsen DW (2001) Albumin thiolate anion is an intermediate in the formation of albumin-S-S-homocysteine. J Biol Chem 276(32):30111–30117. https://doi.org/10.1074/jbc.M104324200
Sudlow G, Birkett DJ, Wade DN (1975) The characterization of two specific drug binding sites on human serum albumin. Mol Pharmacol 11(6):824–832
Terawaki H, Yoshimura K, Hasegawa T, Matsuyama Y, Negawa T, Yamada K, Matsushima M, Nakayama M, Hosoya T, Era S (2004) Oxidative stress is enhanced in correlation with renal dysfunction: examination with the redox state of albumin. Kidney Int 66(5):1988–1993. https://doi.org/10.1111/j.1523-1755.2004.00969.x
van Boekel MA, van den Bergh PJ, Hoenders HJ (1992) Glycation of human serum albumin: inhibition by diclofenac. Biochem Biophys Acta 1120(2):201–204. https://doi.org/10.1016/0167-4838(92)90270-n
van der Vusse GJ, Glatz JF, Stam HC, Reneman RS (1992) Fatty acid homeostasis in the normoxic and ischemic heart. Physiol Rev 72(4):881–940. https://doi.org/10.1152/physrev.1992.72.4.881
van der Vusse GJ, van Bilsen M, Glatz JF, Hasselbaink DM, Luiken JJ (2002) Critical steps in cellular fatty acid uptake and utilization. Mol Cell Biochem 239(1–2):9–15
Walker JE (1976) Lysine residue 199 of human serum albumin is modified by acetylsalicylic acid. FEBS Lett 66(2):173–175
Wanwimolruk S, Birkett DJ, Brooks PM (1982) Protein binding of some non-steroidal anti-inflammatory drugs in rheumatoid arthritis. Clin Pharmacokinet 7(1):85–92. https://doi.org/10.2165/00003088-198207010-00005
Watanabe H, Tanase S, Nakajou K, Maruyama T, Kragh-Hansen U, Otagiri M (2000) Role of arg-410 and tyr-411 in human serum albumin for ligand binding and esterase-like activity. Biochem J 349(Pt 3):813–819. https://doi.org/10.1042/bj3490813
Watanabe H, Kragh-Hansen U, Tanase S, Nakajou K, Mitarai M, Iwao Y, Maruyama T, Otagiri M (2001) Conformational stability and warfarin-binding properties of human serum albumin studied by recombinant mutants. Biochem J 357(Pt 1):269–274. https://doi.org/10.1042/0264-6021:3570269

Watanabe H, Imafuku T, Otagiri M, Maruyama T (2017) Clinical implications associated with the posttranslational modification-induced functional impairment of albumin in oxidative stress-related diseases. J Pharm Sci 106(9):2195–2203. https://doi.org/10.1016/j.xphs.2017.03.002

Weijers RN (1977) Amino acid sequence in bovine serum albumin. Clin Chem 23(7):1361–1362

Yamada K, Yokomaku K, Kureishi M, Akiyama M, Kihira K, Komatsu T (2016) Artificial blood for dogs. Sci Rep 6:36782. https://doi.org/10.1038/srep36782

Yamagishi S, Inagaki Y, Okamoto T, Amano S, Koga K, Takeuchi M (2003) Advanced glycation end products inhibit de novo protein synthesis and induce TGF-beta overexpression in proximal tubular cells. Kidney Int 63(2):464–473. https://doi.org/10.1046/j.1523-1755.2003.00752.x

Yamasaki K, Maruyama T, Kragh-Hansen U, Otagiri M (1996) Characterization of site I on human serum albumin: concept about the structure of a drug binding site. Biochem Biophys Acta 1295(2):147–157. https://doi.org/10.1016/0167-4838(96)00013-1

Yamazaki E, Inagaki M, Kurita O, Inoue T (2005) Kinetics of fatty acid binding ability of glycated human serum albumin. J Biosci 30(4):475–481

Chapter 16
High-Density Lipoproteins and Apolipoprotein A1

Emiel P. C. van der Vorst

Abstract High-density lipoprotein (HDL) and its main protein component apolipoprotein (apo)A-I, play an important role in cholesterol homeostasis. It has been demonstrated that HDLs comprise of a very heterogeneous group of particles, not only regarding size but also composition. HDL's best described function is its role in the reverse cholesterol transport, where lipid-free apoA-I or small HDLs can accept and take up cholesterol from peripheral cells and subsequently transport this to the liver for excretion. However, several other functions have also been described, like anti-oxidant, anti-inflammatory and anti-thrombotic effects. In this article, the general features, synthesis and metabolism of apoA-I and HDLs will be discussed. Additionally, an overview of HDL functions will be given, especially in the context of some major pathologies like cardiovascular disease, cancer and diabetes mellitus. Finally, the therapeutic potential of raising HDL will be discussed, focussing on the difficulties of the past and the promises of the future.

Keywords High-density lipoproteins (HDLs) · Apolipoprotein A1 · Lipid metabolism · Therapy

---

E. P. C. van der Vorst (✉)
Institute for Cardiovascular Prevention (IPEK), Ludwig-Maximilians-University Munich, Pettenkoferstr. 8a und 9, D-80336 Munich, Germany
e-mail: evandervorst@ukaachen.de

Department of Pathology, Cardiovascular Research Institute Maastricht (CARIM), Maastricht University Medical Centre, Maastricht, The Netherlands

Institute for Molecular Cardiovascular Research (IMCAR)/Interdisciplinary Center for Clinical Research (IZKF), RWTH Aachen University, Aachen, Germany

DZHK (German Centre for Cardiovascular Research), Partner Site Munich Heart Alliance, Munich, Germany

U. Hoeger and J. R. Harris (eds.), *Vertebrate and Invertebrate Respiratory Proteins, Lipoproteins and other Body Fluid Proteins*, Subcellular Biochemistry 94,
https://doi.org/10.1007/978-3-030-41769-7_16

Introduction

The discovery of high-density lipoproteins (HDLs) dates back to 1929 when a protein-rich, lipid-poor complex was isolated from equine serum (*for review see* Kingwell et al. 2014). Several years later in the 1950s, HDL was isolated from human serum by using ultracentrifugation (Barr et al. 1951). It took however a further decade before the first biological roles of HDL were described, especially its impact on the cardiovascular system (Gofman 1956). Probably the best studied phenomenon is the inverse correlation between HDL-cholesterol (HDL-C) levels and the risk for developing coronary heart disease, which suggests that HDL-C levels in plasma could be protective against atherosclerosis development (Papachristou et al. 2017). Recent data from experimental mice and clinical trials however indicate that the classical focus on HDL-C might be too limited, and that HDL particle functionality as determined by apolipoprotein (apo) and lipid content is even more important than HDL-C levels alone (Constantinou et al. 2016; Karavia et al. 2014). Therefore, one should realize that the generally used HDL-synonym "good cholesterol" is rather misleading, as HDL should be considered as a complex assembly of proteins (apolipoproteins) and lipids (esterified cholesterol, phospholipids, sphingolipids, ceramides, etc.).

Today, a wide variety of HDL functions has already been identified that play an important role in physiological and pathological processes. Therefore, besides the general principles of HDL biogenesis and metabolism, several HDL functions will be discussed in the context of pathologies like cardiovascular disease (CVD), cancer and diabetes mellitus.

Sub-types of HDL

HDLs are the smallest (7–12 nm diameter), but densest (1.063–1.25 g/ml) population of plasma lipoproteins. This population is very heterogeneous, comprising several distinct subpopulations that vary in shape, size, density, protein/lipid composition and surface charge (Rye et al. 1999). In normal plasma, most of the HDL particles are spherical, consisting of a hydrophobic core (mainly cholesteryl esters with a small amount of triglycerides) surrounded by a surface monolayer consisting of phospholipids, unesterified cholesterol and apolipoproteins (Fig. 16.1). A small HDL population has a discoidal shape, consisting only of a bilayer of phospholipids with some unestrified cholesterol and apolipoproteins, without containing a hydrophobic core (Rye and Barter 2004). Plasma HDL contains two main apolipoproteins, being apoA-I (70% of total HDL protein content) and apoA-II (20% of total HDL protein content). Some HDL particles also contain minor amounts of other apolipoproteins such as apoA-IV, apoA-V, apoC, apoD, apoE, apoJ and apoL. Other proteins, especially enzymes can also be incorporated in HDL particles and play an important role

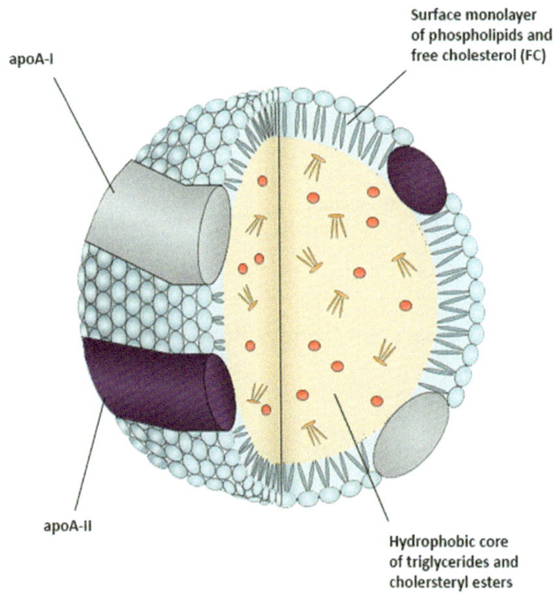

Fig. 16.1 Generalized structure of spherical HDLs

in their metabolism as will be discussed later, for example cholesteryl ester transfer protein (CETP) and lecithin: cholesterol acyltransferase (LCAT).

The different HDL subpopulations can be distinguished based on different parameters (Fig. 16.2). As already mentioned, HDLs can be present in a spherical or discoidal form, making shape one of the distinguishing factors. Additionally, based on the apolipoprotein composition, HDLs can be separated into two major subpopulations. The first population contains apoA-I but not apoA-II (A-I HDLs), while the other population contains both apoA-I and apoA-II (A-I/A-II HDLs) (Cheung and Albers 1982, 1984). Another way of defining HDL subpopulations is based on size and density. Based on density, HDLs can be separated into two main subfractions, being HDL_2 ($1.063 < d < 1.125$ g/ml) and HDL_3 ($1.125 < d < 1.21$ g/ml). These populations can even be further subdivided based on size into HDL_{2b} (diameter 10.6 nm), HDL_{2a} (9.2 nm), HDL_{3a} (8.4 nm), HDL_{3b} (8.0 nm) and HDL_{3c} (7.6 nm) (Blanche et al. 1981). Finally, HDL particles can also vary in surface charge. When separating HDLs using agarose gel electrophoresis, they can exhibit alpha, pre-beta of gamma migration. The alpha-migrating particles are spherical HDLs, including the HDL_2 and HDL_3 subfractions as well as most of the A-I HDLs and A-I/A-II HDLs (Rye and Barter 2004). Pre-beta HDLs consist of the small population of lipid-free ApoA-I and the discoidal particles containing two or three molecules of apoA-I complexed with phospholipids and small amounts of unesterified cholesterol (Barrans et al. 1996). A minor subpopulation of HDLs, consisting of HDLs containing apoE and phospholipids, exhibits gamma migration (Huang et al. 1994).

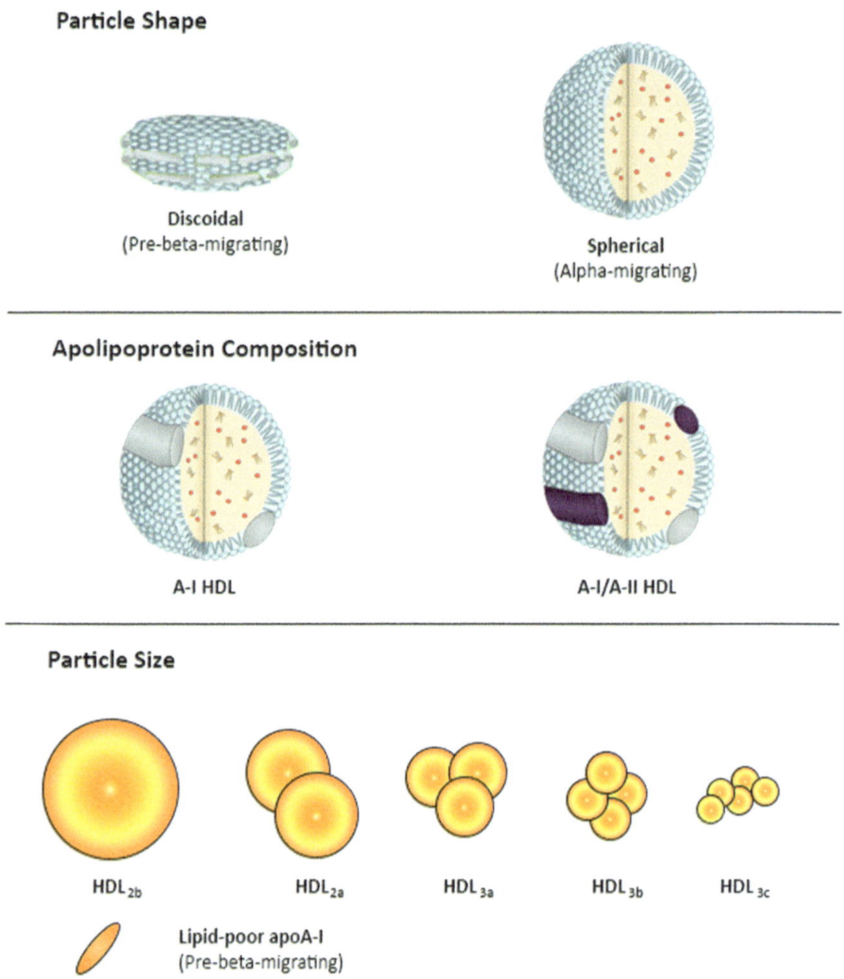

Fig. 16.2 Overview of the different HDL subpopulations

HDL Metabolism

The metabolism of HDL (Fig. 16.3) starts with the production and secretion of ApoA-I, the main protein component of HDLs, by the liver (70%) and in smaller amounts by the intestines (30%) in a lipid-free form (Kingwell et al. 2014). Once in the plasma, this lipid-free apoA-I will rapidly interact with cells and obtain phospholipids and cholesterol from the cell membranes in a reaction mediated by the ATP-binding cassette A1 (ABCA1) transporter. This will result in the formation of small, lipid-poor discoidal pre-beta-migrating HDLs (Camont et al. 2011). Subsequently, the cholesterol present in these discoidal HDLs will be esterified by LCAT, forming cholesteryl

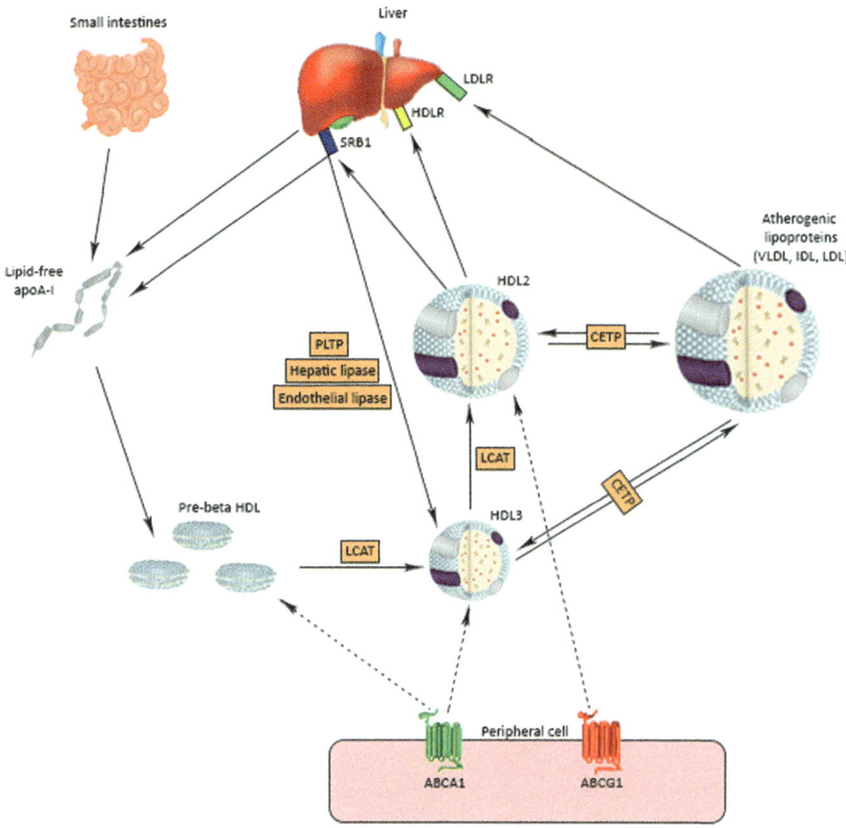

Fig. 16.3 General overview of HDL metabolism. ABCA1 = ATP binding cassette subfamily A member 1; ABCG1 = ATP binding cassette subfamily G member 1; ApoA-I = apolipoproteinA-I; CETP = cholesteryl ester transfer protein; HDL = high density lipoprotein; HDLR = high density lipoprotein receptor; IDL = intermediate density lipoprotein; LCAT = lecithin-cholesterol acyltransferase; LDL = low density lipoprotein; LDLR = LDL receptor; PLTP = phospholipid transfer protein; SR-B1 = scavenger receptor class B member 1; VLDL = very low-density lipoprotein

esters. Since these cholesteryl esters are highly hydrophobic, they cannot remain on the surface of the HDL disc where they would be in direct contact with the plasma water. Therefore, they will move into the inside of the particle and accumulate in the centre, forming a lipid core. This process will transform the discoical pre-beta-migrating HDLs into the alpha-migrating, spherical HDL particles that predominate in the plasma (Barter 2002). Initially, these newly formed spherical HDLs contain only two molecules of apoA-I per particle. However, they are subject to extensive remodelling by various factors, changing their composition and size. For example, small spherical HDLs can further increase in size by taking up more free cholesterol from peripheral cells by cholesterol efflux mediated by the transporters ABCA1 and ABCG1 or the scavenger receptor type B1 (SR-B1) and subsequent esterification of this cholesterol by LCAT. After remodelling, the HDL particles contain three or four molecules of apoA-I, rather than the original two molecules per particle. Additionally, phospholipid transfer protein (PLTP) can transfer phospholipids from very low-density lipoproteins (VLDLs) and chylomicrons to HDL, thereby contributing to the remodelling (Masson et al. 2009). This phospholipid transfer can also lead to apolipoprotein destabilization, resulting in remnant particle fusion and formation of a larger alpha-migrating HDL particle (Settasatian et al. 2001). The latter can undergo constituent-exchange with apoB containing lipoproteins (VLDL and low-density lipoproteins (LDL)) via CETP. The net effect of CETP is the transfer of cholesteryl esters from HDL to apoB containing lipoproteins in exchange for triglycerides, generating triglyceride-rich HDL (Camont et al. 2011; Pirillo et al. 2013). This CETP mediated transfer will result in the transition of large alpha-migrating HDL_2 particles into smaller alpha-migrating HDL_3 particles. Additionally, this transition can be caused by SR-B1 mediated cholesteryl ester uptake by the liver or by hydrolysis of triglycerides by hepatic lipase or endothelial lipase (von Eckardstein et al. 2001). The combined action of CETP and lipases eventually results in size reduction of HDL, accompanied by partial loss of apoA-I. This released, lipid-free, apoA-I is excreted by the kidneys or recycled as it will again be able to interact with the ABCA1 transporter to start a new cycle of lipidation.

General Functions of HDL

Carrier in Reverse Cholesterol Transport

The most important function of HDLs is their involvement in plasma cholesterol transport. Before going into the precise role of HDL in this process it is important to consider the overall metabolism of cholesterol. Dietary cholesterol is absorbed by intestinal cells and becomes esterified by acyl-coenzyme A: cholesterol acyltransferase (ACAT). This esterified cholesterol is subsequently packaged into chylomicrons and via the lymph ducts transported to the blood plasma. In the plasma, the triglycerides which are also present in the chylomicrons are broken down by lipoprotein lipase (LPL), releasing free fatty acids as a source of energy or to be stored in adipose tissues. The remaining remnant chylomicrons, containing dietary cholesterol is finally taken up by the liver. Additionally, the liver itself also produces endogenous cholesterol. From the liver, cholesterol (together with triglycerides) is incorporated into VLDLs which are secreted into the plasma, where the triglycerides are again broken down by LPL in a similar manner as described above. By the loss of triglycerides, VLDL particles are in a complex series of reactions converted into LDL particles which are triglyceride-poor but cholesterol-rich. Via these LDL particles, the cholesterol will be delivered and deposited in peripheral tissues making use of the LDL receptor. Finally, HDLs are involved in the transport from peripheral cholesterol back to the liver in a process called reverse cholesterol transport

(RCT). There are various pathways involved in this RCT. ABCA1 is responsible for the transport of cholesterol, but also phospholipids, from peripheral cells towards lipid-free apoA-I or small lipid-poor discoidal HDLs (Favari et al. 2009; Yokoyama 2006). Larger discoidal HDLs and large spherical HDLs can act as main cholesterol acceptors for ABCG1-mediated efflux (Favari et al. 2009; Rothblat and Phillips 2010; Yokoyama 2006). Additionally, these large spherical HDLs can also bind and interact with SR-B1. This interaction is bidirectional and the rate of this efflux correlates positively with the HDL cholesterol concentration (Ji et al. 1997). The final pathway involves receptor-independent passive diffusion of cholesterol from cell membranes to HDL particles along the concentration gradient. This gradient is formed when LCAT esterifies cholesterol, resulting in the migration of these cholesteryl esters towards the centre of the particle thereby depleting the HDL membrane of cholesterol (Adorni et al. 2007). The relative importance of these different processes varies between different cell types and under different physiological or pathological circumstances as also discussed in greater detail later. Once the cholesterol is incorporated into HDLs, it will be delivered to the liver for excretion which is mainly mediated by SR-B1.

Anti-inflammatory Effects

HDLs have a wide range of anti-inflammatory properties. They can limit endothelial cell activation, resulting in decreased adhesion molecule expression such as vascular cell adhesion molecule 1 (VCAM-1) and intercellular cell adhesion molecule 1 (ICAM-1) (Cockerill et al. 1995). Furthermore, HDLs can suppress chemokine secretion of CCL2, CCL5 and CX_3CL1 from multiple cell types, like endothelial cells, smooth muscle cells and monocytes (Bursill et al. 2010; van der Vorst et al. 2013). It was even shown already that the expression of CCR2 and CX_3CR1, the receptors for the chemokines CCL2 and CX_3CL1, respectively, which play a key role in leukocyte migration, was significantly reduced in mice injected with apoA-I (Bursill et al. 2010). These in vivo anti-inflammatory effects of HDLs were further validated in normo- and hypercholesterolemic rabbit models showing reduced vascular inflammation and adhesion molecule and chemokine expression (Nicholls et al. 2005a, b).

Antioxidant Activity

Another well described function of HDLs is their protection from oxidative damage, thereby also inhibiting the potentially atherogenic oxidised LDL formation (Kontush et al. 2003). Although the precise mechanisms behind these antioxidant effects remain an active field of research, the enzyme paraoxonase-1 (PON-1) which is present on the HDL particles seems to play an important role (Mackness et al. 2004). Several other mechanisms have also been described, for example phospholipid hydroperoxidases that are transferred from LDL to HDL via CETP can subsequently be reduced by methionine residues of apoA-I, forming redox-inactive phospholipid hydroxides (Christison et al. 1995; Zerrad-Saadi et al. 2009). Also other apolipoproteins, like apoA-II, apoA-IV and apoE have anti-oxidant properties that can underlie these effects of different HDLs.

Anti-thrombotic Effects

HDLs can also exert vascular protective effects by upregulating endothelial nitric oxide synthase (eNOS) expression (Mineo et al. 2006). Moreover, HDLs can inhibit thrombosis and platelet activation by inhibiting platelet-activating factor/cyclooxygenase A2. HDLs can also reduce factors that play a crucial role in the coagulation cascade, like thrombin and tissue factor to further exert anti-thrombotic effects (Gordon et al. 2011; Nofer et al. 2010).

Other Activities

Another way in which HDL can have beneficial effects on the endothelium is by enhancing the recruitment of endothelial progenitor cells to damaged areas to ensure proper endothelial repair (Tso et al. 2006). In line with this, it has also been shown that HDL can suppress the proliferation of hematopoietic stem cells, thereby reducing leucocytosis and monocytosis (Yvan-Charvet et al. 2010a). Additionally, HDLs are involved in the transport of various microRNAs. Several studies have already identified that purified HDLs contain a wide variety of small RNAs (Vickers et al. 2011). However, the elucidation of the exact mechanisms behind the loading of microRNAs onto HDL and the exact biological significance of this remains an active field of investigation.

Role of HDL in Pathologies

Cardiovascular Disease

Already in 1964, The Framingham Heart Study was the first study describing compelling evidence that an inverse relationship exists between HDLs and the risk for cardiovascular events (Kannel et al. 1964). This was the start and the driving force behind the concept that HDLs could possibly protect against CVD and its main underlying cause atherosclerosis. Several animal studies during the 1990s supported this hypothesis, showing for example that infusing cholesterol-fed rabbits with HDL plasma fractions resulted in a regression of atherosclerotic lesions (Badimon et al. 1990). Furthermore, mice that overexpress human apoA-I and hence have high circulating HDL levels displayed reduced development of atherosclerosis compared to controls (Rubin et al. 1991). These studies were the foundation of a wide range of pre-clinical and clinical studies over the following decades to further elucidate and validate this association. Various meta-analyses of clinical trials performed over the last two decades have been performed that clearly demonstrate that high levels of HDL are associated with a lower risk of cardiovascular disease (Cordero et al. 2012; Fan et al. 2014).

Several mechanisms have already been described by which HDL can exert atheroprotective effects. First, and probably the best recognized one, is the capacity of HDL to remove cholesterol from arterial wall cells via RCT. As this cholesterol extraction mainly affects macrophages and macrophage-derived foam cells that play a critical role in the development of atherosclerosis, it is conceivable that this cholesterol efflux property provides the basis for the observed inverse association of HDL and CVD risk (von Eckardstein et al. 2001). As discussed previously, there are different pathways for the cholesterol efflux from cells. Specifically for macrophages, the ABCA1-mediated efflux accounts for at least 80% of the efflux stimulated by cellular cholesterol loading (Adorni et al. 2007). The atheroprotective role of ACBA1-mediated cholesterol efflux from macrophages has been extensively described in several animal models (Rader et al. 2009; Yvan-Charvet et al. 2010b; Zhao et al. 2010). The role of the ABCG1-mediated pathway of cholesterol efflux from macrophages appears to be less important compared to ABCA1 (Adorni et al. 2007; Kontush 2014). The passive cholesterol efflux seems to be especially important in non-stimulated conditions in which this pathway is even more potent then SR-B1 (Adorni et al. 2007). However, upon cholesterol loading of cells which is a hallmark for cells present in atherosclerotic lesions, these pathways become of minor importance.

Growing evidence support the role of other anti-atherogenic effects of HDL. These beneficial actions involve for example the vasodilatory activity. HDLs can bind to the extracellular loop of SR-B1, initiating intracellular signalling events mediated by Akt and intracellular Ca^{2+} mobilization and increasing the nitric oxide (NO) production (Drew et al. 2004; Mineo et al. 2006; Nofer et al. 2004). Another pathway contributing to these vasodilatory effects of HDL in endothelial cells is ABCG1-mediated cholesterol efflux, which stimulates the formation of eNOS dimers resulting in decreased reactive oxygen species (ROS) production (Terasaka et al. 2008). The diminished production of ROS, which normally inactivate NO, can consequently lead to a higher bioavailability of NO and improving the vasodilation. Clinical evidence for this beneficial role of HDL on endothelial function is accumulating.

Several studies highlight that the anti-inflammatory effects of HDLs may be critical for the protection against CVD. HDLs have been shown to potently decrease adhesion molecule expression on endothelial cells and thereby also the monocyte adhesion to the endothelium both in vitro and in vivo (Kontush 2014). Besides affecting the vascular cells, HDL also inhibits the stimulation of T lymphocytes by antigen-presenting cells (Yu et al. 2010). Furthermore, neutrophil activation is reduced by HDLs, showing a broad interaction of HDL with both stromal and inflammatory cells (Liao et al. 2005). Interestingly, HDL has also been associated with processes of resolution of inflammation, as HDL stimulates the emigration of macrophages and monocyte-derived dendritic cells from atherosclerotic lesions (Pagler et al. 2011; Yu et al. 2010). Although it is plausible that several mechanisms underlie this wide range of effects of HDLs, it is plausible that cholesterol efflux plays an important part in this. For example, removal of cellular cholesterol down-regulates the inflammatory profile of several cell types, like macrophages, adipocytes and endothelial cells (Kontush 2014). In addition, as described above, protective effect of HDL on LDL oxidation can have a major effect on lesion development as oxidized LDL is known to be a key driver for atherosclerosis development. HDLs also diminish the cellular production of superoxide and hydrogen peroxide in endothelial cells, further explaining the anti-oxidative effects of HDL (de Souza et al. 2010).

As already briefly eluded to before, HDLs carry several microRNAs (Vickers et al. 2011). Several of these microRNAs are important intracellular regulators of gene expression and thereby post-transcriptionally control cholesterol homeostasis/efflux (Chen et al. 2013). For example, microRNA-33 down-regulate the expression of the transporters ABCA1 and ABCG1 and has thereby been implicated in reduced macrophage cholesterol efflux and thereby progression of atherogenesis (Horie et al. 2012; Rayner et al. 2011).

HDLs, therefore, have various properties and mechanisms of action that contribute to their ability to protect against the development of atherosclerosis (Fig. 16.4).

Cancer

Several studies suggest that tumour cells require an increased cholesterol supply and are able to accumulate relatively high amounts of cholesterol (Mooberry et al. 2016). Although, the interaction between tumour cells and cholesterol/lipids is not straightforward, as human cancers exhibit both high and low levels of HDLs and plasma lipids can also change in a bidirectional manner (Chawda et al. 2011; Poorey and Thakur 2016). However, it seems that there is inverse relationship between plasma HDL-cholesterol levels and the risk of developing cancer, which is also confirmed in a large meta-analysis demonstrating that every 10 mg/dL increase in plasma HDL-cholesterol results in a 36% reduction of the risk of cancer (Jafri et al. 2010; Mogilenko et al. 2010). It is still unclear however, whether decreased HDL-cholesterol levels are causal or consequential to cancer progression (Morin et al. 2018). Furthermore, the exact relationship between HDLs and cancer remains complicated as several factors (e.g. type of cancer, age, tobacco exposure) can exert confounding influence (Ahn et al. 2009). In general, the relationship between HDLs, HDL-cholesterol and cancer incidence and mortality rates remain controversial and an interesting field of active research. It might very well be that this relationship is highly tumour-type-dependent as several studies report negative, while others report positive associations (Liu et al. 2016). This specificity has already been shown in

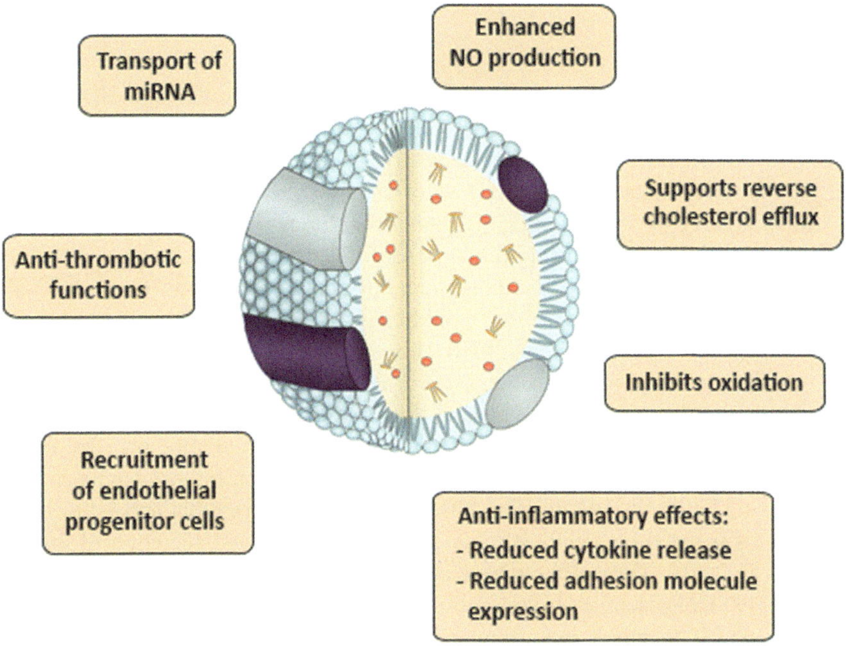

Fig. 16.4 To date identified roles of HDL in the protection against atherosclerosis development/progression

in vitro studies, where HDLs by its anti-oxidant activity could suppress the proliferation of prostate cancer cells, while HDLs stimulate the migration of breast cancer cells (Danilo et al. 2013; Ruscica et al. 2018). Also the effects of apoA-I are rather controversial as some in vitro studies demonstrate a cancer-promoting effect, while other show conflicting anti-tumour effects (Hashemi et al. 2014; Su et al. 2010). Recently, it has been suggested that the variable HDL composition and functionality might better explain the association between HDLs and specific tumour-types (Ganjali et al. 2019). This is supported by the fact that several HDL-associated proteins and enzymes play an important role in cancer biology. For example, cancer cells upregulate the expression of SR-B1, which is likely a survival mechanism to increase HDL-cholesterol uptake needed for proliferation and hormone production (Morin et al. 2018). Additionally, inactivation of the ABCA1 transporter, which plays a crucial role in the cholesterol efflux pathway, also results in the accumulation of cholesterol in cancer cells, which at least partially accounts for the biological aggressiveness of cancer cells (Ganjali et al. 2019). Therefore, at the moment, especially ABCA1 and SR-B1 are being investigated as potential novel diagnostic or prognostic indicators of cancer. Future research should further identify mechanisms behind the interaction between HDL and cancer, and elucidate whether this role is causal or consequential and whether the risk association is positive or negative in specific situations.

Diabetes Mellitus

There is already abundant epidemiological evidence showing that low HDL levels are correlated with an increased risk of Diabetes Mellitus (DM) (Abbasi et al. 2013). For example, in a large multi-centre prospective study of almost 7000 patients, it has been demonstrated that patients with low levels of HDL have double the prevalence of DM and thereby also a significantly higher risk for cardiovascular complications (Ahmed et al. 2016). Supporting the notion of this association between HDL and DM is the fact that there are also consistent correlations between low HDL levels and the manifestation of DM-related complications, like peripheral neuropathy and diabetic nephropathy (Wong et al. 2018). However, similar to cancer, regarding DM it also remains unclear so far whether there is also a causal relationship between low HDL levels and the incidence of DM and its complications.

A wide range of HDL properties can have a beneficial outcome on DM. Especially interesting is its direct role in mediating glycaemic control through its actions on pancreatic beta cells (Wong et al. 2018). It has been shown that apoA-I or HDL treatment of isolated beta cells stimulated the production and secretion of insulin (Fryirs et al. 2010). These effects have been shown to be mediated by both SR-B1 and ABCA1. Consistent with this notion, both apoA-I and HDL have also been shown to inhibit apoptosis in murine and human pancreatic islets, which is probably related to the capacity of HDL to counteract the endoplasmic reticulum stress response (Petremand et al. 2012; Rutti et al. 2009). The beneficial role of HDL in regulating glucose metabolism has also been confirmed in in vivo studies, where *apoA-I* deficient mice showed significantly impaired glucose tolerance compared to control mice (Han et al. 2007). Additionally, when human apoA-I was injected into insulin-resistant mice, these animals showed an acute improvement in insulin secretion demonstrating the potential therapeutic potential (Stenkula et al. 2014). This therapeutic potential was further substantiated by a human study with type 2 DM patients. Infusions of reconstituted HDLs into these patients significantly reduced the plasma glucose levels by increasing the insulin production and secretion (Drew et al. 2009).

Besides these direct effects on glucose metabolism, the effects of HDLs on the RCT, endothelial function, as well as anti-inflammatory, anti-oxidative, anti-thrombotic properties all contribute to the beneficial effects of HDLs on DM (Wong et al. 2018).

Dysfunctional HDL

Originally, the relationship between HDL and several pathologies has only been investigated based on the levels of HDL-cholesterol. However, recent years the function of HDL has also become of greater interest as it has been shown that several pathologies like hyperuricemia, DM, CVD trigger structural and functional alterations in HDLs. This highlights the point that not only the concentration of HDLs or HDL-cholesterol is important, but especially also the composition and thereby its function. The term dysfunctional HDL has been introduced to describe HDLs that have been changed in their composition and functional properties, due to conditions of oxidative stress, infection or inflammation (Estrada-Luna et al. 2018). For example, it has been demonstrated that patients who suffered a coronary event show dysfunctional HDLs, despite having relatively high HDL-cholesterol levels (Asztalos and Schaefer 2003). This dysfunctionality was shown by the fact that patients HDLs have a reduced cholesterol efflux capacity, reduced anti-oxidative effects and impaired anti-inflammatory effects on endothelial cells. Another study, described that patients with atherosclerosis also have altered HDLs, since these HDL particles show an enrichment of inflammatory proteins apoC-III and serum amyloid A (Han et al. 2016). Also in other pathologies like DM, HDL function can be significantly changed as HDL from type 2 DM patients for example exhibit an impaired capacity to stimulate endothelial cell proliferation/migration (Pan et al. 2012). Although the concept of dysfunctional HDL is still rather new and needs to be further explored, it is already clear that it is of paramount importance to not only focus on HDL quantities but also on HDL quality or composition. This concept will be further explored in the future and will open up new avenues for therapeutic of even prognostic/diagnostic approaches.

HDL Therapies

Since HDLs have so much beneficial functions and epidemiological data demonstrate an inverse relationship with various pathologies especially CVD, many efforts have been made to increase plasma HDL in patients. Of course, the most straightforward approach is lifestyle intervention. For example, high levels of aerobic activity are associated with higher levels of HDL-cholesterol. Physical activity, especially of overweight individuals, increases the HDL levels at least partly by stimulating the lipoprotein lipase activity. Furthermore, it has been observed that weight reduction is accompanied by increasing levels of HDLs, although the exact underlying mechanism is unknown this probably coincides with improved diet composition and physical activity. Additionally, alcohol consumption has been shown to increase the plasma HDL levels, probably due to the inhibition of CETP though reports are not consistent about this. Finally, smoking cessation has been associated with a striking 10% increase in HDL-cholesterol levels, further supporting the major effects that lifestyle intervention can have on HDLs.

Besides these interventions, various therapeutic/pharmacological approaches have also been investigated (Fig. 16.5). The main therapeutic strategies that were and are still being developed will be discussed here. Niacin, or nicotinic acid, has already been used for decades to modulate lipids and is an effective agent to raise

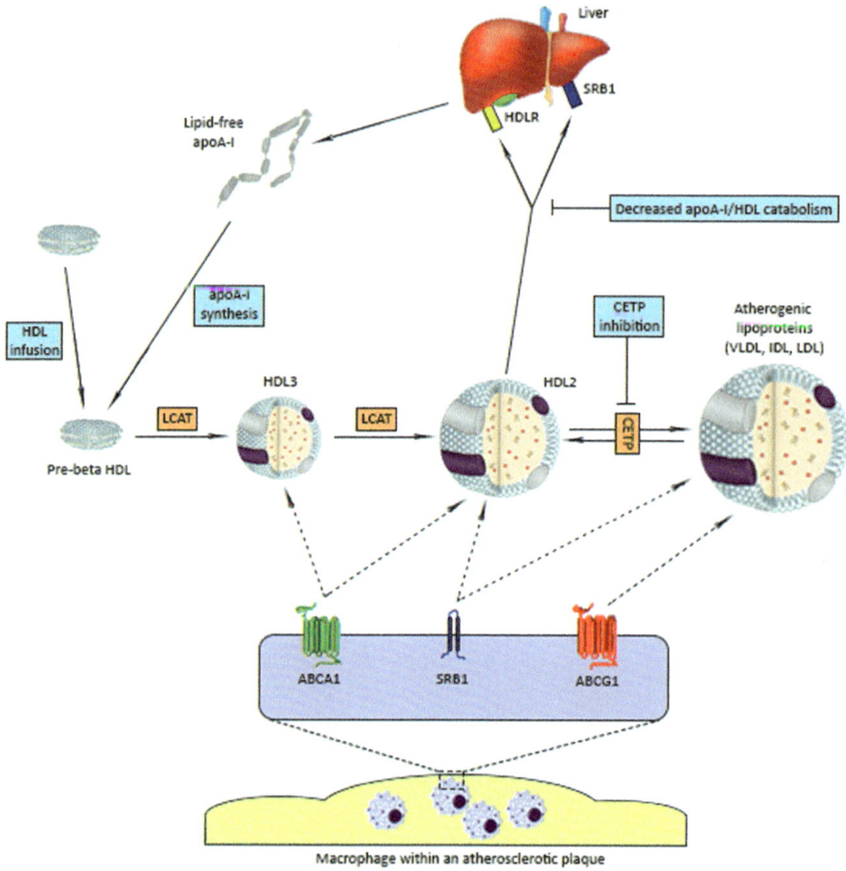

Fig. 16.5 Mode of action of currently available HDL-raising therapies. ABCA1 = ATP binding cassette subfamily A member 1; ABCG1 = ATP binding cassette subfamily G member 1; ApoA-I = apolipoproteinA-I; CETP = cholesteryl ester transfer protein; HDL = high density lipoprotein; HDLR = high density lipoprotein receptor; IDL = intermediate density lipoprotein; LCAT = lecithin-cholesterol acyltransferase; LDL = low density lipoprotein; LDLR = LDL receptor; SR-B1 = scavenger receptor class B member 1; VLDL = very low-density lipoprotein

HDL levels (Carlson 2005). Interestingly though, is the fact that the exact underlying mechanism of action if not very well understood. Some studies show that it can reduce the apoA-I catabolism in a direct manner (Kamanna et al. 2013), while others show that it can inhibit CETP (van der Hoorn et al. 2008). However, the beneficial effects of Niacin on HDL and the lipid profile could not be matched in various clinical studies as no effect on cardiovascular risk could be found (Weber et al. 2017). The most recent trial, the Second Heart Protection Study—Treatment of HDL to Reduce the Incidence of Vascular Events (HPS2-THRIVE), was even terminated early due to a high occurance of serious bleedings and infections in the niacin/laropiprant-treated group, which entailed the withdrawal of niacin from many markets (Group 2013).

Fibrates is another pharmacological drug used to increase HDL levels. It has been shown that the HDL levels increase approximately by 10%, although it seemed to have yet undefined unfavourable effects on the large HDL subclasses. Fibrates increase the synthesis of apoA-I and ABCA1 via the activation of peroxisome proliferator-activated receptor α (PPARα) (Staels et al. 1998). However, fibrates also have a significant effect on the triglycerides level. It is as yet rather unclear what the exact contribution of HDLs is in the general beneficial effect of fibrate treatment.

One of the most promising drug developments in recent years in the field of HDLs is focused on CETP inhibition. As mentioned, CETP normally transfers cholesteryl esters from HDL to apoB-containing lipoproteins (VLDL, LDL), thus resulting in a decreased HDL-cholesterol. Several large clinical trials have been performed to validate the therapeutic potential of CETP inhibition, but this turned out to be more challenging than expected. The first trial using the drug torcetrapib (ILLUMINATE trial) was terminated early because of unforeseen off-target effects resulting in increased mortality rates in the treated group (Barter et al. 2007). In follow-up experiments it was shown that torcetrapib influences the renin-angiotensin-aldosterone system, leading to hypertension and hypokalemia in patients. More recent trials, using dalcetrapib (dal-OUTCOMES), anacetrapib (REALIZE) or evacetrapib (ACCELERATE) were all stopped due to futility (Weber et al. 2017). However, since the REALIZE trial was only performed with a rather small population size; an additional trial has been performed using anacetrapib (REVEAL). This turned out to be the first trial that showed a significant reduction of the incidence of cardiovascular events after anacetrapib treatment, without any major safety concerns. These beneficial effects only became apparent after two years of treatment (Barter 2002). This may hold even bigger promise for CETP inhibition, as for example the evacetrapib was terminated before the two years follow-up point.

Another, more direct way, to influence plasma HDL levels is direct infusion. The ERASE trial is a small randomized clinical trial showing that the infusion of reconstituted HDL significantly reduced atheroma volume (Tardif et al. 2007). Initial studies demonstrated that the infusion of apoA-I Milano containing HDLs also had beneficial effects, although the more recent and larger MILANO-PILOT study showed that five weekly injections of MDCO-216, containing apoA-I Milano, had no favourable effect on atherosclerosis development (Weber et al. 2017). Another interesting approach is the use of HDL mimetic peptides. Already a wide range of peptides have been tested in pre-clinical studies, like peptide 5A that was shown to reduce the expression of adhesion molecules in human endothelial cells (Tabet et al. 2010), or the apoA-I mimetic peptide 4F that was shown to reduce atherosclerosis development in mice (Wool et al. 2011). Two HDL-mimetics, CER-001 and CSL-112, demonstrated to have a substantial impact on HDL levels, are currently being tested in phase II clinical trials. Another method would be to increase the endogenous synthesis of apoA-I in order to increase the levels of HDLs. Apabetalon (RVX-208) is a small molecule that inhibits the bromodomain and extra terminal (BET), thereby epigenetically modulating genes that result in increased apoA-I synthesis. Currently, a phase III trial (BETonMACE) is being conducted to investigate the effects on cardiac events in patients.

Thus there is a lively debate about the previously described HDL hypothesis and about the question whether raising HDL is a suitable and beneficial therapeutic strategy. This discussion is fuelled by the various clinical trials that, although providing very promising pre-clinical data, did not show the expected beneficial effects. Further adding to this discussion are several meta-analysis, for example one that analysed 39 clinical trials, showing that niacin, fibrates and CETP inhibitors did not reduce all-cause mortality or the incidence of MI or stroke (Keene et al. 2014). Additionally, polymorphisms exclusively altering HDL levels did not affect the risk of MI in Mendelian randomisation analyses (Voight et al. 2012). All by all, the debate about the therapeutic potency of HDLs is ongoing and also in this context, focus is more and more shifting to HDL function rather than HDL levels.

Closing Remarks

HDLs play an important role in the cholesterol metabolism in the physiological state. Over the years the appreciation of the complexity of HDLs has increased significantly and researchers realized that HDLs comprise a very heterogeneous group of lipoproteins and that especially the pathophysiological effects of HDLs are very complex and interrelated. Future research will undoubtedly even identify more complexity in this whole system, which will give more insight and potentially also open up new avenues for therapeutic approaches. Though there is an ongoing discussion about the therapeutic potential of raising HDL levels, it seems that understanding HDL function and the therapeutic potency of its modulation is still in the very early stages of research and development, holding a great promise for HDL-based therapies in the future.

Acknowledgements The author would like to express his gratitude to Linsey Peters for preparing the graphics of this book chapter.

References

Abbasi A, Corpeleijn E, Gansevoort RT, Gans RO, Hillege HL, Stolk RP, Navis G, Bakker SJ, Dullaart RP (2013) Role of HDL cholesterol and estimates of HDL particle composition in future development of type 2 diabetes in the general population: the PREVEND study. J Clin Endocrinol Metab 98(8):E1352–E1359. https://doi.org/10.1210/jc.2013-1680

Adorni MP, Zimetti F, Billheimer JT, Wang N, Rader DJ, Phillips MC, Rothblat GH (2007) The roles of different pathways in the release of cholesterol from macrophages. J Lipid Res 48(11):2453–2462. https://doi.org/10.1194/jlr.M700274-JLR200

Ahmed HM, Miller M, Nasir K, McEvoy JW, Herrington D, Blumenthal RS, Blaha MJ (2016) Primary low level of high-density lipoprotein cholesterol and risks of coronary heart disease, cardiovascular disease, and death: results from the multi-ethnic study of atherosclerosis. Am J Epidemiol 183(10):875–883. https://doi.org/10.1093/aje/kwv305

Ahn J, Lim U, Weinstein SJ, Schatzkin A, Hayes RB, Virtamo J, Albanes D (2009) Prediagnostic total and high-density lipoprotein cholesterol and risk of cancer. Cancer Epidemiol Biomark Prev: Publ Am Assoc Cancer Res Cosponsored by the Am Soc Prev Oncol 18(11):2814–2821. https://doi.org/10.1158/1055-9965.EPI-08-1248

Asztalos BF, Schaefer EJ (2003) High-density lipoprotein subpopulations in pathologic conditions. Am J Cardiol 91(7A):12E–17E

Badimon JJ, Badimon L, Fuster V (1990) Regression of atherosclerotic lesions by high density lipoprotein plasma fraction in the cholesterol-fed rabbit. J Clin Investig 85(4):1234–1241. https://doi.org/10.1172/JCI114558

Barr DP, Russ EM, Eder HA (1951) Protein-lipid relationships in human plasma. II. In atherosclerosis and related conditions. Am J Med 11(4):480–493

Barrans A, Jaspard B, Barbaras R, Chap H, Perret B, Collet X (1996) Pre-beta HDL: structure and metabolism. Biochem Biophys Acta 1300(2):73–85

Barter PJ (2002) Hugh sinclair lecture: the regulation and remodelling of HDL by plasma factors. Atheroscler Suppl 3(4):39–47

Barter PJ, Caulfield M, Eriksson M, Grundy SM, Kastelein JJ, Komajda M, Lopez-Sendon J, Mosca L, Tardif JC, Waters DD, Shear CL, Revkin JH, Buhr KA, Fisher MR, Tall AR, Brewer B, Investigators I (2007) Effects of torcetrapib in patients at high risk for coronary events. New Engl J Med 357(21):2109–2122. https://doi.org/10.1056/NEJMoa0706628

Blanche PJ, Gong EL, Forte TM, Nichols AV (1981) Characterization of human high-density lipoproteins by gradient gel electrophoresis. Biochem Biophys Acta 665(3):408–419

Bursill CA, Castro ML, Beattie DT, Nakhla S, van der Vorst E, Heather AK, Barter PJ, Rye KA (2010) High-density lipoproteins suppress chemokines and chemokine receptors in vitro and in vivo. Arterioscler Thromb Vasc Biol 30(9):1773–1778. https://doi.org/10.1161/ATVBAHA.110.211342

Camont L, Chapman MJ, Kontush A (2011) Biological activities of HDL subpopulations and their relevance to cardiovascular disease. Trends Mol Med 17(10):594–603. https://doi.org/10.1016/j.molmed.2011.05.013

Carlson LA (2005) Nicotinic acid: the broad-spectrum lipid drug. A 50th anniversary review. J Intern Med 258(2):94–114. https://doi.org/10.1111/j.1365-2796.2005.01528.x

Chawda JG, Jain SS, Patel HR, Chaduvula N, Patel K (2011) The relationship between serum lipid levels and the risk of oral cancer. Indian J Med Paediatr Oncol: Off J Indian Soc Med Paediatr Oncol 32(1):34–37. https://doi.org/10.4103/0971-5851.81888

Chen WJ, Zhang M, Zhao GJ, Fu Y, Zhang DW, Zhu HB, Tang CK (2013) MicroRNA-33 in atherosclerosis etiology and pathophysiology. Atherosclerosis 227(2):201–208. https://doi.org/10.1016/j.atherosclerosis.2012.11.025

Cheung MC, Albers JJ (1982) Distribution of high density lipoprotein particles with different apoprotein composition: particles with A-I and A-II and particles with A-I but no A-II. J Lipid Res 23(5):747–753

Cheung MC, Albers JJ (1984) Characterization of lipoprotein particles isolated by immunoaffinity chromatography. Particles containing A-I and A-II and particles containing A-I but no A-II. J Biol Chem 259(19):12201–12209

Christison JK, Rye KA, Stocker R (1995) Exchange of oxidized cholesteryl linoleate between LDL and HDL mediated by cholesteryl ester transfer protein. J Lipid Res 36(9):2017–2026

Cockerill GW, Rye KA, Gamble JR, Vadas MA, Barter PJ (1995) High-density lipoproteins inhibit cytokine-induced expression of endothelial cell adhesion molecules. Arterioscler Thromb Vasc Biol 15(11):1987–1994

Constantinou C, Karavia EA, Xepapadaki E, Petropoulou PI, Papakosta E, Karavyraki M, Zvintzou E, Theodoropoulos V, Filou S, Hatziri A, Kalogeropoulou C, Panayiotakopoulos G, Kypreos KE (2016) Advances in high-density lipoprotein physiology: surprises, overturns, and promises. Am J Physiol Endocrinol Metab 310(1):E1–E14. https://doi.org/10.1152/ajpendo.00429.2015

Cordero A, Moreno-Arribas J, Bertomeu-Gonzalez V, Agudo P, Miralles B, Masia MD, Lopez-Palop R, Bertomeu-Martinez V (2012) Low levels of high-density lipoproteins cholesterol are

independently associated with acute coronary heart disease in patients hospitalized for chest pain. Rev Esp Cardiol 65(4):319–325. https://doi.org/10.1016/j.recesp.2011.07.022
Danilo C, Gutierrez-Pajares JL, Mainieri MA, Mercier I, Lisanti MP, Frank PG (2013) Scavenger receptor class B type I regulates cellular cholesterol metabolism and cell signaling associated with breast cancer development. Breast Cancer Res: BCR 15(5):R87. https://doi.org/10.1186/bcr3483
de Souza JA, Vindis C, Negre-Salvayre A, Rye KA, Couturier M, Therond P, Chantepie S, Salvayre R, Chapman MJ, Kontush A (2010) Small, dense HDL 3 particles attenuate apoptosis in endothelial cells: pivotal role of apolipoprotein A-I. J Cell Mol Med 14(3):608–620. https://doi.org/10.1111/j.1582-4934.2009.00713.x
Drew BG, Duffy SJ, Formosa MF, Natoli AK, Henstridge DC, Penfold SA, Thomas WG, Mukhamedova N, de Courten B, Forbes JM, Yap FY, Kaye DM, van Hall G, Febbraio MA, Kemp BE, Sviridov D, Steinberg GR, Kingwell BA (2009) High-density lipoprotein modulates glucose metabolism in patients with type 2 diabetes mellitus. Circulation 119(15):2103–2111. https://doi.org/10.1161/CIRCULATIONAHA.108.843219
Drew BG, Fidge NH, Gallon-Beaumier G, Kemp BE, Kingwell BA (2004) High-density lipoprotein and apolipoprotein AI increase endothelial NO synthase activity by protein association and multisite phosphorylation. Proc Natl Acad Sci USA 101(18):6999–7004. https://doi.org/10.1073/pnas.0306266101
Estrada-Luna D, Ortiz-Rodriguez MA, Medina-Briseno L, Carreon-Torres E, Izquierdo-Vega JA, Sharma A, Cancino-Diaz JC, Perez-Mendez O, Belefant-Miller H, Betanzos-Cabrera G (2018) Current therapies focused on high-density lipoproteins associated with cardiovascular disease. Molecules 23(11). https://doi.org/10.3390/molecules23112730
Fan J, Qi Y, Zhao D (2014) A meta-analysis on the association between high-density lipoprotein particle subfractions and cardiovascular disease events. Zhonghua xin xue guan bing za zhi 42(1):57–61
Favari E, Calabresi L, Adorni MP, Jessup W, Simonelli S, Franceschini G, Bernini F (2009) Small discoidal pre-beta1 HDL particles are efficient acceptors of cell cholesterol via ABCA1 and ABCG1. Biochemistry 48(46):11067–11074. https://doi.org/10.1021/bi901564g
Fryirs MA, Barter PJ, Appavoo M, Tuch BE, Tabet F, Heather AK, Rye KA (2010) Effects of high-density lipoproteins on pancreatic beta-cell insulin secretion. Arterioscler Thromb Vasc Biol 30(8):1642–1648. https://doi.org/10.1161/ATVBAHA.110.207373
Ganjali S, Ricciuti B, Pirro M, Butler AE, Atkin SL, Banach M, Sahebkar A (2019) High-density lipoprotein components and functionality in cancer: state-of-the-art. Trends Endocrinol Metab: TEM 30(1):12–24. https://doi.org/10.1016/j.tem.2018.10.004
Gofman JW (1956) Serum lipoproteins and the evaluation of atherosclerosis. Ann NY Acad Sci 64(4):590–595
Gordon SM, Hofmann S, Askew DS, Davidson WS (2011) High density lipoprotein: it's not just about lipid transport anymore. Trends Endocrinol Metab: TEM 22(1):9–15. https://doi.org/10.1016/j.tem.2010.10.001
Group HTC (2013) HPS2-THRIVE randomized placebo-controlled trial in 25 673 high-risk patients of ER niacin/laropiprant: trial design, pre-specified muscle and liver outcomes, and reasons for stopping study treatment. Eur Heart J 34(17):1279–1291. https://doi.org/10.1093/eurheartj/eht055
Han CY, Tang C, Guevara ME, Wei H, Wietecha T, Shao B, Subramanian S, Omer M, Wang S, O'Brien KD, Marcovina SM, Wight TN, Vaisar T, de Beer MC, de Beer FC, Osborne WR, Elkon KB, Chait A (2016) Serum amyloid A impairs the antiinflammatory properties of HDL. J Clin Investig 126(2):796. https://doi.org/10.1172/JCI86401
Han R, Lai R, Ding Q, Wang Z, Luo X, Zhang Y, Cui G, He J, Liu W, Chen Y (2007) Apolipoprotein A-I stimulates AMP-activated protein kinase and improves glucose metabolism. Diabetologia 50(9):1960–1968. https://doi.org/10.1007/s00125-007-0752-7

Hashemi M, Pooladi M, Razi Abad SK (2014) The investigation of changes in proteins expression (Apolipoprotein A1 and albumin) in malignant astrocytoma brain tumor. J Cancer Res Ther 10(1):107–111. https://doi.org/10.4103/0973-1482.131413

Horie T, Baba O, Kuwabara Y, Chujo Y, Watanabe S, Kinoshita M, Horiguchi M, Nakamura T, Chonabayashi K, Hishizawa M, Hasegawa K, Kume N, Yokode M, Kita T, Kimura T, Ono K (2012) MicroRNA-33 deficiency reduces the progression of atherosclerotic plaque in ApoE-/- mice. J Am Heart Assoc 1(6):e003376. https://doi.org/10.1161/JAHA.112.003376

Huang Y, von Eckardstein A, Wu S, Maeda N, Assmann G (1994) A plasma lipoprotein containing only apolipoprotein E and with gamma mobility on electrophoresis releases cholesterol from cells. Proc Natl Acad Sci USA 91(5):1834–1838

Jafri H, Alsheikh-Ali AA, Karas RH (2010) Baseline and on-treatment high-density lipoprotein cholesterol and the risk of cancer in randomized controlled trials of lipid-altering therapy. J Am Coll Cardiol 55(25):2846–2854. https://doi.org/10.1016/j.jacc.2009.12.069

Ji Y, Jian B, Wang N, Sun Y, Moya ML, Phillips MC, Rothblat GH, Swaney JB, Tall AR (1997) Scavenger receptor BI promotes high density lipoprotein-mediated cellular cholesterol efflux. J Biol Chem 272(34):20982–20985

Kamanna VS, Ganji SH, Kashyap ML (2013) Recent advances in niacin and lipid metabolism. Curr Opin Lipidol 24(3):239–245. https://doi.org/10.1097/MOL.0b013e3283613a68

Kannel WB, Dawber TR, Friedman GD, Glennon WE, McNamara PM (1964) Risk factors in coronary heart disease. an evaluation of several serum lipids as predictors of coronary heart disease; the Framingham study. Ann Intern Med 61:888–899

Karavia EA, Zvintzou E, Petropoulou PI, Xepapadaki E, Constantinou C, Kypreos KE (2014) HDL quality and functionality: what can proteins and genes predict? Expert Rev Cardiovasc Ther 12(4):521–532. https://doi.org/10.1586/14779072.2014.896741

Keene D, Price C, Shun-Shin MJ, Francis DP (2014) Effect on cardiovascular risk of high density lipoprotein targeted drug treatments niacin, fibrates, and CETP inhibitors: meta-analysis of randomised controlled trials including 117,411 patients. BMJ 349:g4379. https://doi.org/10.1136/bmj.g4379

Kingwell BA, Chapman MJ, Kontush A, Miller NE (2014) HDL-targeted therapies: progress, failures and future. Nat Rev Drug Discov 13(6):445–464. https://doi.org/10.1038/nrd4279

Kontush A (2014) HDL-mediated mechanisms of protection in cardiovascular disease. Cardiovasc Res 103(3):341–349. https://doi.org/10.1093/cvr/cvu147

Kontush A, Chantepie S, Chapman MJ (2003) Small, dense HDL particles exert potent protection of atherogenic LDL against oxidative stress. Arterioscler Thromb Vasc Biol 23(10):1881–1888. https://doi.org/10.1161/01.ATV.0000091338.93223.E8

Liao XL, Lou B, Ma J, Wu MP (2005) Neutrophils activation can be diminished by apolipoprotein A-I. Life Sci 77(3):325–335. https://doi.org/10.1016/j.lfs.2004.10.066

Liu YY, Lin SJ, Chen YY, Liu LN, Bao LB, Tang LQ, Ou JS, Liu ZG, Chen XZ, Xu Y, Ma J, Chan AT, Chen M, Xia YF, Liu WL, Zeng YX, Mai HQ, Zeng MS, Pan JJ, Zhang X (2016) High-density lipoprotein cholesterol as a predictor of poor survival in patients with nasopharyngeal carcinoma. Oncotarget 7(28):42978–42987. https://doi.org/10.18632/oncotarget.7160

Mackness MI, Durrington PN, Mackness B (2004) The role of paraoxonase 1 activity in cardiovascular disease: potential for therapeutic intervention. Am J Cardiovasc Drugs: Drugs, Devices, Other Interv 4(4):211–217. https://doi.org/10.2165/00129784-200404040-00002

Masson D, Jiang XC, Lagrost L, Tall AR (2009) The role of plasma lipid transfer proteins in lipoprotein metabolism and atherogenesis. J Lipid Res 50(Suppl):S201–S206. https://doi.org/10.1194/jlr.R800061-JLR200

Mineo C, Deguchi H, Griffin JH, Shaul PW (2006) Endothelial and antithrombotic actions of HDL. Circ Res 98(11):1352–1364. https://doi.org/10.1161/01.RES.0000225982.01988.93

Mogilenko DA, Shavva VS, Dizhe EB, Orlov SV, Perevozchikov AP (2010) PPARgamma activates ABCA1 gene transcription but reduces the level of ABCA1 protein in HepG2 cells. Biochem Biophys Res Commun 402(3):477–482. https://doi.org/10.1016/j.bbrc.2010.10.053

Mooberry LK, Sabnis NA, Panchoo M, Nagarajan B, Lacko AG (2016) Targeting the SR-B1 Receptor as a Gateway for Cancer Therapy and Imaging. Front Pharmacol 7:466. https://doi.org/10.3389/fphar.2016.00466

Morin EE, Li XA, Schwendeman A (2018) HDL in Endocrine Carcinomas: Biomarker, Drug Carrier, and Potential Therapeutic. Front Endocrinol 9:715. https://doi.org/10.3389/fendo.2018.00715

Nicholls SJ, Cutri B, Worthley SG, Kee P, Rye KA, Bao S, Barter PJ (2005a) Impact of short-term administration of high-density lipoproteins and atorvastatin on atherosclerosis in rabbits. Arterioscler Thromb Vasc Biol 25(11):2416–2421. https://doi.org/10.1161/01.ATV.0000184760.95957.d6

Nicholls SJ, Dusting GJ, Cutri B, Bao S, Drummond GR, Rye KA, Barter PJ (2005b) Reconstituted high-density lipoproteins inhibit the acute pro-oxidant and proinflammatory vascular changes induced by a periarterial collar in normocholesterolemic rabbits. Circulation 111(12):1543–1550. https://doi.org/10.1161/01.CIR.0000159351.95399.50

Nofer JR, Brodde MF, Kehrel BE (2010) High-density lipoproteins, platelets and the pathogenesis of atherosclerosis. Clin Exp Pharmacol Physiol 37(7):726–735. https://doi.org/10.1111/j.1440-1681.2010.05377.x

Nofer JR, van der Giet M, Tolle M, Wolinska I, von Wnuck Lipinski K, Baba HA, Tietge UJ, Godecke A, Ishii I, Kleuser B, Schafers M, Fobker M, Zidek W, Assmann G, Chun J, Levkau B (2004) HDL induces NO-dependent vasorelaxation via the lysophospholipid receptor S1P3. J Clin Investig 113(4):569–581. https://doi.org/10.1172/JCI18004

Pagler TA, Wang M, Mondal M, Murphy AJ, Westerterp M, Moore KJ, Maxfield FR, Tall AR (2011) Deletion of ABCA1 and ABCG1 impairs macrophage migration because of increased Rac1 signaling. Circ Res 108(2):194–200. https://doi.org/10.1161/CIRCRESAHA.110.228619

Pan B, Ma Y, Ren H, He Y, Wang Y, Lv X, Liu D, Ji L, Yu B, Wang Y, Chen YE, Pennathur S, Smith JD, Liu G, Zheng L (2012) Diabetic HDL is dysfunctional in stimulating endothelial cell migration and proliferation due to down regulation of SR-BI expression. PLoS ONE 7(11):e48530. https://doi.org/10.1371/journal.pone.0048530

Papachristou NI, Blair HC, Kypreos KE, Papachristou DJ (2017) High-density lipoprotein (HDL) metabolism and bone mass. J Endocrinol 233(2):R95–R107. https://doi.org/10.1530/JOE-16-0657

Petremand J, Puyal J, Chatton JY, Duprez J, Allagnat F, Frias M, James RW, Waeber G, Jonas JC, Widmann C (2012) HDLs protect pancreatic beta-cells against ER stress by restoring protein folding and trafficking. Diabetes 61(5):1100–1111. https://doi.org/10.2337/db11-1221

Pirillo A, Norata GD, Catapano AL (2013) High-density lipoprotein subfractions—what the clinicians need to know. Cardiology 124(2):116–125. https://doi.org/10.1159/000346463

Poorey VK, Thakur P (2016) Alteration of Lipid Profile in Patients with Head and Neck Malignancy. Indian J Otolaryngol Head Neck Surg: Off Publ Assoc Otolaryngol India 68(2):135–140. https://doi.org/10.1007/s12070-015-0829-4

Rader DJ, Alexander ET, Weibel GL, Billheimer J, Rothblat GH (2009) The role of reverse cholesterol transport in animals and humans and relationship to atherosclerosis. J Lipid Res 50(Suppl):S189–S194. https://doi.org/10.1194/jlr.R800088-JLR200

Rayner KJ, Sheedy FJ, Esau CC, Hussain FN, Temel RE, Parathath S, van Gils JM, Rayner AJ, Chang AN, Suarez Y, Fernandez-Hernando C, Fisher EA, Moore KJ (2011) Antagonism of miR-33 in mice promotes reverse cholesterol transport and regression of atherosclerosis. J Clin Investig 121(7):2921–2931. https://doi.org/10.1172/JCI57275

Rothblat GH, Phillips MC (2010) High-density lipoprotein heterogeneity and function in reverse cholesterol transport. Curr Opin Lipidol 21(3):229–238

Rubin EM, Krauss RM, Spangler EA, Verstuyft JG, Clift SM (1991) Inhibition of early atherogenesis in transgenic mice by human apolipoprotein AI. Nature 353(6341):265–267. https://doi.org/10.1038/353265a0

Ruscica M, Botta M, Ferri N, Giorgio E, Macchi C, Franceschini G, Magni P, Calabresi L, Gomaraschi M (2018) High density lipoproteins inhibit oxidative stress-induced prostate cancer cell proliferation. Sci Rep 8(1):2236. https://doi.org/10.1038/s41598-018-19568-8

Rutti S, Ehses JA, Sibler RA, Prazak R, Rohrer L, Georgopoulos S, Meier DT, Niclauss N, Berney T, Donath MY, von Eckardstein A (2009) Low- and high-density lipoproteins modulate function, apoptosis, and proliferation of primary human and murine pancreatic beta-cells. Endocrinology 150(10):4521–4530. https://doi.org/10.1210/en.2009-0252

Rye KA, Barter PJ (2004) Formation and metabolism of prebeta-migrating, lipid-poor apolipoprotein A-I. Arterioscler Thromb Vasc Biol 24(3):421–428. https://doi.org/10.1161/01.ATV.0000104029.74961.f5

Rye KA, Clay MA, Barter PJ (1999) Remodelling of high density lipoproteins by plasma factors. Atherosclerosis 145(2):227–238

Settasatian N, Duong M, Curtiss LK, Ehnholm C, Jauhiainen M, Huuskonen J, Rye KA (2001) The mechanism of the remodeling of high density lipoproteins by phospholipid transfer protein. J Biol Chem 276(29):26898–26905. https://doi.org/10.1074/jbc.M010708200

Staels B, Dallongeville J, Auwerx J, Schoonjans K, Leitersdorf E, Fruchart JC (1998) Mechanism of action of fibrates on lipid and lipoprotein metabolism. Circulation 98(19):2088–2093

Stenkula KG, Lindahl M, Petrlova J, Dalla-Riva J, Goransson O, Cushman SW, Krupinska E, Jones HA, Lagerstedt JO (2014) Single injections of apoA-I acutely improve in vivo glucose tolerance in insulin-resistant mice. Diabetologia 57(4):797–800. https://doi.org/10.1007/s00125-014-3162-7

Su F, Kozak KR, Imaizumi S, Gao F, Amneus MW, Grijàlva V, Ng C, Wagner A, Hough G, Farias-Eisner G, Anantharamaiah GM, Van Lenten BJ, Navab M, Fogelman AM, Reddy ST, Farias-Eisner R (2010) Apolipoprotein A-I (apoA-I) and apoA-I mimetic peptides inhibit tumor development in a mouse model of ovarian cancer. Proc Natl Acad Sci USA 107(46):19997–20002. https://doi.org/10.1073/pnas.1009010107

Tabet F, Remaley AT, Segaliny AI, Millet J, Yan L, Nakhla S, Barter PJ, Rye KA, Lambert G (2010) The 5A apolipoprotein A-I mimetic peptide displays antiinflammatory and antioxidant properties in vivo and in vitro. Arterioscler Thromb Vasc Biol 30(2):246–252. https://doi.org/10.1161/ATVBAHA.109.200196

Tardif JC, Gregoire J, L'Allier PL, Ibrahim R, Lesperance J, Heinonen TM, Kouz S, Berry C, Basser R, Lavoie MA, Guertin MC, Rodes-Cabau J, Effect of r HDLoA-S, Efficacy I (2007) Effects of reconstituted high-density lipoprotein infusions on coronary atherosclerosis: a randomized controlled trial. Jama 297(15):1675–1682. https://doi.org/10.1001/jama.297.15.jpc70004

Terasaka N, Yu S, Yvan-Charvet L, Wang N, Mzhavia N, Langlois R, Pagler T, Li R, Welch CL, Goldberg IJ, Tall AR (2008) ABCG1 and HDL protect against endothelial dysfunction in mice fed a high-cholesterol diet. J Clin Investig 118(11):3701–3713. https://doi.org/10.1172/JCI35470

Tso C, Martinic G, Fan WH, Rogers C, Rye KA, Barter PJ (2006) High density lipoproteins enhance progenitor-mediated endothelium repair in mice. Arterioscler Thromb Vasc Biol 26(5):1144–1149. https://doi.org/10.1161/01.ATV.0000216600.37436.cf

van der Hoorn JW, de Haan W, Berbee JF, Havekes LM, Jukema JW, Rensen PC, Princen HM (2008) Niacin increases HDL by reducing hepatic expression and plasma levels of cholesteryl ester transfer protein in APOE*3Leiden. CETP mice. Arterioscler Thromb Vasc Biol 28(11):2016–2022. https://doi.org/10.1161/atvbaha.108.171363

van der Vorst EP, Vanags LZ, Dunn LL, Prosser HC, Rye KA, Bursill CA (2013) High-density lipoproteins suppress chemokine expression and proliferation in human vascular smooth muscle cells. FASEB J: Off Publ Federation Am Soc Exp Biol 27(4):1413–1425. https://doi.org/10.1096/fj.12-212753

Vickers KC, Palmisano BT, Shoucri BM, Shamburek RD, Remaley AT (2011) MicroRNAs are transported in plasma and delivered to recipient cells by high-density lipoproteins. Nat Cell Biol 13(4):423–433. https://doi.org/10.1038/ncb2210

Voight BF, Peloso GM, Orho-Melander M, Frikke-Schmidt R, Barbalic M, Jensen MK, Hindy G, Holm H, Ding EL, Johnson T, Schunkert H, Samani NJ, Clarke R, Hopewell JC, Thompson JF, Li M, Thorleifsson G, Newton-Cheh C, Musunuru K, Pirruccello JP, Saleheen D, Chen L, Stewart

A, Schillert A, Thorsteinsdottir U, Thorgeirsson G, Anand S, Engert JC, Morgan T, Spertus J, Stoll M, Berger K, Martinelli N, Girelli D, McKeown PP, Patterson CC, Epstein SE, Devaney J, Burnett MS, Mooser V, Ripatti S, Surakka I, Nieminen MS, Sinisalo J, Lokki ML, Perola M, Havulinna A, de Faire U, Gigante B, Ingelsson E, Zeller T, Wild P, de Bakker PI, Klungel OH, Maitland-van der Zee AH, Peters BJ, de Boer A, Grobbee DE, Kamphuisen PW, Deneer VH, Elbers CC, Onland-Moret NC, Hofker MH, Wijmenga C, Verschuren WM, Boer JM, van der Schouw YT, Rasheed A, Frossard P, Demissie S, Willer C, Do R, Ordovas JM, Abecasis GR, Boehnke M, Mohlke KL, Daly MJ, Guiducci C, Burtt NP, Surti A, Gonzalez E, Purcell S, Gabriel S, Marrugat J, Peden J, Erdmann J, Diemert P, Willenborg C, Konig IR, Fischer M, Hengstenberg C, Ziegler A, Buysschaert I, Lambrechts D, Van de Werf F, Fox KA, El Mokhtari NE, Rubin D, Schrezenmeir J, Schreiber S, Schafer A, Danesh J, Blankenberg S, Roberts R, McPherson R, Watkins H, Hall AS, Overvad K, Rimm E, Boerwinkle E, Tybjaerg-Hansen A, Cupples LA, Reilly MP, Melander O, Mannucci PM, Ardissino D, Siscovick D, Elosua R, Stefansson K, O'Donnell CJ, Salomaa V, Rader DJ, Peltonen L, Schwartz SM, Altshuler D, Kathiresan S (2012) Plasma HDL cholesterol and risk of myocardial infarction: a mendelian randomisation study. Lancet 380(9841):572–580. https://doi.org/10.1016/S0140-6736(12)60312-2

von Eckardstein A, Nofer JR, Assmann G (2001) High density lipoproteins and arteriosclerosis. role of cholesterol efflux and reverse cholesterol transport. Arterioscler Thromb Vasc Biol 21(1):13–27

Weber C, Badimon L, Mach F, van der Vorst EPC (2017) Therapeutic strategies for atherosclerosis and atherothrombosis: past, present and future. Thromb Haemost 117(7):1258–1264. https://doi.org/10.1160/TH16-10-0814

Wong NKP, Nicholls SJ, Tan JTM, Bursill CA (2018) The role of high-density lipoproteins in diabetes and its vascular complications. Int J Mol Sci 19(6). https://doi.org/10.3390/ijms19061680

Wool GD, Cabana VG, Lukens J, Shaw PX, Binder CJ, Witztum JL, Reardon CA, Getz GS (2011) 4F Peptide reduces nascent atherosclerosis and induces natural antibody production in apolipoprotein E-null mice. FASEB J: Off publ Federation Am Soc Exp Biol 25(1):290–300. https://doi.org/10.1096/fj.10-165670

Yokoyama S (2006) ABCA1 and biogenesis of HDL. J Atheroscler Thromb 13(1):1–15

Yu BL, Wang SH, Peng DQ, Zhao SP (2010) HDL and immunomodulation: an emerging role of HDL against atherosclerosis. Immunol Cell Biol 88(3):285–290. https://doi.org/10.1038/icb.2009.112

Yvan-Charvet L, Pagler T, Gautier EL, Avagyan S, Siry RL, Han S, Welch CL, Wang N, Randolph GJ, Snoeck HW, Tall AR (2010a) ATP-binding cassette transporters and HDL suppress hematopoietic stem cell proliferation. Science 328(5986):1689–1693. https://doi.org/10.1126/science.1189731

Yvan-Charvet L, Wang N, Tall AR (2010b) Role of HDL, ABCA1, and ABCG1 transporters in cholesterol efflux and immune responses. Arterioscler Thromb Vasc Biol 30(2):139–143. https://doi.org/10.1161/ATVBAHA.108.179283

Zerrad-Saadi A, Therond P, Chantepie S, Couturier M, Rye KA, Chapman MJ, Kontush A (2009) HDL3-mediated inactivation of LDL-associated phospholipid hydroperoxides is determined by the redox status of apolipoprotein A-I and HDL particle surface lipid rigidity: relevance to inflammation and atherogenesis. Arterioscler Thromb Vasc Biol 29(12):2169–2175. https://doi.org/10.1161/ATVBAHA.109.194555

Zhao Y, Van Berkel TJ, Van Eck M (2010) Relative roles of various efflux pathways in net cholesterol efflux from macrophage foam cells in atherosclerotic lesions. Curr Opin Lipidol 21(5):441–453. https://doi.org/10.1097/MOL.0b013e32833dedaa

Chapter 17
Serum Amyloid A (SAA) Proteins

George H. Sack Jr.

Abstract As normal constituents of blood serum, the Serum Amyloid A (SAA) proteins are small (104 amino acids in humans) and remarkably well-conserved in mammalian evolution. They are synthesized prominently, but not exclusively, in the liver. Fragments of SAA can associate into insoluble fibrils (called "amyloid") characteristic of "secondary" amyloid disease in which they can interrupt normal physiology and lead to organ failure. SAA proteins comprise a family of molecules, two members of which (SAA1 and SAA2) are (along with C-reactive protein, CRP) the most prominent members of the acute phase response (APR) during which their serum levels rise dramatically after trauma, infection and other stimuli. Biologic function(s) of SAA are unresolved but features are consistent with a prominent role in primordial host defense (including the APR). SAA proteins are lipophilic and contribute to high density lipoproteins (HDL) and cholesterol transport. SAA proteins interact with specific receptors and have been implicated in tissue remodeling through metalloproteinases, local tissue changes in atherosclerosis, cancer metastasis, lung inflammation, maternal–fetal health and intestinal physiology. Molecular details of some of these are emerging.

Keywords Serum amyloid A · SAA · Acute phase reaction · Lipoproteins · Inflammation · Cytokine · Atherosclerosis · Metastasis · Arthritis · Amyloidosis · APR · HDL

Introduction

Serum amyloid A (SAA) proteins constitute a family of very closely-related, highly conserved serum molecules whose pathophysiology has been studied for nearly 70 years. As their name implies, molecules in this category originally were identified

G. H. Sack Jr. (✉)
Departments of Biological Chemistry and Medicine, The Johns Hopkins University School of Medicine, 725 N. Wolfe Street, Physiology 615, Baltimore, MD 21205, USA
e-mail: gsack@jhmi.edu

U. Hoeger and J. R. Harris (eds.), *Vertebrate and Invertebrate Respiratory Proteins, Lipoproteins and other Body Fluid Proteins*, Subcellular Biochemistry 94,
https://doi.org/10.1007/978-3-030-41769-7_17

as constituents of so-called "amyloid" deposits in both clinical and laboratory studies. "Amyloid" (from Greek, meaning "starch-like") is a well-recognized pathologic finding in which regions appearing amorphous by light microscopy stain characteristically with dyes like Congo Red or Thioflavin T (*see* Fig. 17.1). When viewed by electron microscopy, however, amyloid deposits are seen to be composed of fibrils (*see* Fig. 17.2) (Cohen and Calkins 1959). SAA proteins are related to so-called "reactive" or "secondary" amyloid—where deposits develop in response to chronic or recurrent infections and/or inflammation. Such amyloid deposits often were found in histopathologic studies in the preantibiotic era when chronic infections were particularly common. In "secondary" amyloid, the unique fibrillary proteins (usually ≈76 aa long) were called "Amyloid A" or AA (historically designating the *first* amyloid protein identified). AA proteins were later shown to be derived from partial proteolysis of a larger (104 aa) serum protein—hence the name of this latter species was proposed as "Serum Amyloid A" (SAA). (Thus meaning "the serum protein precursor of the first [hence, "A"] well-characterized protein ["AA"] found in amyloid deposits.) Subsequent analysis has identified other types of "amyloid" and with protein constituents related to their pathophysiologic context(s) (Sack 2009, 2019).

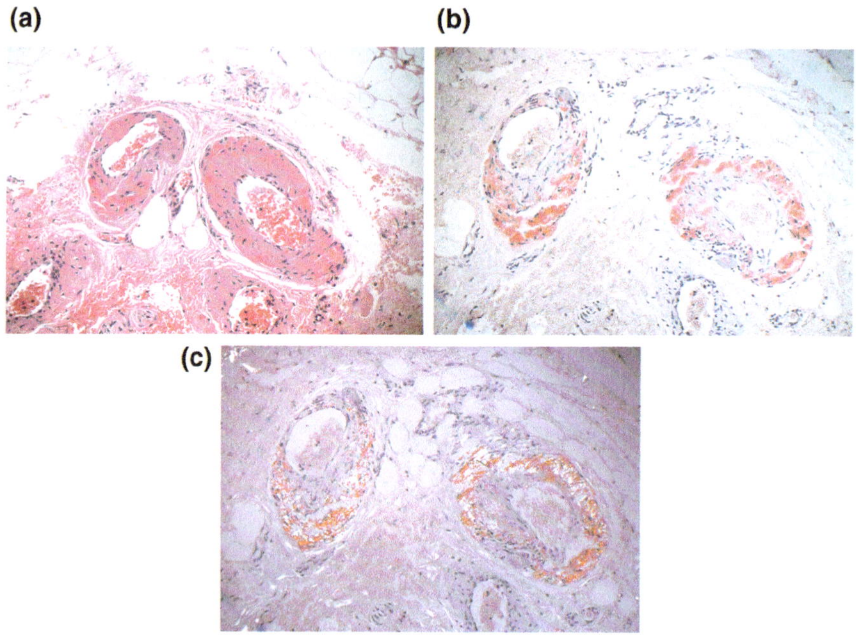

Fig. 17.1 Photomicrographs of serial sections of human tissue showing amyloid deposits in blood vessel walls. **a** Hematoxylin and eosin staining viewed with regular illumination. Note loss of normally visible details. **b** Congo red staining viewed with regular illumination. Note that the amyloid deposits appear pink but without details. **c** Congo red staining viewed with polarized light showing characteristic birefringence of the amyloid deposits (collection of author)

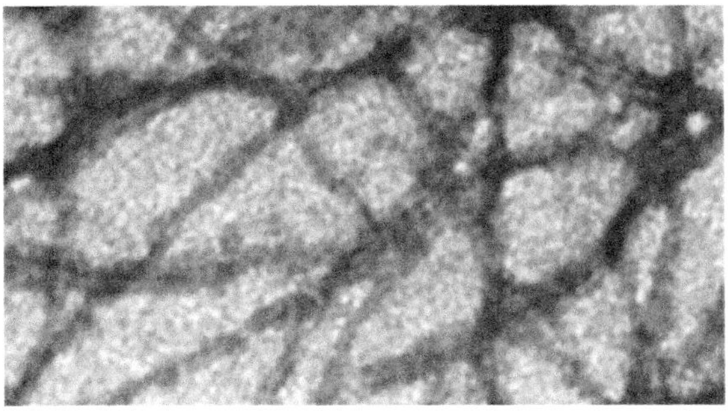

Fig. 17.2 Electron microscopic image of amyloid fibrils (Alzheimer amyloid-β(1–42)) (Lührs et al. 2005) (Copyright [2005] National Academy of Sciences, used by permission)

The unique sequence(s) of SAA proteins were identified relatively early. These have served as basic identifying data in many studies. Particularly notable is the remarkable aa sequence conservation throughout evolution in mammals as well as other vertebrates (for examples, *see* Fig. 17.3). Although human SAA proteins have 104 aa small length variations are found in other species (Fig. 17.3). Amino acid polymorphisms have been identified in large human population studies and a classification has been proposed (*see* Fig. 17.4 for a consensus) (Sipe 1999; Sun and Ye 2016; Benson 2019). However, most of these variants represent conservative aa changes.

Monomer and Fibril Structure

The 3-dimensional structure of the SAA monomer is notable for having 4 α-helices and a relatively unstructured C-terminal domain (Fig. 17.5) (Lu et al. 2014). As predicted from earlier circular dichroism (CD) studies (Meeker and Sack 1998), no β-sheet regions have been identified in the monomer. Because, as noted above, fibrils in tissue amyloid deposits generally do not contain isolated monomers, particular interest has been directed to determining structural features of the fibrils themselves. To date, many details of AA fibril structure remain unknown. However, studies of AA-derived as well as other types of amyloid fibrils have shown at least some common features. The most prominent fibril features are arrays of β-pleated sheets (usually antiparallel but in some cases parallel) stacked above and below one another and stabilized by extensive hydrogen bonding and van der Waals interactions. The general dimensions of amyloid fibrils include a 20 nm diameter with 10 Å between the stacked sheets. As predicted by earlier sequencing, SAA fibrils do not contain the entire 104 aa parent protein. Most fibrils comprise ≈76 N-terminal aa residues although there are reports of both shorter and longer forms. Reference to the monomer model (Fig. 17.5) shows an unstructured region at aa 69–73 that may make nearby regions more susceptible to proteolysis. Given the absence of β-sheet domains in the parent monomer structure (Meeker and Sack 1998; Lu et al. 2014), substantial reorganization of the primary chain must occur prior to fibril assembly.

(A)

R-S-F-F-S-F-L-G-E-A-F-D-G-A-R-D-M-W-R-A-Y-S-D-M-R-E-A-N-Y-I30

G-S-D-K-Y-F-H-A-R-G-N-Y-D-A-A-K-R-G-P-G-G-V-W-A-A-E-A-I-S-D60

A-R-E-N-I-Q-R-F-F-G-H-G-A-E-D-S-L-A-D-Q-A-A-N-E-W-G-R-S-G-K90

D-P-N-H-F-R-P-A-G-L-P-E-K-Y104

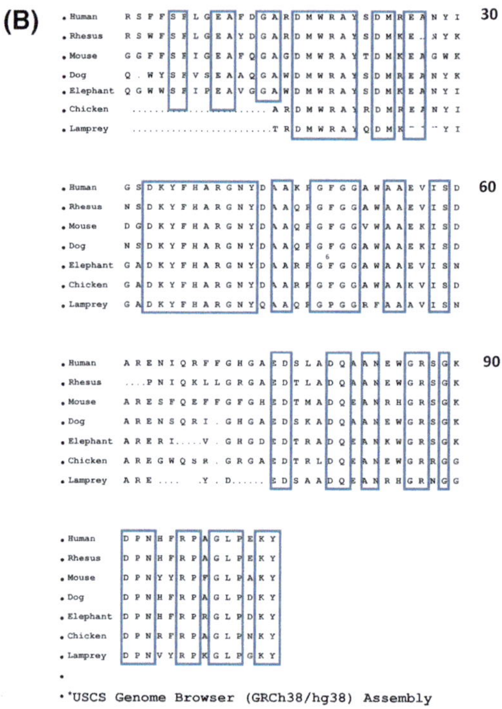

Fig. 17.3 **a** Consensus sequence for human SAA1. Variants recognized in Fig. 17.4. **b** Comparison of SAA sequences in multiple organisms. Preserved amino acids are enclosed. Note length variation(s) (USCS Genome Browser [GRCh38/hg38] Assembly)

Cytologic data indicate that at least initial AA amyloid fibril formation occurs intracellularly, in lysosomal compartments where low pH likely contributes to structural changes (Claus et al. 2017; Sack 2018). Once formed, small fibril "nuclei" likely interact with other N-terminal monomer fragments as "seeds" to extend the longitudinal dimensions of the fibrils into largely insoluble and relatively chemically insert macromolecular arrays. These become the fibrils seen by electron microscopy and are the site(s) of specific planar dye interaction(s) that are responsible for birefringence.

SAA1 :

```
R-S-F-F-S-F-L-G-E-A-F-D-G-A-R-D-M-W-R-A-Y-S-D-M-R-E-A-N-Y-I-G-S-D-K-Y-F-H-A-R-G40
N-Y-D-A-A-K-R-G-P-G-G-V-W-A-A-E-A-I-S-D-A-R-E-N-I-Q-R-F-F-G-H-G-A-E-D-S-L-A-D-Q80   (1.1)
                     A         V                            D                        (1.2)
                     A                                                               (1.3)
                     A         V    N                                                (1.4)
                     A         V                                                     (1.5)
A-A-N-E-W-G-R-S-G-K-D-P-N-H-F-R-P-A-G-L-P-E-K-Y104
```

SAA2:

```
R-S-F-F-S-F-L-G-E-A-F-D-G-A-R-D-M-W-R-A-Y-S-D-M-R-E-A-N-Y-I-G-S-D-K-Y-F-H-A-R-G40
N-Y-D-A-A-K-R-G-P-G-G-V-W-A-A-E-A-I-S-D-A-R-E-N-I-Q-R-F-F-G-H-G-A-E-D-S-L-A-D-Q80
                     A         V    N              L T                               (2.1)
                     A         V    N              L T R                             (2.2)

A-A-N-E-W-G-R-S-G-K-D-P-N-H-F-R-P-A-G-L-P-E-K-Y104
   K           R
   K           R
```

Fig. 17.4 Recognized variants in human SAA1 and SAA2. Underlined regions appear invariant (Sipe 1999; De Buck et al. 2016b; Benson et al. 2019)

Fig. 17.5 Three-dimensional structure of monomeric SAA1.1. Note 4 α-helices (1—aa 1–27; 2—aa 32–47; 3—aa 50–69; 4—aa 73–88). C-terminal tail—aa 89–104. (Lu et al. 2014) (Copyright [2014] National Academy of Sciences, used by permission)

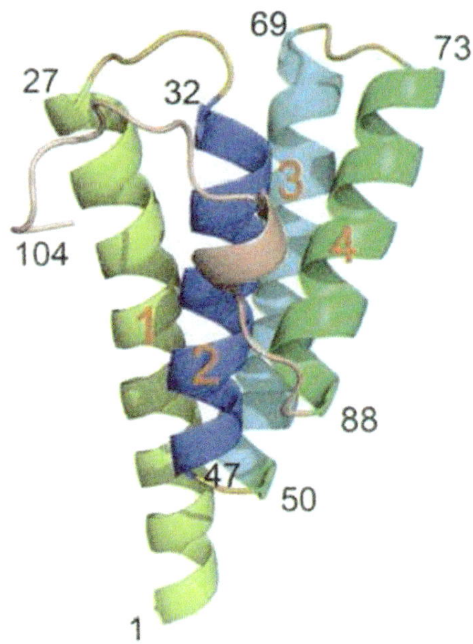

A characteristic feature of AA and other types of amyloid fibrils is very slow spontaneous dissociation and this, among other features, contributes to their cytopathology by interrupting cell–cell communication, nutrient and ion transport and visible organ microstructure. Fibril stability (and the macroscopic fibril deposits) reflects at least several factors. First, formation of individual fibrils involves considerable negative free energy ($-\Delta G$). Second, individual fibril microarchitecture with tightly packed polypeptide chains makes the structure(s) relatively inaccessible to proteases. Third, the fibrils themselves are generally packed together tightly. These features help explain the association of amyloid fibrils (generally of all types) with chronic, often progressive, organ dysfunction and disease and slow (if any) spontaneous resolution.

Molecular Biology

SAA-related proteins and genes constitute a closely-related family. In all organisms studied to date, genes for all members of the SAA "family" are clustered on a single chromosome. Human SAA genes are on chromosome 11p; those of the mouse are on murine chromosome 7. Figure 17.6 shows the human SAA gene family. A low level of aa polymorphism is recognized among family members (corresponding to the variations noted in Fig. 17.4), much of which has been cataloged. In mice there are 4 functional SAA genes and one pseudogene. These correspond to SAA1 and 2 (virtually identical APR proteins) and SAA3. The latter differs from SAA1 and 2 by aa changes in the N-terminal region. Human SAA1 and 2 genes similarly encode nearly identical proteins. The human SAA3 gene closely resembles its murine counterpart. However, although capable of being transcribed, the human SAA3 gene contains a single nucleotide insertion leading to an altered reading frame with early termination of transcription and likely loss of the corresponding mRNA through nonsense-mediated decay (Sack et al. 2018). No protein corresponding to SAA3 has been found in human serum. In mice, most anti SAA antibodies detect all species (SAA1, 2, and 3) but similar use of antibodies in humans reveals only SAA1 and 2. The SAA4 gene is constitutively transcribed in humans and mice and contains an insertion of 8 aa leading to a serum protein of 112 aa (De Buck et al. 2016b). Glycosylation may occur within the octapeptide and both glycosylated and unglycosylated forms have been detected in humans.

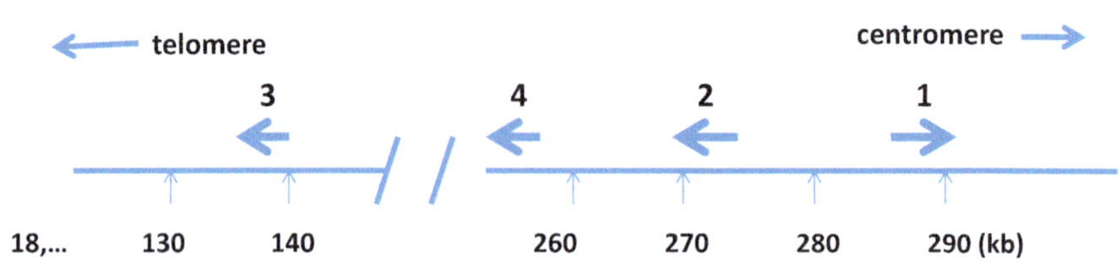

Fig. 17.6 Map of human SAA gene family on chromosome 11

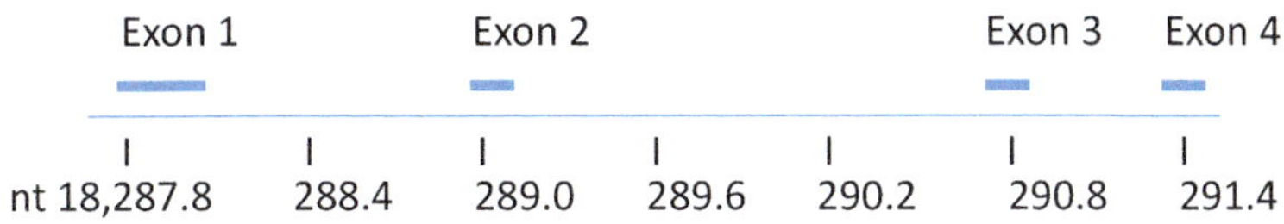

Fig. 17.7 Exon structure of human SAA1 gene. A common structure for all SAA genes

The remarkable evolutionary conservation of SAA protein sequences also is reflected in structural features of their corresponding genes (Fig. 17.6). SAA genes have been characterized in many organisms and all, to date, share a structure of 4 exons (Fig. 17.7). The aa of the mature protein(s) are encoded in exons 2–4. Consistent with SAA's status as a secreted or "serum" protein, the primary translated aa sequence also encodes an 18 aa N-terminal leader. The leader also is well-conserved and is removed from the primary translation product prior to secretion of the 104 aa species (or 112 aa species for SAA4). Consistent with features of some other proteins and their genes, exon/intron junctions correspond reasonably well to turns or unstructured loops in the mature, folded monomer. At least some SAA in the serum associates into hexamers, but these likely are in equilibrium with monomers, particularly in lipid environments (e.g. HDL).

Acute Phase Response

An important feature of SAA proteins, recognized early in their study, is their prominence among the serum constituents of the "acute phase response" (APR). This stereotyped process is well-characterized as a response to experimental as well as clinical conditions—endotoxin, trauma, infections, etc. (Kushner 1982). Among the serum protein changes during the APR are rapid elevations in the levels of SAA and C-reactive protein (CRP, *q.v.*); these can rise 1000-fold in 24–36 h following the initiating stimulus. These impressive changes as recognized in humans and experimental animals reflect new synthesis and secretion of both SAA and CRP largely, although not exclusively, by the liver. In the absence of persistent stimulus (i) the entire APR resolves, usually beginning 48–72 h after induction. Interestingly, prolonged inflammation or other stimuli can cause long-term elevation of SAA and CRP levels. Chronically, or repeatedly, elevated SAA levels set the stage for their cleavage and deposition as pathologic amyloid fibrils as noted above. This helps explain "secondary" (i.e. "AA") amyloid deposition in individuals with chronic or recurrent inflammation.

Note that the C-reactive protein (CRP) (*see* Chap. 20) also is a well-conserved serum protein and specifically binds to the C-polypeptide of *S. pneumoniae* (hence its name). CRP has historically been considered to be a contributor to primordial host defense, generally in the context of the APR.

SAA gene transcription responds to various APR "inducers." Some of these are well-characterized at the molecular level (including IL-1β, IL-6, TNFα and participants in MAPK activation including pERK1/2, pJNK, p38) while other less well-defined agents include sepsis, intraperitoneal casein, and lipopolysaccharides (LPS, bacterial endotoxin). At the onset of the APR levels of SAA mRNA can rise dramatically (in mouse liver SAA mRNA can comprise 5% of mRNA 24 h after APR induction) (Morrow et al. 1981). In most systems studied, elevated SAA and CRP serum levels resolve completely. Viewed phenomenologically this is a circuit with stereotyped "feedback" attenuation. The multiple pathways to APR induction involve different factors that interact with different cells in different organs in different ways, making generalizations difficult. An important, currently unresolved, question relates to details of how the APR process is resolved once the initiating stimulus is withdrawn.

An established protocol in mice—CLP (cecal ligation and puncture)—provides an APR model by producing acute polymicrobial sepsis (in which LPS is prominent). About 75% of normal mice survive at day 5 after CLP. By contrast, mice unable to synthesize characteristic APR SAA due to homozygosity for liver-specific deletion of gp130 (*gp130Δhepa*) showed ≈90% lethality using the same protocol. Treating these mice with exogenous SAA restored normal survival levels. In addition, using this model in normal mice treated with a monoclonal antibody against aa 33–43 region of SAA (helix 2 in model, *see* Fig. 17.5) also led to ≈90% lethality. These studies are all consistent with a critical "survival" function for SAA (and, possibly specifically the helix 2 region containing the epitope(s) recognized by the monoclonal antibody) (Linke et al. 2017).

Another CLP-APR study in mice transgenic for constitutive SAA production evaluated protection against lung injury (Cheng et al. 2018). Using intratracheal TNFα to induce the APR in the lung of the transgenic mice led to tissue damage and a prominent cytokine response that was unaffected by the constitutive presence of SAA. This is consistent with a protective effect of SAA against LPS that did not generalize to TNFα as a stimulus. Further study showed direct molecular interaction between SAA and LPS leading to macrophage uptake (via the SR-B1 receptor—*see* below) of the product and attenuation of serologic APR changes (reduced IL-6 and TNFα, and increased IL-10). The protective effect of SAA (and the cytokine responses) was largely blocked by treating the mice with an aa 32–47 polypeptide (also overlapping the helix 2 domain of SAA), consistent with competitive displacement of full length SAA from LPS and thus reducing macrophage uptake of LPS that would have attenuated the APR.

As further evidence for changes in other study systems, treating HeLa cells transgenic for receptor TLR-2 with SAA induced pERK1/2 and pJNK—a response different from treating with an aa 11–58 SAA polypeptide (overlapping helices 1 and 2) which increased pERK1/2 and p38 but not pJNK, indicating that presumably other domains of SAA could affect the response (Zhou et al. 2017). Attempts to assign specific function(s) to specific parts of the SAA monomer (*recall* Fig. 17.5) have been complicated by not knowing whether these exist as specific domains and/or can function independently in the parent protein. None has been identified as an independent species in vivo.

Thus, SAA1 (and, particularly aa 32–44/47 domain of helix 2) appears to specifically interact with LPS to attenuate lethality. Since SAA production is stimulated in macrophages (as well as liver) during the APR this adds a potential source for direct interaction with, at least, LPS species in a polymicrobial situation. SAA's response to the initiating APR stimulus could thus help to eliminate LPS as a stimulus and reduce the magnitude of the APR.

SAA "Receptors"

SAA interacts with specific cells and, possibly, macromolecules such as LPS. Recalling that CRP was defined (and named) by bacterial binding, it is notable that SAA binds to the outer membrane protein A (ompA) of *E. coli.*, acting as an opsonin and enhancing bacterial uptake by polymorphonuclear leukocytes. *E. coli* not producing ompA were unaffected (Hari-Dass et al. 2005; Shah et al. 2006). *E. coli* growth in culture can be reduced by SAA.

The notion of cellular "receptors" for SAA has proved complicated and, not surprisingly, studies of this problem have led to complicated answers. Association of SAA with various "receptors" has been studied using different systems. The array of reported interactions (*see* Table 17.1) implies that at least some of them likely are primary; others may be incidental. Nevertheless, considering them together illustrates possibilities for multiple physiologic roles for SAA. A particularly well-characterized example is the "scavenger receptor" SR-B1. As discussed above, this is likely involved with SAA/LPS uptake by macrophages. SR-B1 also is involved with cholesterol transport (*see* below). Toll-like receptors likely help mediate inflammatory processes. Others likely are involved both with advancing and, finally, resolving, inflammation.

Table 17.1 Receptors interacting with SAA[a]

Receptor	Function
SR-B1	cholesterol efflux
TLR2	Induce cytokines; NLRP3 inflammasome activation
TLR4	Activate NF-κB; AP-1 → NOS
FPRL1	Stimulates MMP-9
FPR2	Phagocyte migration, chemotaxis
RAGE	Activate NF-κB, cytokine induction
Bacteria	Opsonize Gram negative species

[a]*See also* Ye and Sun (2015), De Buck et al. (2016a), Sack (2018)

SAA Physiology

Many anecdotal reports have implicated SAA in (patho)physiology. Not all observations have been studied in rigorous detail but several important processes clearly involve SAA. Molecular details of SAA participation in some of these have been defined. Important overlaps between the categories indicate that a more inclusive view of SAA biology is needed. Prominent aspects of SAA biology will be considered below. Several features are notable: (1) no enzymatic activity has been identified for SAA (the protein is likely too small). (2) participation in APR and inflammation is compatible with a primordial signaling/communication/defense function (this would be consistent with its impressive evolutionary preservation as well as its clinical/laboratory association with chronic disorders). (3) lipophilicity of the SAA monomer (consistent with its exposed aa residues—*see* Fig. 17.5) makes it likely to be partitioned into lipid environments and also to associate with host and bacterial lipoproteins. (4) participation in aspect(s) of inflammation is likely (based both on the APR as well as clinical /laboratory association with chronic infections). (5) formation of the AA amyloid fibril itself likely represents a biologic "dead end" (since such deposits are long-lived and at least relatively inert).

It is likely important to distinguish acute, short-term activities of SAA from those associated with chronic, high-level, or localized longer-term effect(s). Interpreting different reports involves consideration of the source of SAA studied. Early concern about recombinant SAA produced in *E. coli* was that the material might contain endotoxin and that this could explain some results. This problem was recognized and, although contaminating endotoxin could largely be removed, Burgess et al. (2018) recently examined such preparations and found multiple bacterial proteins, at least some of which were lipoproteins. They emphasized that recombinant SAA proteins should be derived from eukaryotic cells to avoid this important contamination (this was not the case with many previous reports). De Buck (2016a) summarized data showing that SAA derived from blood serum or plasma lacks much of the proinflammatory effect of recombinants derived from bacteria. After nearly 70 years of study no single role can be assigned to SAA; participation in multiple pathways is likely. Unless otherwise noted below, most studies have been done using the nearly-identical SAA1/2 proteins.

Lipophilicity and Lipid Interactions

As already noted, SAA1/2 proteins are remarkably lipophilic and this has led to their characterization as apolipoproteins of high density lipoprotein (HDL). In serum, most SAA is associated with HDL and the concentration of free SAA monomer is quite low. Measuring SAA concentration in serum generally involves dissociating it from HDL particles. In addition to its being partitioned into the HDL fraction in the serum, SAA has been related to cholesterol transport and recycling. Many study systems have been used and despite some controversy [likely reflecting, at least in part, bacterial protein contamination of recombinant SAA used (*see* above)] several consistent features are recognized. As summarized elsewhere (Sack 2018): (1) SAA promotes an increase in cholesterol ester hydrolase (hence increased intracellular free cholesterol). (2) Macrophage acyl-CoA:cholesterol acyl transferase (ACAT) activity is reduced by SAA. (3) HDL containing SAA is targeted to the macrophage within which it can be loaded with free cholesterol for transport (usually back to the liver). (4) Multiple receptors including SR-B1 are involved in recognition and uptake of SAA. (5) Different SAA isoforms may affect this process (but all are not fully characterized).

Kisilevsky and coworkers proposed that SAA is central to mobilizing cholesterol from macrophages and other constituents of sites of tissue injury, trauma, breakdown, etc. (Kisilevsky and Manley 2012). Such injury or tissue damage could, for example, accompany APR induction during which SAA levels would be high as described above. According to this hypothesis, the APR and its role in cholesterol recycling through salvage and transport could be an evolutionarily favored process, helping explain SAA biology and conservation as important for survival.

By contrast to acute lipid mobilization changes due to injury, chronic inflammation also changes local tissue architecture and cellular as well as extracellular contents. Especially when prolonged, this changes endothelial transport leading to extravascular accumulation of monocytes and macrophages and an altered local tissue proteoglycan milieu. SAA is known to induce metalloproteinases (*see* below) and can extend local tissue damage and release more lipid. Macrophages can take up the released LDL and then persist as "foam cells." The entire process can become self-perpetuating with chronic reorganization of the site—as large-scale juxtavascular change(s) accumulate this can become recognized as atherosclerosis (King et al. 2011). Evidence connecting all parts of this process has been gathered from multiple systems. In addition to the proposed (and detected) involvement of SAA, many more studies have been devoted to CRP, possibly because it has been simpler to measure and large data collections are available. As an inflammatory marker, CRP has been used as a "proxy" or biomarker for the inflammation that is central to atherosclerosis. The CANTOS trial in humans provided convincing evidence that reducing chronic inflammation—and CRP levels (*e.g.* reducing IL-1β with the monoclonal antibody Canakinumab)—was associated with reduced vascular complications (Ridker et al. 2017). Presumably, SAA participation is similarly affected (although not yet studied in detail).

Tissue Remodeling

SAA's relation to extracellular metalloproteinase release and function was shown many years ago. Treating rabbit synovial fibroblasts with phorbolmyristate not only released collagenase but also small molecules including β_2-microglobulin and an "SAA-like" protein (later designated SAA3 and featuring a unique TFLK sequence near the N-terminus). Significantly, adding this SAA3 to otherwise untreated, quiescent cells also stimulated collagenase release—consistent with a paracrine circuit capable of prolonging tissue destruction and remodeling (Brinckerhoff et al. 1989). SAA transcripts also were found by in situ hybridization in synovial tissue from caprine retroviral arthritis (Sack and Zink 1992). More recent studies also detected SAA expression in arthritic synovium and related it to metalloproteinase induction (Connolly et al. 2010, 2012).

Cancer

Serum SAA levels have been studied as potential biomarkers for cancer and clinical associations have been described for many tumor types (Table 17.2). This has not uniformly led to utility as a biomarker, however, but chronic inflammation and cytokine participation has been recognized (Landskron et al. 2014). Of more mechanistic interest has been dissecting direct relationships. SAA production *within* the tumor tissue has been found in colorectal, ovarian, uterine and glioblastoma cancers (Moshkovskii 2012).

SAA also has been implicated in tumor growth. Melanoma cells with SAA expression correlated with finding immunosuppressive neutrophils within the tumor. SAA in humans and mice has been considered an effector for metastasis promotion by S100A4. In mice, cells surrounding pancreatic ductal adenocarcinoma were found to secrete IL-6. This cytokine led to SAA synthesis by the liver (via STAT3). After the IL-6/SAA stimulation, metastatic "niches" were formed within the liver that were followed by tumor metastases. This process did not occur in SAA knock-out mice (although metastasis to lung and other areas was unaffected) (Lee et al. 2019). Consistent with the observations relating SAA to metalloproteinases and paracrine effects (*see* above) this is consistent with a central role for SAA in local tissue remodeling that could affect metastasis (i.e. "niche") formation.

Table 17.2 Examples of human cancers associated with elevated SAA levels[a]

Astrocytoma	**Melanoma**
Breast	Nasopharynx
Colon	Ovary
Esophagus	Pancreas
Glioblastoma	Prostate
Kidney	Stomach
Lung	Uterus

[a]*See also* Knebel et al. (2017), Sack (2018)

Intestinal Immunology

Adhesion of segmented filamentous bacteria (sfb) to intestinal epithelial cells leads to specific, IL-17 secreting Th17 cell accumulation in the lamina propria. Within these cells SAA transcripts are the most upregulated species. Actin reorganization within the cells led to C/EBPδ expression which is proposed to interact with two DNase hypersensitive sites 3′ to the SAA1 gene (Ivanov et al. 2009). Further study of this system showed that ILC3 cells in the ileum secreted IL-22 after sfb adhesion and that this then led to SAA production via a STAT-independent mechanism (Atarashi et al. 2015). In another murine system, colonic bacteria (and possibly other bacterial products) led to SAA1/2 expression to modulate Th17 cell differentiation and cytokine production (Sano et al. 2015).

Lung Disease and Sarcoidosis

Exacerbations of symptoms in individuals with chronic obstructive pulmonary disease (COPD) showed better correlation with blood SAA levels than with those of CRP. SAA levels in bronchoalveolar lavage fluid correlated well with IL-8 and neutrophil elastase levels; both related to the pathophysiologic changes (Bozinovski et al. 2008, 2012). Detailed study of noncaseating granulomata, macrophages and multinucleated giant cells from sarcoidosis patients showed prominent SAA. Noting that TLR2 and RAGE receptor polymorphisms are related to disease progression in sarcoidosis, it has been proposed that SAA is a central participant in the characteristic "granulomatous inflammation" seen in these patients (Chen et al. 2010).

Maternal/Fetal Health

Pregnancy and Delivery

Measuring SAA levels during pregnancy has led to important associations. In mice, SAA levels rose with preterm delivery (Yang et al. 2009). In humans, SAA levels correlate with the severity of neonatal encephalopathy and mortality (Aly et al. 2011). Ibrahim et al. (2017) found that SAA levels provided an independent indicator of early pregnancy loss. Elevated cord blood levels of SAA, CRP and haptoglobin (all members of the APR) were associated with early onset neonatal sepsis in preterm infants (Mithal et al. 2017).

Mammary-Derived SAA

Bovine colostrum (early milk) contains an SAA protein with a characteristic TFLK sequence near its N-terminus (in mouse this would be classified as SAA3) (McDonald et al. 2001). Exposing intestinal epithelial cells to this protein led to prominent MUC3 mucin production as well as a corresponding decrease in adherence of enteropathogenic *E. coli* (Mack et al. 2003). Both LPS and prolactin can stimulate transcription of this "SAA3" protein in vitro using human breast epithelial cell cultures (Larson et al. 2005). This process is consistent with a pro-survival transmission of this "SAA3" molecule to the newborn by nursing and, particularly, by early exposure to colostrum, hence reducing the risk of infection for the neonate. In humans, as noted above, the SAA3 protein is not translated, likely due to nonsense-mediated mRNA decay; instead SAA1 is present in colostrum (Sack et al. 2018). Its protective effect(s) have not yet been confirmed.

Perspective

Despite many years of study, all biologic interactions and functions of SAA proteins remain unresolved. Initially discovered due to prominent participation in the acute phase response as well as its contribution to pathologic fibril formation in chronic inflammation, SAA has been implicated in many processes. The hypothesis that SAA proteins have an evolutionarily preserved role in survival appears correct; their widespread distribution across species as well as striking amino acid sequence conservation are consistent with this. The APR itself is recognized as a "primordial" defensive response, serving as a generic front-line for maintaining organismal integrity. With a perspective based on evolution, it is likely that SAA and the APR have served their survival functions well. Over the course of longer organismal lifespans, however, the pathophysiologic roles have had broader implications. As shown, a wide array of "receptors" (some established, others implicated) can serve multiple pathways. In at least some instances, high blood levels of SAA can simply be "biomarkers" that reflect underlying inflammation (infection, trauma, cancer, etc). (CRP can be viewed similarly.) Gathering reports, however, implicate SAA participation in multiple basic processes and disorders. For example, paracrine stimulation of metalloproteinase(s) can be central to arthritic joint destruction as well as remodeling niches for tumor metastasis. Stimulating mucin secretion in the neonatal gut may have been essential for survival, particularly in human evolution and animal communities. Lipid transfer between site(s) for recycling, storage or modification (e.g. atherosclerosis) is consistent with basic chemical features of the SAA monomer as well as its many serum interactions. As noted, there is some irony in having the name of this family of molecules derived from what likely is a "terminal" state of their biology—the AA fibril appears chemically inert, but biologically quite stable.

References

Aly H, Hamed Z, Mohsen L et al (2011) Serum amyloid A protein and hypoxic ischemic encephalopathy in the newborn. J Perinatol 31:263–268. https://doi.org/10.1038/jp.2010.130

Atarashi K, Tanoue T, Ando M et al (2015) Th17 cell induction by adhesion of microbes to intestinal epithelial cells. Cell 163:367–380

Benson MD, Buxbaum JN, Eisenberg DS et al (2019) Amyloid nomenclature 2018: recommendations by the International Society of Amyloidosis (ISA) nomenclature committee. Amyloid 5pp. https://doi.org/10.1080/13506129.2018.1549825

Bozinovski S, Hutchinson A, Thompson M et al (2008) Serum amyloid A is a biomarker of acute exacerbations of chronic obstructive pulmonary disease. Am J Respir Crit Care Med 177:269–278

Bozinovski S, Uddin M, Thompson M et al (2012) Serum amyloid A opposes lipoxin A_4 to mediate glucocorticoid refractory lung inflammation in chronic obstructive pulmonary disease. Proc Natl Acad Sci 109:935–940

Brinckerhoff CE, Mitchell TI, Karmilowicz MJ et al (1989) Autocrine induction of collagenase by serum amyloid A-like and B_2-microglobulin-like proteins. Science 243:655–657

Burgess EJ, Hoyt LR, Randall MJ et al (2018) Bacterial lipoproteins constitute the TLR-stimulating activity of serum amyloid A. J Immunol 201:2377–2384

Chen ES, Song Z, Willett MH et al (2010) Serum Amyloid A regulates granulomatous inflammation in sarcoidosis through toll-like receptor-2. Am J Respir Crit Care Med 181:360–373

Cheng N, Liang Y, Du X et al (2018) Serum amyloid A promotes LPS clearance and suppresses LPS-induced inflammation and tissue injury. EMBO Rep (e45517):14pp. https://doi.org/10.15252/embr.201745517

Claus S, Meinhardt K, Aumuller T et al (2017) Cellular mechanism of fibrila formation from serum amyloid A1 protein. EMBO Rep 18(8):1352–1366

Cohen AS, Calkins E (1959) Electron microscopic observations on a fibrous component in amyloid of diverse origins. Nature 183:1202–1203

Connolly M, Marelli A, Blades M et al (2010) Acute serum amyloid A induces migration, angiogenesis, and inflammation in synovial cells in vitro and in a human rheumatoid arthritis/SCID mouse chimera model. J Immunol 184:6427–6437

Connolly M, Mullan RH, McCormick J et al (2012) Acute=phase serum amyloid A regulates tumor necrosis factor alpha and matric turnover and preficts disease progression in pastients with inflammatory arthritis before and after biologic therapy. Arthritis Rheum 64(4):1035–1045

De Buck M, Gouwy M, Wang JM et al (2016a) The cytokine-serum amyloid A-Chemokine network. Cytokine Growth Factor Rev 30:55–69

De Buck M, Gouwy M, Wang JM et al (2016b) Structure and expression of different serum amyloid A (SAA) variants and their concentration-dependent functions during host insults. Curr Med Chem 23:1725–1755

Hari-Dass R, Shah C, Meyer DJ et al (2005) Serum amyloid A proteins binds to outer membrane protein A of gram-negative bacteria. J Biol Chem 280(19):18562–18567

Ibrahim M, Ramy AR, Abdelhamid A et al (2017) Maternal serum amyloid A level as a novel marker of primary unexplained recurrent early pregnancy loss. Int J Gynecol Obstet 136:298–303

Ivanov II, Atarashi K, Manel N et al (2009) Induction of intestinal Th17 cells by segmented filamentous bacteria. Cell 139:485–498

King VL, Thompson J, Tannock LR (2011) Serum amyloid A in atherosclerosis. Curr Opin Lipidol 22:302–307

Kisilevsky R, Manley PN (2012) Acute-phase serum amyloid A: perspectives on its physiological and pathological roles. Amyloid 19(1):5–14

Knebel FH, Uno M, Galatro TF, Bellé LP, Oba-Shinjo SM, Marie SKN, Campa A (2017) Serum amyloid A1 is upregulated in human glioblastoma. J Neuro-Oncol 132(3):383–391

Kushner I (1982) The phenomenon of the acute phase response. Ann N Y Acad Sci 39–48

Landskron G, De La Fuente M, Thuwajit P et al (2014) Chronic inflammation and cytokines in the tumor microenvironment. J Immunol Res 2014:149185. https://doi.org/10.1155/2014/149185

Larson MA, Weber A, Weber AT et al (2005) Differential expression and secretion of bovine serum amyloid A3 (SAA3) by mammary epithelial cells stimulated with prolactin or lipopolysaccharide. Vet Immunol Immunopathol 107:255–264
Lee JW, Stone ML, Porrett PM et al (2019) Hepatocytes direct the formation of a pro-metastatic niche in the liver. Nature 567:249–252
Linke RP, Meinel A, Chalcroft JP et al (2017) Serum amyloid A (SAA) treatment enhances the recovery of aggravated polymicrobial sepsis in mice, whereas blocking SAA's invariant peptide results in early death. Amyloid 24(S1):149–150
Lu J, Yu Y, Zhu I et al (2014) Structural mechanism of serum amyloid A-mediated inflammatory amyloidosis. Proc Natl Acad Sci 111(14):5189–5194
Luhrs T, Ritter C, Adrian M et al (2005) 3D structure of Alzheimer's amyloid-β(1–42) fibrils. Proc Natl Acad Sci USA 102(48):17341–17347
Mack DR, McDonald TL, Larson MA et al (2003) The conserved TFLK motif of mammary-associated Serum Amyloid A3 is responsible for Up-regulation of intestinal MUC3 mucin expression in vitro. Pediatr Res 53(1):137–142
McDonald TL, Larson MA, Mack DR et al (2001) Elevated extrahepatic expression and secretion of mammary-associated serum amyloid A3 (M-SAA3) into colostrum. Vet Immunol Immunopathol 83:203–211
Meeker AK, Sack GH Jr (1998) A fusion protein between serum amyloid A and staphylococcal nuclease—synthesis, purification, and structural studies. Proteins Struct Funct Genet 30:381–387
Mithal LB, Palac HL, Yogev R et al (2017) Cord blood acute phase reactants predict early onset neonatal sepsis in preterm infants. PLoS ONE 12(1):E0168677. https://doi.org/10.1371/journal.pone.0168677
Morrow JF, Stearman RS, Peltzman CG et al (1981) Induction of hepatic synthesis of serum amyloid A protein and actin. Proc Natl Acad Sci USA 78:4718–4722
Moshkovskii SA (2012) Why do cancer cells produce serum amyloid A acute-phase protein? Biochemistry (Moscow) 77(4):339–341
Ridker PM, Everett BM, Thuren T et al (2017) Antiinflammatory therapy with Canakinumab for atherosclerotic disease. N Engl J Med 377(12):1119–1131
Sack GH Jr (2009) Amyloidosis (Chap. 46). In: Stone JH (ed) A clinician's pearls and myths in rheumatology. Springer, Dordrecht, pp 461–466
Sack GH Jr (2018) Serum amyloid A - a review. Mol Med 24:46. https://doi.org/10.1186/s10020-10018-10047-10020
Sack GH Jr (2019) The pathophysiology of amyloid fibril formation. In: Amyloidosis. IntechOpen, London. https://doi.org/10.5772/intechopen.81965
Sack GH Jr, Zink MC (1992) Serum amyloid A (SAA) gene expression in synovial cells in retroviral arthritis. Am J Pathol 141:525–529
Sack GH Jr, Zachara N, Rosenblum N et al (2018) Serum amyloid A1 (SAA1) protein in human colostrum. FEBS OpenBio 8:435–441. https://doi.org/10.1002/2211-5463.12383
Sano T, Huang W, Hall JA et al (2015) An IL-23/IL-22 circuit regulates epithelial serum amyloid A to promote local effector Th17 responses. Cell 163:381–393
Shah C, Hari-Dass R, Raynes JG (2006) Serum amyloid A is an innate immune opsonin for Gram-negative bacteria. Blood 108:1751–1757
Sipe JD (1999) Revised nomenclature for serum Amyloid A (SAA). Nomenclature committee for the international society of amyloidosis. Part 1. Amyloid 6:67–70
Sun L, Ye RD (2016) Serum amyloid A1: structure, function and gene polymorphism. Gene 583:48–57
Yang Q, Whitin JC, Ling XB et al (2009) Plasma biomarkers in a mouse model of preterm labor. Pediatr Res 66(1):11–16
Ye RD, Sun L (2015) Emerging functions of serum amyloid A in inflammation. J Leukoc Biol 98(6):923–929
Zhou H, Chen M, Zhang G et al (2017) Suppression of lipopolysaccharide-induced inflammatory response by fragments from serum amyloid A. J Immunol 199:1105–1112

Chapter 18
Physiological Roles of the von Willebrand Factor-Factor VIII Interaction

Klytaimnistra Kiouptsi and Christoph Reinhardt

Abstract Von Willebrand factor (VWF) and coagulation factor VIII (FVIII) circulate as a complex in plasma and have a major role in the hemostatic system. VWF has a dual role in hemostasis. It promotes platelet adhesion by anchoring the platelets to the subendothelial matrix of damaged vessels and it protects FVIII from proteolytic degradation. Moreover, VWF is an acute phase protein that has multiple roles in vascular inflammation and is massively secreted from Weibel-Palade bodies upon endothelial cell activation. Activated FVIII on the other hand, together with coagulation factor IX forms the tenase complex, an essential feature of the propagation phase of coagulation on the surface of activated platelets. VWF deficiency, either quantitative or qualitative, results in von Willebrand disease (VWD), the most common bleeding disorder. The deficiency of FVIII is responsible for Hemophilia A, an X-linked bleeding disorder. Here, we provide an overview on the role of the VWF-FVIII interaction in vascular physiology.

Keywords von Willebrand factor (VWF) · Coagulation factor VIII (FVIII) · Platelet · Acute phase protein · Hemophilia · Hemostasis

Introduction

Under physiological conditions, blood circulates in a closed system of vessels. When the vessels are disrupted by trauma, a cascade of events occurs, namely hemostasis, that lead to the repair of the vessel, preventing blood loss into the surrounding tissues.

K. Kiouptsi · C. Reinhardt (✉)
Center for Thrombosis and Hemostasis (CTH), University Medical Center Mainz, Langenbeckstrasse 1, Building 708, 55131 Mainz, Germany
e-mail: Christoph.Reinhardt@unimedizin-mainz.de

K. Kiouptsi
e-mail: Klytaimnistra.Kiouptsi@unimedizin-mainz.de

C. Reinhardt
German Center for Cardiovascular Research (DZHK), Partner Site RheinMain, Mainz, Germany

U. Hoeger and J. R. Harris (eds.), *Vertebrate and Invertebrate Respiratory Proteins, Lipoproteins and other Body Fluid Proteins*, Subcellular Biochemistry 94,
https://doi.org/10.1007/978-3-030-41769-7_18

Platelets are key players in this process, forming the primary hemostatic plug. Proteases of the coagulation cascade then are activated and fibrin polymers are generated to strengthen the primary hemostatic plug, forming a stable, rigid thrombus.

The cell types involved in coagulation vary widely in invertebrates. Many marine invertebrates have a single cell type circulating in the blood, called the hemocyte. Hemocytes are responsible for defense mechanisms and hemostasis (Arneth 2019). Insects on the other hand contain coagulocytes that ensure wound healing (Brehelin et al. 1978). The first cells that evolved to be specifically used in hemostasis were the thrombocytes (Ratnoff 1987). Thrombocytes are nucleated spindle-shaped cells found in fish, avian, amphibians and reptiles and have many similarities to the mammalian platelets (Grant 2015). Platelets are anucleated discoid cells, derived from megakaryocytes, which contain many receptors on their extracellular membrane that after ligand binding, lead to cell activation and aggregation.

The receptors that ensure hemostatic platelet function include integrins, leucine-rich repeat family receptors, seven-transmembrane receptors, receptors of the immunoglobulin superfamily, c-type lectin family receptors, tetraspanins, and many others (Rivera et al. 2009). The primary receptor on the platelet surface that mediates initial binding to subendothelial matrix proteins, exposed in case of vascular injury, is the glycoprotein Ibα (GPIbα). Interestingly, GPIbα can bind to several ligands, including von Willebrand factor (VWF), thrombospondin (TSP), collagen, P-selectin, αMβ2 (Mac-1), thrombin, FXI, kininogen, FXII, and FVIIa (Andrews and Berndt 2013). During hemostasis and thrombosis, two major receptors on the platelet surface, GPIbα and GPIIbIIIa (integrin $\alpha_{IIb}\beta_3$), mediate the deposition of platelets to the injured vessel wall through the interaction with VWF multimers.

VWF multimers vary in size, ranging from 5×10^2 to 5×10^4 kDa. They are composed of monomeric 250 kDa VWF glycoproteins that are trafficking via the secretory pathway of endothelial cells and are present in plasma and subendothelial matrix. VWF is widely known for its vital role in hemostasis, mediating platelet adhesion and aggregation at the site of vascular injury. VWF multimers can be secreted by either endothelial cell Weibel-Palade bodies that are 0.1–0.2 μm wide and up to 4 μm long (Wagner 1990) or by platelet α-granules (Giddings et al. 1982), but in contrast to the generation of α-granules in platelets, the formation of Weibel-Palade bodies in endothelial cells critically depends on VWF (Denis et al. 1998; Haberichter et al. 2005). In response to inflammatory stimuli, endothelial cells release the content of their Weibel-Palade bodies, which increases the circulatory concentration of VWF, making it an acute phase response protein (Hollestelle et al. 2006). In the setting of arterial thrombosis, the mechanical properties of VWF and the affinity of its binding partners are critical, especially in stenosed arteries, where the high shear rates would otherwise prevent platelet capture and activation (Ruggeri et al. 2006). Interestingly, VWF plasma levels increase with age (Coppola et al. 2003; Albanez et al. 2016) and high levels of VWF are associated with an increased risk for cardiovascular and cerebrovascular disease (Frankel et al. 2008; Wieberdink et al. 2010). Apart from its central role in thrombosis and hemostasis, VWF serves a myriad of other biological functions. Amongst others, VWF was revealed to participate in angiogenesis (Starke et al. 2011; Randi et al. 2018), smooth muscle cell proliferation (Ishihara et al.

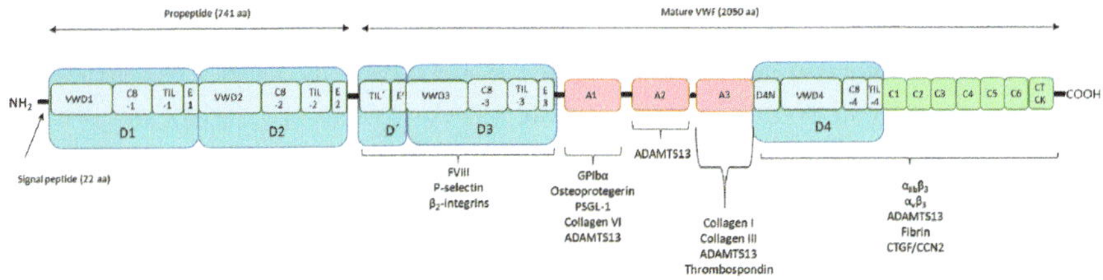

Fig. 18.1 VWF domain architecture according to Zhou et al. (2012) assigned to the interaction with various other plasma proteins and receptors. VWF consists of a 22 amino acid (aa) signal peptide, a 741 aa propeptide and the mature processed molecule consisting of 2050 aa. The D domains are shown in blue, the A domains in pink and the C domains in green

2019), tumor cell metastasis (O'Sullivan et al. 2018), and instructs proper function of cell-based immunity (Chen and Chung 2018; Urisono et al. 2018).

In addition to the various adhesive cell-cell interactions ensured by the domain architecture of VWF (Ruggeri and Mendolicchio 2015) (Fig. 18.1), the interaction with VWF stabilizes coagulation factor VIII (FVIII) in the circulation, the precursor of FVIIIa, constituting the cofactor in the coagulation pathway that together with FIXa forms the Xase complex on the surface of activated platelet membranes (Muntean and Leschnik 1989). This cofactor-protease complex localizes FXa formation to the site of platelet activation and it promotes quantitative thrombin generation and the resulting thrombus stabilization by fibrin polymers. Patients with functional or quantitative deficiencies of VWF suffer from a bleeding phenotype, show impaired thrombus formation and reduced FVIII levels, a deficiency known as von Willebrand factor disease (Leebeek and Eikenboom 2016).

Here, we explain the functional roles of VWF and how the molecular interaction of VWF with FVIII impacts on vascular physiology in health and disease.

Synthesis, Processing and Clearance of Von Willebrand Factor

Due to its covalently linked multimers, VWF is considered the largest blood-borne vertebrate protein (Wagner 1990). It is predominantly expressed in capillary endothelial cells (Pan et al. 2016), but it is also expressed in megakaryocytes and stored in platelet α-granules. Platelet-derived VWF is enriched in ultra-large VWF multimers (Ruggeri and Zimmerman 1980) and due to its glycosylation pattern it does not undergo significant proteolysis (Fernandez et al. 1982; McGrath et al. 2013). In contrast to endothelial VWF, platelet-derived VWF is not essential for hemostasis in mice, but plays an important role in stroke pathology (Dhanesha et al. 2016; Verhenne et al. 2015).

The nucleic acid sequence of human VWF and human pre-pro-VWF was described in 1985 (Ginsburg et al. 1985; Lynch et al. 1985; Sadler et al. 1985; Verweij et al. 1985; Bonthron et al. 1986) and the amino acid sequence of this large protein was determined by Titani and colleagues in 1986 (Titani et al. 1986). The pro-VWF molecule contains 17 N-linked carbohydrate structures that are highly variable, of which 4 are located in the pro-peptide sequence. The cyanogen bromide cleavage sequencing analysis confirmed N-glycosylation at Asn94, Asn384, Asn468, Asn752, Asn811, Asn1460, Asn1527, Asn1594, Asn1637, Asn1783, Asn1822 and Asn2027. O-glycosylations were detected at Thr485, Thr492, Thr493, Thr705, Thr714, Ser723, Thr724, Thr916, and Thr1535. Of note, these glycan structures are capped mostly by two sialic acid residues (Canis et al. 2010, 2012). This is relevant since the clearance of VWF is partially achieved through binding to the asialoglycoprotein receptor (Ashwell-Morell receptor) that is expressed by hepatocytes (Grewal et al. 2008; Lenting et al. 2015). In addition, 10 monosulfated N-glycans were identified in human plasma-derived VWF (Canis et al. 2012). Interestingly, approximately 13% of the N-glycans and 1% of the O-glycans harbour blood group determinants on the mature endothelial VWF subunit (Matsui et al. 1992). While N-glycosylation takes place in the ER, the addition of O-glycan moieties takes place in the Golgi apparatus (Brehm 2017). Overall, VWF is highly glycosylated. In additon, an interesting feature in its primary structure is the high abundance of Pro and Cys residues.

The domain structure of the human VWF is D1-D2-D'-D3-A1-A2-A3-D4-C1-C2-C3-C4-C5-C6-CK, with the D1-D2 domains representing a propeptide and the residual peptide chain the mature VWF (Zhou et al. 2012) (Fig. 18.1). VWF forms three intermolecular disulfide bonds resulting in a reduction-resistant dimeric structure that is very stable (Zhou and Springer 2014). This dimerization step occurs through C-terminal cysteine knot domains (CK) in the endoplasmatic reticulum (ER) (tail-to-tail dimerization) (Marti et al. 1987). It was revealed that the formation of VWF dimers requires protein disulfide isomerase A1 function (Lippok et al. 2016). In the Golgi apparatus, D3 domain-mediated interdimeric cross-linking and disulfide formation between D3 domains, which is most likely catalyzed by the VWF propeptide, link VWF pro-dimers to multimers, consisting of 2–60 subunits (head-to-head multimerization) (Mayadas and Wagner 1989; Wagner 1990). By furin cleavage, the VWF propeptide is removed when reaching the Golgi apparatus (Springer 2011).

In the trans-Golgi network, the VWF multimers are packaged into immature Weibel-Palade bodies, that are subsequently linked to the actin cytoskeleton via a protein complex that consists of Rab27a, Rab27a-interaction proteinand myosin-Va (Rojo Pulido et al. 2011). These storage organelles have a rod-shaped or cigar-like structure and contain many other proteins, e.g. P-selectin, interleukin-8, angiopoietin-2 and coagulation factor FVIII. Weibel-Palade bodies release VWF continuously (Giblin et al. 2008), leading to random fusion of these storage organelles and release of their cargo to the subendothelium and into the blood (Romani de Wit et al. 2003). However, under challenge conditions, when the vascular endothelium is stimulated, these storage organelles release massive amounts of VWF, thus supporting platelet adhesion and aggregate formation (Table 18.1).

Upon secretion, the multimeric size of VWF is tightly regulated by "a disintegrin and metalloproteinase with a thrombospondin type 1 motif, member 13" (ADAMTS13), which cleaves multimeric VWF and thus reduces its pro-adhesive

Table 18.1 Platelet receptor GPIb and $\alpha_{IIb}\beta_3$ ligands

Receptor	Ligand
GPIbα	VWF (Fujimura et al. 1987) TSP (Jurk et al. 2003) $\alpha_M\beta_2$ (Simon et al. 2000) P-selectin (Romo et al. 1999) FVIIa (Weeterings et al. 2008) FXI (Baglia et al. 2002) FXII (Bradford et al. 2000) Kininogen (Joseph et al. 1999)
$\alpha_{IIb}\beta_3$	Fibrinogen (Bennett and Vilaire 1979) Fibronectin (Weisel et al. 1992) Vitronectin (Mohri and Ohkubo 1991) Thrombospondin 1 (Kowalska and Tuszynski 1993) VWF (Weisel et al. 1992)

function, having major implications for thromboembolic diseases (Denorme et al. 2017). If the activity of this metalloprotease is impaired, this can result in the rare thrombotic microangiopathy thrombotic thrombocytopenic purpura (Moschcowitz syndrome) (Furlan and Lammle 1999; Bohm et al. 2005).

Interestingly, clearance of VWF occurs in similar efficacy in the liver and the spleen (van Schooten et al. 2008; Lenting et al. 2015). Histologic analyses revealed that VWF and FVIII are co-localized in macrophages in mouse spleen and liver sections, which was confirmed by in vitro analyses (van Schooten et al. 2008). VWF is bound and transported via the entocytotic pathway into macrophages (Casari et al. 2013a; Castro-Nunez et al. 2012). So far, a number of receptors have been identified, involved in the clearance of VWF (Casari et al. 2013b). Among these receptors are the low-density lipoprotein-related receptor protein-1 (LRP1), sialic acid binding Ig like lectin 5 (Siglec-5), C-type lectin domain family 4 member M (CLEC4M), and asialoglycoprotein receptor (Ashwell-Morell receptor) (Lenting et al. 1999; Rastegarlari et al. 2012; Pegon et al. 2012; Ellies et al. 2002). However, since VWF is strongly sialylated, the asialoglycoprotein receptor is thought to play a minor role in clearance.

In summary, the degradation of VWF multimers and VWF clearance, but also the interaction with its binding partners and receptors, determines the prothrombotic function of VWF.

VWF in Platelet Activation, Adhesion and Inflammation

The Von Willebrand Factor-Glycoprotein Ibα Interaction

GPIbα is part of the leucine-rich repeat receptor GPIb-IX-V complex and is a well conserved platelet specific receptor. VWF release from endothelial cells into the circulation in form of ultra large VWF (ULVWF) multimers, which are prothrombotic and bind spontaneously to the GPIbα, leads to platelet aggregation (Federici et al. 1989). The binding of the A1 domain of VWF to GPIbα requires a conformational change of free floating plasmatic VWF by high shear rates (Reininger et al. 2006; Schneider et al. 2007), whereas the elongation of VWF to the vessel wall occurs at lower shear rates (Pappelbaum et al. 2013). ADAMTS-13 regulates VWF multimer size by cleavage in the A2 domain of VWF, resulting in smaller and less adhesive multimers that require high shear forces to bind to the N-terminal domain of GPIbα (Dong et al. 2002). ADAMTS 13 deficiency can be either congenital, caused by a defect on the *ADAMTS13* gene, resulting in decreased or absent enzyme activity (Levy et al. 2001), or acquired, due to autoantibodies that inhibit the activity or increase the clearance of the molecule (Tsai and Lian 1998). The ULVWF multimers, which bind, on the luminal side, to the endothelial integrin αvβ3 and are anchored to the endothelial glycocalyx and immobilized on collagen fibres of the subendothelium (Huang et al. 2009; Kalagara et al. 2018), are crucial for the initial capture of the flowing platelets (Federici et al. 1989). Consequently, shear stress regulates the initiation of thrombus formation, by modulating the VWF multimer binding to the GPIb-IX-V receptor complex (Dong et al. 2002). This binding is necessary to decrease the platelet velocity and makes it possible for other receptors, like GPVI, to bind to the subendothelial matrix proteins (Baker et al. 2004; Savage et al. 1996).

The signaling, which results from the binding of platelet agonists to their receptors, leads to conformational changes of the integrin $\alpha_{IIb}\beta_3$ (GPIIbIIIa) that enhance its affinity for soluble ligands (inside out signaling) (Bledzka et al. 2013). Vascular injury leads to collagen and VWF exposure from the subendothelium that engage the platelets through the GPVI (Gibbins et al. 1997) and the GPIbα (Savage et al. 1998) receptor. The inside out signaling which is initiated by this binding results in clustering of the GPVI receptor (Poulter et al. 2017) and activates a cascade of events leading to the phosphorylation of intracellular proteins provoking the unclasping of the cytoplasmic tails of the integrin $\alpha_{IIb}\beta_3$. Thus, integrin αIIbβ3 changes its conformation from a closed to an open-active form (Bledzka et al. 2013). This unclasping reveals the extracellular binding site of the integrin for soluble ligands, like fibrinogen, which in turns acts like a bridge connecting adjacent platelets through their $\alpha_{IIb}\beta_3$ integrins (Floyd and Ferro 2012). Extracellular ligand binding to the extended conformation of $\alpha_{IIb}\beta_3$ in turn, triggers the outside in signaling that promotes the binding of cytoskeletal and signaling proteins, which result in actin polymerization, cytoskeletal reorganization and force transmission, all necessary for platelet spreading, thrombus stabilization and retraction (Durrant et al. 2017).

Although platelet capture by VWF at high shear forces is a rapid process, it is rapidly reversible and insufficient for stable adhesion (Savage et al. 1996). The dynamics of GPIbα/VWF-dependent platelet adhesion at the cellular level have been studied using extracellular matrix, collagen-coated surfaces in flow chambers, and intravital microscopy in live mice or rats, at low physiological (<600 s^{-1}) (large veins) (Sakariassen et al. 2015), high physiological (600–900 s^{-1}) (venules and capillaries) and pathological (>1500 s^{-1}) (atherosclerosis) shear rates, with the latter often leading to thrombotic events (Savage et al. 1996; Goto et al. 1998).

The von Willebrand Factor-Integrin $\alpha_{IIb}\beta_3$ Interaction

Another function of VWF in hemostasis is mediated by the binding of VWF to the most abundantly expressed platelet-specific receptor, the integrin $\alpha_{IIb}\beta_3$, through the VWF C1 domain, which contains the RGD motif (Alevriadou et al. 1993; Fressinaud et al. 1990). After platelet activation by different agonists, resulting in inside-out signaling and integrin $\alpha_{IIb}\beta_3$ activation, the platelets start to form aggregates mediated by molecules that serve as bridges between activated platelets. The molecules mediating this cell-cell interaction are fibrinogen and VWF. Platelets bind preferentially to VWF under conditions of high shear (Savage et al. 1996). On the other hand, integrin $\alpha_{IIb}\beta_3$ mediates platelet adhesion under static conditions (Sivaraman and Latour 2011). Although the binding of fibrinogen to integrin $\alpha_{IIb}\beta_3$ is the predominant interaction, studies with VWF mutants, inhibiting binding of VWF to integrin $\alpha_{IIb}\beta_3$, showed delayed thrombus growth and increased vessel occlusion times in a ferric-chloride induced injury model (Marx et al. 2008). In the absence of fibrinogen, binding of VWF to integrin $\alpha_{IIb}\beta_3$ supports platelet aggregation in a multimeric size dependent manner, similarly to GPIbα (Federici et al. 1989; De Marco et al. 1986; Trabold et al. 2019). Although the GPIbα is the main receptor for VWF, VWF has a higher dissociation constant for its interaction with GPIbα than for integrin $\alpha_{IIb}\beta_3$ binding, meaning that it binds tighter to the $\alpha_{IIb}\beta_3$ integrin receptor (Federici et al. 1989).

Willebrand Factor as an Acute Phase Protein and Its Involvement in Vascular Inflammation

Apart from its well-studied role in hemostasis, VWF is also an inflammatory marker, due to its release from endothelial cells upon inflammatory stimuli. Being stored in the Weibel-Palade bodies together with P-selectin, VWF is massively secreted upon activation of endothelial cells by inflammatory stimuli and therefore is considered an acute phase protein during inflammatory processes (Pottinger et al. 1989).

The use of VWD animal models highlighted the importance of the molecule since they proved that VWF is a key molecule involved in Weibel-Palade body biogenesis and size determination (Ferraro et al. 2014). Even though these organelles can show selectivity to their cargo release according to the stimulus (Babich et al. 2008), meaning that Weibel-Palade body exocytosis is not absolutely linked to VWF, the VWF-deficient mice were unable to form Weibel-Palade bodies. As a consequence, they showed defective P-selectin expression and decreased leukocyte rolling and recruitment in models of acute inflammation (Denis et al. 1998). However, atherosclerosis models applied in these animals showed P-selectin expression even in the absence of VWF, implying that there was no storage of the protein mediating synthesis and secretion (Methia et al. 2001). Nonetheless, even in this inflammatory model, there was a considerable reduction in leukocyte recruitment (Methia et al. 2001), an effect that was also observed in studies using anti-VWF antibodies (Hillgruber et al. 2014). Similarly, VWF-deficiency or the use of anti-VWF antibodies led to a decreased vascular permeability in an intracerebral hemorrhage model (Zhu et al. 2016), although it seems that the extent of how VWF mediates vascular permeability is likely to be dependent on the experimental model applied (Suo et al. 2014).

Efforts have been made in humans to use VWF levels as a biomarker for the prognosis for disease severity (Paulus et al. 2011; Reinhart et al. 2002). However, VWF alone is not sufficient as an inflammatory marker. The VWF propeptide (VWFpp)/VWF ratio should be quantified as well in order to distinguish between acute inflammatory responses and vascular dysfunction (Borchiellini et al. 1996; Vischer et al. 1997; van Mourik et al. 1999). Although both molecules are secreted in equal concentrations from endothelial cells, the VWFpp half-life is much shorter, making a low VWFpp/VWF ratio a marker for chronic inflammation (van Mourik et al. 1999). Another way to correlate VWF levels to inflammation was approached by measuring the active VWF, meaning the VWF which is in a GPIb-binding conformation and which is only present in pathological situations (Hyseni et al. 2014). This research was the first effort to use VWF levels as a prediction for mortality (Hyseni et al. 2014; Groot et al. 2007).

Another aspect of VWF's contribution to inflammation is through its ability to bind to neutrophil extracellular traps (NETs). NET formation is an inflammatory response (NETosis) where neutrophils and DNA form extracellular fibers to capture invading pathogens (Brinkmann V Science 2004). Although it has been known for years that the VWF A1 domain contains an interactive site for histones (Ward et al. 1997), VWF binding to NETs was confirmed much later, in a study showing that VWF-NET interaction prevents VWF degradation by ADAMTS-13 (Grassle et al. 2014). VWF, with its large size, provides binding sites for multiple binding partners and hence mediates many inflammatory pathophysiological processes.

Synthesis, Processing and Clearance of Anti-haemophilic Coagulation Factor VIII

Mature FVIII (anti-haemophilic factor) consists of a heavy (domains A1-A2-B) and a light chain (domains A3-C1-C2), circulating as a non-covalently linked heterodimeric glycoprotein, bound to VWF. The quantitative or qualitative deficiency of FVIII is known as hemophilia A, a rare hereditary bleeding disorder, which occurs at 1 in 5,000 individuals. FVIII together with FIX forms the tenase complex, amplifying the production of FX, leading to thrombin generation, a major platelet agonist with pleiotropic functions. FVIII, isolated by Legaz et al. (1973) and cloned by Gitschier et al. (1984), is only present in a selected subset of endothelial cells (Rondaij et al. 2006; Jacquemin et al. 2006). The human F8 gene comprises 26 exons that are transcribed into a single polypeptide chain. Human FVIII consists of 2332 amino acid residues that form 6 domains (A1-A2-B-A3-C1-C2) (Thompson 2003) (Fig. 18.2). The 19 amino acid signal peptide is removed in the ER, where Asn-linked N-glycosylation occurs. The three A domains of FVIII show homology with FV and ceruloplasmin and each bind a single copper ion (Pemberton et al. 1997; Tagliavacca et al. 1997). The sequence residues 558-565 in a loop region in the A2 domain are likely to represent the major interaction site between FVIIIa cofactor and the FIXa protease (Bajaj et al. 2001), but a second binding site for FIXa exists in the sequence Glu1811-Lys1818 near the amino terminus of the A3 domain (Lenting et al. 1996). In addition, FX binding can be achieved in the absence of phospholipid binding via the carboxyterminal regions of the A1 domain and activated protein C is bound between His2007 and Val2016 within the carboxyterminal half of the A3 domain (Lapan and Fay 1997; Walker et al. 1990). The B domain of FVIII contains most of the Asn-linked glycosylations and gets O-glycosylated in the Golgi compartment. Prior to secretion, the B domain is cleaved off after Arg1648 and after Arg1313 and thus a heterodimer, the secretory form of FVIII, is formed (Thompson 2003). It also contains the thrombin activation sites at the C-terminus of the A1 and the A2 domain of FVIII (Arg372 and Arg740) and an additional thrombin cleavage site is located at Arg1689 in the light chain (A3 domain). The C1 and C2 domain show homology with FV and a class of lectins and these domains are engaged in binding of FIX and FX as well as in phospholipid binding of activated FVIII (FVIIIa) (Jacquemin et al. 2006; Pratt et al. 1999). The C2 domain contains a putative lipid binding surface in beta-strands 3 and 4 that is opposite to the amino- and carboxyterminal sequences (Liu et al. 2000) and an important VWF binding site (Saenko and Scandella 1997). In addition, the binding sites for thrombin and FXa are situated in the C2 domain (Nogami et al. 1999, 2000). In summary, the modular domain architecture ensures hemostasis and place FVIIIa cofactor function as a central regulatory element in the coagulation cascade.

The liver is the major source of circulating FVIII, as liver transplantation corrects FVIII levels in haemophilic patients (Marchioro et al. 1969; Bontempo et al. 1987), but detection of FVIII mRNA clearly indicated that the spleen, the lung, lymph nodes and the kidney are additional sources (Jacquemin et al. 2006; Wion et al.

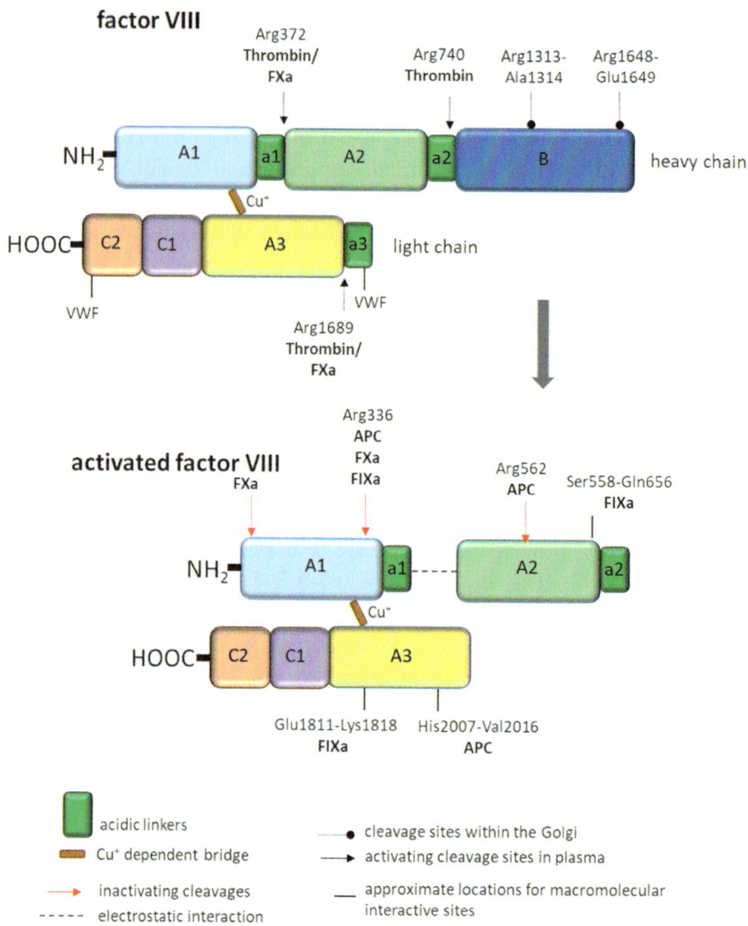

Fig. 18.2 Structure of coagulation FVIII in its inactive and active form. The cleavage sites that are proteolyzed in the Golgi are indicated with lollipops, the activating cleavage sites in plasma are shown with black arrows and the inactivating cleavage sites in plasma with red arrows. A dashed line represents the electrostatic interaction between the A1 and the A2 domain and the orange square represents the copper dependent ion bridge between the A1 and the A3 domain. A black line represents the approximate locations for macromolecular interaction sites. Activation of FVIII requires removal of the B domain and the a3 acidic linker

1985). FVIII expression was shown to be specific for fenestrated endothelial cells in liver sinusoids, renal glomeruli and high endothelial venules (Do et al. 1999; Shahani et al. 2014; Pan et al. 2016; Hollestelle et al. 2001). In contrast, hepatocytes do not significantly contribute to the plasma FVIII pool (Shahani et al. 2014; Everett et al. 2014). Interestingly, for VWF a mosaic expression pattern was revealed (Yuan et al. 2016; Pan et al. 2016). The expression of VWF and FVIII is heterogeneous and strongly dependent on the capillary bed. In case of VWF it underlies epigenetic regulatory mechanisms (Senis et al. 1996; Guan et al. 1999; Yuan et al. 2016; Pan et al. 2016; Everett et al. 2014). FVIII mRNA is highly expressed in the sinusoidal endothelium and recent studies with tissue-specific deletion of FVIII-expression have unequivocally demonstrated that plasma FVIII is of endothelial cell origin (Hollestelle et al. 2001; Shi et al. 2010; Fahs et al. 2014).

FVIII is cleared from the circulation through different routes. The ubiquitously expressed low-density lipoprotein-related receptor protein (LRP) partially contributes to FVIII clearance by binding FVIII, interacting with its light chain (A3-C1-C2) (Schwarz et al. 2000; Lenting et al. 1999; Saenko et al. 1999). In contrast, clearance of full length FVIII by binding via N-glycans of the B domain to the asialoglycoprotein-receptor was suggested to play a minor role (Bovenschen et al. 2005). In addition, the endothelial lectin clearance receptor CLEC4M was recently identified to bind and internalize FVIII in a VWF-dependent and independent manner (Swystun et al. 2019). Furthermore, the half-life and immunogenicity of the VWF-FVIII complex was found to be regulated by the scavenger receptor stabilin-2, expressed on the sinusoidal endothelial cells of the liver, spleen, and lymphatics that binds glycosaminoglycans (Swystun et al. 2018).

In conclusion, FVIII has a vital regulatory role in the coagulation system and specifically its interaction with VWF is influencing many aspects of coagulation physiology.

Physiological Roles of the Von Willebrand Factor-Factor VIII Interaction

When released into the circulation, FVIII binds to the VWF D' (TIL'E') domain with high affinity by interacting with its light chain (Lollar et al. 1988) through an acidic sequence (Glu1649-Arg1689) that connects the B with the A3 domain of FVIII (Przeradzka et al. 2018; Dagil et al. 2019), supported by the sulphated Tyr1680 residue (Foster et al. 1988; Leyte et al. 1991) and the FVIII C2 domain (Saenko and Scandella 1997) (Fig. 18.3). In turn, the VWF interaction site with FVIII is situated within the first 272 amino acid residues of mature VWF (D' domain), mediated by residues 78 and 96 and the sequence around residue 53 (Foster et al. 1987; Bahou et al. 1989; Jorieux et al. 1994).

A population-based patient-control study has revealed that non-0 blood group, plasma VWF concentration and plasma FVIII levels are related to venous thrombosis and link FVIII as a risk factor with a dose-response in relation to the occurrence of venous thrombosis (Koster et al. 1995). Of note, this increased thrombosis risk was not associated with polymorphisms of the VWF and FVIII genes (Kamphuisen et al. 2001). In mouse models it was demonstrated that VWF and FVIII are independently required for the formation of occlusive venous thrombi (Chauhan et al. 2007). Besides the role of the VWF-FVIII complex in thrombosis, many additional functional roles arise or are influenced through the interaction of VWF with FVIII in plasma.

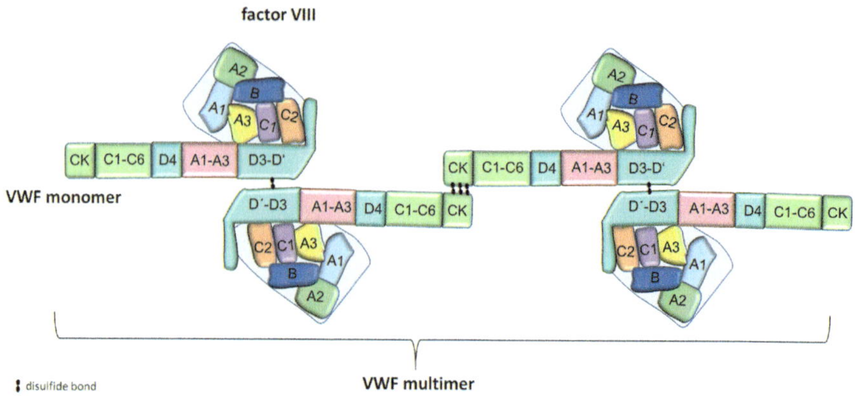

Fig. 18.3 Domain architecture of coagulation factor VIII and VWF. The domain architecture of the molecules is indicated in different colors. The dots respresent disulfide bonds linking two CK or two D3 domains that are responsible for dimer formation and multimerization, respectively

VWF protects FVIII from proteolytic degradation, premature activation and clearance and transfers this cofactor of the coagulation cascade to the site of vascular injury (Nesheim et al. 1991; Weiss et al. 1977; Wise et al. 1991). The stabilizing role of VWF is most apparent in patients with severe von Willebrand disease (VWD), who have a secondary deficiency of FVIII. The molecular ratio of FVIII to VWF monomer is about 1:50 in human plasma (Vlot et al. 1995). Hence FVIII levels in plasma are strongly dependent on VWF plasma levels (Tuddenham et al. 1982), which are regulated through multiple genetic and environmental influences (Jackel et al. 2017; Pinsky et al. 1996). In turn, deficiency of FVIII was linked to elevated VWF plasma levels in a haemophilia A mouse model and vice versa, the infusion of FVIII in patients with haemophilia A resulted in decreased VWF levels (Kiouptsi et al. 2017; van Bladel et al. 2014). The half-life of FVIII is very short in the absence of VWF, but increasing VWF concentration in plasma stabilizes the newly synthetized FVIII (Goudemand et al. 1992). The half-life of FVIII in plasma shows great variability in haemophilia A patients ranging from 10 to 20 h, depending on the ABO blood group and other influences on VWF levels (Vlot et al. 2000). For instance in VWD, the absence of VWF results in rapid proteolysis of FVIII, reducing its half-life from 12 h to 1–2 h (Montgomery and Gill 2000).

Co-expression of FVIII with VWF occurs in the high endothelial venules of lymphatics but not in most other endothelial cell types (Pan et al. 2016). However, when the gene of FVIII is introduced into a cell, which produces and stores VWF, the two molecules have been demonstrated to traffic together and be stored in the secretory granules. This has been proven both in a pituitary tumor cell line (Rosenberg et al. 1998) and in human umbilical vein endothelial cells (HUVECs) (Rosenberg et al. 2000). Interestingly, in the α-granules of platelets, FVIII is stored together with VWF (Shi et al. 2003). Moreover, there is evidence describing the secretion of FVIII bound to ultra-large VWF strings from the Weibel-Palade bodies of human glomerular microvascular endothelial cells and human umbilical vein endothelial cells (Turner and Moake 2015). Interestingly, in previous work, FVIII secretion was demonstrated to depend on the presence of VWF (Weiss et al. 1977; Kaufman et al. 1988).

Relevance of the Von Willebrand Factor-Factor VIII Interaction in Von Willebrand Factor Disease and Haemophilia A

Haemophilia A is an inherited bleeding disorder caused by a deficiency in coagulation FVIII. Its prevalence is approximately 1 in 5,000, usually affecting males because it is X-chromosome-linked (heterosomal). The FVIII-deficiency can occur by a large number of different mutations, either deletions of large portions of the gene, inversions or point mutations causing premature stop codons (Graw et al. 2005). The severity of the disease also varies depending on the FVIII levels in plasma. Severe cases have less than 1% of the normal FVIII levels and need to have prophylactic treatment to prevent joint and muscle bleeding. Moderate cases show 1–5% of normal FVIII levels while mild cases have the required levels for hemostasis (<30%) (White et al. 2001).

Treatment for haemophilia includes either infusion of plasma-derived FVIII (pdFVIII) or recombinant FVIII (rFVIII) (Franchini 2010; Lai et al. 2017). Although the life expectancy of treated haemophilia patients is comparable to healthy individuals, replacement therapy often leads to inhibitory antibodies (alloantibodies) against FVIII, the most common complication of the disease (Graw et al. 2005). Plasma derived products are less immunogenic than the recombinant, but they bear a risk of infectious diseases (Lai et al. 2017). In-depth knowledge on the reason for the immunogenicity of FVIII concentrates would improve the treatment and the life quality of patients. The co-existence of VWF and FVIII in plasma and the protective role of VWF is considered the reason for the lower inhibitor incidence in pdFVIII concentrates, which can be explained by two hypothetical mechanisms. The VWF binding sites on the FVIII molecule are also the inhibitor epitope sites. Therefore, VWF masks these epitopes when in complex with pdFVIII (Gharagozlou et al. 2009; Kallas and Talpsep 2001; Suzuki et al. 1996). The other hypothesis is that VWF protects FVIII from endocytosis by dendritic cells (Dasgupta et al. 2007).

Comparative analyses of commercial FVIII products with varying VWF concentrations pinpointed the best VWF/FVIII ratio and the therapeutic levels resulting in reduced inhibitor occurrence. These studies concluded that products with high VWF levels produced fewer alloantibodies and more thrombin (Salvagno et al. 2007). However, in standard inhibitor assays, low-VWF pdFVIII concentrates showed higher inhibitor titers than both high-VWF pdFVIII concentrates and rFVIII, concluding no correlation between VWF content and inhibitor titer (Verbruggen et al. 2001). Another parameter, confirming the latter finding, is the half-life of FVIII in either pdFVIII or rFVIII. FVIII and VWF have a very low dissociation constant (Kd = 0.4 nM), therefore the complex assembles in 12 s after intra venous injections of FVIII (Vlot et al. 1996). The k_d of affinity-purified natural anti-FVIII antibodies is 10-fold higher (Moreau et al. 2000). Since individuals with blood group 0, who have the lowest levels of VWF, clear FVIII faster than individuals with other blood groups, it is clear that the half-life of FVIII depends on VWF levels (Preston and Barr 1964; Souto et al. 2000). Regarding the second hypothesis, pre-incubation of VWF with FVIII reduced the uptake of FVIII by dendritic cells compared with FVIII alone (Dasgupta et al. 2007). In contrast, work of others showed that neither FVIII alone or in complex with VWF can stimulate an immunogenic response (Pfistershammer et al. 2006) (to the heterogeneous mutations in the FVIII gene, ethnicity, polymorphisms of immune response genes) and environmental factors (the multiple FVIII concentrate products that are available, the age at first exposure to FVIII).

Von Willebrand factor disease is the most common inherited bleeding disorder with a prevalence of 1:1,000 (Bowman et al. 2010). Like in haemophilia A, there are three types of hereditary VWD (Leebeek and Eikenboom 2016). Type I is the most common and patients have a quantitative deficiency in VWF levels, ranging from 20 to 50% of normal levels. However, they exhibit mild symptoms. Type 2 instead is rarer (15–30% of patients) and the deficiency is qualitative. There are four subtypes included in type 2 (2A, 2B, 2M and 2N), depending on the presence and behavior of VWF multimers. The rarest type of VWD is type 3 (5–10% of patients) and like type 1, it is a quantitative deficiency with severe and spontaneous bleeding episodes in the muscles and joints. There is an additional type of VWD, which is not hereditary but acquired, namely the acquired VWD that is usually a result of an autoimmune disease. It is not surprising that severe VWD patients exhibit reduced FVIII levels in plasma. The absence of VWF leads to rapid proteolysis of FVIII, reducing its plasma half-life from 12 h to 1–2 h (Montgomery and Gill 2000).

The VWD treatment depends on the diagnosis and severity. Usually patients are administered with desmopressin (DDAVP), a synthetic derivative of the blood pressure-regulating cyclic peptide hormone vasopressin, which stimulates the release of VWF from endothelial cells. In general, this is sufficient to treat minor bleeding and to prevent bleeding during minor surgery interventions (Federici 2008). DDAVP administration stimulates FVIII release, resulting in a rapid increase of both VWF and FVIII (Kawai and Montgomery 1987; Mannucci et al. 1981). However, DDAVP is only effective if adequate and functionally normal cellular VWF is present (Federici et al. 2004). In severe type 3 VWD there is no release of VWF after DDAVP administration because VWF is not synthetized and no release of FVIII occurs because it is not stored (Haberichter et al. 2006). In contrast, in Hemophilia A where FVIII is absent, DDAVP leads to a normal release of VWF (Haberichter et al. 2006).

Severe VWD cases or patients that do not respond to DDAVP are treated with plasma derived VWF concentrates. These concentrates often contain varying VWF/FVIII ratios (from 0.2 to 2.9) and varying amounts of high molecular weight multimers, which are more effective in restoring hemostasis (Budde et al. 2006; Batlle et al. 2009; Lethagen et al. 2004). It should be noted that VWD patients have normal FVIII synthesis but increased FVIII proteolysis due to reduced FVIII binding to

VWF. This worsens the bleeding phenotype and increases the bleeding risk. Although FVIII levels increase after treatment with VWF/FVIII concentrates, the endogenous FVIII half-life increases as well, which requires the regular quantification of both molecules. Concentrates with high FVIII content may induce high FVIII levels, a thrombosis risk factor (Lethagen et al. 2004; Tracy et al. 1999). Another rare but sever complication of the plasma derived concentrate treatment is the development of alloantibodies against VWF (James et al. 2013; Baaij et al. 2015)

There have been efforts to purify VWF with low or no FVIII content (Goudemand et al. 1992). Recombinant VWF (rVWF) concentrates are also available, reaching a purity of more than 99% (Peyvandi et al. 2019b; Turecek et al. 2009). During the manufacturing process, rVWF is not exposed to ADAMTS13 and therefore contains more large, highly active VWF multimers (Peyvandi et al. 2019a). This might result in a slower clearance in comparison to the plasma derived products (Mannucci et al. 2013). Although administration of VWF concentrates into a patient with VWD type 3 leads to a rapid increase or VWF, the increase of FVIII reached its peak 12–24 h after the infusion (Goudemand et al. 1992). Most patients reach FVIII levels that exceed 60 U/dl at 6h after infusion of pure VWF concentrate (Borel-Derlon et al. 2007). Therefore, additional FVIII administration might be necessary in cases where immediate hemostatic levels of both factors should be achieved, as in spontaneous or post-traumatic bleedings or urgent surgeries (Sadler et al. 2006).

Conclusion

The secretion of VWF and FVIII from endothelial cells is dependent on the respective vascular bed and varies between vascular endothelium of different organs. Weibel-Palade body exocytosis and release of VWF and FVIII is triggered by various physiologic stimuli, constituting a crucial element of vascular adaptation, *e.g.* during blood pressure regulation and in the adjustment of the local inflammatory response. The balance of the VWF FVIII interaction is pivotal for hemostatic function, becoming most apparent in VWD patients. The VWF-FVIII complex also contributes to the increasing risk of arterial thrombosis in the ageing population. Hence, the assembly of the VWF-FVIII complex influences various aspects of vascular homeostasis and the etiology of many vascular diseases. The increasing insights in the regulation and mechanisms defining this macromolecular interaction are likely to result in improved therapeutic intervention strategies to combat the extent of thrombo-inflammation and vascular inflammation, representing a major disease burden.

References

Albanez S, Ogiwara K, Michels A, Hopman W, Grabell J, James P, Lillicrap D (2016) Aging and ABO blood type influence von Willebrand factor and factor VIII levels through interrelated mechanisms. J Thromb Haemost 14(5):953–963. https://doi.org/10.1111/jth.13294

Alevriadou BR, Moake JL, Turner NA, Ruggeri ZM, Folie BJ, Phillips MD, Schreiber AB, Hrinda ME, McIntire LV (1993) Real-time analysis of shear-dependent thrombus formation and its blockade by inhibitors of von Willebrand factor binding to platelets. Blood 81(5):1263–1276

Andrews RK, Berndt MC (2013) The GPIb-IX-V complex. In: Michelson AD (ed) Platelets, 3rd edn. Elsevier, London, pp 195–213

Arneth B (2019) Coevolution of the coagulation and immune systems. Inflamm Res 68(2):117–123. https://doi.org/10.1007/s00011-018-01210-y

Baaij M, van Galen KP, Urbanus RT, Nigten J, Eikenboom JH, Schutgens RE (2015) First report of inhibitory von Willebrand factor alloantibodies in type 2B von Willebrand disease. Br J Haematol 171(3):424–427. https://doi.org/10.1111/bjh.13395

Babich V, Meli A, Knipe L, Dempster JE, Skehel P, Hannah MJ, Carter T (2008) Selective release of molecules from Weibel-Palade bodies during a lingering kiss. Blood 111(11):5282–5290. https://doi.org/10.1182/blood-2007-09-113746

Baglia FA, Badellino KO, Li CQ, Lopez JA, Walsh PN (2002) Factor XI binding to the platelet glycoprotein Ib-IX-V complex promotes factor XI activation by thrombin. J Biol Chem 277(3):1662–1668. https://doi.org/10.1074/jbc.M108319200

Bahou WF, Ginsburg D, Sikkink R, Litwiller R, Fass DN (1989) A monoclonal antibody to von Willebrand factor (vWF) inhibits factor VIII binding. Localization of its antigenic determinant to a nonadecapeptide at the amino terminus of the mature vWF polypeptide. J Clin Invest 84 (1):56–61. https://doi.org/10.1172/jci114169

Bajaj SP, Schmidt AE, Mathur A, Padmanabhan K, Zhong D, Mastri M, Fay PJ (2001) Factor IXa:factor VIIIa interaction. helix 330–338 of factor ixa interacts with residues 558-565 and spatially adjacent regions of the a2 subunit of factor VIIIa. J Biol Chem 276 (19):16302-16309. https://doi.org/10.1074/jbc.m011680200

Baker J, Griggs RK, Falati S, Poole AW (2004) GPIb potentiates GPVI-induced responses in human platelets. Platelets 15(4):207–214. https://doi.org/10.1080/09537100410001701010

Batlle J, Lopez-Fernandez MF, Fraga EL, Trillo AR, Perez-Rodriguez MA (2009) Von Willebrand factor/factor VIII concentrates in the treatment of von Willebrand disease. Blood Coagul Fibrinolysis 20(2):89–100. https://doi.org/10.1097/MBC.0b013e3283254570

Bennett JS, Vilaire G (1979) Exposure of platelet fibrinogen receptors by ADP and epinephrine. J Clin Invest 64(5):1393–1401. https://doi.org/10.1172/JCI109597

Bledzka K, Pesho MM, Ma Y, Plow EE (2013) Integrin αIIbβ3. In: Michelson AD (ed) Platelets, 3rd edn. Elsevier, London, pp 233–248

Bohm M, Betz C, Miesbach W, Krause M, von Auer C, Geiger H, Scharrer I (2005) The course of ADAMTS-13 activity and inhibitor titre in the treatment of thrombotic thrombocytopenic purpura with plasma exchange and vincristine. Br J Haematol 129(5):644–652. https://doi.org/10.1111/j.1365-2141.2005.05512.x

Bontempo FA, Lewis JH, Gorenc TJ, Spero JA, Ragni MV, Scott JP, Starzl TE (1987) Liver transplantation in hemophilia A. Blood 69(6):1721–1724

Bonthron D, Orr EC, Mitsock LM, Ginsburg D, Handin RI, Orkin SH (1986) Nucleotide sequence of pre-pro-von Willebrand factor cDNA. Nucleic Acids Res 14(17):7125–7127. https://doi.org/10.1093/nar/14.17.7125

Borchiellini A, Fijnvandraat K, ten Cate JW, Pajkrt D, van Deventer SJ, Pasterkamp G, Meijer-Huizinga F, Zwart-Huinink L, Voorberg J, van Mourik JA (1996) Quantitative analysis of von Willebrand factor propeptide release in vivo: effect of experimental endotoxemia and administration of 1-deamino-8-D-arginine vasopressin in humans. Blood 88(8):2951–2958

Borel-Derlon A, Federici AB, Roussel-Robert V, Goudemand J, Lee CA, Scharrer I, Rothschild C, Berntorp E, Henriet C, Tellier Z, Bridey F, Mannucci PM (2007) Treatment of severe von

Willebrand disease with a high-purity von Willebrand factor concentrate (Wilfactin): a prospective study of 50 patients. J Thromb Haemost 5(6):1115–1124. https://doi.org/10.1111/j.1538-7836.2007.02562.x
Bovenschen N, Rijken DC, Havekes LM, van Vlijmen BJ, Mertens K (2005) The B domain of coagulation factor VIII interacts with the asialoglycoprotein receptor. J Thromb Haemost 3(6):1257–1265. https://doi.org/10.1111/j.1538-7836.2005.01389.x
Bowman M, Hopman WM, Rapson D, Lillicrap D, James P (2010) The prevalence of symptomatic von Willebrand disease in primary care practice. J Thromb Haemost 8(1):213–216. https://doi.org/10.1111/j.1538-7836.2009.03661.x
Bradford HN, Pixley RA, Colman RW (2000) Human factor XII binding to the glycoprotein Ib-IX-V complex inhibits thrombin-induced platelet aggregation. J Biol Chem 275(30):22756–22763. https://doi.org/10.1074/jbc.M002591200
Brehelin M, Zachary D, Hoffmann JA (1978) A comparative ultrastructural study of blood cells from nine insect orders. Cell Tissue Res 195(1):45–57. https://doi.org/10.1007/bf00233676
Brehm MA (2017) Von Willebrand factor processing. Hamostaseologie 37(1):59–72. https://doi.org/10.5482/HAMO-16-06-0018
Budde U, Metzner HJ, Muller HG (2006) Comparative analysis and classification of von Willebrand factor/factor VIII concentrates: impact on treatment of patients with von Willebrand disease. Semin Thromb Hemost 32(6):626–635. https://doi.org/10.1055/s-2006-949668
Canis K, McKinnon TA, Nowak A, Panico M, Morris HR, Laffan M, Dell A (2010) The plasma von Willebrand factor O-glycome comprises a surprising variety of structures including ABH antigens and disialosyl motifs. J Thromb Haemost 8(1):137–145. https://doi.org/10.1111/j.1538-7836.2009.03665.x
Canis K, McKinnon TA, Nowak A, Haslam SM, Panico M, Morris HR, Laffan MA, Dell A (2012) Mapping the N-glycome of human von Willebrand factor. Biochem J 447(2):217–228. https://doi.org/10.1042/BJ20120810
Casari C, Du V, Wu YP, Kauskot A, de Groot PG, Christophe OD, Denis CV, de Laat B, Lenting PJ (2013a) Accelerated uptake of VWF/platelet complexes in macrophages contributes to VWD type 2B-associated thrombocytopenia. Blood 122(16):2893–2902. https://doi.org/10.1182/blood-2013-03-493312
Casari C, Lenting PJ, Wohner N, Christophe OD, Denis CV (2013b) Clearance of von Willebrand factor. J Thromb Haemost 11(Suppl 1):202–211. https://doi.org/10.1111/jth.12226
Castro-Nunez L, Dienava-Verdoold I, Herczenik E, Mertens K, Meijer AB (2012) Shear stress is required for the endocytic uptake of the factor VIII-von Willebrand factor complex by macrophages. J Thromb Haemost 10(9):1929–1937. https://doi.org/10.1111/j.1538-7836.2012.04860.x
Chauhan AK, Kisucka J, Lamb CB, Bergmeier W, Wagner DD (2007) von Willebrand factor and factor VIII are independently required to form stable occlusive thrombi in injured veins. Blood 109(6):2424–2429. https://doi.org/10.1182/blood-2006-06-028241
Chen J, Chung DW (2018) Inflammation, von Willebrand factor, and ADAMTS13. Blood 132(2):141–147. https://doi.org/10.1182/blood-2018-02-769000
Coppola R, Mari D, Lattuada A, Franceschi C (2003) Von Willebrand factor in Italian centenarians. Haematologica 88(1):39–43
Dagil L, Troelsen KS, Bolt G, Thim L, Wu B, Zhao X, Tuddenham EGD, Nielsen TE, Tanner DA, Faber JH, Breinholt J, Rasmussen JE, Hansen DF (2019) Interaction Between the a3 Region of Factor VIII and the TIL'E' Domains of the von Willebrand Factor. Biophys J 117(3):479–489. https://doi.org/10.1016/j.bpj.2019.07.007
Dasgupta S, Repesse Y, Bayry J, Navarrete AM, Wootla B, Delignat S, Irinopoulou T, Kamate C, Saint-Remy JM, Jacquemin M, Lenting PJ, Borel-Derlon A, Kaveri SV, Lacroix-Desmazes S (2007) VWF protects FVIII from endocytosis by dendritic cells and subsequent presentation to immune effectors. Blood 109(2):610–612. https://doi.org/10.1182/blood-2006-05-022756

De Marco L, Girolami A, Zimmerman TS, Ruggeri ZM (1986) von Willebrand factor interaction with the glycoprotein IIb/IIa complex. Its role in platelet function as demonstrated in patients with congenital afibrinogenemia. J Clin Invest 77 (4):1272–1277. https://doi.org/10.1172/jci112430

Denis C, Methia N, Frenette PS, Rayburn H, Ullman-Cullere M, Hynes RO, Wagner DD (1998) A mouse model of severe von Willebrand disease: defects in hemostasis and thrombosis. Proc Natl Acad Sci U S A 95(16):9524–9529. https://doi.org/10.1073/pnas.95.16.9524

Denorme F, Kraft P, Pareyn I, Drechsler C, Deckmyn H, Vanhoorelbeke K, Kleinschnitz C, De Meyer SF (2017) Reduced ADAMTS13 levels in patients with acute and chronic cerebrovascular disease. PLoS ONE 12(6):e0179258. https://doi.org/10.1371/journal.pone.0179258

Dhanesha N, Prakash P, Doddapattar P, Khanna I, Pollpeter MJ, Nayak MK, Staber JM, Chauhan AK (2016) Endothelial Cell-Derived von Willebrand Factor Is the Major Determinant That Mediates von Willebrand Factor-Dependent Acute Ischemic Stroke by Promoting Postischemic Thrombo-Inflammation. Arterioscler Thromb Vasc Biol 36(9):1829–1837. https://doi.org/10.1161/ATVBAHA.116.307660

Do H, Healey JF, Waller EK, Lollar P (1999) Expression of factor VIII by murine liver sinusoidal endothelial cells. J Biol Chem 274(28):19587–19592. https://doi.org/10.1074/jbc.274.28.19587

Dong JF, Moake JL, Nolasco L, Bernardo A, Arceneaux W, Shrimpton CN, Schade AJ, McIntire LV, Fujikawa K, Lopez JA (2002) ADAMTS-13 rapidly cleaves newly secreted ultralarge von Willebrand factor multimers on the endothelial surface under flowing conditions. Blood 100(12):4033–4039. https://doi.org/10.1182/blood-2002-05-1401

Durrant TN, van den Bosch MT, Hers I (2017) Integrin alphaIIbbeta3 outside-in signaling. Blood 130(14):1607–1619. https://doi.org/10.1182/blood-2017-03-773614

Ellies LG, Ditto D, Levy GG, Wahrenbrock M, Ginsburg D, Varki A, Le DT, Marth JD (2002) Sialyltransferase ST3Gal-IV operates as a dominant modifier of hemostasis by concealing asialoglycoprotein receptor ligands. Proc Natl Acad Sci U S A 99(15):10042–10047. https://doi.org/10.1073/pnas.142005099

Everett LA, Cleuren AC, Khoriaty RN, Ginsburg D (2014) Murine coagulation factor VIII is synthesized in endothelial cells. Blood 123(24):3697–3705. https://doi.org/10.1182/blood-2014-02-554501

Fahs SA, Hille MT, Shi Q, Weiler H, Montgomery RR (2014) A conditional knockout mouse model reveals endothelial cells as the principal and possibly exclusive source of plasma factor VIII. Blood 123(24):3706–3713. https://doi.org/10.1182/blood-2014-02-555151

Federici AB (2008) The use of desmopressin in von Willebrand disease: the experience of the first 30 years (1977-2007). Haemophilia 14(Suppl 1):5–14. https://doi.org/10.1111/j.1365-2516.2007.01610.x

Federici AB, Bader R, Pagani S, Colibretti ML, De Marco L, Mannucci PM (1989) Binding of von Willebrand factor to glycoproteins Ib and IIb/IIIa complex: affinity is related to multimeric size. Br J Haematol 73(1):93–99. https://doi.org/10.1111/j.1365-2141.1989.tb00226.x

Federici AB, Mazurier C, Berntorp E, Lee CA, Scharrer I, Goudemand J, Lethagen S, Nitu I, Ludwig G, Hilbert L, Mannucci PM (2004) Biologic response to desmopressin in patients with severe type 1 and type 2 von Willebrand disease: results of a multicenter European study. Blood 103(6):2032–2038. https://doi.org/10.1182/blood-2003-06-2072

Fernandez MF, Ginsberg MH, Ruggeri ZM, Batlle FJ, Zimmerman TS (1982) Multimeric structure of platelet factor VIII/von Willebrand factor: the presence of larger multimers and their reassociation with thrombin-stimulated platelets. Blood 60(5):1132–1138

Ferraro F, Kriston-Vizi J, Metcalf DJ, Martin-Martin B, Freeman J, Burden JJ, Westmoreland D, Dyer CE, Knight AE, Ketteler R, Cutler DF (2014) A two-tier Golgi-based control of organelle size underpins the functional plasticity of endothelial cells. Dev Cell 29(3):292–304. https://doi.org/10.1016/j.devcel.2014.03.021

Floyd CN, Ferro A (2012) The platelet fibrinogen receptor: from megakaryocyte to the mortuary. JRSM Cardiovasc Dis 1 (2). https://doi.org/10.1258/cvd.2012.012007

Foster PA, Fulcher CA, Marti T, Titani K, Zimmerman TS (1987) A major factor VIII binding domain resides within the amino-terminal 272 amino acid residues of von Willebrand factor. J Biol Chem 262(18):8443–8446

Foster PA, Fulcher CA, Houghten RA, Zimmerman TS (1988) An immunogenic region within residues Val1670-Glu1684 of the factor VIII light chain induces antibodies which inhibit binding of factor VIII to von Willebrand factor. J Biol Chem 263(11):5230–5234

Franchini M (2010) Plasma-derived versus recombinant Factor VIII concentrates for the treatment of haemophilia A: recombinant is better. Blood Transfus 8(4):292–296. https://doi.org/10.2450/2010.0067-10

Frankel DS, Meigs JB, Massaro JM, Wilson PW, O'Donnell CJ, D'Agostino RB, Tofler GH (2008) Von Willebrand factor, type 2 diabetes mellitus, and risk of cardiovascular disease: the framingham offspring study. Circulation 118(24):2533–2539. https://doi.org/10.1161/CIRCULATIONAHA.108.792986

Fressinaud E, Girma JP, Sadler JE, Baumgartner HR, Meyer D (1990) Synthetic RGDS-containing peptides of von Willebrand factor inhibit platelet adhesion to collagen. Thromb Haemost 64(4):589–593

Fujimura Y, Holland LZ, Ruggeri ZM, Zimmerman TS (1987) The von willebrand factor domain-mediating botrocetin-induced binding to glycoprotein IB lies between Val449 and Lys728. Blood 70(4):985–988

Furlan M, Lammle B (1999) von Willebrand factor in thrombotic thrombocytopenic purpura. Thromb Haemost 82(2):592–600

Gharagozlou S, Sharifian RA, Khoshnoodi J, Karimi K, Milani M, Okita DK, Shokri F, Conti-Fine BM (2009) Epitope specificity of anti-factor VIII antibodies from inhibitor positive acquired and congenital haemophilia A patients using synthetic peptides spanning A and C domains. Thromb Haemost 101(5):834–839

Gibbins JM, Okuma M, Farndale R, Barnes M, Watson SP (1997) Glycoprotein VI is the collagen receptor in platelets which underlies tyrosine phosphorylation of the Fc receptor gamma-chain. FEBS Lett 413(2):255–259. https://doi.org/10.1016/s0014-5793(97)00926-5

Giblin JP, Hewlett LJ, Hannah MJ (2008) Basal secretion of von Willebrand factor from human endothelial cells. Blood 112(4):957–964. https://doi.org/10.1182/blood-2007-12-130740

Giddings JC, Brookes LR, Piovella F, Bloom AL (1982) Immunohistological comparison of platelet factor 4 (PF4), fibronectin (Fn) and factor VIII related antigen (VIIIR:Ag) in human platelet granules. Br J Haematol 52(1):79–88. https://doi.org/10.1111/j.1365-2141.1982.tb03863.x

Ginsburg D, Handin RI, Bonthron DT, Donlon TA, Bruns GA, Latt SA, Orkin SH (1985) Human von Willebrand factor (vWF): isolation of complementary DNA (cDNA) clones and chromosomal localization. Science 228(4706):1401–1406. https://doi.org/10.1126/science.3874428

Gitschier J, Wood WI, Goralka TM, Wion KL, Chen EY, Eaton DH, Vehar GA, Capon DJ, Lawn RM (1984) Characterization of the human factor VIII gene. Nature 312(5992):326–330. https://doi.org/10.1038/312326a0

Goto S, Ikeda Y, Saldivar E, Ruggeri ZM (1998) Distinct mechanisms of platelet aggregation as a consequence of different shearing flow conditions. J Clin Invest 101(2):479–486. https://doi.org/10.1172/JCI973

Goudemand J, Mazurier C, Marey A, Caron C, Coupez B, Mizon P, Goudemand M (1992) Clinical and biological evaluation in von Willebrand's disease of a von Willebrand factor concentrate with low factor VIII activity. Br J Haematol 80(2):214–221. https://doi.org/10.1111/j.1365-2141.1992.tb08903.x

Grant KR (2015) Fish Hematology and Associated Disorders. Clin Lab Med 35(3):681–701. https://doi.org/10.1016/j.cll.2015.05.015

Grassle S, Huck V, Pappelbaum KI, Gorzelanny C, Aponte-Santamaria C, Baldauf C, Grater F, Schneppenheim R, Obser T, Schneider SW (2014) von Willebrand factor directly interacts with DNA from neutrophil extracellular traps. Arterioscler Thromb Vasc Biol 34(7):1382–1389. https://doi.org/10.1161/ATVBAHA.113.303016

Graw J, Brackmann HH, Oldenburg J, Schneppenheim R, Spannagl M, Schwaab R (2005) Haemophilia A: from mutation analysis to new therapies. Nat Rev Genet 6(6):488–501. https://doi.org/10.1038/nrg1617

Grewal PK, Uchiyama S, Ditto D, Varki N, Le DT, Nizet V, Marth JD (2008) The Ashwell receptor mitigates the lethal coagulopathy of sepsis. Nat Med 14(6):648–655. https://doi.org/10.1038/nm1760

Groot E, de Groot PG, Fijnheer R, Lenting PJ (2007) The presence of active von Willebrand factor under various pathological conditions. Curr Opin Hematol 14(3):284–289. https://doi.org/10.1097/MOH.0b013e3280dce531

Guan J, Guillot PV, Aird WC (1999) Characterization of the mouse von Willebrand factor promoter. Blood 94(10):3405–3412

Haberichter SL, Merricks EP, Fahs SA, Christopherson PA, Nichols TC, Montgomery RR (2005) Re-establishment of VWF-dependent Weibel-Palade bodies in VWD endothelial cells. Blood 105(1):145–152. https://doi.org/10.1182/blood-2004-02-0464

Haberichter SL, Shi Q, Montgomery RR (2006) Regulated release of VWF and FVIII and the biologic implications. Pediatr Blood Cancer 46(5):547–553. https://doi.org/10.1002/pbc.20658

Hillgruber C, Steingraber AK, Poppelmann B, Denis CV, Ware J, Vestweber D, Nieswandt B, Schneider SW, Goerge T (2014) Blocking von Willebrand factor for treatment of cutaneous inflammation. J Invest Dermatol 134(1):77–86. https://doi.org/10.1038/jid.2013.292

Hollestelle MJ, Thinnes T, Crain K, Stiko A, Kruijt JK, van Berkel TJ, Loskutoff DJ, van Mourik JA (2001) Tissue distribution of factor VIII gene expression in vivo–a closer look. Thromb Haemost 86(3):855–861

Hollestelle MJ, Donkor C, Mantey EA, Chakravorty SJ, Craig A, Akoto AO, O'Donnell J, van Mourik JA, Bunn J (2006) von Willebrand factor propeptide in malaria: evidence of acute endothelial cell activation. Br J Haematol 133(5):562–569. https://doi.org/10.1111/j.1365-2141.2006.06067.x

Huang J, Roth R, Heuser JE, Sadler JE (2009) Integrin alpha(v)beta(3) on human endothelial cells binds von Willebrand factor strings under fluid shear stress. Blood 113(7):1589–1597. https://doi.org/10.1182/blood-2008-05-158584

Hyseni A, Kemperman H, de Lange DW, Kesecioglu J, de Groot PG, Roest M (2014) Active von Willebrand factor predicts 28-day mortality in patients with systemic inflammatory response syndrome. Blood 123(14):2153–2156. https://doi.org/10.1182/blood-2013-08-508093

Ishihara J, Ishihara A, Starke RD, Peghaire CR, Smith KE, McKinnon TAJ, Tabata Y, Sasaki K, White MJV, Fukunaga K, Laffan MA, Lutolf MP, Randi AM, Hubbell JA (2019) The heparin binding domain of von Willebrand factor binds to growth factors and promotes angiogenesis in wound healing. Blood 133(24):2559–2569. https://doi.org/10.1182/blood.2019000510

Jackel S, Kiouptsi K, Lillich M, Hendrikx T, Khandagale A, Kollar B, Hormann N, Reiss C, Subramaniam S, Wilms E, Ebner K, Bruhl MV, Rausch P, Baines JF, Haberichter S, Lammle B, Binder CJ, Jurk K, Ruggeri ZM, Massberg S, Walter U, Ruf W, Reinhardt C (2017) Gut microbiota regulate hepatic von Willebrand factor synthesis and arterial thrombus formation via Toll-like receptor-2. Blood 130(4):542–553. https://doi.org/10.1182/blood-2016-11-754416

Jacquemin M, Neyrinck A, Hermanns MI, Lavend'homme R, Rega F, Saint-Remy JM, Peerlinck K, Van Raemdonck D, Kirkpatrick CJ (2006) FVIII production by human lung microvascular endothelial cells. Blood 108(2):515–517. https://doi.org/10.1182/blood-2005-11-4571

James PD, Lillicrap D, Mannucci PM (2013) Alloantibodies in von Willebrand disease. Blood 122(5):636–640. https://doi.org/10.1182/blood-2012-10-462085

Jorieux S, Gaucher C, Pietu G, Cherel G, Meyer D, Mazurier C (1994) Fine epitope mapping of monoclonal antibodies to the NH2-terminal part of von Willebrand factor (vWF) by using recombinant and synthetic peptides: interest for the localization of the factor VIII binding domain. Br J Haematol 87(1):113–118. https://doi.org/10.1111/j.1365-2141.1994.tb04879.x

Joseph K, Nakazawa Y, Bahou WF, Ghebrehiwet B, Kaplan AP (1999) Platelet glycoprotein Ib: a zinc-dependent binding protein for the heavy chain of high-molecular-weight kininogen. Mol Med 5(8):555–563

Jurk K, Clemetson KJ, de Groot PG, Brodde MF, Steiner M, Savion N, Varon D, Sixma JJ, Van Aken H, Kehrel BE (2003) Thrombospondin-1 mediates platelet adhesion at high shear via glycoprotein Ib (GPIb): an alternative/backup mechanism to von Willebrand factor. FASEB J 17(11):1490–1492. https://doi.org/10.1096/fj.02-0830fje

Kalagara T, Moutsis T, Yang Y, Pappelbaum KI, Farken A, Cladder-Micus L, Vidal YSS, John A, Bauer AT, Moerschbacher BM, Schneider SW, Gorzelanny C (2018) The endothelial glycocalyx anchors von Willebrand factor fibers to the vascular endothelium. Blood Adv 2(18):2347–2357. https://doi.org/10.1182/bloodadvances.2017013995

Kallas A, Talpsep T (2001) von Willebrand factor in factor VIII concentrates protects against neutralization by factor VIII antibodies of haemophilia A patients. Haemophilia 7(4):375–380. https://doi.org/10.1046/j.1365-2516.2001.00530.x

Kamphuisen PW, Eikenboom JC, Rosendaal FR, Koster T, Blann AD, Vos HL, Bertina RM (2001) High factor VIII antigen levels increase the risk of venous thrombosis but are not associated with polymorphisms in the von Willebrand factor and factor VIII gene. Br J Haematol 115(1):156–158. https://doi.org/10.1046/j.1365-2141.2001.03089.x

Kaufman RJ, Wasley LC, Dorner AJ (1988) Synthesis, processing, and secretion of recombinant human factor VIII expressed in mammalian cells. J Biol Chem 263(13):6352–6362

Kawai Y, Montgomery RR (1987) Endothelial cell processing of von Willebrand proteins. Ann N Y Acad Sci 509:60–70. https://doi.org/10.1111/j.1749-6632.1987.tb30984.x

Kiouptsi K, Grill A, Mann A, Dohrmann M, Lillich M, Jackel S, Malinarich F, Formes H, Manukyan D, Subramaniam S, Khandagale A, Karwot C, Thal SC, Bosmann M, Scharrer I, Jurk K, Reinhardt C (2017) Mice deficient in the anti-haemophilic coagulation factor VIII show increased von Willebrand factor plasma levels. PLoS ONE 12(8):e0183590. https://doi.org/10.1371/journal.pone.0183590

Koster T, Blann AD, Briet E, Vandenbroucke JP, Rosendaal FR (1995) Role of clotting factor VIII in effect of von Willebrand factor on occurrence of deep-vein thrombosis. Lancet 345(8943):152–155. https://doi.org/10.1016/s0140-6736(95)90166-3

Kowalska MA, Tuszynski GP (1993) Interaction of thrombospondin with platelet glycoproteins GPIa-IIa and GPIIb-IIIa. Biochem J 295(Pt 3):725–730. https://doi.org/10.1042/bj2950725

Lai J, Hough C, Tarrant J, Lillicrap D (2017) Biological considerations of plasma-derived and recombinant factor VIII immunogenicity. Blood 129(24):3147–3154. https://doi.org/10.1182/blood-2016-11-750885

Lapan KA, Fay PJ (1997) Localization of a factor X interactive site in the A1 subunit of factor VIIIa. J Biol Chem 272(4):2082–2088. https://doi.org/10.1074/jbc.272.4.2082

Leebeek FW, Eikenboom JC (2016) Von Willebrand's Disease. N Engl J Med 375(21):2067–2080. https://doi.org/10.1056/NEJMra1601561

Legaz ME, Schmer G, Counts RB, Davie EW (1973) Isolation and characterization of human Factor VIII (antihemophilic factor). J Biol Chem 248(11):3946–3955

Lenting PJ, van de Loo JW, Donath MJ, van Mourik JA, Mertens K (1996) The sequence Glu 1811-Lys1818 of human blood coagulation factor VIII comprises a binding site for activated factor IX. J Biol Chem 271(4):1935–1940. https://doi.org/10.1074/jbc.271.4.1935

Lenting PJ, Neels JG, van den Berg BM, Clijsters PP, Meijerman DW, Pannekoek H, van Mourik JA, Mertens K, van Zonneveld AJ (1999) The light chain of factor VIII comprises a binding site for low density lipoprotein receptor-related protein. J Biol Chem 274(34):23734–23739. https://doi.org/10.1074/jbc.274.34.23734

Lenting PJ, Christophe OD, Denis CV (2015) von Willebrand factor biosynthesis, secretion, and clearance: connecting the far ends. Blood 125(13):2019–2028. https://doi.org/10.1182/blood-2014-06-528406

Lethagen S, Carlson M, Hillarp A (2004) A comparative in vitro evaluation of six von Willebrand factor concentrates. Haemophilia 10(3):243–249. https://doi.org/10.1111/j.1365-2516.2004.00893.x

Levy GG, Nichols WC, Lian EC, Foroud T, McClintick JN, McGee BM, Yang AY, Siemieniak DR, Stark KR, Gruppo R, Sarode R, Shurin SB, Chandrasekaran V, Stabler SP, Sabio H, Bouhassira

EE, Upshaw JD Jr, Ginsburg D, Tsai HM (2001) Mutations in a member of the ADAMTS gene family cause thrombotic thrombocytopenic purpura. Nature 413(6855):488–494. https://doi.org/10.1038/35097008

Leyte A, van Schijndel HB, Niehrs C, Huttner WB, Verbeet MP, Mertens K, van Mourik JA (1991) Sulfation of Tyr1680 of human blood coagulation factor VIII is essential for the interaction of factor VIII with von Willebrand factor. J Biol Chem 266(2):740–746

Lippok S, Kolsek K, Lof A, Eggert D, Vanderlinden W, Muller JP, Konig G, Obser T, Rohrs K, Schneppenheim S, Budde U, Baldauf C, Aponte-Santamaria C, Grater F, Schneppenheim R, Radler JO, Brehm MA (2016) von Willebrand factor is dimerized by protein disulfide isomerase. Blood 127(9):1183–1191. https://doi.org/10.1182/blood-2015-04-641902

Liu ML, Shen BW, Nakaya S, Pratt KP, Fujikawa K, Davie EW, Stoddard BL, Thompson AR (2000) Hemophilic factor VIII C1- and C2-domain missense mutations and their modeling to the 1.5-angstrom human C2-domain crystal structure. Blood 96 (3):979–987

Lollar P, Hill-Eubanks DC, Parker CG (1988) Association of the factor VIII light chain with von Willebrand factor. J Biol Chem 263(21):10451–10455

Lynch DC, Zimmerman TS, Collins CJ, Brown M, Morin MJ, Ling EH, Livingston DM (1985) Molecular cloning of cDNA for human von Willebrand factor: authentication by a new method. Cell 41(1):49–56. https://doi.org/10.1016/0092-8674(85)90060-1

Mannucci PM, Canciani MT, Rota L, Donovan BS (1981) Response of factor VIII/von Willebrand factor to DDAVP in healthy subjects and patients with haemophilia A and von Willebrand's disease. Br J Haematol 47(2):283–293. https://doi.org/10.1111/j.1365-2141.1981.tb02789.x

Mannucci PM, Kempton C, Millar C, Romond E, Shapiro A, Birschmann I, Ragni MV, Gill JC, Yee TT, Klamroth R, Wong WY, Chapman M, Engl W, Turecek PL, Suiter TM, Ewenstein BM (2013) Pharmacokinetics and safety of a novel recombinant human von Willebrand factor manufactured with a plasma-free method: a prospective clinical trial. Blood 122(5):648–657. https://doi.org/10.1182/blood-2013-01-479527

Marchioro TL, Hougie C, Ragde H, Epstein RB, Thomas ED (1969) Hemophilia: role of organ homografts. Science 163(3863):188–190. https://doi.org/10.1126/science.163.3863.188

Marti T, Rosselet SJ, Titani K, Walsh KA (1987) Identification of disulfide-bridged substructures within human von Willebrand factor. Biochemistry 26(25):8099–8109. https://doi.org/10.1021/bi00399a013

Marx I, Christophe OD, Lenting PJ, Rupin A, Vallez MO, Verbeuren TJ, Denis CV (2008) Altered thrombus formation in von Willebrand factor-deficient mice expressing von Willebrand factor variants with defective binding to collagen or GPIIbIIIa. Blood 112(3):603–609. https://doi.org/10.1182/blood-2008-02-142943

Matsui T, Titani K, Mizuochi T (1992) Structures of the asparagine-linked oligosaccharide chains of human von Willebrand factor. Occurrence of blood group A, B, and H(O) structures. J Biol Chem 267 (13):8723–8731

Mayadas TN, Wagner DD (1989) In vitro multimerization of von Willebrand factor is triggered by low pH. Importance of the propolypeptide and free sulfhydryls. J Biol Chem 264 (23):13497–13503

McGrath RT, van den Biggelaar M, Byrne B, O'Sullivan JM, Rawley O, O'Kennedy R, Voorberg J, Preston RJ, O'Donnell JS (2013) Altered glycosylation of platelet-derived von Willebrand factor confers resistance to ADAMTS13 proteolysis. Blood 122(25):4107–4110. https://doi.org/10.1182/blood-2013-04-496851

Methia N, Andre P, Denis CV, Economopoulos M, Wagner DD (2001) Localized reduction of atherosclerosis in von Willebrand factor-deficient mice. Blood 98(5):1424–1428. https://doi.org/10.1182/blood.v98.5.1424

Mohri H, Ohkubo T (1991) How vitronectin binds to activated glycoprotein IIb-IIIa complex and its function in platelet aggregation. Am J Clin Pathol 96(5):605–609. https://doi.org/10.1093/ajcp/96.5.605

Montgomery RR, Gill JC (2000) Interactions between von Willebrand factor and Factor VIII: where did they first meet. J Pediatr Hematol Oncol 22(3):269–275. https://doi.org/10.1097/00043426-200005000-00017
Moreau A, Lacroix-Desmazes S, Stieltjes N, Saenko E, Kaveri SV, D'Oiron R, Sultan Y, Scandella D, Kazatchkine MD (2000) Antibodies to the FVIII light chain that neutralize FVIII procoagulant activity are present in plasma of nonresponder patients with severe hemophilia A and in normal polyclonal human IgG. Blood 95(11):3435–3441
Muntean W, Leschnik B (1989) Factor VIII influences binding of factor IX and factor X to intact human platelets. Thromb Res 55(5):537–548. https://doi.org/10.1016/0049-3848(89)90386-1
Nesheim M, Pittman DD, Giles AR, Fass DN, Wang JH, Slonosky D, Kaufman RJ (1991) The effect of plasma von Willebrand factor on the binding of human factor VIII to thrombin-activated human platelets. J Biol Chem 266(27):17815–17820
Nogami K, Shima M, Hosokawa K, Suzuki T, Koide T, Saenko EL, Scandella D, Shibata M, Kamisue S, Tanaka I, Yoshioka A (1999) Role of factor VIII C2 domain in factor VIII binding to factor Xa. J Biol Chem 274(43):31000–31007. https://doi.org/10.1074/jbc.274.43.31000
Nogami K, Shima M, Hosokawa K, Nagata M, Koide T, Saenko EL, Tanaka I, Shibata M, Yoshioka A (2000) Factor VIII C2 domain contains the thrombin-binding site responsible for thrombin-catalyzed cleavage at Arg1689. J Biol Chem 275(33):25774–25780. https://doi.org/10.1074/jbc.M002007200
O'Sullivan JM, Preston RJS, Robson T, O'Donnell JS (2018) Emerging Roles for von Willebrand Factor in Cancer Cell Biology. Semin Thromb Hemost 44(2):159–166. https://doi.org/10.1055/s-0037-1607352
Pan J, Dinh TT, Rajaraman A, Lee M, Scholz A, Czupalla CJ, Kiefel H, Zhu L, Xia L, Morser J, Jiang H, Santambrogio L, Butcher EC (2016) Patterns of expression of factor VIII and von Willebrand factor by endothelial cell subsets in vivo. Blood 128(1):104–109. https://doi.org/10.1182/blood-2015-12-684688
Pappelbaum KI, Gorzelanny C, Grassle S, Suckau J, Laschke MW, Bischoff M, Bauer C, Schorpp-Kistner M, Weidenmaier C, Schneppenheim R, Obser T, Sinha B, Schneider SW (2013) Ultralarge von Willebrand factor fibers mediate luminal Staphylococcus aureus adhesion to an intact endothelial cell layer under shear stress. Circulation 128(1):50–59. https://doi.org/10.1161/CIRCULATIONAHA.113.002008
Paulus P, Jennewein C, Zacharowski K (2011) Biomarkers of endothelial dysfunction: can they help us deciphering systemic inflammation and sepsis? Biomarkers 16(Suppl 1):S11–S21. https://doi.org/10.3109/1354750X.2011.587893
Pegon JN, Kurdi M, Casari C, Odouard S, Denis CV, Christophe OD, Lenting PJ (2012) Factor VIII and von Willebrand factor are ligands for the carbohydrate-receptor Siglec-5. Haematologica 97(12):1855–1863. https://doi.org/10.3324/haematol.2012.063297
Pemberton S, Lindley P, Zaitsev V, Card G, Tuddenham EG, Kemball-Cook G (1997) A molecular model for the triplicated A domains of human factor VIII based on the crystal structure of human ceruloplasmin. Blood 89(7):2413–2421
Peyvandi F, Kouides P, Turecek PL, Dow E, Berntorp E (2019a) Evolution of replacement therapy for von Willebrand disease: From plasma fraction to recombinant von Willebrand factor. Blood Rev 38:100572. https://doi.org/10.1016/j.blre.2019.04.001
Peyvandi F, Mamaev A, Wang JD, Stasyshyn O, Timofeeva M, Curry N, Cid AR, Yee TT, Kavakli K, Castaman G, Sytkowski A (2019b) Phase 3 study of recombinant von Willebrand factor in patients with severe von Willebrand disease who are undergoing elective surgery. J Thromb Haemost 17(1):52–62. https://doi.org/10.1111/jth.14313
Pfistershammer K, Stockl J, Siekmann J, Turecek PL, Schwarz HP, Reipert BM (2006) Recombinant factor VIII and factor VIII-von Willebrand factor complex do not present danger signals for human dendritic cells. Thromb Haemost 96(3):309–316. https://doi.org/10.1160/TH05-11-0729
Pinsky DJ, Naka Y, Liao H, Oz MC, Wagner DD, Mayadas TN, Johnson RC, Hynes RO, Heath M, Lawson CA, Stern DM (1996) Hypoxia-induced exocytosis of endothelial cell Weibel-Palade

bodies. A mechanism for rapid neutrophil recruitment after cardiac preservation. J Clin Invest 97 (2):493–500. https://doi.org/10.1172/jci118440

Pottinger BE, Read RC, Paleolog EM, Higgins PG, Pearson JD (1989) von Willebrand factor is an acute phase reactant in man. Thromb Res 53(4):387–394. https://doi.org/10.1016/0049-3848(89)90317-4

Poulter NS, Pollitt AY, Owen DM, Gardiner EE, Andrews RK, Shimizu H, Ishikawa D, Bihan D, Farndale RW, Moroi M, Watson SP, Jung SM (2017) Clustering of glycoprotein VI (GPVI) dimers upon adhesion to collagen as a mechanism to regulate GPVI signaling in platelets. J Thromb Haemost 15(3):549–564. https://doi.org/10.1111/jth.13613

Pratt KP, Shen BW, Takeshima K, Davie EW, Fujikawa K, Stoddard BL (1999) Structure of the C2 domain of human factor VIII at 1.5 A resolution. Nature 402 (6760):439–442. https://doi.org/10.1038/46601

Preston AE, Barr A (1964) The Plasma Concentration of Factor Viii in the Normal Population. Ii. The Effects of Age, Sex and Blood Group. Br J Haematol 10:238–245. https://doi.org/10.1111/j.1365-2141.1964.tb00698.x

Przeradzka MA, Meems H, van der Zwaan C, Ebberink E, van den Biggelaar M, Mertens K, Meijer AB (2018) The D' domain of von Willebrand factor requires the presence of the D3 domain for optimal factor VIII binding. Biochem J 475(17):2819–2830. https://doi.org/10.1042/BCJ20180431

Randi AM, Smith KE, Castaman G (2018) von Willebrand factor regulation of blood vessel formation. Blood 132(2):132–140. https://doi.org/10.1182/blood-2018-01-769018

Rastegarlari G, Pegon JN, Casari C, Odouard S, Navarrete AM, Saint-Lu N, van Vlijmen BJ, Legendre P, Christophe OD, Denis CV, Lenting PJ (2012) Macrophage LRP1 contributes to the clearance of von Willebrand factor. Blood 119(9):2126–2134. https://doi.org/10.1182/blood-2011-08-373605

Ratnoff OD (1987) The evolution of hemostatic mechanisms. Perspect Biol Med 31(1):4–33. https://doi.org/10.1353/pbm.1987.0003

Reinhart K, Bayer O, Brunkhorst F, Meisner M (2002) Markers of endothelial damage in organ dysfunction and sepsis. Crit Care Med 30(5 Suppl):S302–S312. https://doi.org/10.1097/00003246-200205001-00021

Reininger AJ, Heijnen HF, Schumann H, Specht HM, Schramm W, Ruggeri ZM (2006) Mechanism of platelet adhesion to von Willebrand factor and microparticle formation under high shear stress. Blood 107(9):3537–3545. https://doi.org/10.1182/blood-2005-02-0618

Rivera J, Lozano ML, Navarro-Nunez L, Vicente V (2009) Platelet receptors and signaling in the dynamics of thrombus formation. Haematologica 94(5):700–711. https://doi.org/10.3324/haematol.2008.003178

Rojo Pulido I, Nightingale TD, Darchen F, Seabra MC, Cutler DF, Gerke V (2011) Myosin Va acts in concert with Rab27a and MyRIP to regulate acute von-Willebrand factor release from endothelial cells. Traffic 12(10):1371–1382. https://doi.org/10.1111/j.1600-0854.2011.01248.x

Romani de Wit T, Rondaij MG, Hordijk PL, Voorberg J, van Mourik JA (2003) Real-time imaging of the dynamics and secretory behavior of Weibel-Palade bodies. Arterioscler Thromb Vasc Biol 23(5):755–761. https://doi.org/10.1161/01.ATV.0000069847.72001.E8

Romo GM, Dong JF, Schade AJ, Gardiner EE, Kansas GS, Li CQ, McIntire LV, Berndt MC, Lopez JA (1999) The glycoprotein Ib-IX-V complex is a platelet counterreceptor for P-selectin. J Exp Med 190(6):803–814. https://doi.org/10.1084/jem.190.6.803

Rondaij MG, Bierings R, Kragt A, van Mourik JA, Voorberg J (2006) Dynamics and plasticity of Weibel-Palade bodies in endothelial cells. Arterioscler Thromb Vasc Biol 26(5):1002–1007. https://doi.org/10.1161/01.ATV.0000209501.56852.6c

Rosenberg JB, Foster PA, Kaufman RJ, Vokac EA, Moussalli M, Kroner PA, Montgomery RR (1998) Intracellular trafficking of factor VIII to von Willebrand factor storage granules. J Clin Invest 101(3):613–624. https://doi.org/10.1172/JCI1250

Rosenberg JB, Greengard JS, Montgomery RR (2000) Genetic induction of a releasable pool of factor VIII in human endothelial cells. Arterioscler Thromb Vasc Biol 20(12):2689–2695. https://doi.org/10.1161/01.atv.20.12.2689

Ruggeri ZM, Mendolicchio GL (2015) Interaction of von Willebrand factor with platelets and the vessel wall. Hamostaseologie 35(3):211–224. https://doi.org/10.5482/HAMO-14-12-0081

Ruggeri ZM, Zimmerman TS (1980) Variant von Willebrand's disease: characterization of two subtypes by analysis of multimeric composition of factor VIII/von Willebrand factor in plasma and platelets. J Clin Invest 65(6):1318–1325. https://doi.org/10.1172/JCI109795

Ruggeri ZM, Orje JN, Habermann R, Federici AB, Reininger AJ (2006) Activation-independent platelet adhesion and aggregation under elevated shear stress. Blood 108(6):1903–1910. https://doi.org/10.1182/blood-2006-04-011551

Sadler JE, Shelton-Inloes BB, Sorace JM, Harlan JM, Titani K, Davie EW (1985) Cloning and characterization of two cDNAs coding for human von Willebrand factor. Proc Natl Acad Sci U S A 82(19):6394–6398. https://doi.org/10.1073/pnas.82.19.6394

Sadler JE, Budde U, Eikenboom JC, Favaloro EJ, Hill FG, Holmberg L, Ingerslev J, Lee CA, Lillicrap D, Mannucci PM, Mazurier C, Meyer D, Nichols WL, Nishino M, Peake IR, Rodeghiero F, Schneppenheim R, Ruggeri ZM, Srivastava A, Montgomery RR, Federici AB, Party Working on von Willebrand Disease C (2006) Update on the pathophysiology and classification of von Willebrand disease: a report of the Subcommittee on von Willebrand Factor. J Thromb Haemost 4(10):2103–2114. https://doi.org/10.1111/j.1538-7836.2006.02146.x

Saenko EL, Scandella D (1997) The acidic region of the factor VIII light chain and the C2 domain together form the high affinity binding site for von willebrand factor. J Biol Chem 272(29):18007–18014. https://doi.org/10.1074/jbc.272.29.18007

Saenko EL, Yakhyaev AV, Mikhailenko I, Strickland DK, Sarafanov AG (1999) Role of the low density lipoprotein-related protein receptor in mediation of factor VIII catabolism. J Biol Chem 274(53):37685–37692. https://doi.org/10.1074/jbc.274.53.37685

Sakariassen KS, Orning L, Turitto VT (2015) The impact of blood shear rate on arterial thrombus formation. Future Sci OA 1 (4):FSO30. https://doi.org/10.4155/fso.15.28

Salvagno GL, Astermark J, Ekman M, Franchini M, Guidi GC, Lippi G, Poli G, Berntorp E (2007) Impact of different inhibitor reactivities with commercial factor VIII concentrates on thrombin generation. Haemophilia 13(1):51–56. https://doi.org/10.1111/j.1365-2516.2006.01400.x

Savage B, Saldivar E, Ruggeri ZM (1996) Initiation of platelet adhesion by arrest onto fibrinogen or translocation on von Willebrand factor. Cell 84(2):289–297. https://doi.org/10.1016/s0092-8674(00)80983-6

Savage B, Almus-Jacobs F, Ruggeri ZM (1998) Specific synergy of multiple substrate-receptor interactions in platelet thrombus formation under flow. Cell 94(5):657–666. https://doi.org/10.1016/s0092-8674(00)81607-4

Schneider SW, Nuschele S, Wixforth A, Gorzelanny C, Alexander-Katz A, Netz RR, Schneider MF (2007) Shear-induced unfolding triggers adhesion of von Willebrand factor fibers. Proc Natl Acad Sci U S A 104(19):7899–7903. https://doi.org/10.1073/pnas.0608422104

Schwarz HP, Lenting PJ, Binder B, Mihaly J, Denis C, Dorner F, Turecek PL (2000) Involvement of low-density lipoprotein receptor-related protein (LRP) in the clearance of factor VIII in von Willebrand factor-deficient mice. Blood 95(5):1703–1708

Senis YA, Richardson M, Tinlin S, Maurice DH, Giles AR (1996) Changes in the pattern of distribution of von Willebrand factor in rat aortic endothelial cells following thrombin generation in vivo. Br J Haematol 93(1):195–203. https://doi.org/10.1046/j.1365-2141.1996.4661005.x

Shahani T, Covens K, Lavend'homme R, Jazouli N, Sokal E, Peerlinck K, Jacquemin M (2014) Human liver sinusoidal endothelial cells but not hepatocytes contain factor VIII. J Thromb Haemost 12(1):36–42. https://doi.org/10.1111/jth.12412

Shi Q, Wilcox DA, Fahs SA, Kroner PA, Montgomery RR (2003) Expression of human factor VIII under control of the platelet-specific alphaIIb promoter in megakaryocytic cell line as well as storage together with VWF. Mol Genet Metab 79(1):25–33

Shi Q, Fahs SA, Kuether EL, Cooley BC, Weiler H, Montgomery RR (2010) Targeting FVIII expression to endothelial cells regenerates a releasable pool of FVIII and restores hemostasis in a mouse model of hemophilia A. Blood 116(16):3049–3057. https://doi.org/10.1182/blood-2010-03-272419

Simon DI, Chen Z, Xu H, Li CQ, Dong J, McIntire LV, Ballantyne CM, Zhang L, Furman MI, Berndt MC, Lopez JA (2000) Platelet glycoprotein ibalpha is a counterreceptor for the leukocyte integrin Mac-1 (CD11b/CD18). J Exp Med 192(2):193–204. https://doi.org/10.1084/jem.192.2.193

Sivaraman B, Latour RA (2011) Delineating the roles of the GPIIb/IIIa and GP-Ib-IX-V platelet receptors in mediating platelet adhesion to adsorbed fibrinogen and albumin. Biomaterials 32(23):5365–5370. https://doi.org/10.1016/j.biomaterials.2011.04.011

Souto JC, Almasy L, Muniz-Diaz E, Soria JM, Borrell M, Bayen L, Mateo J, Madoz P, Stone W, Blangero J, Fontcuberta J (2000) Functional effects of the ABO locus polymorphism on plasma levels of von Willebrand factor, factor VIII, and activated partial thromboplastin time. Arterioscler Thromb Vasc Biol 20(8):2024–2028. https://doi.org/10.1161/01.atv.20.8.2024

Springer TA (2011) Biology and physics of von Willebrand factor concatamers. J Thromb Haemost 9(Suppl 1):130–143. https://doi.org/10.1111/j.1538-7836.2011.04320.x

Starke RD, Ferraro F, Paschalaki KE, Dryden NH, McKinnon TA, Sutton RE, Payne EM, Haskard DO, Hughes AD, Cutler DF, Laffan MA, Randi AM (2011) Endothelial von Willebrand factor regulates angiogenesis. Blood 117(3):1071–1080. https://doi.org/10.1182/blood-2010-01-264507

Suo J, Linke B, Meyer dos Santos S, Pierre S, Stegner D, Zhang DD, Denis CV, Geisslinger G, Nieswandt B, Scholich K (2014) Neutrophils mediate edema formation but not mechanical allodynia during zymosan-induced inflammation. J Leukoc Biol 96(1):133–142. https://doi.org/10.1189/jlb.3A1213-628R

Suzuki T, Arai M, Amano K, Kagawa K, Fukutake K (1996) Factor VIII inhibitor antibodies with C2 domain specificity are less inhibitory to factor VIII complexed with von Willebrand factor. Thromb Haemost 76(5):749–754

Swystun LL, Lai JD, Notley C, Georgescu I, Paine AS, Mewburn J, Nesbitt K, Schledzewski K, Geraud C, Kzhyshkowska J, Goerdt S, Hopman W, Montgomery RR, James PD, Lillicrap D (2018) The endothelial cell receptor stabilin-2 regulates VWF-FVIII complex half-life and immunogenicity. J Clin Invest 128(9):4057–4073. https://doi.org/10.1172/JCI96400

Swystun LL, Notley C, Georgescu I, Lai JD, Nesbitt K, James PD, Lillicrap D (2019) The endothelial lectin clearance receptor CLEC4M binds and internalizes factor VIII in a VWF-dependent and independent manner. J Thromb Haemost 17(4):681–694. https://doi.org/10.1111/jth.14404

Tagliavacca L, Moon N, Dunham WR, Kaufman RJ (1997) Identification and functional requirement of Cu(I) and its ligands within coagulation factor VIII. J Biol Chem 272(43):27428–27434. https://doi.org/10.1074/jbc.272.43.27428

Thompson AR (2003) Structure and function of the factor VIII gene and protein. Semin Thromb Hemost 29(1):11–22. https://doi.org/10.1055/s-2003-37935

Titani K, Kumar S, Takio K, Ericsson LH, Wade RD, Ashida K, Walsh KA, Chopek MW, Sadler JE, Fujikawa K (1986) Amino acid sequence of human von Willebrand factor. Biochemistry 25(11):3171–3184. https://doi.org/10.1021/bi00359a015

Trabold K, Makhoul S, Gambaryan S, van Ryn J, Walter U, Jurk K (2019) The Direct Thrombin Inhibitors Dabigatran and Lepirudin Inhibit GPIbalpha-Mediated Platelet Aggregation. Thromb Haemost 119(6):916–929. https://doi.org/10.1055/s-0039-1685139

Tracy RP, Arnold AM, Ettinger W, Fried L, Meilahn E, Savage P (1999) The relationship of fibrinogen and factors VII and VIII to incident cardiovascular disease and death in the elderly: results from the cardiovascular health study. Arterioscler Thromb Vasc Biol 19(7):1776–1783. https://doi.org/10.1161/01.atv.19.7.1776

Tsai HM, Lian EC (1998) Antibodies to von Willebrand factor-cleaving protease in acute thrombotic thrombocytopenic purpura. N Engl J Med 339(22):1585–1594. https://doi.org/10.1056/NEJM199811263392203

Tuddenham EG, Lane RS, Rotblat F, Johnson AJ, Snape TJ, Middleton S, Kernoff PB (1982) Response to infusions of polyelectrolyte fractionated human factor VIII concentrate in human haemophilia A and von Willebrand's disease. Br J Haematol 52(2):259–267. https://doi.org/10.1111/j.1365-2141.1982.tb03888.x

Turecek PL, Mitterer A, Matthiessen HP, Gritsch H, Varadi K, Siekmann J, Schnecker K, Plaimauer B, Kaliwoda M, Purtscher M, Woehrer W, Mundt W, Muchitsch EM, Suiter T, Ewenstein B, Ehrlich HJ, Schwarz HP (2009) Development of a plasma- and albumin-free recombinant von Willebrand factor. Hamostaseologie 29(Suppl 1):S32–S38

Turner NA, Moake JL (2015) Factor VIII Is Synthesized in Human Endothelial Cells, Packaged in Weibel-Palade Bodies and Secreted Bound to ULVWF Strings. PLoS ONE 10(10):e0140740. https://doi.org/10.1371/journal.pone.0140740

Urisono Y, Sakata A, Matsui H, Kasuda S, Ono S, Yoshimoto K, Nishio K, Sho M, Akiyama M, Miyata T, Okuchi K, Nishimura S, Sugimoto M (2018) Von Willebrand Factor Aggravates Hepatic Ischemia-Reperfusion Injury by Promoting Neutrophil Recruitment in Mice. Thromb Haemost 118(4):700–708. https://doi.org/10.1055/s-0038-1636529

van Bladel ER, Tuinenburg A, Roest M, de Groot PG, Schutgens RE (2014) Factor VIII concentrate infusion in patients with haemophilia results in decreased von Willebrand factor and ADAMTS-13 activity. Haemophilia 20(1):92–98. https://doi.org/10.1111/hae.12266

van Mourik JA, Boertjes R, Huisveld IA, Fijnvandraat K, Pajkrt D, van Genderen PJ, Fijnheer R (1999) von Willebrand factor propeptide in vascular disorders: A tool to distinguish between acute and chronic endothelial cell perturbation. Blood 94(1):179–185

van Schooten CJ, Shahbazi S, Groot E, Oortwijn BD, van den Berg HM, Denis CV, Lenting PJ (2008) Macrophages contribute to the cellular uptake of von Willebrand factor and factor VIII in vivo. Blood 112(5):1704–1712. https://doi.org/10.1182/blood-2008-01-133181

Verbruggen B, Giles A, Samis J, Verbeek K, Mensink E, Novakova I (2001) The type of factor VIII deficient plasma used influences the performance of the Nijmegen modification of the Bethesda assay for factor VIII inhibitors. Thromb Haemost 86(6):1435–1439

Verhenne S, Denorme F, Libbrecht S, Vandenbulcke A, Pareyn I, Deckmyn H, Lambrecht A, Nieswandt B, Kleinschnitz C, Vanhoorelbeke K, De Meyer SF (2015) Platelet-derived VWF is not essential for normal thrombosis and hemostasis but fosters ischemic stroke injury in mice. Blood 126(14):1715–1722. https://doi.org/10.1182/blood-2015-03-632901

Verweij CL, Hofker M, Quadt R, Briet E, Pannekoek H (1985) RFLP for a human von Willebrand factor (vWF) cDNA clone, pvWF1100. Nucleic Acids Res 13(22):8289. https://doi.org/10.1093/nar/13.22.8289

Vischer UM, Ingerslev J, Wollheim CB, Mestries JC, Tsakiris DA, Haefeli WE, Kruithof EK (1997) Acute von Willebrand factor secretion from the endothelium in vivo: assessment through plasma propeptide (vWf:AgII) Levels. Thromb Haemost 77(2):387–393

Vlot AJ, Koppelman SJ, van den Berg MH, Bouma BN, Sixma JJ (1995) The affinity and stoichiometry of binding of human factor VIII to von Willebrand factor. Blood 85(11):3150–3157

Vlot AJ, Koppelman SJ, Meijers JC, Dama C, van den Berg HM, Bouma BN, Sixma JJ, Willems GM (1996) Kinetics of factor VIII-von Willebrand factor association. Blood 87(5):1809–1816

Vlot AJ, Mauser-Bunschoten EP, Zarkova AG, Haan E, Kruitwagen CL, Sixma JJ, van den Berg HM (2000) The half-life of infused factor VIII is shorter in hemophiliac patients with blood group O than in those with blood group A. Thromb Haemost 83(1):65–69

Wagner DD (1990) Cell biology of von Willebrand factor. Annu Rev Cell Biol 6:217–246. https://doi.org/10.1146/annurev.cb.06.110190.001245

Walker FJ, Scandella D, Fay PJ (1990) Identification of the binding site for activated protein C on the light chain of factors V and VIII. J Biol Chem 265(3):1484–1489

Ward CM, Tetaz TJ, Andrews RK, Berndt MC (1997) Binding of the von Willebrand factor A1 domain to histone. Thromb Res 86(6):469–477. https://doi.org/10.1016/s0049-3848(97)00096-0

Weeterings C, de Groot PG, Adelmeijer J, Lisman T (2008) The glycoprotein Ib-IX-V complex contributes to tissue factor-independent thrombin generation by recombinant factor VIIa on

the activated platelet surface. Blood 112(8):3227–3233. https://doi.org/10.1182/blood-2008-02-139113
Weisel JW, Nagaswami C, Vilaire G, Bennett JS (1992) Examination of the platelet membrane glycoprotein IIb-IIIa complex and its interaction with fibrinogen and other ligands by electron microscopy. J Biol Chem 267(23):16637–16643
Weiss HJ, Sussman, II, Hoyer LW (1977) Stabilization of factor VIII in plasma by the von Willebrand factor. Studies on posttransfusion and dissociated factor VIII and in patients with von Willebrand's disease. J Clin Invest 60 (2):390–404. https://doi.org/10.1172/jci108788
White GC, 2nd, Rosendaal F, Aledort LM, Lusher JM, Rothschild C, Ingerslev J, Factor V, Factor IXS (2001) Definitions in hemophilia. Recommendation of the scientific subcommittee on factor VIII and factor IX of the scientific and standardization committee of the International Society on Thrombosis and Haemostasis. Thromb Haemost 85 (3):560
Wieberdink RG, van Schie MC, Koudstaal PJ, Hofman A, Witteman JC, de Maat MP, Leebeek FW, Breteler MM (2010) High von Willebrand factor levels increase the risk of stroke: the Rotterdam study. Stroke 41(10):2151–2156. https://doi.org/10.1161/STROKEAHA.110.586289
Wion KL, Kelly D, Summerfield JA, Tuddenham EG, Lawn RM (1985) Distribution of factor VIII mRNA and antigen in human liver and other tissues. Nature 317(6039):726–729. https://doi.org/10.1038/317726a0
Wise RJ, Dorner AJ, Krane M, Pittman DD, Kaufman RJ (1991) The role of von Willebrand factor multimers and propeptide cleavage in binding and stabilization of factor VIII. J Biol Chem 266(32):21948–21955
Yuan L, Chan GC, Beeler D, Janes L, Spokes KC, Dharaneeswaran H, Mojiri A, Adams WJ, Sciuto T, Garcia-Cardena G, Molema G, Kang PM, Jahroudi N, Marsden PA, Dvorak A, Regan ER, Aird WC (2016) A role of stochastic phenotype switching in generating mosaic endothelial cell heterogeneity. Nat Commun 7:10160. https://doi.org/10.1038/ncomms10160
Zhou YF, Springer TA (2014) Highly reinforced structure of a C-terminal dimerization domain in von Willebrand factor. Blood 123(12):1785–1793. https://doi.org/10.1182/blood-2013-11-523639
Zhou YF, Eng ET, Zhu J, Lu C, Walz T, Springer TA (2012) Sequence and structure relationships within von Willebrand factor. Blood 120(2):449–458. https://doi.org/10.1182/blood-2012-01-405134
Zhu X, Cao Y, Wei L, Cai P, Xu H, Luo H, Bai X, Lu L, Liu JR, Fan W, Zhao BQ (2016) von Willebrand factor contributes to poor outcome in a mouse model of intracerebral haemorrhage. Sci Rep 6:35901. https://doi.org/10.1038/srep35901

Chapter 19
Antigen–Antibody Complexes

A. Brenda Kapingidza, Krzysztof Kowal and Maksymilian Chruszcz

Abstract In vertebrates, immunoglobulins (Igs), commonly known as antibodies, play an integral role in the armamentarium of immune defense against various pathogens. After an antigenic challenge, antibodies are secreted by differentiated B cells called plasma cells. Antibodies have two predominant roles that involve specific binding to antigens to launch an immune response, along with activation of other components of the immune system to fight pathogens. The ability of immunoglobulins to fight against innumerable and diverse pathogens lies in their intrinsic ability to discriminate between different antigens. Due to this specificity and high affinity for their antigens, antibodies have been a valuable and indispensable tool in research, diagnostics and therapy. Although seemingly a simple maneuver, the association between an antibody and its antigen, to make an antigen–antibody complex, is comprised of myriads of non-covalent interactions. Amino acid residues on the antigen binding site, the epitope, and on the antibody binding site, the paratope, intimately contribute to the energetics needed for the antigen–antibody complex stability. Structural biology methods to study antigen–antibody complexes are extremely valuable tools to visualize antigen–antibody interactions in detail; this helps to elucidate the basis of molecular recognition between an antibody and its specific antigen. The main scope of this chapter is to discuss the structure and function of different classes

Maksymilian Chruszcz—To whom correspondence should be addressed: Maksymilian Chruszcz, Department of Chemistry and Biochemistry, University of South Carolina, Columbia, SC 29208, USA

A. B. Kapingidza · M. Chruszcz (✉)
Department of Chemistry and Biochemistry, University of South Carolina, Columbia, SC 29208, USA
e-mail: chruszcz@mailbox.sc.edu

A. B. Kapingidza
e-mail: anyway@email.sc.edu

K. Kowal
Department of Allergology and Internal Medicine, Medical University of Bialystok, Bialystok, Poland
e-mail: kowalkmd@umb.edu.pl

Department of Experimental Allergology and Immunology, Medical University of Bialystok, Bialystok, Poland

U. Hoeger and J. R. Harris (eds.), *Vertebrate and Invertebrate Respiratory Proteins, Lipoproteins and other Body Fluid Proteins*, Subcellular Biochemistry 94,
https://doi.org/10.1007/978-3-030-41769-7_19

of antibodies and the various aspects of antigen–antibody interactions including antigen–antibody interfaces—with a special focus on paratopes, complementarity determining regions (CDRs) and other non-CDR residues important for antigen binding and recognition. Herein, we also discuss methods used to study antigen–antibody complexes, antigen recognition by antibodies, types of antigens in complexes, and how antigen–antibody complexes play a role in modern day medicine and human health. Understanding the molecular basis of antigen binding and recognition by antibodies helps to facilitate the production of better and more potent antibodies for immunotherapy, vaccines and various other applications.

Keywords Antigen · Antibody · Antigen–antibody interface · Antigen recognition · CDR · Framework region · X-ray crystallography · NMR · CryoEM · Proteins · Superantigens · Allergic diseases

Introduction to Antibodies

Immunoglobulins

Immunoglobulins (Igs), also known as antibodies, play an integral role in the armamentarium of immune defense in vertebrates. The presence of these protective molecules mainly found in the blood stream and other body fluids, was first identified by von Behring and Kitasato in 1890 (Schroeder and Cavacini 2010). The duo reported the existence of an agent in the blood that could neutralize diphtheria toxin. Later, it was noticed that the agent could discriminate between two immune substances. Through the years, it was discovered that antibodies eradicate and neutralize extracellular pathogens including viruses and bacteria (Schroeder and Cavacini 2010). More than 120 years of research and investigations have not only demystified the function, structure and the immunological role of these protective agents, but also emphasized the antibodies' complex nature. A significant fraction of all studies in this area focus on human and murine antibodies, so in this chapter we will mainly concentrate on antibodies from these two organisms.

Antibodies are secreted by B lymphocytes, specifically differentiated B cells called plasma cells. When antibodies are secreted into blood and tissue fluids, they can exist as soluble proteins in the blood plasma, or as membrane bound proteins, adhered to the surface of B cells as B-cell receptors (BCR). These BCRs facilitate the activation of the B cells including their differentiation into memory B cells that recognize "second-time" antigen offenders to launch a better immune response, or antibody producing plasma cells (Borghesi and Milcarek 2006). However, for a full immune response to be achieved, usually B cells should interact with T-helper lymphocytes, upon antigen binding, leading to antibody production (Parker 1993).

Antibodies, except for those observed in sharks and camelids (Stanfield et al. 2004; Hamers-Casterman et al. 1993), are hetero-oligomeric glycoproteins that belong to the immunoglobulin superfamily. Visualized simply, the shape of antibodies basic

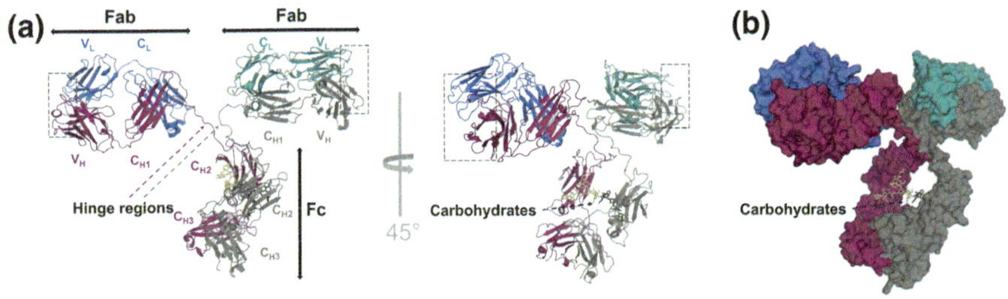

Fig. 19.1 Structure of murine IgG1. **a** Cartoon representation of IgG1 (PDB code: 1IGT (Harris et al. 1997)) with light chains (blue and teal) and heavy chains (purple and gray) marked in different colors. Domains forming both light and heavy chains, as well as Fc and Fab fragments are labeled. Antigen binding regions are marked with boxes. Carbohydrate residues are shown in stick representation. **b** Surface representation of IgG1. Orientation of the molecule is the same as for the neighboring carton representation. The figure indicates that the carbohydrate residues are located in cavity between two C_H2 domains

structure can be compared to the letter "Y" (or "T" or a Y/T hybrid; Fig. 19.1) (Harris et al. 1998; Tian et al. 2015). Both extremities of these Y-shaped proteins are important for their function (Janeway et al. 2001; Huber 1980). The top part of the Y, the antibody fragment (Fab), has the antigen binding site called the paratope which is specific to one part of the antigen referred to as an epitope. The lower part of the Y consists of the crystallizable fragment (Fc), which binds to specific receptors to communicate with other components of the immune system (Maverakis et al. 2015). There are five types of Fc regions that enable antibodies to bind to different Fc receptors thereby initiating various immune responses depending on the receptor bound. These differences in the Fc regions give rise to five antibody classes (Fig. 19.2), (Arnold et al. 2007): IgM, IgG, IgA, IgD and IgE. IgG and IgA are further divided into subclasses IgG1, IgG2, IgG3, IgG4, IgA1 and IgA2 respectively (Spiegelberg 1989; Leder 1982).

More specifically, the structure of all antibodies consists of two heavy (H) and two light (L) polypeptide chains that have an NH_2 terminal variable-domain (V) and a COOH-constant domain (C) (Fig. 19.1). The variable domains bind antigens and the constant domains specify effector functions such as binding to Fc receptors (Spiegelberg 1989). The five different types of Fc regions on the heavy chain, α (alpha), γ (gamma), δ (delta), ε (epsilon), and μ (mu), give rise to IgA, IgG, IgD, IgE, and IgM antibody classes respectively. Both L chains have one identical constant domain; whereas both H chains have three, or four such domains. There are two different light chains, κ (kappa) and λ (lambda), which are characterized by different physicochemical properties (Townsend et al. 2016). There is no evidence, however, that the light chain constant region determines, or affects the effector function of an antibody. Variable domains can be further divided into three regions of sequence variability, complementarity determining regions (CDRs), and four regions that have comparatively constant amino acid sequences named the framework regions (FRs) (Schroeder Jr and Cavacini 2010; Spiegelberg 1989). A paratope is, more often than

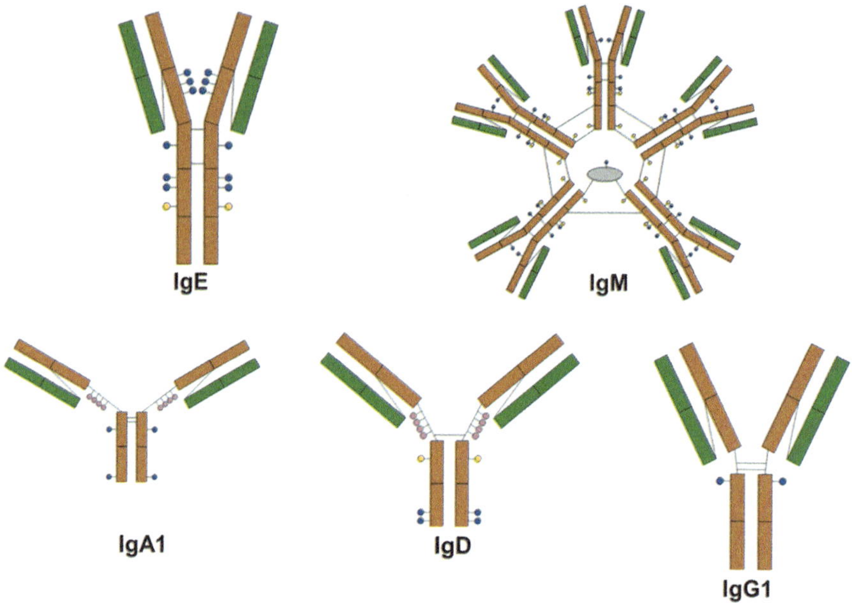

Fig. 19.2 Diagrammatic representations of antibody classes IgE, IgM, IgA1, IgD, and IgG1. Light chains are in green, heavy chains-brown, inter-chain disulfide bridges-black and IgM J chain-grey. N-glycosylation sites are presented in blue, oligomannose glycans in yellow and O-glycans in pink (Arnold et al. 2007)

not, composed of amino acids originating from the three CDRs of the heavy chain and the three CDRs of the light chain (Fig. 19.3). The antigen binding site is extremely variable and this enables antibodies to recognize millions of different antigens even within the same antibody class (Market and Papavasiliou 2003).

Despite their different classes and subclasses, antibodies implement prominent modes of action, to either eradicate or mark the pathogen for devastation by immune cells. Antibody action on pathogens includes, but is not limited to, neutralization, opsonization, antibody dependent cellular cytotoxicity (ADCC) and agglutination (Abbas et al. 2014). Pathogens use molecules in their cell walls, or envelopes to bind the host cell and gain entry into it. Antibodies like IgA and IgG attach to these pathogen molecules thereby blocking entry into the host cell. Antibodies also coat pathogens for degradation in a process called opsonization (Fig. 19.4). These antibody-coated pathogens are marked for phagocytosis by neutrophils and macrophages. In ADCC, antibodies like IgG coat the surface of a cell and generate signals that activate natural killer cells to discharge their toxic granule proteins on an infected cell thereby killing the cell (Abbas et al. 2014). Another way of eradicating pathogens that antibodies employ is agglutination. In this mechanism, antibodies glue pathogens together into clumps. The clumped cells simultaneously become inactive and are susceptible to phagocytosis or, become a target for the complement system (Elek et al. 1964). Antibodies execute their protective roles in the immune system

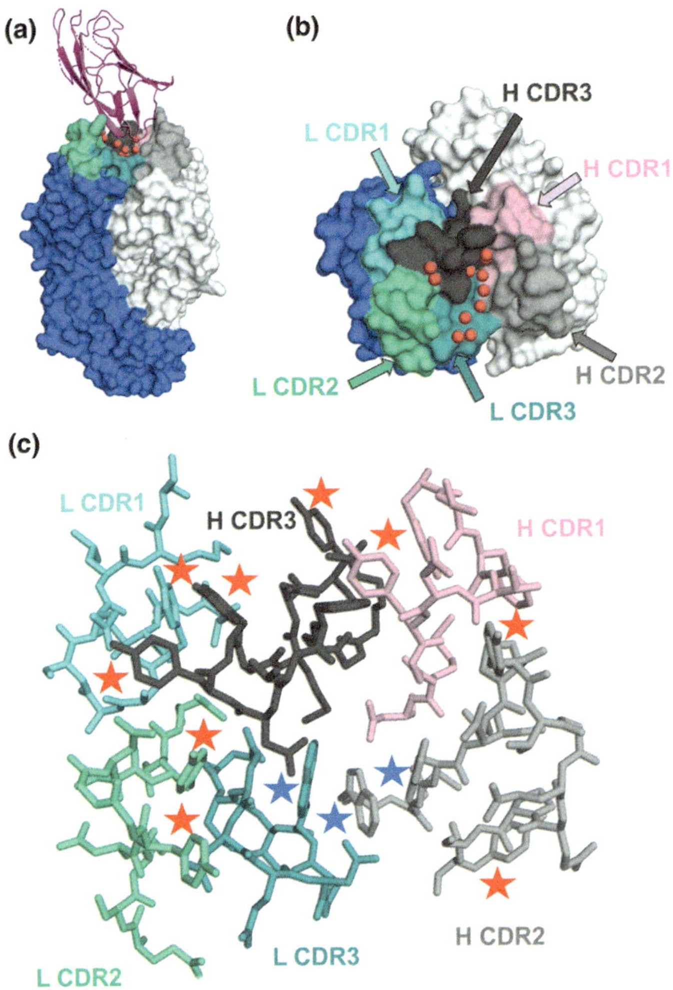

Fig. 19.3 Structure of a complex between Fab fragment of murine antibody 7A1 (IgG) and house dust mite allergen Der p 2 (PDB code: 6OY4 (Glesner et al. 2019)). **a** Overall structure of the complex. Der p 2 is shown in carton representation (magenta) and Fab fragments shown in space filling representation. Water molecules buried between the antibody and antigen are represented by red spheres. **b** CDRs of 7A1 mapped on the antibody surface and shown in different colors. **c** CDRs shown using stick representation. Red stars mark position of tyrosine residues and blue stars mark position of tryptophan residues

by employing various mechanisms, and all of these functions include some form of interaction between an antibody and its antigen. For clarification, there are at least two definitions of an antigen. The first definition is that an antigen is a substance that induces an immune response, and the second states that anything that binds to immunoglobulins, or T-cell receptors is an antigen. For the purpose of this chapter,

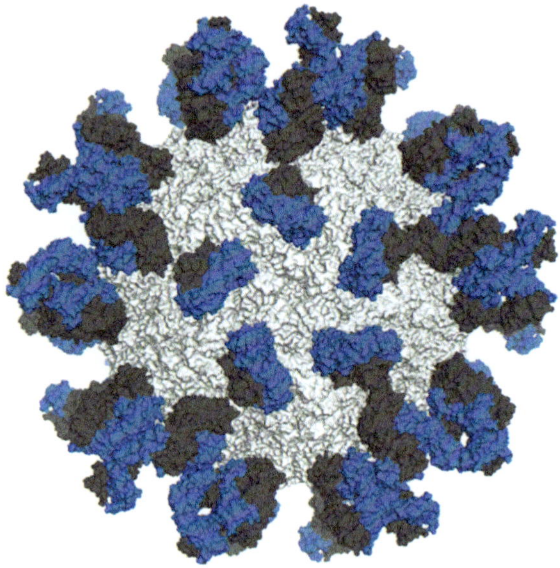

Fig. 19.4 Complex of foot-and-mouth disease virus capsid (light gray) with Fab fragments of neutralizing antibody SD6 (PDB code: 1QGC (Hewat et al. 1997)). Heavy chain of the antibody is presented in dark gray and light chain of the antibody is shown in blue

the second definition will be adopted. Irrespective of the antibody class, the antibody should somehow interact with its antigen to elicit an immune response. For example, in neutralization, the immune response is heavily dependent on the interaction of an antibody's variable region with its antigen. However, other mechanisms like ADCC, depend on the interaction between the antibody's constant domain with immune cells through Fc receptors and/or complement proteins (Pollara and Tay 2019).

Antibody Classes and Functions

Although the general structure of all antibody classes is very similar, each antibody class has unique biological characteristics, functions and elicits specific immune responses. Starting with immunoglobulin M, IgM, is the most conserved antibody class expressed first in immune-development and it is found in all vertebrates (Boes 2000). A naïve B cell, never exposed to an antigen, expresses cell surface monomeric IgM in bound form (here by "monomer" we mean a basic antibody unit composed of two heavy and two light chains). Most circulating IgM is "natural" IgM; thus, IgM is produced spontaneously without a known exogenous antigenic, or microbial stimulus. There are three main known forms of secreted IgM: monomeric, pentameric, and hexameric. In healthy people, circulating IgM exists predominantly in the pentameric form. Pentameric IgM is formed in the endoplasmic reticulum, contains 5 monomers of IgM linked together by a J-(joining) chain (Fig. 19.2). The multimeric IgM has high avidity despite low affinity of its individual components. This makes IgM very efficient in the coating and neutralization of antigens. In addition, upon binding to an antigen, polymeric IgM potently activates the complement system. IgM is also associated with primary immune response and is involved in immunoregulation (Boes 2000; Grönwall et al. 2012). The relatively low affinity IgMs are also known as natural antibodies. These natural antibodies play a role in augmenting the clearance of dying cells, as well as reinforcing the mechanisms that protect the body from the development of autoimmune diseases (Grönwall et al. 2012).

Even though a naïve B cell expresses IgM, it also expresses cell surface immunoglobulin D (IgD). IgD mainly functions as an antigen receptor on naïve B cell surfaces, although it is also believed to play a part in the production of antimicrobial factors by activating mast cells and basophils to release pro-inflammatory cytokines (Chen et al. 2009b). The co-expression of both IgM and IgD renders the B lymphocyte ready to respond to an antigen. When the cell bound antibody binds to an antigen, this activates the B cell causing it to divide and differentiate into an antibody producing "factory", a plasma cell (Geisberger et al. 2006). The antibodies produced by the plasma cell are not only specific for the bound antigen but, are also secreted and no longer membrane bound. Intriguingly, some daughter cells from the activated B cell, are subjected to class switch recombination and somatic hypermutation to help generate a diverse and high-affinity repertoire of antibody classes and subclasses (Durandy 2003). These two processes cease the production of IgM, and/or IgD and start the production of other antibody classes like IgG, IgE and IgA leading to a specific effector function needed for each specific antigenic challenge (Geisberger et al. 2006; Stavnezer and Amemiya 2004).

Immunoglobulin G, IgG, does not only have the longest half-life in the serum, but is the most predominant antibody in the body. In all its four forms, IgG (IgG1, IgG2, IgG3 and IgG4) provides the majority of immunity in antibody-based immunity and is the only antibody class that is able to cross the placenta to give passive immunity to a fetus (Schroeder and Cavacini 2010). IgG subclasses were identified due to functional, structural and antigenic differences in the heavy chain constant region particularly C_H1 and C_H3. These C_H1 regions control the mobility and flexibility of the antibody. Hence, IgG subclasses exhibit different functional activities (Schroeder and Cavacini 2010). Structural differences in their heavy chains determine the ability to activate the complement system and to bind to IgG receptors such as CD64 (FcεRI), CD32 (FcεRII) and CD16 (FcεRIII). For example, it was suggested that IgG1 and IgG3 are mainly induced in response to protein antigens, however, IgG2 is usually activated in response to polysaccharide antigens and IgG4 antibodies are often formed after a long-term, or repeated exposure to an antigen in a non-infectious setting like in allergic diseases (Schroeder and Cavacini 2010; Vidarsson et al. 2014). IgG subclasses also have similar functions such as participation in secondary immune response, trans-placental transport as well as neutralization of toxins and viruses. However, subclasses still impact the outcome of the immune response. For example, IgG3 antibodies were observed to be more effective in neutralizing the HIV virus than IgG1 (Cavacini et al. 2003). IgG antibodies, especially IgG1 and IgG4, have also been shown to influence antigen–antibody complex formation and binding to B cells in food allergies (Meulenbroek et al. 2013).

Immunoglobulin E, IgE, although the least abundant antibody class in the serum with the shortest half-life, it is a very potent antibody which is a troublesome instigator of allergic reactions (Platts-Mills 2001; Burton and Oettgen 2011). However, the major role of IgE involves immunity to parasites through its binding to eosinophils. IgE also binds to mast cells and basophils with high affinity but, it is of rudimentary importance to highlight that the IgE receptor on eosinophils is seen only in some clinical conditions, such as, parasite infestations (Gounni et al. 1994). IgE immune responses are called "immediate hypersensitivity". This term does not only denote the extreme nature of IgE sensitivity to antigens, but also the tremendous speed in which the immune response takes place (Burton and Oettgen 2011). IgE has an extremely high affinity for FcεRI receptors expressed on mast cells, basophils, eosinophils and Langerhans cells (Schroeder and Cavacini 2010). When an allergen enters the human body, it binds IgE on mucosal mast cells, or if administered parenterally, it also binds to basophils circulating in the blood. However, eosinophils and Langerhans cells are not involved in immediate hypersensitivity. They participate in the late phase response which occurs 3–8 h after the initial challenge. IgE strongly binds to FcεRI receptors after which an antigen crosslinks two IgE molecules leading to degranulation of mast cells and the release of histamine and other pro-inflammatory mediators. Immediate hypersensitivity is manifested in one, or several symptoms like urticaria in the skin, food-induced diarrhea and anaphylaxis of the gut, bronchospasms and so on depending on the organ where sensitization occurred (Schroeder and Cavacini 2010; Burton and Oettgen 2011).

Immunoglobulin A, IgA, the most abundant antibody at mucosal surfaces and in secretions, is the first line of defense in the immune system armamentarium against ingested and inhaled antigens like toxins, viruses, and bacteria at mucosal surfaces (Underdown and Schiff 1986). These mucosal surfaces include the gut, the urogenital tract, and the respiratory tract. IgA has also been shown to be found in saliva and breast milk. IgA serum levels are higher than IgM, but lower by considerable magnitudes compared to IgG. Generally, IgA is found in monomeric form in the serum, but can assume different oligomeric states at the mucosa. Termed secretory IgA (sIgA) at the mucosa, the antibody mainly exists in dimeric form. Dimeric sIgA was shown to be more effective in preventing damage (by toxins) to epithelial cells than monomeric (Schroeder and Cavacini 2010). IgA is found in two main subclasses, IgA1 and IgA2. The difference between the two subclasses is mainly in their hinge region, with IgA1 having a longer hinge region. The elongated IgA1 hinge region consists of repeated amino acids that increases the antibody's propensity to bacterial protease degradation. For that reason, 90% of serum sIgA is IgA1, and IgA2 predominates on the mucosal surface. IgA fights against viruses and bacteria by neutralizing, or preventing the binding of the pathogens on the mucosal surface (Schroeder and Cavacini 2010; Underdown and Schiff 1986).

Antigen–Antibody Complexes

Since 1890, when the existence of antibodies was first defined as part of the immune system armament against pathogens, scientists went on a quest to demystify and understand the interactions between an antibody and its antigen. Antigen–antibody complexes became an extremely valuable tool in understanding detailed antigen–antibody interactions and elucidating the basis of molecular recognition between an antibody and its specific antigen (Davies et al. 1988). Antigen–antibody complexes have not only helped to predict the biological function of particular antigens, or proteins, but also explained the mechanisms by which antigens can elicit an immune response. For example, it was found that in allergic diseases some allergens promote sensitization through their proteolytic activity, or by mimicking certain proteins in signaling pathways (Karp 2010). The main aim of this part of the chapter is to discuss various aspects of antigen–antibody interactions including antigen–antibody interfaces, methods used to study these complexes, antigen recognition by antibodies, types of antigens in complexes and how antigen–antibody complexes play a role in modern day medicine and human health.

Structural Methods to Study Antigen–Antibody Complexes

In modern day research, various ways of studying antigen antibody-complexes have been developed. Among all those methods, X-ray crystallography, nuclear magnetic resonance (NMR) and Single Particle Cryogenic Electron Microscopy (CryoEM) have been the most widely used. Of the three methods, X-ray crystallography championed in the Protein Data Bank (PDB) with almost 94% of antigen–antibody complex structures determined using this method. X-ray crystallography is followed by NMR and CryoEM that contributed approximately 6% of the reported structures of antigen–antibody complexes in the PDB (Berman et al. 2000). Structural models of antigen–antibody complexes generated using these methods are not only interesting from the scientific point of view, but their analysis is also performed for various medicinal and other commercial purposes. In studying antigen–antibody complexes, all the afore-mentioned methods have their merits and demerits. A brief summary of the method as well as its advantages and disadvantages will be discussed below.

X-Ray Crystallography

Since most B cell epitopes (fragments of antigens interacting with paratopes) are conformational (Van Regenmortel 2009), to date, X-ray crystallography remains the gold standard that provides a detailed three-dimensional structure of an antigen–antibody complex. X-ray crystallography is able to produce a snapshot of an antigen–antibody complex at high resolution that makes it possible to visualize the specific interactions between the antigen and antibody (Sheriff et al. 1987; Jeffrey et al. 1995). However, due to antibodies large size, flexibility and glycosylation, it is extremely difficult, to determine their structure by X-ray crystallography (King and Brooks 2018). In fact, currently there are only three available crystal structures of intact antibodies (Harris et al. 1997, 1998; Saphire et al. 2001), and there is no single structure of a whole antibody in complex with an antigen. Fortunately, antibody variants like antibody fragments (Fab), single chain variable fragments (ScFv) and other various antibody formats are easier to crystallize and therefore, they are used for determination of antigen–antibody structures using X-ray crystallography.

X-ray crystallography comprises three basic steps (Chruszcz et al. 2008). The first step is to obtain an X-ray diffracting crystal. This step is usually the bottle-neck for using this method in studying antigen–antibody complexes. The second step is obtaining the crystal X-ray diffraction pattern. Structure determination using X-ray crystallography depends exclusively on diffracted beams that are produced when X-rays interact with the crystal (King and Brooks 2018; Wlodawer et al. 2008). The diffraction data collected is combined and processed computationally to produce an electron density map that contains information about the chemical structure and orientation of the atoms in the molecular structure under study. Finally, the atomic model generated from the electron density map is refined using several software programs that employ various parameters important for structure determination, to yield the final crystal structure (Wlodawer et al. 2008).

Nuclear Magnetic Resonance

Although NMR does not give as exhaustive information on antigen–antibody binding as X-ray crystallography, this method is able to map residues forming epitopes and/or paratopes, and shed light on intermolecular interactions (Zuiderweg 2002; Wider and Wüthrich 1999). The biggest advantage of NMR over X-ray crystallography is that it does not need the formation of a crystal, so it is usually used for those proteins that are hard to crystallize. The basic idea employed in NMR epitope mapping and antigen–antibody interactions studies is that when an amino acid on an antigen binds to an antibody, the amino acid's chemical environment changes. The chemical environment of residues on the antigen–antibody interface changes because protein-solvent interactions are replaced by protein-protein interactions (Zuiderweg 2002). The NMR signal that a residue emits unbound and bound to an antibody changes, and it is these changes in NMR signal that are studied to deduce residues that are involved in antibody binding. By studying the differences in NMR spectra of an antigen before and after binding an antibody, epitope residues can be easily identified. Retrospectively, if two different antibodies exhibit the same changes in the NMR signals, then they are regarded as sharing the same epitope (Simonelli et al. 2018). However, for the NMR signals to be generated, the antigen has to be labelled by either ^{15}N, or ^{13}C. Therefore, expression systems like *E. coli,* or yeast are required which might not be practical for some proteins.

^{15}N-Heteronuclear Single Quantum Coherence (^{15}N-HSQC) is usually the most used NMR method for studying antigen–antibody interactions. The N–H of each amino acid residue generates an NMR signal. So sensitive are the positions of these signals to the protein conformation that ^{15}N-HSQC is often referred to as the protein fingerprint. NMR use in structure determination and antigen–antibody interactions is readily applicable to proteins that are 40–50 kDa. However, NMR signal intensity is greatly weakened by the increase in molecular weight of the complex. Intense reduction in signal, for bigger proteins and complexes, from 60–100 kDa, almost renders NMR inapplicable. Consequently, for larger antigen–antibody complexes, Transverse Relaxation Optimized Spectroscopy (TROSY) is used (Zuiderweg 2002; Wider and Wüthrich 1999). Although a very powerful tool, NMR requires high amounts of pure homogeneous sample of 0.5 mM or, more. Detailed and comprehensive discussion on NMR usage in antigen–antibody interactions research is found in these excellent articles (Simonelli et al. 2018; Rosen and Anglister 2009).

Single Particle Cryogenic Electron Microscopy

Although X-ray crystallography and NMR are the fundamental techniques for structure determination at atomic resolution, over the past decade, CryoEM has been intrinsically gaining popularity in the field of structural biology. This is mainly because in CryoEM large multi-subunit complexes of viruses, bacterial appendages, eukaryotic ribosome and cellular organelles are able to be visualized without the need for crystallization (Chen et al. 2009a; Donnarumma et al. 2015). Currently, structures of small proteins, or other antigen of similar size cannot be analyzed using CryoEM, however this problem can be mitigated by forming antigen–antibody complexes. In this manner structures of complexes as small as 65 kDa can be determined. This approach could be the solution to solving antigen–antibody structures with full IgE or IgG, since these proteins are very flexible and hard to crystallize (Donnarumma et al. 2015; Wu et al. 2012).

In CryoEM, a small amount of sample suspended in buffer solution is quickly frozen to form a non-ctystalline glass-like specimen called vitreous ice. This vitrified sample is then exposed to high frequency electrons which scatter through the specimen due to electrostatic interactions with the sample's atoms. Scattered and unscattered electrons form an interference pattern that results in 2D projection images which are recorded. These 2D projections are recorded from viewing the specimen from different angles. For structure determination, the combined 2D projections are further digitally processed and analyzed taking into account all orientation parameters (Skiniotis and Southworth 2016). The final structure is then solved enabling visualization of biological complexes under close to physiological conditions. Unfortunately, quite often this technique provides low-resolution structural information, and it has to be combined with other methods to provide more detailed information on epitopes and paratopes involved in antigen–antibody interactions. For a further thorough discussion of CryoEM in structural biology, see the following articles (Skiniotis and Southworth 2016; Boekema et al. 2009).

Other Methods to Study Antigen–Antibody Interactions

Although the three methods discussed above are the "go-to" techniques for mapping antibody epitopes in studying antigen–antibody interactions, they do possess major restrictions. Hence, several other relatively simpler methods are employed in B-cell epitope mapping (Potocnakova et al. 2016). Examples of such methods are mutagenesis, hydrogen–deuterium exchange coupled with mass spectroscopy, and peptide-based approaches. In the mutagenesis approach, to putatively determine antibody epitopes, binding of the antibody to the antigen mutants is analyzed. A reduction in antibody binding is deciphered as an indication that the mutated amino acid residues form part of the epitope. Using "shotgun mutagenesis", or Ala scanning, thousands of proteins can be screened simultaneously (Abbott et al. 2014). In the other method used, hydrogen-deuterium exchange coupled to mass spectrometry, the rate of deuteration is slowed down if some part of the antigen is interacting with an antibody. After deuteration, the level of deuteration for the pepsin digested antigen is analyzed by mass spectrometry. This technique is applicable to most proteins with a moderate level of purity and even impure antibodies can be used (Wei et al. 2014). One more common method in epitope mapping is the peptide-based approach. A number of overlapping peptides covering the whole primary sequence of an antigen are synthesized. These peptides are immobilized onto a solid matrix and their binding to the antibody of interest is analyzed in an ELISA-like format. The peptide(s) that bind to the antibody are then sequenced and the mimotopes deciphered. This technique is applicable to linear epitopes, quick and easy to perform. Huge peptide libraries can be generated allowing for fast and comprehensive epitope mapping (Potocnakova et al. 2016; Ahmad et al. 2016). There are many more techniques, be it computational and in vitro, or in vivo that this chapter cannot comprehensively cover. Other methods not discussed here are reviewed by Ahmad et al. (2016).

Antigen–Antibody Interactions

Complementarity Determining Regions

Just as their name suggests, the six hypervariable loops of the heavy and light chain known as the complementarity determining regions (CDRs) (Fig. 19.3), have been solely designated to be responsible for direct antigen binding. Hence, in studying the structural and molecular basis for antigen recognition, finding the CDR boundaries and the residues involved in antigen binding have been the focus of many studies. Several methods to decipher CDR boundaries have been developed over the years like the Kabat, Clothia and IMGT numbering nomenclatures (Al-Lazikani et al. 1997; Chothia and Lesk 1987; Chothia et al. 1989; Sela-Culang et al. 2013). However, as mentioned earlier, the general structure of an antibody comprises of the Fab regions and the constant region(s). The Fab region contains two variable domains from the light and heavy chain and two constant domains (Fig. 19.1). Each of the variable domains contains three hypervariable loops (CDRs). Therefore, there are three CDRs from the heavy chain (H1, H2 and H3) and the three from the light chain (L1, L2 and L3) (Sela-Culang et al. 2013; Schroeder and Cavacini 2010). The folding of the light and heavy variable domains brings the hypervariable loops together allowing them to form an antigen binding site, also known as the paratope. The hypervariable loops are supported by framework of beta-sheets that are the basic building blocks of the characteristic immunoglobulin domains (Schroeder and Cavacini 2010).

Paratopes and Epitopes

Identification of CDRs is used as the starting point in identifying paratopes, since the residues that make contact with the antigen are quite often found at the center of the region formed by the CDRs' residues (Sela-Culang et al. 2013). Interestingly though, 3D structure analysis of antibodies showed that only 20–33% of the CDRs' amino acid residues actually take part in antigen binding and these are the most variable within each CDR (Padlan et al. 1995). This also suggests that CDRs of a particular antibody may form several paratopes that are able to recognize various epitopes. For example, such a situation was observed in the case of a bH1 antibody that is able to recognize fragments of two completely different proteins (Fig. 19.5a), (Bostrom et al. 2009). This scenario is an example of the so-called true cross-reactivity, in which an antibody raised against one antigen also binds to another unrelated antigen. True cross-reactivity like this, also shows that the antibody may be poly-specific (Van Regenmortel 2014). Additionally, the presence of such cross-reactivity also highlights the importance of distinguishing between immunogenic and antigenic properties of molecules. A single antibody may recognize two unrelated antigens;

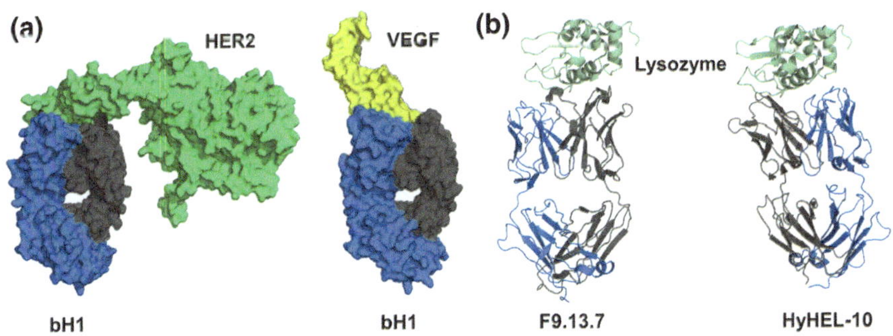

Fig. 19.5 **a** Structures of the same antibody (bH1) with two completely different proteins (PDB codes: 3BE1 (Bostrom et al. 2009) and 3BDY (Bostrom et al. 2009)). **b** Structures of two unrelated antibodies binding to the same epitope on lysozyme surface (PDB codes: 1FBI (Lescar et al. 1995) and 3HFM (Padlan et al. 1989)). Fab fragments of antibodies were used for structural studies. Light chains are shown in blue and heavy chains are shown in dark gray

however, it is also likely that two unrelated antibodies are binding to the same epitope. This is illustrated in Fig. 19.5b (Pons et al. 2002), where the same epitope located on a lysozyme protein molecule is recognized by two different paratopes originating from two different antibodies. The second situation corresponds to the presence of an immunodominant epitope that is recognized by various polyclonal antibodies. This phenomenon was vividly demonstrated in one study whereby allergenic epitopes of bovine α_{s1} casein, a major allergen from cow milk, were identified using 188 overlapping peptides. All sera from bovine α_{s1} casein allergic patients bound to three common regions on the allergen surface. These regions showed the most reactivity to the polyclonal antibodies from the patients' sera, although the antibodies also bound to other regions. Indeed, the three regions were immunodominant epitopes compared to the other epitopes (Spuergin et al. 1996).

True cross-reactivity is observed less frequently than the so-called shared reactivity (Van Regenmortel 2014; Berzofsky and Schechter 1981). In shared reactivity two antigens possess a common fragment in their epitopes that is recognized by the same antibody. The case of two-house dust allergens, Der f 1 and Der p 1, that both bind to a cross-reactive antibody, 4C1, provides an excellent example to illustrate shared reactivity (Fig. 19.6), (Chruszcz et al. 2012). Der p 1 and Der f 1 share approximately 80% sequence identity, and 4C1 (murine IgG) binds to epitopes that are almost identical in both allergens. The residues forming the core of the 4C1 epitopes on Der f 1 and Der p 1 are not only conserved in terms of their sequence, but the side chains of the conserved residues also adopt very similar conformations. The 4C1 epitopes on Der f 1 and Der p 1 are examples of conformational and discontinues epitopes, as they are composed of residues that are distant when considering the proteins' amino acid sequences (primary structure of the proteins), however, the residues forming the epitopes are brought in vicinity of each other when the proteins are folded. Such epitopes can be destroyed by denaturation of the proteins (loss of the tertiary and secondary structure), or by modification of the proteins' surfaces.

A second example of shared reactivity was demonstrated by Stanfield et al. (2006a). The authors elucidated the binding of human immunodeficiency virus type 1 (HIV-1) neutralizing antibody 2219, that is able to bind three peptides derived from V3 fusion protein (Fig. 19.7). The three peptides binding to 2219 have approx-

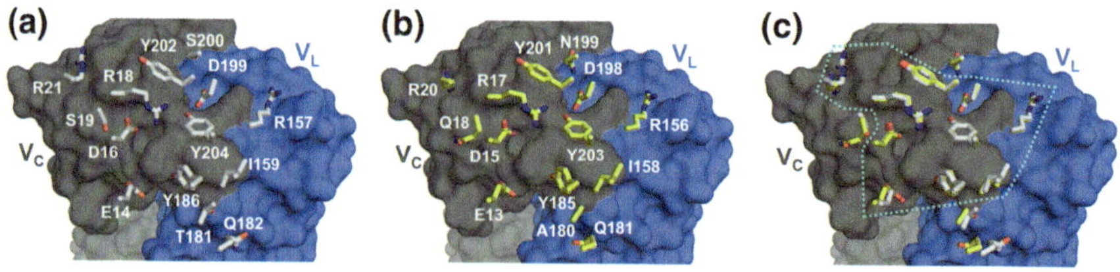

Fig. 19.6 Comparisons of 4C1 binding epitopes on Der f 1 (PDB code: 5VPL (Chruszcz et al. 2012)) (**a**) and Der p 1 (PDB code: 5VPG (Chruszcz et al. 2012)) (**b**). The residues forming the epitopes are shown in stick representation. Antibody is shown in space-filling representation with the heavy chain in gray and light chain in blue. **c** Superposition of Der f 1 and Der p 1 epitopes. The residues inside of the marked regions are identical for both Der f 1 and Der p 1

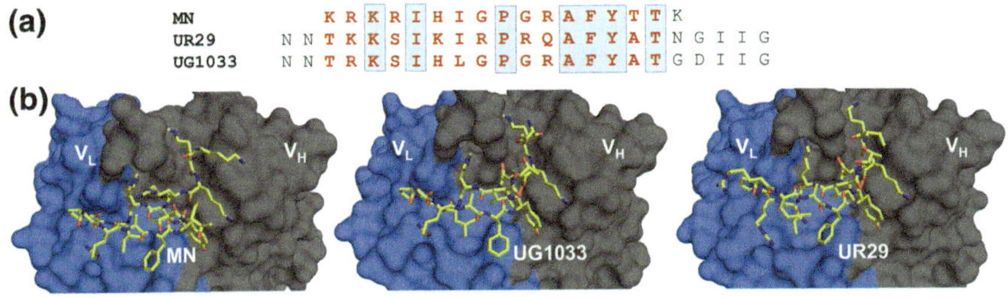

Fig. 19.7 **a** Three different peptides derived from third hypervariable loop of HIV gp120 that bind human monoclonal antibody 2219. Residues in dark red are ordered in the crystal structures shown below. Amino acids highlighted with blue boxes are conserved in all three peptides. **b** Crystal structures of antibody 2219 in complex with peptides MN (left; PDB code: 2B0S (Stanfield et al. 2006a)), UG1033 (middle; PDB code: 2B1A (Stanfield et al. 2006a)) and UR29 (right; PDB code: 2B1H (Stanfield et al. 2006a))

imately 50% sequence identity, however, they all adopt a very similar β-hairpin conformation. Therefore, these peptides can be simultaneously described as both being linear and conformational epitopes. A third example of shared reactivity is related to interactions of antibodies with some polysaccharides that are N- or O-linked to proteins. These alleged cross-reactive carbohydrate determinants (CCDs) are important from the perspective of allergy diagnostics (Mari et al. 1999) because the presence of IgEs directed against CCDs may result in some false positives and suggests a polysensitization.

Structural Characterization of Paratopes and Epitopes

Numerous structural studies of antigen–antibody complexes allowed for more detailed characterization of paratopes and epitopes. It was shown that the five out of the six hypervariable regions adopt a relatively small number of main chain conformations (Al-Lazikani et al. 1997; North et al. 2011), and only H CDR3 displays a pronounced conformational variability (Weitzner et al. 2015). It was also observed that of all the CDR chains, L CDR3 and H CDR3 residues dominated in antigen binding (Padlan et al. 1995; Padlan 1994). A number of research studies revealed that each CDR consists of distinct and unique amino acid residues different from other CDRs (Kunik and Ofran 2013; Zhao and Li 2010; Raghunathan et al. 2012). Furthermore, according to the so-called hotspot hypothesis, only a few very specific residues on the antigen–antibody interface within each paratope and epitope are critical for antigen recognition and binding (Bogan and Thorn 1998). Site directed mutagenesis on epitope residues have supported this hypothesis. Mutations of one, or two epitope residues significantly reduced, or even abrogated antibody binding showing that only a few residues dominated the energetics of the antigen–antibody interactions (Li et al. 2001; Glesner et al. 2017). It was also shown that paratopes are enriched in some residues, like aromatic amino acids (Phe, Trp and Tyr), with tyrosine being the most important (Birtalan et al. 2008; Peng et al. 2014; Robin et al. 2014). This is shown in Fig. 19.3c, which clearly illustrates the enrichment of CDRs in tyrosine residues. While enrichment of paratopes in some amino acids can be easily demonstrated, such effects are not observed for proteinous epitopes, and their composition is generally the same as the composition of the protein surface (Kunik and Ofran 2013; Kringelum et al. 2013).

Structural studies also demonstrated that quite often water molecules are trapped between paratopes and epitopes (Braden et al. 1995; Bhat et al. 1994). Water molecules enhance antigen–antibody interactions and stability by improving charge complementarity through hydrogen bonding, and increasing packing density and paratope-epitope complementarity by filling "empty" cavities at the interface (Braden et al. 1995; Mariuzza and PoIjak 1993). Generally, surfaces corresponding to paratopes may be divided into concave, ridged, planar, and convex (MacCallum et al. 1996). Research showed that the concave paratopes are mainly observed for complexes of antibodies with haptens (Fig. 19.8), ridged for complexes with peptides (Fig. 19.7) and planar for proteinous antigens (Fig. 19.6). It was also observed that

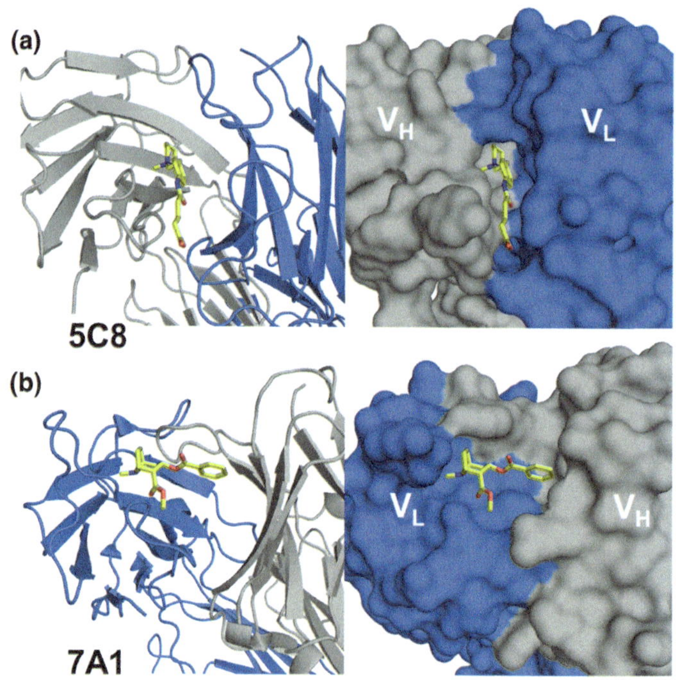

Fig. 19.8 Crystal structure of Fab fragments with haptens. The binding sites are shown in two representations—carton (left) and surface (right). Haptens are shown in stick representation. **a** Complex of catalytical antibody 5C8 with a transition state analog (PDB code: 25C8 (Gruber et al. 1999)). **b** Complex of 7A1 antibody with cocaine (PDB code: 2AJV (Zhu et al. 2006))

the concave paratope topology is characteristic for antibodies binding disordered antigens (MacRaild et al. 2016). Convex paratopes were perceived for dromedary heavy-chain antibodies (Fig. 19.9a), (De Genst et al. 2006). Carbohydrate antigens are mainly recognized by ridged paratopes (Dingjan et al. 2015; Haji-Ghassemi et al. 2015). However, structural studies revealed an interesting mode of antibody binding for carbohydrates. Namely, in the case of an anti-carbohydrate HIV neutralizing antibody in complex with a hexasaccharide, the antibody Fab forms an unusual dimer with heavy chains swapped (Fig. 19.10) (Stanfield et al. 2006a). Usually the affinities of anti-carbohydrate antibodies are lower than those observed for protein or, peptide antigens (Haji-Ghassemi et al. 2015), therefore, the unusual mode of binding observed for the mentioned anti-carbohydrate HIV neutralizing antibody may compensate for the lower affinity by forming contacts with the two antigens. In this situation, instead of affinity, the term avidity should be used to describe the overall strength of the binding between the antibody with multiple binding sites and antigen(s).

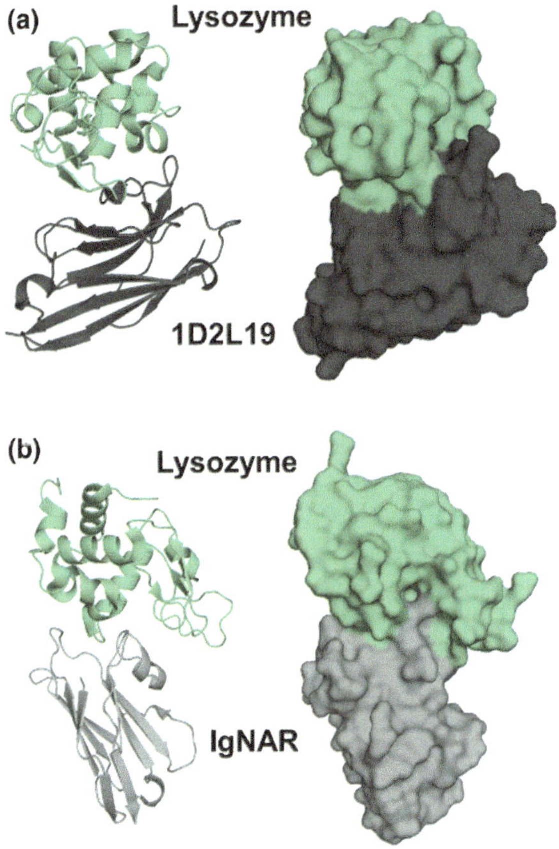

Fig. 19.9 Structures of lysozyme with variable domains of single chain antibodies. The complexes are shown in cartoon (left) and surface representation (right). **a** Complex with camelid antibody 1D2L19 (PDB code: 1RI8 (De Genst et al. 2005)). **b** Complex with shark IgNAR (PDB code: 1SQ2 (Stanfield et al. 2004))

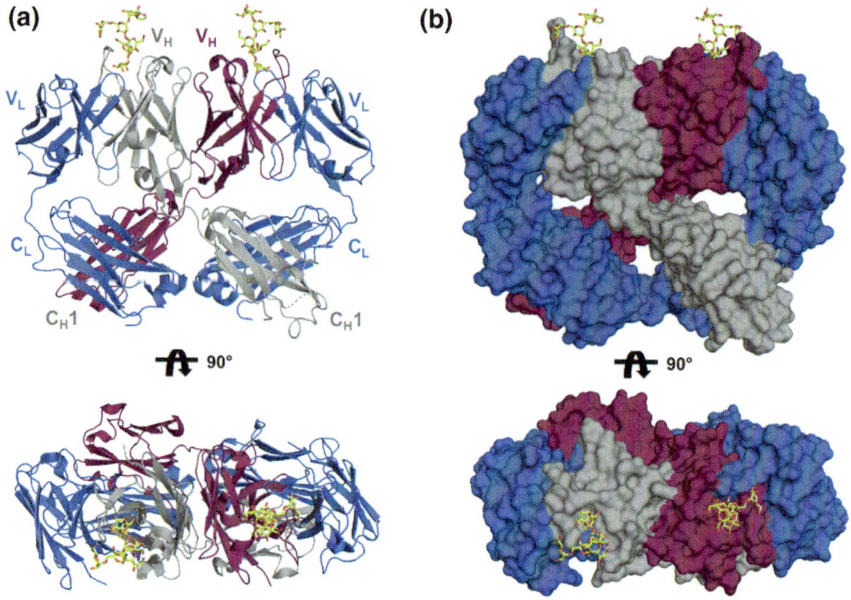

Fig. 19.10 Structure of a complex between a dimeric Fab 2G12 and Rv3 hexasaccharide (PDB code: 4RBP (Stanfield et al. 2015)) in cartoon (**a**) and surface (**b**) representations. The swapped heavy chains are shown in gray and purple, while light chains are shown in blue. The antigen is shown in stick representation (carbon atoms in yellow and oxygen atoms in red)

As implied above, the stability of macromolecular assemblies, like antigen–antibody complexes, is also facilitated by the same molecular interactions that stabilize correctly folded native proteins. Hence, just as the chemical environment affects the three-dimensional state of a protein, the pH, ionic strength and solvent also affect antigen–antibody complexes (Torres and Casadevall 2008). These chemical environmental effects can influence the distribution of charges and conformational effects thereby affecting binding of antibodies to their antigen. The major contributions to the free energy of an antigen–antibody binding comes from hydrophobic and hydrogen bonding interactions. (Torres and Casadevall 2008; Webster et al. 1994). However, in the fine tuning of antibody specificities charge–charge interactions, like salt bridges or cation-π interactions, play very important roles (Dalkas et al. 2014).

Other Non-CDR Components Important for Antigen Binding

As discussed previously, the constant domains have long since been believed to be exclusively responsible for mediating effector functions. However, there is a growing body of evidence suggesting that constant regions, framework regions (FRs) and other residues that are not part of the CDRs play a crucial role in antigen binding (Torres and

Casadevall 2008; Sela-Culang et al. 2013). Framework regions are part of the variable regions providing support to the CDRs to assume the right conformation for antigen binding (Sela-Culang et al. 2013). They are mainly believed to be scaffolds for the CDRs, but during murine antibody humanization when FR residues were omitted, substantial decrease, or complete abrogation of antigen binding was observed. When some of the FR residues were mutated back into the murine antibody sequence antigen binding was retained (Kettleborough et al. 1991). This highlighted the importance of FR residues in antigen binding. In fact, some of the FR residues due to their proximity to the CDRs can actually bind the antigen (Xiang et al. 1995, 1999). Interestingly, some FR residues that are further from the CDRs in sequence, but closer in 3D structure were also found to be involved in antigen binding especially heavy chain FR-3 residues, that account for 1.3% of human antigen–antibody contacts (Raghunathan et al. 2012; Capra and Kehoe 1974). FR residues that do not make contact with the antigen help maintain the CDR structural confirmation and orientation needed for antigen binding (Haidar et al. 2012).

Constant regions mainly deemed to be responsible for antibody effector functions, like Fc receptor binding and avidity, have been recently proven to be involved in antibody specificity and affinity for its antigen (Torres and Casadevall 2008). Although antibody classes have identical variable domains, but different constant domains, it was observed that they bind the same antigen with different affinities (Casadevall and Janda 2012; Tudor et al. 2012; Adachi et al. 2003; Xia et al. 2012). This phenomenon has also been shown in antigen–antibody complexes, whereby the dissociation constant of a single chain variable fragment antibody in complex with its antigen was different from a Fab format of the same antibody. Consequently, the importance of constant domains in antigen binding is undeniable (Adachi et al. 2003). Allosteric effects of the constant domains on the variable domain conformation have been attributed to be the cause of the differences perceived in antibody class affinity and specificity (Xia et al. 2012). Conclusively, constant regions are indeed important players in antigen–antibody complexes. Hence, the antibody class in antibody engineering for either research, vaccines or, immunotherapy should be carefully chosen, because the affinity, specificity and the epitope bound can be easily affected by the class used.

Other than constant and framework regions, research shows that the glycosylation of antibodies plays a pivotal role in antibody function and antigen binding. Although all immunoglobulins are glycosylated, they exhibit substantial diversity both in the number and location of the conserved glycosylation sites be it on the variable, or constant regions (Fig. 19.2). Glycosylation accounts for 12–14% of the total molecular weight of IgE, IgM and IgD. Interestingly, IgG glycosylation is only 2–3% of its total molecular weight (Arnold et al. 2007). Antibody glycans play multiple roles from maintaining antibody effector functions by enhancing the binding of the Fc regions to its receptors (Mimura et al. 2001) to facilitating subcellular transport, secretion and clearance (Gala and Morrison 2002). Most significantly, antibody glycans play important structural roles by maintaining antibody conformation, as well as stability (Mimura et al. 2000), and they also participate in binding events (Malhotra et al. 1995). For example, in one study the removal of glycans from IgG antibodies led to the reduction, or complete abolition of IgG binding to its receptors, FcγRs, demonstrating the importance of glycans in IgG antibody function (Schroeder and Cavacini 2010). Not only that, both antibody-dependent cellular cytotoxicity and complement dependent cytotoxicity were shown to be heavily influenced by the glycosylation and glycan composition of IgG antibody (Schroeder and Cavacini 2010). For an exhaustive discussion on the importance of antibody glycosylation, please consult the review by Arnold et al. (2007).

Conformational Changes Associated with Antigen–Antibody Binding

Perspectives on antigen–antibody interactions have changed over time. Initially it was believed that these interactions could be described using the "lock and key" model proposed by Fisher. However, as more and more structural data became available it became clear that in many cases the "lock and key" model did not properly describe the antigen–antibody interaction (Wilson and Stanfield 1994), and the so-called "induced fit" model provided a significantly better picture of antigen and antibody binding. As demonstrated by structural studies, this induced fit model was made possible due to antibody flexibility (Stanfield et al. 2006b), that allows for profound conformational changes in paratope regions upon antibody–antigen complex formation. It was also stressed that proteins (including antibodies) are highly dynamic molecules, which in normal physiological conditions undergo continuous conformational changes involving both main and side chain domains. This dynamic nature of proteins suggests that in solution immunoglobulins adopt various conformations and there is a pre-existing equilibrium between various (sometimes very similar) conformations of these molecules (Keskin 2007). This pre-existing equilibrium model explains how antibody flexibility may lead to promiscuity, or cross-reactivity (Van Regenmortel 2014); the same set of amino acids may adopt various conformations enabling them to recognize and interact with various epitopes. In fact, the antigens are also dynamic, hence, epitopes may also undergo various conformational changes (Liang et al. 2016). The plasticity of the epitopes and paratopes can be illustrated using the idea of adjustable locks and flexible keys that was proposed by Khan and Salunke (2014).

The analysis of antigen–antibody complexes and flexibility of the molecules also provides an interesting insight into antibody maturation process. Namely, structural studies have proposed that for antibodies that exhibit higher affinity for their antigens, "lock and key" type of binding is observed and that the binding is mainly driven by ionic and polar interactions. These interactions contribute towards higher binding energy between the antibody and its antigen by maintaining the rigidity of the antigen binding site (Wedemayer et al. 1997; Sinha and Smith-Gill 2002; Chong et al. 1999). Therefore, it was suggested that during the process of maturation, the flexibility of the paratope forming region of the antibody is reduced, and the reduction of this region's plasticity is accompanied by changes that allow the formation of more specific polar interactions, like hydrogen bonds, or charge–charge interactions (Pons et al. 2002; LeBrasseur 2003; Cauerhff et al. 2004; Sinha et al. 2002; DeKosky et al. 2016). At the same time, the increase in antibody affinity is driven by the expansion of the paratope area and the formation of additional non-polar interaction (Li et al. 2003).

The orientation of the CDRs and paratope forming residues is affected by the orientation of variable regions of an antibody. One may expect that the same variable regions may have a different relative orientation dependent on the constant region of a particular antibody (Torres and Casadevall 2008; Narayanan et al. 2009). Therefore, class-switching combined with flexibility of the antigen binding site may results in an improved fit to the antigen.

Antigen Recognition and Interactions in Heavy Chain-Only Antibodies

As discussed earlier, the conventional antibody paratope is made up of six CDRs, three from the variable domain heavy chain and another three from the light chain. Interestingly though, there are species that deviate from this norm (Hamers-Casterman et al. 1993). Camelid and shark antibodies lack the light chain domain making the antigen binding site composed of only the heavy chain CDRs (Hamers-Casterman et al. 1993; Greenberg et al. 1995). The variable domains (antigen binding domains) of these heavy chain-only antibodies are usually referred to as single-domain antibodies (sdAbs). As warranted by their lack of light chain CDRs, sdAbs have smaller paratope diameter and surface area. Compared to conventional antibodies, the size of sdAbs paratopes is about half (Henry and MacKenzie 2018). On the sdAbs paratopes, the heavy chain CDR3 is unusually long, folding over the area supposedly to have been for the light chain interface, thereby interacting more with the cognate antigen. Despite these differences, the heavy chain CDRs of sdAbs are made up of similar amino acids compositions, like conventional antibodies (Henry and MacKenzie 2018; Harmsen et al. 2000; Muyldermans et al. 1994), although it was also suggested that these paratopes display a greater diversity in comparison with conventional antibodies (Mitchell and Colwell 2018). Consequently, sdAbs also implement the same antigen–antibody interactions and can bury similar solvent accessible areas as conventional antibodies although the energetics are more pronounced on the sdAbs small epitopes producing high affinity interactions (Henry and MacKenzie 2018). Merited by their small molecular size, sdAbs are claimed to be able to access recessed sites on antigen interfaces like enzyme active sites and viral glycoproteins (Rouet et al. 2015; Jahnichen et al. 2010). However, the dominant mechanism employed by conventional antibodies in antigen–antibody complexes with haptens, small molecule lipids as well as oligosaccharides, is the same as sdAbs exhibit in antigen–antibody complexes (Henry and MacKenzie 2018; Fanning and Horn 2011).

Antigen–Antibody Complexes in Human Health

The binding of an antigen to its specific antibody, a presumably simple immunological event, dramatically changes the properties of the bound antigen leading to various immune responses including antigen presentation and processing, inflammatory responses and receptor signaling (Wen et al. 2016). Modern day medicine has exploited this immunologically simple event to innovate new drugs, immunotherapy and biomedical technologies in the treatment of not only cancer, but autoimmune diseases, antibody deficiencies and pediatric infections (Wen et al. 2016; Kholodenko et al. 2019). Although full antibodies are used in immunotherapy (Jolles et al. 2005; Wood 2012), due to the relatively large size of antibody molecules, for example IgG~150 kDa, they cannot effectively penetrate into some tissues like tumors and cross the blood–brain barrier to gain access to the central nervous system (Kholodenko et al. 2019). Hence, in antigen–antibody complexes used in health care today, antibody fragments are usually preferred (Nelson 2010; Strohl and Strohl 2013). Antibody fragments still possess the same antigen specificity as full-length antibodies. They also have other unique characteristics that are important for both diagnostic and therapeutic purposes: antibody fragments have a shorter serum circulating half-life, are easy to manufacture and manipulate, can be engineered to be multi-specific and due to the absence of constant regions, they do not trigger cytotoxic immune responses, or antibody dependent cell-mediated cytotoxicity which can lead to adverse side effects (Strohl and Strohl 2013). However, in some cases the presence of the Fc fragment is indispensable. This is the case when an antibody employs the complement system, or enhanced phagocytosis, both processes are dependent on the presence of the Fc fragment. A good example of this scenario is in anti-CD20 therapy (Freeman and Sehn 2018).

In human health, the association of an antibody with its specific antigen leads to a cascade of events that either result in a beneficial, or sometimes detrimental immune response to the host. Taking HIV and cancer as examples, use of antibody complexes in vaccine development and immunotherapy for these two incurable diseases has been promising. For example, in one study on HIV in mice, antibodies recognizing the V3 region of the envelope glycoprotein, gp120, elicited production of strong neutralizing antibodies against the virus (Visciano et al. 2008). This immune response, caused by an antibody binding to its epitope, was merited to the exposure of a concealed V3 region which is usually inaccessible without antibody binding (Pan et al. 2015). Other antibodies generated for binding gp120, but outside the CD4 binding region, also induced production of neutralizing antibodies against several HIV strains (Liao et al. 2004). These and other findings are crucial in the design of an HIV vaccine. In cancer and tumor immunotherapy, one common method of treatment is targeting surface factors crucial for tumor cell growth (Nimmerjahn and Ravetch 2005). When antibodies recognize and bind to these factors, like epidermal growth factor receptor (EGFR) and vascular endothelial growth factor (VEGF), forming an antigen–antibody complex, ADCC is triggered which leads to tumor lysis (Moalli et al. 2010). Apart from triggering ADCC, binding of antibodies to antigens on the surface of tumor cells can also induce tumor cell death mainly via apoptosis (Ludwig et al. 2003). Additionally, another study showed that the formation of these antigen–antibody complexes facilitates tumor suppression by employing therapeutic, prophylactic and memory-inducing methods (Rafiq et al. 2002). All these efforts and many more not mentioned here, have led to several antibodies approved by the FDA for use in cancer therapy (Kholodenko et al. 2019). A comprehensive discussion of antigen–antibody complexes use in the study and treatment of other diseases is found here (Wen et al. 2016).

On the negative side, the formation of antigen–antibody complexes has been linked to the development of various autoimmune and allergic diseases (Theofilopoulos and Dixon 1980). One prominent example is the major role played by immunoglobulin E, IgE, in what has been termed type 1 hypersensitivity. IgE is implicated in hypersensitivity reactions stemming from food allergies, allergic asthma, atopic dermatitis, allergic rhinitis to some types of drugs, and sting allergies (Platts-Mills 2001). In allergic diseases, antigen presenting cells present the antigen to T helper cells, Th2 cells. The activated Th2 cells releases interleukins, like interleukin 4 and 13, which leads to the upregulation of IgE production. This antigen specific IgE attaches itself to the high affinity FcεRI receptors on mast cells, or basophils through CH3 domain of its constant regions. At this juncture, IgE is geared for attack upon a subsequent antigen exposure (Presta et al. 1994). When an antigen binds to IgE variable regions of two adjacent antibodies, formation of this antigen–antibody complex cross-links the receptors. This crosslinking together with the uptake of calcium ions into the mast cell, or basophil results in cell activation with rapid release of mediators such as histamine, or tryptase from cell granules as well as newly-synthesized such as leukotrienes, prostaglandins and cytokines. These mediators are indicted to be responsible for both the immediate and late allergic reactions that can be as severe as anaphylaxis (Oettgen and Geha 1999; Williams and Galli 2000).

Superantigens

Of all the antigens discussed in this chapter, superantigens are one of the most, if not the most, potent antigens capable of inducing massive immune responses and wreaking havoc in the human immune system (Fraser 2011). In comparison to a normal antigen-induced T-cell response where only 1 in 100,000 (0.001%) of the body's T-cells are activated, superantigens are capable of activating up to 1 in 5 (20%) of the body's T-cells (Van Kaer 2018). Attributed to this tremendous immune response, superantigens have been implicated in diverse human diseases ranging from food poisoning, toxic shock syndrome, autoimmune diseases, allergic diseases to HIV-1 (Van Kaer 2018; Marone et al. 2006).

The ability of superantigens to evoke such a strong immune response compared to conventional antigens, lies in their capability to bind both the Major Histocompatibility complex (MHC) class II molecules on antigen presenting cells and T-cell receptors on T-cells (Fraser 2011). In a normal setting, microbial antigens are engulfed and digested by antigen presenting cells (APC) to yield small peptides that are presented on the APC cell surface bound together MHC class II molecules. The MHC and peptide complex are consequently recognized by T-cell receptors on T cells, and an immune response specific to that peptide antigen is launched (Fraser 2011). Research has shown that superantigens, unlike conventional antigens, by-pass antigen presentation stage and bind both MHC and T-cell receptors as an intact protein (Van Kaer 2018). This T-cell activation results in the excessive production of cytokines, interleukin-2 and tumor necrosis factor alpha, which elicits the massive immune responses associated with different diseases (Fraser 2011).

Other antigens that do not solely elicit T-cell mediated immune response also exhibit similar binding characteristics (Marone et al. 2006). For example, Protein Fv that is both a superallergen and superantigen, is produced in the human liver and is associated with Hepatitis A, B, C and E. It binds with high affinity to the human antibody heavy chain variable domain regardless of the antibody class, or subclass. Protein Fv is a superantigen because it binds to the VH3 fragment of different human antibodies. The protein also acts as a superallergen because it binds to a variable region of IgE (but not IgM or IgG) present on mast cells and basophils through its high-affinity interaction with IgE VH3+ (Bouvet and Marone 2007). This binding is done outside of the conventional antigen binding region. Intriguingly, one Fv protein molecule can bind up to 12 antibody fragments of human IgG, IgE and IgM (Bouvet et al. 1990). Other superallergens like Protein A associated with *Staphylococcus aureus* also possess two binding sites, one that binds to the Fab region of IgG, IgM, IgA and IgE and another binding site that interacts with receptors. For IgE, binding of its receptors lead to mast cells and basophils degranulation which is a hallmark of allergic diseases. HIV-1 gp-120 superallergen also cause basophil and mast cells degranulation in the same manner (Marone et al. 2006).

Interestingly though, not only superallergens can possess these dual binding sites. For example, Phl p 7, an important calcium binding allergen from *Phleum Pratense* (Timothy grass), can bind a human IgG antibody in a canonical way by interacting with five of the hypervariable loops, but could simultaneously interact in a non-canonical way with CDR-L2 and framework residues (Fig. 19.11). This phenomenon, technically makes Phl p 7 a superantigen (Mitropoulou et al. 2018). Additionally, two molecules of Phl p 7 could interact with two antibody molecules contradicting the general dogma that only dimeric antigens can interact with two antibody identical molecules at a specific time. When anti-Phl p 7 IgG antibody was switched to IgE, only one molecule of Phl p 7 cross-linked and signaled effector functions. These findings further confirmed the unusual antibody binding for Phl p 7 (Mitropoulou et al. 2018). For a detailed discussion of superantigens and superallergens and their role in human diseases please consult this excellent book by Marone (2007).

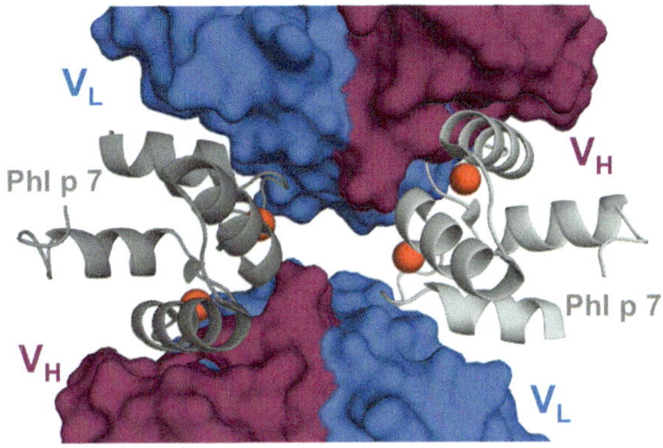

Fig. 19.11 Crystal structure of the complex between two Fab fragments of a human antibody and two molecules of timothy grass pollen allergen—Phl p 7 (PDB code: 5OTJ (Mitropoulou et al. 2018)). The antibody is binding two molecules of the monomeric allergen using conventional and superantigen-like recognition. Phl p 7 molecules are shown in cartoon representation (gray) and the antibodies are shown in surface representation (light chains in blue and heavy chains in purple). Calcium ions bound by Phl p 7 are presents as red spheres

Closing Remarks

As one of the most versatile natural biological sensors, antibodies have been extensively used in numerous settings in research, disease diagnostics and therapy. Antibodies protect vertebrates against invasion by harmful microbes by binding to antigens to launch an immune response; also by activating other components of the immune system to annihilate the invaders. More than a century of research and investigations have greatly improved our understanding of the function, structure and the immunological role of these protective agents, and also emphasized the complex nature of antibodies.

To launch an immune response after encountering an antigen, antibodies recognize and bind to their specific antigens making antigen–antibody complexes. Although the central dogma states that in antigen–antibody complexes CDR amino acid residues hold the keys to antigen recognition, research has shown that non-contacting residues, on both constant and framework regions, are responsible for preserving the structural conformations needed for the CDRs to make antigen binding possible. By affecting the molecular flexibility of the antibody, non-contacting residues can induce substantial impact in antigen binding especially in the affinity and specificity of the antibody. Moreover, some residues on the framework regions actually make contact with the antigen and substantially contribute to the energetics needed for antigen binding. Binding of an antigen to its specific antibody dramatically changes the properties of the bound antigen leading to various immune responses including antigen presentation and processing. The binding of superantigens to antibodies compared to conventional antigens irrefutably illustrates the importance of antigen–antibody complexes in immunity and explains the molecular basis of diseases in human health.

Modern day medicine, has exploited antigen–antibody complexes to innovate new drugs, immunotherapy and biomedical technologies in the treatment of various diseases like cancer, autoimmune diseases, antibody deficiencies and pediatric infections. Studying the structural and molecular mechanisms by which antibodies recognize and bind to their specific antigens helps to understand the basis of immunity, facilitate disease diagnostics and enable better design of vaccines and immunotherapy.

Acknowledgements The authors would like to thank Dr. Tapiwa Pekoyo for insightful comments and discussions. This work was supported by grant R01AI077653 from National Institute of Allergy and Infectious Diseases. The content is solely the responsibility of the authors and does not necessarily represent the official views of the National Institutes of Health.

References

Abbas AK, Lichtman AH, Pillai S (2014) Basic immunology: functions and disorders of the immune system. Elsevier Health Sciences

Abbott WM, Damschroder MM, Lowe DC (2014) Current approaches to fine mapping of antigen–antibody interactions. Immunology 142(4):526–535

Adachi M, Kurihara Y, Nojima H, Takeda-Shitaka M, Kamiya K, Umeyama H (2003) Interaction between the antigen and antibody is controlled by the constant domains: normal mode dynamics of the HEL–hyHEL-10 complex. Protein Sci 12(10):2125–2131

Ahmad TA, Eweida AE, Sheweita SA (2016) B-cell epitope mapping for the design of vaccines and effective diagnostics. Trials Vaccinol 5:71–83

Al-Lazikani B, Lesk AM, Chothia C (1997) Standard conformations for the canonical structures of immunoglobulins. J Mol Biol 273(4):927–948

Arnold JN, Wormald MR, Sim RB, Rudd PM, Dwek RA (2007) The impact of glycosylation on the biological function and structure of human immunoglobulins. Annu Rev Immunol 25:21–50

Berman HM, Westbrook J, Feng Z, Gilliland G, Bhat TN, Weissig H, Shindyalov IN, Bourne PE (2000) The protein data bank. Nucleic Acids Res 28(1):235–242

Berzofsky JA, Schechter AN (1981) The concepts of crossreactivity and specificity in immunology. Mol Immunol 18(8):751–763

Bhat TN, Bentley GA, Boulot G, Greene MI, Tello D, Dall'Acqua W, Souchon H, Schwarz FP, Mariuzza RA, Poljak RJ (1994) Bound water molecules and conformational stabilization help mediate an antigen-antibody association. Proc Natl Acad Sci USA 91(3):1089–1093

Birtalan S, Zhang Y, Fellouse FA, Shao L, Schaefer G, Sidhu SS (2008) The intrinsic contributions of tyrosine, serine, glycine and arginine to the affinity and specificity of antibodies. J Mol Biol 377(5):1518–1528

Boekema EJ, Folea M, Kouřil R (2009) Single particle electron microscopy. Photosynth Res 102(2–3):189

Boes M (2000) Role of natural and immune IgM antibodies in immune responses. Mol Immunol 37(18):1141–1149

Bogan AA, Thorn KS (1998) Anatomy of hot spots in protein interfaces. J Mol Biol 280(1):1–9

Borghesi L, Milcarek C (2006) From B cell to plasma cell. Immunol Res 36(1–3):27–32

Bostrom J, Yu S-F, Kan D, Appleton BA, Lee CV, Billeci K, Man W, Peale F, Ross S, Wiesmann C (2009) Variants of the antibody herceptin that interact with HER2 and VEGF at the antigen binding site. Science 323(5921):1610–1614

Bouvet J-P, Marone G (2007) Protein Fv: an endogenous immunoglobulin superantigen and superallergen. Chem Immunol Allergy 93:58–76. Karger Publishers

Bouvet JP, Pires R, Lunel-Fabiani F, Crescenzo-Chaigne B, Maillard P, Valla D, Opolon P, Pillot J (1990) Protein F. A novel F(ab)-binding factor, present in normal liver, and largely released in the digestive tract during hepatitis. J Immunol 145(4):1176–1180

Braden BC, Fields BA, Poljak RJ (1995) Conservation of water molecules in an antibody–antigen interaction. J Mol Recognit 8(5):317–325

Burton OT, Oettgen HC (2011) Beyond immediate hypersensitivity: evolving roles for IgE antibodies in immune homeostasis and allergic diseases. Immunol Rev 242(1):128–143

Capra JD, Kehoe JM (1974) Variable region sequences of five human immunoglobulin heavy chains of the VHIII subgroup: definitive identification of four heavy chain hypervariable regions. Proc Natl Acad Sci USA 71(3):845–848

Casadevall A, Janda A (2012) Immunoglobulin isotype influences affinity and specificity. Proc Natl Acad Sci USA 109(31):12272–12273

Cauerhff A, Goldbaum FA, Braden BC (2004) Structural mechanism for affinity maturation of an anti-lysozyme antibody. Proc Natl Acad Sci 101(10):3539–3544

Cavacini LA, Kuhrt D, Duval M, Mayer K, Posner MR (2003) Binding and neutralization activity of human IgG1 and IgG3 from serum of HIV-infected individuals. AIDS Res Hum Retroviruses 19(9):785–792

Chen JZ, Settembre EC, Aoki ST, Zhang X, Bellamy AR, Dormitzer PR, Harrison SC, Grigorieff N (2009a) Molecular interactions in rotavirus assembly and uncoating seen by high-resolution cryo-EM. Proc Natl Acad Sci USA 106(26):10644–10648

Chen K, Xu W, Wilson M, He B, Miller NW, Bengten E, Edholm E-S, Santini PA, Rath P, Chiu A (2009b) Immunoglobulin D enhances immune surveillance by activating antimicrobial, proinflammatory and B cell–stimulating programs in basophils. Nat Immunol 10(8):889

Chong LT, Duan Y, Wang L, Massova I, Kollman PA (1999) Molecular dynamics and free-energy calculations applied to affinity maturation in antibody 48G7. Proc Natl Acad Sci USA 96(25):14330–14335

Chothia C, Lesk AM (1987) Canonical structures for the hypervariable regions of immunoglobulins. J Mol Biol 196(4):901–917

Chothia C, Lesk AM, Tramontano A, Levitt M, Smith-Gill SJ, Air G, Sheriff S, Padlan EA, Davies D, Tulip WR (1989) Conformations of immunoglobulin hypervariable regions. Nature 342(6252):877

Chruszcz M, Wlodawer A, Minor W (2008) Determination of protein structures—a series of fortunate events. Biophys J 95(1):1–9

Chruszcz M, Pomes A, Glesner J, Vailes LD, Osinski T, Porebski PJ, Majorek KA, Heymann PW, Platts-Mills TA, Minor W, Chapman MD (2012) Molecular determinants for antibody binding on group 1 house dust mite allergens. J Biol Chem 287(10):7388–7398

Dalkas GA, Teheux F, Kwasigroch JM, Rooman M (2014) Cation–π, amino–π, π–π, and H-bond interactions stabilize antigen–antibody interfaces. Proteins: Struct Funct Bioinform 82(9):1734–1746

Davies DR, Sheriff S, Padlan EA (1988) Antibody-antigen complexes. J Biol Chem 263(22):10541–10544

De Genst E, Silence K, Ghahroudi MA, Decanniere K, Loris R, Kinne J, Wyns L, Muyldermans S (2005) Strong in vivo maturation compensates for structurally restricted H3 loops in antibody repertoires. J Biol Chem 280(14):14114–14121

De Genst E, Silence K, Decanniere K, Conrath K, Loris R, Kinne J, Muyldermans S, Wyns L (2006) Molecular basis for the preferential cleft recognition by dromedary heavy-chain antibodies. Proc Natl Acad Sci USA 103(12):4586–4591

DeKosky BJ, Lungu OI, Park D, Johnson EL, Charab W, Chrysostomou C, Kuroda D, Ellington AD, Ippolito GC, Gray JJ, Georgiou G (2016) Large-scale sequence and structural comparisons of human naive and antigen-experienced antibody repertoires. Proc Natl Acad Sci 113(19):E2636–2645

Dingjan T, Spendlove I, Durrant LG, Scott AM, Yuriev E, Ramsland PA (2015) Structural biology of antibody recognition of carbohydrate epitopes and potential uses for targeted cancer immunotherapies. Mol Immunol 67(2):75–88

Donnarumma D, Bottomley MJ, Malito E, Settembre E, Ferlenghi I, Cozzi R (2015) Structural biology in vaccine research. In: Advanced vaccine research methods for the decade of vaccines, p 103

Durandy A (2003) Mini-review Activation-induced cytidine deaminase: a dual role in class-switch recombination and somatic hypermutation. Eur J Immunol 33(8):2069–2073

Elek S, Smith BK, Highman W (1964) The interaction of antigen and antibody in agglutination: a study by electron microscopy. Immunology 7(5):570

Fanning SW, Horn JR (2011) An anti-hapten camelid antibody reveals a cryptic binding site with significant energetic contributions from a nonhypervariable loop. Protein Sci 20(7):1196–1207

Fraser JD (2011) Clarifying the mechanism of superantigen toxicity. PLoS Biol 9(9):e1001145

Freeman CL, Sehn L (2018) Anti-CD20 directed therapy of B cell lymphomas: are new agents really better? Curr Oncol Rep 20(12):103

Gala FA, Morrison SL (2002) The role of constant region carbohydrate in the assembly and secretion of human IgD and IgA1. J Biol Chem 277(32):29005–29011

Geisberger R, Lamers M, Achatz G (2006) The riddle of the dual expression of IgM and IgD. Immunology 118(4):429–437

Glesner J, Vailes LD, Schlachter C, Mank N, Minor W, Osinski T, Chruszcz M, Chapman MD, Pomes A (2017) Antigenic determinants of Der p 1: specificity and cross-reactivity associated with IgE antibody recognition. J Immunol 198(3):1334–1344

Glesner J, Kapingidza AB, Godzwon M, Offermann LR, Mueller GA, DeRose EF, Wright P, Richardson CM, Woodfolk JA, Vailes LD, Wünschmann S, London RE, Chapman MD, Ohlin M, Chruszcz M, Pomés A (2019) Human IgE antibody binding site on Der p 2 for the design of a recombinant allergen for immunotherapy. J Immunol 203(9):2545–2556

Gounni AS, Lamkhioued B, Ochiai K, Tanaka Y, Delaporte E, Capron A, Kinet J-P, Capron M (1994) High-affinity IgE receptor on eosinophils is involved in defence against parasites. Nature 367(6459):183

Greenberg AS, Avila D, Hughes M, Hughes A, Mckinney EC, Flajnik MF (1995) A new antigen receptor gene family that undergoes rearrangement and extensive somatic diversification in sharks. Nature 374(6518):168–173

Grönwall C, Vas J, Silverman GJ (2012) Protective roles of natural IgM antibodies. Front Immunol 3:66

Gruber K, Zhou B, Houk KN, Lerner RA, Shevlin CG, Wilson IA (1999) Structural basis for antibody catalysis of a disfavored ring closure reaction. Biochemistry 38(22):7062–7074

Haidar JN, Yuan QA, Zeng L, Snavely M, Luna X, Zhang H, Zhu W, Ludwig DL, Zhu Z (2012) A universal combinatorial design of antibody framework to graft distinct cdr sequences: a bioinformatics approach. Proteins: Struct Funct Bioinform 80(3):896–912

Haji-Ghassemi O, Blackler RJ, Martin Young N, Evans SV (2015) Antibody recognition of carbohydrate epitopes. Glycobiology 25(9):920–952

Hamers-Casterman C, Atarhouch T, Muyldermans S, Robinson G, Hamers C, Songa EB, Bendahman N, Hamers R (1993) Naturally occurring antibodies devoid of light chains. Nature 363(6428):446–448

Harmsen MM, Ruuls RC, Nijman IJ, Niewold TA, Frenken LGJ, de Geus B (2000) Llama heavy-chain V regions consist of at least four distinct subfamilies revealing novel sequence features. Mol Immunol 37(10):579–590

Harris LJ, Larson SB, Hasel KW, McPherson A (1997) Refined structure of an intact IgG2a monoclonal antibody. Biochemistry 36(7):1581–1597

Harris LJ, Skaletsky E, McPherson A (1998) Crystallographic structure of an intact IgG1 monoclonal antibody. J Mol Biol 275(5):861–872

Henry KA, MacKenzie CR (2018) Antigen recognition by single-domain antibodies: structural latitudes and constraints. MAbs 10(6):815–826

Hewat EA, Verdaguer N, Fita I, Blakemore W, Brookes S, King A, Newman J, Domingo E, Mateu MG, Stuart DI (1997) Structure of the complex of an Fab fragment of a neutralizing antibody with foot-and-mouth disease virus: positioning of a highly mobile antigenic loop. EMBO J 16(7):1492–1500
Huber R (1980) Spatial structure of immunoglobulin molecules. Klin Wochenschr 58(22):1217–1231
Jahnichen S, Blanchetot C, Maussang D, Gonzalez-Pajuelo M, Chow KY, Bosch L, De Vrieze S, Serruys B, Ulrichts H, Vandevelde W, Saunders M, De Haard HJ, Schols D, Leurs R, Vanlandschoot P, Verrips T, Smit MJ (2010) CXCR4 nanobodies (VHH-based single variable domains) potently inhibit chemotaxis and HIV-1 replication and mobilize stem cells. Proc Natl Acad Sci 107(47):20565–20570
Janeway CA Jr, Travers P, Walport M, Shlomchik MJ (2001) The structure of a typical antibody molecule. In: Immunobiology: the immune system in health and disease, 5th edn. Garland Science
Jeffrey PD, Bajorath J, Chang CY, Yelton D, Hellström I, Hellström KE, Sheriff S (1995) The X-ray structure of an anti-tumour antibody in complex with antigen. Nat Struct Biol 2(6):466
Jolles S, Sewell WA, Misbah SA (2005) Clinical uses of intravenous immunoglobulin. Clin Exp Immunol 142(1):1–11
Karp CL (2010) Guilt by intimate association: what makes an allergen an allergen? J Allergy Clin Immunol 125(5):955–960
Keskin O (2007) Binding induced conformational changes of proteins correlate with their intrinsic fluctuations: a case study of antibodies. BMC Struct Biol 7(1):31
Kettleborough CA, Saldanha J, Heath VJ, Morrison CJ, Bendig MM (1991) Humanization of a mouse monoclonal antibody by CDR–grafting: the importance of framework residues on loop conformation. Protein Eng Des Sel 4(7):773–783
Khan T, Salunke DM (2014) Adjustable locks and flexible keys: plasticity of epitope–paratope interactions in germline antibodies. J Immunol 192(11):5398–5405
Kholodenko RV, Kalinovsky DV, Doronin II, Ponomarev ED, Kholodenko IV (2019) Antibody fragments as potential biopharmaceuticals for cancer therapy: success and limitations. Curr Med Chem 26(3):396–426
King MT, Brooks CL (2018) Epitope mapping of antibody-antigen interactions with x-ray crystallography. In: Epitope Mapping Protocols. Springer, Switzerland, pp 13–27
Kringelum JV, Nielsen M, Padkjær SB, Lund O (2013) Structural analysis of B-cell epitopes in antibody: protein complexes. Mol Immunol 53(1–2):24–34
Kunik V, Ofran Y (2013) The indistinguishability of epitopes from protein surface is explained by the distinct binding preferences of each of the six antigen-binding loops. Protein Eng Des Sel 26(10):599–609
LeBrasseur N (2003) A view of antibody maturation. J Cell Biol 161(5):834
Leder P (1982) The genetics of antibody diversity. Sci Am 246(5):102–115
Lescar J, Pellegrini M, Souchon H, Tello D, Poljak RJ, Peterson N, Greene M, Alzari PM (1995) Crystal-structure of a cross-reaction complex between Fab F9.13.7 and guinea-fowl lysozyme. J Biol Chem 270(30):18067–18076.
Li Y, Lipschultz CA, Mohan S, Smith-Gill SJ (2001) Mutations of an epitope hot-spot residue alter rate limiting steps of antigen–antibody protein–protein associations. Biochemistry 40(7):2011–2022
Li Y, Li H, Yang F, Smith-Gill SJ, Mariuzza RA (2003) X-ray snapshots of the maturation of an antibody response to a protein antigen. Nat Struct Biol 10(6):482–488
Liang Y, Guttman M, Davenport TM, Hu S-L, Lee KK (2016) Probing the impact of local structural dynamics of conformational epitopes on antibody recognition. Biochemistry 55(15):2197–2213
Liao HX, Alam SM, Mascola JR, Robinson J, Ma BJ, Montefiori DC, Rhein M, Sutherland LL, Scearce R, Haynes BF (2004) Immunogenicity of constrained monoclonal antibody A32-human immunodeficiency virus (HIV) env gp120 complexes compared to that of recombinant HIV type 1 gp120 envelope glycoproteins. J Virol 78(10):5270–5278

Ludwig DL, Pereira DS, Zhu Z, Hicklin DJ, Bohlen P (2003) Monoclonal antibody therapeutics and apoptosis. Oncogene 22(56):9097–9106
MacCallum RM, Martin AC, Thornton JM (1996) Antibody-antigen interactions: contact analysis and binding site topography. J Mol Biol 262(5):732–745
MacRaild CA, Richards JS, Anders RF, Norton RS (2016) Antibody recognition of disordered antigens. Structure 24(1):148–157
Malhotra R, Wormald MR, Rudd PM, Fischer PB, Dwek RA, Sim RB (1995) Glycosylation changes of IgG associated with rheumatooid arthritis can activate complement via the mannose-binding protein. Nat Med 1(3):237
Mari A, Iacovacci P, Afferni C, Barletta B, Tinghino R, Di Felice G, Pini C (1999) Specific IgE to cross-reactive carbohydrate determinants strongly affect the in vitro diagnosis of allergic diseases. J Allergy Clin Immunol 103(6):1005–1011
Mariuzza RA, PoIjak RJ (1993) The basics of binding: mechanisms of antigen recognition and mimicry by antibodies. Curr Opin Immunol 5(1):50–55
Market E, Papavasiliou FN (2003) V(D)J recombination and the evolution of the adaptive immune system. PLoS Biol 1(1):e16
Marone G (2007) Superantigens and superallergens, vol 93. Karger Medical and Scientific Publishers, Basel
Marone G, Spadaro G, Liccardo B, Rossi FW, D'Orio C, Detoraki A (2006) Superallergens: a new mechanism of immunologic activation of human basophils and mast cells. Inflamm Res 55(Suppl 1):S25–27
Maverakis E, Kim K, Shimoda M, Gershwin ME, Patel F, Wilken R, Raychaudhuri S, Ruhaak LR, Lebrilla CB (2015) Glycans in the immune system and the altered glycan theory of autoimmunity: a critical review. J Autoimmun 57:1–13
Meulenbroek LA, de Jong RJ, den Hartog Jager CF, Monsuur HN, Wouters D, Nauta AJ, Knippels LM, van Neerven RJ, Ruiter B, Leusen JH (2013) IgG antibodies in food allergy influence allergen–antibody complex formation and binding to B cells: a role for complement receptors. J Immunol 191(7):3526–3533
Mimura Y, Church S, Ghirlando R, Ashton P, Dong S, Goodall M, Lund J, Jefferis R (2000) The influence of glycosylation on the thermal stability and effector function expression of human IgG1-Fc: properties of a series of truncated glycoforms. Mol Immunol 37(12–13):697–706
Mimura Y, Sondermann P, Ghirlando R, Lund J, Young SP, Goodall M, Jefferis R (2001) Role of oligosaccharide residues of IgG1-Fc in FcγRIIb binding. J Biol Chem 276(49):45539–45547
Mitchell LS, Colwell LJ (2018) Analysis of nanobody paratopes reveals greater diversity than classical antibodies. Protein Eng Des Sel 31(7–8):267–275
Mitropoulou AN, Bowen H, Dodev TS, Davies AM, Bax HJ, Beavil RL, Beavil AJ, Gould HJ, James LK, Sutton BJ (2018) Structure of a patient-derived antibody in complex with allergen reveals simultaneous conventional and superantigen-like recognition. Proc Natl Acad Sci 115(37):E8707–E8716
Moalli F, Doni A, Deban L, Zelante T, Zagarella S, Bottazzi B, Romani L, Mantovani A, Garlanda C (2010) Role of complement and Fcγ receptors in the protective activity of the long pentraxin PTX3 against aspergillus fumigatus. Blood 116(24):5170–5180
Muyldermans S, Atarhouch T, Saldanha J, Barbosa JA, Hamers R (1994) Sequence and structure of VH domain from naturally occurring camel heavy chain immunoglobulins lacking light chains. Protein Eng 7(9):1129–1135
Narayanan A, Sellers BD, Jacobson MP (2009) Energy-based analysis and prediction of the orientation between light- and heavy-chain antibody variable domains. J Mol Biol 388(5):941–953
Nelson AL Antibody fragments: hope and hype. In: MAbs, 2010. vol 1. Taylor & Francis, pp 77–83
Nimmerjahn F, Ravetch JV (2005) Divergent immunoglobulin G subclass activity through selective Fc receptor binding. Science 310(5753):1510–1512
North B, Lehmann A, Dunbrack RL Jr (2011) A new clustering of antibody CDR loop conformations. J Mol Biol 406(2):228–256

Oettgen HC, Geha RS (1999) IgE in asthma and atopy: cellular and molecular connections. J Clin Invest 104(7):829–835
Padlan EA (1994) Anatomy of the antibody molecule. Mol Immunol 31(3):169–217
Padlan EA, Silverton EW, Sheriff S, Cohen GH, Smith-Gill SJ, Davies DR (1989) Structure of an antibody-antigen complex: crystal structure of the HyHEL-10 Fab-lysozyme complex. Proc Natl Acad Sci 86(15):5938–5942
Padlan EA, Abergel C, Tipper J (1995) Identification of specificity-determining residues in antibodies. FASEB J. 9(1):133–139
Pan R, Chen Y, Vaine M, Hu G, Wang S, Lu S, Kong XP (2015) Structural analysis of a novel rabbit monoclonal antibody R53 targeting an epitope in HIV-1 gp120 C4 region critical for receptor and co-receptor binding. Emerg Microbes Infect 4(7):e44
Parker DC (1993) T cell-dependent B cell activation. Annu Rev Immunol 11(1):331–360
Peng H-P, Lee KH, Jian J-W, Yang A-S (2014) Origins of specificity and affinity in antibody–protein interactions. Proc Natl Acad Sci USA 111(26):E2656–E2665
Platts-Mills TA (2001) The role of immunoglobulin E in allergy and asthma. Am J Respir Crit Care Med 164 (supplement_1):S1-S5
Pollara J, Tay MZ (2019) Antibody-dependent cellular phagocytosis in antiviral immune responses. Front Immunol 10:332
Pons J, Stratton JR, Kirsch JF (2002) How do two unrelated antibodies, HyHEL-10 and F9. 13.7, recognize the same epitope of hen egg-white lysozyme? Protein Sci 11(10):2308–2315
Potocnakova L, Bhide M, Pulzova LB (2016) An introduction to B-cell epitope mapping and in silico epitope prediction. J Immunol Res 2016:6760830
Presta L, Shields R, O'Connell L, Lahr S, Porter J, Gorman C, Jardieu P (1994) The binding site on human immunoglobulin E for its high affinity receptor. J Biol Chem 269(42):26368–26373
Rafiq K, Bergtold A, Clynes R (2002) Immune complex-mediated antigen presentation induces tumor immunity. J Clin Invest 110(1):71–79
Raghunathan G, Smart J, Williams J, Almagro JC (2012) Antigen-binding site anatomy and somatic mutations in antibodies that recognize different types of antigens. J Mol Recognit 25(3):103–113
Robin G, Sato Y, Desplancq D, Rochel N, Weiss E, Martineau P (2014) Restricted diversity of antigen binding residues of antibodies revealed by computational alanine scanning of 227 antibody–antigen complexes. J Mol Biol 426(22):3729–3743
Rosen O, Anglister J (2009) Epitope mapping of antibody–antigen complexes by nuclear magnetic resonance spectroscopy. In: Epitope mapping protocols. Springer, Switzerland, pp 37–57
Rouet R, Dudgeon K, Christie M, Langley D, Christ D (2015) Fully human VH single domains that rival the stability and cleft recognition of camelid antibodies. J Biol Chem 290(19):11905–11917
Saphire EO, Parren PW, Pantophlet R, Zwick MB, Morris GM, Rudd PM, Dwek RA, Stanfield RL, Burton DR, Wilson IA (2001) Crystal structure of a neutralizing human IGG against HIV-1: a template for vaccine design. Science 293(5532):1155–1159
Schroeder HW Jr, Cavacini L (2010) Structure and function of immunoglobulins. J Allergy Clin Immunol 125(2):S41–S52
Sela-Culang I, Kunik V, Ofran Y (2013) The structural basis of antibody-antigen recognition. Front Immunol 4:302
Sheriff S, Silverton EW, Padlan EA, Cohen GH, Smith-Gill SJ, Finzel BC, Davies DR (1987) Three-dimensional structure of an antibody-antigen complex. Proc Natl Acad Sci USA 84(22):8075–8079
Simonelli L, Pedotti M, Bardelli M, Jurt S, Zerbe O, Varani L (2018) Mapping antibody epitopes by solution NMR spectroscopy: practical considerations. In: Epitope mapping protocols. Springer, Switzerland, pp 29–51
Sinha N, Smith-Gill SJ (2002) Electrostatics in protein binding and function. Curr Protein Pept Sci 3(6):601–614
Sinha N, Mohan S, Lipschultz CA, Smith-Gill SJ (2002) Differences in electrostatic properties at antibody-antigen binding sites: Implications for specificity and cross-reactivity. Biophys J 83(6):2946–2968

Skiniotis G, Southworth DR (2016) Single-particle cryo-electron microscopy of macromolecular complexes. Microscopy 65(1):9–22
Spiegelberg HL (1989) Biological role of different antibody classes. Int Arch Allergy Immunol 90(Suppl. 1):22–27
Spuergin P, Mueller H, Walter M, Schiltz E, Forster J (1996) Allergenic epitopes of bovine αs1-casein recognized by human IgE and IiG. Allergy 51(5):306–312
Stanfield RL, Dooley H, Flajnik MF, Wilson IA (2004) Crystal structure of a shark single-domain antibody V region in complex with lysozyme. Science 305(5691):1770–1773
Stanfield RL, Gorny MK, Zolla-Pazner S, Wilson IA (2006a) Crystal structures of human immunodeficiency virus type 1 (HIV-1) neutralizing antibody 2219 in complex with three different V3 peptides reveal a new binding mode for HIV-1 cross-reactivity. J Virol 80(12):6093–6105
Stanfield RL, Zemla A, Wilson IA, Rupp B (2006b) Antibody elbow angles are influenced by their light chain class. J Mol Biol 357(5):1566–1574
Stanfield RL, De Castro C, Marzaioli AM, Wilson IA, Pantophlet R (2015) Crystal structure of the HIV neutralizing antibody 2G12 in complex with a bacterial oligosaccharide analog of mammalian oligomannose. Glycobiology 25(4):412–419
Stavnezer J, Amemiya CT (2004) Evolution of isotype switching. In: Seminars in immunology, vol 4. Elsevier, Amsterdam, pp 257–275
Strohl W, Strohl L (2013) Antibody fragments as therapeutics. Therap Anti Eng: Curr Fut Adv Driv Strong Growth Area Pharm Ind 265–299
Theofilopoulos AN, Dixon FJ (1980) Immune complexes in human diseases: a review. Am J Pathol 100(2):529–594
Tian X, Vestergaard B, Thorolfsson M, Yang Z, Rasmussen HB, Langkilde AE (2015) In-depth analysis of subclass-specific conformational preferences of IgG antibodies. IUCrJ 2(1):9–18
Torres M, Casadevall A (2008) The immunoglobulin constant region contributes to affinity and specificity. Trends Immunol 29(2):91–97
Townsend CL, Laffy JM, Wu Y-CB, Silva O'Hare J, Martin V, Kipling D, Fraternali F, Dunn-Walters DK (2016) Significant differences in physicochemical properties of human immunoglobulin kappa and lambda CDR3 regions. Front Immunol 7:388
Tudor D, Yu H, Maupetit J, Drillet A-S, Bouceba T, Schwartz-Cornil I, Lopalco L, Tuffery P, Bomsel M (2012) Isotype modulates epitope specificity, affinity, and antiviral activities of anti–HIV-1 human broadly neutralizing 2F5 antibody. Proc Natl Acad Sci USA 109(31):12680–12685
Underdown BJ, Schiff JM (1986) Immunoglobulin A: strategic defense initiative at the mucosal surface. Annu Rev Immunol 4(1):389–417
Van Kaer L (2018) How superantigens bind MHC. J Immunol 201(7):1817–1818
Van Regenmortel MH (2009) What is a B-cell epitope? In: Epitope mapping protocols. Springer, Switzerland, pp 3–20
Van Regenmortel MH (2014) Specificity, polyspecificity, and heterospecificity of antibody-antigen recognition. J Mol Recognit 27(11):627–639
Vidarsson G, Dekkers G, Rispens T (2014) IgG subclasses and allotypes: from structure to effector functions. Front Immunol 5:520
Visciano ML, Tuen M, Gorny MK, Hioe CE (2008) In vivo alteration of humoral responses to HIV-1 envelope glycoprotein gp120 by antibodies to the CD4-binding site of gp120. Virology 372(2):409–420
Webster DM, Henry AH, Rees AR (1994) Antibody-antigen interactions. Curr Opin Struct Biol 4(1):123–129
Wedemayer GJ, Patten PA, Wang LH, Schultz PG, Stevens RC (1997) Structural insights into the evolution of an antibody combining site. Science 276(5319):1665–1669
Wei H, Mo J, Tao L, Russell RJ, Tymiak AA, Chen G, Iacob RE, Engen JR (2014) Hydrogen/deuterium exchange mass spectrometry for probing higher order structure of protein therapeutics: methodology and applications. Drug Discov Today 19(1):95–102
Weitzner BD, Dunbrack RL Jr, Gray JJ (2015) The origin of CDR H3 structural diversity. Structure 23(2):302–311

Wen YM, Mu L, Shi Y (2016) Immunoregulatory functions of immune complexes in vaccine and therapy. EMBO Mol Med 8(10):1120–1133
Wider G, Wüthrich K (1999) NMR spectroscopy of large molecules and multimolecular assemblies in solution. Curr Opin Struct Biol 9(5):594–601
Williams CM, Galli SJ (2000) The diverse potential effector and immunoregulatory roles of mast cells in allergic disease. J Allergy Clin Immunol 105(5):847–859
Wilson IA, Stanfield RL (1994) Antibody-antigen interactions: new structures and new conformational changes. Curr Opin Struct Biol 4(6):857–867
Wlodawer A, Minor W, Dauter Z, Jaskolski M (2008) Protein crystallography for non-crystallographers, or how to get the best (but not more) from published macromolecular structures. FEBS J 275(1):1–21
Wood P (2012) Human normal immunoglobulin in the treatment of primary immunodeficiency diseases. Ther Clin Risk Manag 8:157–167
Wu S, Avila-Sakar A, Kim J, Booth DS, Greenberg CH, Rossi A, Liao M, Li X, Alian A, Griner SL (2012) Fabs enable single particle cryoEM studies of small proteins. Structure 20(4):582–592
Xia Y, Pawar RD, Nakouzi AS, Herlitz L, Broder A, Liu K, Goilav B, Fan M, Wang L, Li Q-Z (2012) The constant region contributes to the antigenic specificity and renal pathogenicity of murine anti-DNA antibodies. J Autoimmun 39(4):398–411
Xiang J, Sha Y, Jia Z, Prasad L, Delbaere LT (1995) Framework residues 71 and 93 of the chimeric B72. 3 antibody are major determinants of the conformation of heavy-chain hypervariable loops. Elsevier
Xiang J, Prasad L, Delbaere LT, Jia Z (1999) Light-chain framework region residue Tyr71 of chimeric B72. 3 antibody plays an important role in influencing the TAG72 antigen binding. Protein Eng 12(5):417–421
Zhao L, Li J (2010) Mining for the antibody-antigen interacting associations that predict the B cell epitopes. BMC Struct Biol 10(1):S6
Zhu XY, Dickerson TJ, Rogers CJ, Kaufmann GF, Mee JM, McKenzie KM, Janda KD, Wilson IA (2006) Complete reaction cycle of a cocaine catalytic antibody at atomic resolution. Structure 14(2):205–216
Zuiderweg ER (2002) Mapping protein–protein interactions in solution by NMR spectroscopy. Biochemistry 41(1):1–7

Chapter 20
C-Reactive Protein and Its Structural Isoforms: An Evolutionary Conserved Marker and Central Player in Inflammatory Diseases and Beyond

James D. McFadyen, Johannes Zeller, Lawrence A. Potempa, Geoffrey A. Pietersz, Steffen U. Eisenhardt and Karlheinz Peter

Abstract C-reactive protein (CRP) is an evolutionary highly conserved member of the pentraxin superfamily of proteins. CRP is widely used as a marker of inflammation, infection and for risk stratification of cardiovascular events. However, there is now a large body of evidence, that continues to evolve, detailing that CRP directly

---

J. D. McFadyen (✉) · G. A. Pietersz · K. Peter (✉)
Atherothrombosis and Vascular Biology Laboratory, Baker Heart and Diabetes Institute, Melbourne, VIC, Australia
e-mail: James.McFadyen@monash.edu

K. Peter
e-mail: Karlheinz.Peter@baker.edu.au

G. A. Pietersz
e-mail: pietersz.geoff@gmail.com

J. D. McFadyen · K. Peter
Department of Medicine, Monash University, Melbourne, VIC, Australia

J. D. McFadyen
Department of Clinical Haematology, The Alfred Hospital, Melbourne, VIC, Australia

Australian Centre for Blood Diseases, Monash University, Melbourne, VIC, Australia

J. Zeller · S. U. Eisenhardt
Department of Plastic and Hand Surgery, Medical Faculty of the University of Freiburg, University of Freiburg Medical Centre, Freiburg, Germany
e-mail: johannes.zeller@uniklinik-freiburg.de

S. U. Eisenhardt
e-mail: steffen.eisenhardt@uniklinik-freiburg.de

L. A. Potempa
College of Pharmacy, Roosevelt University, Schaumburg, IL, USA
e-mail: lpotempa01@roosevelt.edu

G. A. Pietersz · K. Peter
Department of Immunology, Monash University, Melbourne, VIC, Australia

G. A. Pietersz
Burnet Institute, Melbourne, VIC, Australia

U. Hoeger and J. R. Harris (eds.), *Vertebrate and Invertebrate Respiratory Proteins, Lipoproteins and other Body Fluid Proteins*, Subcellular Biochemistry 94,
https://doi.org/10.1007/978-3-030-41769-7_20

mediates inflammatory reactions and the innate immune response in the context of localised tissue injury. These data support the concept that the pentameric conformation of CRP dissociates into pro-inflammatory CRP isoforms termed pCRP* and monomeric CRP. These pro-inflammatory CRP isoforms undergo conformational changes that facilitate complement binding and immune cell activation and therefore demonstrate the ability to trigger complement activation, activate platelets, monocytes and endothelial cells. The dissociation of pCRP occurs on the surface of necrotic, apoptotic, and ischaemic cells, regular β-sheet structures such as β-amyloid, the membranes of activated cells (e.g., platelets, monocytes, and endothelial cells), and/or the surface of microparticles, the latter by binding to phosphocholine. Therefore, the deposition and localisation of these pro-inflammatory isoforms of CRP have been demonstrated to amplify inflammation and tissue damage in a broad range of clinical conditions including ischaemia/reperfusion injury, Alzheimer's disease, age-related macular degeneration and immune thrombocytopaenia. Given the potentially broad relevance of CRP to disease pathology, the development of inhibitors of CRP remains an area of active investigation, which may pave the way for novel therapeutics for a diverse range of inflammatory diseases.

Keywords C-reactive protein · CRP · Inflammation · Complement · CRP structure · Ischaemia/reperfusion injury · Alzheimer's disease · Atherosclerosis

Introduction

C-reactive protein (CRP) is a member of the highly conserved and phylogenetically ancient protein family, known as pentraxins, according to their five subunit structure (Pepys et al. 1978; Baltz et al. 1982). Indeed, CRP is a constitutively expressed protein in the horseshoe crab, which has been termed a living fossil given its existence for over 250 million years (Armstrong 2015). Moreover, CRP has been found to be expressed in every species where its presence has been investigated (Pathak and Agrawal 2019). CRP from humans and all other animal species all share the classical cyclical pentameric structure. The exception being the horseshoe crab which exhibits a hexameric structure (Nguyen et al. 1986). Nevertheless, CRP across all species exhibits a high degree of sequence homology and calcium dependent ligand binding (Pathak and Agrawal 2019). Despite the highly conserved nature of CRP, the biological role of CRP in humans remained elusive for many years and continues to evoke controversy. However, a large body of evidence now supports the concept that CRP is part of the innate immune system where it acts as a pattern recognition molecule that binds to damaged and inflamed cells in vivo thus serving to amplify inflammation. As such, the pathogenic role of CRP in mediating a range of important human diseases and the therapeutic targeting of CRP is now of particular interest.

K. Peter
Heart Centre, The Alfred Hospital, Melbourne, VIC, Australia

The Discovery of CRP

Despite the evolutionary conserved nature of CRP, the identification and elucidation of CRP function has remained obscure until relatively recently. Indeed, the first characterization of CRP was heralded by the discovery of Tillett and Francis in 1930 whilst investigating the immune responses of patients with pneumococcal pneumonia, where they observed that a distinct third fraction from the sera of patients with pneumo-coccus infection could precipitate the "C" polysaccharide derived from the pneumococcus cell wall (Tillett and Francis 1930). Although not known at the time, these experiments detailed the acute phase response since the precipitation reaction began to diminish or disappear as patients recovered from infection whilst remaining positive in fatal cases. Further investigation of patients with subacute bacterial endocarditis, rheumatic fever, lung abscesses and osteomyelitis revealed that the sera from these patients also produced strong precipitation reactions in the presence of the "C" polysaccharide (McCarty 1982). These observations lead to the notion that the reaction was produced in response to gram positive bacteria, however this concept was dispelled by work by Rachel Ash at the Children's Hospital of Philadelphia where sera from patients suffering gram negative infection were also shown to precipitate upon exposure to fraction C (Ash 1933). It was soon realized that the production of CRP was due to a nonspecific reaction to bacterial infection as opposed to a humoral antibody response directed against a specific bacterial antigen. Shortly after, an early CRP assay was developed as a means to monitor the clinical course of patients with rheumatic fever (Anderson and McCarty 1950; McCarty 1982). These seminal, early studies lead to the subsequent investigation of CRP in other species and the eventual purification of the CRP and resolution of the crystal structure.

CRP Clinical Use

The routine analysis of serum CRP concentrations as a measure of inflammation and disease activity remains one of the most widely utilised assays in medicine. Indeed, the modern CRP immunoassay pioneered by Mark Pepys has been described by the European Union Science Hub as one of the most important analytes in clinical chemistry. A fundamental reason that CRP measurement holds such clinical utility is due to the fact the CRP production is swiftly upregulated in a broad range of important clinical conditions, including infection, cardiovascular disease, inflammatory diseases, trauma and malignancy (Pepys and Hirschfield 2003). Moreover, in addition to serving as a marker of disease activity CRP has prognostic value in myocardial infarction, and importantly, is a robust predictor of incident cardiovascular events to the extent that assessment of high sensitivity CRP is now used as a part of global cardiovascular risk assessment (de Beer et al. 1982; Liuzzo et al. 1994; Ridker 2009). The observation that CRP helps predict cardiovascular risk, combined with the data that therapies that reduce CRP levels result in the reduction of cardiovascular risk, has spurred the controversy regarding whether CRP is merely a biomarker of the hallmark of atherosclerosis, inflammation, or is an active player in the pathogenesis of cardiovascular disease. Whilst Mendelian randomization studies have not supported a causal role, at least for the pentameric form, of CRP in cardio-vascular disease, there is a wealth of experimental data that are consistent with a pro-inflammatory, immunomodulating and platelet-activating effects (Zacho et al. 2008; McFadyen et al. 2018).

CRP Production

CRP is principally produced in the liver (Hurlimann et al. 1966), and whilst there is evidence of CRP production by a range of other cells types, including leukocytes (Kuta and Baum 1986), adipocytes (Calabro et al. 2005), neuronal cells (Yasojima et al. 2000), renal cells (Jabs et al. 2003), and respiratory epithelial cells (Ramage et al. 2004), the extra-hepatic production is not thought to significantly influence plasma CRP concen-trations. In keeping with its role as an acute phase reactant, the hepatic production of CRP is largely influenced by the pro-inflammatory cytokines interleukin-6 (IL-6) and interleukin-1β (IL-1β) (Zhang et al. 1996, p. 3). IL-6 appears to play a more significant role in regulating CRP expression, while IL-1β enhances the effects of IL-6 on CRP production (Ganter et al. 1989). The translation of the CRP gene, located on the short arm of chromosome 1, is under the control of the transcription factors STAT3, C/EBP, and NF-κB (Agrawal et al. 2003a, b). Thus, in the context of inflammation, IL-6 and IL-1β result in the recruitment and activation of the C/EBP family members, C/EBPδ and C/EBPβ, and the binding of STAT3 and Rel proteins to the proximal promoter region of the CRP gene (Young et al. 2008). These interactions enhance and stabilise the binding of the C/EBP transcription factor to the CRP gene yielding maximal gene expression and hence CRP production. The mechanisms by which the extrahepatic synthesis of CRP is regulated remains to be elucidated. However, given CRP production from these sites does not significantly influence plasma levels the biological significance of this remains unknown. In this regard, it has been proposed that extrahepatic CRP may produce autocrine and paracrine inflammatory function given the demonstration of local CRP production in atherosclerotic plaques (Kobayashi et al. 2003; Venugopal et al. 2005).

The mean serum concentration of CRP, under physiological conditions, is 0.8 mg/L. However, in the context of inflammation, CRP levels rise dramatically up to 10,000 fold over 24–72 h. Once secreted, CRP has a half life of approximately 19 h, and given the clearance of CRP follows first order kinetics, the serum level of CRP is principally a result of any ongoing hepatic synthesis due to inflammation. These properties make CRP an ideal marker of tissue inflammation and infection and has led to the widespread adoption of CRP assays in clinical medicine.

CRP Structure

CRP has a pentameric structure, composed of five identical non covalently linked 23 kDa protomers (Osmand et al. 1977; Shrive et al. 1996) (Fig. 20.1). Each protomer consists of 206 amino acids, with X ray crystallography studies revealing that each protomer is folded into two antiparallel beta sheets with a flattened jellyroll topology (Osmand et al. 1977; Shrive et al. 1996). The ability of CRP to trigger inflammatory responses is structurally related to the fact that each protomer contains a binding face and effector face. The binding face contains a phosphocholine binding site, which facilitates the calcium dependent interactions with PC exposed on apoptotic and inflamed cell membranes and bacterial cell walls (Thompson et al. 1999) (Fig. 20.1). Co-crystallisation, and site-directed mutagenesis studies of CRP with PC suggest that the Phe-66 and Glu-81 located in a hydrophobic pocket are the two critical residues that mediate PC binding to CRP (Thompson et al. 1999; Agrawal et al. 2002; Black et al. 2003). The exposed face of the Phe-66 residue within the hydrophobic pocket provides hydrophobic interactions with the methyl groups of PC. The Glu-81 residue is located at the opposite end of the hydrophobic pocket where it interacts with positively charged nitrogen. These residues are highly conserved across a large number of species underscoring the importance of these residues in mediating PC interactions with CRP (Shrive et al. 1996).

The opposite face of the CRP pentamer is the effector face, which can activate the innate immune system, by binding to the globular head of complement C1q and the Fcγ receptor (Agrawal and Volanakis 1994; Agrawal et al. 2001; Gaboriaud et al. 2003). Based on the crystal structure of the globular head of C1q, it is proposed that the predominantly negatively charged head of C1q binds to the negatively charged

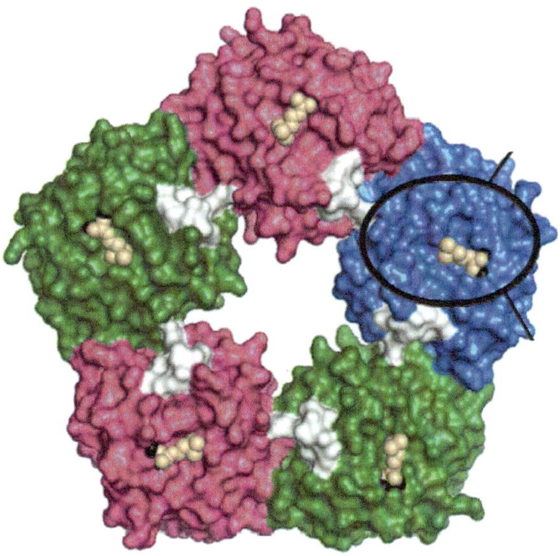

Fig. 20.1 pCRP structure. The structure of the pCRP pentamer with each individual monomer highlighted in a different colour (purple, green, blue) and phosphocholine (cream spheres) and calcium ions (black spheres) located in the ligand binding pocket (highlighted in black circle) with the CHO binding areas shown in white. Adapted from Braig et al., Nature comm. (2018)

central pore of the CRP pentamer. Interestingly, the interaction of CRP with the C1q head is sterically hindered in this proposed model thus suggesting that optimal interactions between CRP and C1q require a degree of conformation change in the CRP pentamer (Agrawal and Volanakis 1994, p. 1; Agrawal et al. 2001; Gaboriaud et al. 2003). This has led to the concept that CRP undergoes a conformational change at sites of tissue damage to exert its pro-inflammatory effects (Lv and Wang 2018).

CRP Isoforms

The exposure of pCRP to heat, urea, or an acidic environment can lead to the dissociation into monomeric CRP (mCRP) (Potempa et al. 1983; Kresl et al. 1998). The dissociation of the pentameric structure yields monomers that display previously cryptic epitopes that are thought to account for the distinct pro-inflammatory effects of mCRP (Ji et al. 2007; Braig et al. 2017). Our group, and others, have demonstrated that CRP can dissociate on the membranes of activated platelets (Eisenhardt et al. 2009), monocytes (Braig et al. 2017), endothelial cells (Thiele et al. 2014) and microparticles (Ji et al. 2007), all of which are abundant in phosphocholine thus providing the requisite binding site for CRP (Volanakis and Wirtz 1979). Interestingly, the dissociation of pCRP to mCRP appears to be dependent upon phospholipase A2 (PLA2) given its role in lipid metabolism and generation of LPC, which serves to facilitate CRP binding (Thiele et al. 2014). Structurally, the dissociation of pCRP to the monomers is associated with the marked shift from the β-sheet tertiary protein conformation of pCRP to individual protomers with an α-helical tertiary structure, thus leading to the exposure of previously cryptic interprotomer contacts (Braig et al. 2017). Not surprisingly, these conformational changes produce a marked change in the solubility of pCRP from a soluble protein to the highly insoluble mCRP. This observation, coupled with the observation that mCRP is generated locally within inflamed tissues, has led many to speculate that mCRP represents a tissue- or cell-bound isoform of CRP found in the context of inflammation.

Intriguingly, our group has recently demonstrated the existence of a distinct isoform of CRP, termed pCRP*, which appears to possess the required structural changes necessary to bind and activate complement (Braig et al. 2017). Indeed, the binding of pCRP to microparticles (MPs), which have a curved surface and contain requisite PC for CRP binding, leads to structural changes in CRP without the frank disruption of its overall pentameric symmetry (Braig et al. 2017) (Fig. 20.2). However, importantly, this conformational change from pCRP to pCRP* is associated with the exposure of neoepitopes and allows the binding of complement C1q. Consequently, pCRP* is generated on the surface of MPs released from activated cells and can amplify tissue inflammation in vivo and comprises a significant portion of CRP species in injured human tissue, including inflamed human muscle, burn wounds, and human atherosclerotic plaques (Braig et al. 2017). Thus, in contrast to mCRP, which is likely rapidly cleared in vivo due to its insolubility and highly disordered structure, MPs appear to act as chaperone to transport bound pCRP* to

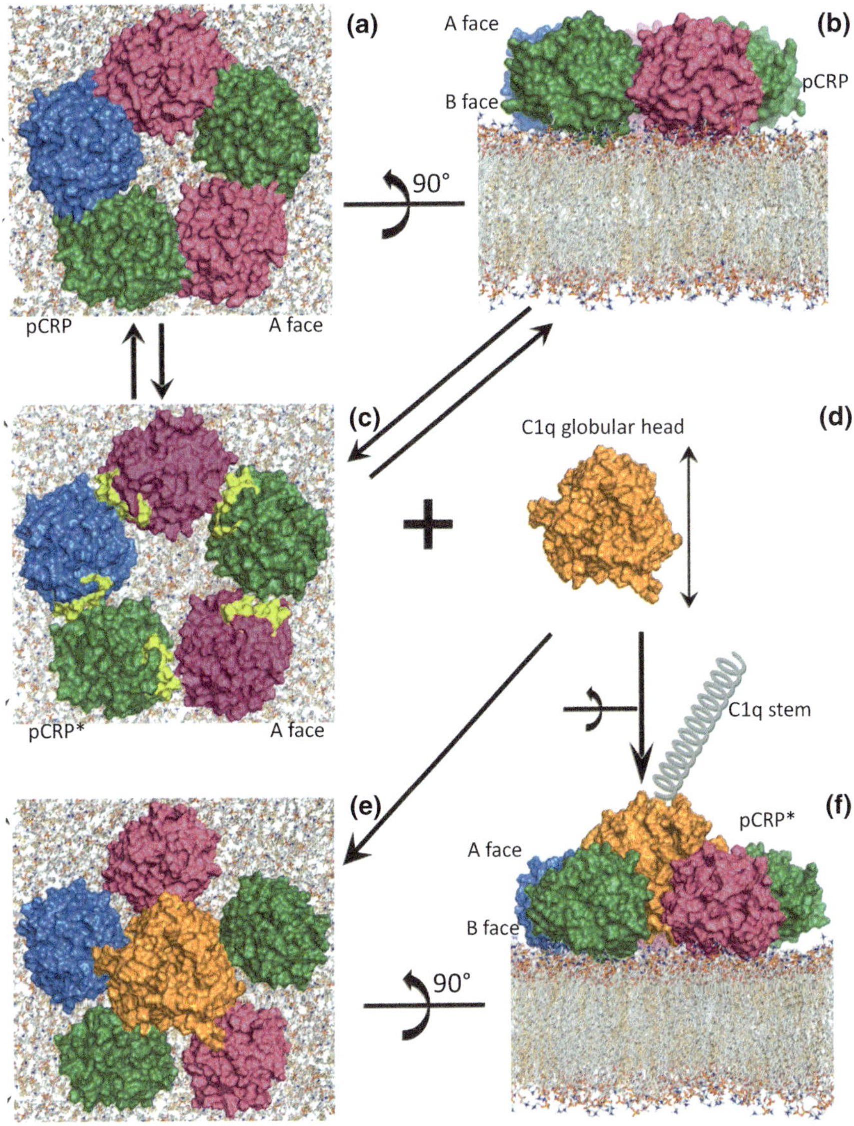

Fig. 20.2 Model of pCRP dissociation. **a** The binding of pCRP pentamer to a PC containing lipid bilayer occurs via the PC binding domain of each CRP subunit on the B face of pCRP. **b** A side view of the B face of pCRP bound to a PC containing lipid bilayer. **c** This leads to the moving apart of the pCRP subunits such that neoepitopes are exposed (yellow) which are recognized by the conformation specific antibodies 1D6 and 3H12. **d**, **e** The globular head of the C1q domain binds into the void of the pCRP* molecule, which creates further separation of the CRP subunits. **f** A cross section view of the pCRP*-C1q complex bound to a PC containing lipid bilayer. Adapted from Braig et al., Nature comm. (2018)

distant sites. In this regard, dissociated CRP has been demonstrated on the surface of MPs from patients with myocardial infarction (Habersberger et al. 2012).

These data have led to the concept that CRP regulates the innate immune response in a conformation-dependent fashion. Thus, whilst pCRP does not display any intrinsic pro-inflammatory properties, the structural isoforms, mCRP and pCRP*, possess potent pro-inflammatory effects which can mediate immune, inflammatory, and prothrombotic responses in a range of diseases.

The Pro-inflammatory and Prothrombotic Properties of CRP

As discussed, the mCRP and pCRP* isoforms of CRP are considered to mediate the pro-inflammatory effects of the CRP system (Fig. 20.3). To date, the majority of the

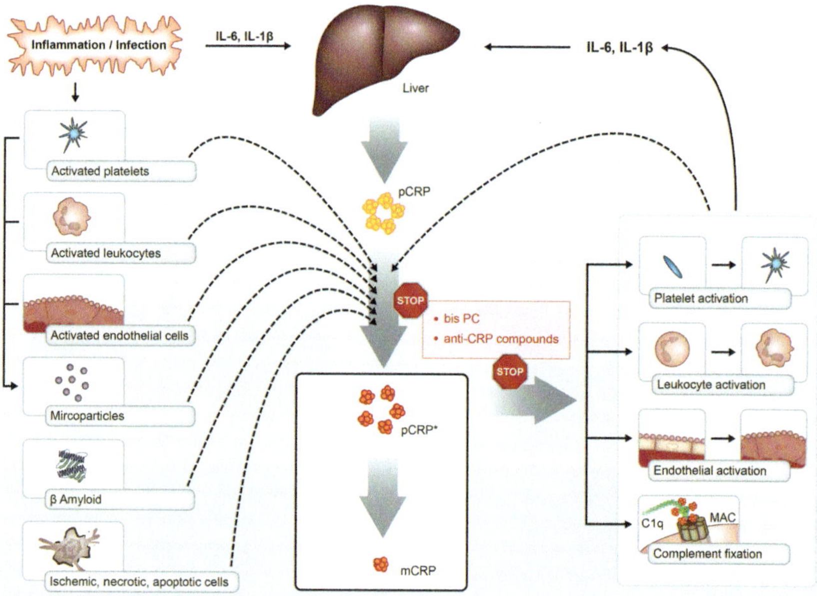

Fig. 20.3 CRP synthesis and isoform generation. CRP is synthesized by the liver in the context of inflammation in response to the inflammatory cytokines IL-6 and IL-1β. CRP is secreted as a pentamer (pCRP) which can bind to the surface of activated platelets, leukocytes, endothelial cells, PC-rich microvesicles, dying cells and β amyloid. The binding of pCRP to these substrates leads to the dissociation of pCRP into its pro-inflammatory structural isoforms—pCRP* and mCRP. pCRP* and mCRP can induce the activation of platelets, leukocytes and endothelial cells in addition to binding complement C1q. The pro-inflammatory effects of pCRP* and mCRP can be inhibited with compounds such as bisPC that prevent the dissociation of pCRP and inhibit the PC binding pocket. Figure adapted from McFadyen et al., Frontiers Immunology (2018)

data has shown the pro-inflammatory role of mCRP, however, it appears that pCRP* exhibits similar effects in vitro and in vivo.

CRP and Monocytes

mCRP activates monocytes leading to the activation of the major leukocyte integrin, Mac-1, and resultant monocyte adhesion (Eisenhardt et al. 2009). In addition, mCRP induces monocyte reactive oxygen species (ROS) production (Singh et al. 2017), which forms an important part of the innate immune response to invading pathogens. Additionally, unquenched monocyte ROS generation has been linked to the process of monocyte to macrophage differentiation and enhanced inflammation within atherosclerotic plaques. Whilst the precise mechanism(s) by which mCRP activates monocytes remains to be fully elucidated, the binding of mCRP to the Fcγ receptors (CD64, CD32, CD16), integrin $\alpha_4\beta_1$ and lipid raft domains appear to play an important role in mediating monocyte activation (Bang et al. 2005; Lu et al. 2008; Ji et al. 2009; Fujita et al. 2014). Accordingly, blocking antibodies to the Fcγ receptors, and disruption of the lipid raft domains with nystatin attenuates the effects of mCRP. mCRP has also been demonstrated to induce Mac-1 activation and enhance neutrophil binding to endothelial cells in vitro in addition to stimulating the production of IL-8 from neutrophils in a process linked to increased neutrophil peroxynitrate production (Zouki et al. 2001; Khreiss et al. 2005). The binding of mCRP to integrin $\alpha_4\beta_1$ induces integrin-dependent chemotaxis and induces PI3K/AKT signaling pathways.

The effect of pCRP* on leukocyte function has principally been assessed in vivo, where in the context of vascular inflammation, pCRP* enhances leukocyte recruitment and transmigration (Braig et al. 2017). Importantly, the use of bis PC an antagonist of pCRP abrogates the pro-inflammatory effects of mCRP and pCRP* in vivo since pretreatment of animals in various animal models inhibits leukocyte adhesion and migration.

CRP and Endothelial Function

The resting endothelium normally provides a quiescent environment that prevents the unwanted adhesion of platelets and leukocytes. However, in the context of inflammation, endothelial cells upregulate the expression of a number of adhesion receptors, including intercellular adhesion molecule-1 (ICAM-1) and vascular cell adhesion molecule-1 (VCAM-1)—both of which serve to facilitate platelet and leukocyte adhesion. Indeed, ICAM-1 acts as a counter receptor for the leukocyte integrin, lymphocyte function-associated antigen 1 (LFA-1), and thereby enhances leukocyte adhesion at sites of endothelial inflammation (Springer 1994). In addition, ICAM-1 has been demonstrated to bind fibrinogen and therefore plays an important role in mediating stable platelet-endothelial adhesion via platelet glycoprotein (GP)IIb/IIIa binding to the ICAM-1/fibrinogen complex (Massberg et al. 1999). VCAM-1 also serves to facilitate leukocyte adhesion to the endothelium by acting as the cognate receptor for the leukocyte integrin, very late antigen 4 (VLA-4) (Elices et al. 1990; Springer 1994). Stimulation of endothelial cells with mCRP upregulates the expression of ICAM-1 and VCAM-1 (Khreiss et al. 2004; Ji et al. 2009). Likewise, pCRP* also induces the expression of ICAM-1 and VCAM-1 in endothelial cells (Braig et al. 2017). Contrastingly, native pCRP exhibits no effects on the endothelial expression of VCAM-1 or ICAM-1 (Khreiss et al. 2004).

CRP and Platelet Activation

Platelets are central mediators of thrombosis and haemostasis in addition to having important immune and inflammatory functions. Accordingly, mCRP has been demonstrated to activate platelets. Indeed, stimulation of platelets ex vivo with mCRP results in the activation of the major platelet adhesion receptor, GPIIb/IIIa, platelet aggregation in addition to P-selectin expression (a marker of alpha granule exocytosis) (Molins et al. 2011; de la Torre et al. 2013). Activated platelets have also been demonstrated to provide a substrate for pCRP dissociation, which in turn leads to enhanced platelet activation and platelet thrombus formation under shear stress in vitro. Whilst the platelet receptors responsible for the prothrombotic effects of mCRP have yet to be fully delineated, the platelet scavenger receptor, CD36, appears to play an important role given CD36 inhibition partially inhibits some of the platelet activating effects of mCRP (de la Torre et al. 2013). Consistent with the role of CD36 in regulating mCRP mediated platelet activation, inhibition of the downstream signaling partners of the CD36 receptor, JNK and MAPK, also partially abrogates the prothrombotic effects of mCRP (de la Torre et al. 2013). Given the effects of CD36 inhibition are only partial, there are no doubt other mechanisms by which mCRP activates platelets. In this regard, whether mCRP interacts with lipid raft domains of platelets, as has been demonstrated in leukocytes, has not been investigated.

CRP and the Complement Cascade

The complement system represents a major arm of the innate immune response. As discussed above, pCRP* binds the globular domain of complement C1q leading to activation of the complement cascade (Ji et al. 2006). Activation of the complement cascade ultimately leads to the formation of the membrane attack complex (MAC) and generates the anaphylatoxins, C3a and C5a, and the opsonins C3b and C4b which results in cell lysis, enhanced leukocyte recruitment and phagocytosis, respectively (Holers 2014). Interesting, CRP induced activation of the complement system is inefficient at generating MAC, however, it efficiently cleaves C3, a crucial step for generation of the C3a anaphylatoxin and C3b opsonin (Sjöberg et al. 2006; Mihlan et al. 2009; Thiele et al. 2014). Accordingly, microvesicles with bound pCRP* generates C3b in vitro, whilst in vivo, CRP co-localises with complement staining in various models of inflammation (Griselli et al. 1999; Braig et al. 2017). These observations, combined with the fact that inhibitors of CRP diminish tissue injury and complement deposition suggest that the activation of complement represents a major pro-inflammatory mechanism in vivo.

Intriguingly, CRP also appears to play a role in regulating the complement system. Indeed, recent data suggests that CRP can bind the complement regulatory protein, Factor H (Mihlan et al. 2009; O'Flynn et al. 2016). As such, CRP can recruit factor H to sites of cellular damage, which acts to facilitate the inactivation of C3b thereby dampening further inflammation and aiding in the opsonisation and subsequent clearance of damaged cells (Mihlan et al. 2009, 2011). However, in pathological conditions, these processes are likely to become maladaptive, and as will be discussed, promote the pathogenesis of a range of inflammatory diseases.

CRP and Ischaemia/Reperfusion Injury

Ischaemia/reperfusion injury refers to the exacerbation of injury after restoration of blood flow to a previously ischemic organ (Eltzschig and Eckle 2011). Clinically, the treatment of IRI remains a major unmet clinical need in cardiovascular medicine where the adverse manifestations of IRI significantly contributes to the morbidity and mortality of ischaemic stroke and myocardial infarction (Hausenloy and Yellon 2013). Indeed, in the context of myocardial infarction, it is estimated that up to 50% of the infarct size is the result of IRI (Hausenloy and Yellon 2013). However, to date, no effective therapy is available to prevent IRI. A cardinal feature of IRI is the initiation of a maladaptive immune response that leads to widespread leukocyte infiltration, microvascular dysfunction and organ injury (Eltzschig and Eckle 2011). There is now a significant body of data implicating an important role of CRP in mediating IRI. Indeed, seminal work from Pepys and colleagues demonstrated that the administration of CRP to rats prior to the onset of myocardial ischaemia increased the infarct size by 40% (Griselli et al. 1999). Significantly, depletion of C3 completely abrogated the effects of CRP. Accordingly, CRP and C3 staining is abundant in the post-ischaemic myocardium (Griselli et al. 1999). Our group has subsequently demonstrated that the vast majority of CRP deposited in ischaemic myocardium is mCRP, with the degree of mCRP deposition directly correlating with the degree of leukocyte infiltration, apoptosis, IL-6 and tissue necrosis factor (TNF)-alpha expression (Thiele et al. 2014) (Fig. 20.4). Critically, mCRP staining co-localises with complement in infarcted human myocardial tissue whilst the potential relevance of CRP in a clinical context has been underscored by data highlighting that elevated CRP levels in the context of myocardial infarction are associated with increased mortality and is an independent predictor of the development of heart failure (Suleiman et al. 2003; Thiele et al. 2014).

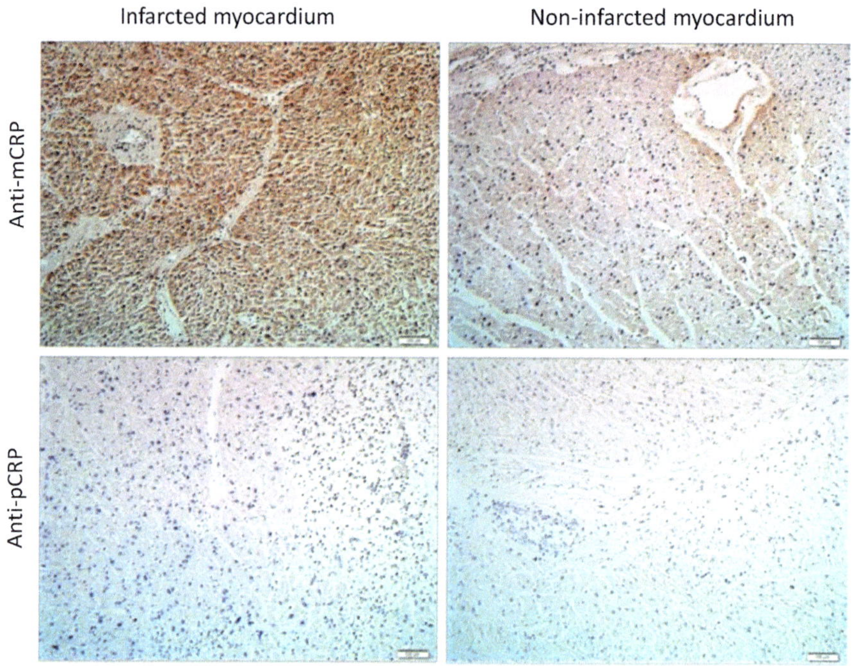

Fig. 20.4 mCRP abundance in infarcted human myocardial tissue. Photomicro-graph of sections of infarcted and non-infarcted human myocardium, stained with conformation specific CRP antibodies. Upper panel, mCRP is widely distributed in infarcted myocardium, but largely absent from non-infarcted tissue. Conversely, no pCRP is seen within infarcted or non-infarcted myocardium. Figure adapted from Thiele et al., Circulation (2014)

These observations have recently been recapitulated in a rat model of renal IRI where rats administered pCRP display more severe renal tubular damage. Importantly, the degree of renal injury correlates with mCRP deposition, leukocyte infiltration, and caspase-3 activation after IRI (Thiele et al. 2018). The deleterious effects of CRP in this model of renal IRI appear primarily dependent upon enhancing endothelial–leukocyte interactions and leukocyte ROS generation in a process linked to mCRP binding to leukocyte lipid raft domains (Thiele et al. 2018). Accordingly, transgenic mice that express human CRP exhibit more severe renal injury after renal IRI, which is associated with a pro-inflammatory M1 monocyte profile (Pegues et al. 2013).

Our group has used intravital microscopy to delineate the spatiotemporal aspects and effects of CRP conformation changes in the context of tissue inflammation. These studies have demonstrated that pCRP binds to microparticles and inflamed endothelial cells in the inflamed microcirculation which provides the necessary structure for dissociation of pCRP into its pro-inflammatory isoforms, pCRP* and mCRP (Thiele et al. 2014, 2018; Braig et al. 2017). These pro-inflammatory isoforms subsequently become deposited at sites of tissue damage leading to enhanced leukocyte

ROS generation in addition to leukocyte adhesion and transmigration (Thiele et al. 2014, 2018; Braig et al. 2017). CRP can also bind to activated, trans-migrating monocytes which in turn facilitates conformational changes in pCRP and thus complement activation (Braig et al. 2017). The importance of complement in mediating the pro-inflammatory effects of CRP in vivo are highlighted by the observation that depletion of complement with cobra venom factor diminish the effects of CRP on leukocyte adhesion and transmigration (Griselli et al. 1999; Braig et al. 2017).

CRP and Burns Wounds

Consistent with the data and concept that CRP deposits at sites of tissue damage to amplify the local inflammatory response, dissociated isoforms have been demonstrated to be deposited, and co-localise with complement C3d and leukocytes in tissue from patients with burns (van de Goot et al. 2009). Interestingly, mCRP is only observed in areas of tissue damage and this co-localises with CD68 positive monocytes/macrophages and correlates with the depth of tissue injury (Braig et al. 2014). Correlative in vitro assays suggest that the dissociation of pCRP into mCRP in the context of burns plays a role in opsonising necrotic cells thereby enhancing phagocytosis and removal of dead tissue by macrophages (Braig et al. 2014). Moreover, mCRP has been shown to facilitate keratinocyte migration, and therefore given its proposed role in clearing necrotic tissue, CRP may also play a role in aiding wound healing in the context of burns injury.

CRP and Atherosclerosis

It is now well appreciated that atherosclerosis is a chronic inflammatory disease (Ross 1999). This has led to significant interest, and controversy, regarding the role of CRP as a marker, and potentially pathogenic factor in atherosclerosis (Libby and Ridker 2004; Ridker 2004). Indeed, both the Women's Health Study and Physician's Health Study have demonstrated that CRP is a strong predictor of cardiovascular risk (Ridker et al. 1997, 1998). In addition, a large randomized trial (JUPITER trial) has shown that rosuvastatin treatment of asymptomatic patients with elevated CRP levels provides benefit in the reduction of cardiovascular events and mortality (Ridker et al. 2008). More recently, the CANTOS trial demonstrated that anti-inflammatory treatment with a monoclonal antibody (canakinumab) targeting IL-1β reduces recurrent cardiovascular events in patients with a high sensitive CRP and history of cardiovascular disease (Ridker et al. 2017). Interestingly, the extent of the reduction of CRP achieved with canakinumab directly correlated with the reduction of cardiovascular outcomes (Ridker et al. 2018).

Despite the evidence demonstrating the association between raised CRP levels and cardiovascular disease, whether CRP plays a direct role in the pathogenesis of atherosclerosis continues to evoke some controversy (Yousuf et al. 2013). Early studies suggesting that CRP possessed proatherogenic properties seem to be flawed, since it later emerged that the source of recombinant CRP produced in *E. coli* used in many of these studies were potentially contaminated with LPS, thus potentially accounting for the pro-inflammatory properties observed (Pepys 2005; Pepys et al. 2005). Indeed, the infusion of pharmaceutical grade human CRP into healthy human volunteers was not associated with any inflammatory response (Lane et al. 2014). However, as discussed above, the pro-inflammatory functions of CRP are associated with conformational changes. Therefore, it is possible that many of the discrepancies related to the role of CRP in atherosclerosis may stem from the fact that standard CRP assays measure all isoforms of CRP rather than specifically detecting the pro-inflammatory mCRP and pCRP* isoforms. In this regard, our laboratory has shown that mCRP, but not pCRP, is deposited in human atherosclerotic plaques, and that mCRP associated microparticles are significantly upregulated in patients with acute coronary syndromes (Habersberger et al. 2012). However, further studies are required to elucidate whether the pro-inflammatory isoforms of CRP directly contribute to the pathogenesis of atherosclerosis and plaque instability.

CRP and Alzheimer's Disease

Alzheimer's disease (AD) is a leading cause of dementia (Scheltens et al. 2016). With the ageing of the population in Western societies, the health burden of AD and economic consequences are likely to significantly increase. Given current therapies display limited efficacy, there remains significant interest in developing novel therapeutic targets that may be able to inhibit the pathological events underlying the development of AD. Through this lens, recent experimental data suggests CRP may play a direct role in the pathogenesis of AD. The cardinal feature of AD is the accumulation of amyloid beta (Aβ) plaques in extracellular spaces and within the walls of the vasculature (Masters et al. 2015; Wang et al. 2017). As AD progresses, neurofibrillary tangles (comprised of Tau protein aggregates) accumulate in neurons (Masters et al. 2015; Wang et al. 2017). Histological studies have demonstrated that neuronal tissue from patients with AD exhibit higher levels of CRP compared to controls and is co-localised with senile plaques (Wood et al. 1993; Iwamoto et al. 1994). Suggesting a pathogenic role of CRP, in vitro, Aβ plaques can dissociate pCRP, and accordingly mCRP and C1q co-localise with Aβ plaques in human AD sections (Strang et al. 2012).

Evidence that the pro-inflammatory isoforms of CRP directly contribute to AD pathogenesis has been provided by data from a mouse model of AD where the direct injection of mCRP into the hippocampal region leads to cognitive decline and behavioral changes (Slevin et al. 2015). Importantly, in addition to cognitive decline, mCRP treated mice exhibit abnormal neuronal morphology in addition to enhanced p-Tau and beta amyloid plaque production (Slevin et al. 2015). Mechanistically, in vitro, mCRP has been shown to induce Tau filament polymerisation and directly trigger signaling pathways linked to neurodegeneration including p-ERK1/2 and p-Tau (Slevin et al. 2015). An interesting observation from these studies is that the microvasculature of patients with AD also exhibits an abundance of mCRP deposition where it co-localises with beta amyloid and CD105 (Slevin et al. 2010, 2015). Interestingly, these findings are most prominent in patients with a history of ischaemic stroke, suggesting that mCRP deposition in these regions may promote dysfunctional vascular development thus contributing to the small vessel vasculopathy characteristic of AD (Slevin et al. 2010).

CRP and Age-Related Macular Degeneration

Recent evidence has emerged implicating a role for CRP in the development of age-related macular degeneration (AMD), which is one of the leading causes of blindness globally (Friedman et al. 2004). It has previously been demonstrated that elevated serum CRP levels are an independent risk factor for AMD development and progression (Seddon et al. 2005). More recently, it has emerged that mCRP may have a direct pathogenic role in AMD development in patients harbouring the high-risk CFH SNP (Y402H), which is considered one of the major risk factors for AMD (Edwards et al. 2005; Chirco et al. 2016). In this regard, the choroid of patients with the CFH SNP (Y402H) display elevated mCRP levels (Chirco et al. 2016). In vitro, the treatment of choroid endothelial cells with mCRP induces endothelial cell permeability and migration, which are both important prerequisites for AMD development (Chirco et al. 2016). Moreover, the treatment of human choroidal tissue with mCRP leads to an altered transcriptome where both the expression of ICAM-1 and carbonic anhydrase 4 (CA4) is altered at the mRNA and protein level, with mCRP resulting in increased ICAM-1 expression and decreased CA4 (Chirco et al. 2016). Importantly, these changes are observed in patients with AMD and appear central to AMD pathogenesis (Yuan et al. 2010; Skeie et al. 2010).

CRP and Immune Thrombocytopaenic Purpura

Immune-mediated thrombocytopaenia (ITP) leads to the destruction of platelets via the binding of allo- or auto-antibodies to platelets that ultimately leads to Fc gamma receptor-mediated phagocytosis in the reticuloendothelial system (Semple et al. 2011). Consistent with the role of CRP in amplifying immune responses, CRP has been demonstrated to enhance the IgG-mediated phagocytosis of platelets (Kapur et al. 2015). In the context of ITP, the binding of anti-platelet antibodies to platelets leads to platelet activation via oxidative damage and exposure of the CRP ligand, phosphocholine (Kapur et al. 2015). This, in turn, allows the binding of CRP to the activated platelets, which acts to facilitate the Fc mediated platelet phagocytosis.

Intriguingly, neither complement nor the platelet FcγIIa receptor appears to play a role in this process. Rather, CRP bound to the activated platelets leads to enhanced FcR mediated functions such as phagocytosis and respiratory burst. Importantly, consistent with other data, CRP did not display any effects on resting platelets. Clinically, higher CRP levels in patients with ITP are linked to the slower recovery of platelet counts (Kapur et al. 2015). These data may help explain the observation that ITP is often triggered by infections and suggest that therapeutic targeting of CRP may be a useful adjunctive treatment for ITP.

CRP as a Therapeutic Target

Although the role of CRP in atherosclerosis remains a matter of debate, the ability of CRP to bind to sites of tissue damage and amplify inflammation is universally accepted, having been demonstrated and recapitulated by a number of groups independently. Therefore, given the large number of conditions in which CRP is elevated, the therapeutic inhibition of CRP is potentially a highly desirable target for a range of inflammatory, ischaemic and malignant conditions. In this regard, a number of different agents have been developed with the aim to inhibit the pro-inflammatory function of CRP. Foremost amongst these is the bis(phosphocholine) alkane, bis(phosphocholine) hexane, a small molecule inhibitor that binds and cross-links two CRP pentamers thereby stabilising the pentameric structure, blocking PC binding and abrogating its ability to bind complement C1q (Pepys et al. 2006). This compound has shown significant ability to abrogate the deleterious effects of CRP in a rat model of myocardial infarction (Pepys et al. 2006). However, the pharmacokinetic profile of bis(phospho-choline) has significant shortcomings since it exhibits a relatively low affinity for CRP (Kd = 150 nM) and short half-life of 90 min in mice. Therefore, other therapeutic approaches that have been developed including antisense oligonucleotides to down-regulate hepatic CRP synthesis as well as therapeutic apheresis to remove circulating CRP (Slagman et al. 2011, p. 200; Jones et al. 2012). Thus, the development of novel, more potent inhibitors of CRP remains an active source of investigation given the potentially wide scope of therapeutic indications and potential benefits in the clinic.

Acknowledgements We thank Eliana Stanziano for her excellent manuscript editing and formatting.

References

Agrawal A, Volanakis JE (1994) Probing the C1q-binding site on human C-reactive protein by site-directed mutagenesis. J Immunol Baltim Md 1950 152:5404–5410

Agrawal A, Shrive AK, Greenhough TJ, Volanakis JE (2001) Topology and structure of the C1q-binding site on C-reactive protein. J Immunol Baltim Md 1950 166:3998–4004. https://doi.org/10.4049/jimmunol.166.6.3998

Agrawal A, Simpson MJ, Black S et al (2002) A C-reactive protein mutant that does not bind to phosphocholine and pneumococcal C-polysaccharide. J Immunol Baltim Md 1950 169:3217–3222. https://doi.org/10.4049/jimmunol.169.6.3217

Agrawal A, Cha-Molstad H, Samols D, Kushner I (2003a) Overexpressed nuclear factor-kappaB can participate in endogenous C-reactive protein induction, and enhances the effects of C/EBPbeta and signal transducer and activator of transcription-3. Immunology 108:539–547. https://doi.org/10.1046/j.1365-2567.2003.01608.x

Agrawal A, Samols D, Kushner I (2003b) Transcription factor c-Rel enhances C-reactive protein expression by facilitating the binding of C/EBPbeta to the promoter. Mol Immunol 40:373–380. https://doi.org/10.1016/s0161-5890(03)00148-2

Anderson HC, McCarty M (1950) Determination of C-reactive protein in the blood as a measure of the activity of the disease process in acute rheumatic fever. Am J Med 8:445–455. https://doi.org/10.1016/0002-9343(50)90226-9

Armstrong PB (2015) Comparative biology of the pentraxin protein family: evolutionarily conserved component of innate immune system. Int Rev Cell Mol Biol 316:1–47. https://doi.org/10.1016/bs.ircmb.2015.01.002

Ash R (1933) Nonspecific precipitins for pneumococcic fraction C in acute infections. J Infect Dis 53:89–97

Baltz ML, de Beer FC, Feinstein A et al (1982) Phylogenetic aspects of C-reactive protein and related proteins. Ann N Y Acad Sci 389:49–75. https://doi.org/10.1111/j.1749-6632.1982.tb22125.x

Bang R, Marnell L, Mold C et al (2005) Analysis of binding sites in human C-reactive protein for FcγRI, FcγRIIA, and C1q by site-directed mutagenesis. J Biol Chem 280:25095–25102. https://doi.org/10.1074/jbc.M504782200

Black S, Agrawal A, Samols D (2003) The phosphocholine and the polycation-binding sites on rabbit C-reactive protein are structurally and functionally distinct. Mol Immunol 39:1045–1054. https://doi.org/10.1016/s0161-5890(03)00031-2

Braig D, Kaiser B, Thiele JR et al (2014) A conformational change of C-reactive protein in burn wounds unmasks its proinflammatory properties. Int Immunol 26:467–478. https://doi.org/10.1093/intimm/dxu056

Braig D, Nero TL, Koch H-G et al (2017) Transitional changes in the CRP structure lead to the exposure of proinflammatory binding sites. Nat Commun 8:14188. https://doi.org/10.1038/ncomms14188

Calabro P, Chang DW, Willerson JT, Yeh ETH (2005) Release of C-reactive protein in response to inflammatory cytokines by human adipocytes: linking obesity to vascular inflammation. J Am Coll Cardiol 46:1112–1113. https://doi.org/10.1016/j.jacc.2005.06.017

Chirco KR, Whitmore SS, Wang K et al (2016) Monomeric C-reactive protein and inflammation in age-related macular degeneration. J Pathol 240:173–183. https://doi.org/10.1002/path.4766

de Beer FC, Hind CR, Fox KM et al (1982) Measurement of serum C-reactive protein concentration in myocardial ischaemia and infarction. Br Heart J 47:239–243. https://doi.org/10.1136/hrt.47.3.239

de la Torre R, Peña E, Vilahur G et al (2013) Monomerization of C-reactive protein requires glycoprotein IIb-IIIa activation: pentraxins and platelet deposition. J Thromb Haemost JTH 11:2048–2058. https://doi.org/10.1111/jth.12415

Edwards AO, Ritter R, Abel KJ et al (2005) Complement factor H polymorphism and age-related macular degeneration. Science 308:421–424. https://doi.org/10.1126/science.1110189

Eisenhardt SU, Habersberger J, Murphy A et al (2009) Dissociation of pentameric to monomeric C-reactive protein on activated platelets localizes inflammation to atherosclerotic plaques. Circ Res 105:128–137. https://doi.org/10.1161/CIRCRESAHA.108.190611

Elices MJ, Osborn L, Takada Y et al (1990) VCAM-1 on activated endothelium interacts with the leukocyte integrin VLA-4 at a site distinct from the VLA-4/fibronectin binding site. Cell 60:577–584. https://doi.org/10.1016/0092-8674(90)90661-w

Eltzschig HK, Eckle T (2011) Ischemia and reperfusion–from mechanism to translation. Nat Med 17:1391–1401. https://doi.org/10.1038/nm.2507

Friedman DS, O'Colmain BJ, Muñoz B et al (2004) Prevalence of age-related macular degeneration in the United States. Arch Ophthalmol Chic Ill 1960 122:564–572. https://doi.org/10.1001/archopht.122.4.564

Fujita M, Takada YK, Izumiya Y, Takada Y (2014) The binding of monomeric C-reactive protein (mCRP) to integrins $\alpha v\beta 3$ and $\alpha 4\beta 1$ is related to its pro-inflammatory action. PLoS ONE 9:e93738. https://doi.org/10.1371/journal.pone.0093738

Gaboriaud C, Juanhuix J, Gruez A et al (2003) The crystal structure of the globular head of complement protein C1q provides a basis for its versatile recognition properties. J Biol Chem 278:46974–46982. https://doi.org/10.1074/jbc.M307764200

Ganter U, Arcone R, Toniatti C et al (1989) Dual control of C-reactive protein gene expression by interleukin-1 and interleukin-6. EMBO J 8:3773–3779

Griselli M, Herbert J, Hutchinson WL et al (1999) C-reactive protein and complement are important mediators of tissue damage in acute myocardial infarction. J Exp Med 190:1733–1740. https://doi.org/10.1084/jem.190.12.1733

Habersberger J, Strang F, Scheichl A et al (2012) Circulating microparticles generate and transport monomeric C-reactive protein in patients with myocardial infarction. Cardiovasc Res 96:64–72. https://doi.org/10.1093/cvr/cvs237

Hausenloy DJ, Yellon DM (2013) Myocardial ischemia-reperfusion injury: a neglected therapeutic target. J Clin Invest 123:92–100. https://doi.org/10.1172/JCI62874

Holers VM (2014) Complement and its receptors: new insights into human disease. Annu Rev Immunol 32:433–459. https://doi.org/10.1146/annurev-immunol-032713-120154

Hurlimann J, Thorbecke GJ, Hochwald GM (1966) The liver as the site of C-reactive protein formation. J Exp Med 123:365–378. https://doi.org/10.1084/jem.123.2.365

Iwamoto N, Nishiyama E, Ohwada J, Arai H (1994) Demonstration of CRP immunoreactivity in brains of Alzheimer's disease: immunohistochemical study using formic acid pretreatment of tissue sections. Neurosci Lett 177:23–26. https://doi.org/10.1016/0304-3940(94)90035-3

Jabs WJ, Lögering BA, Gerke P et al (2003) The kidney as a second site of human C-reactive protein formation in vivo. Eur J Immunol 33:152–161. https://doi.org/10.1002/immu.200390018

Ji S-R, Wu Y, Potempa LA et al (2006) Effect of modified C-reactive protein on complement activation: a possible complement regulatory role of modified or monomeric C-reactive protein in atherosclerotic lesions. Arterioscler Thromb Vasc Biol 26:935–941. https://doi.org/10.1161/01.ATV.0000206211.21895.73

Ji S-R, Wu Y, Zhu L et al (2007) Cell membranes and liposomes dissociate C-reactive protein (CRP) to form a new, biologically active structural intermediate: mCRP(m). FASEB J Off Publ Fed Am Soc Exp Biol 21:284–294. https://doi.org/10.1096/fj.06-6722com

Jones NR, Pegues MA, McCrory MA et al (2012) A selective inhibitor of human C-reactive protein translation is efficacious in vitro and in C-reactive protein Transgenic Mice and Humans. Mol Ther Nucleic Acids 1:e52. https://doi.org/10.1038/mtna.2012.44

Ji S-R, Ma L, Bai C-J et al (2009) Monomeric C-reactive protein activates endothelial cells via interaction with lipid raft microdomains. FASEB J Off Publ Fed Am Soc Exp Biol 23:1806–1816. https://doi.org/10.1096/fj.08-116962

Kapur R, Heitink-Pollé KMJ, Porcelijn L et al (2015) C-reactive protein enhances IgG-mediated phagocyte responses and thrombocytopenia. Blood 125:1793–1802. https://doi.org/10.1182/blood-2014-05-579110

Khreiss T, József L, Potempa LA, Filep JG (2004) Conformational rearrangement in C-reactive protein is required for proinflammatory actions on human endothelial cells. Circulation 109:2016–2022. https://doi.org/10.1161/01.CIR.0000125527.41598.68

Khreiss T, József L, Potempa LA, Filep JG (2005) Loss of pentameric symmetry in C-reactive protein induces interleukin-8 secretion through peroxynitrite signaling in human neutrophils. Circ Res 97:690–697. https://doi.org/10.1161/01.RES.0000183881.11739.CB

Kobayashi S, Inoue N, Ohashi Y et al (2003) Interaction of oxidative stress and inflammatory response in coronary plaque instability: important role of C-reactive protein. Arterioscler Thromb Vasc Biol 23:1398–1404. https://doi.org/10.1161/01.ATV.0000081637.36475.BC

Kresl JJ, Potempa LA, Anderson BE (1998) Conversion of native oligomeric to a modified monomeric form of human C-reactive protein. Int J Biochem Cell Biol 30:1415–1426. https://doi.org/10.1016/s1357-2725(98)00078-8

Kuta AE, Baum LL (1986) C-reactive protein is produced by a small number of normal human peripheral blood lymphocytes. J Exp Med 164:321–326. https://doi.org/10.1084/jem.164.1.321

Lane T, Wassef N, Poole S et al (2014) Infusion of pharmaceutical-grade natural human C-reactive protein is not proinflammatory in healthy adult human volunteers. Circ Res 114:672–676. https://doi.org/10.1161/CIRCRESAHA.114.302770

Libby P, Ridker PM (2004) Inflammation and atherosclerosis: role of C-reactive protein in risk assessment. Am J Med 116(Suppl 6A):9S–16S. https://doi.org/10.1016/j.amjmed.2004.02.006

Liuzzo G, Biasucci LM, Gallimore JR et al (1994) The prognostic value of C-reactive protein and serum amyloid a protein in severe unstable angina. N Engl J Med 331:417–424. https://doi.org/10.1056/NEJM199408183310701

Lu J, Marnell LL, Marjon KD et al (2008) Structural recognition and functional activation of FcgammaR by innate pentraxins. Nature 456:989–992. https://doi.org/10.1038/nature07468

Lv J-M, Wang M-Y (2018) In vitro generation and bioactivity evaluation of C-reactive protein intermediate. PLoS ONE 13:e0198375. https://doi.org/10.1371/journal.pone.0198375

Massberg S, Enders G, Matos FC et al (1999) Fibrinogen deposition at the postischemic vessel wall promotes platelet adhesion during ischemia-reperfusion in vivo. Blood 94:3829–3838

Masters CL, Bateman R, Blennow K et al (2015) Alzheimer's disease. Nat Rev Dis Primer 1:15056. https://doi.org/10.1038/nrdp.2015.56

McCarty M (1982) Historical perspective on C-reactive protein. Ann N Y Acad Sci 389:1–10. https://doi.org/10.1111/j.1749-6632.1982.tb22121.x

McFadyen JD, Kiefer J, Braig D et al (2018) Dissociation of C-reactive protein localizes and amplifies inflammation: evidence for a direct biological role of C-reactive protein and its conformational changes. Front Immunol 9:1351. https://doi.org/10.3389/fimmu.2018.01351

Mihlan M, Stippa S, Józsi M, Zipfel PF (2009) Monomeric CRP contributes to complement control in fluid phase and on cellular surfaces and increases phagocytosis by recruiting factor H. Cell Death Differ 16:1630–1640. https://doi.org/10.1038/cdd.2009.103

Mihlan M, Blom AM, Kupreishvili K et al (2011) Monomeric C-reactive protein modulates classic complement activation on necrotic cells. FASEB J Off Publ Fed Am Soc Exp Biol 25:4198–4210. https://doi.org/10.1096/fj.11-186460

Molins B, Peña E, de la Torre R, Badimon L (2011) Monomeric C-reactive protein is prothrombotic and dissociates from circulating pentameric C-reactive protein on adhered activated platelets under flow. Cardiovasc Res 92:328–337. https://doi.org/10.1093/cvr/cvr226

Nguyen NY, Suzuki A, Cheng SM et al (1986) Isolation and characterization of Limulus C-reactive protein genes. J Biol Chem 261:10450–10455

O'Flynn J, van der Pol P, Dixon KO et al (2016) Monomeric C-reactive protein inhibits renal cell-directed complement activation mediated by properdin. Am J Physiol Renal Physiol 310:F1308–F1316. https://doi.org/10.1152/ajprenal.00645.2014

Osmand AP, Friedenson B, Gewurz H et al (1977) Characterization of C-reactive protein and the complement subcomponent C1t as homologous proteins displaying cyclic pentameric symmetry (pentraxins). Proc Natl Acad Sci U S A 74:739–743. https://doi.org/10.1073/pnas.74.2.739

Pathak A, Agrawal A (2019) Evolution of C-reactive protein. Front Immunol 10:943. https://doi.org/10.3389/fimmu.2019.00943

Pegues MA, McCrory MA, Zarjou A, Szalai AJ (2013) C-reactive protein exacerbates renal ischemia-reperfusion injury. Am J Physiol Renal Physiol 304:F1358–F1365. https://doi.org/10.1152/ajprenal.00476.2012

Pepys MB (2005) CRP or not CRP? That is the question. Arterioscler Thromb Vasc Biol 25:1091–1094. https://doi.org/10.1161/01.ATV.0000169644.88847.28

Pepys MB, Hirschfield GM (2003) C-reactive protein: a critical update. J Clin Invest 111:1805–1812. https://doi.org/10.1172/JCI18921

Pepys MB, Dash AC, Fletcher TC et al (1978) Analogues in other mammals and in fish of human plasma proteins, C-reactive protein and amyloid P component. Nature 273:168–170. https://doi.org/10.1038/273168a0

Pepys MB, Hawkins PN, Kahan MC et al (2005) Proinflammatory effects of bacterial recombinant human C-reactive protein are caused by contamination with bacterial products, not by C-reactive protein itself. Circ Res 97:e97–103. https://doi.org/10.1161/01.RES.0000193595.03608.08

Pepys MB, Hirschfield GM, Tennent GA et al (2006) Targeting C-reactive protein for the treatment of cardiovascular disease. Nature 440:1217–1221. https://doi.org/10.1038/nature04672

Potempa LA, Maldonado BA, Laurent P et al (1983) Antigenic, electrophoretic and binding alterations of human C-reactive protein modified selectively in the absence of calcium. Mol Immunol 20:1165–1175. https://doi.org/10.1016/0161-5890(83)90140-2

Ramage L, Proudfoot L, Guy K (2004) Expression of C-reactive protein in human lung epithelial cells and upregulation by cytokines and carbon particles. Inhal Toxicol 16:607–613. https://doi.org/10.1080/08958370490464599

Ridker PM (2004) High-sensitivity C-reactive protein, inflammation, and cardiovascular risk: from concept to clinical practice to clinical benefit. Am Heart J 148:S19–S26. https://doi.org/10.1016/j.ahj.2004.04.028

Ridker PM (2009) C-reactive protein: eighty years from discovery to emergence as a major risk marker for cardiovascular disease. Clin Chem 55:209–215. https://doi.org/10.1373/clinchem.2008.119214

Ridker PM, Cushman M, Stampfer MJ et al (1997) Inflammation, aspirin, and the risk of cardiovascular disease in apparently healthy men. N Engl J Med 336:973–979. https://doi.org/10.1056/NEJM199704033361401

Ridker PM, Buring JE, Shih J et al (1998) Prospective study of C-reactive protein and the risk of future cardiovascular events among apparently healthy women. Circulation 98:731–733. https://doi.org/10.1161/01.cir.98.8.731

Ridker PM, Danielson E, Fonseca FAH et al (2008) Rosuvastatin to prevent vascular events in men and women with elevated C-reactive protein. N Engl J Med 359:2195–2207. https://doi.org/10.1056/NEJMoa0807646

Ridker PM, Everett BM, Thuren T et al (2017) Antiinflammatory therapy with canakinumab for atherosclerotic disease. N Engl J Med 377:1119–1131. https://doi.org/10.1056/NEJMoa1707914

Ridker PM, MacFadyen JG, Everett BM et al (2018) Relationship of C-reactive protein reduction to cardiovascular event reduction following treatment with canakinumab: a secondary analysis from the CANTOS randomised controlled trial. Lancet Lond Engl 391:319–328. https://doi.org/10.1016/S0140-6736(17)32814-3

Ross R (1999) Atherosclerosis–an inflammatory disease. N Engl J Med 340:115–126. https://doi.org/10.1056/NEJM199901143400207

Scheltens P, Blennow K, Breteler MMB et al (2016) Alzheimer's disease. Lancet Lond Engl 388:505–517. https://doi.org/10.1016/S0140-6736(15)01124-1

Seddon JM, George S, Rosner B, Rifai N (2005) Progression of age-related macular degeneration: prospective assessment of C-reactive protein, interleukin 6, and other cardiovascular biomarkers. Arch Ophthalmol Chic Ill 1960 123:774–782. https://doi.org/10.1001/archopht.123.6.774

Semple JW, Italiano JE, Freedman J (2011) Platelets and the immune continuum. Nat Rev Immunol 11:264–274. https://doi.org/10.1038/nri2956

Shrive AK, Cheetham GM, Holden D et al (1996) Three dimensional structure of human C-reactive protein. Nat Struct Biol 3:346–354. https://doi.org/10.1038/nsb0496-346
Singh SK, Thirumalai A, Pathak A et al (2017) Functional transformation of C-reactive protein by hydrogen peroxide. J Biol Chem 292:3129–3136. https://doi.org/10.1074/jbc.M116.773176
Sjöberg AP, Trouw LA, McGrath FDG et al (2006) Regulation of complement activation by C-reactive protein: targeting of the inhibitory activity of C4b-binding protein. J Immunol Baltim Md 1950 176:7612–7620. https://doi.org/10.4049/jimmunol.176.12.7612
Skeie JM, Fingert JH, Russell SR et al (2010) Complement component C5a activates ICAM-1 expression on human choroidal endothelial cells. Invest Ophthalmol Vis Sci 51:5336–5342. https://doi.org/10.1167/iovs.10-5322
Slagman AC, Bock C, Abdel-Aty H et al (2011) Specific removal of C-reactive protein by apheresis in a porcine cardiac infarction model. Blood Purif 31:9–17. https://doi.org/10.1159/000320763
Slevin M, Matou-Nasri S, Turu M et al (2010) Modified C-reactive protein is expressed by stroke neovessels and is a potent activator of angiogenesis in vitro. Brain Pathol Zurich Switz 20:151–165. https://doi.org/10.1111/j.1750-3639.2008.00256.x
Slevin M, Matou S, Zeinolabediny Y et al (2015) Monomeric C-reactive protein–a key molecule driving development of Alzheimer's disease associated with brain ischaemia? Sci Rep 5:13281. https://doi.org/10.1038/srep13281
Springer TA (1994) Traffic signals for lymphocyte recirculation and leukocyte emigration: the multistep paradigm. Cell 76:301–314. https://doi.org/10.1016/0092-8674(94)90337-9
Strang F, Scheichl A, Chen Y-C et al (2012) Amyloid plaques dissociate pentameric to monomeric C-reactive protein: a novel pathomechanism driving cortical inflammation in Alzheimer's disease? Brain Pathol Zurich Switz 22:337–346. https://doi.org/10.1111/j.1750-3639.2011.00539.x
Suleiman M, Aronson D, Reisner SA et al (2003) Admission C-reactive protein levels and 30-day mortality in patients with acute myocardial infarction. Am J Med 115:695–701. https://doi.org/10.1016/j.amjmed.2003.06.008
Thiele JR, Habersberger J, Braig D et al (2014) Dissociation of pentameric to monomeric C-reactive protein localizes and aggravates inflammation. in vivo proof of a powerful proinflammatory mechanism and a new anti-inflammatory strategy. Circulation 130:35–50. https://doi.org/10.1161/CIRCULATIONAHA.113.007124
Thiele JR, Zeller J, Kiefer J et al (2018) A conformational change in C-reactive protein enhances leukocyte recruitment and reactive oxygen species generation in ischemia/reperfusion injury. Front Immunol 9:675. https://doi.org/10.3389/fimmu.2018.00675
Thompson D, Pepys MB, Wood SP (1999) The physiological structure of human C-reactive protein and its complex with phosphocholine. Struct Lond Engl 1993 7:169–177. https://doi.org/10.1016/S0969-2126(99)80023-9
Tillett WS, Francis T (1930) Serological reactions in pneumonia with a non-protein somatic fraction of pneumococcus. J Exp Med 52:561–571. https://doi.org/10.1084/jem.52.4.561
van de Goot F, Krijnen PAJ, Begieneman MPV et al (2009) Acute inflammation is persistent locally in burn wounds: a pivotal role for complement and C-reactive protein. J Burn Care Res Off Publ Am Burn Assoc 30:274–280. https://doi.org/10.1097/BCR.0b013e318198a252
Venugopal SK, Devaraj S, Jialal I (2005) Macrophage conditioned medium induces the expression of C-reactive protein in human aortic endothelial cells: potential for paracrine/autocrine effects. Am J Pathol 166:1265–1271. https://doi.org/10.1016/S0002-9440(10)62345-0
Volanakis JE, Wirtz KW (1979) Interaction of C-reactive protein with artificial phosphatidylcholine bilayers. Nature 281:155–157. https://doi.org/10.1038/281155a0
Wang J, Gu BJ, Masters CL, Wang Y-J (2017) A systemic view of Alzheimer disease - insights from amyloid-β metabolism beyond the brain. Nat Rev Neurol 13:612–623. https://doi.org/10.1038/nrneurol.2017.111
Wood JA, Wood PL, Ryan R et al (1993) Cytokine indices in Alzheimer's temporal cortex: no changes in mature IL-1 beta or IL-1RA but increases in the associated acute phase proteins IL-6, alpha 2-macroglobulin and C-reactive protein. Brain Res 629:245–252. https://doi.org/10.1016/0006-8993(93)91327-o

Yasojima K, Schwab C, McGeer EG, McGeer PL (2000) Human neurons generate C-reactive protein and amyloid P: upregulation in Alzheimer's disease. Brain Res 887:80–89. https://doi.org/10.1016/s0006-8993(00)02970-x

Young DP, Kushner I, Samols D (2008) Binding of C/EBPbeta to the C-reactive protein (CRP) promoter in Hep3B cells is associated with transcription of CRP mRNA. J Immunol Baltim Md 1950 181:2420–2427. https://doi.org/10.4049/jimmunol.181.4.2420

Yousuf O, Mohanty BD, Martin SS et al (2013) High-sensitivity C-reactive protein and cardiovascular disease: a resolute belief or an elusive link? J Am Coll Cardiol 62:397–408. https://doi.org/10.1016/j.jacc.2013.05.016

Yuan X, Gu X, Crabb JS et al (2010) Quantitative proteomics: comparison of the macular Bruch membrane/choroid complex from age-related macular degeneration and normal eyes. Mol Cell Proteomics MCP 9:1031–1046. https://doi.org/10.1074/mcp.M900523-MCP200

Zacho J, Tybjaerg-Hansen A, Jensen JS et al (2008) Genetically elevated C-reactive protein and ischemic vascular disease. N Engl J Med 359:1897–1908. https://doi.org/10.1056/NEJMoa0707402

Zhang D, Sun M, Samols D, Kushner I (1996) STAT3 participates in transcriptional activation of the C-reactive protein gene by interleukin-6. J Biol Chem 271:9503–9509. https://doi.org/10.1074/jbc.271.16.9503

Zouki C, Haas B, Chan JS et al (2001) Loss of pentameric symmetry of C-reactive protein is associated with promotion of neutrophil-endothelial cell adhesion. J Immunol Baltim Md 1950 167:5355–5361. https://doi.org/10.4049/jimmunol.167.9.5355

Index

U. Hoeger and J. R. Harris (eds.), *Vertebrate and Invertebrate Respiratory Proteins, Lipoproteins and other Body Fluid Proteins*, Subcellular Biochemistry 94,
https://doi.org/10.1007/978-3-030-41769-7

CPI Antony Rowe
Eastbourne, UK
March 20, 2020